FOREWORD

This book by the late Dr. Ning Chien and his student Dr. Zhaohui Wan, is based on their experience of teaching and practicing in the field of engineering sedimentation accumulated mainly in China for about forty years. The first draft of this book was completed in 1951 by the senior author while he was a staff of the sedimentation laboratory led by the late Professor H.A. Einstein at the University of California at Berkeley. Between 1955 and 1986, Dr. Chien conducted numerous classes on mechanics of sediment transport for graduate students and hydraulic engineers with these notes in China. Up until 1982, about 830 engineers and graduate students attended these courses. Intensive discussions were conducted in the classes and, based on these discussions, revisions of the notes were carefully made. The result is a book culminating in logic and clearness of presentation.

Research engineers and university professors in China are often asked to do consultant work for engineering projects. The authors are no exceptions. Thus Dr. Chien had conducted reconnaissance of many major rivers in China, especially the Yellow river. For several years, he was stationed at Zhengzhou on the right bank of the lower Yellow River, from which he made numerous inspection tours to different parts of the River, ranging far upstream and all the way downstream to the estuary. For closer observation, in most cases these trips were made on foot under primitive and rough conditions. His first-hand and in-depth knowledge of the river gained in this down-to-the earth approach eventually earned him the reputation of modern-time authority on the Yellow River. Practical knowledge plus a strong theoretical background enabled Dr. Chien to offer valuable advises to many important river-training projects in China. This practical spirit also impregnated the writing of the book. Thus, although the book contains a wealth of theoretical material, the selection and the presentation are by no means academic.

The book is a comprehensive treatise on the mechanics of sediment movement. It covers every essential phase of the subject and is now the standard textbook in Chinese universities and the main reference used by practicing engineers in China. The edition in Chinese was published by the Science Press in Beijing. Over five thousand copies of three printings have been sold. To readers outside China, it is perhaps worthwhile to mention that the book also incorporates Chinese developments on the subject or references thereof otherwise not available to the outside world. In the last decades, out of necessity, large-scale hydraulic constructions have been carried out in China on large streams that are mostly sediment-laden and present many sedimentation problems. To solve these problems, a great deal of research has been conducted. Results obtained prior to 1983 have been reviewed in this book, while references to many later developments are appended in the end.

Last but not least, it is indeed fortunate that the quality of the edition in English was immensely enhanced by Professor John S. McNown, who patiently and thoroughly revised all draft translations performed by Chinese engineers to render them correct, accurate and readable. Professor McNown is an authority in sedimentation himself. He is also a member of the American Engineering Academy and an honorary member of the International Association of Hydraulic Research (IAHR). The contribution to the English edition made by such a distinguished scholar is greatly appreciated.

Bingnan Lin, Ph.D.

Chairman, Advisory Council, IRTCES

Chief of Sedimentation Panel, Three Gorges Project

Member, The Chinese Academy of Sciences

ACKNOWLEDGMENTS

American Society of Civil Engineers,

Figs. 14-21, 14-22 (Alger, G. R.and B. Simons, 1968), Fig.3-25 (McLaughlin, R.T. Jr., 1959), 4-23 (McQuivey, R.S. and E.V. Richardson, 1969), 6-4 (Liu, H.K., 1957), 6-12 (Simons, D.B. and E.V. Richardson, 1960), 6-16 (Karahan, M. E. and A. W. Peterson, 1980), 6-25 (Yalin, M. S., 1973), 6-36 (Albertson, M.L., D.B. Simons, E.V. Richardson. 1958), 6-47 (Yalin, M. S., 1977), 6-48 (Yalin, M. S. and E. Karahan, 1979), 6-54 (Carey, W.C. and M.D. Keller,1957), 7-7 (Liu, H.K.and S. Y. Hwang, 1959), 7-9 (Culbertson, J.K. and C.F. Nordin, Jr., 1960), 7-29, 7-30, (Lovera, F. and J. F. Kennedy, 1969), 7-35 (Alam, A. M. Z. and J. F. Kennedy 1969), 7-38, 7-39 (Kouwen, N. and T. E. Unny, 1973), 7-43 (Rouse, H.., 1965), 8-5 (Grass, A.J., 1970), 8-8 (Yang Chih -Ted, 1973), 8-15 (Kartha, V.C. and H.J. Leutheusser, 1970), 8-18 (Ippen, A.T. and P.A. Drinker, 1962), 8-28 (Task committee on erosion of cohesive sediments, 1968), 8-29 (Flaxman E.M., 1963), 10-18 (Jobson, H.E. and W.W. Sayre, 1970), 10-30 (Hjelmfelt, A.T. and C.W. Lenau, 1970), 10-33. 10-34 (McQuivey R.S. and T.K. Keefer, 1976), 10-36 (Lau, Y. Lam and B.G. Krishnappen, 1977), 10-38 (El-Hadi N.D. Abd. and K.S. Davar, 1976), 11-1 (Grigg, N.S., 1970), 11-3 (Bishop, A.A., D.B. Simmons and E.V. Richardson, 1965), 11-10 (Ackers, P. and W.R. White, 1973), 11-15 (Cooper, R.H., A.W. Petersen and T. Blench, 1972), 13-24 (Takahashi, T., 1978), 14-28 (Wood, I.R., 1967), 14-30 (Fietz, T.R. and I.R. Wood, 1967) 14-33 (Harleman, D.R.F., R.S.Gooch and A.T. Ippen, 1958), 14-39 (French, R.H., 1979)16-4 (Silvester, R, 1959), 16-44 (Eagleson, P.S. and R.G.Dean, 1959), 8-19, 8-20 (Nece, R.E. and J.D.Smith, 1970), 11-25 (Franco, J.J., 1965),12-17 (Vanoni, V.A. and G.N.Nomicos, 1960), 14-11, 14-12 (Albertson, M.L., Y.B. Dai, R.A.Jensen and H.Rouse, 1950), 14-21, 14-22 (Macagno, E.S. and H. Rouse , 1961), 16-15, 16-46 (Iwagaki, Y. and H. Noda, 1963), 16-17 (Dyhr-Nielsen, M. and T. Sorensen, 1970, 16-21 (Bijker, E.W., J.P.Th. Kalkwijk and T. Pieters, 1974), 16-34,16-35 (Madsen, O. S. and W. D. Grant, 1976), 16-36 (Nakato,T. et al., 1977, 16-38 and 16-42 (Das M.M., 1972), 16-48 (Dalrymple, R.A.and W.W. Thompson, 1976), 16-54 (Tubman, N.W. and J.N. Suhayda, 1976), 16-57 (Mogridge, G.R. and J.W. Kamphuis, 1972), 16-58 (Dingler, J.R. and D. L. Inman, 1976),16-62 and 16-63 (Masashi Hom-ma and C.Sonu, 1963), 17-46 (Zandi, I. and G. Govatos, 1967),.8-18 (Ippen, A.T. and P.A. Drinker, 1962), 16-10 (Kamphuis, J.W., 1975), Tables, 6-5 (Simons, D.B. and E.V. Richarson,1960), 7-2 (Shen, H.W.,1962),11-4 (Cooper, R.H., A.W. Petersen and T. Blench, 1972), 16-2 (Einstein, H.A., 1972),

American Chemical Society, 1155 Sixteenth Street, N.W., Washington, D.C. 20036, USA,   Table 3-5 (Steinour, 1944).

American Institute of Chemical Engineers, 345 East 47 Street, New York, NY 10017-2395, USA, Figures 3-11 (Christiansen and Barker, 1965), 17-6 (Metzner and Reed, 1955), 17-8 (Hanks, 1963), 17-10 (Dodge and Metzner, 1959), 17-48 (Virk, 1975), 17-51 (Radin, Zakin and Patterson, 1975), 17-43, 17-44 and Tables 17-5, 17-6 (Turian and Yuan, 1977).

Annual Reviews Inc., 4139 EL Camino Way, PO Box 10139, Palo Alto, CA 94303-0139, USA, Figures 6-22, 6-23 and Table 6-3 (Kennedy, 1969), Figure 10-37 and Table 10-5 (Fischer, 1973).

BHR Group Limited, The Fluid Engineering Centre, Cranfield, Bedfordshire MK43 0AJ, UK, Figures 12-3, 12-29 (Bruhl and Kazanskij, 1976), 17-21 (Sasic and Marjamovic, 1978), 17-22 (Duckworth and Argyros, 1972), 17-25 (Novak and Nalluri, 1974), 17-28, 17-29 (Bantin and Street, 1972), 17-37, 17-38 (Wiedenroth and Kirchner, 1972), 17-39 (Weber and Godde, 1976), 17-41 (Kazanskij, 1980), 17-45 (Hisamitsu, Shoji and Kosugi, 1978), 17-47 (Kao and Hwang, 1979), 17-52, 17-53 (Heywood and Richardson, 1978), 17-55 (Kazanskij, Bruhl and Hinsch, 1974), 17-56 (Wilson, 1976).

Blackwell Science Ltd, Osney Mead, Oxford, OX2 0EL, UK,   Figures 15-13, 15-26 (Wilson, 1972).

Professor F. Bo Peterson, Intitute of Hydrodynamics and Hydraulic Engineering, Technical University of Denmark, Building 115, DK-2800 Lyngby, Denmark,   Figures 14-10 (1980), 14-35 (1980).

Cambridge University Press, 40 West 20th Street, New York, NY 10011-4211, USA,   Figure 14-37 (Turner,1973).

Cambridge University Press, The Edinburgh Building, CB2 2RU, Cambridge,UK,   Figures 4-7, 4-22 (Grass, 1971), 4-13 (Offen and Kline, 1975), 6-21, 6-24 (Kennedy, 1963), 6-26, 6-27 (Engelund, 1970), 15-6 (Owen, 1964).

Elsevier Science - NL, Sara Burgehartstraat 25, 1055 KV Amsterdam, The Netherlands,   Figure 6-10 (Coleman, 1969).

Professor F. Engelund, Intitute of Hydrodynamics and Hydraulic Engineering, Technical University of Denmark, Building 115, DK-2800 Lyngby, Denmark,   Figures 6-28 (1974), 6-29 (1974), 6-45 (1969).

Professors V. Fidleris and R.L. Whitemore, Department of Mining and Fuels, University of Notinggham, LE12 5RD, UK,   Figure 3-23 (1961)

Professor W.H. Graf, LRH/DGC/EPFL, CH-1015 Lausanne, Switzerland,   Figure 8-30 (1971).

Institution of Chemical Engineers, Davis Building, 165-189 Railway Terrace, Rugby CV21 3HQ,UK,   Figures 17-2 (Newitt, Richardson, Abbott and Turtle, 1955), 17-12 (Newitt, Richardson and Shook, 1962), 17-35, 17-36 (Sinclair,1962).

The Institution of Engineers, 8 Gokhale Road, Calcutta-700 020, India,   Figure 4-24 (Rao, Govinda and Swamy, 1964).

Intellectual Property Counsel, MIT, NE25-230, Five Cambridge Center, Kendall Square, Cambridge, MA 02142-1493, USA,   Figures 12-4, 12-5 (Elata and Ippen, 1961), 12-27 (Montes and Ippen, 1973), 12-30(Daily and Chu, 1961), 12-31 (Roberts, Kennedy and Ippen, 1967).

International Research and Training Centre on Erosion and Sedimentation (IRTCES), PO Box 366, Beijing, China,   Figure 11-14 (Ackers and White, 1980).

John Wiley & Sons, Inc., 605 Third Avenue, New York, NY10158-0012, USA,   Figures 3-4, 4-4 (Rouse, 1946).

La Houille Blanche, 48, rue de la Procession, 75724 Paris Cedex 15, France,   Figures 2-19, 3-29, 3-30 (Migniot, 1968), 8-23 (Migniot, 1968), 8-24, 16-25 (Migniot, 1977), 16-52 (Goddet, 1960).

Professor B.N.Lin, IRTCES, PO Box 366, Beijing 100044, China,   Figure 12-19 (1955).

The MIT Press, 55 Hayward Street, Cambridge, Massachusetts 02142-1399, USA,   Figure 15-20 (Mabbutt, 1977).

Oxford University Press, Great Clarendon Street, Oxford OX2 6DP, UK,   Figure 14-7 (Lock, 1951).

Princeton University Press, 41 William Street, Princeton, New Jersey 08540-5237, USA,   Figure 4-6 (Bakhmeteff, 1936).

Professor A.J.Raudkivi, 7 Coates Road, Howick, Auckland 1705, New Zealand,   Figure 6-17 (1976).

The Royal Society, 6 Carlton House Terrace, London, SW1Y 5A9, UK,   Figures 7-12 (Bagnold, 1956), 8-10 (White, 1940), 15-2 (Sheppard, 1947), 15-10 (Bagnold, 1956), 16-20 (Longuet-Higgins, 1953).

Professor H.W. Shen, Department of Civil Engineering, University of California at Berkeley, CA 94720-1712, USA,   Figures 2-25, 8-11 (Sedimentation, 1972), 8-1, 8-13, 8-27 (River Mechanics, 1971), 6-41, 6-42, 6-43, 6-44 (Stochastic Approaches to Water Resources, 1976).

Soil Science Society of America, 677 South Segoe Road, Madison WI 53711, USA,   Figure 5-7 (Chepil, 1961).

Professor V.L. Streeter, 1035 Heatherway, Ann Arbor, MI 48104, USA,   Figure 14-32 (1961).

Professor B.M. Sumer, Intitute of Hydrodynamics and Hydraulic Engineering, Technical University of Denmark, Building 115, DK-2800 Lyngby, Denmark,   Figure 10-1 (1979), Table 10-1 (1979).

Professor A. Sundborg, Institute of Hydraulics, Royal Institute of Technology, Stockhom, Sweden,   Figure 6-8 (1956).

Teknisk Forlag a/s, DK-1780 Kφbenhavn V, Telefon 31 21 68 01, Denmark,   Figures 7-33, 11-9 and Table 6-1 (Engelund and Hansen,1972)

Thomas Telford Services Limited, Thomas Telford House, 1 Heron Quay, London E14 4JD, UK, Figure 16-29 (Bagnold, 1940)

The University of Chicago Press, 5801 Ellis Avenue, Chicago, Illinois 60637, USA,   Figures 3-16 (Komar and Reimers, 1978), 15-17 (Sharp, 1963), Table 2-7 (Graton and Fraser, 1935).

Professor K.C. Wilson, Department of Civil Engineering, Queen扭 University, Kingston, Ontario, K7L 3N6, Canada,   Figure 17-27 (1976).

Professor M. S. Yalin, Department of Civil Engineering, Queen扭 University, Kingston, Ontario, K7L 3N6, Canada,   Figures 6-31, 6-32, 11-2 (1972).

Cover photo provided by Mr. Yin, Hexian.

TRANSLATORS

Chapter	Translator
Preface	Yuqian Long
1	Yuqian Long
2	Lianzhen Ding
3	Zhaohui Wan
4	Zhaohui Wan
5	Zhaoyin Wang
6	Siow-Yong Lim
7	Xiaoqing Yang
8	Ren Zhang
9	Zhaoyin Wang
10	Jinren Ni
11	Lianzhen Ding
12	Zhaohui Wan
13	Zhaohui Wan
14	Renshou Fu
15	Lianzhen Ding
16	Zhide Zhou
17	Zhaohui Wan
Remarks	Yuqian Long

PREFACE

Mechanics of sediment transport is a branch of basic technical science in which the processes of erosion, transportation, and deposition of sediment particles take place under action of gravity, flowing water, wave, and wind. It is an independent discipline of science. Although a number of specialized writings have already been published, this book is the first attempt to unify in detail the movement of sediment under a variety of dynamic actions and boundary conditions. In the late 1940's, I started to collect relevant material abroad and prepared lecture notes from them; I used them in lectures at universities, research institutes, and engineering departments after returning to China. The manuscript had been revised several times on the basis of comments from the audiences and students, and also by the addition of new research results on developments in this branch of learning. By the early 1960's, the first 16 chapters had been completed. The first 12 chapters were distributed in a mimeographed manuscript entitled "Basic Law of Sediment Transport," and it has been reproduced several times by different institutions. In the latter part of the 1970's, I revised the original manuscript thoroughly and amended it by adopting new achievements in this field of learning both at home and abroad; I also expanded the volume into the present book by adding five more chapters. Dr. Zhaohui Wan assisted in the writing of Chapters 2, 3, 4 and, 12.

The writing of this book was begun while I was still at an age in the prime of life and ended at the age with greying temples. The manuscript was finally sent for publication only after many trials and vicissitudes of life over a span of 30-odd years. I devote this book to the people who are working assiduously in the scientific and engineering field of sedimentation in the construction of modernization of our motherland. I will be greatly rewarded if this book is of some help in their work.

It was under the guidance of Prof. Hans A. Einstein that I started to work in the field of sediment science. I worked with him for seven years with sincere and deeply friendly feelings. His thorough and inspiring instruction is still lingering in my ears. It was deeply regretted that he passed away just before his planned visit to China so that he was unable to see for himself the flourishing development of sediment science in China. Allow me to express a few words here both my grief and fond memories of my dearest Professor.

The completion of this book cannot be separated from the great support and assistance of my wife Wei-yao over the past several decades, especially the difficult times we together had gone through. I wish to express my sincere gratitude for the valuable comments on the manuscript from Professors Shunong Zhang, Jiahua Fan, and Guoxiang Hua. I wish also to express my gratitude for the careful proof reading by Meiqing Yang and Baoyu Chen, and the extensive work in the drafting of the figures by Tianjin Jiang.

Ning Chien

POSTSCRIPT

The book, *Mechanics of Sediment Transport*, was translated from the Chinese version, originally published by China Science Press in 1983. Soon after publication, it was awarded a national first class prize as one of the top scientific and technical books.

We are grateful for the encouragement and support received from many friends at home and abroad for the translation of the book.

Acknowledgment should go to the **Sediment Research Laboratory** of **Tsinghua University**, **International Research and Training Center on Erosion and Sedimentation, Department of Sedimentation Research** of **China Institute of Water Resources and Hydro-Power Research** and **China Talent Fund** for their sponsoring and financing the translation.

We wish to express our profound gratitude to Professor John Stephenson McNown, a former Kansas University Engineering College Dean, for his great contribution to the work. He came a long way to China and made every effort to help us. He thoroughly revised and polished the English manuscript. Being an authority in sedimentation himself, he also checked the technical accuracy of the translations. Unfortunately, Professor McNown passed away on February 17, 1998. It is much regrettable that he cannot see the English edition published.

Our heartfelt thanks are also to the 10 translators, all being our friends and experts in sediment engineering, for their enthusiastically taking part in the translation.

We should also extend our thanks to many colleagues and students of the three institutes mentioned above for their enormous work of editing the manuscript into camera-ready form. Among them, Lichun Zhang, Zhaosong Qu, Danxun Li, Dianchang Wang, and Huimei Li should be specially mentioned.

Finally, our special thanks are due to Professor Bingnan Lin, a long-time friend of Professor Ning Chien, for writing a foreword to the book and also for the instructions he gave us all the time during the course of the translation.

Weiyao Gong (Mrs. Ning Chien)

Zhaohui Wan

China Institute of Water Resources
and Hydro-Power Research,
P.O. Box 366
Beijing, China

CONTENTS

CHAPTER 1

INTRODUCTION

Mechanics of sediment transport is the study of the laws of sediment movement in fluids and of the processes of erosion, transportation, and deposition. Various types of sediment movement occur in nature and are encountered in engineering practice; these include sediment movement in rivers and canals, in reservoirs, along the seashore, in deserts, and in pipelines, and they take place as the result of stream flow, wind, and waves. This book is the first attempt to unify and to present these topics systematically. Hence it signifies the growth and development of a new discipline in science, and reflects also the requirements of those engaged in practice.

1.1 IMPORTANCE OF SEDIMENTATION PROBLEMS IN PRACTICE

Parts of the territory of China are overlaid with loess; these include the southern part of northeast China and the southeast part of northwest China. The loess widely spreads over the Yellow River basin extending from east of Qinghai Province in the west, relics of the Great Wall in the north, Qinling Mountain in the south, and the coastal region in the east. The loess is quite uniform in its textural lack of granular structure, and it is bound together mainly by calcium sulphate that is highly soluble and apt to be leached and eroded by rainfall. In addition, with a porosity ratio as high as 40 %, loess is characterized by well-developed vertical joints that are susceptible to erosion and weathering. Over history, most of the vegetation in the basin has been destroyed and the erosion process aggravated. Since the founding of the People's Republic of China, the severe soil erosion has not been brought entirely under control over the basin, although great efforts have been devoted to soil conservation. According to preliminary statistics, annual soil loss in the middle Yellow River basin amounts to 3700 t/km^2, on average, which is about 27.5 times the average annual rate in the world, 134 t/km^2. Enormous amounts of sediment are eroded from the basin and flow through mountain creeks and streams to the river, and they produce a sediment concentration that is higher than in almost any other part of the world. The water runoff and sediment loads of some of the world's major rivers are listed in Table 1.1 [1].

Statistics show that 13 of the large rivers in the world carry annual sediment loads of over 5.8 billion tons. Among these, the Yellow River plays a leading role insofar as total load and average sediment concentration are concerned. Next comes the Bramaputra River of Bangladesh, which has an annual sediment load of 499 million tons, even though its average sediment concentration is only 0.768 kg/m^3. The Indus River in Pakistan ranks third with an annual sediment load of 435 million tons and an average sediment concentration 2.49 kg/m^3. About 30% of the sediment load carried by the 13 rivers mentioned above comes from the Yellow River and the Yangtze River. Actually, sediment concentrations of some of the tributaries in the middle

1

Yellow River basin are still higher than that of the main stem. For instance, the average annual sediment concentration of the Zuli River, a tributary of the Yellow River in the Gansu Province, is close to 600 kg/m³. In several tributaries of the Yellow River, the maximum sediment concentrations have been about 1,600 kg/m³. That amount indicates that about 60% of the volume of the water body is occupied by the sediment. An old saying tells us that several deciliters of mud are contained in a hectoliter of water in the Jinhe River.

Table 1.1 Comparison of annual runoff and sediment load
for the major rivers in the world

In China

Drainage basin	River	Drainage area	Length	Station	An. Mean runoff	An. mean sediment	Av. conc.	Max. conc.	Modulus
		1000km²	km		10⁸m³	10⁸ t	kg/m³	kg/m³	t/ km²/yr
Yellow	Yellow	752.4	5464	Samenxia	432	16.4	37.6	911	2480
Yangtze	Yangtze	1807.2	6300	Datong	9211	4.78	0.54	3.24	280
Haihe	Yunding	50.8	650	Guanting	14	0.81	60.8	436	1944
Huaihe	Huaihe	261.5	1000	Banpu	261	0.14	0.46	11.0	153
Liaohe	Liaohe	166.3	1404	Tieling	56	0.41	6.86	46.6	240
	Dalinhe	23.2	360	Dalinhe	21	0.36	21.9	142	1490
Pearl	Xijiang	355.0	2055	Wuzhou	2526	0.69	0.35	4.08	260

In other countries

Country	River	Drainage area	An. mean runoff	An. mean sediment	Average concentration
		1000 km²	10⁹m³	10⁹ t	kg/m³
USA	Colorado	637	4.9	0.135	27.5
India, Bangladesh	Ganges	955	344.0	0.196	0.57
USA	Missouri	1370	616.0	0.218	3.54
Pakistan	Indus	969	175.0	0.435	2.49
Bangladesh, India	Bramaputra	560	650.0	0.499	0.768
Egypt, Sudan	Nile	2978	89.2	0.111	1.25
Vietnam	Red	119	123.0	0.130	1.06
Burma	Irrawady	430	427.0	0.299	0.70

A river that is heavily laden with sediment has peculiarities that cause it to differ extensively from rivers that carry much less sediment. These differences introduce the following problems in engineering practice:

1. Flood control

Floods in the rivers of North China are formed by heavy rains that are characterized by both high peak flows and large volumes; the discharge varies dramatically during both the ascending and descending periods. Vast areas of land adjacent to the river would be inundated if the dike were breached. Flood hazards are strongly affected by the excessive deposition along the river course of the enormous amount of the sediment the river carries. On the one hand, flood conveyance capacity can be reduced as a result of aggradation of the river bed; hence, the occurrence of an extraordinary flood would put the dikes in great danger of being overtopped or breached. On the other hand, the main flow path meanders over the river bed in the broad and shallow reaches. Once the main current impinges on the main levee, the floodwater could break through the dike. For example, in some 2,500 years (from 602 BC~1911), the dikes of the lower Yellow River were breached 1593 times; during some of these incidents, 26 large avulsions took place. In period 420~1911, dike breachings were recorded almost every year. The area that suffered inundation and damages extended over the vast North China Plain from Tianjin in the north to the Huaihe River basin in the south.

With the founding of the People's Republic of China in 1949, the outlook for the Yellow River began to change. No breachings have taken place during the summer and autumn floods in the past four decades. Miserable situations, in which vast expanses of fertile land were inundated and houses, human beings, and animals were totally washed away, no longer took place. However, the river is still in a state of aggradation. In the present course, the bottom has raised so that it is much higher than the adjacent ground; as a consequence of this sedimentation, the river, for conveyance of the sediment-laden flow over this long period, has become a divide in the broad North China Plain. Since even an ant hole might cause the collapse of a dike extending more than a thousand kilometers, the ensuring of safety against the hazard of flooding is still an urgent need and a matter of great concern that must continue to have a high priority.

In the vigorous development of water resources in the Yellow River basin, those in areas of relatively clear water were first exploited, but the sediment yield from the area subject to severe soil erosion either remains largely unchanged or has not been greatly reduced. As a result, sediment concentration of floods entering the lower Yellow River has become higher than ever. In August 1977, a flood with a hyperconcentration of sediment took place in the lower Yellow River. The maximum concentration observed at Xiaolangdi station amounted to 898 kg/m^3. As the hyperconcentrated flood propagated farther downstream, it overflowed the floodplain; the floodwater could then no longer maintain its flow and stagnated, due to the shallow depth and low velocity. In the lower reaches, the temporal retention of a part of the floodwater on the floodplain upstream was equivalent to a reduction in flood volume. As a consequence, for the downstream reaches stretching nearly 100 km above Huayuankou station, the stages along the river course suddenly dropped by 0.7-1.2 m during the ascending period of the flood. Later on, as the flood discharge continued to increase, the depth over the floodplain increased accordingly, leading to an increase of the boundary shear stress and the stagnated slurries resumed their movement; this process was equivalent to an addition of floodwater to the existing flow, and it induced a sudden rise in stage along the river. The stage at Jiabu, one of the gauging stations, rose by 2.84 m in one and one-half hours. The peak discharge of the flood at Huayuankou was 700 m^3/s greater than that observed at the upstream stations, even though no additional floodwater had entered the intermediate drainage area. The abnormal variations of flood stages during hyperconcentrated floods introduce a series of new problems in flood prevention.

2. Reservoir sedimentation

Reservoirs built in the upper and middle parts of a river basin can be used for multiple purposes, including flood control, irrigation, and power generation. However, along with the impoundment of water in the reservoir, sediment transported by water is also retained. If this process were allowed to continue, the reservoir capacity would gradually reduce and eventually be lost. In addition, deposition in the reservoir may extend sometimes to a large distance upstream from the original backwater deposits occurred in the initial period of impoundment. More and more of the valley upstream would thus be subjected to inundation and salinization.

The seriousness of capacity loss introduced by reservoir sedimentation was convinced by people only in the recent three or four decades when the finding of suitable dam-sites for exploitation of water resources became increasingly difficult. In the United States, the total annual amount of deposition in reservoirs had reached 1.2 billion tons. In Japan, up to 1979, from statistics on 425 reservoirs with a combined capacity exceeding 1 million m^3, 6.3% of the reservoir capacity had been lost due to deposition. In India, according to statistics presented in 1969, the annual rate of loss of reservoir capacity was 0.5 to 1.0% for 21 reservoirs with a combined capacity greater than 1.1 billion m^3 [2] Disturbed by the alarming rate of capacity loss, people began to consider the feasibility of replacing the process of sediment retention in large reservoirs constructed on the main stem or major tributaries of a river by introducing engineering measures like soil conservation and the construction of medium- or small-sized reservoirs in the upland areas of the basin. [3]

These problems are more acute and complex for the more heavily sediment-laden rivers. According to the preliminary statistics from Shaanxi Province, the amount of deposition in reservoirs exceeding 1 million m^3 in capacity had been 512 million m^3, thus constituting 15.3 % of the original capacity. In recent years, the capacity of newly built reservoirs, each exceeding 1 million m^3 in capacity, was increased by 260 million m^3 annually; however, more than 80 million m^3 of that annual increased capacity was lost due to deposition in the reservoirs; that is, about one-third of the increase in capacity was lost each year [4] Observations of 20 key reservoirs, organized directly by the Ministry of Water Resources and shown in Table 1.2, most of which had been operated less than 20 years, show that the total amount of deposition was 7.8 billion m^3, or 18.6% of their original capacity.

In the initial stage of planning of the Sanmenxia Project, due to lack of experience and the inadequate attention paid to the sedimentation problems, the target for multi-purpose development was set much too high. For instance, the maximum outlet discharge was scheduled to drop to 6,000 m^3/s from a design flood of 32,500 m^3/s with a recurrence period of 1,000 years, for purpose of flood control. Eight units of generators with a total capacity of 1,160 mw were to be installed. Impoundment in the reservoir was planned to irrigate an area of 4.14 million hectares in the lower reaches and to maintain a water depth of no less than one meter in the lower reaches for navigation. To attain these targets, the normal pool level proposed in the preliminary study was set at an elevation of 360 m, with a corresponding capacity of 64.7 billion m^3 and an inundated area of 3,500 km^2 from which 870,000 inhabitants had to be resettled. It was decided later to construct the dam in steps. The first step was to construct the dam up to an elevation of 350 m and to operate the reservoir below elevation 340 m. Closure of the dam was accomplished in 1958 and the construction work was essentially completed by September 1960.

Table 1.2 Deposition in Some Reservoirs in China

No.	Reservoir	River	Drainage area	Dam height	Design capacity	Periods of statistics	Total deposition	Dep./capa.
			1000 km^2	m	$10^8 m^3$		$10^8 m^3$	%
1	Liujiaxia	Yellow	181.7	147	57.2	1968~78	5.8	10.1
2	Yangouxia	Yellow	182.8	57	2.2	1961~78	1.6	72.7
3	Bapanxia	Yellow	204.7	43	0.49	1975~77	0.18	35.7
4	Qintongxia	Yellow	285.0	42.7	6.20	1966~77	4.85	78.2
5	Sanshengong	Yellow	314.0	gate	0.80	1961~77	0.40	50
6	Tianqiao	Yellow	388.0	42	0.68	1976~78	0.075	11
7	Sanmenxia	Yellow	688.4	106	96.4	1960~78	37.6	39
8	Bajiazui	Puhe	3.52	74	5.25	1960~78	1.94	37
9	Fengjiashan	Qianhe	3.23	73	3.89	1974~78	0.23	5.9
10	Hesonling	Yeyuhe	0.37	45.5	0.086	1961~77	0.034	39
11	Fenhe	Fenhe	5.27	60	7.00	1959~77	2.60	37.1
12	Guanting	Yongding	47.6	45	22.7	1953~77	5.52	24.3
13	Hongshan	Xiliaohe	24.5	31	25.6	1960~77	4.75	18.5
14	Laodehai	Liuhe	4.50	41.5	1.96	1942~	0.38	19.5
15	Yeyuan	Mihe	0.79	23.7	1.68	1959~72	0.12	7.2
16	Gangnan	Hutuohe	15.9	63	15.58	1960~76	2.35	15.1
17	Gongzui	Daduhe	76.4	88	3.51	1967~78	1.33	38
18	Bikou	Beilong	27.6	101	5.21	1976~78	0.28	5.4
19	Danjiankou	Hanjiang	95.2	110	160.5	1968~74	6.25	3.9
20	Xingqiao	Hongliuhe	1.33	47	2.00	14 yrs	1.56	78

During the initial period of impoundment, deposition in the reservoir was so serious that it amounted to nearly 1.5 billion tons by March 1962, retaining 93% of the oncoming sediment load. By 1964, the total amount of deposition had reached 4.4 billion tons, and the backwater deposits had extended upstream to the tributary arm of Weihe River, thus endangering both the agricultural and industrial development in the vicinity of the city of Xian and the vast alluvial plain on both banks of the Weihe River. Afterwards, the mode of operation had to be changed from impoundment on an annual basis to flood detention only. Reconstruction was carried to enlarge the outlet discharge capacity, and the turbine-generator unit already installed had to be dismantled. Not only were the indexes of benefits greatly reduced, but the construction period was also extended by more than ten years, with the additional construction to be carried out in two phases. This experience was indeed a lesson not to be soon forgotten.

3. Sedimentation problems in irrigation canal systems

One of the major measures introduced to ensure an increase in agricultural production in China was to enlarge the area under irrigation. In North China where the climate is relatively arid, the demand for irrigation is urgent. However, sediment concentration is in general quite high in most rivers located in this area. Thus, sediment would be withdrawn together with the water diverted for irrigation. The great North China Plain is the alluvial plain formed by the Yellow River. Siltation in the canal system is a serious problem because the sediment transport capacity is limited by the small gradient inherent in the flatness of the ground surface. In order to reduce the deposition in the irrigation area, scientists and engineers involved in water conservancy works all over the world have studied how to reduce the amount of sediment entering the canal system, and they have achieved some promising results. In the layout of intakes, principle of circulation currents is followed to minimize the coarse sediment from entering into the canal system. Also, desilting basins are widely used. In Chinese rivers heavily laden with sediments, however, due to the extreme fineness of the sediment and relative uniformity in its vertical distribution, the difference in the amount of sediment extracted with water from the top layer and from the bottom layer is not large. Also, density currents in the form of turbidity underflows may take place in the settling basin, leading to a reduction in the effectiveness of desilting. These are special problems that occur in engineering practice. On the other hand, over a long period of time, Chinese people have adopted a traditional method of diverting sediment-laden floodwater for warping and for using the sediment as fertilizer along some mountain streams or tributaries. The fruitful experiences inherent in such a practice remain to be analyzed.

4. Sedimentation problems in harbors and estuaries

Sediment rapidly settles out where a river empties into the sea. Sediments deposited on shoals or beaches are later agitated and re-suspended by wind-induced waves. Under the actions of tidal and littoral currents, the sediment may be transported offshore initially and later deposited in harbors and estuaries, thus jeopardizing navigation and drainage. The magnificent Chinese sea coast composed of silty deposits was originally formed and developed by the deposition of enormous amounts of fine particles originated from the loess area. Silty materials brought into suspension by wave action can move in density currents even with no external force.

Evolution of the estuarine process along a silty coast differ in many respects from that of a sand or gravel coast.

In summary, on account of high sediment content and fineness of particles in many rivers in North China, a series of difficult confrontations has been encountered in the exploitation of water resources and the construction of water conservancy measures. It is imperative to conduct basic research on the behavior of sediment movement over time. Moreover, sedimentation problems are the concern of many departments and branches of government, not limited to those dealing with water resources.

Vast desert areas are scattered over northwest China. Movement of eolian sand endangers both agricultural production and railway communication. Stabilization of moving sand by afforestation, interception of wind-blown sediment by sediment barriers present major challenges to production and construction in these areas. The mechanism of the movement of the wind-blown sediment is quite similar to that of sediment transport in stream flow. Movement of eolian sand occurs primarily in the form of bed load; it is both relatively simple to deal with and is less difficult to observe and to measure.

Since the 1950s, the transportation of granular solids in pipelines has become more and more wide spread. Slurries to be transported include sand and gravel, coal, paper pulp, syrup, ore products, and various chemical raw materials. The distance of transportation was once limited to construction sites and manufacturing plants. Today, however, coal powder is conveyed by pipeline over long distances directly from the mining area to the steam plant. In recent years, for the sake of savings on transportation costs, fine particles such as clay or high polymers have been added to modify the viscosity of slurries, thus changing the conveyance of the granular material to a state of hyperconcentrated flow [5]. The two-phase heterogeneous flow in pipes is a form of sediment movement with special boundary conditions, and it moves in a way that is basically similar to the movement of fluvial sediment.

Techniques of fluidization, rapidly developed in recent years, are used to bring solid granular material into a state of suspension by vertical currents of water or air so that mixing or heat transfer can be carried out. For instance, raw materials of ore can be reduced to metal in a fluidized bed process without a blast furnace, and cereal or other plant seeds can be dried in this way. The fluidization technique, in combination with pipe transport of granular material, plays an important role in petroleum and chemical engineering, in cereal processing industries, and in the production of atomic energy. Obviously, movement of granular material within a fluidized bed has some similarities with the suspension and settling of sediment in rivers.

Environmental pollution is a critical problem confronting particularly all countries. The water quality can decline because of the disposal of waste water released from industrial plants or agricultural fields. The pollution problem could be further complicated by fluvial sediment that is polluted by the waste water. The contamination can exist a long time and accumulate in nature and thus become a long-term source of pollution. As a result, the diffusion and dispersion process of sediment under different conditions is an important topic in environmental science.

In addition, some understanding of the fundamentals of the mechanics of sediment transport is indispensable to the development of some disciplines of science, even though sediment movement is not normally their major theme of study. For example, the geomorphologic evolution of the earth surface under endogenic and exogenic forces is the major theme of study in geomorphology. In it, the exogenic dynamic process reflects, in fact, exactly the erosion, transportation, and deposition process of the material of the earth surface under the action of gravity, running water, wind, wave action, and glacier movement. The inherent mechanism of formation and development of the morphologic features of the earth surface could not be well formulated if the movement of materials under different dynamic actions were not thoroughly studied. In a book entitled *Theoretical Geomorphology* published in 1960s[6], extended coverage was devoted to the discussion of problems related to sediment movement.

Another excellent example is the book entitled *The Physics of Blown Sand and Desert Dunes*. The book was authored by R.A. Bagnold in the early 1940s, in which the growth and development process of ground configuration under wind action was clearly explained in terms of its dynamics[7]. In the study of sedimentology and paleogeography, the paleogeographical environment may be demonstrated by analysis of the characteristics of grain size and its distribution, shape, roundness, orientation, structure and texture of the continental deposits[8]. The study should have played an important role in the search for petroleum and gas resources as well as in the study of paleobiology in earth strata. Study of the sorting process of sediment grains under various dynamic actions will no doubt be helpful in obtaining a deeper understanding of the meaning of formation of the various deposits. For instance, for a long time, people believed that the interlaced layers of coarse and fine materials in deposits were the result of actions under different dynamic intensities. However, it was proved in a flume experiment conducted by Einstein and the senior author that deposits interlaced with coarse and fine materials could be formed even in a state of steady flow for highly intensive deposition with large ranges of grain sizes [9] Since then, other phenomena of sedimentology have been studied more extensively in laboratories [10-12] Additional experimental studies of the deposition process under various complicated dynamic conditions will be conducted along with the development of the techniques of physical modeling and the use of light-weight modeling materials. Possibly, a new frontier and an independent discipline of science--Dynamic Sedimentology--may evolve in this way. The mechanics of sediment transport also have close links with many branches of the science of geologic geography. The authors firmly believe that the study of the law of movement of heavy minerals would promote the development of mineralogy. The mechanics of sediment transport should be part of a basic course of study for people engaged in the field of geology or geography. For this reason, the authors wish to contribute to this process by writing this book.

Development of modern scientific research indicates that a frontier discipline of science may occasionally arise from several existing disciplines; or a new discipline in basic technical science observing common objective laws may be formulated. Study in these frontier disciplines of science, or in these technical sciences, would, in turn, promote the development of various disciplines of science that already exist. The mechanics of sediment transport is such a technically oriented science, one that is in the process of growth and development. It is still an immature one, but it is full of vigor and has bright prospects for development.

1.2 PRESENT STATUS AND THE NATURE OF THE DISCIPLINE OF SCIENCE

The authors assume, perhaps prematurely, that the mechanics of sediment transport should be a component in the science of sediment with its extensive scope of study. This component should involve primarily the following four aspects:

1. Formation of sediment and its properties--to study the function of weathering and its products; characteristics of sediment grains and sediment groups;

2. Mechanics of sediment transport--to study the law of sediment movement in the processes of erosion, transportation, and deposition;

3. Field measurements and laboratory experimentation—to study the methodology and instrumentation for taking measurements and statistical samples of deposits; laboratory analysis of sediment properties and the ways of presenting results; techniques of experimentation in flumes; wind tunnel experiments; physical modeling;

4. Applied science of sedimentation--application of knowledge of sediment science in the three foregoing respects so as to solve sedimentation problems encountered in practice.

Up to now, no book has been published that covers comprehensively all of these four aspects of sediment science. In the book entitled *An Introduction to Movement of Sediment* authored by Sha Yuqing in 1965 in China, mechanics of sediment movement is the major topic presented, and it includes also problems related to experimentation with physical models and the design of stable canals. However, it is restricted to sediment movement in open channel flow [13]. Nearly half of the volume in a book entitled *River Dynamics*, written by the staff of the Wuhan Institute of Hydraulic and Electrical Engineering (WIHEE) in 1961, discusses the law of sediment movement under the action of flowing water, including density currents. [14] No comprehensive introduction or comments were made in either of these two books concerning the huge volume of literature available.

The senior author and Fan Jiahua made a thorough and systematic review of the literature on density currents that was published prior to 1958 [15], but quite a few research results have been published since then. In the former USSR., many monographs or books in the field of river dynamics and ocean dynamics have been published in which the mechanism of sediment movement in running water or with wave action were treated [16-20]. In western countries, the publication of numerous monographs and writings related to mechanics of sediment movement have been presented, but only in the most recent decades, except for the afore mentioned book on movement of eolian sediment written by Bagnold. In 1971, three volumes of *River Mechanics*, edited by S.W. Shen, were published, in which many papers were devoted to discussions of sediment movement [21]. A systematic introduction of sediment transport in almost all its aspects was presented by W. Graf in his book entitled *Hydraulics of Sediment Transport* [22]. M.S. Yalin describes various aspects of the mechanics of sediment transport on the basis of dimensional analysis (first edition in 1972, second in 1977) [23]. Early in the 1950s, the American Society of Civil Engineers edited *Sedimentation Engineering*, in which all of the chapters were authored by knowledgeable experts on aspects of sedimentation engineering. Each chapter was presented first in the preliminary proceedings and then amended on the basis of comments received. The book was finally edited and published in 1979 [24]. J. Bogardi presented many of the results of research and experimentation in European countries in his book *Sediment Transport in Alluvial Rivers* [25] (published in Hungarian in 1955 and later revised and published in English in 1974). The book, *Sediment Transport Technology*, co-authored by D.B. Simons and F. Senturk, presented many examples of calculations that help to provide beginners with an understanding of applications of the law of sediment transport [26].

All these writings have doubtless made valuable contributions and provided reference material for professionals, but they are still not quite complete in regard to the degree of integrated approach. For example, the book *Sedimentation Engineering* included almost all the four aspects of sediment science and may reflect the level of achievements in the western countries, but it was too limited in size to contain the material in detail. In the part dealing with the mechanics of sediment movement, the movement of suspended sediment was treated in a separate section, but the treatment of the related bed load movement was brief; also it was not integrated into a presentation of a theoretical system. In addition to the publications in the field of mechanics of sediment transport in streams, quite a few additional writings have been published in recent years regarding sediment movement in coastal zones [27-31] and transport of granular materials in pipelines [32-34].

Mechanics of sediment transport is developing into a new discipline of science. Even more important are the many papers in periodicals and proceedings, in addition to the publication of the afore mentioned monographs and books. Publication of these papers is scattered widely in engineering journals and in periodicals of mathematics and physics. In addition, a number of papers have appeared in publications of geologic geography, rheology, and chemistry. An urgent need exists to assemble and compare the disparate research results and to present them comprehensively and systematically.

Furthermore, although the amount of literature on the mechanics of sediment transport that has appeared is as vast as an open sea, its content is often rather immature. This may be attributed to the following two reasons:

In the first place, the phenomenon of sediment movement is quite complicated. In general, the movement of sediment is a topic in two-phase flow. Sediment moves under the action of a flow, and its presence influences the flow in turn. The two are mutually affected and interactive, and together they possess properties different from ordinary water flow. Sediment concentrations in rivers in other countries than China are generally not high, so that only the law of sediment movement under the action of flow is taken into consideration. Functions of the feedback of sediment upon the flow characteristics are either neglected or not even considered. It would naturally be impossible to provide a deep understanding of the mechanical properties of sediment-laden flows if the two parts of the complex process were not treated integrally. The effects of sediment on the flow in the sediment-laden rivers in China are quite pronounced. Large errors would be introduced if they were neglected in a simplified treatment. If the water contains a high concentration of fine particles (less than 0.01 mm in diameter), the physical and chemical properties on the surface of the granular material play an important role. The fluid is no longer even quasi-Newtonian. Studies of this type of fluid belong to the field of rheology. Besides, the movement of sediment is generally on movable beds, except for those in mountainous streams or in pipes; therefore, the boundary of the flow, composed of movable sedimentary particles, is deformable under the action of flow, and it influences the flow in turn. These form a dual feedback system. For flow on a rigid bed with a stable boundary, roughness is usually taken to be a constant; a condition that does not properly represent the flow on movable beds. Not only the flow structure but also the intensity and distribution of the sediment movement are closely related to the bed configuration. The latter can vary continuously with the intensity of flow, thus further complicating the problem.

Secondly, practical difficulties occur in taking direct measurements. The degree of turbidity can be high in sediment-laden flows, so that direct observation of sediment movement is difficult. In recent years, the trend toward using plastic granular materials in the flume experiment in lieu of natural sediment is an attempt to circumvent this difficulty. In the study of sediment movement, the most important region is in the vicinity of the bed; there, both the velocity and the concentration gradient are relatively large, and the potential energy is actively transformed into kinetic energy of turbulence. Exchanges between the bed material and the bed load and between the bed load and the suspended load also take place in this region. The thickness of this layer is not large, well under one tenth of the depth of flow; it is a small region that often cannot be reached by conventional instruments for measuring velocity or for taking samples. Near a movable bed, the boundary and flow conditions can be altered due to local disturbances created by the presence of instruments. Although the use of a hot wire anemometer to measure turbulence structure has already become conventional in wind tunnel experiments, the techniques of taking measurements in a sediment-laden flow have not yet been resolved. It is still more difficult to take measurements in natural rivers that are characterized by unsteady flow and three-dimensional properties.

Because of the extreme complications involved in the movement of sediment and difficulties in taking measurements, no breakthroughs have yet been made on some of the key problems, even though many scientists and engineers have made great efforts to resolve them. The understanding of these problems is still limited to the perceptual stage of cognition, a fact that is reflected in the subjective and often controversial ideas concerning the basic concepts and in the variety of formulae proposed for use in solving related problems. The sedimentation research at present should be continuously directed to solve problems that arise in practice, and, an effort should be made to acquire a deeper understanding and to clarify the inherent mechanism of sediment transport. The necessary steps are (a) acquiring more reliable experimental data and (b) analyzing these data thoroughly. In addition, efforts should be made to compare the existing formulae and to make a synthesis of them. By eliminating the false and retaining the true, and by selecting the essential and discarding the useless, some formulae or computational methods that are theoretically sound and have a more reliable basis in experiment should be obtainable for use in practice. The present state of uncertainty and confusion in the use of these many formulae and methods may then be overcome.

1.3 GUIDELINE FOR WRITING AND ORGANIZATION OF THIS BOOK

This book has been written to reflect the following guidelines:

1. Because the tremendous amount of literature related to the mechanics of sediment transport is scattered world-wide, it may not be generally accessible to the readers. Therefore, the authors have made an effort to collect those materials that are the most relevant to the study, and they have synthesized, appraised, and commented on them whenever possible in order to provide the readers with a better understanding of the present status of development in the discipline. Also, selected references out of the extensive literature are presented at the end of each chapter to guide anyone wishing to go more deeply into the subject.

2. Unification of the selected material on both history and logic was sought in the preparation of this book. The main lines of approach in the development of our understanding has to be made clear and also the logical system of theory should be described as well as possible. Also, different theoretical systems for the interpretation of the mechanics of sediment movement exist at present; if all these theoretic systems

were introduced indiscriminately, no line of thought could be followed all the way through this book. In that way, mixed and misleading concepts would be conveyed. Therefore, while a variety of research results were collected, the materials presented in this book were organized according to a consistent concept in which the most promising theoretical systems were included. From this point of view, some theories have not been fully introduced, nor does the presentation in this book reflect their content.

3. Mathematical developments were considerably simplified because an effort was made to satisfy the demands of both the engineering professionals and people engaged in the field of geologic geography. From the experience of the senior author in giving lectures based on this book, readers from the field of geologic geography, possessing as they do the fundamentals of mathematics and physics, can understand most of the contents of this book. Furthermore, since the deeper aspects of the mathematical analyses in some research work had to be simplified or omitted, this book should be considered at this stage as the *Elementary Mechanics of Sediment Transport*.

4. To be helpful in providing an understanding of the fundamentals, simplified models were used to describe the rather complicated physics. In so doing, some aspects may have been over-emphasized and the others under-emphasized or distorted. To provide the reader with a better and clearer presentation of some topics, a number of graphs and tables with many experimental data are included because the original literature may not be readily accessible to all.

5. Attention has been paid to the fact that the theoretical structure is growing. A new idea may not yet be mature in itself and remains to be developed and studied, but it can still provide an avenue of approach for exploring new avenues of research. As research deepens our knowledge, the ideas may be improved and supplemented, or they may be deleted if found to be misleading.

6. Research results both in China and abroad have been introduced systematically. However, there are still many questions in these works, whether in the method of thought or in the degree of close combination with problems encountered in practice. The authors have worked to discard the useless and select the essential in presenting these research results, and also to make appropriate comments on the theories that have the greater relevance. However, limited by our ability, we did not do enough work of this kind. In applications of these research results, the reader should adopt an attitude of inheritance and criticism in conformity with practical conditions.

This book has 17 chapters. In Chapter 1, the background for the mechanics of sediment transport as a new discipline of science is explained. Chapter 2 contains the fundamental properties of sediment; its contents do not belong within the scope of a study of the mechanics of sediment transport. However, a brief explanation of the nomenclature and definitions is required at the start, as are the sediment properties that are used frequently in later chapters. Chapter 3 is devoted to the settling velocity, which is one of the important sediment properties because it is related to the processes of settling that occur in the two-phase medium.

Movement of sediment in running water is introduced progressively in Chapters 4 through 13. Flow structures in open channel flow with respect to turbulence and resistance are described in Chapter 4 and Chapter 7. Only after obtaining a thorough understanding of the mechanical properties of the flow is it possible to study the sediment movement caused by the flow. The topic to be dealt with is the sediment-laden flow on movable bed. It would be difficult to explain the properties of the flow as a whole if the movement of sediment were not considered in its entirety. For this purpose, fundamental concepts related to sediment movement and movement of sand waves are inserted and described in Chapters 5 and 6. In many books, movement of sand waves is treated as a part of bed load movement. From the point of view of the movement of sand waves considered to be a group movement of the bed load, this approach is reasonable. However, the sand wave is a major component in the resistance. It would be impossible to explore the inherent mechanism of the resistance in alluvial rivers if the development and decay of sand waves, as well as their properties, were not thoroughly explained initially. Hence, the movement of sand waves is introduced in the chapters before those on resistance. These two arrangements may have both advantages and disadvantages. In Chapters 8 to 10, incipient motion, bed load and suspended load movement are discussed. In Chapter 11, sediment transport capacity and its related problems are treated with the incorporation of the combined movement of bed load and suspended load. In the foregoing chapters, the structure of flow in clear water medium is considered. The effect of sediment on the flow is specially treated in Chapter 12. Since the state of flow is sometimes a hyperconcentrated flow, which is quite different from ordinary sediment-laden flow in a number of rivers in China, Chapter 13 is devoted to the discussion of the mechanism of the hyperconcentrated flow.

Chapters 14 through 17 form, somewhat independently, a set of treatises on the movement of sediment under different dynamic actions or boundary conditions; these include the density current, eolian sediment movement, sediment movement under wave action and transport in pipelines. The law of sediment movement for these conditions has some similarities with that in running water, but it also has some features peculiar to these special fields of study. The present status of the mechanics of sediment transport as a whole is reviewed and prospects for future development are put forward in the concluding remarks as a summary of this book.

This book is the first attempt to introduce systematically and integrally the mechanics of sediment transport that belong to the many different categories. Therefore, many aspects remain to be developed more fully. Limited by the present level of understanding, our approach is bound to contain errors. Criticism and comments will be cordially welcomed to help overcome the inevitable mistakes and omissions.

REFERENCES

[1] Chien, Ning and Dingzhong Dai. "River Sediment Problems and Status of Research in China," *Proceeding of International Symposium on River Sedimentation*, Vol. 1. Guanghua Press, 1981, pp. 3-49.

[2] Chien, Ning. "Reservoir Sedimentation and Slope Stability--Technical and Environmental Effects, " *General Report, Proceeding of 14th Congress of International Committee on Large Dams. Brazil*, 1982, pp. 639-690.

[3] Peterson, E.T. *Big Dam Foolishness, The Problem of Modern Flood Control and Water Storage*. The Devin-Adair Co., 1954, pp. 224.

[4] River and Canal Laboratory, Northwest Hydrotechnical Scientific Research Institute (NWHRI). "Study on the Planning and Design of Reservoirs Built on Sediment-laden Rivers." *Compilation of Research Reports on Reservoir Sedimentation Studies*, 1972, pp. 238-246

[5] Virk, P.S. "Drag Reduction Fundamentals," *Journal of American Institute of Chemical Engineers*, Vol. 21, No. 4, 1975, pp. 625-656.

[6] Scheidegger, A.E. *Theoretical Geomorphology*. Springer-Verlag, Berlin Göttinggen-Heidelberg, 1961, pp. 333.

[7] Bagnold, R.A. *The Physics of Blown sand and Desert Dunes*. Methuen and Co., Ltd., London, 1941, pp. 242.

[8] Chendu Institute of Geology. *Treatise on Sedimentary Phase and Paleo-sedimentology*. China Industrial Press, 1961, pp.142.

[9] Einstein, H. A., and Ning Chien. "Transport of Sediment Mixtures with Large Ranges of Grain Sizes," *M.R.D. Sediment Series No. 2*, Missouri River Div., U.S. Corps of Engrs., 1953, pp. 49.

[10] Mckee, E.D. "Flume Experiments on the Production of Stratification and Cross Stratification," *Journal of Sediment Petrology*, Vol. 27, No. 2, 1957, pp. 129-134.

[11] Jopling, A.V. "Interpreting the Concept of the Sedimentation Unit," *Journal of Sediment Petrology*, Vol. 34, No. 1, 1964, pp. 165-172.

[12] Middleton, G.V. *Primary Sediment Structures and their Hydrodynamic Interpretation*. Sp. Pub. No. 12, Soc. Economic Paleontologists and Mineralogists, 1965, pp. 265.

[13] Sha, Yuqing. *Introduction to Mechanics of Sediment Movement*. China Industrial Press, 1965, pp. 302.

[14] Wuhan Institute of Hydraulic and Electric Engineering. *River Dynamic*. China Industrial Press, 1961, pp. 288.

[15] Chien, Ning, Jiahua Fan et al. *Density Current*. Water Conservancy Press, 1958, pp. 215.

[16] Shamov, G.E. *Fluvial Sediment*. Hydro-meteorological Press (in Russian), 1954, pp. 378.

[17] Levi, E.E. *River Dynamics*. National Energy Press (in Russian), 1957.

[18] Goncharov, M.A. *River Dynamics*. Hydrology Press (in Russian), 1962.

[19] Velikanov, M.A. *Fluvial Process*. Hydro-meteorological Press (in Russian), 1958.

[20] Zenkovich, V.P. *Fundamentals of Coastal Fluvial Process*. Science Press (in Russian), 1962, pp. 710.

[21] Shen, H.W. *River Mechanics*. 3 Vols, Water Resource Pub., Fort Collins, Colo., U.S.A., 1971, pp. 1323.

[22] Graf, W. *Hydraulics of Sediment Transport*. McGraw Hill Book Co., 1971

[23] Yalin, M.S. *Mechanics of Sediment Transport*. Pergamon Press, Oxford, 1972 and 1977, pp. 290.

[24] Vanoni, V.A. *Sedimentation Engineering*. Manual No. 54, American Society of Civil Engineers, 1975, pp. 745.

[25] Bogardi, J. *Sediment Transport in Alluvial Rivers*. Akademiai Kiodo, Budapest, 1974, pp. 826.

[26] Simons, D.B., and F. Sentürk. *Sediment Transport Technology*. Water Resource Pub., Fort Collins, U.S.A., 1977, pp. 807.

[27] Ingle, J.L. *The Movement of Beach Sand*. Elsevier, 1966.

[28] Swift, D.J.P. and O.H. Pilkey. *Shelf Sediment Transport, Process and Pattern*. Dowden, Hutchinson and Ross, Inc., 1972, pp. 656.

[29] Hails, J. and A. Carr. *Nearshore Sediment Dynamics and Sedimentation.* John Wiley and Sons, 1975.

[30] Stanley, D.J. and D.J.P. Swift. *Marine Sediment Transport and Environmental Management.* John Wiley and Sons, 1976, pp. 602.

[31] Komar, P.D. *Beach Processes and Sedimentation.* Prentice Hall Inc., 1976, pp. 429.

[32] Zandi, I. *Advances in Solid-Liquid Flow in Pipes and its Application.* Pergamon Press, 1971.

[33] Govier, G.W. and K. Aziz. *The Flow of Complex Mixtures in Pipes.* Van Nostrand Reinhold Co., 1972.

[34] Wasp, E.J., J.P. Kenny, and B.L. Gandhi. *Solid-Liquid Flow Slurry Pipeline Transportation.* Trans. Tech.Pub., 1977.

CHAPTER 2

ORIGIN AND FORMATION OF SEDIMENT AND ITS PROPERTIES

This chapter deals with the definition and origin of sediment, the properties of sediment and muddy water, and the commonly adopted methods of classification.

2.1 ORIGIN OF SEDIMENT AND ROCK WEATHERING

Sediment is defined to be solid particles or debris transported in fluid media or found in deposit after transportation by flowing water, wind, wave, glacier, and gravitational action.

Rock weathering is the main origin of sediment. In addition to weathering, skeletons and shells of living creatures, volcanic ash, scoria, air-borne objects ejected during volcanic eruptions, magma on sea bed or magma flowing from hot springs, and disintegrated pieces of acrolite passing through the atmosphere can also turn into sediment.

Rock weathering is a unified process that includes mechanical detachment and chemical decomposition [1]. The relative importance of these two aspects varies both in time and in space. They occur simultaneously and supplement each other. For example, crevices in rocks caused by mechanical action expose the interior of the rock to the atmosphere, accelerate chemical reactions, and promote further weathering. Generally, the role of chemical decomposition is more important than that of mechanical detachment. Fine sediment, in particular, is produced primarily by chemical decomposition.

2.1.1 Mechanical detachment of rocks

Rock can be broken into pieces or even into particles by mechanical action (Fig. 2.1). Three main types of mechanical detachment occur:

1. Detachment to blocks. Cracks first appear on the surface of the rock, then cracking proceeds along further, and extends the crevice; a large mass changes into smaller blocks.

2. Disintegration into grains. Due to the absence of coagulation between individual mineral grains, rocks can disintegrate into small grains or sediment. Such disintegration is limited to coarse-grained rocks, mostly to coarse-grained granite.

3. Stripping of surface layers. The outer layer can be detached from the inner layer if the rock is acted on by certain forces. As time passes, the outer layer is removed and

the inner surface is exposed to the atmosphere. The process of denudation thus goes on layer by layer.

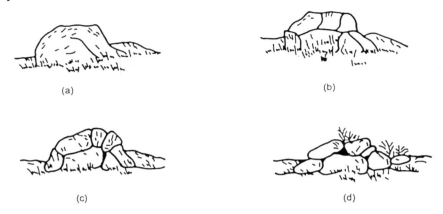

(a)

(b)

(c)

(d)

Fig. 2.1 Process of mechanical detachment

The causes of detachment of rocks are diverse. The most important ones are the following:

1. Unloading. The massive rock under the earth's crust may become exposed at the ground surface due to crustal uplift or erosion. As the pressure decreases, dilation fissures form and result in detachment.

2. Change in temperature. Heat conductivity in rocks is small. When the temperature rises, the upper strata are hotter than the lower strata. In this way, the upper strata of the rock body are constrained by forces that link them to lower strata during the process of unequal expansion. Tiny fissures form that are parallel to strata. In the opposite sense, when the temperature drops, the upper strata become cooler than the lower strata; then the rock in upper strata is constrained by the lower strata as they shrink. This process causes fissures perpendicular to the rock surface. As a result of repeated cooling and· heating, criss-crossed fissures appear on the rock surface and eventually cause rock blocks of various sizes to form. Also, the coefficients of expansion for various mineral compounds are different. Even for the same mineral, the coefficient of expansion varies with the orientation of its crystals. Because of the variability of the amplitudes of expansion and contraction, the rock body is nearly always in a state of stress somewhere. After a long time, detachment of rock takes place.

Research results show that the stresses induced by the variable expansion and contraction are usually within the rock's elastic strength. In combination with other factors, changes in temperature can accelerate rock weathering. Thus, change in temperature does not in itself play a decisive role [2]. Even in a dry or desert climate, or in high mountains where the amplitude of temperature variation is large, the duration of heating and cooling is short; also the rate of change in temperature is more rapid if the moisture content is low. Still, the destruction of rocks due to exclusively changes in temperature is rarely observed nonetheless [3].

3. Freezing. At high altitudes, freezing and thawing frequently alternate. Water in rock crevices expands when its freezes (the volume of water can be 9% greater when frozen). Hence it causes large destructive forces. Porosity of sandstone amounts to 10 to 30%, so sandstone is the most affected by freezing. For igneous rocks, the volume of the water crevices is negligible, and freezing can cause only some detachment into blocks.

4. Growing of crystals. When rainfall infiltrates into rock crevices, it occasionally carries along minerals in solution or as dissolved soluble salts from rocks. If the crevice water evaporates when heated, mineral crystals can grow from the deposit in the crevices. The expansion that occurs as crystals grow can disintegrate the rock in a manner similar to the way ice crystals disintegrate rocks during freezing. Weathering of rocks in the Egyptian desert is often caused by the growth of salt crystals.

5. Effect of external forces due to animals, plants, and human activities. Roots of plants that extend deeply into cracks and fissures of rocks can intensify the splitting. Fissure habitat animals can enlarge fissures so that actions due to atmospheric changes can penetrate the interior of rocks. Road construction, tunnel excavation, and cultivation practices can lead to the breakage of rocks by mechanical crushing or dynamite explosion, and the result is more exposure of the parent soil at the ground surface, an occurrence that promotes weathering.

6. Abrasion. As rocks are transported by flowing water or glaciers, they rub and impinge on each other. Often they break into smaller pieces in the process. Rocks along the shores of seas and lakes are under the continuous attack by wave and they are easily broken and dislodged. Also, within a fault, rocks are subjected to crushing and grinding and thus become smaller and smaller.

Among these several mechanical actions, unloading, freezing, and growing of crystals are the more important ones. Mechanical detachment is not usually a major agent in rock weathering. Only in frigid zones or arid regions where chemical actions are not so strong, is the role played by mechanical actions dominant in the formation of sediment through weathering of rocks.

2.1.2 Chemical decomposition of rocks

The mineral or chemical changes of rocks under atmospheric action are called decomposition. Generally, the atmosphere consists of nitrogen, oxygen, carbon dioxide, inert gases, and water vapour. Under certain conditions or in specific areas, it also contains acidic material resulting from volcanic ash or industrial dust. Oxygen, carbon dioxide, and acidic material become attached to raindrops and fall with them to the ground; they then infiltrate the top soil and come into contact with rocks. Then oxidation, hydration, hydrolysis, and dissolution take place, leading to the decomposition of rocks. Besides, bio-chemical action induced by organic matter can also result in rock weathering. This action usually comes after one of the following process:

1. Oxidation of rocks

The patterns of oxidation of rocks can differ greatly. Some of them are induced by organic matter (microbial activity), some by non-organic matter. The most commonly encountered ones are the decomposition of ferrosilicate in augite, amphibole and olivine due to oxidation, and the conversion of ferrous to ferric oxide — either Fe_2O_3 or $Fe_2O_3 \cdot H_2O$. Because ferrous sulfide can produce sulfuric acid after oxidation, it possesses the highest potential force of the erosion of rocks.

2. Hydration and hydrolysis

Hydration refers to the changes in minerals that adsorb moisture as they become a hydrous mineral, but the basic properties of the rock are not changed. Common examples are the change between hematite and limonite and that between anhydride and gypsum.

$$CaSO_4 + 2H_2O \Leftrightarrow CaSO_4 \cdot 2H_2O$$

$$2Fe_2O_3 + 3H_2O \Leftrightarrow 2FeO_4 \cdot 3H_2O$$

Hydrolysis forms the hydroxide radical. It indicates that a certain chemical reaction of the following type is taking place

$$KAISi_3O_8 + H_2O \longrightarrow HAISi_3O_8 + KOH$$

Alumina silicate is unstable; under certain conditions it transforms into a clayey mineral. Carbon dioxide plays an extremely important role. When dissolved in water, it forms carbonic acid. As soon as potassium hydroxide contacts carbonic acid, it changes into potassium carbonate, which is soluble in water.

$$2KOH + H_2CO_3 \longrightarrow K_2CO_3 + 2H_2O$$

Such a chemical reaction is called carbonatization.

An important consideration is that various weathering processes are often interrelated in nature. The weathering of feldspar is represented by the following chemical reaction

$$2KAISi_3O_8 + 2H_2O + CO_2 \longrightarrow AI_2Si_2O_5(OH)_4 \text{ (Kaolinite) } + 4SiO_2 + K_2CO_3$$

Thus after hydrolysis and carbonatization, feldspar becomes soluble carbonate, silicon oxide, and colloidal clay.

Mineral grains in rocks, after absorbing water, expand in volume and extrude neighboring mineral grains or other solid, further promoting the mechanical detachment of rocks. Since the stratified denudation of rocks is mostly the result of unloading and cracking induced by expansion, the hydration-affected seam may also exert an influence. The disintegration of coarse-grained igneous rock into grains was once considered to be mainly the consequence of unequal expansion and contraction of rocks. However, recent research reveals that it can also result from hydration.

3. Dissolution of rocks

Dissolution of rocks is one of most important factors in the rock weathering. The afore-mentioned formation of dissoluble carbonate by carbonatization of rocks due to the carbon dioxide dissolved in water serves as an example. Movement of a solution continually changes the chemical environment of weathering regions. Moreover, the chemical actions of a solution are extremely strong. Ionization of dissolved matter further promotes exchanges between a solution and salts in a solid.

The weathering of rocks generally increases the surface area of the rock mass. As chemical reactions take place primarily at the interfaces between objects, the smaller the mineral grains are, the more intensive can be the chemical action they are subjected to. If the diameter of particles is less than 0.005 to 0.20 μ, mineral grains will contact the solution directly to form a colloid. Colloids can also be formed as the result of decaying induced by microbes. Chemical action of colloid is less important than that of solution; colloidal particles usually adsorb water and various ions; the adsorbed matter can act directly on rock surfaces and promote further weathering.

4. Biochemical action

Biochemical action of bacteria and microbes greatly accelerates rock weathering. Their sizes are so small that they can easily enter the tiniest fissures in rocks. Saprophytic bacteria can decompose the organic matter of their food supply into water and carbonic acid, which in turn is capable of dissolving limestone and destroying mica and feldspar. Making use of carbon dioxide, saprophytic bacteria can also decompose organic matter into organic acid, a process that facilitates the dissolving of calcium carbonate and magnesium carbonate. Thus saprophytic bacteria too results in the weathering of marlite, dominate and limestone.

2.1.3 Rate of rock weathering

The rate of weathering depends upon the balance between the effectiveness of weathering agents and the resistance of rocks to weathering.

The intensity of chemical decomposition is closely related to local temperature and humidity. In arid regions, chemical reactions can hardly take place. Also in frigid regions, the temperature is too low, and moisture exists only in its solid state, which is also unfavorable to chemical weathering. At the North Pole, for example, scarcely any weathering can occur. Mechanical detachment of rocks is also related to moisture. If fissure water is continuously subjected to freezing and thawing, the weathering proceeds more rapidly. One can readily conclude that a strong rock cannot be expected to weather in two regions: one where the temperature is so high that no freezing occurs and the other where the temperature is so low that no thawing occurs. Fig. 2.2 shows the interrelationship among annual mean temperature, annual mean precipitation, and the relative intensities of chemical and physical weathering [4]. These two climatic factors determine not only the rate of weathering but also the relative importance of mechanical detachment and chemical decomposition.

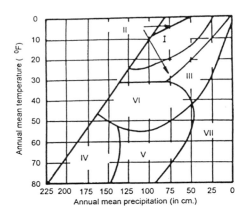

mechanical detachment : I intensive, II moderate, III weak
chemical decomposition: IV intensive,V moderate, VI weak
VII rarely weathering

Fig. 2.2 Relative importance of different patterns of weathering of rocks with respect to precipitation and temperature (after Peltier, L.)

The capability of rocks to resist weathering depends upon whether rocks have defects, and also upon the mineral composition of rocks. If rocks contain many fissures, beddings, and faults, then water can penetrate the interior of the rock through them, and thus accelerate the weathering. As already mentioned, weathering of rock is a process of readjustment induced by changes in the physical and chemical environment of the rock as it is lifted up from a deep strata in the interior of the earth's crust. The more the mineral constitution and chemical composition of the rock differ from the physico-chemical conditions at the surface of the crust, the more intensive the weathering will be. The basic igneous rocks possess the property of mono-oxide, which is the mostly contradictory with the oxidative atmosphere of the ground surface, hence destruction of such rocks is most rapid. The chemical decomposition of acidic igneous rocks proceeds much more slowly. Sedimentary rocks have experienced weathering once or more often, and their rate of weathering is even slower. In addition to the afore-mentioned factors, the rate of displacement of weathering products determines whether a newly exposed rock face will appear. Therefore it affects the rate of weathering consequently. On cliffs and steep slopes, all weathered products move down rapidly and rocks in the upper strata are continuously exposed. Consequently, the weathering process is more intensive.

2.1.4 Products of rock weathering

During the weathering of rock, some minerals behave stably, so that no major changes occur; but other minerals may completely lose their original nature. Based on their relative stability with regard to rock weathering, various rocks can be ranked as follows [5]: quartz, white mica, potassium feldspar, biotite, alkali anorthosite, hornblende, orthopyroxene, calcium-bearing plagioclase, olivine. The preceding minerals have a higher stability. Clearly, the relative stability during the weathering of minerals depends upon both the contrast of the initial environment of the minerals and the environment for weathering.

The final step along the way as minerals in rocks change is the formation of diversified clay minerals. The main components of clay minerals are hydrous alumina silicate. According to the different physico-chemical properties and different infrastructures, they can be classified into three groups: montmorillonite, hydromica, and kaolinite.

The clay mineral montmorillonite is the primitive product of chemical decomposition; its crystal consists of many flat cells parallel to each other. An indefinite quantity of water molecules can be absorbed among cells so the crystal lattice (structure) is both active and hydrophilic. Sometimes when too many water molecules are absorbed, the neighboring cells begin to lose their binding force and smaller particulate soil is formed. The Al^{+++} in cells can be replaced by Fe^{+++}, Fe^{++}, Ca^{++}, and Mg^{++}, and then diversified minerals of the montmorillonite family form. Bentonite is composed of various montmorillonite minerals.

The clay mineral family of hydromica has a crystal lattice similar to that of the montmorillonite family, but the binding force among its crystal cells is much stronger. Hence, the crystal lattice of hydromica is not as active as in that of mortmorillonite, and so is its hydrophilicity. The most common mineral in the hydromica family is illite hydromica, and potassium is one of the main elements in its mineral compound.

The binding force between crystal cells of the kaolinite mineral family is rather strong. Water molecules can enter the spaces in a crystal lattice, and its hydrophylicity is even weaker than hydromica. Kaolinite is coarse-grained, but sometimes it can exist in the form of powder-like substance. Insofar as the broken face of crystal cells of kaolinite along the lateral direction can absorb water molecules to form a hydrated film, the particulate kaolinite is a hydrophylic material.

Generally, alkaline or neutral ground water that is poorly drained under bad drainage conditions can promote the formation of montmorillonite, whereas acidic ground water with good drainage favours the formation of kaolinite.

In addition to clay minerals, ferric oxide and aluminum oxide are also important products of weathering. These two newly formed minerals appear mostly in those regions where the temperature is high during the summer season. Besides, if silicate minerals are subject to chemical reaction, silica changes into a colloid or dissolves in water, but no quartz forms. Calcium carbonate and magnesium carbonate are easy to dissolve in water, so that these minerals crystalize and then deposit elsewhere. Sodium compounds exist in the form of salts, thus they can be dissolved in water in large quantities finally carried to the open sea.

Except for the portion of sediment originating from rock weathering that remains in place and forms soil, called eluvium, the soil is either dissolved in water or transported by flowing water, waves, glaciers, wind, or gravitational force. This module of the sediment redeposits after moving some distance, and it can consolidate so as to form sedimentary rocks; these could, in turn, be changed back into sediment particles by the weathering process.

2.2 BASIC PROPERTIES OF SEDIMENT PARTICLES [6]

The basic properties of sediment particles are studied in a broad range of disciplines. The concepts described herein are limited to those aspects that are related to the process of erosion, transport and deposition of sediment. These can be grouped into two categories: properties of individual sediment particles and properties of a mixture composed of many sediment particles of various sizes.

2.2.1 Properties of an individual sediment particle

The properties of an individual sediment particle reflect directly or indirectly the history of the particle. For example, sediment size is related to the medium in which the particle is transported and to the velocity of its movement. Shape and roundness of a sediment particle involve the medium, and so do the distance and intensity of the movement. The mineral composition of a sediment particle indicates the possible source area and the distance it has been transported. The orientation of a sediment particle is determined by the flow direction and the various forces acting on it during deposition. The surface texture of a sediment particle reflects its history of abrasion and changes induced by dissolution. From such properties, one can obtain a schematic or an outline of the factors pertaining to the source area and to the process of erosion, transport and deposition of the particle.

2.2.1.1 Sediment size

The size of a geometrical solid with a regular shape can be simply defined (e.g. diameter of a sphere, side length of a cube), and there is no difficulty in its measurement. However, for all irregularly-shaped objects such as a sediment particle, to give only its size is not enough unless the measuring method and the definition of the size must also be fully explained. Only then can the measured size be used in practical applications.

The terminology used to define the size of sediment particles starts from a boulder and descends to clay particles; their sizes vary by a factor of more than one million times. Naturally, more than one method of measuring size is needed. For particles larger than cobbles, three axial lengths are taken as the representative dimensions, and a mean value is then be obtained. For particle sizes ranging between cobbles and fine sand, sieving is the most convenient method of measuring size. Results obtained from sieving indicate only that the particle size is between two sieve mesh openings, D_1 and D_2. It gives a range for the size $D_2>D>D_1$, rather than an absolute value. The mean size can be expressed as the algebraic average $(\dfrac{D_1+D_2}{2})$, the geometric average $(\sqrt{D_1 D_2})$, or more generally $\dfrac{1}{3}(D_1+D_2+\sqrt{D_1 D_2})$. Sediment size obtained by sieving is neither the maximum, nor the minimum, but an intermediate one. For sediment finer than fine sand, either a microscopic method or a settling method is adopted. The former refers to the sediment size projected onto a certain plane, while the latter refers to the diameter of a fictitious sphere with the same fall velocity and specific gravity, often called the effective diameter or settling diameter of the particle. The effective diameter is related not only to the size of sediment particle, but also to other factors such as shape and specific gravity. Sometimes, the diameter of a sphere of equal volume is used, and it is called the nominal diameter.

Settling velocity, or fall velocity, of a sphere can be obtained from

$$\omega^2 = \frac{4}{3C_D}\,\frac{\rho_s-\rho}{\rho}\,gD \qquad (2.1)$$

$$C_D = f(\mathrm{Re}) = f(\omega D/v) \qquad (2.2)$$

in which:

ω — fall velocity of the sphere;

C_D — drag coefficient;

$\rho,\ \rho_s$ — densities of fluid and sphere, respectively;

g — gravitational acceleration;

D — diameter;

Re — Reynolds number;

v — kinetic viscosity of the liquid.

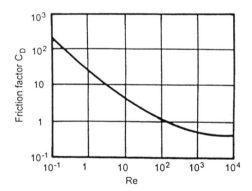

Fig. 2.3 C_D vs. Re for a sphere

In view of the wide range of sediment sizes, Krumbein suggested the use the parameter of Φ to express particle diameter:

$$\Phi = -\log_2 D \qquad (2.3)$$

The particle diameter D is in mm.

This expression has the advantage of being able to cope with a wide range of sediment size and is convenient for statistical methods of processing grain size data.

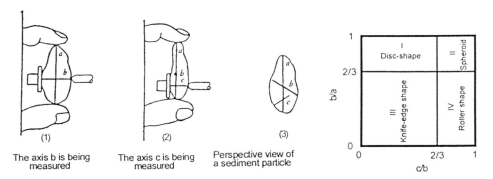

(1)	(2)	(3)
The axis b is being measured	The axis c is being measured	Perspective view of a sediment particle

Fig. 2.4 The maximum, intermediate, and minimum diameter of a sediment particle

Fig. 2.5 Zingg's classification of sediment particle shape (after Zing, Th.)

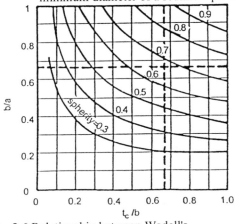

Fig. 2.6 Relationship between Wadell's sphericity and Zingg's axis ratios (after Krumbein, W.C.)

2.2.1.2 Shape and roundness

Shape and roundness are two different properties of sediment particles. Roundness refers to the sharpness of edges, or their radius of curvature, and shape describes the geometric form of the particle as a whole.

Wadell suggested the use of the sphericity Λ to describe shape [7]:

$$\Lambda = A' / A \qquad (2.4)$$

in which:

A--sphericity of a sediment particle;
A'--surface area of a sphere with equal volume of the sediment particle;
A--surface area of the sediment particle.

Because it is difficult to measure the surface area of an irregular particle, sphericity is usually approximated by

$$\Lambda = D_n / D_s \qquad (2.5)$$

in which:

D_n--the nominal diameter;
D_s--diameter of circumscribed circle and is equivalent to the maximum dimension of the particle.

Many researchers have also used the maximum, intermediate, and minimum diameters, indicated in Fig. 2.4 by a, b, and c, respectively, to describe the shape of a particle. For example, Zingg classified the shape of sediment particles into four groups based on c/b and b/a [8] (Fig. 2.5). These four shapes are disc, spheroid, roller, and blade edge [8]. Fig. 2.6 indicates the relationship between Zingg's classification of shape and Wadell's sphericity [9]. It shows that a disc-shaped particle and a roller-shaped particle may have the same sphericity, even though they behave quite differently in transport and in deposition. Therefore, as a shape factor, sphericity is far from perfect. Many other suggestions have been made for the use of various axis ratios to describe shape of sediment particles [10]. Some representative parameters of axis ratios are listed in Table 2.1. Some of these parameters will be discussed later.

Table 2.1 Shape parameters of sediment particles

Author	Expression of parameter	Physical meaning
Curry (1949) Fork, R.L. (1958)	$c/\sqrt{ab}$	shape factor
Wentworth, C.K. (1922) Cailleux (1945)	$\dfrac{a+b}{2c}$	degree of oblateness
Krumbein W.C. (1942)	$[(b/a)^2(c/b)]^{1/3}$	sphericity
Aschenbrenner (1956)	$\dfrac{12.8[(c/b)^2(b/a)]^{1/3}}{1+\dfrac{c}{b}(1+\dfrac{b}{a})+6\{1+(\dfrac{c}{b})^2[1+(\dfrac{b}{a})^2]\}^{1/2}}$	sphericity
McNown J.S. Malaika J. (1950)	$a_1/\sqrt{b_1c_1}$ $\begin{cases} a_1\text{—semi - axes in direction of movement} \\ b_1 \text{ and } c_1 \text{—semi - axes perpendicular to direction of movement} \end{cases}$	shape factor

Roundness is a parameter that is used to indicate the sharpness of edges of an irregular sediment particle, and it is defined by Wadell as follows

$$\Pi = \dfrac{\sum\limits_N r/R}{N} \tag{2.6}$$

in which:

R — the radius of the maximum inscribed circle measured on the plane that shows the maximum projection of the particle (Fig. 2.7);
r — radius of curvature of edges on the plane;
N — number of angles formed by edges.

27

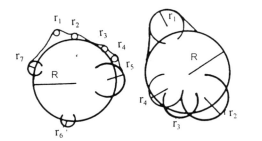
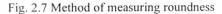

Fig. 2.7 Method of measuring roundness

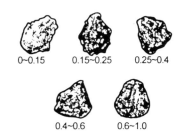

0~0.15	0.15~0.25	0.25~0.4

0.4~0.6	0.6~1.0

Fig. 2.8 Appearances of sediment particles with different roundnesses

Actual measurement of roundness requires a lot of work. Krumbein has calculated roundness of some typical shapes of sediment particles by Wadell's method as shown in Fig. 2.8. With applications of Fig. 2.8 as a standard, one can obtain the roundness of a sciencific sediment particle by comparison. Application of Fig. 2.8 shows that results thus obtained are rather close to the results calculated by Wadell's method. However, the precision of measurement drops appreciably with the decrease in sediment size and radius of curvature.

The roundness of a sediment particle reflects its history of abrasion. The farther the distance it has been transported, the more likely that it has experienced abrasion. The sharp edges are thus ground off and its roundness more closely approaches that of a spheroid. Sphericity of a sediment particle is also affected by abrasion, but it depends even more on its original shape. In another sense, the shape of a particle can greatly influence its transport and deposition. Sorting according to sediment shape can help people to judge the paleogeographical environment of the accumulational relief of sediment deposits.

2.2.1.3 Mineral composition

Sediment originates from the weathering of rocks that are mostly a mixture of various minerals. Naturally, these mineral compositions are found in the sediments. Nine principal minerals are found in the rocks. Their properties are listed in Table 2.2.

The mineral composition of common igneous rocks and sedimentary rocks are given in Table 2.3.

Quartz and feldspar are the two substances that occur most commonly in the mineral composition of sediment. The bed material of the Yangtze River at Jingjiang reach consists of quartz, feldspar, hornblende, calcite biolite, ferric oxide, pyroxene, chlorite, and others. Among these, quartz constitutes 79~80 %, and feldspar 5~10 % [11]. Because quartz and feldspar form the major part of sediment, despite the diversity and complexity of the mineral constitution of sediment, the specific gravity of sediment particles usually varies within the narrow range of 2.60~2.70.

For coarse sediment particles, with sizes greater than 2mm, the mineral composition may include more than one mineral; those with a size less than 2mm are usually composed of a single mineral.

Table 2.2 Principal minerals in the composition of rocks

Name	Nomen-clature	Molecular formula	Color	Joint	Luster	Hardness	Specific gravity
Feldspar	Orthoclase	$KAlSiO_3$	Red, pink	2 sets that intersect orthogonally	like glass-or pearl-like	6.0	2.56
	Plagioclase	$NaAlSi_3O_3$ $CaAl_2Si_2O_3$	white, grey	2 sets, intersect at 86°	like glass and pearl	6.0	2.6~2.8
Quartz	——	SiO_2	white, transparent	no	like glass	7.0	2.66
Pyroxene	Diopside	$CaMg(SiO_3)_2$	green, black	2 sets intersecting at 87°-93°	like silk	5 6	3.2~3.6
	Hypersth-ene	$(Mg, Fe)SiO_3$					3.3~3.5
	Orthopy-roxene	$(Al, Fe)SiO_3$					3.2~3.4
Hornb-lende	Tremolite	$Ca_2Mg_5Si_8O_{22}(OH)_2$	green, black	2 sets intersect at 12.4°	like glass and pearl	5 6	2.9~3.1
	Actinolite	$Ca_2(Mg, Fe)_5Si_8O_{22}(OH)_2$					3.0~3.2
Mica	Muscovite	$KAl_2(Si_3Al)O_{10}(OH)_2$	transparent	1 sets extremely fragile and easy to	like pearl	2.5 3.0	2.8~2.9
	2-Biolite	$K_2(Mg, Fe)_6(SiAl)_8O_{20}(OH)_2$	black, brown, dark green	be split into sheets			3.0~3.1
Olivine	——	$(Mg, Fe)_2SiO_4$	green, yellow	not clear	like glass	5.6 7.0	3.2~3.6
Carbonate	Calcite	$CaCO_3$	transparent, white, grey, brown, yellowish grey	3 sets intersect at 74°	like glass	3.0	2.72
	Dolomite	$CaMg(CO_3)$				3.5 4.0	2.87
	Siderite	$FeCO_3$				3.5 4.0	3.8~3.9
Kaolinite	——	$Al_2Si_2O_5(OH)_4$	white, grey, brown, black	no	muddy soil	1 2.5	2.6
Ferric Oxide	Hemilite	Fe_2O_3	red, brown, grey	no	close to metal or like muddy soil	5.5 6.0	4.3±
	Limolite	$2Fe_2O_3 \cdot 3H_2O$	yellow, brown	no	like muddy soil or like silk	5 5.5	3.8

Table 2.3 Mineral compositions of igneous rocks and sedimentary rocks

Igneous rocks		Sedimentary rocks		
Nomenclature	%	Nomenclature	% of shale	% of sandstone
Quartz	12.0	Quartz	22.3	66.8
Feldspar	59.5	Feldspar	30.0	11.5
Pyroxene & Hornblende	16.8	Clay minerals	25.0	6.6
Mica	3.8	Limonite	5.6	1.8
others	7.9	Carbonate	5.7	11.1
		others	11.4	2.2

Some heavy minerals are found in the sediments; common ones are magnetite (specific gravity 5.2), iron core (specific gravity 4.7), garnet (specific gravity 3.8), and hornblende (specific gravity 3.2). For example, at the Nanjing reaches of the Yangtze River, ironshot constitutes 1.1% in the deposits. The total amount of such heavy minerals is not large, but their occurrence serves as an excellent indicator of the source of sediment and of the relative amounts of sediment from various regions within the river basin.

2.2.1.4 Surface texture

Surface texture of sediment particle varies greatly, but nonetheless it can be grouped into the two categories of shiny or dark. Another factor is whether there are marks on the particle's surface. William summarized the feature of surface texture as follows [12]:

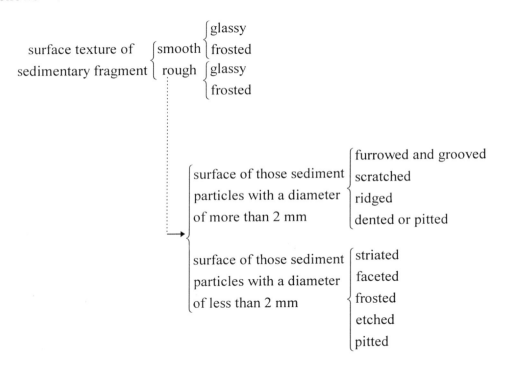

Like shape and roundness of sediment particles, surface texture inherits, on one side, its original nature from parent rocks, and on the other, changes occurring during its transport. Existing data indicate that surface texture of sediment particles changes noticeably as it travels some distance. Hence, surface texture reflects more nearly the circumstances of its recent transportation process.

The same result can be induced in various ways. For example, a glassy surface may form due to gentle and slow rubbing, to solution in the liquid, or to the deposit of a thin bright layer of other elements. Moreover, measuring techniques have not yet been standardized. Thus, study of this aspect remains in an early stage of development.

2.2.1.5 Orientation

Orientation of a sediment particle refers to the angle between the flow direction and either the long axis or the plane of the maximum projected area of the particle, either during transport or after deposition. Orientation can be used to investigate flow direction and vectorial properties of the permeability of sediment deposits.

Orientation of a sediment particle can be expressed by two angles: the angle between the horizontal plane (or plane parallel to flow direction) and the long axis (i.e., dip angle of the long axis) and the azimuth of the long axis with respect to some fixed direction on the horizontal plane (i.e., direction of the long axis). For disc-shaped particles, the orientation is not expressed by the long axis but by the plane of disc, then the dip angle of the orthogonal line to the plane of maximum projected area can be used to express the orientation.

Fig. 2.9 shows a typical imbricate texture that occurs frequently on gravel river beds. Experimental results reveal that once the natural orientation of a gravel river bed is destroyed by external forces, the river bed will lose its equilibrium until the gravel particles restore their original orientation.

Flow direction

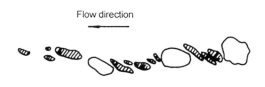

Fig. 2.9 Imbricate texture of gravel bed

Among the afore-mentioned variety of properties of sediment particles, the most important one is the particle size (grain diameter). The reasons for this are the following:

1. During the process of sediment movement and deposition, the weight (or submerged weight) of particles, which acts all the time, is proportional to the cube of its diameter; also, areal forces caused by the fluid acting on the particle, such as uplift, drag, and skin friction, are generally proportional to the square of the diameter. Because sediment size varies over a very wide range, effects of size variation on the acting forces are much greater than those of shape, specific weight, orientation, etc.

2. The effect of physico-chemical action on the surface of a sediment particle is directly related to its size. As discussed elsewhere, a thin water film is often adsorbed on a particle surface, and its thickness is about 0.1μ, varying slightly with the mineral composition and particle size. For coarse particles, the volume occupied by the water film is far less than that occupied by the particle; hence the water film has little effect on sediment movement and deposition. For fine particles, the situation is quite different. Especially for particles finer than 1μ, the water film is not only inseparable

from the particle, but it also plays a more dominant role in sediment movement and deposition.

3. Other properties of sediment particles (such as mineral composition, shape, etc.) may be directly or indirectly related to particle diameter.

Table 2.4 shows the relationship between particle size and mineral composition[13]. As shown, particles coarser than 2 mm are mostly poly-mineralic in composition, and sand particles are mostly mono-mineralic; they are mainly composed of quartz, feldspar, mica, and other rock-forming minerals. Silt is usually composed of minerals quartz that strongly withstand weathering. Clay is exclusively composed of secondary minerals (including secondary silicon dioxide, clay minerals, etc) and humus.

Table 2.4 Relationship between particle size and mineral composition

Mineral composition \ Groups / Name / Size(mm)	Gravel Pebble Cobble Boulder >2	Sand 2~0.05	Silt 0.05~0.005	Clay Coarse 0.005~0.001	Clay Medi 0.001~0.0001	Clay Fine <0.0001
Minerals / Mono-mineral — Ploy-mineralic						
Quartz						
Feldspar						
Mica						
Secondary minerals / Clay mineral — SiO$_2$						
Kaoline						
Hydromica						
Montmorillo-nite						
Al$_2$O$_3$, Fe$_2$O$_3$						
CaCO$_3$, MgCO$_3$						
Humus						

Roundness and sphericity of sediment particles are also closely related to particle size. Fig. 2.10 shows the relationships among the three parameters for lacustrine beach sand and riverine beach sand. It shows that coarse sediment particles are more nearly spherical than are fine ones, and their edges and corners are less sharp.

For the before-mentioned reasons, sediment size is generally the principal parameter in the study of sediment movement in the following chapters. In the rest of this chapter the classification of sediment is based on sediment size groups.

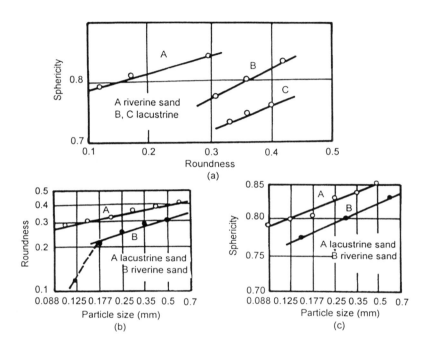

Fig. 2.10 Relationship between particle size, roundness, and sphericity for sediment particles

2.2.2 Properties of sediment mixtures

Sediment refers usually not to individual particles but rather to a mixture with innumerable particles of different sizes, shapes, and mineral compositions. Although sediment particles are not cemented together to form an entity, many properties pertinent to sediment manifest themselves mainly through the existence of an ensemble of sediment particles.

2.2.2.1 Size distribution

Size distribution of sediment particles, including the degree of uniformity, reflects directly the properties of parent rocks and the intensity of the sorting process by river flow; it is also closely related to the amount of sediment transported.

Several methods are used to express the size gradation, and the following three are the conventional ones.

1. Frequency histogram

In a frequency histogram, particle diameter (or its logarithm) is taken as the abscissa and percentage of frequency (by weight or by number of particles) as the ordinate. A sediment sample is first divided into groups according the size. The width of each bar in the frequency histogram represents the interval for the successive sizes, and the height represents the percentage of the total falling within that group. Table 2.5 shows the results from the sieving of a sediment sample. If the data from the second and the fourth columns are plotted against each other on a semi-logarithmic scale, the frequency histogram shown like that in Fig. 2.11 is obtained.

Table 2.5 Results of a sieve analysis

Sieve no.	Sieve opening (mm)	Median size of sieves (mm) $D = \sqrt{D_1 D_2}$	Percent by weight between sieves Δp	Percent finer by weight	$\Delta R = \log(D_1/D_2)$	$N = \Delta p/\Delta R$
14	1.168			99.75		
		0.991	0.48		0.15	3.20
20	0.833			99.27		
		0.701	5.29		0.15	35.25
28	0.589			93.98		
		0.495	21.68		0.15	145.00
35	0.417			72.30		
		0.351	52.23		0.15	350.00
48	0.295			20.07		
		0.246	18.39		0.15	122.50
65	0.208			1.68		
		0.175	1.48		0.15	9.87
100	0.147			0.20		
		0.124	0.1		0.15	0.67
150	0.104			0.10		
200	0.074	0.088	0.05	0.05	0.15	0.33

The shape of a frequency histogram depends strongly on the number and spacing intervals of the size groups. Fig. 2.12 shows three frequency histograms for the same sediment sample, but with different spacing the intervals for the size groups. Small differences are evident in the results. If one spacing is adopted for the size group, the curve seems symmetric; but for another spacing, the appearance is different. These facts reveal a serious considerable limitation on the application of frequency histograms. Generally, the more subdivisions and the smaller the spacing interval, the

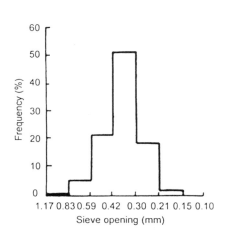

Fig. 2.11 Frequency histogram

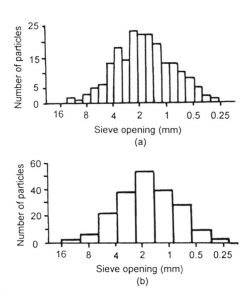

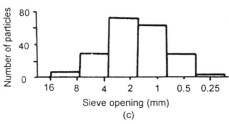

Fig. 2.12 Frequency histograms with different width of interval of sieve openings

more exactly the curve represents the size distribution. If the interval of spacing diminishes, for an unchanged size of sample, then in the extreme case, the frequency histogram approaches a continuous curve called the frequency distribution. Fig. 2.13 shows the size-frequency distribution for the sediment sample presented in Table 2.5. Such a frequency distribution cannot be plotted directly from the measured data.

2. Cumulative size-frequency curve

In a cumulative size-frequency curve, sediment size (or its logarithm) is taken as the ordinate, and the percentage by weight (or by number) of sediment particles that is smaller than the given size is taken as the abscissa. A cumulative size-frequency curve is the integral of the size-frequency curve. A plot of the data from the third and fifth columns of Table 2.5 on semi-logarithmic paper is the cumulative frequency curve obtained for the sediment size distribution, as shown in Fig. 2.14.

The best way to plot cumulative curve for size-frequency distribution is to use a special kind of graph paper on which the ordinate scale is logarithmic and the abscissa probabilistic. Fig. 2.15 presents the data of Fig. 2.14 on a logarithmic-probability scale and is the cumulative size-probability curve. If the sediment size distribution plotted on a logarithmic scale follows the normal error distribution, the curve is a straight line. The curves for natural sediment are generally close to straight lines, so this diagram is convenient for comparing size compositions of different sediment mixtures. Moreover, the coarsest and finest particles occupy only small portions of the total, yet these parts

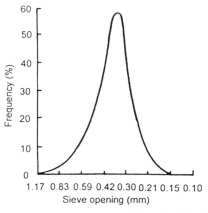

Fig. 2.13 Frequency curve of size distribution

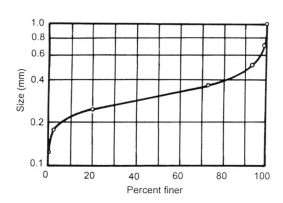

Fig. 2.14 Cumulative frequency curve of size distribution

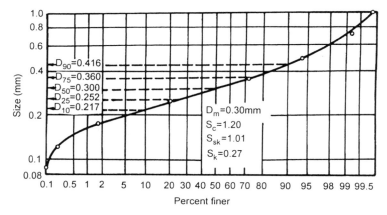

Fig. 2.15 Cumulative size-frequency curve

have a considerable influence on the properties of a sediment mixture. If a common logarithmic algebraic scale is used (Fig. 2.14), the curve changes abruptly on both ends and approaches the vertical, hence it is nearly impossible to read the percentage of very coarse and very fine particles. If a logarithmic-probabilistic scale is used, this difficulty is usually avoided.

3. Differential frequency curve

Bagnold suggested the use of the logarithm of the differential of the cumulative size-frequency curve and the logarithm of particle size for the two axes of the curve, i.e., the data from the third and the seventh columns in Table 2.5; these are plotted in Fig. 2.16. The curve is called a differential frequency curve [14]. The height of the peak for the differential frequency

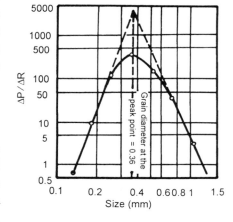

Fig. 2.16 Differential frequency curve (after Bagnold, R.A.)

curve corresponds to the maximum slope of the cumulative size-frequency curve, and the corresponding particle size has the highest frequency of occurrence. For ordinary sediment samples, constant rates of reduction form nearly straight lines to the right and left of the peak. Gradients of the two straight lines are called the coefficients of the coarse and fine sand groups. The size corresponding to the intersection of the two straight lines is called the particle size of the peak point. Bagnold proved that during the transport and deposition process of wind-brown sediment, the particle size at the peak point and the coefficients of the coarse and fine sand grades change and follow a certain rule.

Size distribution curves described in the foregoing paragraphs can reflect distinguishing features of sediment samples, but such curves may not be convenient to use in the quantitative descriptions and comparisons of one example with another. Therefore scholars have proposed the use of a variety of characteristic parameters to describe features of sediment size distribution and to carry out statistical analyses based on such parameters. Table 2.6 contains some of those that are more commonly used.

Table 2.6 Commonly used statistical parameters pertinent to sediment size distribution (after Graton, L.C., and H.J. Fraser)

Authors	Representative value	Extent of dispersion	Symmetry	Degree of concentration	Remarks
W.C. Krumbein [15] (1938)	median diameter D_m, D_{50}	sorting coefficient S_c, $\sqrt{D_{75}D_{25}}$	skewness S_{sk}, $\dfrac{D_{75}D_{25}}{D_{50}^2}$	kurtosis, S_k, $\dfrac{D_{75}-D_{25}}{2(D_{90}-D_{10})}$	—
D.L. Inman [16] (1952)	(1) median ø value, ϕ_{50} (2) mean ø value, $\frac{1}{2}(\phi_{16}+\phi_{84})$	deviation of measurement, $\frac{1}{2}(\phi_{84}-\phi_{16})$	(1) skewness of measured ø, $\dfrac{\frac{1}{2}(\phi_{16}+\phi_{84})-\phi_{50}}{\frac{1}{2}(\phi_{84}-\phi_{16})}$, (2) second skewness of measured ø, $\dfrac{\frac{1}{2}(\phi_5+\phi_{95})-\phi_{50}}{\frac{1}{2}(\phi_{84}-\phi_{16})}$	kurtosis of measured ø, $\dfrac{\frac{1}{2}(\phi_{95}-\phi_5)-\phi_{50}}{\frac{1}{2}(\phi_{84}-\phi_{16})}$	measured by graphics
D. Fork & C. Ward [17] (1957)	mean value by graphics, $\frac{1}{3}(\phi_{16}+\phi_{50}+\phi_{84})$	sum of standard deviation measured by graphs, $\frac{1}{4}(\phi_{16}-\phi_{50})+\frac{1}{6.6}(\phi_{95}-\phi_5)$	sum of deviation measured by graphs, $\dfrac{\phi_{16}+\phi_{84}-2\phi_{50}}{2(\phi_{84}-\phi_{16})}+\dfrac{\phi_5+\phi_{95}-2\phi_{50}}{2(\phi_{95}-\phi_5)}$	kurtosis by graphs, $\dfrac{\phi_{95}-\phi_5}{2.44(\phi_{75}-\phi_{25})}$	measured by graphs
M. Friedman [18] (1962)	mean value ø, $\frac{1}{100}\Sigma f \cdot m\phi$	standard deviation, $\sigma\phi$ $\frac{1}{100}\Sigma f \cdot (m\phi - \bar{x\phi})^2)^{1/2}$	skewness $\alpha_3\phi$, $\frac{1}{100}\cdot\sigma\phi^{-3}\cdot\Sigma f(m\phi-\bar{x\phi})^3$	kurtosis, $\sigma_4\phi$, $\frac{1}{100}\cdot\sigma\phi^{-4}\cdot\Sigma f(m\phi-\bar{x\phi})^4$	moment measure

In order to explain the physical meaning of these parameters, six size frequency histograms of different sediment samples are presented in Fig. 2.17. Among them, the medians of samples (a), (b), and (c) are the same. Curve (b) has a wider range of distribution than does curve (a); curve (c) has the same median diameter and the same sorting coefficient as curve (b) but it has a higher percentage at the central part; i.e., its kurtosis is larger than that of curve (b).

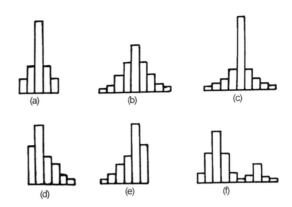

Fig. 2.17 Different types of frequency histograms of sediment size distribution

Samples (d) and (e) are both asymmetric; in curve (d), coarse particles constitute a higher percentage, a trend that is called positive skewness; in curve (e), fine particles constitute a higher percentage and that is called negative skewness. Each of the five sediment samples from (a) to (e) has only one with a higher than normal percentage. In curve (f), two groups have percentages that are higher than the neighboring ones, i.e. curve (f) is a double-peak curve that is sometimes called a bimodal distribution.

Applications of the parameters used to describe the characteristics of sediment samples are simple and convenient. They can be used for statistical analyses and usually give satisfactory results. However, in the statistical sense, these parameters are still not strictly defined. In some cases, such as at the bimodal distribution, the parameters do not present a complete picture.

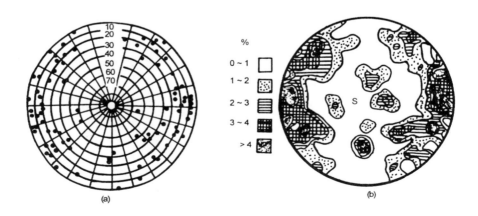

Fig. 2.18 Graphs from an analysis of orientation distribution

2.2.2.2 Analysis of orientation distribution

Just as sediment particles can have different sizes, they can also have different orientations. Investigation of this property is called analysis of orientation distribution. As mentioned before, the orientation of sediment particles is defined by both the dip angle and the azimuth of the long axis (or other axis). Conceptually a sediment particle is placed at the center of a hollow sphere with its natural orientation. The prolonged long axis intersects the surface of the hollow sphere at points in the southern and northern hemispheres. If a map of southern hemisphere is plotted in polar coordinates, then the position of the intersection point on that map in polar coordinates represents both the orientation angle and the dip angle of the sediment particle. Such a map is called a graph of orientation distribution (Fig. 2.18a). Each point in the map represents the orientation of the long axis of one sediment particle. If the points are concentrated at one place, the sediment particles have some dominant orientation during the processes of movement or of deposition. In order to demonstrate the degree of concentration of a group of points, isolines of the number of intersection points per unit area are used, and these are usually expressed in relative values. A typical set of isolines is given in Fig. 2.18b.

2.2.2.3 Porosity

Porosity is the percentage of voids in the total space occupied by a number of sediment particles. The porosity depends upon the size, uniformity, and shape of particles, and also on the pattern of deposition; hence, it depends on the external forces acting on the deposits and the time over which they act.

Theoretically, sediment size should have nothing to do with porosity. But in fact, fine sediment particles have more voids than coarse particles because the surface area of fine particles is relatively larger. Therefore, the friction between particles is larger and so also are both the adsorption action and the formation of texture. For example, the porosity of coarse sand usually falls between 39% and 41%, that of medium sand 41 to 48%, and that of fine sand 44 to 49%. If a small amount of clay particles is mixed with sandy soils, the porosity may be 50 to 54%.

Uniform sediment particles have the highest porosity, because in a mixture of non-uniform sediment, the voids between coarse particles can be filled with fine ones. Experimental results show that the porosity of a mixture of different-sized spheres can be as small as 15%. The porosity of natural sediment usually falls between 25% and 50%. For sediment finer than 5 μm, and if floc structures form during their settling process, the porosity can be as high as 90%. Together with the rate of oncoming sediment flow, the porosity of the deposits is an important factor that affects the speed of reservoir siltation. In Table 2.7 the porosity of various sediment samples is given for diameters ranging between 0.5 and 1.0 mm [19].

Table 2.7 Porosity of various sediment samples

Sample	Specific gravity	Dry sample		Wet sample	
		loosely piled	compact	loosely piled	compact
Lead stearate	11.21	41	37	42	30
Oceanic sand	2.68	39	35	43	35
Lake beach sand	2.66	41	37	47	38
Lacustrine sand	2.68	41	38	45	39
Crushed calcite	2.67	51	41	55	43
Crushed quartz block	2.65	48	41	54	44
Broken mica	2.84	94	87	92	87

Table 2.7 shows that undisturbed sediment (crushed calcite and debris of quartz blocks) has a higher porosity than does transported sediment. Here, the effect of sediment shape on porosity is involved; the abrasion during sediment movement makes the particle shapes close to spheroidal. Actually, the relationship between particle size and porosity is not the sole cause because the shape factor is also involved. The coarser the sediment particle, the more intensive is the abrasion it experiences during transportation. In the process, its shape approaches spheroid, and consequently, the porosity tends to be less. The effect of sediment shape on the porosity can be illustrated by the high porosity of broken mica, also given in Table 2.7.

The way in which sediment particles settle greatly affects the porosity after deposition. Bagnold pointed out that sediment particles sliding down along the surface of collapse on sand dunes in the desert have a large porosity, so that deposits thus formed possess properties of the moving sand. In contrast, if sediment settles out because of a decrease of sediment carrying capacity, due to increased friction or changes in ground surface structure, saltation of the particles still takes place. Impingements then bring about some shaking and cause the settled particles to be laid down selectively in their most stable position. The texture of deposits thus formed is quite compact and capable of bearing heavy loads [14]. For alluvial deposits, similar phenomena exist. Deposits on the flood plains of the Yellow River have different porosities and they are called "loosely sunken sand" and "well consolidated sand." Formation of a given kind of deposit may have similar causes, such as the afore-mentioned wind-drift sand.

If a bed of fine sand is loaded, the porosity of particles gradually decreases. The consolidation process will be discussed later in this chapter.

2.2.2.4 Permeability

Because the voids among sediment particles are interconnected, water can permeate through them. The flow that occurs obeys the well-known Darcy law

$$\frac{Q}{A} = kJ \tag{2.7}$$

in which:

  Q—discharge passing through a sediment column with cross-section A;
  J—hydraulic gradient;
  K—coefficient of permeability.

The coefficient of permeability of sediment depends upon particle size and size distribution. The following expression was obtained from the experimental results of Krumbein and Monk.

$$k \sim D_m^2 e^{-a\sigma} \tag{2.8}$$

in which:

  D_m — geometric mean diameter;
  σ — standand deviation;
  a — constant.

2.2.2.5 Angle of repose

Once the surface of a pile of sediment particles reaches a certain slope, particles begin to slide down it. The specific angle is called the angle of repose and it is designated by α.

Migniot found that the angle of repose for submerged sand and small gravel varies between 31° and 40° (Fig. 2.19), whereas that for bakelite powder with a specific weight of 1.40 varies between 34° and 46°. Also, tanα varies inversely with the square root of particle size. If the specific weight of particles is less, α is larger [21].

A large amount of experimental data pertaining to angle of repose for submerged sediment particles has been published, and the resulting divergence is significant. The differences are mainly due to the different methodologies used, but they also depend somewhat on the different properties of sediment samples. For example, if sediment particles are angular and have sharp edges, then the

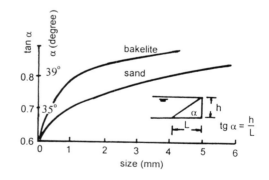

Fig. 2.19 Relationship between particle diameter and angle of repose for submerged cohesionless sediment (after Migniot, C.)

particles can interlock, and have a higher angle of repose, some 5° to 10° higher than samples composed of rounded particles. Until more reliable data become available, the angle of repose for cohesionless sediment particles greater than 0.5 mm in size can be taken from Fig. 2.20 [22].

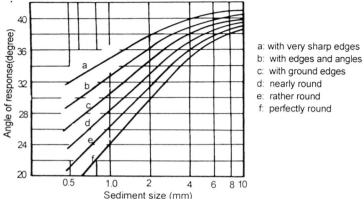

Fig. 2.20 Angle of repose of cohesionless sediment particle (after Lane, E.W.)

2.2.3 Physico-chemical action on the surface of fine sediment particles

The finer the sediment particle, the larger is its surface area per unit volume (i.e. specific surface area). Physico-chemical effects on a particle surface often play a significant role in the erosion, transport, and deposition of fine sediment particles particularly.

2.2.3.1 Double electric layer and the adsorbed water film

Water in nature is usually not pure. It often contains a few dielectrics and sometimes many. If a sediment particle is surrounded by water with dielectrics, one of two cases is bound to occur; either ions of the dielectrics are adsorbed on the surface of the particle, or molecules of surface sediment particles dissociate and ions of one charge leave and go into the water, while ions of opposite charge remain on the surface. Either occurrence results in the formation of an electric charge on the surface of the particles. The surface of most sediment particles is charged negatively. Because of the attraction due to static electricity, ions with opposite electric charges are distributed in the sediment particle of surrounding water, forming double electric layers, or layers of ions with opposite charge, as shown in Fig. 2.21.

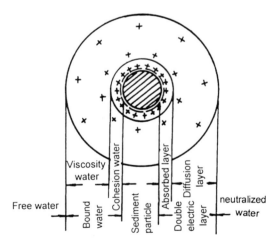

Fig. 2.21 Sketch of double electric layer and the adsorbed water film

Electric charges on a particle surface attracts not only ions of opposite sign but also water molecules. A water molecule is an even electrode consisting of an isosceles triangle with two hydrogen atoms (H^+) at the base angles and one oxygen atom (O^-) at the vertex. Because of negative charge on a particle surface, water molecules close to the surface lose their ability to move freely and are compacted tightly in an arrangement that is called cohesion water. The attraction force acting on cohesion water is about ten thousand bars. Under such a tremendous pressure, the density of cohesion water reaches 1.2 to 2.4 g/cm^3 with an average of 2.0 g/cm^3. This cohesion water layer consisting of highly compressed water molecules has the mechanical properties of solid matter. Its viscosity, elasticity, and shearing strength are all large, and it cannot transmit static pressure. Water molecules surrounding the cohesion layer are also subject to an attraction caused by static electricity. However, the greater the distance, the smaller the intensity of attraction, and the more the water molecules are able to move freely. These water molecules have a relatively loose order and may be only slightly oriented. This region is called viscosity water. The density of viscosity water, being 1.3 to 1.74 g/m^3, is not as high as that of the cohesion water, but is still much higher than that of ordinary liquid water. This water also has a high viscosity and high shearing strength, and it, too, cannot transmit static pressure.

Cohesion water and viscosity water together form what is called the bound water layer. The difference between the two parts lies in whether a relative movement to the sediment particle can be induced by hydraulic action or an electric field. Water molecules outside the bound water layer are almost unaffected by the attraction of static electricity and keep their original ability to move freely, is called free water. The cohesion water layer including the ions together with the particle surface forms the inner part of the double electric layer (adsorbed layer), and the viscosity water layer including the ions in the layer forms the outer layer (diffusion layer) of the double electric layer. In the layer of cohesion water and viscosity water, the water molecules partially or totally lose the ability to move freely, and they form a water film around the sediment particle, called the adsorbed water film. It is the product of an interaction between the sediment particle and water; from the viewpoint of mechanical properties, it is the transitional pattern between solid phase and liquid phase.

The thickness of the adsorbed water film depends mainly on the mineral composition of the sediment particle and the chemical composition of water (pH value, kind and concentration of ions). The thickness of a water film is on the order of 0.1μ. For coarse sediment particles, the volume of the film is only a tiny percentage of that of the sediment particle. The effect of the water film is accordingly limited. Therefore, the properties of coarse sediment particles are only those of the sediment itself. For fine sediment particles, especially particles less than 1μm in size, the situation is quite different. The water film that is inseparable from the sediment particle plays a more important role than does the particle itself. Therefore properties of very fine sediment particles often vary greatly because they depend on the mineral composition and the dielectrics of the water.

The distribution of electric potential in the double electric layer is shown in Fig. 2.22. The electric potential between the solid surface and free water is called the thermoelectric potential and is designated by ψ_0. Its value depends on the total number of ions adsorbed on the particle surface. The potential drops considerably across the adsorbed layer (or the inner layer) due to the neutralization of ions with opposite charge in that layer. The residual potential is called the electro-dynamic potential, or the ζ-potential. It is the potential difference between the inside and outside of the diffusion layer. The ζ-potential determines directly the thickness of the diffusion layer and hence the latter is often expressed by the ζ-potential, which depends on the difference between the total number of ions adsorbed on the particle surface and the number of ions with opposite charge in the adsorbed layer.

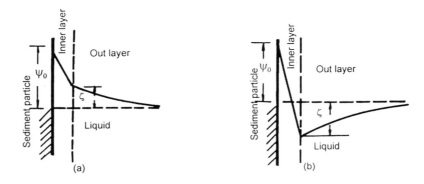

Fig. 2.22 Distribution of electric potential in double electric layer

Factors that cause the variation of the thickness of the double electric layer and of the electrodynamic potential are mainly the mineral composition of the sediment particle and the chemical composition of water. The mineral composition of the particle determines the manner in which the double electric layer forms. Consequently, it determines which chemical components in the water control the thickness of the diffusion layer. Changes in ion concentration or type of dielectrics in the water can change the thickness of the double electric layer and the electrodynamic potential. If the ion concentration increases, ions with opposite charge are pressed toward the solid surface, so that more ions move from the diffusion layer to the adsorbed layer; in this way the thickness of the double electric layer decreases, and the ζ-potential drops correspondingly from ζ_1 to ζ_2 as shown in Fig. 2.23. In the opposite sense, if the ion concentration decreases, ions with opposite charge diffuse toward the liquid phase; some ions enter into the diffusion layer from the adsorbed layer, the thickness of double electric layer increases and the ζ-potential increases correspondingly. If the dielectric of the ions, different from the ion with opposite charge in the double electric layer, is added, then ion-exchange takes place. Ions of higher valence, whose exchange capacity is stronger, are adsorbed more easily by the solid surface, and the original ions with opposite charge in the double electric layer are replaced. The sequence of capacity for ion exchange is as follows:

44

$$Fe^{+++} > Al^{+++} > H^+ > Ba^{++} > Ca^{++} > Mg^{++} > K^+ > Na^+ > Li^+;$$

Generally, if ion-exchange takes place, the valence of the ion with opposite charge in the double electric layer becomes higher, and the attraction to ions on the particle surface becomes stronger. This results in decrease of the thickness of the double electric layer and sharp drop of the ζ-potential. If the ion concentration increases further and surpasses the original charge on the solid surface, it causes a reversal of the electric potential at the outer boundary of the adsorbed layer of double electric layer, the normal ζ_1-potential then changes to the abnormal potential of ζ_3 (Fig. 2.23).

Properties and variations of the double electric layer and the adsorbed water film play important roles in the movement of fine sediment particles and in their properties. Details are presented in the following sections.

2.2.3.2 Flocculation and dispersion

The finer the sediment particles, the less important is the role played by gravity and the more important the role of the interactions among particles. If two sediment particles with adsorbed water films approach each other, a common adsorbed water film forms. If the particle is fine enough, the common adsorbed water film may connect them together. If ions exist in the film, the phenomenon becomes more evident and more complex. The

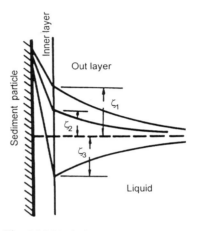

Fig. 2.23 Variation of electro-dynamic potential induced by changes of ion concentration or type of dielectrics in the solution

adsorbed water film and the layer of ions with opposite charge are more or less the same thing. If a common adsorbed water film exists, so does a common layer of ions with opposite charge, more specifically, a common diffusion layer. Although electric charges on the surfaces of neighboring particles dispel each other because they have the same electric charges, these particles may link together closely because of the attraction of ions with opposite charge and form an aggregate. The phenomenon of neighboring particles binding together and forming a group under certain conditions is called coagulation. The strength of the attraction is related to the thickness of the diffusion layer. The thinner the diffusion layer is, the stronger the attraction and the more tightly the soil particles coagulate. On the contrary, if the diffusion layer is thick, ions in the diffusion layer have a higher degree of mobility. Instead of attracting each other, the soil particles repel each other and disperse. The change from the state of a coagulated aggregate to a state of separated particles, due to the increase of the thickness of diffusion layer, is called dispersion.

As mentioned, properties of a dielectric in water and its concentration directly affect the ζ-potential and the thickness of the diffusion layer; consequently, they also directly affect the flocculation and dispersion of the sediment particles. In order to accelerate the settling of sediment particles, one should lower the ζ-potential on the particle surfaces to make the diffusion layer thinner and to promote flocculation. Thus, one should add a small amount of flocculating agent, such as alum, green vitriol, calcium chloride. However, one should not use too much of flocculating agent. Otherwise, the ζ-potential may change to the opposite sign and result in stabilization of sediment particles in water. In the reverse sense, to destroy the flocculation of sediment particles and prevent them from settling rapidly, one should add an anti-flocculating agent such as sodium carbonate, amonium hydroxide and sodium metaphosphate.

Flocculated sediment forms large granules (with a large proportion of voids filled with water), called floc. These generally settle more rapidly than do single particles. However, if the sediment concentration exceeds a certain limit, flocs link together to form a honeycomb-like framework anywhere, then the settling tendency is obstructed. More details about this issue are contained in Chapter 3.

2.2.3.3 Cohesion

The relationship between shear resistance per unit area τ and pressure per unit area p is expressed as follows

$$\tau = p \tan \phi + c \qquad (2.9)$$

in which:

ϕ — angle of internal friction;

c — cohesion force.

From Eq. (2.9), the shear resistance of fine sediment particles (or clay) consists of two parts: one is a friction force related to pressure and the other is a cohesive force that is not related to the pressure. The difference between the two forces can be illustrated by a simple example. Cohesion force is similar to the case when two sheets of fly paper are put together with their sticky sides face-to-face. Even if no vertical pressure is exerted on them, some force is still required to pull them apart along the contact surface. The shear resistance comes from the adsorption of stickiness. On the other hand, if two sheets of sand papers are put together with the rough sides face-to-face. If no vertical pressure acts on sand papers, one can easily slide one of them along the contact surface. If a normal force is applied to press the sheets together, the shear resistance due to friction increases with that force.

If fine sediment particles are rather densely concentrated, a texture with a certain strength forms, as already mentioned. The cohesion of the fine sediment particles is a reflection of the strength of the texture. The following example of the transmission of pressure by the bound water film on a particle surface shows how the additional pressure is produced and how results from the existence of a thin water film on the contact surface of fine sediment particles [23]. If two thin plates at depth h are located a distance apart (Fig. 2.24), a thin water film around the plates exists, and it does not behave like a liquid. The distribution of pressure within the water film does not obey Pascal's law of static hydraulic pressure; but the pressure can be transmitted along a straight line. If the gap between two plates is large, besides the thin water film close the plate, the space is filled with free water capable of transmitting pressure in all the directions. The water pressures on the two sides of each plate are balanced, as shown in Fig. 2.24a, and hence no other external force is needed to lift the upper plate, other than one equal to its submerged weight. On the contrary, if the gap is small, there is no free water between the two plates; then only one side of each plate is subject to hydraulic pressure and atmospheric pressure, as shown in Fig. 2.24b. In order to lift the upper plate one has to overcome a pressure force in addition to the submerged weight of the plate. For coarse sediment particles, most of the voids are filled with free water, and the forces acting are analogous to those shown in Fig. 2.24a. For fine sediment particles, the voids are small and are filled mostly by thin water films. An additional pressure analogous to those shown in Fig. 2.24b acts so as to hold the particles together; consequently it is not easy to separate one from another.

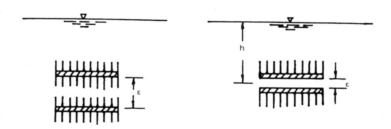

(a) The gap is filled with both water film and free water if ε is large.

(b) The gap is filled only with water film if ε is small.

Fig. 2.24 Sketch showing additional pressure due to the water film between two plates

The concept that cohesion is a product of specific features of pressure transmission in a thin water film is far from exact. The cohesion produced has the form of an additional pressure, and it differs from the shear resistance given in Eq. (2.9).

The cohesion of sediment is closely related to its erosivity. As the physical mechanism of cohesion is still not thoroughly understood, no proper physical index that can properly relate cohesion to erosivity quantitatively has been developed.

2.2.3.4 Consolidation process and phenomena of thixotropy

Coarse sediment particles like sand, gravel, and cobbles do not further consolidate once they settle down onto the river bed. But for fine sediment, especially clay, the situation is different. Because the latter flocculate while settling, they are linked to one another and form flocs, and these, in turn, connect with other flocs to form an aggregate. Aggregates can then further conbine to form texture [24]. The new deposits have a loose honeycomb-like texture with high water content and low density, as shown in Fig. 2.25a. The shear resistance or cohesion of flocculated particles is quite low.

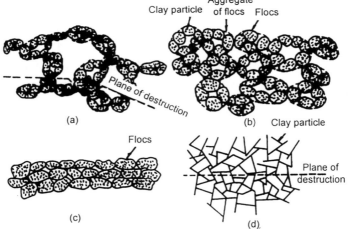

Fig. 2.25 Various textures of cohesive soil that formed under pressure during deposition (after Partheniades, E.)

Under the action of its own weight or another external force, the weakest connection among aggregates breaks first. The texture of deposits then changes progressively into a denser state of equilibrium, as shown in Fig. 2.25b, and they have a stronger cohesion.

If the pressure increases further, the connections among the flocs are broken, aggregates of flocs disappear, and innumerable particles pile up strata, as shown in Fig. 2.25c.

Still further increase of pressure causes the deformation of flocs, and a decrease of their voids in flocs; deposits then acquire a uniform texture composed of densely arranged particles, as shown in Fig. 2.25d.

Evidently, properties like density, cohesion, and shear resistance of deposits formed of the same fine sediment particles but with different textures can differ greatly; their ability to resist erosion differs accordingly.

As mentioned, a newly deposited flocculate has a loose honeycomb-like texture with high water content and low density; it behaves somewhat like a semi-solid material. If such deposits undergo by shaking, stirring, supersonic wave action, or an electric current, deposits often liquefy and become a sol or suspension. As soon as the external force is removed, they recoagulate. The phenomenon of such abrupt change is called thixotropy. Experiments reveal that thixotropic sediment deposits are characterized by their size, shape, and mineral composition; the water in the deposits also has some specific chemical composition. The size of thixotropic sediment particles should be less than 0.01 mm, and the deposit must contain a sufficient proportion of the fine particles with sizes less than 0.001 mm. The shape of the particles should be disc-like or strip-like so as to facilitate the formation of a net-work texture. Montmoriollonite minerals are highly hydrophilic and disc-like shaped and their particles are small. Hence, they form deposits that readily become thixotropic. Besides, the pH-value of water and the concentration of dielectrics have larger effects on the thixotropy of sediment deposits.

2.3 PROPERTIES OF MUDDY WATER

Water containing a suspension of fine sediment particles is called muddy water, and it differs from the clear water in significant ways.

2.3.1 Unit weight and sediment concentration of muddy water implied

The weight of muddy water per unit volume, its unit weight, is usually expressed in kg/m^3. The relative amount of sediment in muddy water is its sediment concentration. Commonly used expressions for sediment concentration are:

Percent by volume $\quad S_v = \dfrac{\text{volume occupied by sediment}}{\text{volume of muddy water}}$

Percent by weight $\quad S_\omega = \dfrac{\text{weight of dry sediment}}{\text{weight of muddy water}}$

Mixed expression $\quad S = \dfrac{\text{weight of dry sediment}}{\text{volume of muddy water}}$

The unit weight of muddy water γ_m and sediment concentration are interrelated by the following relationship:

$$\gamma_m = \gamma + (\gamma_s - \gamma)S_v = \gamma + (1 - \frac{\gamma}{\gamma_s})S \qquad (2.10)$$

in which γ_s and γ represent the unit weights of sediment and of water, respectively.

The three different ways of expressing sediment concentration can be converted from one to another as follows:

$$S = \gamma_s S_v \qquad (2.11)$$

$$S_\omega = \frac{\gamma_s S_v}{\gamma + (\gamma_s - \gamma)S_v} = \frac{S}{\gamma + (1 - \frac{\gamma}{\gamma_s})S} \qquad (2.12)$$

2.3.2 Viscosity of muddy water

2.3.2.1 Classification of rheological patterns

In two-dimensional flow with a velocity gradient, a unit body of flow abcd (Fig. 2.26), has height Δy, length Δx, and thickness 1. If Δy is extremely small, the velocity varies approximately linearly over that distance. Because of the difference in velocity Δu between the two planes ad and bc, the unit body of flow will deform into $ab'c'd$ with an angular deviation after time interval of Δt. From experimental results, the rate of deformation of the unit body of flow $\Delta\theta/\Delta t$ is proportional to the shear stress on a unit area, τ,

$$\tau = \mu\frac{\Delta\theta}{\Delta t} = \mu\frac{1}{\Delta t}(\frac{\Delta u}{\Delta y}\cdot\Delta t) = \mu\frac{\Delta u}{\Delta y}$$

As Δy approaches zero,

$$\tau = \mu\frac{du}{dy} \tag{2.13}$$

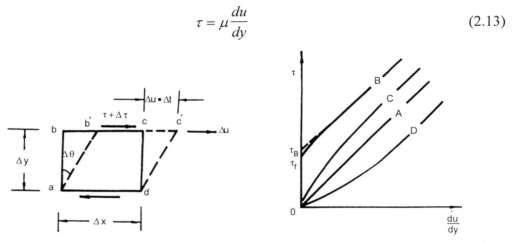

Fig. 2.26 Deformation of a free body under an external force in two-dimensional flow

Fig.2.27 Rheograms for time-independent fluids

This relationship is Newton's law for shear in a fluid. Fluid that obeys this law is called a Newtonian fluid, and μ in the equation is called the dynamic viscosity, or simply the viscosity. Gases, water, and most other liquids are Newtonian fluids. In rheology, the mathematical equation for the relationship between shear stress and shear rate is called the rheological equation, and its graphical relationship is called the rheogram. Fig. 2.27 shows rheograms for time-independent fluids. Among them, curve A represents a Newtonian fluid.

If sediment particles are carried by the flow, the viscosity increases because of the deformation of stream-lines around solid particles and the whirling of asymmetric particles that occur in a flow field with a velocity gradient; an additional factor is the increased friction due to flocculation. Moreover, if the sediment concentration (especially the content of fine sediment particles) exceeds a certain limit, the relationship between shear stress and shear rate no longer follows the Newtonian relationship. Such fluids are called non-Newtonian.

Non-Newtonian fluids are further subdivided into three categories:

1. Purely viscous non-Newtonian fluids. The fluid is characterized by the fact that the rate of deformation at any point is a function only of the shear stress; thus it is not affected by other factors. Commonly encountered fluids of this type includes Bingham fluids, pseudo-plastic fluids, and dilatant fluids. The corresponding rheological equations are given in Table 2.8 and the rheograms in Fig. 2.27.

Table 2.8 Rheological properties of Newtonian fluid and pure viscous non-Newtonian fluid

Rheological pattern		Curve	Rheological eq.	Definition of parameter and meaning	Remarks
Newtonian fluid		A	$\tau = \mu \dfrac{du}{dy}$	μ—coefficient of viscosity, slope of straight line A	
Purely viscous non-Newtonian fluid	Bingham fluid	B	$\tau = \tau_B + \eta \dfrac{du}{dy}$	τ_B—critical Bingham shear stress, intersection of straight line B η—rigidity coefficient, slope of straight line B	
	Pseudoel-plastic fluid	C	$\tau = K(\dfrac{du}{dy})^m$	K—stickiness coefficient m—plasticity coefficient, $m<1$	slope of straight line B a straight line on log-log plot
	Dilitant fluids	D	$\tau = K(\dfrac{du}{dy})^m$	K—stickiness coefficient, a straight line m—plasticity coefficient, $m>1$	slope of straight line B a straight line on log-log plot

2. Time-dependent non-Newtonian fluids. If the relationship between shear stress and shear rate of a fluid depends upon the time a shear is applied or varies with the history of the motion, it is called a time-dependent non-Newtonian. If shear stress decreases with time for a constant rate of deformation, it is called thixotropic fluid; if the shear stress increases with time, it is called antithixotropic fluid.

3. Visco-elastic fluids are characterized by a dual nature of both viscosity and elasticity. After deformation, the fluid partially recovers its original form because of elasticity.

2.3.2.2 Effects of suspended sediment particles on viscosity of muddy water and its rheological patterns

The presence of suspended sediment particles in muddy water increases the viscosity and can even change its rheological pattern from Newtonian to non-Newtonian. The causes are as follows [25]:

1. Deformation of stream-lines near the solid particles.

For non-spherical particles, especially disc-like or cylinder-like particles, the shear action in muddy water causes the particles to reorient themselves with their long-axis parallel to the shear direction.

2. Formation of flocs, aggregates, and textures.

Such flocs and textures are fragile and sensitive to shear stress. On one hand, they are easily destroyed, and on the other, they can readily reform their texture.

3. A certain degree of elasticity of the texture or chain formed by particles (especially for cylindrical ones) and the adsorbed water film due to flocculation.

If the shear rate of muddy water increases or decreases, the arrangement of sediment particles and the destruction and restoration of flocs and texture also change. However, the readjustment takes time. Therefore, for muddy water containing non-spherical particles, especially fine particles, the relationship between shear stress and shear rate varies with time, the shear can react as well with the history of the motion. In other words, muddy water behaves like a time-dependent non-Newtonian fluid but not like a time independent purely viscous one. Besides, as the adsorbed water film of particles and the chain or texture formed due to flocculation are somewhat elastic, the muddy water containing fine sediment particles can also be visco-elastic. The mathematical description of these two kinds of fluid is quite complicated. Fortunately in the case of steady uniform flow, some simplification can be made. First, in such case the elastic force does not appear and can be neglected. The condition is like a spring moving with constant speed but not subject to any temporal change, the length of spring does not change but acts as though it were an iron bar [25]. Second, if the arrangement of sediment particles reaches a certain condition, the destruction and restoration of flocs compensate to an acceptable degree. For these reasons, muddy water can be treated approximately as a purely viscous non-Newtonian fluid. Of course, for each other unsteady motion with rapid variation or if the muddy water passes cross-sections with abrupt contraction or enlargement, the simplified treatment may deviate significantly from the actual situation.

2.3.2.3 Theoretical analysis and experimental results of the rheology of muddy water

For low concentrations of coarse sediment, muddy water retains the features of Newtonian fluids. The relative viscosity μ_r, which can be used to assess the influence of sediment, is the ratio of the viscosity of muddy water to that of clear water at the same temperature.

If the sediment concentration is low, the distances between particles are rather large. Each particle has a certain effect on the flow near it. The greater the distance, the less the effect is. If the effect of the first particle does not extend to the position of a second particle, its effect is negligible. In other words, for flows with very low sediment concentrations, the interaction among particles is negligible. At any point in the muddy water, the effect induced by the existence of other sediment particles is the algebraic sum of all of their independent influences on neighboring particles. From this viewpoint, Einstein deduced the following well known formula [26].

$$\mu_r = 1 + 2.5 S_\upsilon \qquad (2.14)$$

in which S_υ is sediment concentration by volume.

For somewhat higher concentrations, neighboring particles have some effect on any given particle; i.e., the interaction of forces among particles becomes significant. In such circumstances, Eq. (2.14) needs to be modified. Usually, a polynomial expression is used

$$\mu_r = 1 + k_1 S_\upsilon + k_2 S_\upsilon^2 + k_3 S_\upsilon^3 + \cdots$$

The higher the concentration is, the larger the number of terms needed in the equation. If the polynomial extends to S_υ^2, it has the form

$$\mu_r = 1 + k_1 S_\upsilon + k_2 S_\upsilon^2 \qquad (2.15)$$

Most authors use the value of 2.5 for k_1 as used in Eq. (2.14). The values used for k_2, however, diverge considerably [26]. Among them, three proposed values are given in Table 2.9, and the resulting curves are compared in Fig. 2.28.

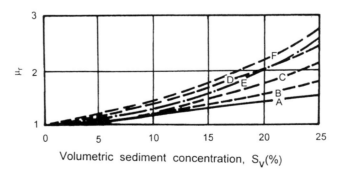

Fig. 2.28 Various expressions for μ_r versus S_υ

Table 2.9 Relationships between relative viscosity and sediment concentration for Newtonian fluids

Sediment conc.	Author	Expression	Curve in Fig. 2.28
Extremely low	A. Einstein	$\mu_r = 1 + 2.5S_\upsilon$	A
Low	H. Debruijin, J.M. Burgers& N. Saite	$\mu_r = 1 + 2.5S_\upsilon + 2.5S_\upsilon^2$	B
	V.Vand	$\mu_r = 1 + 2.5S_\upsilon + 7.35S_\upsilon^2$	C
	E. Guth, R. Simha, O. Gold	$\mu_r = 1 + 2.5S_\upsilon + 14.1S_\upsilon^2$	D
Relatively high	M. Mooney	$\mu_r = \exp(\dfrac{2.5S_\upsilon}{1 - kS_\upsilon})$, for suspension of glass ball k=1.43	E
	R. Roscoe	$\mu_r = \dfrac{1}{(1 - 1.35S_\upsilon)^{2.5}}$	F

For muddy water containing fine sediment particles, in which the sediment concentration is high enough, particles link together and form flocs and then textures. Under such conditions, muddy water no longer behaves as a Newtonian fluid, nor can it be analyzed like a simple problem of mechanics. The changes in rheological properties cannot be deduced theoretically. Over a long period of time, many experimental studies have been conducted in China and other countries on the rheological properties of muddy water. The principal achievements are summarized in Table 2.10, in which the relative viscosity μ_r has the forms:

$$\mu_r = \begin{cases} \mu / \mu_0 & \text{for Newtonian fluid} \\ \eta / \mu_0 & \text{for Bingham fluid} \\ K / \mu_0 & \text{for pseudoplastic fluid} \end{cases}$$

In which μ, η and K are viscosity, rigidity, and stickiness, respectively, and μ_0 is the viscosity of clear water at the same temperature.

For muddy water with the pattern of a Bingham fluid, the Bingham yield stress is proportional to the sediment concentration raised to n.

$$\tau_B \infty S_v^n$$

The value of n ranges between 3 and 5.4 in the results obtained by various authors. As sediment concentration increases, τ_B increases much more. In some formulas the expression differs somewhat, but the trends are similar.

The variation of rigidity (or stickiness) with sediment concentration is shown in Fig. 2.29; several empirical formula are plotted for sediment particles with three different median diameters.

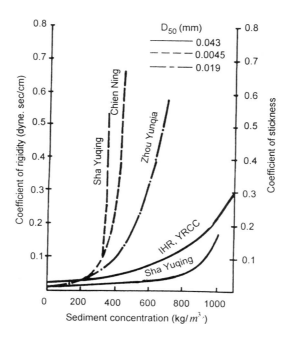

Fig. 2.29 Relationship between rigidity (or stickness) and the concentration of muddy water

Within the range of moderate sediment concentrations, rigidity (or stickiness) increases slowly with the increase of sediment concentration. If the concentration exceeds a certain critical value, the rigidity (or stickiness) increases rapidly with increasing concentration. If the median diameter of the particle is small, the critical value is also small. However, a large divergence in empirical formula by various authors is evident, especially for high concentrations. This divergence is primarily due to differences in the sediment samples used in the experiments, and secondly, differences in instrumentation and measuring techniques. In fact, the median diameter alone is far from being a sufficient representation of the properties of the sediment particles. The content of fine particles ($D<0.02\sim0.03$ mm) in the non-uniform sediment sample is probably a more important factor. Experiments conducted by the authors indicate that muddy water still behaves as a Newtonian fluid even at concentrations as high as $300\sim400$ kg/m³ if the sediment sample is entirely composed of particles with sizes greater than 0.02 to 0.03 mm. Migniot conducted an experiment by adding fine sediment of sizes 0.025 to 0.03 mm from the estuarine deposits of a sediment with a concentration of 385 kg/m³ so as to increase the concentration up to 456 kg/m³, and he found that the Bingham yield stress remained unchanged, only the rigidity η had increased [21]. Recent experiments show that for muddy water containing high concentration of coarse sediment. a Bingham yield stress exists, but the absolute value of τ_B is comparatively small. For example, Bagnold conducted a rheological experiment using a mixture of lead powder and paraffin wax ($d = 1.39$ mm, and the specific weight close to that of the fluid) in a rotational viscometer. He paid special attention to preventing any sorting of the solid particles. Different fluids of various viscosities were used (water, mixtures of glycerine-water-ethyl alcohol). Sediment concentrations by volume varied in the range from 13 % to 62 %. The experiment showed that only if the volume concentration of coarse particles exceeded 60 %, would the residual shear stress exit at the zero shear rate [30,31].

As mentioned above, the formation of flocs and texture due to flocculation affects the viscosity of muddy water; also the kind of ions and the ion concentrations in water affect the flocculation to some degree. Consequently, the water quality should also affect the viscosity of muddy water. Migniot pointed out that, for a given concentration, the Bingham yield stress and the relative viscosity of silt deposits in sea water are larger than those in river water (fresh water), the ratio of the two is in the range 1.4~1.7. For salinity within the range of 0~0.005, the Bingham yield stress increases rapidly with the increase of salinity. Above this range, it varies only slightly. The Bingham yield stress would be appreciably lower if an anti-flocculating agent were added to the silty deposits [21]. Elliott et al. found in experiments on the hydraulic transport of coal slurry that one can make the slurry of fine coal powder either more or less sticky by regulating its pH value [34]. Besides, the mineral composition of sediment particles is a factor that significantly affect the flocculation, consequently, it affects also the viscosity of muddy water. For example, the behavior of bentonite differs greatly from that of kaolinite. However, up to now, the research on this aspect does not yet give the whole story.

Table 2.10 Scope of experimental data pertaining to viscosity of muddy water and the empirical formulae of best fit

Author	source of sample	\multicolumn{3}{c}{Data scope of experimental}		\multicolumn{4}{c}{Empirical formula of best fit}	
		\multicolumn{2}{c}{size (mm)}	max.conc.(% by vol.)	rheological pattern	relative coefficient of viscosity	limiting sed.conc.(% by vol.)	Bingham yielding stress	Ref.
		D_{50}	D_{90}						
Sha, Yuqing	loess from Wugong	0.041		32	Bingham fluid	$\mu_r = \dfrac{r_m}{r} \cdot \dfrac{1}{1 - \dfrac{S_v}{2\sqrt{D_{50}}}}$	0.405		[27, 28]
		0.0205		23			0.286		
		0.0064		7.5			0.160		
Chien	Baotou	0.0058	0.021	11.2	Bingham fluid	$\mu_r = (1 - 2.9S_v)^{-2.5}$	0.345	$\tau_B = 6 \times 10^{-5} e^{19.1S_v}$	
Ning & Ma, Huimin	Guanting	0.0090	0.020	17.4		$\mu_r = (1 - 2.4S_v)^{-2.5}$	0.417		[29]
	old Tanggu	0.0080	0.047	14.6		$\mu_r = (1 - 3.1S_v)^{-2.5}$	0.323		
	new Tanggu	0.0045	0.013	8.7		$\mu_r = (1 - 4.9S_v)^{-2.5}$	0.204	$\tau_B = 6 \times 10^{-5} e^{32.7S_v}$	
	Zhengzhou	0.014	nearly uniform	15.5	Newtonian fluid	——			
		0.035		14.0		$\mu_r = 1 + 2.55S_v$			
		0.049		14.1					

Table 2.10 Continued

Zhou Yunqia	Floodplain deposits at Nanhechuan sta. We0ihe River	0.019	0.076	15.4	Pseudo-plastic fluid	$\mu_r = 10^{6.75S_v}$		1)
IHR, YRCC	loess from Zhenzhou	0.056	0.12				$\tau_B = 26.4 S_v^{5.4}$	2)
		0.043	0.074		Bingham fluid	$\mu_r = 1 + e^{8.24 S_v}$	$\tau_B = 101 S_v^{5.4}$	
		0.061	0.087				——	
		0.087	0.098	34.7	Newton fluid		——	
IGIFDR, CAS	slurry in debris flow at Dongchuan			75.2	Bingham fluid	$\mu_r = k_1 e^{k_2 r_m}$	$\tau_B = k_3 e^{k_4 r_m}$	3)
C. Migniot	deposits in harbors and reservoirs, Kaolinite, lime powder				Bingham fluid	——	$\tau_B = k_5 S_v^{(4\sim5)}$	[21]
H.E. Babbitt, D.H. Caldwell	clay				Bingham fluid	——	$\tau_B = 2.3 \times 10^{-3} e^{18 S_w}$	[32]
D.G. Tomas	Kaolinite				Bingham fluid	$\mu_r = e^{k_6 S_v}$	$\tau_B = 16.7 S_v^3$	[33]
Govier and Winning	clay				Bingham fluid	$\mu_r - 1 + \varepsilon_1 = k_7 e^{S_v}$	$\tau_B + \varepsilon_2 = k_8 S_v$	[25]

Note: Limiting sediment concentration refers to sediment concentration by volume as viscosity approaches infinity. k is a coefficient, ε_1—small amount, τ_B in units of g/cm^2.

IHR,YRCC refers to the Institute of Hydraulic Research, Yellow River Conservancy Commission.

IGFSDR,CAS refers to the Institute of Glacial, Frozen Soil and Desert Research, Chinese Academy of Sciences.

1) Zhou, Yongqia. "Rheological Properties of Muddy Water." Dept. of Sedimentation Engineering. Institute of Water Conservancy and Hydroelectric Power Research, 1963.

2) Institute of Hydraulic Research, Yellow River Conservancy Commission, 11th Bureau of Hydraulic Engineering Construction, Ministry of Water Conservancy and Electric Power, Administration Bureau of Luohui Irrigation District, Shaanxi Province, "Experimental Studies of Hyperconcentration of Sediment," 1978, PP 5-6.

3) Institute of Glacial, Frozen Soil and Desert Research, Chinese Academy of Sciences, "Preliminary Study on Rheological Properties of Debris Slurry ," 1977.

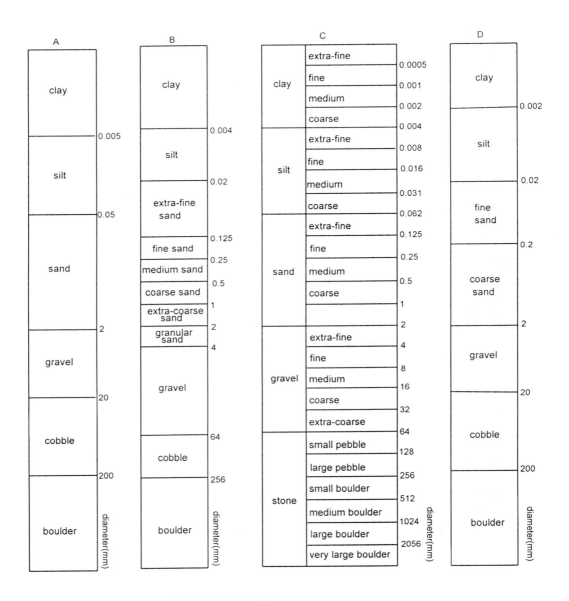

A: adopted by Chinese Hydrological Institutions
B: Wentworth's classification
C: adopted by AGU in USA
D: suggested by Atterberg and approved by International Association of Soil Sciences

Fig. 2.30 Classifications of sediment

2.4 CLASSIFICATION OF SEDIMENT

To differentiate various types of particles, one must subdivide them into groups. For a long time, various kinds of terminology have been used for the different sizes of sediment particles. In ancient Chinese river engineering, the three kinds of soils were named loessial, sticky, and plain. Different sizes of sediment can have different physical properties and different sources. Recent tendencies in sediment classification define narrower intervals and hence more groups, and the classification process has become more detailed.

Fig. 2.30 shows the classification of sediment that is commonly used. Atterberg made his classification at the beginning of the nineteenth century. It was approved by the International Association for Soil Sciences in 1927 as the standard in soil analysis, and it has been widely adopted in European countries. Most American geologists use Wentworth's classification [35] . In 1947 the American Geophysical Union drew up a new standard for sediment classification [36]. This standard is based on the same groups that Wentworth used. The only difference is that each group was subdivided, so that the denomination of classes is more complete. Classification of sediment in the hydraulic engineering of China followed that adopted by the former Soviet Union. Some divergence remains between this classification and those in European countries and the United States.

Despite the several classifications of sediment used in various countries, they have some points in common. Most of the intervals for the sediment groups are unequal as shown in Fig. 2.30, because the sizes cover such a wide range. Stone blocks and fine clay particles differ by a factor of more than a million. If an algebraic scale were used (with equal spaces between size groups), a scale suitable for coarse particle cannot be used for fine ones, and vice versa. For example, fine sediment particles with diameters of 0.01 mm and 0.06 mm behave rather differently, so that an interval of not more than 0.05 mm should be used in the classification. However, coarse sediment particles of sizes $d = 50$ mm and $d = 49.95$ mm, do not differ at all; one cannot even distinguish one from the other by the measuring techniques in common use. On the contrary, if an interval of 5 mm were used to suit coarse particles, then all sand, silt, and clay would belong to the same group; such a classification would be meaningless for fine sediment. Evidently, the classification of sediment size must follow a geometric scale; i.e., an appropriate ratio between neighboring groups must be taken. For the Atterberg and the Chinese classification, the ratio is 10, and for the AGU classification the ratio is 2. In connection with this aspect, sieve openings for granulometry also follow a certain ratio. For example, the sizes of the openings in the Tailor sieves follow the ratio of $2^{0.25}$.

Fig. 2.30 shows also that grain sizes such as 0.005 (or 0.004), 0.05 (or 0.06) and 2 mm are common to all classification as demarcations of sediment groups. These standard sizes one used because sediment particles larger and finer than these tend to show quite different characteristics, as shown in Fig. 2.31 [37].

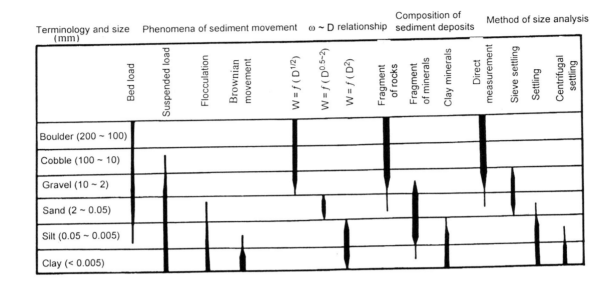

Fig. 2.31 Relationship between particle diameter and sediment properties

REFERENCES

[1] Reiche, P. *A Survey of Weathering Processes and Product.* Univ. New Mexico Pub. in Geology, No.3, Revised ed., 1950, p. 95.

[2] Griggs, D.T. "The Factor of Fatigue in Rock Exfoliation," *Journal of Geology,* Vol. 44, 1936, pp. 783-796.

[3] Roth, E.S. "Temperature and Water Content as Factors in Desert Weathering," *Journal of Geology,* Vol. 73, No. 3, 1965, pp. 454-468.

[4] Peltier, L. "The Geographical Cycle in Periglacial Regions as It is Related to Climatic Geomorphology," Ann., Association of American Geology, Vol. 40, 1950, pp. 214-236.

[5] Goldich, S.S. "A Study of Rock Weathering," *Journal of Geology,* Vol. 46, 1938, pp.17-58.

[6] Krumbein, W. C. "Fundamental Attributes of Sedimentary Particles," *Proceedings of 2nd Hydraulic Conference,* State Univ. Iowa, 1943, pp. 318-331.

[7] Wadell, H. "Volume, Shape and Roundness of Rock Particles," *Journal of Geology,* Vol. 40, 1932, pp. 443-451.

[8] Zing, Th. "Beitrage Zur Schotteranalyse," *Schweiz. Min. u. Pet. Mitt.,* Vol. 15, 1935, pp. 39-140.

[9] Krumbein, W.C. "Measurement and Geological Significance of Shape and Roundness of Sedimentary Particles," *Journal of Sediment Petrology,* Vol. 11, 1941, pp.64-72.

[10] Flemming, N.C. "Form and Function of Sedimentary Particles," *Journal of Sediment Petrology,* Vol. 35, No. 2, 1965, pp.381-390.

[11] Section of River Engineering, Dept. of Hydrology, Yangtze River Planning Office. "Preliminary Analysis of Bed Material Load for the Jinjiang Reach," *Journal of Sediment Research* , Vol. 3, No. 1, 1958, pp.1-14 (in Chinese)

[12] William, L. "Classification and Selected Biblography of the Surface Textures of Sedimentary Fragments," *Rept. Comm. Sedim.* 1936-1937, National Research Council, 1937, pp.114-128.

[13] Beijing Institute of Geology. *Soil Science.* China Industrial Press, 1961. (in Chinese)

[14] Bagnold, R.A. *The Physics of Blown Sand and Desert Dunes.* Methuen & Co London, 1941.

[15] Krumbein, W.C., and Petijohn, F.J. *Manual of Sedimentary Petrography.* Appleton-Century-Croffs, Inc., N.Y., 1938, pp.228-268.

[16] Inman, D.L. "Measures for Describing the Size Distribution of Sediments," *Journal of Sediment Petrology,* Vol. 22, 1952, pp.125-145.

[17] Fork, R.l. and Ward, W.C. "Brazoo River Bar: A Study in the Significance of Grain Size Paramenters," *Journal of Sediment Petrology,* Vol. 27, 1957, pp.3-27.

[18] Fridman, G.M. "Comparison of Moment Measures for Sieving and Thin-Section Data in Sedimentary Petrological Studies," *Journal of Sediment Petrology,* Vol. 32, 1962, pp.15-25.

[19] Gratio, L.C. and H.J. Fraser, "Systematic Packing of Spheres, with Particular Relation to Porosity and Permeability," *Journal of Geology,* Vol. 43, 1935, pp.785-909.

[20] Krumbein, E.C. and G.D. Monk. "Permeability as A Function of the Size Parameters of Unconsolidated Sands." *Amer. Inst. Mining Met.Engrs.* Tech. Pub. No. 1492, 1942.

[21] Migniot, C. 1968, *Etude des Proprietes Physiques de Differents Sediments Tres Fins et de Leur Comportement Sous des Actions Hydrodynamiques,* Houille Blanche, No.7, p.591-620.

[22] Lane, E.W. "Progress Report on Studies on the Design of Stable Channels by the Bureau of Reclamation." *Proceeding of American Society of Civil Engineers,* Vol.79, Separate No.280, 1953.

[23] Wuhan Institute of Hydraulic and Electric Engineering. *River Dynamics.* China Industrial Press, 1961, pp. 24-25 (in Chinese).

[24] Partheniades, E. "Results of Recent Investigation on Erosion and Deposition of Cohesive Sediments," In Sedimentation, by H.W. Shen (Ed.), 1972, p.39.

[25] Govier, G.W. and K. Aziz. *The Flow of Complex Mixtures in Pipes.* Van Nostrand Reinhold Co., 1972.

[26] Frisch, H.L. and R. Simha. "The Viscosity of Colloidal Suspensions and Macromolecular Solution," *In Rheology, Theory and Applications,* by F.R. Eirich (Ed.), Academic Press Inc., 1956, pp. 525-613.

[27] Li, Hanru. "Preliminary Measurement of Viscosity of Loessial Soils," *Voice of Northwestern Hydraulic Engineers*, 1945, Vol. 6, No. 4. pp. 57-68 (in Chinese).

[28] Sha, Yuqing. "On the Classification and Nomenclature of Sediment Particles," *Journal of Water Conservancy*, 1947, Vol.15, No.1, pp.152-178 (in Chinese).

[29] Chien, Ning and Huimin Ma "Viscosity and Flow Pattern of Muddy Water." *Journal of Sediment Research*, 1958, Vol. 3, No. 3, pp. 52-77 (in Chinese).

[30] Bagnold, R.A. "Experiments on A Gravity Free Dispersion of Large Solid Spheres in A Newtonian Fluid under Shear." *Proceeding of Royal Society*, London, Ser.A, Vol. 225, 1954, pp. 49-63.

[31] Bagnold, R.A. "The Flow of Cohesionless Grains in Fluids." *Philosophy Transition of Royal Society*, London, Ser. A, Vol. 249, 1956, pp.235-297.

[32] Babbitt, H.E. and D.H. Caldwell. "Turbulent Flow of Sludges in Pipes." *Bulletin*, Univ. Illinois, Vol. 38, No.13, 1940, p.44.

[33] Thomas, D.G. "Transport Characteristics of Suspensions: III. Laminar Flow Properties of Flocculated Suspensions." *Journal of American Institute of Chemical Engineers.*, Vol. 7, No. 3, 1961, pp. 431-437.

[34] Elliott, D.E. and B.J. Gliddon. "Hydraulic Transport of Coal at High Concentrations." *Proceeding of Hydrotransport 1, 1st. International Conference on the Hydro-Transport of Solids in Pipes*, Paper G2, 1970.

[35] Wentworth, C.K. "A Scale of Grade and Class Terms for Clastic Sediments," Journal of Geology, Vol. 30, 1922, pp. 373-392.

[36] Subcommittee on Sediment Terminology, American Geophysic Union. "Report on the Subcommittee on Sediment Terminology, American Geophysical Union." *Transition of American Geophysic Union*, Vol. 28, No. 6, 1947, pp. 936-938.

[37] Ruhin, A.B. "On the Properties and Classification of Clastic Sediment." Selected Papers in Geology (translated from Russian), 1957, Nov. pp. 33-48 (in Chinese).

CHAPTER 3

FALL VELOCITY OF SEDIMENT PARTICLES

The terminal velocity of sediment particles settling in liquids, often called the fall velocity, is an important physical quantity, one that is used in characterizing sediment transport. This chapter contains the principles governing the settling of sediment particles and the various factors that affect their fall velocity.

3.1 MECHANICS OF SPHERES SETTLING IN QUIESCENT WATER

The simplest case is a single sphere falling with a constant velocity in quiescent water of large extent. The force of gravity W that acts on a sphere with diameter D as it falls in water is

$$W = (\gamma_s - \gamma) \frac{\pi D^3}{6}$$

The resistance to the motion F is

$$F = C_D \frac{\pi D^2}{4} \cdot \frac{\rho \omega^2}{2} \tag{3.1}$$

in which ω is the fall velocity of the sphere and C_D is the drag coefficient.

At the beginning of the settling process, the velocity of the sphere is small, and the force of gravity is greater than the resistance. Hence, the sphere moves with acceleration, and the resistance to the motion increases with the velocity. After a certain distance of travel, the resistance equals the force of gravity; the sphere then falls with a constant velocity, called its fall velocity. That is, W and F are equal, and the equation for the fall velocity is

$$\omega^2 = \frac{4}{3} \cdot \frac{1}{C_D} \cdot \frac{\gamma_s - \gamma}{\gamma} gD \tag{3.2}$$

in which the drag coefficient is a function of the Reynolds number $(\omega D/\nu)$,(Fig. 3.1).

During the settling process, the motion of a sediment particle causes the surrounding fluid to move also. If the Reynolds number is less than about 0.4, the effect of inertia forces induced in the fluid by the motion is much less than those due to the fluid viscosity; the motion of the fluid surrounding the sediment particle is then as shown in Fig. 3.2. Settling of a single particle affects a large region of the surrounding fluid if the Reynolds number is low. If the inertia forces in the fluid are negligible, the Navier-Stokes equations can be linearized and solved. Early in 1851, Stokes obtained the following relationship in this way [1]:

$$F = 3\pi D \mu \omega \qquad (3.3)$$

and it is known as Stokes law. In this case, the drag coefficient is inversely proportional to the Reynolds number,

$$C_D = \frac{24}{R_e} = \frac{24}{\dfrac{\omega D}{v}} \qquad (3.4)$$

and it therefore follows a straight line with a slope of -1 in Fig. 3.1. By substituting the above equation into Eq.(3.2), one obtains the fall velocity of the sphere in the form:

$$\omega = \frac{1}{18} \cdot \frac{\gamma_s - \gamma}{\gamma} \cdot \frac{g D^2}{v} \qquad (3.5)$$

and it is proportional to the square of the sphere diameter. The condition for Stokes

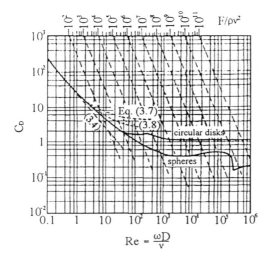

Fig. 3.1 C$_D$ vs. Re for spheres
and circular disks

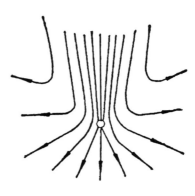

Fig. 3.2 Streamlines for settling of
a sphere in Stokes region

law to apply ($Re < 0.4$), in water at normal temperatures, corresponds to a limiting diameter of

$$\sqrt[3]{\frac{0.4 \times 18}{g} \cdot \frac{\gamma}{\gamma_s - \gamma} v^2} = 0.076 mm$$

Thus, the fall velocity of spheres that pass through the mesh opening of sieve No. 200 can be evaluated from the Stokes formula.

For Reynolds numbers larger than 0.4, fluid inertia becomes more and more important as Re increases. The region in which the viscous deformations are significant

64

is then confined to the immediate vicinity of the sphere, and the flow tends to separate. The separation is hardly noticable for $Re = 3$, but for $Re = 20$, it is well developed; a wake forms above the sphere and vortices are produced, as shown in Fig. 3.3. The inertia force and the flow separation make the motion quite different from that characterized by Stokes law. If the Reynolds number is 2×10^4, the viscous force at the spherical surface is so small that compared with the form resistance, it can be neglected. The drag coefficient is then essentially constant and thus independent of the Reynolds number. Since the change of the drag coefficient is small for $Re > 10^3$, that limiting value is usually taken as a critical value. In the range for Re between 0.4 and 10^3 , both inertia forces and the viscous forces have significant effects. However, if Re exceeds 10^3, viscous forces are negligible and

$$C_D = 0.45$$

$$\omega = 1.72 \sqrt{\frac{\gamma_s - \gamma}{\gamma} gD} \qquad (3.6)$$

In this range ($Re > 10^3$), the fall velocity of a sphere is linearly proportional to the square root of its diameter.

For values of Re up to 2×10^5, the boundary layer on the spherical surface is laminar, and the distribution of the pressure force on the spherical surface is as shown

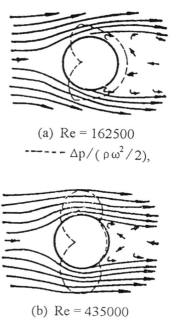

(a) Re = 162500

-- - - - $\Delta p / (\rho \omega^2 / 2)$,

(b) Re = 435000

Fig. 3.3 Wake and vortices above sphere during settling at large Re

Fig. 3.4 Distribution of pressure on spherical surface

in Fig. 3.4a. For larger Reynolds numbers, the flow in the boundary layer is turbulent. The separation point on the sphere is then further downstream, the separation zone is smaller, as shown in the Fig. 3.4b, and the pressure inside the separation zone is higher; therefore, the drag coefficient is much less. The critical Reynolds number at which this phenomenon takes place is related to the properties of the spherical surface. If the spherical surface is smooth, it occurs at about $Re = 2 \times 10^5$, but if the surface is rough, the critical Reynolds number is less [2].

The proper fall-velocity formula is Eq. (3.5) if the surrounding flow is laminar and Eq. (3.6) if it is turbulent.

A number of scholars have derived mathematical equations for the fall velocity in the transition region. If the Reynolds number exceeds only somewhat the limiting value of 0.4, one can use equations of motion that include some effects of inertia. In an extension of the Stokes analysis, Oseen derived the following approximate solution [3]:

$$C_D = \frac{24}{Re}\left(1 + \frac{3}{16}Re\right) \tag{3.7}$$

Goldstein extended Oseen's work and obtained the following more rigorous solution [4]:

$$C_D = \frac{24}{Re}\left(1 + \frac{3}{16}Re - \frac{19}{1280}Re^2 + \frac{71}{20480}Re^3 - \cdots\right) \tag{3.8}$$

These two formulas are also shown in Fig. 3.1. If Re is less than about 2, Goldstein's formula agrees well with the experimental results. However, for somewhat larger Reynolds numbers, form drag due to flow separation becomes important, and then neither the Oseen formula nor the Goldstein formula properly represents the actual situation. Rigorous mathematical analysis of this case is impossible, and one must rely on empirical or semi-empirical formulas.

A simple treatment is based on the assumptions that viscous drag and form drag coexist in the transition region, and that these two drag components can be represented by Eqs. (3.3) and (3.1) but with slightly different coefficients. For fall at a constant velocity,

$$(\gamma_s - \gamma)\frac{\pi D^3}{6} = k_1 \frac{\pi D^2}{4} \cdot \frac{\rho \omega^2}{2} + k_2 \pi D \mu \omega$$

where k_1 and k_2 are constants that must be determined. After some manipulation, the above equation takes the form

$$\omega = -4\frac{k_2}{k_1} \cdot \frac{v}{D} + \sqrt{\left(4\frac{k_2}{k_1}\frac{v}{D}\right)^2 + \frac{4}{3k_1}\frac{\gamma_s - \gamma}{\gamma}gD} \tag{3.9}$$

Rubey [5] and WIHEE [6] independently obtained the above formula. They applied it to determine the fall velocity of natural sediment particles in the transition region. The method for evaluating the coefficients in the formula from experimental results is introduced in section 3.2, along with the discussion of the settling process of natural sediment particles.

The foregoing treatment is, of course, only approximate; even if one can represent the viscous drag and the form drag by Eqs. (3.3) and (3.1), the coefficients in the formulas should not be constants, but rather functions of the Reynolds number. The work of Dou [7] is a partial remedy for this deficiency.

Dou first assumed that, as the Reynolds number increases, the separation zone above the sediment particle increases gradually, and the separation angle increases accordingly, as shown in Fig. 3.5. He obtained the following relationship between the separation angle θ and the Reynolds number Re

$$\frac{d\theta}{d\,Re} = \frac{a_1}{Re}$$

The boundary conditions are

$$\begin{cases} Re = 0.25, & \theta = 0 \\ Re = 850, & \theta = 2\pi \end{cases}$$

In this way, he derived the following equation

$$\theta = 1.78\log 4\,Re \tag{3.10}$$

The projected area of the separation region normal to the fall direction of the sediment particle is

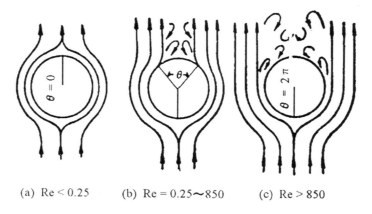

(a) $Re < 0.25$ (b) $Re = 0.25 \sim 850$ (c) $Re > 850$

Fig. 3.5 Sketch of flow pattern

$$\frac{\pi D^2}{4} \sin^2 \frac{\theta}{2}$$

Hence, the form drag due to flow separation in the wake of the sediment particle is given by

$$F_1 = C_{D_1} \frac{\pi D^2}{4} \sin^2 \frac{\theta}{2} \frac{\rho \omega^2}{2} \tag{3.11}$$

in which C_{DI} is the coefficient of form drag, taken to be a constant in the Dou analysis.

For the viscous drag outside the separation region, Dou used the Oseen formula. However, because of the separation zone, the area upon which the viscous drag acts is less. In this way, he obtained the expression

$$F_2 = 3\pi\mu D\omega \left(1 + \frac{3}{16}\text{Re}\right) \frac{1 + \cos\frac{\theta}{2}}{2} \tag{3.12}$$

Hence, the total drag force on the sphere is

$$F = C_D \frac{\pi D^2}{4} \frac{\rho \omega^2}{2} = C_{D1} \frac{\pi D^2}{4} \sin^2 \frac{\theta}{2} \frac{\rho \omega^2}{2} + 3\pi\mu D\omega\left(1 + \frac{3}{16}\text{Re}\right) \frac{1 + \cos\frac{\theta}{2}}{2} \tag{3.13}$$

in which C_D is the total drag coefficient. Since for $Re = 850$

$$\begin{cases} \theta = 2\pi, \ \sin^2 \dfrac{\theta}{2} = 1 \\ F_2 = 0 \\ C_D = 0.45 \end{cases}$$

the corresponding value of C_{DI} is 0.45. By substituting this value into Eq. (3.13), one obtains the following equation for the drag coefficient in the transition region

$$C_D = 0.45 \sin^2 \frac{\theta}{2} + \frac{24}{\text{Re}}\left(1 + \frac{3}{16}\text{Re}\right) \frac{1 + \cos\frac{\theta}{2}}{2} \tag{3.14}$$

in which the separation angle θ is the function of the Reynolds number given in Eq. (3.10).

In all the above formulas for the fall velocity, a drag coefficient is involved, and it is in general a function of the Reynolds number; hence, so is the fall velocity. Therefore, in computing the fall velocity of sediment particles, one must use a trial and error method. In order to avoid repeated trials, one can eliminate the fall velocity ω from the formulas for the drag coefficient and the Reynolds number and obtain the useful parameter

$$C_D = \frac{1}{\mathrm{Re}^2} \frac{8}{\pi} \frac{F}{\rho v^2} \qquad (3.15)$$

For any given value of $F/\rho v^2$, a plot of C_D vs. Re is a straight line on a log-log plot, and these are shown as broken lines in Fig. 3.1; these lines are the locus of points for which the drag is a constant. The intersection of the straight line and the C_D vs. $\omega D/v$ curve gives the Reynolds number and the drag coefficient that correspond to given values of F, ρ, v. For the fall velocity of a sphere

$$\frac{F}{\rho v^2} = \frac{\pi D^3}{6} \frac{\gamma_s - \gamma}{\gamma} \frac{g}{v^2} \qquad (3.16)$$

the corresponding C_D value is available from Fig. 3.1, and by substituting it into Eq. (3.2), one obtains the fall velocity of a sphere with a given particle diameter and the specific gravity of the given liquid [2]. For a quartz sphere with specific gravity 2.65 settling in water, the fall velocities at different temperatures are presented in Fig. 3.6.

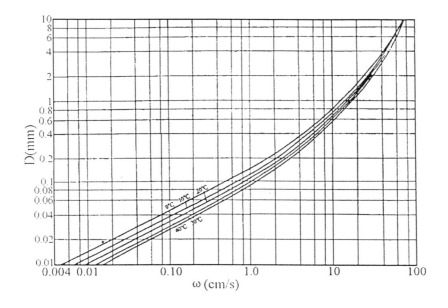

Fig. 3.6 Fall velocity of a quartz sphere settling in clear water

In treatments of sediment, a particle is usually considered to settle at a constant velocity. In reality, however, when a sediment particle starts falling at a given elevation, it takes some distance, and time, before it reaches steady state. Cai conducted research on the change of fall velocity of small spheres during the settling process [8], and obtained the theoretical relationship shown in Fig. 3.7, in which ω_t is the particle velocity at a time t, measured from the beginning of fall, and ω is the fall velocity under steady-state conditions (the Stokes fall velocity). The figure shows that, although the fall velocity of a sphere increases rapidly as it begins to fall, the rate of increase diminishes as the drag increases, and it approaches the fall velocity for steady-state conditions asymptotically. From the beginning of fall, the time required for a particle to reach its steady-state varies with the particle sizes. At normal temperatures, the time is only a few percent of a second for a 0.1 mm sediment particle; for smaller particles, this time is still less. Moreover, the higher the concentration of spherical particles in the flow, the less the time required to reach a constant fall velocity.

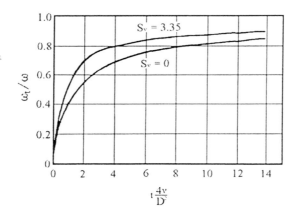

Fig. 3.7 Variation of fall velocity with time

3.2. EFFECT OF SHAPE OF PARTICLE ON FALL VELOCITY ——SETTLING OF NATURAL SEDIMENT PARTICLES

3.2.1 Orientation of settling sediment particles

The settling of natural sediments differs from that of a sphere mainly because the shape of a particle has a significant effect on the fall velocity. For a nonspherical particle, the orientation of the particle can change the projected area normal to the direction of fall. In addition, if the flow separates in the wake of the particle, the location of the point of separation and the size of the wake may also depend on the particle shape. And it can affect the drag force on the particle as well.

From experimental observations, the stable orientation reached by a sediment particle in the process of falling changes markedly with the Reynolds number. If the Reynolds number is less than 0.1, no matter what the initial orientation of the spherical particle is, the particle maintains this orientation and falls steadily. At Reynolds numbers above this value, a particle can oscillate and rotate as it falls, and it may never achieve a fixed orientation. At intermediate Reynolds numbers, the particle often adjusts its orientation so that its maximum cross section is normal to the fall direction: in that case, if a, b, and c represent the major, median, and minor axes of a sediment particle, the c axis will eventually be parallel to the direction the sediment particle falls [9]. The critical Reynolds number, which determines whether a particle falls steadily or not, is different for different particle shapes. Stringham et al. made tests with spheres, circular disks, circular cylinders, oblate spheroids, and prolate spheroids in liquids of different viscosities, and investigated this aspect systematically [10]. Using stroboscopic lighting, they took movies of settling particles to obtain instantaneous pictures of particles in the process of falling. The experiments revealed that oblate spheroids fall stably along the c axis if the Reynolds number is in the range from 10 to 10^5; only a few of the oblate spheroids rotated or oscillated, and they did so only slightly. For Re < 400, circular cylinders fell stably along the short axis; in the range of Re = 400 - 8,000, particles oscillated in a vertical plane as they fell; for Re > 8,000, they also oscillated in the horizontal direction. Prolate spheroids are similar to circular cylinders. The fall of circular disks, which differ the most from spheres, also displayed the most complicated pattern of falling. For Reynolds numbers less than 100, the disks fell stably along the c axis, Fig. 3.8a. For Reynolds numbers somewhat greater than 100, they oscillated regularly to the right and left, Fig. 3.8b, the magnitude of the oscillatory motion increased as the Reynolds number increased, and the direction of fall is almost parallel to the disc surface. For still higher Reynolds numbers, the particle both rotated and oscillated, the magnitude of oscillation was larger but the frequency was smaller, Fig. 3.8c. At still higher Reynolds numbers, a lead circular disc with a large specific gravity rotated 360^0 with an angular velocity that was almost constant, and the direction of fall was an inclined straight line, Fig. 3.8d; in contrast, an aluminum circular disc maintained the oscillatory motion.

Within the range of steady fall at low Reynolds number, the drag coefficients for different body shapes versus the Reynolds number all fall on the same curve. If the flow behind the sediment particle separates, the effect of particle shape is important. Fig. 3.1 compares the drag coefficients for a sphere (as a function of the Reynolds number) with those for a circular disk that falls with its flat surface horizontal. A value of Re of 100 is more or less the point of demarcation. Above this value, the drag force on a circular disk is much larger than that on a sphere, whereas below it, the difference between the two is not significant. The presence of a wake causes the position of the separation point on the upper part of a spherical surface to depend greatly on the Reynolds number. But any separation on a circular disk occurs along its circumference, so that the Reynolds number dependence is relatively small. Hence the drag forces on a sphere and on a circular disk can be quite different. If the flow does not separate or if the form drag due to separation is small, then for a given projected area normal to the fall direction, the difference between the drag forces on a sphere and on a circular disk is not large.

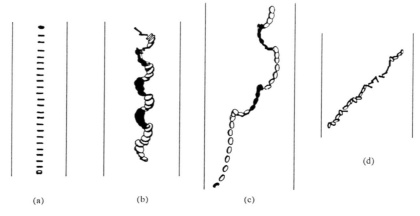

(a)									
(b)									
(c)									
(d)									

Figure	settling pattern	percentage of glycerine in water-glycerin mixture (%)	material of circular disks	diameter of circular disks (cm)	time interval of photo-graphing (s)	ω (cm/s)	Re	C_D
(a)	steady settling	98	lead	2.54	0.075	42	38	2.00
(b)	regular oscillation	70	lead	3.81	0.055	25.1	538	1.63
(c)	oscillation and rotation	0	lead	2.54	0.051	41.6	13400	0.48
(d)	rotation	0	lead	2.54	0.03	47.3	11900	2.02

Fig. 3.8 Settling of a circular disk at different Re

Particles whose shapes depart significantly from the spherical may take various stable orientations while settling. For example, if two equal cylinders with different orientations fall in a liquid but do not start at the same time (with the separation distance not more than 50 diameters), the later one gradually overtakes the earlier as they settle. Finally, they fall together in the form of a cross. If several cylinders fall in an arbitrary pattern, either two of them form a cross, or three form a double cross [11].

3.2.2 Settling of bodies with regular geometric shapes

Within the range of viscous resistance, the drag of several bodies with regular geometric shapes can be obtained analytically. As early as 1876, Oberbeck [12] presented basic solutions for ellipsoids falling slowly in a liquid. McNown and Malaika obtained general solutions for ellipsoids settling along the directions of the three principal axes; their results are shown in Fig. 3.9 [9,13]. In the figure, A denotes the length of the principal axis in the settling direction; B and C are the lengths of the two principal axes in the transverse plane, and K is a correction coefficient for the resistance given by Stokes law

$$F = K(3\pi\mu D\omega) \qquad (3.17)$$

K is related to the drag coefficient C_D in the following way

$$K = \frac{C_D \operatorname{Re}}{24} \qquad (3.18)$$

From the physical point of view, K represents the ratio of the settling velocity of a sphere with the same volume to the settling velocity of the given body. For an ellipsoid

$$K = \frac{16}{3D\zeta} \qquad (3.19)$$

where ζ is, in general, an elliptic function and the effective particle diameter D is given by

$$D = 2\sqrt[3]{ABC} \qquad (3.20)$$

Fig. 3.9 shows that the drag on a spheroid can be even less than that on a sphere. If a spheroid falls in the direction of its major axis, and that axis is twice the length of its minor axes, it has the minimum drag; it is then 95.5% of the drag on a sphere. McNown et al. also gave theoretical solutions for simple bodies with such other shapes as elliptic and circular disks [13].

To further determine the effect of particle shape on the settling velocity, McNown et al. [9,14] conducted experiments on the settling of five bodies with different shapes: prisms, spheroids, double pyramids, double cones, and cylinders in different fluids. From these experiments, they concluded that the best way to represent particle shape is to use the two parameters $A/(BC)^{0.5}$ and B/C. If the diameter D_n of a sphere that has the same volume as a given body is used as the reference diameter, the relation between the coefficient K and the two parameters is as shown in Fig. 3.10. The curves in this diagram represent the theoretical solutions of the resistance-correction coefficient for a spheroid. The points denote the experimental results, the different symbols denote bodies of different shapes, and the numbers beside each point denote the ratio B/C of a body. From the figure, no matter what the shape of the body, the difference

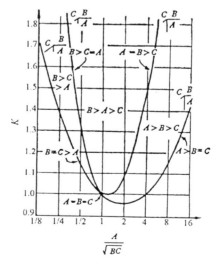

Fig. 3.9 K for an ellipsoid settling along one of three principle axes

between its resistance-correction coefficient and that of a corresponding spheroid is less than 10 percent; hence, the ratios of the principal axes of a nonspherical body can effectively represent the principal fluid-dynamic characteristics of the body. This conclusion is supported by experimental results at higher Reynolds numbers as well.

As already mentioned, a body may oscillate or rotate as it falls if the Reynolds numbers is large because of effects due to its wake. Since the rotation of a body is affected by the moment of inertia, the relationship between the drag coefficient and the Reynolds number is not as simple as that shown in Fig. 3.1; it is also affected by the specific gravity of the body [15]. At a given Reynolds number, the drag coefficient decreases as the specific gravity increases (Fig. 3.11). In their experiments, Stringham et al. found that the relationship between the

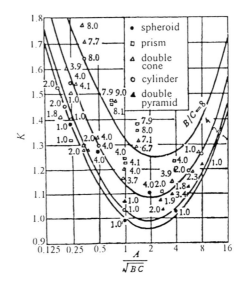

Fig. 3.10 Comparison of K for bodies of various shapes in Stokes range

drag coefficient and the Reynolds number also falls on different curves for circular disks made of tin and aluminum. The relationship between the drag coefficient and the Reynolds number must be modified accordingly:

$$C'_D = C_D \left(\frac{\gamma_s - \gamma}{\gamma} \right)^{1+\alpha} \left(\frac{c}{a} \right)^{\beta} \tag{3.21}$$

In the range of Re between 300 and 10^5, the values of α and β for different body shapes are as given in Table 3.1.

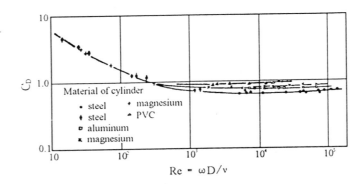

Fig. 3.11 Effect of specific weight on C_D

Table 3.1 Values of α and β

Body shape	α	β
Cylinder	0.05	2.3
Prism	0.05	2.3
Circular disk	0.05	0.315
Body with nearly equal principal axes	0.03	0

3.2.3 Settling of natural sediment

For natural sediments also, one can measure the lengths of the major intermediate, and minor axes, a, b, c, and establish experimentally the relationship between ratios of these axes and the settling velocity. Kira [16] conducted a large number of such experiments, and his results are shown in Fig. 3.12; the corresponding formula is

$$\omega = 15\left(\frac{a}{c}\right)^{-1/3}(abc)^{1/6} \qquad (3.22)$$

Kira used the major-to-minor axis ratio a/c to represent particle shape, and introduced the quantity

$$D = (6abc/\pi)^{1/3} \qquad (3.23)$$

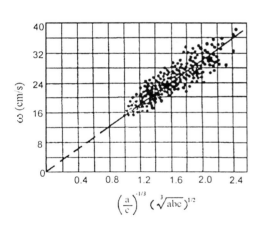

Fig. 3.12 Effect of grain shape on ω

as the effective diameter of a particle, a value that is quite close to the diameter of an equi-volume sphere. Alger and Simons [17] used pebble-sized particles of nine different shapes (spheres, cylinders, wedges, and disks) and conducted experiments in various mixtures of glycerine and water with liquid viscosities varying by a factor of 1,000; they obtained a family of curves relating the drag coefficient to the Reynolds number that is shown in Fig. 3.13, with $\left(\frac{c}{\sqrt{ab}}\right)\left(\frac{D_A}{D_n}\right)$ as a shape parameter; D_n is the equi-volume diameter, and D_A is the diameter of a sphere whose surface area is the same as that of the given particle. In this figure, D_A was used as the length term in the Reynolds number and the drag coefficient because it gave better results than the equi-volume diameter. Unfortunately, in order to use D_A, one must measure the surface area of a particle, and as pointed out in Chapter 2, this is a difficult task to perform. Schulz et al.[18], who use $c/(ab)^{0.5}$ as the shape parameter and Eq. (3.23) to determine the particle diameter, established the relationship between the

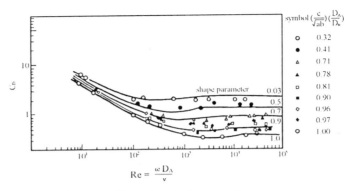

Fig. 3.13 Effect of grain shape on C_D

drag coefficient and the Reynolds number for various values of the shape parameter shown in Fig. 3.14. In addition to natural sediment particles, they used crushed rock fragments in their experiments. The drag coefficients for the latter are clearly higher than those for the former, about 30% higher over much of the range. The difference occurs because the settling velocity of a nonspherical particle is influenced not only by its shape but also by its curvature; surface roughness also has some effect. If $(ab)^{0.5}$ is used as the effective particle diameter, and if the projected area in the direction of fall is used in the drag formula Eq. (3.1), the curves for different values of shape parameter in Fig. 3.14 coalesce for Reynolds numbers less than 100, and they are close to the curve for a sphere. This result is similar to the coalescence of the curves for a sphere and a circular disk in Fig. 3.1, which also occurs for Reynolds numbers less than 100.

Just as many scholars have tried to establish a formula for the settling velocity of a sphere, so have many researchers investigated the mathematical representation of the

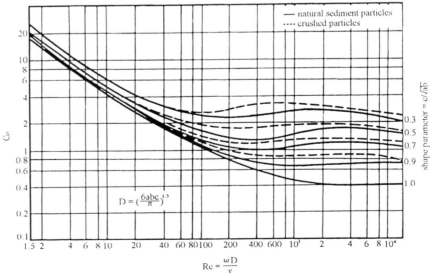

Fig. 3.14 Effect of grain shape on ω

fall velocity of a natural particle. As already mentioned, Rubey and WIHEE obtained the following formula for the settling velocity:

$$\omega = -4\frac{k_2}{k_1}\frac{v}{D} + \sqrt{\left(4\frac{k_2}{k_1}\frac{v}{D}\right)^2 + \frac{4}{3k_1}\frac{\gamma_s - \gamma}{\gamma}gD} \qquad (3.9)$$

For natural sediment, values of k_1 and k_2 are given in Table 3.2.

Table 3.2 Values of coefficients in Eq. (3.9)

Author	k_1	k_2
Rubey	2	3
WIHEE	1.22	4.27

Fig. 3.15 is a comparison between theoretical and experimental results. The figure shows that most of the experimental points fall between the theoretical curves of Rubey and WIHEE.

From an analysis of experimental results, Komar and Reimers [19] gave empirical relationships between the equi-volume diameter D_n of natural sediment particles of

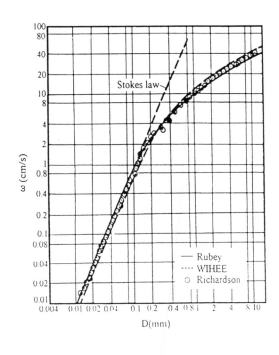

Fig. 3.15 Comparison of theoretical ω with experimental data

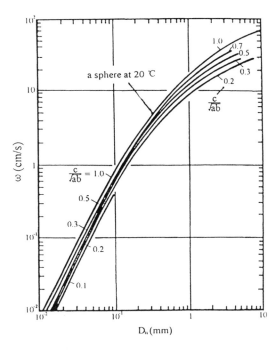

Fig. 3.16 Fall velocities of natural particles with different shapes in water at 20°C

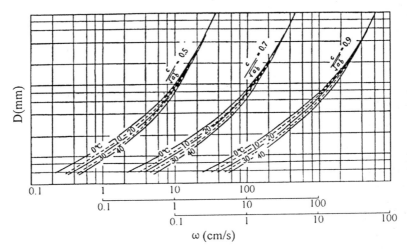

Fig. 3.17 ω of quartz particles of different shapes in clear water

different shapes (using $c/\sqrt{ab}$ as the shape parameter) and the settling velocity ω for water at a temperature of 20^0C, (Fig. 3.16). The family of curves diverges for large values of D_n and ω. Hence, the effect of particle shape on the settling velocity is larger for high Reynolds numbers.

Dou suggested that the drag coefficient should be determined from the following formula, instead of from Eq. (3.14),

$$C_D = 1.2 \sin^2 \frac{\theta}{2} + \frac{32}{Re}\left(1 + \frac{3}{16}Re\right)\frac{1 + \cos\dfrac{\theta}{2}}{2} \qquad (3.24)$$

in which

$$\theta = log4Re$$

The above formula is valid if the Reynolds number is in the range between 0.25 and 350; at the two limits

$$Re = 0.25, \qquad\qquad \theta = 0$$

$$Re = 850, \qquad\qquad \theta = 2\pi$$

To simplify computations, Fig. 3.17 shows the settling velocities of quartz particles of various shapes in clear water. For common natural sediment particles the shape parameter is about 0.7.

3.3 EFFECT OF BOUNDARY ON THE FALL VELOCITY

In the derivations of fall velocity formulas, the domain is usually taken to be unbounded, a condition that is of course not always approached in reality. Especially in laboratory investigations of settling velocities, or in granulometry based on settling velocities, the container size is often relatively small; thus one needs to know the effect of such a boundary on the fall velocity.

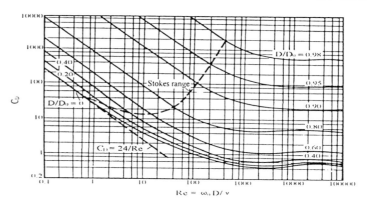

Fig. 3.18 C_D vs. Re for sphere settling in cylindrical containers with different diameters

Within the Stokes law range, the effect of boundary on the fall velocity of a sphere can be studied theoretically, and many scholars have conducted research on this topic. Table 3.3 summarizes the fall velocity formulas for a sphere moving under various boundary conditions [13]; in it, ω_0 is the fall velocity of a sphere in an unbounded fluid, ω is the fall velocity of the same sphere as affected by the boundaries of the flow.

For common sediment problems, Type-1, Type-2, and Type-4 boundary conditions listed in Table 3.3 are the most important. Type-1 is similar to the effect of sediment concentration on the fall velocity, a topic that is discussed in some detail in Section 3.4. In experimental studies of sediment transport in open channel flows, the fall velocity of a particle near the channel wall is smaller than that of particles in more central regions. The effect is equivalent to Type-2, as given in Table 3.3. As a sediment particle approaches the river bed during settling, the fall velocity of the sediment particle decreases because of the presence of the bed; this Type-4 effect can be estimated by the Lorentz formula [20] in Table 3.3.

In addition to investigating the fall velocity of a sphere inside a cylindrical container at low Reynolds numbers [21], McNown and others conducted experiments for a range of higher Reynolds numbers. Their results are shown in Fig. 3.18 [22]. The fall velocity of a sphere is clearly smaller because of the presence of cylindrical boundary. If the ratio of the diameter of a sediment particle to that of the cylindrical container, D/D_0, is small, the resistance effect of the cylindrical container decreases as the Reynolds number increases. As D/D_0 increases, the upper limiting Reynolds number for the Stokes region also increases.

Lorentz gave the following formulas for the settling velocity as affected by both a vertical boundary and a horizontal boundary:

Table 3.3 Effect of boundary condition on the fall (after Lorentz, H. A.)

	Boundary condition				Author
1	sphere settling in a cylindrical container			(1)	Ladenburg
			$\dfrac{D}{D_0} < 1$	(2)	Faxen
				(3)	McNown et al.
		$\dfrac{D}{D_0} \to 1$	sphere settling along axis of cylindrical container (4)		McNown et al.
			rotating, sphere settling near the wall of a container (5)		
2	sphere settling near a vertical wall			(6)	Lorentz
				(7)	Stock
				(8)	
					Oseen
				(9)	Faxen
3	sphere settling between two vertical walls		sphere settling at midway between two walls (10)		
			sphere settling at quarter point(L/4) (11)		Faxen
4	sphere approaching a horizontal bed (12)				Lorentz

Effect of boundary condition on the fall

Formula of fall velocity	Remark
(1) $\dfrac{\omega}{\omega_0} = 1 + 2.4\,\dfrac{D}{D_0}$	——
(2) $\dfrac{\omega}{\omega_0} = 1 - \dfrac{3}{16}\,\mathrm{Re} - \dfrac{D}{D_0}\,f\!\left(\dfrac{\omega_0 D}{\nu}\right) + 2.09\left(\dfrac{D}{D_0}\right)^{3}$ $- 0.95\left(\dfrac{D}{D_0}\right)^{5}\left(\text{approximately } \dfrac{\omega_0}{\omega} = 1 + 2.1\,\dfrac{D}{D_0}\right)$	<table><tr><td>$\dfrac{\omega_0 D}{\nu}$</td><td>0</td><td>2</td><td>4</td><td>8</td><td>20</td></tr><tr><td>f</td><td>2.10</td><td>1.76</td><td>1.48</td><td>1.04</td><td>0.46</td></tr></table>
(3) $\dfrac{\omega_0}{\omega} = 1 + \dfrac{9}{4}\,\dfrac{D}{D_0} + \left(\dfrac{9}{4}\,\dfrac{D}{D_0}\right)^{2}$	——
(4) $\dfrac{\omega_0}{\omega} = 1.66\left(1 - \dfrac{D}{D_0}\right)^{-5/2}$	——
(5) $\dfrac{\omega_0}{\omega} = 0.868\left(1 - \dfrac{D}{D_0}\right)^{-5/2}$	——
(6) $\dfrac{\omega_0}{\omega} = 1 + \dfrac{9}{32}\,\dfrac{D}{s}$	——
(7) $\dfrac{\omega_0}{\omega} = \dfrac{1}{1 - \dfrac{9}{32}\,\dfrac{D}{s}} - \left(\dfrac{D}{4s}\right)^{3}\left(1 + \dfrac{9}{32}\,\dfrac{D}{s}\right)$	——
(8) $\dfrac{\omega}{\omega_0} = 1 - \dfrac{9}{8}\,\dfrac{D}{4s} + \left(\dfrac{D}{4s}\right)^{3}$ $- \dfrac{45}{16}\left(\dfrac{D}{4s}\right)^{4} - 2\left(\dfrac{D}{4s}\right)^{5}$	——
(9) $\dfrac{\omega}{\omega_0} = 1 - \dfrac{3}{16}\,\mathrm{Re} + \dfrac{9}{8}\,\dfrac{D}{4s}\,f\!\left(\dfrac{\omega_0 s}{2\nu}\right)$ $+ \left(\dfrac{D}{4s}\right)^{3} - \dfrac{45}{16}\left(\dfrac{D}{4s}\right)^{4} - 2\left(\dfrac{D}{4s}\right)^{5}$	$f\!\left(\dfrac{\omega_0 s}{2\nu}\right) = f(x) = 1 - \dfrac{4}{3}x + \dfrac{23}{16}x^{2} - \dfrac{16}{9}x^{3}$ $+ \dfrac{317}{864}x^{4} + \dfrac{8}{9}x^{5} - \left[\dfrac{25}{24}x^{4} + \cdots\right]\log 1.78\,\dfrac{x}{2}$
(10) $\dfrac{\omega}{\omega} = 1 - 1.004\,\dfrac{D}{L} + 0.418\left(\dfrac{D}{L}\right)^{3} - 0.169\left(\dfrac{D}{L}\right)^{5}$	——
(11) $\dfrac{\omega}{\omega_0} = 1 - 0.6525\,\dfrac{4D}{L} + 0.1475\left(\dfrac{4D}{L}\right)^{3}$ $- 0.131\left(\dfrac{4D}{L}\right)^{4} - 0.0644\left(\dfrac{4D}{L}\right)^{5}$	——
(12) $\dfrac{\omega_0}{\omega} = 1 + \dfrac{9}{16}\,\dfrac{D}{s}$	——

Vertical boundary

$$\frac{\omega_0}{\omega} = 1 + \frac{9}{32}\frac{D}{s}$$ (3.25)

Horizontal boundary

$$\frac{\omega_0}{\omega} = 1 + \frac{9}{16}\frac{D}{s}$$ (3.26)

In which s is the distance between the center of the sphere and the boundary. These results differ somewhat from the experimental data. For $s/D < 10$, Eq. (3.25) gives results generally higher than the experimental results. Eq. (3.26) is comparatively reliable for $s/D > 15$, but it gives smaller values than the experimental ones for $s/D < 15$ [13].

In practice, for small particles settling within the range of the Stokes law, the boundary effect can be neglected if the distance between the particle and the container wall or the fluid boundary is greater than about 2 to 3 cm.

3.4 EFFECT OF SEDIMENT CONCENTATION ON FALL VELOCITY

If a fluid contains many sediment particles, the settling of any one of them is affected by the presence of the others.

First, as already mentioned, a single sediment particle induces motion of the surrounding fluid as it settles (Fig. 3.2). In the presence of other sediment particles, since a particle is solid and does not deform as does the fluid, the surrounding fluid can not move as freely. The effect of this change is equivalent to an increase in fluid viscosity; hence the fall velocity of the sediment particle decreases. For fine sediment particles that flocculate, the effective viscosity of the suspension changes drastically; the fall velocity of the particles is also affected significantly. The influence of flocculation on the fall velocity is discussed in section 3.6.

Second, a sediment particle entrains some of the nearby liquid as it settles so that some liquid also moves downward. If the liquid extends a large distance, the velocity of any other sediment particle would increase simply because of this induced motion. In contrast, if the liquid is restricted by a boundary, the downward motion of the nearby fluid induces an equal upward flow in more distant fluid, in accordance with the law continuity. Therefore, depending on the spatial arrangement of sediment particles, the fall velocity of a sediment particle may increase or decrease because of the motion of other particles. If a group of sediment particles falls in water and the water far away from them is clear, the fall velocity is greater than that of a single sediment particle falling in an unbounded fluid. On the contrary, if sediment particles are distributed more or less uniformly in the fluid, the fall velocity of each particle is less. The higher the concentration is, the more the effect and the smaller the fall velocity.

Moreover, the presence of many of sediment particles increases both the specific gravity of the suspension and the buoyant force on every sediment particles; consequently, it also affects their fall velocities. For high concentrations of particles, this influence is not negligible.

In the following section, the sediment particles are considered to be randomly but homogeneously distributed throughout the water. The simple case of a uniform sediment is treated first, and then that of non-uniform sediment. In the discussion of uniform sediment, since the mechanism of settling for low concentrations is different from that for high concentrations, the two are treated separately.

3.4.1 Effect of low concentrations of uniform sediment on the fall velocity

Within the Stokes range, two types of analyses have been made of the settling of sediment particles at low concentrations. In the first type of analyses, which was developed by Cunningham [23], McNown [21] and Ushida [24], the settling of sediment particles in a homogeneous suspension was treated as being much the same as the settling of a single sediment particle in a small container. The container size is about the same as the distance between the particles. The treatment of Cai [25] is similar. The second approach was developed by Smoluchowski [26] and Burgers [27]. From the flow pattern around a single sediment particle as it falls in clear water, they treated the settling of a suspension of sediment particles in such a way that flow at any single particle is the sum of the flows induced by all other particles. This approach gave them the fall velocity of a representative particle. If higher-order terms are negligible, both types of analysis produce essentially the following type of formula:

$$\frac{\omega_0}{\omega} = 1 + k \frac{D}{s} \qquad (3.27)$$

in which ω is the fall velocities of sediment particles at concentration S_v and ω_0 that at zero concentration, and s is the effective distance between neighbouring particles; the ratio D/s is related to the volume concentration S_v:

$$D / s = 1.24 S_v^{1/3} \qquad (3.28)$$

The different investigators obtained various values for the coefficient k in Eq. (3.27) shown in Table 3.4.

Table 3.4 Values of coefficient k in Eq.(3.27)

Author	Cunninghain	McNown	Uchida	Cai	Smoluchowski	Burgers
k	1.7~2.25	0.7	0.835	0.75	1.16	1.4

The above analyses are for viscous resistance (Stokes law range). McNown and Lin considered their formula for low concentrations to be applicable for Reynolds numbers less than 2 [28]. They also conducted a series of experiments that essentially verified the theoretical results for volume concentrations in the range of 0.35 to 2.25 percent. Their results and those of other investigators are listed in Table 3.4 and, with the inclusion of higher order terms, are shown in Fig. 3.19. The figure shows that the effect of concentration on the fall velocity is less for large Reynolds numbers. A comparison of the various formulas indicates that Cunningham's result appears to be too high, whereas within the range for Stokes law, the formulas of Smoluchowski, Uchida, and Cai are satisfactory.

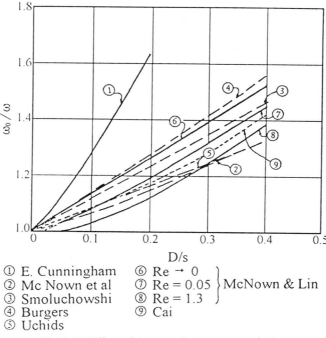

① E. Cunningham ⑥ Re → 0
② Mc Nown et al   ⑦ Re = 0.05 } McNown & Lin
③ Smoluchowshi   ⑧ Re = 1.3
④ Burgers   ⑨ Cai
⑤ Uchids

Fig. 3.19 Effect of S_v on ω (low concentration)

3.4.2 Effect of high concentrations of uniform sediment on the fall velocity

The following analysis is for fine particles settling within the Stokes range [29]. For a single sediment particle settling in clear water

$$\left(\gamma_s - \gamma\right)\frac{\pi D^3}{6} = 3\pi\mu_0 D\omega_0 \qquad (3.29)$$

in which γ and μ_0 are the specific weight and the dynamic viscosity of clear water, respectively, and ω_0 is the fall velocity of a single sediment particle in clear water.

The settling of a particle in sediment-laden water with volumetric concentration S_v has the following characteristics:

1. μ_0 in Eq. (3.29) should be replaced by the dynamic viscosity μ of the suspension;

2. Because of the reverse flow caused by falling particles, the relative velocity between a sediment particle and its surrounding fluid is not the fall velocity of the particle ω but is rather $\omega/(1-S_v)$;

3. In the calculation of the effective weight of a particle by means of Eq. (3.29), γ should be replaced by the specific weight of the suspension γ_m:

$$\gamma_m = \gamma_s S_v + \gamma(1 - S_v) \tag{3.30}$$

but

$$\gamma_s - \gamma_m = (\gamma_s - \gamma)(1 - S_v) \tag{3.31}$$

hence

$$(\gamma_s - \gamma_m)\frac{\pi D^3}{6} = 3\pi\mu\, D\frac{\omega}{1 - S_v} \tag{3.32}$$

From a comparison of equations (3.29) and (3.32), one obtains:

$$\frac{\omega}{\omega_0} = \frac{\mu_0}{\mu}(1 - S_v)^2 \tag{3.33}$$

The relative viscosity μ/μ_0 can be expressed by the Moritake formula

$$\frac{\mu}{\mu_0} = 1 + \frac{3}{\dfrac{1}{S_v} - \dfrac{1}{0.52}} \tag{3.34}$$

Then, one obtains

$$\frac{\omega}{\omega_0} = \frac{(1 - S_v)^2}{1 + \dfrac{3}{\dfrac{1}{S_v} - \dfrac{1}{0.52}}} \tag{3.35}$$

Using Stokes law, Hawksley included the increase of viscosity, the specific weight of the suspension, and the flow reversal due to particle settling, and he obtained the following formula to express the effect of concentration on the fall velocity [30]:

$$\frac{\omega}{\omega_0} = \zeta(1 - S_v)^2 \exp\left(\frac{-k_1 S_v}{1 - k_2 S_v}\right) \tag{3.36}$$

in which:

$\zeta = 1$ if no flocculation occurs, and $\zeta = 2/3$ if the particles are spherical, and flocculation does occur;

k_1 is a shape coefficient, and it equals 5/2 for spheres, and $5\Lambda/2$ for nonspheres; Λ is the sphericity of a sediment particle as defined in Eq. (2.4);

k_2 is a coefficient reflecting the mutual interaction of sediment particles on the viscosity of a suspension, it is 39/64 for spheres.

As pointed out in Section 3.6, the fall velocities obtained from Eq. (3.36) appear to be too high for the condition of flocculation.

Hawksley also pointed out that the mechanics of sedimentation at high concentrations is similar to that of a flow moving upwards through numerous capillary tubes between the sediment particles. If the concentration is high, the settling motion of sediment particles no longer differs much from seepage flow, and the two processes are governed by the same physical laws.

The chemist Steinour also considered the changes in viscosity and specific weight, and the flow reversal induced by concentration. He obtained a fall velocity formula for high concentrations in the form [31]:

$$\frac{\omega}{\omega_0} = [1 - (1 + \varepsilon)S_v]^2 10^{-1.82(1+\varepsilon)S_v} \qquad (3.37)$$

A comparison of this equation with Eqs. (3.35) and (3.36) shows that in addition to his use of different empirical formulas to reflect the change of viscosity, Steinour included a new coefficient ε in his formula. He argued that, if the sediment particles are not only non-spherical but also have rough corners and sharp edges, a part of the clear water will be effectively trapped by the sediment particles and fall with them. The consequence is an increase of the effective concentration of a suspension. The volume of the clear water that moves with falling sediment particles is proportional to the volume of the sediment particles. If flocculation of fine sediment particles occurs at high concentrations, the volume of trapped water is still greater and its effect is equivalent to a large increase in effective concentration. Hence, one should use a large value of ε. The values of ε for various conditions are given in Table 3.5.

Table 3.5 Coefficient ε in Eq.(3.37) (after Steinour, H. H.)

Body	Diameter (mm)	Flocculation phenomenon	ε
Glass balls	0.0135	non-flocculation	0
Emery	0.0096~0.0122	non-flocculation	0.200
Emery	0.0122	flocculation	0.366
	0.0096		0.404
	0.0046		0.538

Table 3.6 Empirical formulas for the effect of concentration
on fall velocity of uniform sediment

Author	Formula	m	Experiments & data	Remark	Reference
Richardson & Zaki	$\dfrac{\omega}{\omega_0}=(1-S_v)^m$	Function of Re	----	m=4.65 in laminar flow region, m=2.39 for Re larger than 500, m varies with Re in transition region	[32]
Guo & Zhuang	"	4.91	----	---	[33]
Field & Gay	"	4.50	Coarse particles settling in fine particle slurry, laminar flow region	---	
Mintz & Schubelt	"	2.25 ~ 4.45	Sand and pebble settling in clear water	For coarse particles m=2.25 and does not vary with diameter. For fine particles m increases rapidly as diameter decreases.	[34]
BMC	"	Function of D	---	(see table below)	
Sha, Y. Q.	$\dfrac{\omega}{\omega_0}=\left(1-\dfrac{S_v}{2\sqrt{D_{50}}}\right)^m$	3.0	Medium diameter of 0.01mm	---	[35]
Wang, S. Y.	$\dfrac{\omega}{\omega_0}=(1-\beta S_v)^m$	2.5	Silt and Steinour experimental data	(see table below)	[36]
Xia & Wang	$\dfrac{\omega}{\omega_0}=(1-S_v)^m$	7.0	Medium diameter of 0.067mm	---	[37]
Wang, Z. Y.	"	7.0	Medium diameter of 0.15mm	---	

BMC Remark table:

D(mm)	2	1.4	0.9	0.5	0.3	0.2	0.15	0.08
m	2.7	3.2	3.8	4.6	5.4	6.0	6.6	7.5

Wang, S. Y. Remark table:

Sediment	Spherical	Vitriol	Silt
β	1.77	3.92	5.0

Particles in suspension at different separation distances are, in effect, equivalent to the fall velocities of solid particles at different concentrations. From dimensional analysis, Richardson and Zaki [32] gave a simple exponential relation between the limiting velocity and the separation distance:

$$\frac{\omega}{\omega_0} = (1 - S_v)^m \tag{3.38}$$

in which m is an exponent to be determined. Several formulas of this type are presented in Table 3.6.

Most of the formulas have the same form — only the values of the exponent m are different. Generally, the values of m are small for coarse particles and large for fine particles; m is always larger than 2. In a comparison of this table with Eq. (3.33), one notes that the factor $(1-S_v)^2$ reflects the effect of both flow reversal and density change of a suspension on the fall velocity, regardless of the size of the sediment particles. The remaining factor, m, reflects the effect of concentration on the viscosity. For coarse particles, the effect is small, and hence the appropriate values of m are small; for fine particles the effect is large, and the values of m should be large. The formulas given by Sha and Wang are slightly different. They introduced the effect of particle size by modifying the value of S_v.

In the Stokes region an abundance of experimental data fall on the same straight line on a log-log plot (Fig. 3.20); thus the exponent m is a constant. Several scholars recommended a value of 4.65 for m, but Guo and Zhuang [33] gave the higher value of 4.91. For flow in the transition region, the exponent m is a function of the Reynolds

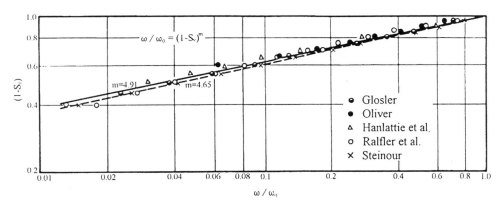

Fig. 3.20 Effect of S_v on ω (high concentration)

number. As the Reynolds number increases, m decreases, thus, the effect of concentration on the fall velocity decreases as the Reynolds number increases (Fig. 3.21). For Reynolds numbers over 500, m has another constant value; Richardson and Zaki recommend a value of 2.39, and Zaminia a value of 2.65.

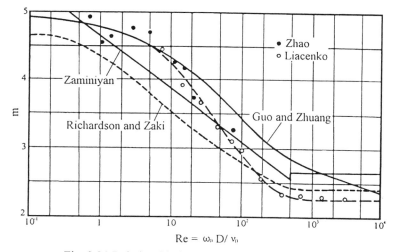

Fig. 3.21 Relationship between m in Eq.(3.38) and Re

An analysis of the experimental data of Riashenko [34] and Zhao[1] showed that they also follow Eq. (3.38), but with the relationship between m and the Reynolds number as shown in Fig. 3.21. The experimental points fall essentially within the same region as the curves of Guo and Zhuang and Richardson and Zaki. For Reynolds numbers less than 5, they are closer to the curve of Guo and Zhuang. For Reynolds number larger than 500, they approach the rather small constant value of 2.25. In general, one can adopt a simple exponential formula such as Eq. (3.38) and use the empirical results in Fig. 3.21 to account for the effect of concentration on the fall velocity. From Fig. 3.21, the curve for small Reynolds numbers could be determined according to the limiting value given in Fig.3.20. However, the experimental results of Beijing Mining

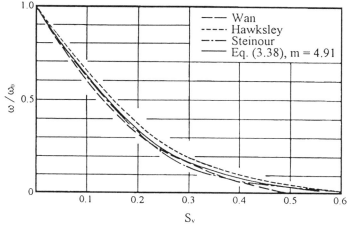

Fig. 3.22 Comparison of formulas for effect of S_v on ω (high concentrations)

[1] Zhao, Naixiong. *The Effect of Sediment Concentration on Fall Velocity of Sediment Particles.* Shaanxi Industrial University, 1964.

College, Xia and Wang, and Wang given in Table 3.6 indicate that the limiting value may be too small. This topic needs further investigation.

Finally, Fig. 3.22 is a comparison of the several equations for spherical particles with no flocculation: Eq. (3.35), Hawksley's, Steinour's and Eq. (3.38) with $m=4.91$. Although the formulas differ in mathematical forms, Eq. (3.35) is close to Eq. (3.37); they cross the curve of Steinour and fall below the curve of Hawksley. For volumetric concentrations of more than 40%, the fall velocity of a particle decreases markedly as the concentration increases. The experimental results of Wilson show that the Steinour formula is more reliable than the Hawksley formula if S_v is greater than 0.1 [38].

3.4.3 Settling of non-uniform sediments

The above discussion is for uniform sediment. The mutual interaction of non-uniform sediments during the settling process is much more complicated. The first case discussed here is one in which uniform coarse particles fall in a suspension composed of uniform fine particles.

3.4.3.1 Settling of uniform coarse particles in a suspension composed of uniform fine particles

Fidleris and Whitemore [39] conducted a series of experiments on the settling velocity of metal spheres with various densities and sizes in a nonflocculated, stable, Newtonian suspension of plastic spheres whose density was the same as or close to that of the liquid. The properties of suspended particles, suspension liquid, and metal spheres used in the experiments are listed in Table 3.7. Experimental results show that, within the Stokes region, the relationship between the drag coefficient and the Reynolds number for metal spheres settling in a suspension is the same as that in a pure liquid, but one must use the density and viscosity of the suspension to calculate both the resistance coefficient and the Reynolds number. Beyond the Stokes region, the situation is the same for a suspension of plastic spheres (I). However, the situation is different for that of coarse plastic spheres (II). For the latter case, the relationship between drag coefficient and Reynolds number is as shown in Fig. 3.23. If the flow is turbulent, the experimental data deviate from the curve for a pure liquid, giving higher drag coefficient for a given Reynolds number. The amount of the deviation depends on the size of the metal spheres, but it is independent of their specific weight. As the size of a metal sphere approaches that of the suspended particles, the drag coefficient reaches a maximum, one that can be nearly ten times that for a pure liquid. For metal spheres of any size, the critical Reynolds number at which the drag coefficient approaches a constant is much smaller for a suspension than it is for a pure liquid.

Table 3.7 Material used in the experiments of fall velocity
conducted by Fidleris and Whitemore

Material			Diameter (mm)	Specific weight
Suspension	Suspended substance	Plastic spheres (I)	0.05~0.18	1.055
			0.12~0.15	
			0.15~0.18	
			0.35~0.42	
		Plastic spheres (II)	1.68~2.06	1.185
			2.41~2.81	
	Suspension medium	Glycerine and lead nitrate solution	——	1.185
		Glycerine, lead nitrate and alcohol solution	——	1.056
			——	1.185
		Glycerine	——	1.248
		Calcium chloride solution	——	1.056
		Olive oil	——	0.912
Metal spheres used in fall velocity exp.		Plastic spheres covered with iron powder	4.95~5.54	1.28
		Magnesium spheres	2.56~10.18	1.80
		Aluminum-magnesium alloy spheres	2.56~7.63	2.57
		Aluminum spheres	2.56~10.18	2.81
		Steel spheres	0.79~25.4	7.72
		Lead spheres	2.51~3.74	11.25
		Tantalum spheres	1.91~5.09	16.60
		Gold spheres	1.00~10.00	19.36

In understanding and applying correctly the experimental results of Fidleris and Whitemore, one should note that they used a stable suspension--one without relative motion between the suspended particles and the suspending liquid. They studied the settling of metal spheres, with a higher specific weight in such a suspension. In natural rivers, a suspension of sediment particles is unstable, relative motions occur between sediment particles of different sizes and with the surrounding liquid. However, if the particle sizes differ a lot in such a non-uniform suspension, the fall velocity of the large particles is much greater than that of the small ones. For the settling of the large particles, the suspension of the small particles is thus effectively stable. For this situation, an important deduction based on the experimental results is the following. for either a fine-particle suspension or a coarse-particle suspension within the Stokes region, the settling of relatively large particles in a non-uniform suspension is the same as that in a pure liquid with a density and viscosity equivalent to those of the suspension; the particle in the suspension must, of course, be considerably smaller than the particle in question.

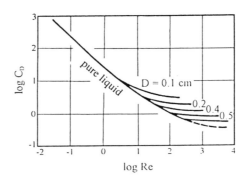

Fig. 3.23 C_D vs. R_e for metal sphere setting in suspension

Simons [40] investigated the validity of this concept experimentally. Fig. 3.24 shows the average settling velocities of three different sediment particles in suspensions of different concentrations of bentonite clay and kaolin clay. If the densities and viscosities of suspensions are used in the calculation, the computed settling velocities

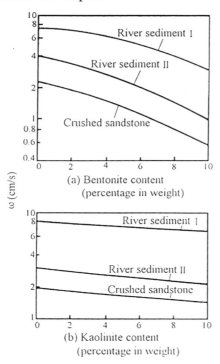

Fig. 3.24 ω of sediment particles in clay suspensions at different concentrations

agree well with the experimental results in the figure. The difference between this result and that of Fidleris and Whitemore is that the size of suspended particles in muddy water is so small that the physico-chemical properties of the particle surface become important. Hence, viscosities of suspensions depend not only on the sizes and concentrations of the suspended particles, but also on their mineral compositions. The viscosity of the bentonite clay suspension is much greater than that of kaolin clay, hence the resistance to settling is also greater.

Another factor is that heavy particles in a stable suspension collide with the suspended particles as they settle. This process increases the effective cross-sectional area of a particle. If the settling particle is much larger than the suspended particles, the increase in cross-sectional area of the former due to collision with the latter is small, and the effect is negligible. If the settling particles and the suspended particles are about the same size, the increase in cross-sectional area is relatively large. Consequently, the resistance force on the particle will be large. That is why the drag coefficient increase as the size of a metal sphere approaches the size of the suspended particles, as shown in Fig. 3.23. In natural unstable suspensions, if the densities of suspended particles are the same and the particle sizes are close to each other, then almost no collisions occur during settling. An increase like that shown in Fig. 3.23 does not occur.

If the concentration of fine particles in a suspension is high, the suspension may no longer be Newtonian. The experimental results of Ansley and Smith [41] indicate that the relationship between the drag coefficient C_D of a Bingham fluid and the following Reynolds number

$$\frac{\rho \omega^2}{\dfrac{\eta \omega}{D} + \dfrac{7\pi}{24}\tau_B}$$

is similar to the usual relationship between drag coefficient and Reynolds number (Fig. 3.1). For a fluid following an exponential law, Brea's study [42] indicates that the above conclusion still applies if the Reynolds number is defined as:

$$\frac{D^m \omega^{2-m}\rho}{\dfrac{K}{8}\left(\dfrac{6m+2}{m}\right)^m}$$

3.4.3.2 Settling of natural sediments with a continuous size distribution

The foregoing discussion is for a simple case of uniform coarse particles settling in fine particle suspension. For natural nonuniform sediments with a continuous size distribution, even if initially particles are evenly distributed in suspension, coarse particles settle faster and continuously leave the suspension as they deposit on the bottom. Thus, the concentration of sediment in the fluid surrounding the fine particles varies continuously. If flocculation occurs, the coarse particles attract fine sediment particles that adhere to them as they settle, and both their effective volumes and their fall velocities increase. The process of settling is thus an unsteady phenomenon. The mean settling velocity at any horizontal cross-section is a function of time. McLaughlin [43] provided a physical explanation for this phenomenon based on his experiments.

At the beginning of the experiments, sediment particles were evenly distributed in the water. Subsequently, samples were taken at different heights periodically. The horizontal axis was set at the initial top level of sediment and the vertical distance was measured downwards from there; the concentration distributions $S_v(y,t)$ obtained at different times and at different heights are shown in Fig. 3.25. For continuity of sediment,

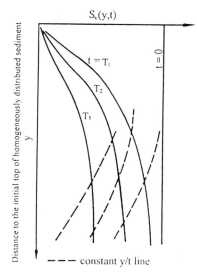

S$_v$(y,t)

t = 0

t = T$_1$

T$_2$

T$_3$

Distance to the initial top of homogeneously distributed sediment

y

— — — constant y/t line

Fig. 3.25 Variation of vertical profile of concentration during the process of setting of non-uniform sediment, initially homogeneously distributed

$$\frac{\partial S_v}{\partial t} + \frac{\partial (\varpi S_v)}{\partial y} = 0 \qquad (3.39)$$

in which ϖ is the mean settling velocity at time t and at a distance y below the horizontal axis. Integrating the above equation, one obtains the expression

$$(\varpi S_v)_{y=y_1} = -\frac{\partial}{\partial t}\int_0^{y_1} S_v dy \qquad (3.40)$$

Applying Eq. (3.40) and using the vertical concentration distributions at different times in Fig. 3.25, one can obtain the variation of ϖ with time and at location y by graphical integration. Fig. 3.26 shows the experimental results for bentonite clay and alum settling in water. The figure shows that, at any level, the mean settling velocity has a maximum. The curve to the left of the maximum reflects the accelerating process of coarse particles due to increasing flocculation as they settled. The curve to the right reflects the fact that coarse particles continually left the suspension and deposited on the bottom; thus, as the sediment remaining in suspension became progressively finer, the mean fall velocity decreased.

In order to better demonstrate the effects on the settling velocity of acceleration due to flocculation and that of retardation due to concentration, dotted lines were drawn in Fig. 3.25 for constant values of y/t. These dotted lines represent the change of concentration as seen by an observer starting from $y = 0$ at $t = 0$ and moving at a constant velocity y/t downwards. Without the effects of acceleration and retardation, the concentration seen by the observer at different heights would be the concentration of those particles, for which the fall velocity is equal to or less

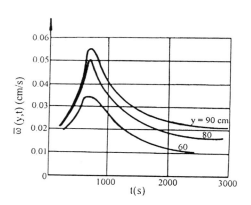

Fig. 3.26 Variation of $\overline{\omega}(y,t)$

than y/t. This concentration would not change with y, that is, a constant y/t line should be a straight line parallel to the y-axis. If the retardation effect due to concentration exceeds the acceleration effect due to flocculation, an observer would see an increase in concentration as he moves downwards, that is, a constant y/t line would be further away from the *y*-axis. If, on the contrary, the acceleration effect due to flocculation exceeds the deceleration effect due to concentration, the concentration would decrease as the depth increases; that is, a constant y/t line would move closer to the *y*-axis. Obviously, as shown in Fig. 3.25, the real situation indicates that the effect of flocculation is larger and dominates.

Recently, the IHR, YRCC [44] and others conducted many tests on the settling of natural nonuniform sediments in quiescent water and obtained useful results. They found two different types of results depending on whether or not the nonuniform sediments contain particles finer than 0.01 mm. If they did not, no flocculation occurred, and the variation of mean settling velocity with concentration was basically the same as that for uniform sediment, as shown in Fig. 3.27 for sediment samples III and IV. If they did contain fine particles, flocculation occurred, and the settling phenomenon was more complex. This case is discussed in more detail in section 3.6.

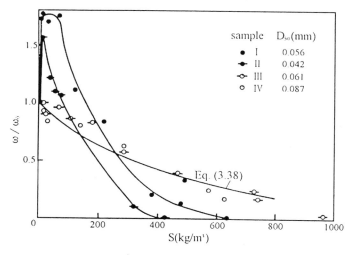

Fig. 3.27 ω / ω_0 vs. S for natural non-uniform sediment

3.4.4 Force on settling particles with different arrangements

In investigations of the effect of concentration on the fall velocity, the velocities at various concentrations are usually either measured directly or the relationship between drag coefficient and Reynolds number is compared with that for a single settling particle. If the force on settling sediment particles with different arrangements could be measured directly, it would help in obtaining an understanding of the mechanism of the entire process.

It is difficult to measure the force on a moving body. An alternative is to fix the sediment particles and let the fluid move with a uniform velocity U through the particles. Thus the unsteady motion is changed into a steady one, and one can measure the force on a fixed particle. Of course, this approach applies only to the case in which all particles settle with the same velocity — to particles all the same size. The flow velocity U then takes the place of the settling velocity of the particle. The flow condition produced in such an experiment conforms to that in a fluidized bed.

Rowe [45,46] conducted experiments on 12.7 mm plastic spheres that were fixed in a flume with various geometric arrangements and using a small flow velocity. He measured the drag force on a representative sphere and compared it with the force acting on a single sphere setting in quiescent water at the same velocity. He kept the Reynolds number $Re=UD/v$ constant at one of the two values — 32.64 or 96. The arrangements of the spheres are shown in Table 3.8; in the tests, the drag force was measured for the black sphere and the white spheres served to create various flows. He found that, in most cases, the force F on a sphere can be expressed in the form

$$\frac{F}{F_0} = 1 + \eta \frac{D}{s} \tag{3.41}$$

in which the separation distance between the spheres is s (as defined for the different arrangements shown in Table 3.8), F_0 is the comparable drag on a single sphere settling in quiescent water at the same velocity, and the coefficient η varies with the sphere arrangement, as shown in Table 3.8. The values of this coefficient were the same for the two Reynolds numbers used.

Table 3.8 Force on fixed sphere, which has different permutations and combinations with others, in a flow (after Rowe, P. N., and G. A. Henwood)

No	Arrangement of spheres			η	Remark
1	Two spheres arranged in a line parallel to the flow direction	Trailing sphere		-0.846	Does not apply for $s/D<2$. Drag ratio approaches 0.4 when two spheres touch each other.
		Leading sphere		-0.151	Does not apply for $s/D<2$. Drag ratio approaches 0.8 when two spheres touch each other.
2	The line linking two spheres forms an angle with the flow direction			—	Experimental results are shown in Fig. 3.28. Generally the drag is reduced, but the drag ratio is larger than 1 in a small region.
3	The line linking two spheres is perpendicular to the flow direction			—	The drag is increased by less than 15%.

4	Four layers of closely packed spheres nearby	Trailing sphere		9	Applies for $s/D > 6$. If s/D is less than 6, the effect is reversed. More layers do not effect the force on the sphere.
		Leading sphere		-1.02	
5	Four layers of closely packed spheres nearby, but the position just opposite to the black ball is empty	Trailing sphere		9	Same condition as for case without empty position. Drag rapidly increases as black sphere approaches the empty position. Drag increases by 100 times.
		Leading sphere		6.6	The drag is larger. Result is completely different from the case without empty position.
6	Spheres touch each other along direction perpendicular to the flow. Around the black sphere are spheres forming a hexagon.			0.68	The drag ratio is 69 if spheres are closely packed ($s/D = 0.01$).

Rowe's experiments on the interference effect for uniform spheres led to the following conclusions, and they are valid for the Reynolds numbers tested:

1. For two spheres settling in tandem, the velocity of the spheres is smaller due to shielding. This effect exists even if the separation distance is as much as 10 times the sphere diameter. Because shielding effects on the two spheres differ, the two spheres move towards each other as they settle.

2. For two spheres settling in parallel, their fall velocities are less. The closer they are to each other, the more the retardation. Nevertheless, the drag force does not differ from that on a single sphere by more than 15%.

3. Two spheres settling with a diagonal separation rarely maintain their relative positions. Usually, the sphere further back catches up with the one in front, and they then settle in parallel. However, other times, the separation distance tends to increase.

4. If many particles are densely packed together, their settling velocity is much less.

5. If a group of densely packed, uniformly distributed particles settle, and one of the spheres lags behind for some reason, the empty space left by that sphere seems to induce a force that pulls the sphere back to its original position. Hence, if uniform sediments settle, a sharp demarcation between sediment-laden and clear water is maintained.

6. Without flocculation, the additional resistance force on a settling sphere due to many surrounding spheres can be quite large, the maximum observed value was not more than 68 times that on a single sphere settling with the same velocity.

3.5 THE EFFECT OF TURBULENCE ON THE FALL VELOCITY

Turbulence in a sediment-laden flow affects the settling of the sediment particles in three ways.

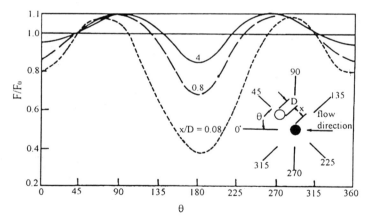

Fig. 3.28 F / F_0 vs. θ

First, flow fluctuations cause the external forces on sediment particles to vary continuously. Moreover, because the eddies rotate, the particles rotate as they settle so that they do not maintain a stable settling orientation. In fact, a sediment particle can rotate as it settles even in quiescent water because of the vortices in its wake, as discussed in section 3.2.

Second, the magnitude and the direction of the fluctuating flow velocity vary continuously in time and in space (as discussed in Chapter 4). Thus, a particle is sometimes accelerated and sometimes decelerated by the flow. Beside the usual drag force on a particle, an additional force results from acceleration or deceleration. In the latter part of the 18th century, DuBuat proved that if a body oscillates in quiescent water, there is an additional effective mass (called virtual mass). An additional mass of fluid must be added to the actual mass of the body in calculations of the forces acting on an accelerated body. The expression for virtual mass is as follows:

virtual mass = coefficient × mass of fluid having same volume as the body

in which the coefficient depends on the size, shape, and motion of the body and on the viscosity and density of the fluid.

Finally, the location of the separation point on the surface of a particle and the surface pressure distribution are affected by turbulence. As a consequence, the drag on a particle can be either less or more than that for steady flow.

The first two of these three effects decrease the settling velocity of a sediment particle.

3.5.1 Analytical study of the effect of turbulence on the fall velocity

The effect of turbulence on settling velocity can be analyzed in several ways. In one, the motion of fine particles in water is viewed as being similar to the motion of water molecules. Then by applying the diffusion equation, one can solve for the actual pattern of flow for the given boundary conditions [47]. A more common way is to use the dynamic balance of the forces acting on a particle from Newton's second law, these include gravitational force, drag force, additional resistance due to virtual mass, fluctuating flow force, etc. The result is

$$\sum F = M d\omega / dt$$

in which M is the particle mass. One can use this equation to study mathematically the trajectory of a moving sediment particle [48,49]. The difficulty with this method lies in how to express the force on a particle due to flow fluctuation. Different authors, using different assumptions and simplifications, have obtained quite different results.

Kada and Hanratty [50] included only the force due to virtual mass in the force-balance equation, and used this term to reflect all of the effects of acceleration. They found that, within the Stokes region, the settling velocity of sediment particles in a turbulent flow and that in a quiescent water do not differ. Beyond the Stokes region, the effect of turbulence is to reduce the settling velocity. In an investigation of the motion of sediment particles in a turbulent flow by altering the form of the drag coefficient, Lhermitte [51] found also that the settling velocity in turbulent flow is smaller than that in a quiescent water. Tang did not solve the equation of motion directly. Instead he assumed that the drag on a sediment particle settling in a turbulent flow is the sum of the drag in a quiescent water and a force due to turbulence. The force of turbulence stress is the component of the Reynolds stress in the vertical direction. Since the Reynolds stresses on the upper and lower sides of a particle are different, they produce a "pressure difference", that gives rise to a buoyancy effect on a particle in turbulent flow. Therefore, the settling velocity of a sediment particle in a turbulent flow is smaller than that in quiescent water[1].

Because of mathematical difficulties encountered in theoretical analyses, sometimes one has to rely on semi-empirical methods. For example, one can assume the settling velocity of a particle in a turbulent flow is equal to the difference of the settling velocity in quiescent water and the fluctuating vertical component of the velocity. Then from the distribution of the fluctuation velocity, with certain averages, one can obtain the mean settling velocity of a particle in moving water, and hence the drag on a sediment particle as it settles. This type of approach was used in the work of Meyer [52] and Bouvard [53]. Their analytical results indicate also that the effect of turbulence is to reduce the settling velocity of a sediment particle. However, the physical explanations provided by Meyer and Bouvard are not convincing. If one follows the same process of deduction, but uses another method of averaging or makes another assumption as to the relationship between drag and settling velocity, one can obtain a result in which the settling velocity in quiescent water is equal to that in moving water.

3.5.2 Experimental study of the effect of turbulence on the fall velocity

Many people have conducted experiments on how turbulence affects the fall velocity. They generally conducted one of three types of experiments.

1. Some subjected sediment particles in a water column to simple harmonic motion, recorded their motions and settling times, and then calculated the settling velocities of the particles. Field, Murray, and Fan conducted experiments of this type independently. The experimental conditions and conclusions are briefly summarized in Table 3.9, in which ω and ω_0 are the settling velocities in moving water and in quiescent water, respectively.

---

[1] Tang, Yunji. *Study on the settling of sediment particles in flow*. Shaanxi Industrial University, 1963.

Table 3.9 Experiments with sediment particles in a water column
undergoing simple harmonic motion

Author	Particles used in experiments	Characteristics of oscillation	Liquid used	$\frac{\omega_0 D}{v}$	Results	Ref.
Lhermitte	Plastic spheres with specific weight of 1.45 and diameter of 0.19-1.38 mm	—	Water	2 ~ 140	$\frac{\omega}{\omega_0} = 1$	[51]
Field	—	Amplitude 2.54 cm, frequency 540/min	Glycerine	140~ 400	$\frac{\omega}{\omega_0} < 1$, ω is reduced by less than 20%	[54]
Murray	Spheres of 2 mm with different specific weight, their fall velocities are 1,2,3, and 4 cm/s respectively	Frequency 3300/min	Water	20 ~80	$\frac{\omega}{\omega_0} < 1$, ω is reduced by less than 30%	[55]
Fan et al.	Plastic sphere of 3mm with a fall velocity in quiescent water 0.25-3.5 cm/s	Amplitude 0.6~6.0cm, frequency 40~200/min	Water	7.5 ~ 105	$\frac{\omega}{\omega_0} = 1$	*

* Fan, Jiahua, Wu, Deyi and Chen, Ming. "Settling of sediment particles in turbulent flow." *Report of Institute of Water Conservancy & Hydroelectric Power Research*, 1964.

The conclusions from their experiments, as shown in the table, differ from each other; perhaps because the experimental conditions were different. Murray and Field stated that the decrease of settling velocity is due to the nonlinearity of the drag term in the equation of motion. High frequency oscillation has a large effect on reducing the settling velocity. The frequency used in the experiments of Fan was comparatively low, and hence it had only a small effect on the settling velocity. Murray also measured the turbulent velocity of a particle, and showed that it followed a Gaussian distribution. He pointed out that the decrease in the settling velocity of a particle is more if both the settling velocity in quiescent water and the intensity of turbulence are large.

2. Others put sediment particles into a turbulent open-channel flow and then calculated their settling velocities from the trajectories of the sediment particles, from the concentration distribution, or from the distribution of sediment particles deposited on the channel bed. Conditions and conclusions from this type of experiments are listed in Table 3.10.

Table 3.10 Settling experiments with sediment particles in turbulent open channel flow

Author	Particles used in experiments	$\omega_0 D / v$	Type of experiments	Results	Ref.
Lhermitte	Plastic spheres with specific weight of 1.45 and diameter of 0.19 to 1.38 mm	2 ~140	—	$\dfrac{\omega}{\omega_0} = 0.72$	[51]
Fan et al.	Plastic spheres, diameter 3 mm	33 ~171	At a fixed point adding spheres into a turbulent flow, determining fall velocity according to the distribution of particles settled on bed.	$\dfrac{\omega}{\omega_0} \approx 1$	* in Table 3.9
Tang	Spheres made of the mixture of talcum powder and tung oil, specific weight 1.56 to 1.61, diameter 2.88 to 9.15 mm	555 ~3350		$\dfrac{\omega}{\omega_0} = 0.72$ ~0.80	1)
Jobson & Sayra	Glass spheres, diameter 0.123 mm with specific weight of 2.42. natural sand with median diameter of 0.39 mm and specific weight 2.65	1.3 ~24.6	Adding sediment particles along the whole width of the flume, determining fall velocity according to concentration distribution	$\dfrac{\omega}{\omega_0} = 1.05$ for coarse sand. no definite conclusion for fine particles	[56]

1) Tang, Yunji. *Study on settling of sediment particles in flow*, Shaangxi Industry University, 1963.

3. Still others fixed a body in a wind tunnel, varied the turbulence intensity of the air flow in the wind tunnel, and studied the resultant variation of the drag coefficient C_D.

Robertson and McLaren [58] conducted experiments of this type. The objects used in their experiments were blunt (non-streamlined) bodies with sharp edges, such as cylinders with square cross sections, circular cylinders (the axis aligned in the direction of flow). Experimental results indicated that, depending on such variables as the body shape, the angle of attack, and the Reynolds number, turbulence can either increase or decrease the drag coefficient . If the length and the width of a body are about the same, the difference in C_D due to turbulence is the greatest. For example, for a circular cylinder with the length equal to the diameter, C_D was 1.4 without turbulence, and 1.0 with turbulence.

For a circular disk or a square plate perpendicular to the flow direction, or a cylinder with a square cross section at an incidence angle of 45°, the streamline, after separating from the body, does not intersect the body again. An increase in turbulence then increases the Reynolds stresses on the separation surface. If the separation region is isolated, one can show that, as the Reynolds stresses on the separation surface increase, the pressure on the lee side of the body must decrease (otherwise the isolated region cannot maintain a balance); therefore C_D increases.

If a body is long enough, a streamline that separates at the edge of the upstream face will intersect the body again. Under this circumstance, an increase in turbulence changes both the flow pattern and the pressure distribution around the body. The two effects caused by the increase in turbulence--the increase of the Reynolds stresses and the change of pressure distribution--act simultaneously to affect the drag force. For a circular cylinder with $L/D = 1$, an increase in turbulence increases the pressure on the lee side of the body, so that C_D decreases.

In applying the various experimental results to the settling of particles, one must keep in mind the following factors. First, the particles in the experiments were large (and the Reynolds number is large); suspended sediment particles in natural rivers are usually not so large. Second, a particle can adjust its orientation during settling (as explained in Section 3.2.1), unlike the fixed bodies in the air flow experiment.

In summary, although there are no definite conclusions on the effect of turbulence on the settling velocity, for the turbulence intensities commonly encountered in rivers, the effect of turbulence on the settling velocity is not significant in comparison with the effect of concentration. However, if fine particles flocculate, the situation is different, as discussed in the next section.

3.6 EFFECT OF FLOCCULATION ON THE FALL VELOCITY

As mentioned in Chapter 2, fine sediment particles have relatively large ratios of surface area to volume, and the physico-chemical effect on the particle surface often produces micro-structural changes between particles. As the number of fine particles increases, the following changes may occur (Fig. 2.25):

1. In homogeneous suspensions of fine sediment particles in water, a film of bound water adheres to the surface of every particle.

2. If several fine particles form a floc, in addition to the bound water film on the particle surface, some free water is confined within the floc. This confined water cannot be separated from the floc by gravitational force. Therefore the effective diameter of the sediment particles is larger. At this stage, the flocs are suspended in water homogeneously.

3. A network structure is formed by the connection of flocs. At first, the structure is loose and the spaces within the structure are relatively large. These spaces are filled with free water that can be squeezed out by gravitational force (called gravitational free water).

4. If the structure is dense and the flocs are close to each other, the spaces within the structure and the corresponding amount of gravitational free water are less. If the spaces are small enough, the gravitational free water within the spaces may change to confined free water.

The range of particle sizes for which flocculation occurs is discussed first, then the formation of flocs and their structures is discussed.

3.6.1 Range of sediment sizes for which flocculation occurs

The mineral composition of sediment particles and the quality of the water affect the existence and degree of flocculation. In addition, particle size is an important factor. The finer the particle is, the stronger the physico-chemical effect on a particle surface, and hence the stronger the flocculation.

Migniot [59], from his long-time research experience on coastal sediment, called sediment with particle sizes less than 0.03 mm silt, and pointed out that silt clearly displays the flocculation.

In the aforementioned settling experiments in quiescent water, conducted by the IHR, YRCC, the effect of concentration on the mean settling velocity for mixed sediment with no particles finer than 0.01 mm is the same as that for uniform sediment; however, the situation is entirely different for mixed sediments that include finer particles. The explanation is that if a certain amount of particles finer than 0.01mm is included in the sample, flocculation will occur.

On the basis of available research results, the limiting particle size for the existence of flocculation is considered to be 0.01 mm.

3.6.2 Settling velocity of flocs and factors affecting their formation

At low concentrations, the effect of flocculation is that flocs, which are made of many sediment particles, have a large fall velocity. As mentioned in Chapter 2, the size distribution and mineral composition of sediment particles, water salinity, etc., all can affect the flocculation of sediment particles, and hence the settling velocity of flocs. Migniot [59] studied the effects of these factors.

3.6.2.1 Effect of sediment particle size

The finer the sediment particles are and the larger the specific surface area, the stronger the flocculation effect will be and larger flocs relative to a basic sediment particle will form. Migniot defined the flocculation factor F to indicate the magnitude of the flocculation effect as follows:

$$F = \omega_{F50} / \omega_{D50} \qquad (3.42)$$

in which ω_{F50} and ω_{D50} are the settling velocities of a floc and a basic sediment particle, respectively, both being represented by their median values. From experiments, he obtained a relationship between the flocculation factor F and the size of a basic particle, as shown in Fig. 3.29. This figure shows that, with flocculation, the settling velocity of the flocs can be a thousand, even 10 thousand times that of a single particle.

The effect of flocculation is less for larger particles. If the particle size is greater than 0.03 mm, flocculation has no effect. Also, the effect of flocculation is small if the particle size is between 0.01 mm and 0.03 mm. Therefore, 0.01 mm is considered as the limiting particle size for flocculation to have any effect.

The finer the sediment, the larger the flocculation factor, i.e., the larger the settling velocity of a floc is in comparison with that of a basic particle. Therefore, flocculation makes the settling velocities of flocs much less variable than are the fall velocities of basic particles. Migniot found that, for silt suspensions with flocs, the mean settling velocity was within the range of 0.15 and 0.6 mm/s, a result that was independent of the size of the basic particles.

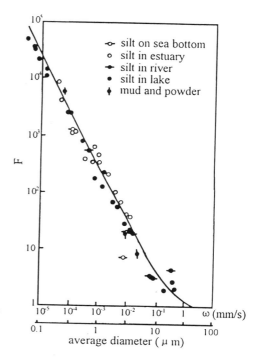

Fig. 3.29 F vs. size of a basic particle

3.6.2.2 Effect of salinity

Fig. 3.30 is a plot of the mean settling velocity of a floc versus the salinity of the water for several different concentrations. This family of curves has a common feature: if the salinity is low, the mean settling velocity of the floc increases steeply as the salinity increases; if the salinity exceeds a certain value, further increase in salinity does not have much effect on the mean settling velocity. The larger the concentration, the smaller the corresponding salinity at the turning point of a curve. If the sediment concentration in water is relatively high (more than 10 kg/m^3), a flocculation structure may appear at high salinity, and it will decrease the settling velocity. This situation is discussed in detail as part of the following point.

3.6.2.3 Effect of turbulence

The effect of turbulence on the fall velocity of sediment, already discussed, is usually not large. However, if flocs form, turbulence can have an effect on their fall velocities. Owen [60] designed a sampler that can be used to take undisturbed suspended sediment samples in situ and to make settling analyses. He used it to take samples

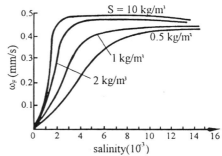

Fig. 3.30 Effect of salinity on ω_F

in a gulf in England and to make settling analyses for different conditions (high tide, low tide), and then compared the results. Since the particle size distribution of coastal sediment changes little with time or sampling location, any difference in the fall velocities is caused by the sizes of the flocs. His results are shown in Fig. 3.31 for the following four types of sediment samples:

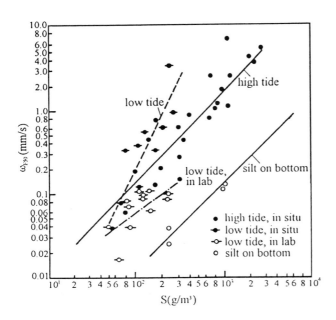

Fig. 3.31 Effect of turbulence on ω_F

1. Suspended sediment samples taken at high tide, on which the settling analysis was made in situ;

2. Suspended sediment samples taken at low tide, for which the settling analysis was made in situ;

3. After taking the suspended sediment samples, settling analysis performed in the laboratory according to conventional procedures (salinity unchanged);

4. Sediment samples taken on the bed surface and the settling analysis made in the laboratory according to conventional procedures (salinity was not changed).

As shown in the figure, the settling velocity of the floc at high tide tends to vary linearly with concentration on a log-log plot; the slope of the straight line is 1.08, or quite close to one. The fall velocity of a floc at low tide is roughly proportional to the square of the concentration; the slope is 2.2. If the concentration is greater than 70 g/m³, the fall velocity of the floc at high tide is smaller than that at low tide. From the result of settling analyses in the laboratory, the fall velocity was found to vary almost linearly with concentration, but its magnitude was much smaller than that in situ.

Owen thought there were two opposing effects of turbulence on the settling velocity of flocs. On the one hand, flocculation of particles occurs after particles collide with each other. Since the turbulence increases the probability of particle collision, it increases flocculation. On the other hand, if the connecting force between particles in a floc cannot overcome high shearing stresses in the field, flocs break up. Since the turbulent eddies increase the shearing rate of the fluid, they may break the flocs. Therefore, the turbulence with median intensity (as in low tide) increases the probability of particle collision and the shear rate is not high enough to break the flocs, thus it enhances the flocculation process. Strong turbulence (as during high tide), also increases the probability of particle collision, but more significantly, the high shear rate causes flocs to break up, so that the net effect is to hinder the flocculation process.

Although flocculation of sediment particles with turbulence and the fall velocity associated with it are obviously an important concern, no additional data are available. Although Owen explained his measured data, the reliability of his explanation should be verified by more field data.

3.6.2.4 Effect of concentration

The process of flocculation and its effect on the settling velocity of flocs are quite sensitive to the sediment concentration. In river water containing salts or in sea water, the fall velocity of a floc increases as the concentration increases until the concentration attains a critical value of about 15 kg/m³ (Fig. 3.32). If the concentration exceeds this level, the fall velocity decreases as increasing concentration. In fact, the microstructure in the flocculation process forms at this stage, as discussed in section 3.4.3.

3.6.3 Formation of flocculation structure and its effect on the fall velocity

A further development in the flocculation process is for flocs to connect with each other and form a skeleton (flocculation structure). With this development the fall velocity decreases greatly. The phenomenon occurs at high concentrations, as shown in Fig. 3.32.

3.6.3.1 Physical description of the formation of the flocculation structure

Recently, IHR, YRCC [44], and NWHRI have conducted a number of experiments on the settling of natural nonuniform sediment in quiescent water. If the nonuniform sediments do not contain fine particles, no flocculation occurs. That is, the relationship between the mean fall velocity and the concentration is the same as that for uniform sediment. If fine particles are present, however, flocculation takes place. One obtains a diagram similar to that of Migniot in Fig. 3.27, samples I and II. For low concentrations, particles form flocs, and the fall velocity is therefore greater than the fall velocity of a single sediment particle. If the concentration is larger, a continuous spatial structure connects the flocs and attains a certain rigidity. In this instance, the fall velocity of the sediment conglomeration is much less. Initially, the spatial structure network is made of fine sediment

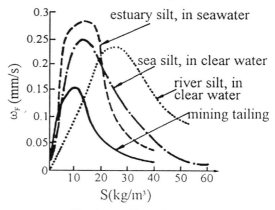

Fig. 3.32 Effect of S on ω_F

particles. Together with clear water, they form a homogeneous mixture and settle down slowly. Such a slurry in a settling tube has an interface separating it from the clear water above it; in the settling tube this interface moves down slowly. Coarse particles, although their fall velocities are less because of the flocculation structure, still maintain their individual entities in the dispersive system and fall freely. During the settling process, coarse particles and fine particles fall separately and sorting takes place. For larger concentrations, more coarse particles are connected to the flocculation structure, and fewer of them settle freely. At some limiting value of the concentration, all of the sediment particles are part of the homogeneous slurry and the sorting phenomenon no longer occurs. Table 3.11 shows the experimental results of NWHRI on nonuniform sediment with a median size of 0.035 mm. The larger the initial concentration is, the coarser the sediment particles forming a homogeneous mixture and the less the settling velocity. For an initial concentration of 800 kg/m^3, the sediment particles settle as an entirety regardless of the particle sizes, and thus no sorting occurs.

Table 3.11 Segregation of a non-uniform sediment, median diameter of 0.035 mm in processes of settling

Initial concentration(kg/m^3)		200	450	600	800
Characteristics of fine sediment particles, comprising the homogeneous slurry	Median diameter (mm)	0.0089	0.0122	0.0173	0.035
	Concentration (kg/m^3)	73	211	297	800
Settling velocity of the interface between the clear water and the slurry (cm/s)		0.00345	0.00085	0.00083	0.00067

3.6.3.2 Settling velocity after the formation of a homogeneous slurry

Once all sediment particles have formed a homogeneous slurry, coarse particles and fine particles fall with the same low velocity, which is the velocity at which the interface between the mixture and the clear water falls. This velocity is hundreds or even thousands of times smaller than the mean fall velocity of the same nonuniform sediment particles in clear water. In this instance, the settling process of sediment is in fact the dehydration process of the whole flocculation structure. Table 3.12 shows the experimental results for samples I and II in Fig. 3.27 [44].

As mentioned in section 3.4.2, Hawksley's finding that flocculation only reduces the fall velocity by a third (that is $\zeta = 2/3$ in equation 3.36), obviously applies only to the special conditions adopted in his experiments; it therefore lacks general validity.

Table 3.12 The limiting concentration for the formation of a homogeneous slurry and its settling velocity as an entity

Sample number	I	II
D_{50} of sediment used in experiments (mm)	0.042	0.056
Limiting concentration for formation of a homogeneous slurry (kg/m^3)	420	630
ω / ω_0	1/250	1/920

3.6.3.3 Determination of the limiting concentration for the formation of a homogeneous slurry

The limiting concentration above which all sediment particles would form a homogeneous slurry obviously depends on the sizes of the sediment particles, particularly on the content of fine particles; the finer the particles, the smaller the limiting concentration is. As mentioned in Chapter 2, a mixture containing fine sediment particles has the characteristics of a Bingham fluid. In a Bingham fluid, even particles with a density higher than the surrounding medium remain stationary do not fall as long as the particles are small enough. In this circumstance, the force-balance condition is that $(\gamma_s-\gamma_m)D$ is proportional to τ_B, where τ_B is the Bingham yield stress [62]. From a synthesis of the experimental results of NWHRI and IHR, YRCC on the settling of sediment in quiescent water, and an analysis the data regarding the limiting concentration, one obtains the following concept: if the Bingham yield stress of a slurry is large enough to prevent most coarse particles (represented by D_{95}) in the nonuniform sediment from settling, all of the nonuniform sediment particles form a homogeneous slurry and no sorting occurs (Fig. 3.33). This occurrence is treated further in Chapter 13.

In the presence of flocculation, the settling of sediments is complex. In the four stages of a flocculation process already mentioned, the distributions of particles, flocs, or structures in the fluid are homogeneous for stages 1, 2, and 4, and heterogeneous for stage 3. If a coarse particle settles in such a fluid and frequently encounters flocs, its fall velocity is very small. If it can pass through the gravitational free water in the gaps of the structure, its settling velocity is much larger. Xia and Wang [37] conducted a series of experiments in which a certain amount of uniform coarse particles were added to a slurry containing fine sediment particles of various concentrations. They discovered that if the concentration of fine particles was either quite low or extremely high, coarse particles had the settling characteristics of uniform sediment, the only difference being that the fall velocity was somewhat smaller. Within a certain range of the concentrations of fine particles, uniform coarse particles exhibited characteristics of nonuniform sediments. This extraordinary result clearly reflects the unusual consequences of the presence of the large amount of fine sediment particles.

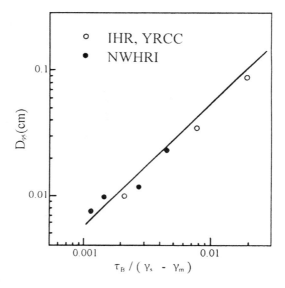

Fig. 3.33 Limiting condition

REFERENCES

[1] Stokes, G. G. "On the Effect of the Internal Friction of Fluids on the Motion of Pendulums." *Transition of Cambridge Philosophy Society*, Vol. 9, Pt. 2, 1851, pp. 8-106.

[2] Rouse, H. *Elementary Mechanics of Fluids*, John Wiley and Sons, 1946, pp. 235-250.

[3] Ossen, C.W. "Neuere Methoden and Ergebnisse in der Hydrodynamik." Akademische Verlagsgesellschaft, Leipzig, 1927.

[4] Goldstein, S. "The Steady Flow of Viscous Fluid Past a Fixed Spherical Obstacle at Small Reynolds Numbers." Proceeding of Royal Society, London, Ser. A, Vol. 123, 1929.

[5] Rubey, W.W. "Settling Velocities of Gravel, Sand, and Silt Particles." *American Journal of Science*, Vol. 25, No. 148, 1933, pp. 325-338.

[6] Wuhan Institute of Hydraulic and Electric Engineering (WIHEE). *River Dynamics*. China Industry Press, 1961, pp. 13-21. (in Chinese)

[7] Dou, Guoren. *Theory of Sediment Transport*. Nanjing Research Institute of Water Conservancy, 1963, pp. 5-1⁻ 5-38. (in Chinese)

[8] Cai, Shutang. "Settling of Sediment Particles on Quiescent Water--(2)The Variation of Fall Velocity with Time." *Journal of Physics*, Vol. 12, No. 5, 1956, pp. 409-418. (in Chinese)

[9] McNown, J. S. and J. Malaiek. "Effects of Particle Shape on Settling Velocity at Low Reynolds Numbers." *Transition of American Geophysical Union*, Vol. 31, No. 1, 1950, pp. 74-82.

[10] Stringham, G. E.; D. B. Simon; and H. P. Guy. "The Behavior of Large Particles Falling in Quiescent Liquids." *U. S. Geological Survey*, Prof. Paper 526-C, 1969, pp. 36.

[11] Jayaweera, K. O. L. F. and B. J. Mason. "The Behavior of Freely Falling Cylinders and Cones on a Viscous Fluid." *Journal of Fluid Mechanics*, Vol. 22, Pt. 4, 1965, pp. 709-720.

[12] Oberbeck, A. "Ueber Stationare Flussigkeitsbewegungen mit Berucksichtigung der Inneren Reibung." *Creiles J.*, Vol. 81, 1876, p. 62.

[13] McNown, J. S. "Particles in Slow Motion." *La Houille Blanche*, Vol. 6, No. 5, 1951, pp. 701-722.

[14] McNown, J. S.; J. Malaika; and R. Pramenik. "Particle Shape and Settling Velocity." *Transition of 4th Meeting of International Association for Hydraulic Research*, Bombay, India, 1951, pp. 511-522.

[15] Christiansen, E. B., and D. H. Barker. "The Effect of Shape and Density on the Free Settling of Particles at High Reynolds Numbers." *Journal of American Institute of Chemical Engineers*, Vol. 11, No. 1, 1965, pp. 145-151.

[16] Kira, Hachiro. "Hydraulical Studies on the Sedimentation in Reservoirs." *Memoirs of Faculty of Agriculture*, Kahawa Univ., No. 12, 1963, p. 191.

[17] Alger, G. R., and B. Simons. "Fall Velocity of Irregular Shaped Particles." *Journal of Hydraulic Division, Proceeding of American Society of Civil Engineers*, Vol. 94, No. HY3, 1968, pp. 721-737.

[18] Schulz, S. F., R. H. Wilde, and M. L. Albertson. "Influence of Shape on the Fall Velocity of Sedimentary Particles." *M.R.D. Sediment Series*, No. 5, Missouri River Division, U.S. Corps of Engineers, 1954, p. 161.

[19] Komar, P. D., and C. E. Reimers. "Grain Shape Effects on Settling Rates." *J. Geol.*, Vol. 86, No. 2, 1978, pp. 193-210.

[20]Lorentz, H. A. "Ein Allgemeiner Satz die Bewegung Einer Reibenden Flussigkeit Betreffend, Nebst Einigen Anwendungen Desselben." *Abhand. uber Theoret. Physik*, Vol. 1, Leipzig, 1907.

[21] McNown, J. S., H. M. Lee, N. B. McPherson, and S. M. Engez. "Influence of Boundary Proximity on the Drag of Spheres." *Proceeding, 7th International Congress for Applied Mechanics*, London, Vol. 2, Pt. 1, 1948, pp. 17-29.

[22] McNown, J. S., and J. T. Newlin. "Drag of Spheres Within Cylindrical Boundaries." *Proceeding of 1st Congress of Applied Mechanics, American Society of Mechanical Engineers*, 1951, pp. 801-806.

[23] Cunningham, E. "On the Velocity of Steady Fall of Spherical Particles Through Fluid Medium." *Proceeding of Royal Society*, London Ser. A, Vol. 83, 1910, pp. 357-365.

[24] Uchida, S. "Slow Viscous Flow Past Closely Spaced Spherical Particles." *Japanese Institute of Science & Technology*, Vol.3, 1949, pp. 97-104.

[25] Cai, Shutang. "Settling of Sediment Particles in Quiescent Water" — (1) The Effect of Concentration on Fall Velocity." *Journal of Physics*, Vol. 12, No. 5, 1965, pp. 402-408. (in Chinese)

[26] Smoluchowski, N. S. "On the Practical Applicability of Stokes Law of Resistance, and the Modification of It Required in Certain Cases." *Proceeding of 5th. International Congress of Mathematics*, Cambridge Press, 1913.

[27] Burgers, J. M. "On the Influence of the Concentration of Suspension Upon the Sedimentation Velocity (In Particular for A Suspension of Spherical Particles)." *Proc., Ned. Akak. Wet.*, Amsterdam, Vol. 45, 1942, p. 126.

[28] McNown, J. A., and P. N. Lin. "Sediment Concentration and Fall Velocity." *Proceeding of 2nd Midwestern Conference on Fluid Mechanics*, 1952, pp. 401-411.

[29] Wan, Zhaohui, and Shuobai Shen. "Hyperconcentrated Flow on The Stem and Tributaries of the Yellow-River." *Selected Papers of Research on The Yellow-River*, Vol. 1, Part. 2, 1976, p. 141-159. (in Chinese)

[30] Hawksley, P. G. W. "The Effect of Concentration on the Settling of Suspensions and Flow Through Porous Media." *In Some Aspects of Fluid Flow*. Edward Arnold and Co., London, 1951, pp. 114-135.

[31] Steinour, H. H. "Rate of Sedimentation--(1) Nonflocculated Suspensions of Uniform Spheres; (2) Suspensions of Uniform Size Angular Particles; (3) Concentrated Flocculated Suspensions of Powders." *Industrial and Engineering Chemistry*, Vol. 36, No. 7, 9 and 10, 1944, pp. 618-624, 840-847, 901-907.

[32] Richardson, J. F., and W.N. Zaki. "Sedimentation and Fluidisation, Part 1." *Transition of Institute of Chemical Engineering*, Vol. 32, No. 1, 1954, pp. 35-53.

[33] Guo, Musun, and Yi'an Zhuang. *The Motion of Uniform Spheres and Fluid on A Vertical System*, Science Press, 1963, pp.97.(in Chinese)

[34] Mintz, D. M., and Shubelt, C. A. *Hydraulics of Granular Material, Press of Ministry of Public Affairs*, Federal Republic of Russia, 1955. (in Russian)

[35] Sha, Yuqing. *Elementary Theory of Sediment Transport*. China Industry Press, 1965, pp. 302. (in Chinese)

[36] Wang, Shangyi. "Settling of Fine Particles in Quiescent Water." *Journal of Hydraulic Engineering*, 1964, Vol. 5. (in Chinese)

[37] Xia, Zenhuan, and Gang Wang. "Settling of Cohesionless Uniform Particles in Fine Particle Suspension." *Journal of Sediment Research*, Vol. 1, 1982. p. 14-23. (in Chinese)

[38] Wilson, B. W. "Sedimentation of Dense Suspensions of Microscopic Spheres." *Australian Journal of Applied Science*, Vol. 4, No. 2, 1953, pp. 274-299.

[39] Fidleris, V., and R. L. Whitemore. "The Physical Interaction of Spherical Particles in Suspension." *Rheologica Acta*, Band 1, No. 4/6, 1961, pp. 573-580.

[40] Simons, D. B., E. V. Richardson, and W. L. Haushild. "Some Effects of Fine Sediment of Flow Phenomena." *U.S. Geological Survey*, Water Supply Paper 1498-G, 1963, p. 47.

[41] Ansley, R. W., and T. N. Smith. "Motion of Spherical Particles in A Bingham Plastic." *Journal of American Institute of Chemical Engineering*, Vol. 134, 1967, p. 1193.

[42] Brea, F. N. "Fluidization of Particles with Non-Newtonian Liquids." *Ph. D. Dissertation*, Univ. of Bradford, 1974.

[43] McLaughlin, R. T. Jr. "The Settling Properties of Suspensions." *Journal of Hydraulic Division, Proceeding of American Society of Civil Engineers*, Vol. 85, No. HY12, 1959, pp 9-41.

[44] Qian, Yiying, Wenhai Yang, et.al. "Basic Characteristics of Flow with Hyperconcentration of Sediment." *Proceeding of The International Symposium on River Sedimentation*, Vol. 1, Guanghua Press, 1981, p. 175-184. (in Chinese)

[45] Rowe, P. N., and G. A. Henwood. "Drag Forces in A Hydraulic Model of A Fluidised Bed, Part 1." *Transition of Institute of Chemical Engineering*, Vol. 39, No. 1, 1961, pp. 41-54.

[46] Rowe, P. N. "Drag Forces in A Hydraulic Model of A Fluidised Bed, Part 2." *Transition of Institute of Chemical Engineering*, Vol.39, No. 3, 1961, pp. 175-180.

[47] Velikanov, M. A. "Application of Gravity Theory to Calculating Sediment Setting in Turbulent Flow." *Bulletin of VNEEG*, Vol. 18, 1936. (in Russian)

[48] Velikanov, M. A. "Fluvial Processes." *National Press of Physical-Meteology Literature*, Moscow, 1985, pp. 241-245. (in Russian)

[49] Tchen, C. M. "Mean Value and Correlation Problems Connected with the Motion of Small Particles Suspended in A Turbulent Flow." *Delft Technical Hoogeschool*, Laboratory of vool Aero. en Hydro-dynamical, Medede 51, 1947, p.125.

[50] Kada, F., and T. J. Hanratty. "Effects of Solids on Turbulence in a Fluid." *Journal of American Institute of Chemical Engineering*, Vol. 6, No.4, 1960, pp. 624-630.

[51] Lhermitte, P. "Influence de la Turbulence sur la Vitesse de Chute des Particules Solides dans les Fluides Pesants." *Annales des Ponts et Chaussees*, Vol. 132, No. 3, 1962, pp. 245-273.

[52] Meyer, R. "A Propos des l'influence de la Turbulence sur la Chute des Particules Solides dan l'eau." *La Houille Blanche*, Vol. 6, No. 2, 1952, p. 285.

[53] Bouvard, M. "Etude de l'influence de la Turbulence sur la Chute des Particules Solides." *La Houille Blanche*, Nov.-Dec. 1951, pp. 862-864.

[54] Field, W. G. "Effects of Density Ratio on Sedimentary Similitude." *Journal of Hydraulic Division, Proceeding of American Society of Civil Engineers*, Vol. 94, No. HY3, 1968, pp. 705-719.

[55] Murray, S.P. "Settling Velocities and Vertical Diffusion of Particles in Turbulent Water." *Journal of Geophysical Research*, Vol. 75, No. 9, 1970.

[56] Jobson, H. E., and W. W. Sayre. "Vertical Transfer in Open Channel Flow." *Journal of Hydraulic Division, Proceeding of American Society of Civil Engineers*, Vol. 96, No. HY3, 1970, pp. 703-722.

[57] Robertson, J. A., C. Y. Lin, and M. D. Stine. "Turbulence Effects on Drag of Sharp-Edged Bodies." *Journal of Hydraulic Division, Proceeding of American Society of Civil Engineers*, Vol. 98, No. HY7, 1972, pp. 1187-1203.

[58] McLaren, F. G., A. F. C. Sherratt, and A. S. Morton. "Effect of Free Stream Turbulence on the Drag Coefficient of Bluff Sharp-Edged Cylinders." *Nature*, Vol. 223, No. 5208, pp. 828-829, Vol. 224, No. 5222, pp. 908-909.

[59] Migniot, C. "Study on Physical Properties of Various Silt and Its Properties under Flow Dynamics." *La Houille Blanche*, No. 7, 1968. (in French)

[60] Owen, M. W. "The Effect of Turbulence on the Settling Velocities of Silt Flocs." *Proceeding of 14th Congress, International Association for Hydraulic Research*, 1971, pp. 27-32.

[61] Zhang, Hao, Zenhai Ren et al. "Settling of Sediment and The Resistance to Flow at Hyperconcentrations." *Proceeding of The International Symposium on River Sedimentation*, Vol. 1, Guanhua Press, 1981, p. 185-194.(in Chinese)

[62] Shishenko, R. E. *Hydraulics of Slurry*. Oil Industry Press, 1957, p. 21-23. (in Russian)

CHAPTER 4

TURBULENCE

Sediment motion is closely related to the phenomenon of flow turbulence. Only because turbulence causes the interchange of eddies in the vertical direction can sediment be maintained in suspension against the action of gravitational force. Incipient motion and bed load motion are also closely related to velocity and pressure fluctuations caused by turbulence near the bed surface. Therefore, before discussing the law of sediment motion, the physical characteristics of turbulence and its analytical method need to be introduced.

Turbulence phenomena have been studied over many decades. The developments can be divided into several stages. In 1925 to 1930, Prandtl and Karman introduced what is now viewed as the classical theory, one that people used at the beginning of the quantitative analysis of turbulence. In 1935 to 1937, Taylor published his statistical theory of turbulence. By his analysis of turbulence as a statistical phenomenon, he provided a more acceptable physical framework. In the meantime, some significant measurements of turbulence in air currents had been obtained. After the second World War, the theoretical work of Kolmogorov, on the one hand, was widely circulated and had a broad influence. On the other hand, many experimental results from studies conducted secretly during the war were gradually disclosed in published articles. They helped people gain a better understanding of the properties of turbulence. Since the sixties, the use of modern measurement techniques such as the hydrogen-bubble method, high-speed photography, and laser have revealed a variety of turbulence patterns, and the use of computers to process the data made possible the application of a new statistical theory. In consequence, a deeper understanding of the origin and the interior structure of turbulence was obtained, and the "bursting" pattern for explaining the origin of turbulence was proposed. Although scientists of theoretical fluid mechanics of the eighties have already discarded the concept of mixing length and momentum exchange, from the practical point of view, there is still no better conceptual basis to replace the classical theory proposed by Prandtl and von Karman.

For a long time, the statistical theory of turbulence was applied only to isotropic homogeneous turbulence. Not until the decades of the seventies and eighties was the statistical theory of turbulence applied to turbulent shear flows. Furthermore, other more complex theories that have been proposed are difficult to apply to practical problems. Therefore, physical patterns of turbulence are illustrated in this chapter with special emphases. And great attention is paid to theories that may be outdated from a strict point of view. In order to avoid a mathematical treatment that is too complicated and to make the illustrations clear and simple, some problems are over-simplified; thus they may deviate somewhat from reality. Readers should therefore consider some of the various points of view introduced here as tools for solving practical problems, and not as a strict theory or a confirmed state of knowledge about turbulence.

4.1 CHARACTERISTICS OF TURBULENCE

4.1.1 Laminar flow and turbulent flow

The difference between laminar flow and turbulent flow can still be best illustrated by the classical experiment of Reynolds. Fig. 4.1 shows the equipment he used in 1883, more than a century ago, to demonstrate the phenomenon of turbulence [1]. Similar methods have been adopted in many laboratories over the years. In the experiment, a glass tube with a rounded inlet is used. The tube terminates in a valve by which the rate of flow can be controlled. At the beginning of the experiment, the tank is full of clear water that had been allowed to stand completely still for some time; the valve is then opened slightly and water begins to flow through the tube. At the same time a dye solution is injected at the entrance to reveal the pattern of flow. At low velocities the dye solution follows a distinct straight line as it passes through the pipe. It does not mix with surrounding flow layers, as shown in Fig. 4.2a. If the valve is gradually opened further, the velocity increases accordingly, and at a certain velocity, the line of dye begins to waver, Fig. 4.2b. With still further increase, the line breaks up and the dye solution diffuses over the entire cross section of the tube, completely losing its original appearance, Fig.4.2c. The flow in the first stage is called laminar, that in the third stage is called turbulent, and that in the second stage is transitional between laminar flow and turbulent flow.

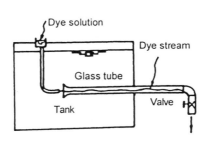

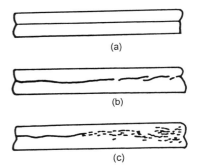

Fig. 4.1 Equipment used by Reynolds for displaying the phenomenon of turbulence

Fig. 4.2 Variation of a dye stream caused by an increase in flow velocity

The experiment reveals that whenever the velocity exceeds a certain value, part of the flow loses its stability so eddies are abruptly induced and spread throughout the flow region. Elements that have formed into eddies rotate vigorously, but no clearly defined picture can be detected for the whole eddy system. Although the flow as a whole moves downstream towards the outlet, the trajectory of each water particle is chaotic and the motion varies randomly both in time and space.

Nevertheless, if one observes the turbulent and laminar flow phenomena at greatly different scales, one finds some similarity between the eddy motion in turbulent flow and the molecular motion in laminar flow. Conceptually, one could measure the velocity at a point in a steady flow by means of a set of sensitive Pitot tubes of different diameters;

$$d_1, \quad d_2, \cdots, \quad d_{k-1}, \quad d_k$$

in which
$$d_n < d_{n-1} \quad (n=1,2,......k)$$

Fig. 4.3 shows the variation of dynamic pressure with time as measured with the different Pitot tubes. The results show that dynamic pressures measured with large Pitot tubes are constant, they do not vary with time. But the dynamic pressures measured with Pitot tubes smaller than some limiting size fluctuate. Although the average dynamic pressure measured with a Pitot tube with diameter d_k is no different from that measured with the others, yet, the measured dynamic

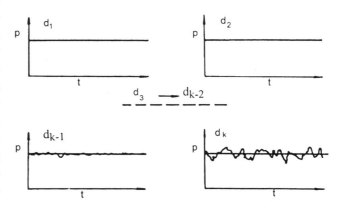

Fig. 4.3 Variation of dynamic pressure as measured with Pitot tubes of various sizes

pressure at different moments varies over a rather sizable range. If d_k corresponds to the average molecular free path of water particles, the flow is laminar. If d_k corresponds to the mixing length of flow, the flow is turbulent. The concept of the mixing length of a flow is illustrated in some detail in the following section. The mixing length of a flow has the same order of size as the scale determining the characteristics of the average motion of the flow (for instance, the size of the conduit or vessel). Usually the mixing length is more than a million times greater than the average molecular free path of water particles. In the Reynolds experiment, even though the dye solution in laminar flow appeared to be a straight line, if the scale of observation were reduced to less than 10^{-6} cm, fluctuation and diffusion would be revealed. At two extremely different scales, phenomena with similar properties exist. Only because the Pitot tube and other commonly used current meters cannot be realistically made as small as the size that characterizes the average molecular free path of water particles, do the usual observations of velocity in a steady laminar flow appear to have a constant value, one that does not fluctuate randomly with time.

The difference between a laminar and a turbulent flow in the foregoing situation can be well illustrated by a vivid example. Let us imagine a disciplined, well-trained troop of soldiers, marching in orderly ranks along a long lane. Each column of soldiers marches in step along a straight line. But if the scale of observation were reduced and only the heads of the soldiers were observed, some fluctuations would be observed. Such marching would correspond to laminar flow. In contrast, suppose a group of drunken men are ordered to move forward. As they move, they might continuously depart from the group, go faster or slower, to the left or the right, and their locations and velocities would vary randomly with time. Still, they would be restricted by the walls, and those in front would be driven along by those behind. As a whole, the drunken men would move forward, and over time, they would maintain a certain forward velocity. Such a group would correspond to turbulent flow. From this illustration the characteristics of laminar and turbulent flows differ in three essential ways:

1. At one point in a long lane, every soldier in an orderly marching troop passing the point does so at a velocity that does not vary with time. In a group of drunken men going forward, every person passing this point can have a different velocity and a different direction, although over a long time period their average velocity would be the same as the average velocity of the group. Therefore, in a steady laminar flow, the velocity components $\bar{u}, \bar{\upsilon}, \bar{\omega}$ at any point in x, y, z directions do not vary with time (the example of marching troop mentioned above corresponds to one-dimensional flow in the x-direction, $\bar{\upsilon} = \bar{\omega} = 0$), In a turbulent flow, the velocity components of a water molecule in three directions are

$$\bar{u} + u', \ \bar{\upsilon} + \upsilon', \ \bar{\omega} + \omega',$$

in which $\bar{u}, \bar{\upsilon}, \bar{\omega}$ are time-averaged velocities, and u', υ', ω' are instantaneous fluctuating velocities; the average values of the latter group over a long time period are equal to zero.

2. Even soldiers that march in step, shoulder next to shoulder, heel next to heel, experience a certain amount of friction. However, a group of drunken men that wander here and there push and press against each other, so that the resistance is much greater. Resistance in turbulent flow is usually some 300 times that in laminar flow.

3. If at the beginning, in each of these two groups, one row of soldiers wears white clothes, a different color from that of the others. In an orderly marching troop, this row of soldiers in white will maintain their relative positions at all times, marching as they do in a clearly visible line. But in a group of drunken men, not long after their start, the row of persons in white will have mixed with others around them so that they become dispersed throughout the group. Because of this type of diffusion in turbulent flow[1], such properties as momentum, heat and mass (including sediment particles) are transferred among flow regions.

---

[1] A turbulent flow possesses both diffusion and mixing effects. In this example, only the diffusion effect is involved, and the effect of mixing is not correctly reflected.

4.1.2 Emergence of turbulence

How does a laminar flow turn into a turbulent one? The reason that a laminar flow loses its stability and becomes turbulent is primarily that some disturbance is inevitable in the course of flow. If the disturbance exceeds a certain limit, the laminar flow loses its stability and eddies form--much as a wave breaks and splashes if its height exceeds a certain value. The disturbance can originate either from the enlargement of a small interior perturbation or from some external action [2]. Whether an interior disturbance is amplified or damped depends on the relative magnitudes of inertial force and viscous force. Flowing water consists of molecules with large fluidity. The weight and speed of these molecules tend to destroy regular motion and induce various disturbances. Thus flow inertia becomes the major force promoting irregular motion and intensifying any instability. Inertia force acting on a unit water volume is proportional to $\rho U^2/L$, in which ρ is water density, L a representative dimension that usually taken to be the diameter of pipe (in pipe flow). The viscous force, behaving as an internal force binding the water molecules, can reduce their fluidity and attenuate the disturbance. The viscous force acting on a unit water volume is proportional to $\mu U^2/L^2$, in which μ is viscosity of water. The turbulence phenomenon essentially depends on the balance between these two forces. The ratio of the inertia force to the viscous force acting on a unit volume of water constitutes a dimensionless number called the Reynolds number, Re:

$$Re = Ud / v \tag{4.1}$$

in which:

$v = \mu/\rho$ — kinematic viscosity of flow;
d — diameter of pipe.

A small Reynolds number indicates that the stabilizing effect of viscous forces is far greater than the disturbing effect of inertia, and the flow is therefore laminar. As the Reynolds number increases, the flow ultimately becomes turbulent. In the experiments conducted by Reynolds, the flow was laminar if the Reynolds number was less than about 2,000, and it always became turbulent if the Reynolds number exceeded a value in the range 10,000-12,000. Later experiments revealed that if the curvature of the entrance was gradual and the experiment was started after the water in tank had become completely quiescent, the critical Reynolds number for the transformation was much higher. This development clearly indicates the effect of external disturbances on the stability of laminar flow. Still, no matter how carefully the experiment is conducted, so as to reduce external disturbances as much as possible, the flow can not remain laminar if the Reynolds number exceeds 40,000 [3]. That is, in addition to any external disturbances, interior instability is an inherent factor promoting the transformation of a laminar flow into a turbulent flow. For flow in rivers, internal disturbances are the causative factor. Such disturbances can have different forms; for instance, eddies can be produced at boundary roughness and in areas of local flow separation.

In order to explain how local disturbances can induce eddies, one can envision a surface of separation in a non-viscous fluid with different velocities on the two sides of the surface, as shown in Fig. 4.4a. If the streamlines on the separation surface are deformed or bent for some reason, as shown in Fig. 4.4b, the velocity is higher, and the

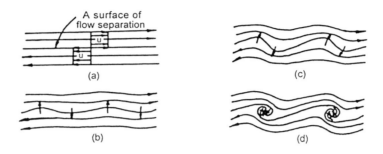

Fig. 4.4 Sketch of vortices evolving at a surface of flow separation (after Rouse, H.)

pressure lower, at places where the streamlines are more concentrated. The situation is just the opposite where streamlines are more widely dispersed. As a consequence, the bending of streamlines is intensified, as shown in Fig. 4.4c. Ultimately, eddies are produced, as shown in Fig. 4.4d [4]. Because of the viscosity of a fluid, such completely opposing flows on the two sides of the separation surface cannot occur in a real flow. Nevertheless, at a place where the velocity gradient is large, the production of eddies is similar to that in Fig. 4.4. In alluvial rivers, not only does the flow separate from dune crests, but it also produces local small-scale separations in the flow around individual protruding sediment particles, and eddies can form at any of these separation surfaces. Furthermore, the velocity gradient in open channel flow is generally large near the boundary. Hence, the entire perimeter of a river bed is a source of turbulence.

Once an eddy forms close to the boundary, the relationship between this eddy and the surrounding fluid is as shown in Fig. 4.5. At the top of the eddy, the velocity component due to the rotation of the eddy u_r and the local velocity component u_1 are in the same direction, but at the bottom they are opposed. Thus, a pressure difference

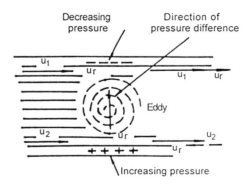

Fig. 4.5 Mechanism of lift for a vortex near the bed and its entrance into the main flow

occurs in the vertical direction, due to the difference in velocity heads, and it causes the eddy to move outward from the bed. Meanwhile the eddy is also being carried downsteam by the main flow. In this manner, eddies originating at the bed gradually spread throughout the flow region and make the entire flow turbulent.

The foregoing illustration is somewhat schematic. Since "bursting phenomenon" was discovered, our understanding of the formation and lifting of eddies has been expanded. It is discussed in detail in section 4.1.4.

4.1.3 Momentum exchange and turbulent shear

The vertical movement of eddies in turbulent flow induces continual exchanges between fluid layers of such fluid properties as momentum, heat, concentration. Among them, the exchange of momentum results in a turbulent shear stress and thus determines the time-averaged velocity field. Through the interpretation of the processes of momentum exchange, the mechanism of sediment suspension can be better understood.

The following visualization [5] shows how the concept of momentum exchange, so important a function of turbulent shear flows, was firstly introduced in hydraulics. As shown in Fig. 4.6a, wagon A loaded with wood and passenger train B move on parallel tracks at the speeds of u_a and u_b. Inasmuch as only the relative velocity $u'=u_b-u_a$ is involved in momentum exchange, for convenience one can consider wagon A to be at rest and passenger train B to move on a parallel track with velocity u'. Machine guns placed in the windows of the passenger train B fire at a right angle to the direction of the train's motion at a constant rate of N bullets per second as the train passes by wagon. All the bullets hit the wagon and penetrate the wood loaded in the wagon. If the stationary wagon is attached to a dynamometer D, a force F exerted on the wagon will show on the scale of the dynamometer. The cause of the force is that each bullet has a momentum in the direction of the train's motion that is equal to the mass of the bullet multiplied by the relative velocity of the passenger train u'. As the bullets become embedded and are brought into rest, that momentum is lost, and force F is thus produced. Its magnitude and direction are equal to those of the momentum change per second in accordance with Newton's Law. If the weight of each bullet is W kg, the force indicated on the dynamometer is

$$F = \frac{NW}{g}\left(u' - 0\right) = \frac{NW}{g}u' \qquad (4.2)$$

If the speed of the passenger train is higher than that of the wagon, firing from the passenger train exerts an accelerative force on the wagon. If it is lower, firing from the passenger train decelerates the wagon, as shown in Fig. 4.6b.

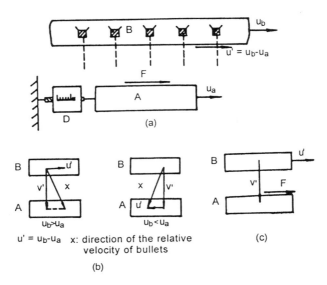

Fig. 4.6 Mechanism of the formation of shear by momentum exchange (after Bakhmeteff, B.A.)

The above reasoning helps us understand the mechanism of momentum exchange between adjacent layers in a moving fluid. In Fig. 4.6c, the adjacent surface area is A_0, the relative velocity between adjacent layers is u' and the fluctuating cross-current velocity is v'. If this cross-current velocity were uniformly distributed over the surface A_0, then the mass transported from layer B to layer A per unit of time would be $\rho v' A_0$. If the particles thus transported into layer A become thoroughly mixed with particles already in layer A and are brought to move axially with the velocity of layer A, the momentum exchange exerts a shear force F on the surface of layer A in the direction of the motion:

$$F = \rho A_0 u' v' \tag{4.3}$$

The shear stress per unit area would then be

$$\tau = F / A_0 = \rho u' v' \tag{4.4}$$

In reality u' and v' vary continuously at any point within the flow, and the shear stress τ is proportional to the average value of the product of u' and v',

$$\tau = \rho \, \overline{u' v'} \tag{4.5}$$

In a river, the velocity generally increases with distance from the bed. If the direction of u' is taken as positive when it coincides with flow direction, and the direction of v' is positive when it is directed upward from the bed, then the sign of u' is always opposite to that of v'. For such a system, the preceding equation should be rewritten as:

$$\tau = -\rho \, \overline{u' v'} \tag{4.6}$$

This is the conventional form for turbulent shear stress. Since the nineteenth century, much of the study concerning turbulence has been an attempt to transform the pattern of velocity fluctuation into a function of the time-averaged velocity and the consequent establishment of a relationship between the distribution of velocity and stress.

4.1.4 Bursting phenomenon

4.1.4.1 Historical review of the study of bursting phenomenon

In early studies of fluid mechanics, flow near a smooth boundary was thought to have a stable laminar layer called the laminar sublayer. Sediment particles on the bed smaller than the thickness of the laminar sublayer would then be sheltered within it, and turbulence would not penetrate the laminar sublayer; hence it could not affect sediment particles on the bed. Early in fifties, Einstein and Li raised a question about this concept. They performed a simple experiment [6]. On the bottom of a flume they drilled a small hole and injected a dye solution into the laminar sublayer. If the flow in the laminar sublayer were to maintain its laminar pattern, the dye solution would move along the bottom in a straight line parallel to flow. However, the experiment showed that the dye solution suddenly lost its stability after flowing near the bed for a short distance, and it then mixed with the turbulent flow in the main flow region and could no longer be detected. The distance along which the dye solution kept its linear pattern was longer at some times and shorter at others. The study revealed a periodic instability in the laminar sublayer. From this experiment, Einstein and Li decided that the laminar sublayer was not really stable, but was rather a region that experienced periodic variations of growth, collapse, and re-growth. With such variations, turbulence could penetrate the laminar sublayer onto the boundary and affect the sediment particles there.

Einstein and Li's work did not attract much attention at the time. Later, in the 1960s, Kline and others at Stanford University developed the hydrogen-bubble technique. They used a platinum wire with a diameter in the range of 10 to 50 μm mounted perpendicular to the flow direction; the wire was the negative electrode of a direct current and the flume walls were the positive electrode. When a burst of electricity passed through the wire, it generated very fine fog-like hydrogen bubbles (with diameters of 0.5 to 1.0 times the wire diameter) around the platinum wire that then moved downstream with the flow. In the experiments, the platinum wire was coated with insulating paint over certain intervals, and a pulsing current was then fed to the wire, also with a certain period. With the intermittent production of hydrogen bubbles from the pulsing current and their separation in space due to the insulating paint, the bubbles formed networks of a kind of fog. With the help of intense illumination and high-speed photography, they were then able to record the turbulence pattern near the platinum wire [7]. Two pictures taken by using hydrogen bubble technique that exemplify their results are shown in Fig. 4.7. In the pictures, the background that appears in patches of dark and light is due to the fog formed by that intermittent release of hydrogen bubbles [8]. From each picture, the profile of instantaneous velocities were obtained. The local profiles of the mean longitudinal velocity and the fluctuating velocities at different points were deduced by synthesizing these pictures. The plot in the middle of Fig. 4.7 shows the profile of the mean longitudinal velocity and the profiles of the longitudinal and vertical instantaneous velocities obtained from such an analysis. Sometimes two or more platinum wires were placed perpendicular to each other, or a dye solution was injected. Such visualizations revealed the spatial and temporal structure of turbulence, and they help to provide a deeper understanding of the production and development of turbulence — in particular, the physical pattern of turbulence near the boundary and the bursting characteristics of turbulence.

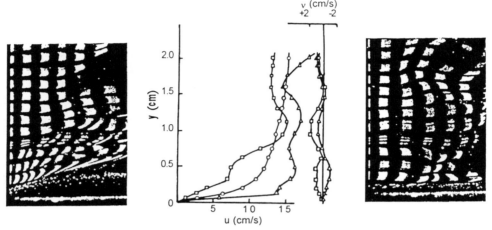

□   u,v velocity profiles associated with low boundary velocity
    and a fluid ejection phase(photograph on the left)
Δ   u,v velocity profiles and a fluid inrush phase(photograph on right)
o   mean longitudinal velocity profile for comparison

Fig. 4.7 Pictures obtained using hydrogen bubble technique, and longitudinal
and vertical velocity profiles obtained from them (after Grass, A. J.)

4.1.4.2 Results of observations of the bursting phenomenon

Recent developments indicate that turbulence is not as random as was initially believed. The concept of simply adding a random velocity field to the time-average motion is not an adequate one. Space related and time-related orderly motions do exist in turbulent flows. These motions can be called a quasi-cyclic process. They result from events that are repeated in time and in space but are not strictly periodical, either in space or in time. The two most striking events, which are observed near the boundary, are (1) the lift-up of low-speed streaks from near the boundary, and (2) the "sweep" of high-speed fluid toward the boundary.

Most of the experiments were conducted for flows with hydraulically smooth boundaries (i.e., flows with laminar sublayers). The observations reveal that the "laminar sublayer" is not two-dimensional and stable as had been assumed in earlier studies, but three-dimensional and unstable, and that it has local velocities perpendicular to the boundary. In the region of the laminar sublayer zone high-speed zones and low-speed streaks are alternatively distributed in both transverse and longitudinal directions; also the lateral velocity ω varies with the longitudinal velocity u. In Fig. 4.8, transverse distributions of instantaneous velocities u and ω measured by means of the hydrogen bubble technique are shown. Statistical results indicate that although transverse intervals between individual low-speed streaks may vary (with a relative standard deviation of 30 to 40%), their average value l_Z is rather stable, and it corresponds to a value of the parameter $\lambda_Z U*/\nu$ that is roughly 100.

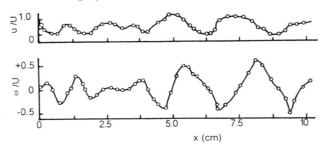

Fig. 4.8 Instantaneous spanwise variation in u and ω (for $(yU*/\nu)=5$)
(after Kim, H. T., S. J. Kline, and W. C. Reynolds)

The intermittently lifted low-speed streaks leave the boundary and penetrate the main flow region. Fig. 4.9 is a sketch of the ejection of low-speed streaks as observed by dye injection [9]. The main part of the observed low-speed streak is indicated by the arrows shown in the figure. Initially, the low-speed streak migrates slowly downstream as a whole, and drifts slowly outward. This stage persists over some distance. But once the low-speed streak attains a critical distance from the boundary, it moves rapidly

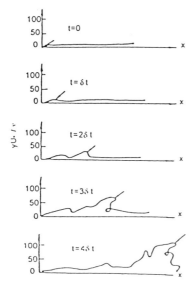

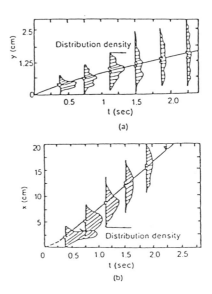

Fig. 4.9 Dye streak breakup as seen from the side (after Kim, H. T., S. J. Kline, and W. C. Reynolds)

Fig. 4.10 Trajectory of ejected eddies (after Kim, H.T., S.J. Kline, and W.C. Reynolds)

125

outward as it moves downstream. The critical distance is not a fixed one as it varies in the statistical sense. The trajectory of the ejected fluid, as measured by Kline, is shown in Fig. 4.10. In the figure, x is in the flow direction, and y is perpendicular to the boundary. The shaded area denotes the distribution density of the fluid ejected at a time t and reaching the location x (or y). Although the individual trajectories vary considerably, the average trajectory is rather stable; also, its average value coincides with its mode. The low-speed streak enters the region of main flow with a longitudinal velocity that is much lower than that of the surrounding fluid. Hence, on the plot of instantaneous longitudinal velocity profile, an inflection point appears at the place reached by the low-speed streak. Velocity profiles with inflection points are often unstable, and can induce flow oscillations further downstream. Such oscillations can be detected in the third picture in Fig. 4.9. The region where the oscillation first appears is in the range of $yU*/v=8$ to 12 (in which y is the distance from the boundary). The oscillation amplifies quickly. After some 3 to 10 periods, this flow structure collapses, and an even more chaotic motion appears. The collapse usually happens in the region $10<yU*/v<30$. Some researchers have suggested that the low-speed streaks rise even to the region $yU*/v>100$. The fourth and fifth pictures in Fig. 4.9 show the twisting and ultimate collapse. The frequency of uplift of low-speed streaks and of bursting per unit width of boundary are proportional to U_*^3.

Another striking feature near the boundary, in addition to the ejection of low-speed streaks, is the "sweep" of high-speed fluid from the main flow region onto the boundary. Grass demonstrated the "sweep" phenomenon by means of sediment particles (diameter 0.1 mm) moving on a smooth bed [8]. In the experiment, the bed was painted black and the sediment particles were white. Coming from the main flow region and reaching the boundary, the high-speed fluid carried away all the sediment particles, and uncovered a path along the black bed. The sediment particles moved downstream with a little lateral diffusion.

Offen and Kline conducted other experiments focusing attention on the inherent correlation between uplift of low-speed streak from the wall and sweep of high-speed fluid. Not only did they inject a dye solution and place a platinum wire close to the boundary, they also injected a dye solution into the main flow region where the velocity profile was logarithmic [10]. They observed two characteristic features: (a) Before almost every uplift of a low-speed streak and the appearance of an oscillation at the boundary, a disturbance originating in the main flow region occurred just upstream. Originating in the logarithmic velocity distribution region, i.e., usually in the range of $20 \leq yU*/v \leq 200$, the disturbance possessed an eddy-like flow pattern with a mean motion toward the wall. The oscillation in the boundary region was always located downstream of that disturbance. (b) The low-speed streak grew and was gradually lifted up. At the end of the growth stage of the oscillating low-speed streak, the mutual action of the fluid in the streak and the fluid in the logarithmic region induced another large eddy-like structure. This eddy system grew downward toward the boundary, and a disturbance in the main flow region formed that moved toward the boundary thus inducing another uplift and oscillation of a low-speed streak at a location further downstream.

4.1.4.3 Model of the formation and development of bursting phenomenon [7,11,12]

Observations of the bursting phenomena suggest the following simplified model to describe its formation and development.

In a flow region that is not directly affected by a boundary in a turbulent shear flow, a vortex with a linear axis has formed downstream from a cylinder, shown as (1) in Fig. 4.11a. Such a vortex is unstable if a certain type of disturbance acts on it; any deformation thus produced develops further because of self-induction and because the velocity is higher in the upper region and lower in the lower region. In stage (2) of the figure, a slight disturbance in the x-z plane of the vortex produces an upward vertical velocity component v at its peak (marked T) and downward velocity at its low point (V). It causes the peak of the vortex to enter a relatively high-speed region and the trough a relatively low-speed one. Thus, as the vortex is carried downstream by the flow in stage (3), it is stretched in the x-direction and forms a U-shaped vortex. At the two wings of the vortex along their axes, slightly inclined to x-direction, a pair of vortices rotating in opposite directions forms.

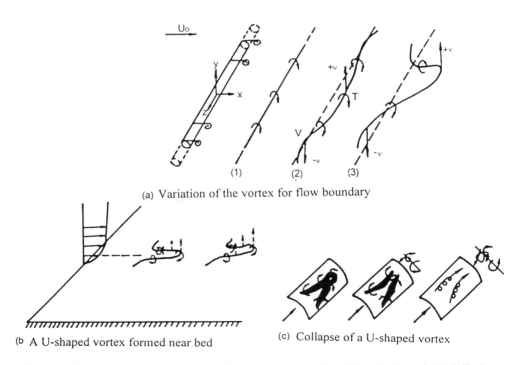

(a) Variation of the vortex for flow boundary

(b A U-shaped vortex formed near bed

(c) Collapse of a U-shaped vortex

Fig. 4.11 Formation and development of U-shaped vortex (after Offen, G. R., and S. J. Kline)

From the same mechanism, a U-shaped vortex may form near the boundary as shown in Fig. 4.11b. Moving downstream and distorting its shape, the U-shaped vortex leaves the boundary and is lifted into the main flow region. Rotation of the vortex intensifies with an increase of curvature. It reaches a maximum at the tip of the U-shaped vortex, where both longitudinal and vertical velocities are high. Once the vortex has developed to a certain stage, an inflection point appears on the vertical profile of the longitudinal velocity (the profile of instantaneous longitudinal velocity at the moment when the low-speed streak is being lifted in Fig. 4.7). There, both the velocity gradient and the shear are large. Finally the vortex collapses and breaks up, forming eddies of various sizes as shown in Fig. 4.11c.

Fig. 4.12 is a sketch of the bursting phenomenon based on the figure. In a quasi-cyclic process, turbulence bursting phenomena can start at any stage of the cycle. In Fig. 4.12 a horseshoe vortex induced by large-scale disturbance outside the boundary region forms near the boundary, then the vortex develops as it moves with the flow.

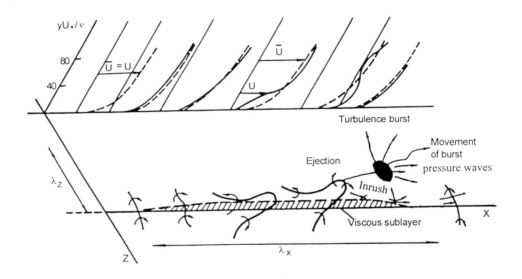

Fig. 4.12 Sketch of turbulence bursting phenomenon (after Offen, G. R., and S. J. Kline)

Furthermore, after the horseshoe vortex is formed, it stretches gradually in flow direction and transforms into a U-shaped vortex as it is altered by the flow. The tip of the vortex then leaves the boundary and progressively penetrates the flow region where it attains a higher and higher velocity. As it stretches, the horse-shoe vortex continuously increases its rate of rotation. In the process, a streak with low momentum is lifted between the two arms of the ring and also enters the region of higher flow. Finally, in the region of $yU*/\nu$=10-30, a horizontal layer with intensive shear forms, it appears as a sharp gradient on the longitudinal velocity profile. Such a velocity profile is unstable. It causes the vortex to break up, thus forming what is called turbulence bursting and producing a pressure wave that propagates outward. The space left as the low momentum streak leaves the boundary is filled by water nearby, part of which comes from the high velocity region. Accelerated by the pressure wave, the water element filling the space drives the high momentum fluid toward the boundary, and sweep the bed making a rather flat angle of 5° to 15° with the boundary. The flow sweeping the boundary is quickly retarded by the strong viscous forces near the boundary and is affected by disturbances in the flow some distance from the boundary, a new horseshoe vortex forms. Thus a new cycle starts.

The process of growth and breakup of the laminar sublayer discovered by Einstein and Li can also be seen clearly in Fig. 4.12. As a horseshoe vortex starts to form and lifts off from the boundary, a laminar sublayer also forms and grows gradually. When a part of the fluid with high momentum sweeps the bed, the laminar sublayer breaks up, and turbulence penetrates far enough to reach the bed. Further downstream, the laminar sublayer has reformed and another horseshoe vortex forms.

The foregoing sequence of events occurs within a narrow strip on the x-z plane. Other horseshoe vortices induced by other disturbances develop independently along the z-direction with a spacing of approximately λ_z. Having experienced stages of uplift and growth, a pair of neighboring low-speed streaks break up independently. After the break-up, part of the low-speed fluid returns to the wall. When it reaches the wall, it spreads sideways and is quickly further retarded. Thus a new low-speed streak further downstream is formed in the region between the locations of the pair of original low speed streaks, as indicated in Fig. 4.13.

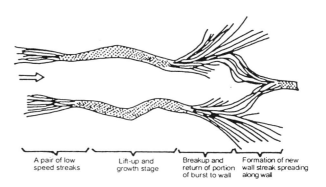

A pair of low speed streaks   Lift-up and growth stage   Breakup and return of portion of burst to wall   Formation of new wall streak spreading along wall

Fig. 4.13 Processes during the formation of a new low-speed streak between two adjacent bursts (after Grass, A.J.)

Liftup and breakup of low-speed streaks produce locally instantaneous stresses due to turbulent shear that are extremely high. They play a dominant role in the transfer of turbulent kinetic energy near the wall. According to the Grass experiments [8], when low-speed streaks pass, the peak value of the Reynolds stress at these points may be five times the local average turbulence shearing stress. High-speed fluid penetrating to the wall can also produce high turbulent shearing stresses, but their effect is limited to the region adjacent to the wall.

The Grass experiments were conducted for three different boundary conditions: hydraulically smooth, transitional and hydraulically rough. The differences between the results for the different boundary conditions are as follows:

First, for the rough boundary, the lifted fluid is ejected almost vertically from the intervals among protruding rough elements. For a smooth boundary, the angle between the ejected fluid and the boundary is smaller.

Second, when high-speed fluid approaches a smooth boundary, it is retarded by viscous resistance and decelerates slowly, whereas for a transitional or rough boundary, form resistance causes a more rapid retardation. Hence, as the degree of boundary roughness increases, the longitudinal velocity decreases in a much shorter distance. In contrast, the vertical fluctuating velocity is greater.

Bursting characteristics of turbulence naturally have far-reaching effects on sediment motion along a bed, as discussed in Chapter 10.

4.1.5 Structure and composition of turbulent eddies

Restricted by the boundary, an eddy is small as it begins to form. In the process of uplift, the diameter of the eddy gradually increases. But such eddies with high speeds of rotation are unstable. They tend to decompose into eddies of second order in size. These eddies are also unstable and decompose further until the Reynolds number characterizing them is so low that further decomposition is unlikely. At that level, energy that has been transferred to these smallest eddies is transformed into heat through viscous effects. Eddies of various sizes have different effects on turbulence phenomena. Pulsations caused by large eddies have long periods and low frequencies, those induced by small eddies have short periods and high frequencies. Because eddies of various sizes exist simultaneously, the pulse frequency displayed is not a single or dual one; it has a continuous frequency spectrum instead.

The exchange phenomenon caused by turbulent flow is mainly the result of large-scale turbulence. The size of large-scale eddies have the same order of magnitude as the size of a conduit (pipe diameter, water depth, etc.). The distribution and orientation of large eddies depend on the way they form. They are not isotropic and homogeneous, i.e., turbulence intensities at different places in space are different, and at any point the root mean squares of fluctuating velocity components in the three directions are also different. However, in any turbulent flow the intensive mixing creates a strong tendency for the root mean squares of the three fluctuating velocity components to approach equality. Therefore, while large eddies decompose in a progression into eddies of smaller size, and as energy is transferred to smaller and smaller eddies, the consequences of the way they were formed originally becomes weaker and weaker, geometrical orientation gradually disappears, and the small eddies tend to be isotropic and homogeneous. Hence, no matter how one turns or reflects the coordinates, their root-mean-squares remain unchanged.

Fig. 4.14 indicates the distribution of energy among eddies of various sizes with the reciprocal of the eddy size as the abscissa [12]. The largest eddies (with a long life) contain only about 20% of the total energy of turbulence. Their distribution and orientation depend on the way they formed. The two curves shown in the figure represent two different patterns of formation. Eddies of medium sizes contain the most kinetic energy. Thus they are often called "energy-containing" eddies and are denoted by the size symbol l_e. The mixing length l proposed by Prandtl falls somewhere between the container size and l_e. With the decrease of eddy size, the energy dissipation caused by viscosity increases. The total energy of the small eddies is limited, but they take energy continuously from larger eddies through momentum transfer. For the small eddies, which

are isotropic and homogeneous, only one curve is required, and it is called the universal curve. At a certain eddy size l_d, the energy dissipation caused by viscous forces is a maximum. Their eddy Reynolds numbers are rather small (of the order of 8 or 10). Eddies can be still smaller, down to an eddy Reynolds number of about 1. For these eddies, viscous forces and inertial forces are equally important. The size of these smallest of all the eddies is designated as l_k.

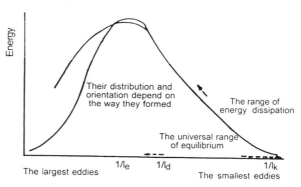

Fig. 4.14 The distribution of energy among eddies with different sizes (after Davis, J.T.)

For isotropic and homogeneous turbulence, as the energy distribution among eddies of various sizes approaches equilibrium, Kolmogoroff defined the size of small eddy l_k and the fluctuating velocity $\sqrt{\upsilon_k'^2}$ in terms of energy dissipation rate per unit mass of water e and kinematic viscosity of the flow v. From dimensional analysis he obtained

$$\sqrt{\upsilon_k'^2} = (\upsilon\varepsilon)^{\frac{1}{4}} \tag{4.7}$$

$$l_k = (\upsilon^3 / \varepsilon)^{\frac{1}{4}} \tag{4.8}$$

The eddy Reynolds number is then equal to 1. These two equations yield:

$$\varepsilon = \left(\sqrt{\upsilon_k'^2}\right)^3 / l_k \tag{4.9}$$

At equilibrium, the rate of continuous transfer of energy from medium-scale eddies to small-scale eddies is constant. The Reynolds number for these medium-scale eddies is rather large. Therefore their average properties depend only on ε; they are unaffected by viscosity. For this category of eddies, one can write a formula, similar to Eq.(4.9),

$$\varepsilon = \left(\sqrt{\overline{v'^2}}\right)^3 / l_e \qquad (4.10)$$

According to Kolmogoroff's assumption, eddies of various sizes in turbulent flow consist of three main parts. Among them, large-scale eddies of the highest order take energy from time-average flow, then transmit energy to eddies of the next smaller order. The large-scale eddies serve as the store house from which energy is supplied, and in the smallest eddies, the energy is consumed. The latter accept energy from eddies of higher order and consume and transmit that energy into heat through viscous effects. Between these two categories of eddies are the eddies of medium order; they are a kind of intermediary between the storehouse and the consumers, their properties depend on the rate of energy transmitted from the storehouse, but they have nothing to do with how the consumers consume the energy.

4.2 CLASSICAL TURBULENCE THEORY

Due to the complexity of the turbulence phenomenon, no complete theory is available that serves to describe and explain fully each of the links of turbulence phenomena. All existing works are based on semi-empirical hypotheses to establish relationships between turbulence stresses caused in the exchanges and the time-averaged velocities. Following the historical development of turbulence theory, basic hypotheses proposed by various classical schools and deduced principal results are presented herein.

4.2.1 Mixing length theory

The following relationship between shear and velocity gradient for laminar flow is well-known.

$$\tau = \mu \frac{du}{dy} = v\rho \frac{du}{dy} \qquad (4.11)$$

in which μ and v are the coefficients of viscosity and of kinematic viscosity for the fluid, respectively. In air currents, air molecules move in a completely random way and velocity components in different directions are equal; thus in a stable flow field with velocity gradient $u = u(y)$, an air molecule at height y will possess a certain momentum $mu(y)$, in which m is the mass of the air molecule. In the course of random motion, the air molecule maintains its momentum unchanged until it travels a distance, on an average, equal to the mean free path l before it collides with another air molecule. Therefore the air molecule, after leaving the y plane vertically, reaches the $y+l$ plane, and no energy is dissipated in the process. Thus in flow layer with a thickness of l, the momentum exchange is

$$mu(y+l) - mu(y) \simeq ml\frac{du}{dy} \qquad (4.12)$$

132

The opportunity for air molecules to move in various directions is equal; therefore, the number of air molecules normally passing through a plane should be one third of the total number of air molecules passing through the plane. If N is the number of air molecules in a unit volume and c the average velocity of air molecules, then the rate of momentum exchange per unit area between layers is

$$\frac{1}{3} Nmcl \frac{du}{dy}$$

The shear stress thus produced at the interface is

$$\tau = \frac{1}{3} Nmcl \frac{du}{dy} \tag{4.13}$$

in which Nm equals the density of air; therefore,

$$\tau = \frac{1}{3} \rho \, cl \frac{du}{dy} \tag{4.14}$$

from the foregoing equation and Eq.(4.11), it can be seen that for air

$$\mu = \frac{1}{3} \rho \, cl \tag{4.15}$$

$$v = \frac{1}{3} cl \tag{4.16}$$

i.e., the kinematic viscosity of air is actually equal to one-third of the product of the velocity of an air molecule and its mean free path.

In 1925 Prandtl first simulated momentum exchange on a macro-scale in a way equivalent to that of the molecular motion of a gas in an effort to explain the mixing phenomenon induced by turbulence in water flow; and he thus established the mixing-length theory for turbulent flow [14]. He replaced the motion of a single air molecule by the motion of a water element called an "eddy," and he replaced the average free path of air molecule by a quantity that he called the mixing length of turbulent flow.

If a small body of water moves randomly in a flow field, it has various properties, like momentum, heat and concentration, that are similar to those of the general flow at its point of origin. These properties do not change until the water element has moved a distance l perpendicular to the flow direction and has become mixed with the local fluid. As it moves, it changes its properties, losing those it started with and gaining those of the fluid at the new location, Fig. 4.15. The difference between the average values of any property at the starting point and at the end point is a measure of the fluctuation of that property at its end point. The distance l, the familiar mixing length of the turbulent flow,

133

is the length over which an eddy moves perpendicularly to the flow during its life span. Mixing length for turbulent flow is a length that characterizes turbulence interchanges at any plane. It was conceived in an analogy with the mean free path of air molecules, but the two concepts diverge somewhat, the mean free path of an air molecule is unaffected by the position and average velocity of the air molecule, but the mixing length of turbulent flow is related to these quantities.

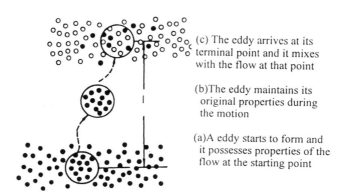

(c) The eddy arrives at its terminal point and it mixes with the flow at that point

(b)The eddy maintains its original properties during the motion

(a)A eddy starts to form and it possesses properties of the flow at the starting point

Fig. 4.15 Sketch showing mixing length in turbulent flow

If x denotes the flow direction and y is perpendicular to it, then according to the definition of mixing length, the fluctuation velocity in the x-direction at point y is

$$u' = l\frac{du}{dy}\bigg|_{y=y} \tag{4.17}$$

The origin of the fluctuation velocity v' in the y direction can be illustrated in the following way. Two elements of water start to move from $y+l$ and $y-l$ toward each other, and they finally meet at y. If the element originating at $y-l$ has a low velocity and arrives at y just ahead of an element originating from $y+l$ with a high velocity, these two elements will collide with a relative velocity $2u'$ at layer y; they are then deflected and depart along the normal to the flow direction and produce a fluctuation velocity v' outward from layer y. Or, if the low velocity element arrives at y later and the high velocity one arrives sooner, then the two elements will depart outward from layer y with velocity $2u'$, and thus the space left behind will be filled by surrounding water. It would result in a fluctuation velocity v' toward layer y. Based on this pattern, u' and v' are not only related to each other, but they also are of the same order of magnitude, i.e.

$$v' \sim u' = l\frac{du}{dy} \tag{4.18}$$

If the element of water reaching layer y comes from the lower part of a flow where the velocity is low, v' is positive (upward) and the exchange will result in a decrease of velocity at y, i.e., the direction of u' is negative. On the contrary, if the element comes from the upper part where the velocity is high, then the direction of v' is negative and the induced u' is positive. The directions of u' and v' usually are opposite, so that

$$\overline{u'v'} \sim -\overline{|u'|} \cdot \overline{|v'|} \tag{4.19}$$

Substituting Eq.(4.18) and Eq.(4.19) into Eq.(4.5), one obtains:

$$\tau = -\rho\, \overline{u'v'} = \text{constant} \times \rho\, l^2 \left(\frac{du}{dy} \right)^2 \tag{4.20}$$

If the constant is absorbed into the mixing length term and the signs are kept consistent, Eq.(4.20) can be written as

$$\tau = \rho\, l^2 \frac{du}{dy} \left| \frac{du}{dy} \right| \tag{4.21}$$

in which $\left| \dfrac{du}{dy} \right|$ denotes the absolute value of the temporal mean velocity gradient.

In analogy to the equation for laminar flow, Eq.(4.21) can be rewritten as

$$\tau = \eta \frac{du}{dy} \tag{4.22}$$

in which

$$\eta = \rho\, l^2 \left| \frac{du}{dy} \right| \tag{4.23}$$

is called eddy viscosity. If the Reynolds number of the flow is not much larger than its critical value, so that turbulent and viscous effects are both important, the shear between fluid layers is the sum of the shears due to turbulence and viscosity, i.e.

$$\tau = (\mu + \eta) \frac{du}{dy} \tag{4.24}$$

Just as there is a difference in concept between the average free path of air molecules and the mixing length of turbulent flow, so is there between the eddy viscosity and the dynamic viscosity; the latter is a property of the fluid and has nothing to do with location in the flow, whereas the former is not a constant, but rather varies with both location and local velocity.

In most turbulent flows, the shear stress results mainly from momentum exchange, and Eq.(4.22) can be written as

$$\tau = \varepsilon_m \frac{d(\rho u)}{dy} \tag{4.25}$$

in which

$$\varepsilon_m = l^2 \left| \frac{du}{dy} \right| \tag{4.26}$$

is called the momentum exchange coefficient. According to Eq.(4.18), the momentum exchange coefficient is actually the product of the mixing length and the root-mean-square of the fluctuating velocity.

$$\varepsilon_m = l\sqrt{\overline{u'^2}} \sim l\sqrt{\overline{v'^2}} \tag{4.27}$$

4.2.2 Similarity hypothesis of turbulent flow

The next step is to determine the mixing length of turbulent flow and then to deduce velocity profile by means of Eq.(4.21). Experimental results show that the shear in turbulent flow is proportional to the square of the velocity. Thus, one can deduce that the mixing length should depend not on the velocity but on the location in the flow. This function can be determined by means of the similarity hypothesis proposed by Von Karman [15].

Von Karman's hypotheses are as follows: 1. except in a region near the boundary, turbulence phenomena are not affected by viscosity; 2. the basic patterns of turbulence at different positions are similar. Differences are only in the scales of time and length. The characteristic length of turbulence is the mixing length. The time-scale, on the other hand, is implicitly included in the velocity scale, and the features that shape the velocity curve at each point are the successive derivatives: du/dy; d^2u/dy^2; d^3u/dy^3

If the velocities at y_1 and y_2 are u_1 and u_2 respectively, the variation of the velocity in the vicinity of y_1 and y_2 may be expressed by means of a Taylor series as follows:

$$\left. \begin{aligned} u(y_1 + dy) &= u_1 + \frac{du_1}{dy}dy + \frac{1}{2}\frac{d^2u_1}{dy^2}dy^2 + \frac{1}{6}\frac{d^3u_1}{dy^3}dy^3 + \cdots \\ u(y_2 + dy) &= u_2 + \frac{du_2}{dy}dy + \frac{1}{2}\frac{d^2u_2}{dy^2}dy^2 + \frac{1}{6}\frac{d^3u_2}{dy^3}dy^3 + \cdots \end{aligned} \right\} \tag{4.28}$$

The idea of similarity of pattern would suggest that there must be some regular proportionality relation between the different factors affecting the change of the velocity curve. In other words, for the factors operating at different points there should obtain proportionality relations:

$$\frac{du_2 / dy}{du_1 / dy} \sim \frac{d^2 u_2 / dy^2}{d^2 u_1 / dy^2} \sim \frac{d^3 u_2 / dy^3}{d^3 u_1 / dy^3} \sim \cdots \quad (4.29)$$

Hence

$$\frac{du_1 / dy}{d^2 u_1 / dy^2} \sim \frac{du_2 / dy}{d^2 u_2 / dy^2}; \quad \frac{d^2 u_1 / dy^2}{d^3 u_1 / dy^3} \sim \frac{d^2 u_2 / dy^2}{d^3 u_2 / dy^3}; \quad \cdots \quad (4.30)$$

Each term in the preceding equation has a length scale that should be proportional to the characteristic length of turbulence (mixing length), i.e.

$$\left. \begin{array}{l} \dfrac{du / dy}{d^2 u / dy^2} = \dfrac{1}{\kappa_1} l; \\[4mm] \dfrac{d^2 u / dy^2}{d^3 u / dy^3} = \dfrac{1}{\kappa_2} l; \end{array} \right\} \quad (4.31)$$

in which k_1, k_2 .... are constants of proportionality. In his analysis, Karman used only the first term in their expression. Hence

$$\frac{du / dy}{d^2 u / dy^2} = \frac{1}{\kappa} l$$

$$l = \kappa \frac{du / dy}{d^2 u / dy^2} \quad (4.32)$$

in which κ, called the Karman constant, does not vary with discharge, average velocity or boundary conditions — neither size nor roughness. From Nikuradse's experiments with clear water flow in pipe, the Karman constant is 0.4 [16]. In an analysis of Basin's experiment data, Keulegan showed that the Karman constant is also 0.4 for clear water flow in open channels of different shapes and with different roughnesses [17]. Only in sediment-laden flows does the Karman constant vary, and it then depends on the concentration and vertical distribution of the sediment. This point is discussed in detail in Chapter 12.

According to Eqs.(4.21) and (4.32), the following relationship between the velocity field and the shear field exists for turbulent flow:

$$\tau = \rho \kappa^2 \frac{(du / dy)^4}{(d^2 u / dy^2)^2} \quad (4.33)$$

This formula can be used to derive the velocity distribution for flow over a fixed bed, as discussed in detail in Chapter 7.

By means of the turbulent mixing length theory proposed by Prandtl and Karman, one can derive a velocity profile for turbulent shear flow, one that is logical to some degree. However, the physical pattern of the turbulence exchange phenomenon hypothecated in such a theory is intermittent. That is, turbulence diffusion is accomplished through a process that involves: motion → mixing → then motion again → mixing again, etc. In fact, the velocity profile and motion of particles of a dye solution are continuous, and the process of diffusion must be continuous too. Therefore, the mixing length theory is at best only a useful approximation. It is more reliable for pulsations of small sizes, and it is considerably less reliable for eddies of large sizes. Experience shows that momentum exchange is mostly accomplished through small eddies [18], but that sediment diffusion is mainly the result of exchanges caused by large eddies. Thus, although both the velocity profile and concentration profile are based on the mixing-length theory, the former result is more reliable than the latter.

4.2.3 Statistical theory of turbulence

In the study presented in the preceding sections, the primary concerns are the relationship between the temporal mean velocity profiles and the resistance caused by turbulence. At the beginning of the analysis, the exchange coefficient, or the mixing length, was introduced to represent the fluctuating properties of the flow. Another, more plausible way is to study the characteristics of the fluctuating velocity components and their relationships directly — to start with the fluctuating phenomena that are inherent in turbulent flow, and then to introduce statistical functions for deducing properties of the time-averaged flow. In the following paragraphs the concepts of statistical theory are used to describe the phenomenon of diffusion.

Diffusion in turbulent flow is similar to molecular diffusion in air and the Browning motion of solid particles in liquids. The process called diffusion denotes the following phenomenon: molecules or other matter concentrated initially at one point are carried along and distributed throughout the flow region through either molecular motion or eddy motion. For instance, dyed particles of the same specific weight as water, injected into the flow at some point, move with the flow and diffuse in directions perpendicular to the flow simultaneously (Fig. 4.16). This process is the result of diffusion in turbulent flow.

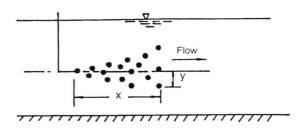

Fig. 4.16 Diffusion in turbulent flow

In aerodynamics, the formula for the molecular diffusion of a flow is

$$n = -\varepsilon_y \frac{dN}{dy} \tag{4.34}$$

in which

    n—number of molecules passing through a unit area, perpendicular to y direction, per unit of time;

    N—concentration of gas molecules at y;

    ε_y—diffusion coefficient in y direction.

In Brownian motion,

$$\varepsilon_y = \frac{1}{2}\overline{Y^2}/t \tag{4.35}$$

in which Y is the displacement of small particles in the y direction in a time t(Fig. 4.16).

The concept of correlation is introduced before presenting the turbulence diffusion theory [19]. Correlation denotes the mutual interdependence of different variables in a statistical sense. The correlation coefficient is a parameter used to express this interdependence. If the velocities at points along a streamline are measured simultaneously, the fluctuating velocity at location x is $u_x{}'$ and that at location $x+\zeta$ is $u_{x+\zeta}{}'$. Then the types of correlations between $u_x{}'$ and $u_{x+\zeta}{}'$ are shown in Fig. 4.17. If $\zeta=0$, then

$$u_x{}' = u_{x+\zeta}{}'$$

and all the data points are distributed along a straight line with a slope of 1(Fig.4.17a). The correlation is perfect. If ξ is small but not zero, the points do not all fall on the straight line with a slope of 1, but they are concentrated near the straight line, as shown in Fig. 4.17b. The values $u_x{}'$ and $u_{x+\zeta}{}'$ are still closely correlated. If ξ is still larger, the data points scatter more, as shown in Fig.4.17c, and the two variables are less well correlated. If ζ is quite large, u_x and $u_{x+\zeta}{}'$ have no interrelationship, and the points

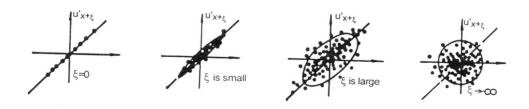

Fig. 4.17 Correlation of fluctuating velocities at two points a distance ζ apart

scatter randomly, as shown in Fig. 4.17d.

The correlation coefficient R_ζ between the fluctuating velocities at two points a distance ζ apart is as follows:

$$R_\zeta = \frac{\overline{u'_x u'_{x+\zeta}}}{\sqrt{\overline{u'^2_x}} \sqrt{\overline{u'^2_{x+\zeta}}}} \qquad (4.36)$$

in which the bar indicates the average value. In uniform flow, the root-mean-square of fluctuating velocity at any point does not vary with location, therefore

$$\sqrt{\overline{u'^2_x}} = \sqrt{\overline{u'^2_{x+\zeta}}} = \sqrt{\overline{u'^2_{x-\zeta}}} = \sqrt{\overline{u'^2}}$$

Substituting these into the preceding equation, one obtains:

$$R_\zeta = \frac{\overline{u'_x u'_{x+\zeta}}}{\overline{u'^2}} = \frac{\overline{u'_{x-\zeta} u'_x}}{\overline{u'^2}} \qquad (4.37)$$

The correlation coefficient R_ζ for this type of flow is normally between 0 and 1, with extreme values of $R=1$ for perfect agreement and $R = 0$ for no correlation.

As it moves, a water element mixes with the surrounding flow; it loses some of the properties that characterized it initially and acquires some of the properties of local flows encountered along its way. Only if it keeps some of its original properties, does it have some correlation with those at the starting point. The degree of correlation thus decreases with the distance traveled. After a certain distance, no correlation exists, i.e. the element has lost its original properties, and reaches the end of that life span, in a sense. From this point of view, the R_ζ—ζ curve (Fig. 4.18) reflects the diffusion process within the flow. Theoretically, R_ζ approaches zero only as ζ approaches infinity. But from a practical point of view, R_ζ is effectively zero if ζ exceeds some finite value, which can be represented by X. Hence, the turbulent mixing length can be defined as follows:

$$l_1 = \int_0^X R_\zeta d\zeta \qquad (4.38)$$

In the foregoing approach, the correlation of properties at two different places are observed at the same time. One can also consider the correlation of properties observed at different times but at the same place. For a sediment-laden flow, a representative correlation coefficient in turbulent diffusion is the time correlation coefficient for the

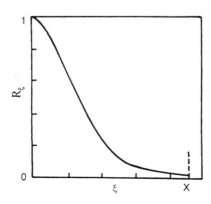

Fig. 4.18 Relationship between R_ζ and ζ

fluctuating velocities in the direction perpendicular to the main flow:

$$R_\zeta = \frac{\overline{\upsilon_t' \upsilon_{t+\zeta}'}}{\overline{\upsilon'^2}} = \frac{\overline{\upsilon_{t-\zeta}' \upsilon_t'}}{\overline{\upsilon'^2}} \tag{4.39}$$

The following equation can be deduced:

$$\int_0^t \overline{\upsilon_t' \upsilon_{t-\zeta}'} d\zeta = \int_0^t R_\zeta \overline{\upsilon'}^2 d\zeta$$

Since $\sqrt{\overline{\upsilon_k'^2}}$ does not vary with ζ, the preceding equation may be written as

$$\int_0^t \overline{\upsilon_t' \upsilon_{t-\zeta}'} d\zeta = \overline{\upsilon'^2} \int_0^t R_\zeta d\zeta \tag{4.40}$$

Furthermore,

$$\overline{\int_0^t \upsilon'_t \, \upsilon'_{t-\zeta} d\zeta} = \overline{\upsilon'_t \int_0^t \upsilon'_{t-\zeta} d\zeta} = \overline{\upsilon'_t \, Y} = \frac{1}{2} \frac{d\overline{Y^2}}{dt} \tag{4.41}$$

in which Y is the displacement of a water molecule or other material (for instance, the dyed particles in Fig. 4.19) in the direction perpendicular to flow in the time interval t. According to Eq.(4.40) and Eq.(4.41)

$$\frac{d\overline{Y^2}}{dt} = 2\overline{\upsilon'^2} \int_0^t R_\zeta d\zeta$$

Integrating, one obtains:

$$\overline{Y^2} = 2\overline{\upsilon'^2} \int_0^T \int_0^t R_\zeta d\zeta dt \tag{4.42}$$

i.e. the diffusion effect varies with the kinetic energy of turbulence and the correlation coefficient R_ζ.

If the displacement of dyed particles near the point of ejection shown in Fig. 4.16 is recorded photographically, or in some other way, then when T is small, near that point

$$R_\zeta \to 1$$

Also,

$$\overline{Y^2} = \overline{\upsilon'^2} T^2$$

and

$$\sqrt{\overline{\upsilon'^2}} = \sqrt{\overline{Y^2}} \, / \, T \tag{4.43}$$

If $\overline{u}$ is the average velocity of flow at the plane where dyed particles are introduced, and X is the distance they have moved in the flow direction, then

$$T = \frac{X}{\bar{u}}$$

If this relationship is substituted into Eq.(4.43),

$$\sqrt{\overline{v'^2}} = \frac{\sqrt{\overline{Y^2}}}{X}\bar{u} \tag{4.44}$$

Now the turbulence intensity can be deduced from the displacement of dyed particles caused by diffusion.

If ζ is larger than some values T_1, R_ζ is quite small, and there is almost no correlation between v'_t and $v'_{t+\zeta}$, thus

$$\int_0^{T_1} R_\zeta \, d\zeta$$

is a constant. That is, if $T > T_1$,

$$\overline{v'^2}\int_0^{T_1} R_\zeta d\zeta = \overline{v'_t Y} = \text{constant} \tag{4.45}$$

T_1 represents, on average, the time required for water molecules to mix sufficiently with the ambient flow that they have lost their original velocity. Hence, one can define another length scale in the following way;

$$l_2 = \sqrt{\overline{v'^2}} \int_0^{T_1} R_\zeta \, d\zeta \tag{4.46}$$

From Eq.(4.45) and Eq.(4.41), one obtains

$$\sqrt{\overline{v'^2}}\, l_2 = \overline{v'_t Y} = \frac{1}{2}\frac{\overline{dY^2}}{dt} = \frac{\bar{u}}{2}\frac{\overline{dY^2}}{dx} \tag{4.47}$$

Integrating this equation, one gets

$$\overline{Y^2} = 2\sqrt{\overline{v'^2}}\, l_2 t \tag{4.48}$$

If T exceeds a certain value, $\overline{Y^2}$ is directly proportional to time t, and

$$\sqrt{\overline{v'^2}}\, l_2 = \frac{1}{2}\frac{\overline{Y^2}}{t} \tag{4.49}$$

A comparison of the preceding equation with Eq.(4.35) shows that $\sqrt{\overline{v'^2}}\, l_2$ is the turbulent diffusion coefficient. Thus in turbulent diffusion, the amount of material passing through a unit area perpendicular to the flow in unit time can be determined from Eq.(4.34), in which

$$\varepsilon_y = \sqrt{\overline{v'^2}}\, l_2 = \frac{1}{2}\left[\frac{\overline{dY^2}}{dt}\right]_{max} = \frac{\bar{u}}{2}\left[\frac{\overline{dY^2}}{dx}\right]_{max} \tag{4.50}$$

142

A comparison of Eq.(4.27) and Eq.(4.50) shows that the diffusion coefficient ε_y in diffusion phenomenon is equivalent to the momentum exchange coefficient ε_m in momentum exchange. Although both of these are products of the fluctuating velocity and a characteristic length of turbulence, the physical process described by mixing length in momentum exchange is intermittent, whereas the pattern envisaged in diffusion theory is continuous.

4.3 MEASUREMENTS OF TURBULENCE CHARACTERISTICS IN OPEN CHANNEL FLOW

4.3.1 Methods for measuring turbulence characteristics in open channel flow

The turbulence characteristics in air currents are measured primarily by means of a hot-wire anemometer. The technique evolved rapidly in the late thirties. However, many difficulties arose in applying this technique to the flow of water. Air bubbles often form around the hot wire and distort the measured values. Also, a hot wire is easily contaminated by the water, so that the apparatus becomes unreliable; the experimenter must then use highly purified water. As a result, although data on turbulence characteristics in air flows are abundant, whereas comparable data on flows of water are sparse.

In the early stages, the technique for measuring turbulence characteristics in water flow was as follows: a dye solution, or solid particles that had same specific gravity as water but did not mix with it, were injected into water; the displacement of these particles was then observed in order to follow the flow pattern, and from those observations the turbulence intensity and diffusion coefficient were deduced. The application of high-speed photography made such measurement more refined and more accurate. The hydrogen bubble technique was another advance on this visualization technique.

Since the sixties, a special type of anemometer has been used for measurements of turbulence in flowing water. A glass cylinder covered by a platinum film, upon which a thin layer of quartz is sprayed, is used for the sensor, instead of the platinum wire. In some cases the sensor is cone-shaped. These improved sensors produce fewer air bubbles and were less likely to become contaminated. Even so, the hot-film anemometer is still not entirely satisfactory for measuring turbulence in water flow.

Still more recently, the laser velocity meter was developed for the measurement of turbulence characteristics in water flow. It is usually used with a computerised data-processing system. In this way, the temporal mean velocity, the fluctuating velocity, and its root-mean-square are all measured and recorded simultaneously. For flows of water with a low concentration of sediment, the laser velocity meter is a promising instrument for the non-intrusive measurement of turbulence.

A few other types of measuring apparatus are introduced in other sections.

4.3.2 Primary results from measurements of turbulence characteristics in open channel flow

4.3.2.1 Longitudinal and vertical velocities fluctuations

The fluctuating velocity is clearly one of the most important properties of turbulence, and in consequence, its measurement is given top priority in studies of turbulence. In fact, most of the available data deal with this quantity.

Usually, the vertical profiles of the root-mean-square values of the longitudinal and

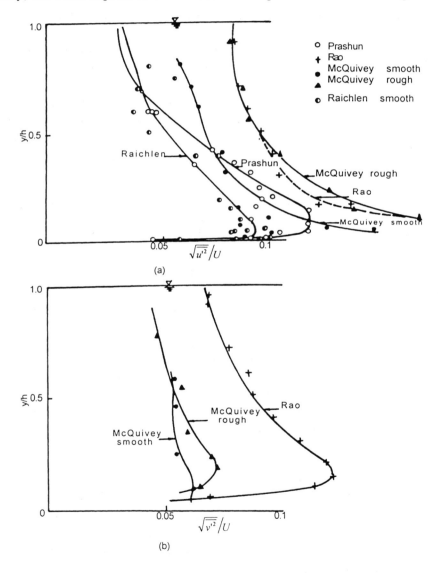

Fig. 4.19 Vertical distribution of $\sqrt{\overline{u'^2}}\,/U$ and $\sqrt{\overline{v'^2}}\,/U$

vertical velocities, $\sqrt{\overline{u'^2}}$ and $\sqrt{\overline{v'^2}}$, are measured. The measured results are written in dimensionless form with either the mean velocity in that vertical U or the shear velocity $U*$ as the normalizing factor. Dimensionless results for the two components of the fluctuating velocity divided by U are shown in Table 4.1 and Fig. 4.19. The measured results shown in the figure indicate a certain degree of scatter.

In Japan, Minami Isao conducted systematic measurements of turbulence in a concrete canal with a cross sectional area of 2.41 m^2 and a mean velocity of 0.78 m/s. The results are shown in Fig. 4.20 [24]. The intensity of the fluctuations at the point close to boundary is about 45 percent of the local temporal mean velocity .

Table 4.1 Measured values of longitudinal and vertical fluctuating velocities

Author	Instrument	Measured parameters	Conditions of experiments	Ref.
McQuivey & Richardson	Hot-film meter	$\sqrt{\overline{u'^2}}$ & $\sqrt{\overline{v'^2}}$	Flume 10 × 0.2 × 0.2 m, smooth and rough boundaries	[20]
Prashun	Hot-film meter	$\sqrt{\overline{u'^2}}$	Flume of 6 m long, 0.3 m wide with smooth and rough walls and sand bed	[21]
Raichlen	Hot-film meter	$\sqrt{\overline{u'^2}}$	Flume 12.2 × 0.25 × 0.267 m	[22]
Rao	Hot-film meter	$\sqrt{\overline{u'^2}}$ & $\sqrt{\overline{v'^2}}$	Flume 15 × 0.6 × 0.9 m	[23]
Isao	Thermo-meter	$\sqrt{\overline{u'^2}}$	Concrete canal with a wetted cross section of 2.41 m^2	[24]

In an analysis of data obtained from five field canals and from laboratory experiments conducted by Minski and himself, Nikichin obtained a vertical profile of relative fluctuation intensity $\sqrt{\overline{v'^2}}/U_*$, as shown in Fig. 4.21 [25]. In the experiments, solid

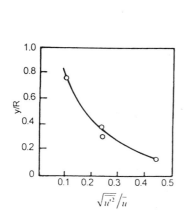

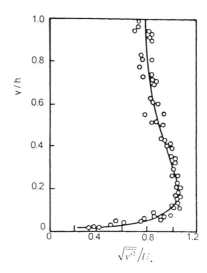

Fig. 4.20 Vertical distribution of relative turbulence intensity in a concrete canal (after Isao, Minami)

Fig. 4.21 Vertical distribution of fluctuating velocity as summarized by Nikichin

145

spheres with a specific gravity the same as water were injected into the flow and their diffusion was recorded by the use of high-speed photography. The results are expressed in dimensionless form with U_* as the denominator. The data follow a well defined trend.

By means of the hydrogen bubble technique and high-speed photography, Grass measured profiles of longitudinal and vertical fluctuation intensity, expressed in dimensionless form with U_* as denominator, for three bed conditions: hydraulically smooth, transitional, and rough, in a flume with glass side walls that is 10 m long, and 0.25 m wide, and 0.05 m deep. The results are shown in Fig. 4.22 [8]. In it, Nikichin's curve, which is quite close to that obtained by Grass for most water depths, is also plotted. Grass also compared the data obtained for the smooth boundary condition with Laufer's results for air flow in a smooth pipe. They also match each other well.

Although the experimental results scatter somewhat, some conclusions can be drawn from them:

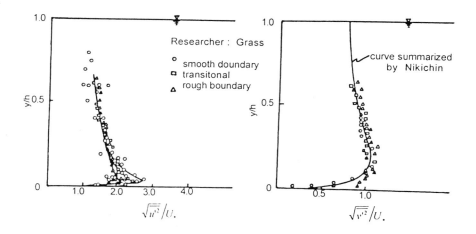

Fig. 4.22 Experimental results of Grass (after Grass, A.J.)

1. Restricted in space, the turbulence intensity at a boundary is zero. Within a slight distance from the boundary, the turbulence intensity increases rapidly and reaches its peak value. Far from the boundary, in the main flow region, the turbulence intensity is somewhat less and essentially constant.

2. In the main flow region, the intensity of the vertical fluctuations approaches the shear velocity, $(\sqrt{\overline{v'^2}}/U_*) \approx 1$. The longitudinal fluctuation intensity is slightly larger than shear velocity. Rao's experimental results for the main flow region show that the longitudinal fluctuation intensity is nearly equal to the vertical fluctuation intensity. Thus, the flow in the main flow region approaches an isotropic homogeneous one. This result is quite different from those of other experiments, perhaps because of experimental errors.

3. The Grass measurements show that the type of boundary has no effect on the intensity of the fluctuations in the main flow region. The boundary has an effect only in the region close to the boundary; there, the rougher the boundary is, the less the longitudinal fluctuations and the more the vertical fluctuations.

4.3.2.2 Turbulent shear stress

The turbulent shear stress ($\rho \overline{u'v'}$) can be obtained directly from measurements of longitudinal and vertical fluctuating velocities. Results obtained by Grass and McQuivey for $\overline{u'v'} / U_*^2$ are shown in Fig. 4.24 [23]. They reveal that in the main flow region, turbulent shearing stresses $\rho \overline{u'v'}$ are distributed linearly; that is, the shearing stress between flow layers is mainly the Reynolds stress induced by turbulence. The results obtained by means of the hydrogen bubble technique showed that for a hydraulically smooth boundary, the turbulent shearing stress decreases rapidly towards the boundary and the shearing stress caused by viscosity increases rapidly and plays a dominant role. For a rough boundary, towards the boundary no decrease in the turbulent shearing stress and no increase in the shearing stress caused by viscosity could be detected. Thus, right up to the boundary, almost the entire shearing stress is transferred through turbulent shearing stress. In the McQuivey experimental results, no matter what the boundary was, no decrease of turbulent shearing stress towards the boundary was detected. Obviously, this result was due to the lack of data near the boundary. Rao's results also reflected the decrease of turbulent shearing stress towards the bed. And he found that the shearing stress started to decrease at a point rather far from the boundary; however, such a situation is improbable.

4.3.2.3 Turbulence spectrum

An energy spectrum is the distribution of turbulent kinetic energy among eddies with different frequencies (corresponding to different scales). Results measured for open channel flow by McQuivey [20], Raichlan [21] and others agree well with each other. As an example, Fig. 4.25 contains the spectrums measured by McQuivey at three different depths for flow with a smooth boundary. In the figure, n is the frequency of eddies; the probability density of the occurrence of such eddies, $F(n)$, is defined from

$$\int_0^\infty F(n)dn = 1 \tag{4.51}$$

The figures show that the main energy of turbulence occurs in its low frequency motion ($n<5$ $1/s$). In the range of $3 < n < 30$ 1/s, the curve has a slope of approximately -5/3. In the high frequency range, the slope is about -7. The figure shows little difference for the data for the three different depths. Hence, the boundary has no great influence on the turbulence spectrum. Only near the boundary is the energy of the high frequency portion slightly larger and that of the low frequency portion (corresponding to large-scale eddies) a little less. Evidently, the presence of a boundary restricts the development of large scale eddies. McQuivey also measured the spectrum function for flow near a rough

boundary, and his results differ little with those shown in Fig. 4.25. He also found that
the type of boundary has no significant effect on the spectrum.

4.3.2.4 Sizes of large-scale and small-scale eddies

Turbulent flow has two characteristic lengths, one for the large-scale eddies

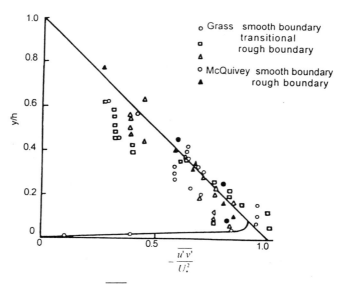

Fig. 4.23 Vertical profile of $\overline{u'v'}/U_*^2$ (after McQuivey, R.S., and E.V. Richardson)

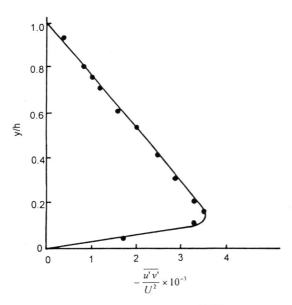

Fig. 4.24 Vertical profile of $\overline{u'v'}/U^2$ (after McQuivey, R.S., and E.V.Richardson)

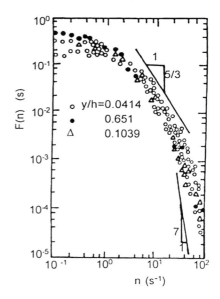

Fig. 4.25 Turbulence spectrum (smooth boundary) (after Raichlen, F.)

148

designated as L_x , and one for small-scale eddies called λ_x. The size of the large-scale eddies is

$$L_x = \int_0^\infty R_\zeta d\zeta \tag{4.52}$$

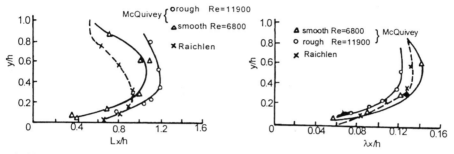

(a) Vertical profiles of sizes of large-scale eddies (b) Vertical profiles of sizes of small-scale eddies

Fig. 4.26 Vertical profiles of sizes of large-scale and small-scale eddies

A comparison of the preceding equation with Eq.(4.38) shows that L_x is actually just the turbulent mixing length l_1, the average length of the region affected by an eddy. McQuivey and Raichlen independently measured the vertical profile for the size of the large-scale eddies with the results shown in Fig. 4.26a [20,22]. The figure shows that there is only a slight difference between their measured results. It may have been caused by such factors as the Reynolds number, boundary, etc. The overall trend indicates that large-scale eddies are restricted near the boundary but are much larger at some distance from it. The size of the large-scale eddies reaches a maximum at a relative depth of about 0.45. Also, the size of the large-scale eddies and the water depth have the same order of magnitude. The eddy size decreases gradually above this level.

The size of small-scale eddies λ_x is defined as:

$$\frac{1}{\lambda_x^2} = \frac{4\pi^2}{U^2} \int_0^\infty n^2 F(n)dn \tag{4.53}$$

It is a length that is related to energy dissipation. The results obtained by McQuivey and Raichlen are shown in Fig. 4.26b. The maximum size occurs for a value of y/h near 0.45, also a little below mid-depth.

4.3.2.5 Other parameters

Other characteristic parameters of turbulence can be indirectly deduced from measured data. Kalinske observed the displacement of dyed liquid particles, and then

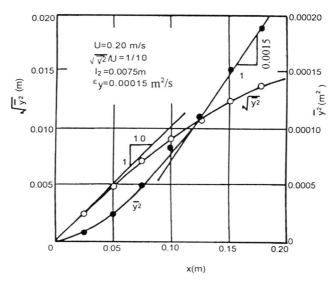

Fig. 4.27 Displacement of dyed liquid particles in turbulent flow as a result of diffusion
(after Kalinske, A.A., and E.R.Van Driest)

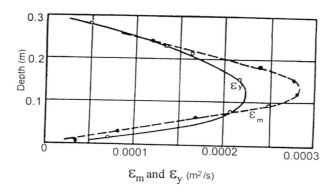

Fig. 4.28 Vertical profiles of diffusion and momentum exchange coefficients in open channel flow
(after Kalinske, A.A.)

deduced the turbulence intensity and diffusion coefficient by means of Eq.(4.44) and (4.50). Fig. 4.27 shows the displacements along the flow direction of dyed particles as the result of diffusion in an open channel, with a cross section of 0.3m×0.3m and a velocity of 0.2 m/s [26]. From the displacements, one readily obtains the following values: fluctuation intensity $\sqrt{v'^2}$ is about one tenth of the average velocity, characteristic length l_2 about 7.5 mm, diffusion coefficient 0.00015 m²/s. Vertical profiles of the diffusion coefficient and momentum exchange coefficient in an open channel, with a cross section 0.76 m wide and 0.3 m deep and a velocity of 0.26 m/s, are shown in Fig. 4.28 [27]. The variations of the two coefficients have a similar trend. Rao measured vertical profiles of eddy viscosity η and mixing length l in a flume 0.6 m by 0.9 m and 15 m long, as shown in Fig.4.29 [23]. He obtained the following results: (1) In the region $y/h < 0.2$, the eddy viscosity increases linearly with the depth; for $y/h > 0.2$, the profile of eddy viscosity has a nearly elliptical shape; the eddy viscosity reaches its maximum at $y = 0.45h$. (2) A comparison of eddy viscosity with the kinematic viscosity shows that except in the

150

region close to the bed, the kinematic viscosity v is negligible in comparison with the eddy viscosity h. (3) In the region of $y/h < 0.3$, the mixing length is indeed linearly related to the distance from the bed, and the proportionality factor is the Karman constant (0.40). Farther from the bed, the mixing length deviates from a linear relation and grows less rapidly.

As for profiles of the velocity and pressure of fluctuation, Nikichin made a summary of the experimental data of Minski and others, and he showed that the distribution of the fluctuation velocity at different heights, even at the bed surface, follows a normal distribution [25]. Einstein and El-Samni measured the fluctuating pressure at the tops of spheres lying on the bed surface. They found that it also follows a normal distribution [28] Since the pressure is proportional to the square of velocity, some contradiction is indicated by both variables following normal distributions. This point is discussed in Chapter 5.

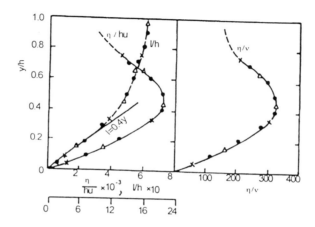

Fig. 4.29 Vertical profiles of eddy viscosity and mixing length

REFERENCES

[1] Reynolds, O. "An Experimental Investigation of Circumstances which Determine Whether the Motion of Water Shall Be Direct or Sinuous, and of the Laws of Resistance in Parallel Channels." *Philosophy Transition of Royal Society*, Vol. 174, 1883, pp. 51-105.
[2] Dryden, H.L. "Recent Advances in the Mechanics of Boundary Layer Flow." *Advances in Applied Mechanics*, Vol. 1, Academic Press, 1948, pp. 1-40
[3] Exman, V.W. "On the Change from Steady to Turbulent Motion of Liquids." *Arkiv. Mat., Astron. Fysik*, Vol. 6, No. 12, 1911.
[4] Rouse, H. *Elementary Mechanics of Fluids*. John Wiley and Sons, 1946, pp. 170-172.
[5] Bakhmeteff, B.A. *The Mechanics of Turbulent Flow*. Princeton Univ. Press, 1936.
[6] Einstein, H.A., and Huon Li. "The Viscous Sublayer Along A Smooth Boundary." *Journal of Engineering Mechanic Division, Proceeding of American Society of Civil Engineers*, Vol. 82, No. EM2, 1956, pp. 27.

[7] Hintz, J.O. *Turbulence*. 2nd ed., McGraw-Hill Inc., 1975, pp. 155-156, 560-561, 607-609, 682-684.

[8] Grass, A.J. "Structural Factors of Turbulent Flow Over Smooth and Rough Boundaries." *Journal of Fluid Mechanics*, Vol. 50, Pt. 2, 1971, pp. 233-255.

[9] Kim, H.T., S.J.Kline, and W.C.Reynolds. "The Production of Turbulence Near a Smooth Wall in a Turbulent Boundary Layer." *Journal of Fluid Mechanics*, Vol. 50, Pt.1, 1971, pp. 133-160.

[10] Offen, G.R., and S.J. Kline. "Combined Dye-Streak and Hydrogen-Bubble Visual Observation of a Turbulent Boundary Layer." *Journal of Fluid Mechanics*, Vol. 62, Pt. 2, 1974, pp. 223-239.

[11] Offen, G.R., and S.J. Kline. "A Proposed Model of the Bursting Process in Turbulent Boundary Layers." *Journal of Fluid Mechanics*, Vol. 70, Pt. 2, 1975, pp. 209-228.

[12] Yalin, M.S. *Mechanics of Sediment Transport*. 2nd ed. Pergamon Press, 1977, pp. 204-206.

[13] Davis, J.T. *Turbulence Phenomena*. Academic Press, 1972, pp. 51-52.

[14] Prandtl, L. "Bericht Uber Untersuchungen Zur Aubgebildeten Turbulenz." *Z. Angew. Math. Mech.*, Vol. 5, No. 2, 1925, p. 136.

[15] Von Karman, Th. "Turbulence and Skin Friction." *Journal of Aeronautical Science*, Vol. 1, No. 1,1934.

[16] Nikuradse, J. "Gesetzmaessigkeiten der Turbulenten Stroemung in Glatten Rohren." *Vereines Deutscher Ingenieur*, Forschungscheft 356, 1932.

[17] Keulegan, G.H. "Laws of Turbulent Flow in Open Channels." *Journal of Research*, U.S. National Bureau of Standards, Vol. 21, 1938, pp. 701-741.

[18] Sutton, O.G. *Micrometeorology*. McGraw Hill Book Co., 1953, pp. 56-104.

[19] Taylor, G.I. "Diffusion by Continuous Movements." *Proceeding of London Mathematics Society*, Ser. 2, Vol. 20, Pt.1, 1921, pp. 196-212.

[20] McQuivey, R.S., and E.V. Richardson. "Some Turbulence Measurements in Open Channel Flow." *Journal of Hydraulic Division, Proceeding of American Society of Civil Engineers*, Vol. 95, No. HY1, 1969, pp. 209-223

[21] Prashun, A.L. "Turbulent Measurements over Sand Beds." *Proceeding of International Symposium on River Mechanics*, International Association for Hydraulic Research, Vol.1, 1973, pp.325-336

[22] Raichlen, F. "Some Turbulence Measurements in Water." *Journal of Engineering Mechanic Division, Proceeding of American Society of Civil Engineers*, Vol. 93, No. EM2, 1967

[23] Rao, N.S. Govinda, and N.V.C. Swamy. "Turbulence Characteristics of Open Channel Flows." *J. Inst. Engrs.(India)*, Vol. 44, No. 5, Pt. CI3, 1964, pp. 341-355.

[24] Minami Isao. "Study on Turbulence and Its Application." *Journal of Sediment Research*, Vol.3, No.4, 1958, pp.73-100 (in Chinese).

[25] Nikichin, E.K. "Turbulent Flow in Near Bed Region and Its Process of Development." *Bulletin of Akademy of Ukraine Socialist Soviet Republic*, Kiev, 1963, pp. 142. (in Russian)

[26] Kalinske, A.A., and E.R. Van Driest. "Application of Statistical Theory of Turbulence to Hydraulic Problems." *Proceeding of 5th International Congress of Applied Mechanics*, 1939, pp. 416-421.

[27] Kalinske, A.A. "Investigation of Liquid Turbulence and Suspended Material Transportation." *Univ. Pennsylvania Bicentennial Conference on Fluid Mechanics and Statistical Methods in Engineering*, 1941, pp. 41-54.

[28] Einstein, H.A., and E.A. El-Samni. "Hydrodynamic Forces on A Rough Wall." *Rev. Modern Physics*, Vol. 21, 1949.

CHAPTER 5

BASIC CONCEPTIONS OF SEDIMENT MOVEMENT

Some terms and concepts need to be introduced prior to a basic discussion of the laws of sediment motion; these include the forces acting on the particles resting on the bed surface, basic patterns of sediment motion, the physical interpretation that distinguishes suspended load from bed load, the difference between sediment motion on rigid and movable beds, and the concepts of wash load and bed material load.

5.1 FORCES ACTING ON PARTICLES RESTING ON THE BED

As water flows over a river bed composed of loose sediment particles, the flow exerts on each a lift force and a drag force. For fine sediment, a cohesive force between particles, in addition to the gravitational force, resists the active forces. If sediment particles move as bed load, a dispersive force between particles exists, and it exerts a pressure on the bed and enhances the stability of the bed particles. If the underground water table at the banks is much different from the water stage and seepage is active, the bed particles are acted by a force resulting from seepage pressure.

5.1.1 Drag force and lift force

As water flows over a channel bed, a frictional force F_1 is exerted on the rough surface represented by the bed of particles. As shown in Fig. 5.1, the frictional force F_1, in the direction of flow, does not act through the center of the particle because only the upper part of the particle is acted on by the flowing water. If the grain Reynolds number ($U_* / D v$, in which U_* is the shear velocity and D the diameter of the particle) is less than 3.5, surface friction is the main force acting on the particle. If it is larger than 3.5, however, separation of streamline in the form of a small wake occurs behind the top of the particles and vortexes form there. Hence, a pressure difference between the front and back surface of the particle exists, and it causes the form resistance F_2. If the particle is spherical, F_2 acts through the center of the particle. The resultant of F_1 and F_2 is called drag force and is expressed as F_D.

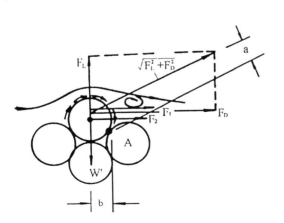

Fig. 5.1 Drag and lift forces acting or particles resting on the bed surface

The velocities in the flow at the top and the bottom of a bed particle are different. The one at the top is the velocity of the flowing water near the bed surface. The one at the bottom is the seepage velocity and is much smaller than the former. The top pressure is lower than the hydrostatic pressure because of the high velocity, and the bottom one is just hydrostatic. The pressure difference results in a lift force, F_L. For spherical particles F_L acts through the center and is directed upward.

The fact that a particle on a bed surface is acted on by a lift force was not widely recognized in the past. Only the drag force was taken into account in the majority of early studies of sediment movement. The existence of this lift force was first demonstrated by Jeffreys in the 1920s [1]. For ideal flow, he analyzed the flow pattern around a cylinder of infinitive length and radius r, lying on the bed surface with its axis perpendicular to the flow; the flow velocity was U, and the complex potential is

$$F = \pi r U \coth \frac{\pi r}{z} \tag{5.1}$$

in which

$$z = x + iy$$

From Eq. (5.1) the cylinder will start to rise if

$$(\frac{1}{3} + \frac{1}{9}\pi^2)U^2 > \frac{\gamma_s - \gamma}{\gamma} gD \tag{5.2}$$

If $(\gamma_s - \gamma)/\gamma = 1.65$, the critical velocity for the cylinder to be lifted is then

$$U = 3.35 \text{ cm/s,} \qquad \text{for } D=0.1\text{mm}$$
$$U_c = 10.6 \text{ cm/s,} \qquad \text{for } D=1\text{mm}$$

In case the bed surface is composed of sediment particles that contact the bed at only a few points, rather than along a line as in the case of a cylinder, some flow passes underneath them. The lift force acting on the particles is smaller than in the case of a cylinder. Still, under ordinary flow conditions, the lift force exerted on sediment particles is appreciable.

The drag force and the lift force can be expressed in general form as follows:

$$F_D = C_D A \frac{\rho u_0^2}{2} \tag{5.3}$$

$$F_L = C_L A \frac{\rho u_0^2}{2} \tag{5.4}$$

in which C_D and C_L are the drag and lift coefficients, respectively, and u_0 is the effective velocity near the bed particle. The drag and lift coefficients depend on the flow pattern around the bed particle and the method of estimating u_0. A number of experimental studies on the drag and lift forces have been made, and the main results are shown in Table 5.1.

The designs of the experiments listed in Table 5.1 differ. The results by Jemianchiev have been widely accepted by Russian scientists. He used two isolated spheres on the bottom plate in a wind tunnel to model sediment particles. His approach is, of course, only an approximation to the reality. Chepil studied the forces acting on a particle as the particle starts to move [6]. He observed that as wind blew over the ground the particles protruding above the ground surface were more likely to be removed. These particles occupied about 10% of the ground surface area. He designed a wind tunnel experiment according to such a scenario. He arranged particles on the bottom plate and measured the forces exerted on particles located at various positions.

The effective velocity u_0 is defined as the velocity at the elevation of kD, in which k is a coefficient. As shown in Table 5.1, researchers used different values of k. Some used the average velocity for u_0 and others the shear velocity.

The following conclusions can be drawn from Table 5.1:

1. The discrepancies of the results presented by different authors is great because the diameter of the particles and the grain Reynolds number selected by the different researchers were in quite different ranges, and the ways of dealing with the data were also different. For instance, different resistance and lift coefficients were obtained because of the different approaches for determining u_0. A basically correct way of determining the drag force and the lift force is to measure the pressure distribution on the surface of the particle and then to integrate it to obtain the resultant components in the horizontal and vertical directions. A rough estimation was provided by Einstein and El-Samni; they took the difference between the pressures at the bottom and the top of a particle as the lift force [4]. Chepil compared his measured data with the estimated lift force and found that the estimates were 1.85 times greater than the measured values [7].

2. The force exerted on a particle depends greatly on its position in the bed. Clearly, the particles in Fig. 5.2 that exposed on the bed surface and the particles in the front row are most likely to move.

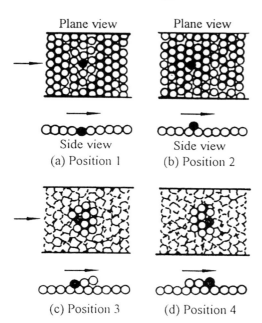

Fig. 5.2 Positions of particles resting
on the bed surface

3. The relationship between the drag coefficient and the Reynolds number is, to some extent, similar to that for a particle falling in a quiet fluid, shown in Fig. 5.3. This similarity is significant. The relationship in Fig. 5.3 is valid only for the particles fully exposed to the flow over the bed surface. For particles at other positions, the C_D vs. Re curve would be different. From the trend shown in Fig. 5.3, the drag coefficient approaches a constant for large Reynolds numbers. For small Reynolds numbers, however, the coefficient may be inversely proportional to the Reynolds number. A major difficulty in the experiments was that the pressure distribution could not be measured if the particle was small, whereas the Reynolds number was too large if the particle was big.

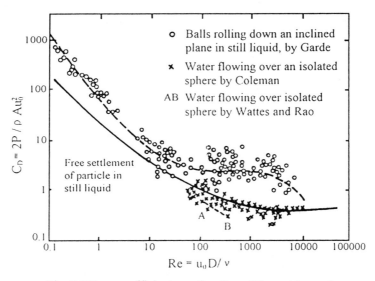

Fig. 5.3 Drag coefficient as a function of Reynolds number

4. A relationship between the lift coefficient and the Reynolds number should exist by inference. Unfortunately, no such results have been reported. A negative lift coefficient is possible for some Reynolds numbers if the particle is placed at an unfavorable position. The reason that all the lift forces in the experiments of Watters and Rao were negative is not clear [11]. Furthermore, little knowledge exists on the forces for non-uniform particles. From his theoretical system, Einstein deduced the lift force on non-uniform particles indirectly. However, the results have not yet been verified by measured data.

5. The results measured by Einstein and El-Samni and by Cheng and Glyde showed that the fluctuation of forces acting on particles exhibits a normal distribution. Gessler also found that the distribution of shear stress on the bed surface is normal and the root mean square of the frequency distribution of the fluctuating shear stress is 0.56 times that of the mean shear stress [13]. In contrast, many researchers found that the fluctuating velocity, rather than shear stress, follows the Gaussian distribution as indicated in Chapter 4. Christensen reanalyzed the data of Einstein and El-Samni and

Table 5. 1 Result of various studies for drag and lift forces acting on bed surface particles

Researchers	Experiment description	Particle diameter (mm)	Grain Re, U.D/v	Determination of u_0	Main experimental results	References
Jemianchiev & Yegiazaov	Gas flowing over two spheres of diameter D and distance l	----	----	Velocity at D/2 apart from the bed	The lift coefficent C was up to 0.88; C_L decreased quickly with increasing l and was sometimes negative. At l/D>0.13 C_L began to rise. For l/D>0.15 C_L approached 0.2 and C_L/C_D approached 0.25	[2, 3]
Einstein & El-Samni	Water flowing over compactly arranged hemispheres of diameter 68.5 mm and gravel of 19~76.5mm	19 ~ 77	3,300~5,600	The theoretical bed was 0.2D below the top of the spheres, u_0 was taken as velocity 0.35D above theoretical bed	C_L=0.178, and the distribution of lift force followed the Gaussian law, $\sigma_L/\overline{F_L}$ =0.364	[4]
Cheng & Clyde	Water flowing over compactly arranged hemispheres of diameter 305mm	305	35,800~63,200	The theoretical bed was 0.15D below the top of spheres, u_0 was taken as the average velocity of the cross section	The distributions of lift force and drag force followed the Gaussian law, $\sigma_L/\overline{F_L}$ =0.18, $\sigma_D/\overline{F_D}$ =0.4~0.8 (depending on water depth)	[5]
Chepil	Gas flowing over hexagonally arranged hemispheres; distance between centers of the hemispheres was 3 diameters	3~102	16~13,680	----	The ratio of the lift force to the drag force varied in the range of 0.53~1.32 with an average value 0.83; C_L=0.068	[6, 7]

Table 5.1 Continued

Author	Experimental condition	Size	Re range	Velocity definition	Remarks	Ref.
Coleman	Water flowing over a isolated sphere on compactly arranged spheres	13	10–1,500	Velocity at 0.5D up from the top of spheres	Relationship between the drag coefficient and u_0D/v was the same as that in case of settlement of a sphere in still water as shown in Fig. 5.3	[8]
		0.6–20	6.5–1,500		The ratio of lift force to submerged weight of the particle K is shown in Fig. 5.5 as a function of the grain Re	[9]
Garde & Sethuraman	Balls rolling down an inclined plane in still liquid	---	---	The same as above	The drag coefficient was larger than the drag coefficient in case of settlement of a particle in still water, as shown in Fig. 5.3	[10]
Wattes & Rao	Oil flowing over shperes arranged in four different ways shown in Fig. 5.2	95.5mm in oil simulating 0.5mm in water	15–100	Velocity was taken as the shear velocity	The relationship between the drag coefficient and the grain Re depended on the position of the sphere, as shown in Fig. 5.6	[11]
Sedi. Dept., WIHEE	Water flowing over gravel arranged in different ways	62×50×39	---	---	The magnitude of the lift force was in the same order of that of the drag force	[12]

concluded that the fluctuating velocity follows the Gaussian distribution more closely than does the shear stress [14]. In fact, if the fluctuating velocity is not too large compared with the mean velocity and the velocity near the bed follows the normal distribution, the distribution of shear stress will be nearly normal [15].

6. Researchers of WIHEE (Wuhan Institute of Hydraulic and Electric Engineering) measured the forces acting on gravel at different positions on the bed surface (Fig. 5.2), and their results are shown in Fig. 5.4; they found that the magnitudes of the lift and drag forces are roughly the same [12]. Still, the relative

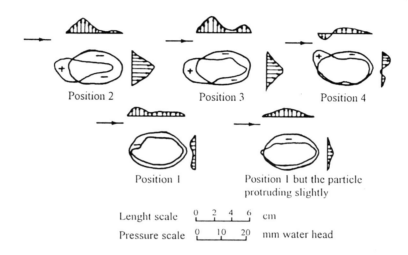

Fig. 5.4 Drag and lift forces acting on the particles at different positions

importance of lift and drag forces does not depend on the magnitude of the forces, but on the effectivity for initiation of motion of the particles. If the incipient motion of a particle protruding on the bed surface at position 2 in Fig. 5.2b is to be determined, both drag and lift have to be considered, but, in general, drag is more important than lift for starting the particle to move. Although particles at position 2 are most liable to move, moving particles hardly ever stop at such a position. Fig. 5.2c and Fig. 5.2d show particles at positions 3 and 4, and these do not represent a general case either. Motion of most particles initiates from position 1 in Fig. 5.2a. In this case, the drag force is not able to affect the stability of the particle because the surrounding particles block any movement. Initiation of such particles depends mainly on the lift force, and therefore, the lift force is more important than the drag force.

The expressions for the drag and lift forces are the same except for the coefficients, and the coefficients are often merged into a composite coefficient in formulas used. Thus, as long as the composite coefficient is accurately based on measured data, it is not important whether drag or lift, or both, are introduced into expressions of the force due to the flow.

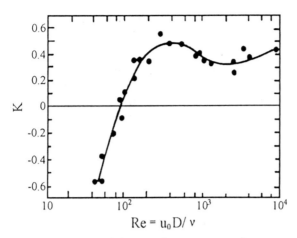

Fig. 5.5 Ratio of lift force acting on compactly arranged spheres to submerged weight of the sphere, K, as a function of Reynolds number

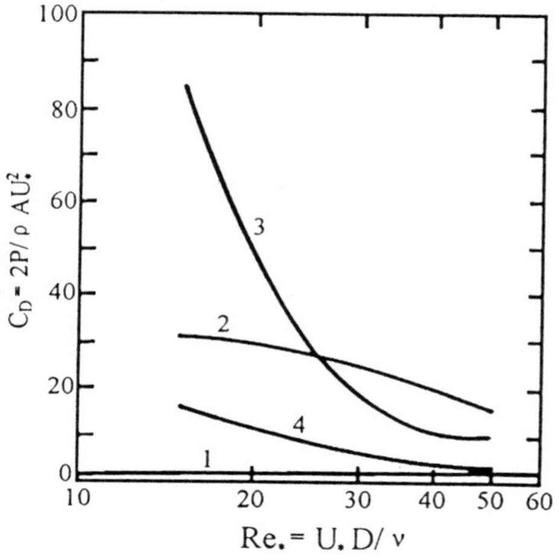

Fig. 5.6 Drag coefficient for a sphere at different positions as a function of Reynolds number

5.1.2 Cohesive force

The submerged weight of the sediment particles resists the drag force and the lift force for cohesionless particles. For fine sediment particles, a cohesive force is also significant in the balancing of the active forces due to flowing water.

The cohesive force is discussed briefly in Chapter 2. As shown in Fig. 2.24, the film water pressure can be illustrated with a model using two parallel plates. If the two plates are close together, an additional force N is needed to lift the upper plate, and this is in addition to the force needed to overcome the submerged weight. It can be expressed as follows:

$$N = \gamma(h_a + h)A \tag{5.5}$$

in which h is water depth; h_a is the head of water equivalent to the atmospheric pressure; A is the contact area of the two plates.

For sediment particles the effective contact area between upper and lower particles should be proportional to the diameter of the particle and inversely proportional to the gap between the particles. The latter can be taken to be proportional to the m-th power of the diameter. Researchers of WIHEE obtained the following expression for the additional pressure of the thin water film [16]:

$$N \sim D^2 \frac{1}{D^m} \gamma \, (h + h_a) \tag{5.6}$$

The value of m was determined from measured data to be 0.72, thus

$$N \sim \gamma (h+h_a)^{1.28} \qquad (5.7)$$

In another study, Dou obtained the following expression of effective contact area of particles from an experiment on friction with an instrument based on a quartz filament [17]:

$$A = (k_1 + k_2 \tanh\frac{h}{h_a})\frac{(\gamma_s - \gamma)D^3}{\gamma h_a} + (k_3 + k_4 \tanh\frac{h}{h_a})\delta D \qquad (5.8)$$

in which k_1, k_2, k_3, and k_4 are constants and can be determined from measured data, δ is the diameter of a water molecule (=3×10^{-8} cm). Substituting Eq. (5.8) into Eq. (5.5), one obtains another expression for the force N.

It is not clear whether the cohesive force can be simply represented by the additional force N. In other words, are there any other forces that have not been taken into consideration yet? The researchers of WIHEE took Eq. (5.7) as the expression of cohesive force between fine sediment particles. Tang gathered from the experiments by Deriaguin that the cohesive force results from the molecular pressure of the thin water film and is independent of water depth and atmospheric pressure [18.19]. He suggested an expression of the cohesive force proportional to the diameter of the particles:

$$N' = \xi D \qquad (5.9)$$

The coefficient ξ is closely related to the surface properties of the particles, the properties of the liquid and the compactness of the contact area of the particles. If the particles are in close contact, the coefficient ξ is a constant. Dou suggested both the additional forces N expressed as in Eq. (5.5) and N' expressed in Eq. (5.9) should be taken into consideration [20]. Obviously, there are still more questions to be clarified concerning the cohesive force among fine particles.

5.1.3 Dispersive force

For a particle that starts to move as bed load, the drag force acting on it increases considerably as it rises into a region of faster moving water; at the same time, the lift force decreases sharply owing to the decrease of the pressure difference on the top and the bottom surfaces of the particle. Fig. 5.7 shows the distribution of the pressure acting on the surface of a spherical particle 0.8 mm in diameter at different distances from the bed [7]. At a distance of 2.5 cm, about three times the particle diameter, the uplift component is almost zero.

However, more than one particle is moving, and relative movement occurs between the particles and the fluid. The movement of particles is mutually affected by the flows around them. As a result, a force between the particles is created which is

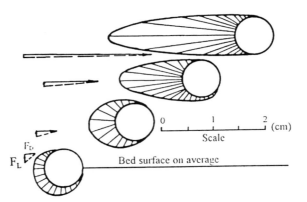

Fig. 5.7 Distributions of force acting on the surface of a spherical particle at different distances from the bed

perpendicular to the flow. To differentiate it from uplift, the force is called a dispersive force. Rowe's experiment is discussed in Chapter 3. He allowed water to flow through the gaps between two rows of balls and found that the balls were acted on not only by a drag F_D in the flow direction, but also by a dispersive force P in the direction perpendicular to the flow [21]. The closer the particles are to each other, the larger the dispersive force. The relationship between P and F_D follows the formula:

$$\frac{P}{F_D} = \frac{0.15D}{s} \tag{5.10}$$

in which F_D is the drag on a single ball and s is the distance between the balls, as shown in No. 3 in Table 3.8.

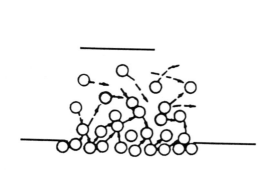

Fig. 5.8 Collisions between particles
during bed load movement

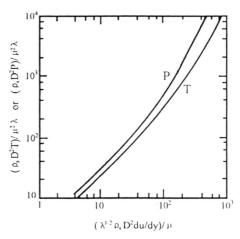

Fig. 5.9 Dispersive shear stress
and dispersive force

In case many bed load particles are moving at the same time, a dispersive force may be generated by the interaction of particles. Some particles move at low speed, while some particles have gained more energy and move much faster. Collisions between the particles are inevitable, and they increase rapidly with increase in bed load movement, as shown in Fig. 5.8. Momentum exchange occurs during the collisions and, therefore, results in a force. Its component in the flow direction is called dispersive shear stress T and the component in the vertical direction is the dispersive force P. Of course, T and P are correlated with each other.

Bagnold mixed spherical particles with a liquid of the same density, sheared the mixture in a rotational co-axial cylinder viscometer, and measured the shear stress at various shear rates. A part of the surface of the inner cylinder was replaced by a rubber film. The film would deform if a dispersive force were produced among particles, and the deformation was proportional to the dispersive force. In this way, Bagnold measured the dispersive shear stress and dispersive force and expressed them in dimensionless numbers, as shown in Fig. 5.9 [22]. In the figure, ρ_s is the density of the particles, and λ is the ratio of the diameter of the particle to the average distance between the particles, a function of the concentration of the particles. As the drag on falling particles, a viscous zone and an inertia regions can be identified, zones in which viscosity or inertia dominates corresponding to different regions of shear rate.

The dispersive force between the bed load particles is ultimately transferred to the bed and acts on the bed as an additional weight, thus it enhances the stability of the bed, a point that is discussed in a following section.

5.1.4 Seepage pressure

Seepage occurs in a river bed if the riparian ground is permeable and water can flow between the river and the ground. The velocity of the seepage flow in the direction perpendicular to the river, υ_S, is given by

$$\upsilon_s = -K\frac{\partial h}{\partial y} \tag{5.11}$$

in which K is the permeability coefficient of the material composing the bed and $\partial h / \partial y$ is the gradient of piezometric head perpendicular to the river bed. Thus the seepage force on a unit volume of sediment in the bed, F_S, is

$$F_s = -C(1+e)\gamma\frac{\partial h}{\partial y} \tag{5.12}$$

in which γ is the specific weight of water, e the porosity of bed material, C a coefficient with value in the range of 0.35-0.5 [23]. If F_S is positive, the force is directed upward.

In a discussion of forces acting on the bed particles, the hydrostatic pressure is usually not taken into consideration, although it exists, because the pressure can be transmitted deep into the bed through the gaps between particles. Sediment particles contact each other usually at one point except for flat particles. Hence, almost the whole surface of every particle is acted upon by hydrostatic pressure. The resultant of the hydrostatic pressure is nothing more than the buoyancy force. Except for the particles so fine that the thin film of water surrounding the particles governs their interaction, the hydrostatic pressure has to be taken into account in the analysis of incipient motion because the hydrostatic pressure cannot be transmitted through the water film and acts only on the particles on the bed surface.

5.2 PATTERN OF SEDIMENT MOTION

Sediment particles in motion can be classified according to their patterns of motion as contact load, saltation load, suspended load, and laminated load; in these, contact load, saltation load, and laminated load all belong to the category of bed load.

In the following, the sediment is considered to be coarse enough that the cohesive force is negligible so as to simplify the analysis.

If an experiment is conducted in a flume with sediment particles on the bed, the state of sediment motion in the flume goes through the following stages as the flow velocity increases. At the beginning of the experiment, the velocity is small and all particles are motionless. As the velocity increases to a threshold value, some particles vibrate because of the fluctuating velocity of the turbulent flow but do not move from their original positions. In this case the force acting on a particle is at some instants high enough to move the particle, but the particle cannot move from its position immediately because of its inertia and the short duration of the force. Therefore the particle sways but does not move.

Following a further increase in velocity, the bed particles begin to move. The pattern of movement depends greatly on the position of the particle on the bed.

5.2.1 Contact load

A particle at position 2, the one that protrudes the furthest of the four arrangements shown in Fig. 5.2, is subjected to the greatest drag force. If

$$F_D > f(W' - F_L) \qquad (5.13)$$

the particle slides forward; W' is the submerged weight of the particle in water, and f is a friction coefficient. The bed surface is not smooth, the particle may still roll along it. The particles, as they slide or roll, make frequently contacts with the bed and therefore are called a contact load (Fig. 5.10b).

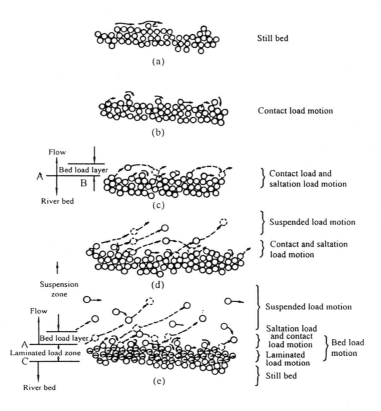

Fig. 5.10 Patterns of sediment motion

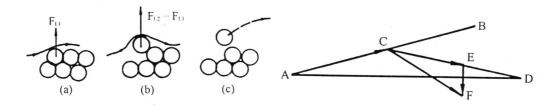

Fig. 5.11 Particle changes from
rolling to saltation

Fig. 5.12 Acceleration of a saltation
particle at elevation A

165

5.2.2 Saltation load

As shown in Fig. 5.11a, a particle at the surface and on the front row, position 3 in Fig. 5.2, begins to roll around its point of contact with the particle just downstream (Point A in Fig. 5.1) if

$$(F_L^2 + F_D^2)^{1/2}a > W'b \qquad (5.14)$$

in which a is the distance of the resultant force of F_L and F_D from point A, b the distance of W' from point A. Once the particle moves to the position shown in Fig. 5.11b, however, the curvature of the streamline at the top of the particle increases, resulting in a local increase of the velocity and a reduction in the pressure. The force acting on the bottom surface of the particle is enlarged because the particle has moved a small distance away from the bed particle underneath it, and the bottom surface area on which the static pressure acts increases. Therefore, the lift force is much greater. In other words, the particle experiences an abrupt increase in lift force at the instant of rolling away from the bed. Consequently, it may jump into the flow from the bed, as shown in Fig. 5.11c [24].

The particle moves away from the bed, enters the flow, which has a higher velocity, and is carried downstream. The combination of movements in the two directions leads the particle to move along the line AB in Fig. 5.12. If AC is the velocity of the particle and AD the velocity of the flow, then CD is the velocity of the particle relative to the flow. The relative movement results in an acceleration represented by CE. If EF is the gravitational acceleration, then CF is the resultant acceleration of the particle.

As the particle rises, the ambient velocity becomes higher and higher, and the horizontal velocity component of the particle larger and larger. The two affect the relative movement between the particle and the flow in opposite ways. As the particle rises to some height, the effect of the increasing horizontal velocity component of the particle is greater than the effect of the flow velocity, and the relative movement between the particle and the flow begins to reduce. In general, as the particle reaches its highest point, its velocity is close

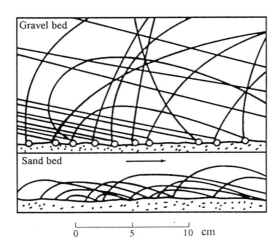

Fig. 5.13 Trajectory of saltation load in wind

to the local flow velocity. From this point, the particle descends. Fig. 5.13 shows the trajectories of particles carried by wind. Particles moving in such a way produce what is called saltation load, as shown in Fig. 5.10c.

A falling particle transmits a shock to the bed particles when it falls down to the bed. The intensity of the shock depends on the height of saltation and the velocity of the flow. If the saltation height is small, the particle gains less momentum because the velocity near the bed is small, and it will probably not rebound. If the saltation height of a particle is large, the particle gains more momentum, and it may rebound after it strikes the bed. If the flow velocity is still higher and the particle gains much more momentum, it will not only rebound, but it may also induce other bed particles that it hits to jump into flow.

The height of particle saltation is inversely proportional to the density of the fluid [25]. Since the density of water is more than 800 times larger than that of air at room temperature, the height of saltation in wind is about 800 times larger than that in water flow if the particles jump from the bed at same initial velocity. The difference is of profound significance. For movement of wind-blown sand, particles jump quite high and gain much more energy from the wind. And when they fall on the bed they splash more particles into the flow. The chain reaction results in a sharp increase in rate of sediment transport soon after sand motion is initiated in the desert. Although the saltation load in flowing water is more important than the contact load, the height of saltation is usually only a few times the particle diameter in water flow, and the kinetic energy it possess falling back down on to the bed is not enough to induce such chain reactions. A particle at position 1 of Fig. 5.2 is constrained by the neighboring particles. Only if

$$F_L > W'$$ (5.15)

can the particle be lifted away from the bed so that it rolls or saltates.

A particle at position 4 of Fig. 5.2 usually rolls on the bed once it moves. If it is hit by a falling particle, however, it may jump up and go into saltation. Eqs. (5.13)-(5.15) are the critical conditions for the initiation of motion of particle.

5.2.3 Suspended load

Flows at high velocity are turbulent and have eddies of various sizes. If a particle jumping from the bed enters such an eddy, it may be carried far away from the bed. For carrying a particle, the size of the eddy must be much larger than the particle and its upward velocity component must be higher than the fall velocity of the particle. If an eddy is of about the same size as a particle, the latter is liable to fall out of the eddy; hence, the eddy would no longer affect the movement of the particle. On the contrary, if an eddy is much greater than a particle, the eddy may carry the particle for a long time. And by the time the particle falls out of the eddy, it may already have been carried into the region of the main flow. Obviously, the transport of suspended particles is mainly the effect of large-scale eddies.

In the main flow zone, even if a particle falls out of an eddy, it may enter another eddy and be carried further along. Therefore, the trajectory of a suspended particle is quite irregular and depends almost completely on the movement of the eddies surrounding it. If a particle is carried by an eddy close to the bed, it may fall on to the bed. These particles, carried by eddies and moving downstream at the same velocity as the flow, are called the suspended load, as shown in Fig. 5.10d. Suspension of particles takes a certain amount of energy from the turbulent flow. Hence, on the one hand, flow turbulence carries sediment particles into suspension, and on the other hand, the existence of suspended load reduces the turbulence intensity.

A significant deduction from the effect of large scale eddies in bringing sediment particles into suspension is as follows: the eddies generated in the zone close to the bed are small because they are constrained by the boundary; thus they can not cause the suspension of particles. Only at a certain distance from the boundary are the eddies big enough to carry particles. Therefore, bed particles must go through the saltation process before becoming a suspended load.

Nevertheless, the following possibility should not be ruled out: during a bursting process of turbulent flow, a big eddy may develop from a small eddy close to the bed or come from the main flow zone, and it can sweep the bed surface and pick up bed sediment and bring elements of it directly into suspension. In general, however, sediment entering directly from the bed into suspension accounts for only a small amount of suspended load. On the one hand, most of eddies move upward and the few large eddies that sweep over the bed surface can only affect part of the bed surface at any instant. On the other hand, as these eddies move upward, most of sediment carried by them is the sediment moving in the vicinity of the bed as bed load. Therefore, the major part of the suspended load should be the sediment being suspended through the transition into saltation. For given hydraulic conditions, the finer the sediment, the more sediment is brought directly from bed into suspension.

5.2.4 Laminated load

As has been mentioned, the sediment on a river bed is subjected to a shear stress induced by the flow. As the river bed is composed of loose granular material, which is not a solid entirety, the shear stress of the flow is transmited to the bed. If the velocity is small, some of the particles on the bed surface--the layer A-B in Fig. 5.10c--slides, rolls, or moves in saltation, depending on the flow velocity. The rest of the particles remain stationary because the friction caused by their effective weight plus the extra pressure exerted on them by the dispersive force is large enough to balance the drag force. With a large shear stress, however, not only the particles on the bed surface, but also the second layer of the bed sediment can enter into motion. And the motion penetrates further into the bed in response to further increase in shear stress. The real river bed in this case is not at the A-B plane in Fig. 5.10c but the C-D plane in Fig. 5.10e. The particles between the two planes are closely packed and can move only in layers. In the process of movement, the moving bed dilates somewhat so as to attain freedom to move. The velocity of the moving sediment is much smaller in a deeper bed. The sediment that moves in such a way is called the laminated load.

Discussions on patterns of sediment movement have been mostly based on the experiments of Gilbert [26]. Since only a few experiments with high shear stress were conducted and details of sediment movement could not be observed because of the turbidity of the flow, Gilbert did not notice the motion of the laminated load. The phenomenon of laminated load was clearly observed by the author in a flume experiment with colorless plastic particles. Bagnold had also conducted detailed experimental studies and theoretical analyses on laminated load motion [27, 28].

5.2.5 Relative importance of bed load and suspended load

Contact load, saltation load, and laminated load belong to the category of bed load and thus clearly differ from suspended load. The relative importance of bed load and suspended load depends on the sediment size and flow velocity. For the same composition of bed sediment, sediment slides, rolls, or moves in saltation if the flow velocity is low, and the movement occurs only in a zone close to the bed surface with a thickness of 1 to 3 times the particle diameter. The zone is called the bed surface layer. At a higher velocity, a part of the sediment is carried into the main flow zone and becomes suspended load. The rest remains in the bed surface layer and moves as bed load, but the thickness of the bed surface layer is augmented. Following still further increase in flow velocity, the suspended load is greater, and it exceeds the bed load. If the velocity is more than some threshold value, however, laminated load motion underneath the bed surface comes into existence and the thickness of the laminated load layer becomes larger and larger with further increases in flow velocity. As a consequence, the relative importance of suspended load motion reduces gradually and becomes secondary due to a weakening of the turbulent intensity.

In his experiments with plastic particles, Bagnold found that the phenomenon of turbulence characterized by randomly occurring eddies had been replaced by rather regularly occurring secondary flows once the volume concentration of plastic sediment was over 25%. As the concentration reached 30%, the turbulence of the flow was greatly attenuated. The turbulence and the secondary flows disappeared altogether and laminated load motion developed for a concentration of 35%. All particles moved forward in layers and the concentration became uniformly distributed. At a concentration of 60%, the distances between particles were so small that the particles might clog the flume and the whole flow might come to standstill as though frozen if a small disturbance was exerted on the flow [27]. The critical concentration for all suspended sediment to transform itself into laminated load depends on the characteristics of both flow and sediment.

In general and for ordinary flows in rivers, sediment coarser than a certain diameter moves mainly as bed load, and sediment finer than that diameter moves mostly in suspension. Kresser concluded from data from four European rivers that the critical diameter that differentiates bed load and suspended load was defined as follows [29]:

$$\frac{U^2}{gD} = 360 \tag{5.16}$$

in which U is the average velocity of flow. Nevertheless, Eq. (5.16) is valid only for European rivers and may not be correctly applied elsewhere.

If the critical conditions for incipient motion, the fall velocity of the sediment, and the nature of the turbulence of flow are known, the patterns of sediment motion in the flow can be roughly predicted. The incipient motion of sediment is discussed in Chapter 8. There, critical conditions for the incipient motion are expressed in different ways. For the curve COD in Fig. 5.14, the condition for initiation of sediment motion is shown with the shear velocity as the main parameter. A conclusion from the data by Nikijin (Fig. 4.21) is that the shear velocity in most zones of the flow equals roughly the root mean square of the vertical component of the fluctuating velocity except within a zone close to the boundary. The curve EOF in Fig. 5.14 represents the fall velocities for sediment of various diameters. The curves COD and EOF divide Fig. 5.14 into several zones, and each of them is characterized by a different kind of sediment movement:

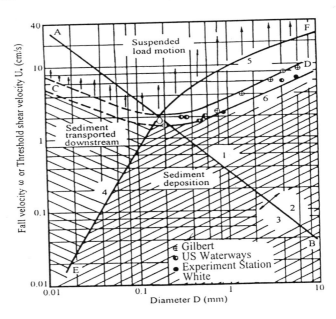

Fig. 5.14 Zoning of sediment movements

1—Grain Reynolds number $U_*D/\nu = 3.5$, 2—Form resistance dominates, 3—Skin friction dominates, 4—Fall velocity ω, 5—Sliding, rolling and saltating, 6—Threshold shear velocity U_*

170

1. In the zone below the curve *DOE*, the fall velocity of the sediment is larger than the vertical component of the fluctuating velocity and therefore sediment will settle out. Because the shear velocity of the flow is lower than the critical value for sediment initiation, i.e. $U_* < U_{*c}$, sediment from upstream will accumulate on the bed.

2. In the zone between *CO* and *OE*, sediment carried by the flow can remain in suspension because the fall velocity of the sediment is less than the upward component of fluctuating velocity. However, sediment on the bed of the same size cannot be picked up by the flow because of the influence of the laminar sublayer and cohesive forces. Of course, since the fluctuating velocity and shear stress vary stochastically, the upward fluctuating velocity may sometimes be less than the fall velocity of sediment and the shear stress on the bed sediment may sometimes be over the threshold value for incipient motion. But such scenarios are rare and one can say for simplicity that sediment coming from upstream is transported through the river channel without any exchange with the bed sediment.

3. In the zone between *DO* and *OF*, the shear stress of the flow is over the threshold value for initiation of motion but the turbulence is not strong enough for the suspension of sediment. Sediment moves in this zone as contact load and saltation load. This zone is smaller than the others.

4. In the zone above *CO* and *OF*, sediment can not resist movement by the flow and is likely to be suspended once it begins to move. Bed load and suspended load co-exist in this region. The higher the shear velocity, the more the suspended load will be.

The mechanism of the transformation from laminated load into suspended load is not yet clear, the laminated load is not shown in Fig. 5.14. The *AB* line in Fig. 5.14 characterizes the resistance regime of the flow near the bed, and it is represented by the following formula:

$$U_*D/\upsilon = 3.5$$

In the zone above the line *AB*, the shear stress on the bed particles consists mainly of form resistance. In the zone below the line, the shear stress on the bed particles consists mainly of skin friction.

5.3. SIGNIFICANCE OF THE DISTINCTION BETWEEN BED LOAD AND SUSPENDED LOAD MOTION

5.3.1 Continuity of sediment movement

Sediment motion, from the river bed to the water surface, is continuous even though the sediment is classified in catagories like contact load, saltation load, suspended load, and laminated load according to its mode of movement. There exist continuous exchanges between these loads as well as between material in the bed and that being transported.

The phenomenon of exchange was observed in two experiments. In the first experiment, water flowed in a flume with a layer of sediment on the bed, and the flow velocity was controlled so that it initiated motion but did not suspend the sediment. At the entrance of the flume, colored sediment was fed into the flume at a rate equal to the sediment transport capacity of the flow. After a period of operation one found that not all the colored sediment went through the flume to the sediment collection tank at downstream of the flume; a part of the colored sediment was distributed over the flume bed. In the second experiment, the bed sediment in a section of the flume was replaced with colored sediment and water was allowed to flow over the bed in equilibrium conditions for a period of time. Later, a part of the colored sediment had been removed from the section and its original positions were filled by uncolored sediment. The two experiments proved that on the one hand bed sediment can be picked up and transported downstream by the flow, and on the other hand sediment carried by the flow can stop moving and become a part of the bed.

Not only do exchanges occur between bed material and sediment in the bed surface layer, but also exchanges occur between suspended load and saltation load or contact load if the flow velocity is high enough to suspend sediment. Moreover, exchanges through bed surface layer occur between suspended load and bed material. The exchanges can be summarized as follows:

Suspension zone $\leftrightarrow$ Bed surface layer $\leftrightarrow$ River bed
(Suspended load)   (Contact and saltation loads)   (Bed material)

With laminated loads, exchanges between the bed material and the bed surface layer are transmitted through the laminated load zone:

Suspension zone $\leftrightarrow$ Bed surface layer $\leftrightarrow$ Laminated load zone $\leftrightarrow$ River bed
(Suspended load)   (Contact and saltation loads)   (Laminated load)   (Bed material)

Whenever a large eddy sweeps over the channel bed, direct exchange between suspended load and bed material can occur. In the zone between CO and OE in Fig. 5.14, no exchange takes place between bed material and moving sediment.

As long as the exchanges occur between different loads and bed material, the concentration distribution of sediment is a continuous curve. For fine sediment, strong turbulence makes the distribution quite uniform. But coarse sediment and weak turbulence cause the distribution to be non-uniform so that more of the sediment is concentrated near the bed.

5.3.2 Essential differences between suspended load and bed load

Although exchanges occur between bed load and suspended load and a sediment particle may sometimes move as bed load and sometimes move as suspended load, the difference between the bed load motion and suspended load motion lies not only in their locations but also in their physical features. For a thorough discussion on the mechanism of sediment movement, essential differences between suspended load and bed load must be understood.

Differences between the suspended load and bed load are described as follows:

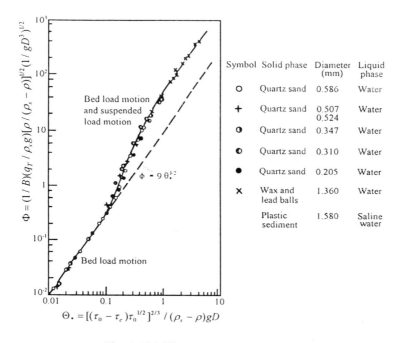

Fig. 5.15 Different laws of sediment movement

1. Different laws of movement. The laws of movement of bed load are different from those of suspended load, as discussed in detail in Chapters 9 and 10. The plot in Fig. 5.15 shows a dimensionless sediment transport parameter, ϕ, to be a function of a dimensionless shear stress, Θ_*, with definitions shown in the figure. It is based on six groups of flume experiments by Bagnold [28], for which g_T is the rate of sediment transport per unit width, in weight per second per unit width; τ_0[1] is the shear acting on a unit area of the bed surface and is normally called the drag force of the flow; τ_c is the critical shear stress for incipient motion of bed sediment; and B is a coefficient. Besides the data for quartz sand transported by flowing water, data of lightweight sediment carried by water or salty water are also included in the figure. In the latter case the difference in the specific gravity of the liquid and solid phases was only 0.004. The figure shows that for the two extremely different cases, the transport of sediment followed two different laws: For $\Theta_* < 0.1$, sediment moved mainly as bed load, and the measured points followed a straight line

---

[1] The shear exerted by the flow on a unit area of bed is τ_0. Strictly speaking, the drag should be only part of the shear, which is in respect to grain friction.

$$\phi = 9\,\Theta_*^{3/2} \tag{5.17}$$

For Θ_* larger than 0.1, a part of sediment was suspended, and the points deviate from the previous straight line and fall on another curve. The study indicated that the law of movement for suspended load is different from that for bed load.

2. Different source of energy supply. Movement of sediment requires energy. The energy must come from the potential energy of flow, kinetic energy of fluid turbulence, and potential energy of sediment. The origin of energy for suspended load is different from that for bed load.

For contact load and saltation load, the velocity component in the flow direction is small as motion begins but grows quickly due to the drag exerted by the flow. Therefore the flow loses energy in carrying the sediment, and the energy is transferred into kinetic energy of sediment motion. For laminated load motion, collisions among particles result in a large resistance to the flow, and a large energy slope is needed to maintain the movement.

For suspended load, sediment particles are carried by eddies, and their velocity in the flow direction is essentially the same as that of the flow. Therefore, suspended load motion does not consume flow energy directly. Nevertheless, suspended sediment is heavier than water and would settle out if there were no turbulent diffusion. Therefore, to maintain its motion, sediment has to draw energy from the turbulence to keep it in suspension.

The fact that suspended load and bed load draw energy from different origins has fundamental and far-reaching consequences. For contact load and saltation load that directly consume energy from the flow, energy consumption and flow resistance increase in accordance with the increase of the load, and the flow velocity decreases. However, suspended load does not draw energy directly from the flow in the same way but extracts it from the turbulence. From the viewpoint of energy transformation the turbulence energy has already taken from the energy of the main flow of water and is a transitional form of energy between potential energy and heat (Section 7.1). Further, the structure of the turbulence in the flow is also affected by the existence of suspended load. As a result, the distribution of the mean velocity is modified; this in turn affects the energy consumption of the flow. The effect, however, is indirect and the mechanism is quite complicated. Some of the data available show that the existence of suspended load does not affect the loss of energy and resistance to the flow, others show that the resistance can be either higher or lower as a result of the existence of a suspended load. This complex question is discussed further in Chapter 12.

3. Different action on the river bed. Sediment has a larger specific weight than does water and tends to settle down onto the bed. Therefore, a force is needed to balance the submerged weight of sediment if its movement is to be maintained.

Sediment is suspended because of momentum received from turbulent eddies. Momentum exchange of eddies in the vertical direction results in a force that balances the submerged weight of suspended particles. Because suspended sediment mixes with water, the specific weight of the suspension is larger than that of pure water and the hydrostatic pressure is also larger in consequence. The static pressure is transmitted to the water in the interstices of the bed material, and it can be measured by means of sophisticated pressure sensors embedded in the sediment on the channel bottom. The concentration of suspended sediment can then be calculated from the measured result [27].

Bed load is supported by a dispersive force. The dispersive force among particles is transmitted to the static particles on the bed and exerts a downward pressure on them. The pressure is equal to the submerged weight of bed load moving above the bed.

Suspended load increases both the specific weight of a suspension and the static pressure of water in the interstices of the bed. Bed load increases pressure on the bed surface and enhances the stability of the bed. The former acts on the water in interstices of the bed particles but the latter acts on the particles themselves.

In some literature the essential differences between the suspended load and bed load have been discussed from other viewpoints. For instance, some authors considered that the movement of bed load was intermittent and the movement of suspended load was continuous; bed load concentrated on river bed could be called "bottom sediment," whereas suspended load could be called correspondingly "suspended sediment."

The intermittence of sediment movement reflects essentially an exchange between different forms of sediment motion. For example, the intermittent movement of a bed load particle is essentially that the particle is rolling or saltating part of the time and is bed material during the rest of the time. Since the particle transforms from bed material to moving material, its motion is intermittent. Suspended particles also exchange with bed load and bed sediment; therefore, the movement of suspended load is not always continuous but is also intermittent. The frequency of intermittency of movement of contact load and saltation load is much larger than that of suspended load, and those for laminated load and suspended load are about the same.

The concept of "bottom sediment" comes from the traditional concept. In fact, if bed load motion occurs only in the forms of sliding, rolling, and saltation, all bed load move within a zone close to the bed with a thickness of a few times of particle diameter. It can, of course, be called as bottom sediment in such a case. Nevertheless, if laminated load motion is considered as a kind of bed load motion, as indicated in Section 2.4, the bed load zone may expand to the surface of the flow. In this case, it would be inappropriate to call such bed load "bottom sediment."

5.3.3 Sediment movement on rigid beds and movable beds

If the channel bed is composed of loose and mobile sediment and the supply of sediment from the bed material to the flow is sufficient, a steady flow can reach its sediment-carrying capacity after flowing for some distance. It is not necessary to analyze whether the sediment concentration reaches the carrying capacity of the flow. Nevertheless, for flow over a rigid bed, sediment carried by the flow comes entirely from upstream reaches of a river or canal, or is fed artificially into a laboratory flume. If the sediment content in the flow is less than the carrying capacity of the flow, no compensation is available from the bed; consequently, the amount of sediment load carried by the flow depends on the oncoming sediment or on the feed-rate of sediment into the flume.

As mentioned earlier, only bed load can enhance the stability of the bed by the dispersive force acting on the bed particles. For a movable bed, suspended load motion cannot occur without bed load motion. If the flow velocity increases, rates of both suspended load and bed load transport increase, as long as the sediment concentration is not so high as to suppress turbulence considerably. The increase of the rate of suspended load transport may be much faster than that for bed load, and the suspended load may dominate the flow, but it is impossible for all moving sediment to become suspended load. If there were no bed load motion and no dispersive force, stability of the bed would depend completely on the submerged weight of the bed particles. If the shear stress of the flow is over the threshold shear stress for incipient motion of bed particles, the first layer of bed particles is removed and the second layer is exposed to the flow. Because the particles eroded from the bed surface would enter directly into suspension and would not strengthen the stability of the bed, the submerged weight of the particles of the second layer would not be enough to withstand the shear of the flow and the second layer would also be eroded as a consequence. Therefore, the movable bed would be eroded layer by layer, and thus is obviously not the real case. On the contrary, if the bed is rigid or immovable, the stability of the bed does not rely on the dispersive force of bed load motion. All sediment can be eroded and move in suspension if the flow velocity and turbulence of the flow are large enough. Wind blowing over a highway is such a scenario; all of the sediment can be transported in suspension.

For a high intensity of bed load motion, granular shear and dispersive shear stress occur in the zone near the boundary, and the distributions of velocity and sediment concentration are affected by the shear stress. If the boundary is rigid, however, the bed load motion near the bed is weak or disappears altogether, and the distributions would be different. Bagnold found from an experiment with movable bed and plastic sediment that if the sediment supply at the entrance of the flume is not sufficient, a part of bottom boundary is eroded to the rigid bottom of the flume. The local mean velocity then increases sharply and the centroid of concentration distribution rises; as a result, the rate of sediment transport increases [27].

The phenomenon illustrates that sediment movement over a rigid bed and over a movable bed are essentially different. In laboratory studies, a rigid bed is often employed to simplify the operation. The results of such tests must be properly analyzed before they can be applied to natural rivers flows.

5.3.4 Practical significance of differentiating bed load and suspended load

Differentiation of bed load and suspended load is essential not only because of their different laws of movement, but also to be able to deal correctly with processes of sediment deposition and river morphology [30].

As one instance, the development of sand waves is closely related to the phase position difference between sediment movement and flow velocity near the bed (Chapter 6, section 6.3.2). The phase position difference is positive if bed load motion dominates; the river bed is unstable in this case and a series of sand dunes develops. The phase difference is negative if suspended load motion dominates; the river bed is then stable and no sand waves can develop. The transition from sand dune to flat bed is closely related to the ratio of the rate of suspended load transport to the rate of bed load transport.

Another example is the effect of helical flow at bends on the direction of sediment movement. Bed load motion is largely affected by the transverse slope whereas suspended load motion is little affected by the helical flow and follows the main current. The difference results directly in the sorting of sediment, the formation of a point bar in the convex bank, and the development of river patterns.

If sediment deposition occurs, suspended load and bed load cause deposits at different locations. As a sediment-laden flow pours into a reservoir, bed load accumulates at the upstream end and suspended load mainly deposits on the top-set and fore-set of delta. Resiltation of dredged channels usually results from bed load and density currents rather than from suspended load.

5.4. BED MATERIAL LOAD AND WASH LOAD

Sediment is classified as either bed load or suspended load according to the patterns and laws of movement. It can also be classified as bed material load and wash load according to the size of the particles and their origins. The characteristics of bed material load and wash load, the criteria for differentiating the two kinds of loads and the significance of the two loads on theory and engineering practice are discussed in this section.

5.4.1 Concept of bed material load and wash load

As early as 1940, Einstein, Anderson, and Johnson analyzed a number of size distribution curves of sediment samples from channel beds and flowing water and found that the ratios of fine to coarse sediment in the channel bed and that in motion are quite different [31]. Sediment in the channel bed is composed of much more coarse and much less fine sediment than is the moving sediment. Moreover, fine fractions of the sediment existing in the moving sediment either do not exist or hardly exist in the channel bed. This fraction of fine sediment hardly exchanges with bed material at all during movement and behaves more like sediment moving over a rigid bed. The content of fine sediment in the flow is not saturated and the rate of transport depends only on the amount contained in the oncoming flow. Thus the amount of coarse sediment carried by the flow depends on the sediment transport capacity and exhibits a well-defined relationship with the discharge of water. In contrast, the concentration of fine sediment depends only on the supply of the sediment from an upstream reach and no obvious correlations with the discharge is found. Fig. 5.16 shows the relationships of transport rate of sediment of various sizes with flow discharge and illustrates this phenomenon.

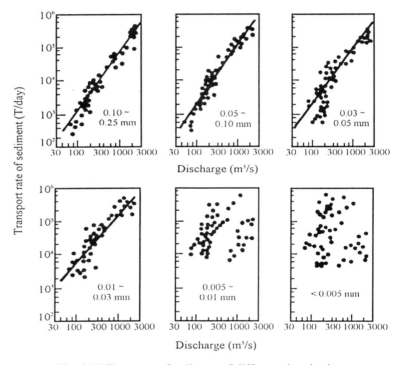

Fig. 5.16 Transport of sediment of different sizes in river

Because coarse sediment always exchanges with bed material during transport, incoming coarse sediment may be thought of as originating directly from the channel bed of the upstream reach. It is directly supplied from the bed and, therefore, is called "bed material load." In contrast, fine sediment, eroded and washed from upland watersheds, that has been transported through the channel over a long distance and is scarcely ever deposited in the channel, is called "wash load." Sediment can be classified as bed material load and wash load, or bed load and suspended load. The two sets of classification of sediment are distinct and should not be intermingled. Bed material load may move as bed load and also may move as suspended load and the same is true for wash load. Of course, wash load is fine and mainly moves as suspended load. It is not correct to identify the bed material load with bed load and wash load with suspended load.

5.4.2 Identity of laws of motion of bed material load and wash load

In some cases, moving sediment does not exchange with bed sediment; one such case is shown as in the area between CO and OE in Fig. 5.14. Sediment in that zone is likely to be suspended. But once a particle falls on the bed it can be protected by the laminar sublayer, or the cohesive force may withstand the shear induced by the flow. Another example occurs in mountain rivers where the flow is torrential. The river bed is often composed of large gravel and boulders. Only when a torrential flood occurs is the bed material moved. Nonetheless, from observations in such watersheds, mountain rivers can also carry some fine material. Fine sediment may also be found in the river bed but the particles are hidden in the interstices of the large particles, and hence they are not included as part of the effective bed material that can exchange frequently with the moving sediment. The movement of the fine sediment in mountain rivers is, moreover, like that in flows over a rigid bed.

The situation in alluvial rivers is much more complicated because these movements of both fine and coarse sediment can be caused by the flow. Does fine sediment exchange with bed material during transport over a long distance? Although the amount of fine sediment in the river bed is much less, does this part of sediment interrelate with the sediment of the same size in the flow? In other words, is there any essential difference between the laws of motion for coarse and fine sediment? Without correct answers to these questions, the concepts of bed material load and wash load cannot be clarified. To answer them, a series of experiments were conducted [32].

In the first group of experiments, non-uniform sediment of sizes ranging from 0.005 mm to 4 mm was employed, and the flow rate was held constant. Sediment was fed continuously into the flow, also at a controlled rate. During the course of the experiment, a part of the sediment that was supplied to the flow deposited on the bed, while another part, especially the fine sediment, circulated continuously; in this way, the rate of transport increased progressively. After some time of continuous operation, a maximum total-load concentration of 15% by weight was reached. At that stage, the feeding of material was discontinued and the test reach was kept in equilibrium. Next, some of the water with its suspended sediment was siphoned off from the system at the entrance of the flume and replaced by an equal amount of clear water. The bed then scoured continuously as the sediment concentration decreased. Thus, during the experiment, the oncoming sediment varied with time. The transport rate of particles of different sizes at the flume outlet are shown in Fig. 5.17. The figure clearly indicates that, for particles coarser than 0.1 mm, the rate of transport for each grain size fluctuated around its average value and was independent of the availability of particles. This part of material, therefore, behaved like bed material load. In contrast, particles

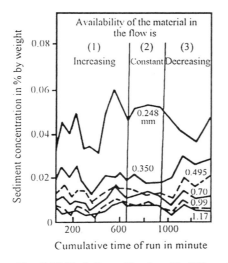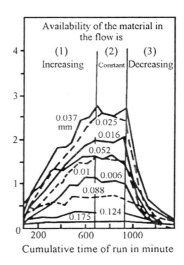

Fig. 5.17 Variation of loads with different availability of material in the flow

smaller than 0.06 mm behaved exactly like wash load; their rates of transport depended entirely on their availability in the oncoming flow. Particles between 0.06 mm and 0.1 mm are intermediate and show characteristics of both the bed material load and the wash load.

The bed composition adjusted itself in accordance with the variation of the oncoming sediment load. Fig. 5.18 shows the compositions of the oncoming sediment, of the bed material, and of the sediment at the outlet, while sediment was accumulated on the bed. The topmost layer of the bed was distinctly finer than the bulk of the bed deposit because of the sorting effect caused by the flow during the development of sand dunes. Although the wash load (<0.06 mm) constituted only about 5% of the volume of the bed deposits, it was more in the surface layer, and it increased with an increase in the incoming sediment concentration. When the concentration reached 15% by weight, the fine material constituted as much as 17% in the surface layer. Because the adjustment of bed composition involved only fine material, the effective roughness of the bed did not change, although the supply of sediment in different

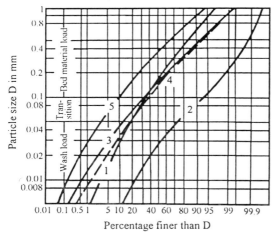

Fig. 5.18 Compositions of feed material, of the total load in the flow and of different bed layers during sediment accumulation on the bed

1—Feed material, 2—Total load in the flow, 3—Bed surface layer (6.1 mm thick), 4—Bed surface layer after increase of feeding rate, 5—Bulk of the bed deposit (65.5 mm thick)

size groups did change. Therefore, the flow maintained its original characteristics during deposition and erosion.

In the second group of experiments, the flume bed was covered with loose sediment. With clear water flowing in the channel, sediment was scoured from the bed; all of the sediment in the flow moved as bed material load. As an equilibrium state was established after a period of running, sediment finer than the bed material was fed into the flow at the entrance. The fine sediment represented wash load. It was found that a small amount of wash load deposited on the bed, although most of the wash load was carried on downstream. Thus the composition of the sediment at the bed surface contained more fine material because of the deposition of the wash load (Fig. 5.19).

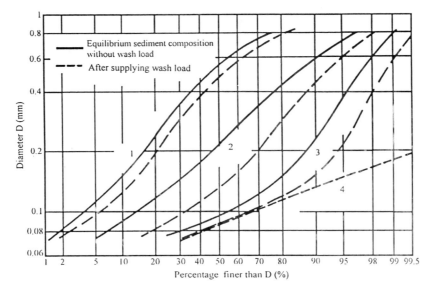

1—Bed sediment underneath the surface layer, 2—Bed surface layer,
3—Total load in the flow, 4—Feed wash load
Fig. 5.19 Change of the bed composition after supplying wash load

The two groups of experiments indicate that the composition of bed material, particularly the part of surface bed material corresponding to the wash load, adjusts itself according to variations of the rate of wash load transport. In the discussion of continuity of sediment movement (Paragraph 5.3.1), bed sediment contributes to the resistance of the flow and represents the boundary condition. Its size composition reflects

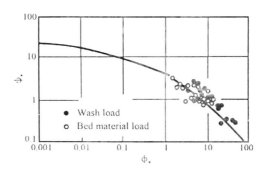

Fig. 5.20 Similarity of laws of movement of wash load and bed material load

181

upstream sediment supply and is an important link in the sediment movement chain. In alluvial rivers, sediment particles can neither be continually transported nor can they rest on the river bed forever. Even in the equilibrium state, settlement always occurs at some places and scour at others. The condition called "equilibrium" implies the amount of settling sediment equal to the amount of sediment scoured from the bed. Sediment transport capacity is the amount of sediment a flow can carry in an alluvial river with a certain composition of bed material; it involves the following three elements: (1) amount of sediment of various sizes in the effective bed material; (2) intensity of flow; (3) rates of transport of various sizes of sediment.

Taking the variation of the bed composition into account and analyzing the transport rate of sediment of various sizes, Einstein and Chien found that, for alluvial rivers composed of movable sediment, the laws of movement of bed material load and wash load are identical; that is, there is no essential difference between them. Fig. 5.20 shows the results calculated with the Einstein bed load function, in which data of the 5% finest sediment in the surface layer is included. The comparison proved that wash load and bed material load have the same law of movement.

5.4.3 Relationship between discharge and rate of sediment transport

If the laws of movement of wash load and bed material load are identical, why are the relationships between the discharge of water and rate of sediment transport for wash load and bed material load different? In fact, the findings in the preceding section do not imply that a definite relationship between the discharge and the rate of sediment transport exists, and that is the same for both the wash load in a river reach and bed material load. The purpose of the study of sediment transport capacity is essentially to determine the following function:

$$f (\text{flow, bed material composition, transport rate}) = 0$$

The function involves three variables. A relationship established between two variables that neglects the third would certainly show scatter of the measured points unless the third variable does not change significantly during the measurement. The key question is how the fractions of different sizes of the bed material adjust themselves to variations in the rate of water and sediment flows.

The concept can be elucidated with the following example of a river reach 30 km long and 1000 m wide, with bed material that is effectively 25 cm thick [33]. Only the sediment of size of 0.05 to 0.1 mm (fine sediment) and 0.5 to 1 mm (coarse sediment) are discussed. The fine sediment comprises 2% and the coarse sediment 38% of the bed material, whereas the former comprises 30% and the latter 3.6% of the sediment in motion. During a flood period, the flow can transport about 16 million m³ of sediment. Nevertheless, the incoming sediment load in a flood is changeable because of a random combination of rainstorms and sediment yields in the upstream watersheds. Now, if the incoming sediment load in a flood is 20% more than the usual amount, how much sediment can be transported downstream through the reach? If the ratio of transport capacity of sediment of a certain size group to the percentage of the sediment in the bed does not change with a slight adjustment of bed material, then the amount of sediment deposition and transportation through the channel can be estimated. Table 5.2 shows the calculated results. From the table, the ratio of the amount of sediment transported through the channel to the increment of incoming sediment is quite different for fine sediment from it is for coarse sediment.

Table 5.2 Variations of the amounts of effective bed sediment and transported sediment owing to 20% increase in incoming sediment load

No.	D	Percentage in bed	Percentage in moving	Effective amount of bed sediment	Amount of sediment transport in a normal flood	Amount of incoming sediment in the flood	Increment of incoming sediment load	Variation induced by increase of incoming sediment load					Ratio of the sediment transport capacity to the amount of bed sediment
								Amount of sediment deposition	Percentage of deposited sediment	Ratio of deposited sediment to volume of bed sediment	Amount of sediment transported	Increment of transported sediment	
	(mm)	(%)	(%)	(10^4 m^3)	(10^4m^3)	(10^4m^3)	(10^4m^3)	(10^4 m^3)	(%)	(%)	(10^4 m^3)	(%)	
1	0.05~0.10	2	30.0	15	480	576	96	2.9	3.0	19.3	573	19.4	32
	0.5~1.0	38	3.6	285	58	70	11.6	9.6	82.8	3.4	60	3.4	0.2
2	0.05~0.01	2	30.0	15	4800	5760	960	3.0	0.3	20.0	5757	20.0	320
	0.5~1.0	38	3.6	285	580	696	116	38.0	32.8	6.5	658	13.5	2

Since the coarse sediment is bed material load, 82.8% of the increment of coarse sediment will deposit on the bed. This fraction of sediment (0.5 to 1 mm), however, is plentiful in the bed, the deposit of 100,000 m^3 of coarse sediment increases the fraction in the bed by only 3.4%. Flow depth and velocity do not change as long as the bed morphology and slope remain unchanged. Therefore, the transport capacity for the coarse sediment and the amount transported downstream through the channel is only 3.4% more than those in the usual case.

The fine sediment in the wash load is mostly transported downstream through the channel; only 3% of it deposits in the channel. Since the fraction of fine sediment in the original bed material is rather small, however, the deposit of 30,000 m^3 of fine sediment increases the amount of that fraction in the bed by a surprising 19.3%. The size composition of the bed material changes little owing to the small amount of deposition. Therefore, the roughness and flow velocity remain unchanged. However, due to the variation of the condition of supply from the bed, the amount of wash load transported downstream through the channel will increase by 19.4%, about the same as for the incoming wash load.

183

The results reveal that as incoming sediment load varies, the amount of wash load transported through the reach equals roughly the amount of incoming wash load, and it is supplied from the watershed. For bed material load, however, the amount of transport varies within a small range, and it is equivalent to the normal load carried under same hydraulic conditions in spite of the change in the incoming sediment. Therefore, the relationship between flow discharge and rate of sediment transport is maintained for bed material load but changed for wash load.

Moreover, even the relationship for bed material load is not absolutely definite. The calculated results in Table 5.2 show that, as the incoming sediment varies, the variation of the sediment outflow depends mainly on the ratio of the sediment transported in normal cases (in general, it is the transport capacity under the same flow conditions) to the amount of the bed material that is available for the flow. For the example shown in Table 5.2, if the amount of bed material remains unchanged and sediment transported in a flood in normal cases is 160 million m^3 rather than 16 million m^3, the transport rate for coarse sediment is much higher. The calculated results are shown in row II for the case of a 20% increase in incoming sediment load. Even for the coarse sediment, the transport rate increases by 13.5%.

In ordinary alluvial rivers, the measured transport rates of bed material load correlate well with flow discharge. In rivers heavily laden with sediment, such as the Yellow and Weihe Rivers in China, even for the bed material load, the measured points for the flow discharge-sediment transport rate show considerable scatter. Fig. 5.21 shows the correlation of the discharge-sediment transport rate measured at Sunkou Gauging Station on the lower Yellow River. The points distribute about straight lines with different concentrations of bed material load as measured at the upstream station [34]. This figure reflects the automatic adjustment of the sediment transport capacity in alluvial rivers. The mechanism of the adjustment was discussed in detail by Chien et al.[35].

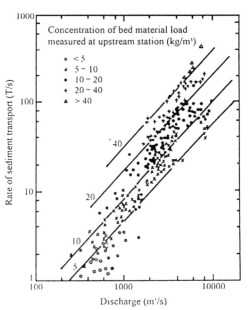

Fig. 5.21 Relationship between the rate of sediment transport and discharge measured at Sunkou Gauging Station of the lower Yellow River

Finally, if the increment of the incoming sediment load is permanent rather than temporary, the amount of sediment transported may not be equal to the incoming amount initially. After a period of time, however, a new equilibrium is reached and the transport rate and incoming sediment load have become identical. For bed material load, if the change with incoming sediment load is permanent, flow will pick up a part of sediment from the bed or a part of sediment will deposit on the bed depending on the direction of the variation. After a long period, the composition of bed material then becomes coarser or finer, and the sediment transport capacity decreases, or increases, accordingly. For rivers composed of gravel and sand, the adjustment of the sediment transport capacity can be completed through the adjustment of the bed composition with no or almost no change of bed form and slope. In most cases, however, the variation of the bed material composition is accompanied by an evident change in bed form and slope. For wash load, a slight adjustment of the bed composition may match the variation of the incoming sediment load, and no change in bed morphology is required. Therefore, bed material load is sometimes referred to as bed-forming load and wash load as non-bed-forming load.

5.4.4 Criterion for distinguishing bed material load and wash load

What is the criterion for distinguishing bed material load and wash load? Partheniades correctly indicated that the laws of movement of wash load and bed material load are identical from the view point of the mechanics of sediment transport. He employed 0.06 mm as a critical diameter to differentiate bed material load and wash load, because for sediment finer than 0.06 mm, the cohesive force among particles is an important factor, and the critical flow intensities for the suspended fine sediment to deposit and the sediment on the bed surface to be scoured are quite different. For flow between the two critical conditions there was no exchange between the bed material and the moving sediment, all the fine sediment coming from the watershed could be transported without deposition [36]. In fact, the cohesive force is important only for sediment much finer than 0.06 mm, as discussed in Chapter 2. To use a fixed value for the critical diameter or assign a given range of sizes as the criterion for distinguishing bed material load and wash load is not appropriate in a discussion of the mechanism of the two type of loads.

From the foregoing discussion, especially the example in Table 5.2, a suitable approach for differentiating bed material load from wash load may be through the ratios of the rate of sediment transport at dominant discharge to the amount of the bed material for various sizes of sediment. If the ratio is over a critical value, the sediment belongs to the category of wash load, whereas if the ratio is less than the critical value, it is bed material load. Two critical values may exist, sediment between the two values is in a transitional region between bed material load and wash load. Further study is required.

An empirical method, which is quite simple and is often used in engineering practice, is to employ D_5 (or D_{10} in some cases) as the criterion. Sediment finer than D_5 is wash load and coarser than D_5 is bed material load, in which D_5 is the grain size for which 5% by weight of the bed material is finer. Strictly speaking, bed material at a bed surface can not be sampled by the samplers in current use. Therefore, the bed material referred to here denotes the material within a certain depth of the bed. Another practical approach is to overlap the accumulative size distribution curves of

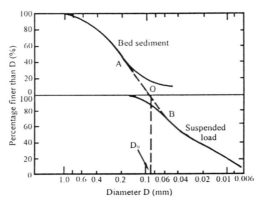

Fig. 5.22 Critical diameter for distinguishing bed material load and wash load

bed sediment and suspended load, as shown in Fig. 5.22. The straight line AB is tangent to and connects the two curves. The intercept of this line with the coordinate axis is at point *o*, the diameter corresponding to point *o* is the critical diameter that distinguishes bed material load from wash load.

No matter what method is employed, the critical diameter for differentiating bed material load and wash load is different for different localities. For a mountain river, gravel is the bed material load and sand and finer materials are wash load. In the middle reaches of a sand-bed river, sand is bed material load and silt and finer material are wash load. And for estuaries, silt may be bed material load and only clay is wash load. Not simply is the critical diameter different in different reaches of a river, but it may also be different in the same reach for different flow discharges. For instance, before the closure of the Sanmenxia Dam, the Yellow River bed at Shanxian was covered by a layer of sand during low flow season. In flood season, the sand layer was scoured away and the gravel underneath was exposed to the flow. Sand in that reach was bed material load in the low flow season and then became wash load during flood season. Furthermore, if the bed composition changes significantly owing to deformation of the river bed, the differentiation of wash load and bed material load may change. For instance, river bed scour usually occurs in the reach directly downstream of a dam after impoundment in a reservoir. After the development of an armoring layer in the process of scour, a part of bed material load becomes wash load. Or, in the upstream reach of a reservoir, a part of the wash load may convert into bed material load as the bed material becomes finer.

Some researchers have suggested that the criterion for differentiating bed material load and wash load should include not only sediment size but also flow intensity.

Wash load is transported mainly as suspended load. Flow that is carrying suspended load, on one hand, has to spend a part of its turbulent energy for suspending sediment, and on the other hand, its potential energy is enhanced due to the larger density of the suspension.

If the concentration does not vary in flow along a river course, the sediment can be thought as moving parallel to the river bed. The sediment particles move down a vertical distance of UJ in unit time. Therefore, the sediment per unit volume contributes an energy to the flow:

$$E_1 = S \frac{\gamma_s - \gamma}{\gamma_s} UJ \qquad (5.18)$$

in which U stands for average velocity, J for the river slope, S for the dry weight of the sediment in unit volume, and ω for the fall velocity of the sediment. Because sediment particles fall in the fluid at some velocity ω, to maintain the particles in suspension the turbulent eddies must create a net transport of sediment upward at the same velocity. Hence, if the concentration of sediment is low, sediment draws energy from turbulence at a unit rate of

$$E_2 = S \frac{\gamma_s - \gamma}{\gamma_s} \omega \qquad (5.19)$$

If

$$E_1 \geq E_2$$

that is

$$\omega \leq UJ \qquad (5.20)$$

the turbulent energy for suspending sediment can be compensated for from the potential energy contributed by the sediment to the flow. Such sediment is wash load. No matter how much of the sediment comes from a watershed, it can be transported without deposition. This concept was put forward by Wang[1] and Bagnold [37] independently.

5.4.5 Implications of distinguishing bed material load and wash load

The advantages of distinguishing bed material load and wash load are the follows:

1. Calculation of the transport rate in different ways   Theoretically, the transport rate of both wash load and bed material load can be calculated if the flow and bed material composition are known because the laws of motion for wash load and bed material load are identical. In practice, however, the two cases are approached differently.

---

[1] Wang, Shangyi. "Wash Load Motion in Sediment-Laden Flows." *Report Series of Tianjin Society of Waterways*, 1959.

The fraction of bed material corresponding to wash load varies with the incoming flow and sediment and its variation cannot be predicted. Therefore, even if the laws of wash load motion are known, the transport rate of wash load is difficult to determine from the relationships established for the mechanics of wash load transport; thus people are obliged to use a statistical approach or results of hydrologic measurements, as discussed in Chapter 11.

For bed material load, the corresponding fraction in the bed material can be dealt with as a constant in ordinary alluvial rivers. If the percentages of various sizes in the bed material are determined by sampling and size analysis, the rate of sediment transport can be calculated by means of the formulas of sediment transport capacity. In rivers heavily laden with sediment, the problem is much more complicated because the composition of the bed material also varies with the incoming flow and sediment. If the average bed composition is known from repetitive sampling and size analyses, the average rate of sediment transport can be calculated.

Formulas of sediment transport capacity derived from flume experiments are valid only for bed material load because of limitations on the experimental conditions. Obviously, the formulas derived from them cannot be used directly for the calculation of total sediment transport. Furthermore, empirical relations of sediment transport capacity derived from measured data in specific rivers include a part that is wash load. They cannot be used for the calculation of sediment transport rate of other rivers with different watershed characteristics. Even for the same river, the construction of a reservoir upstream of a flow region affects the incoming flow, rate of sediment transport, and bed material composition; thus any formula must modified accordingly before application.

2. Provide a way to study variation of sediment input to the downstream reaches after construction of dams. For reaches downstream of places where river water is impounded, two questions have to be answered: can the flow carry sediment, and if so, how much [38]? Although capacity of flow for carrying wash load is quite high, availability of wash load is limited to the floodplain and tributaries because the dam traps most of the sediment coming from the watershed upstream. In most cases the wash load coming from erosion of the floodplain and the tributaries is not abundant. For bed material load, however, the situation is just the reverse. Sufficient sediment can be supplied from the river bed, but the sediment transport capacity of the flow is limited; following adjustment of the bed composition, the sediment transport capacity is accordingly less. Hence, after construction of a large dam on a heavily sediment-laden river, the sediment load entering downstream reaches is bound to be much less.

3. Lay emphasis on appropriate items in the analysis of fluvial processes. Bed material load is the main constituent of sediment carried during the process of the formation of river bed. Wash load mainly affects the channel deformation in the following two aspects:

First, sediment on floodplains is composed mainly by wash load, although it is rare in the main channel. If the deposits from wash load are cohesive, the main current is restrained by the bank composed of such deposits and the development of the river pattern is affected in consequence.

Second, if the concentration of wash load in the flow is high, the physical properties of the flow are considerably changed, and the sediment transport capacity of bed material load is also affected.

The two effects are minor, compared with the effect of bed material load. For studies of the fluvial process, especially of the channel deformation due to variation of incoming flow and sediment in response to human activities, primary attention should be paid to bed material load.

The other arises in the study of reservoir siltation, especially the problem of life span and reduction of reservoir capacity; both bed material load and wash load should then be considered. Wash load often plays the major role in the rate of reservoir siltation because the volume of wash load transported by the flow is often more than bed material load.

Finally, the main features of bed material load and wash load are summarized in Table 5.3, providing a conclusion to the section.

Table 5.3 Main features of bed material load and wash load

Features	Bed material load	Wash load
Origin	Soil erosion in watershed	
Direct origin	River bed upstream	Sediment yield in watershed
Composition of bed material	Main part of bed material usually not changing except for heavily sediment-laden rivers	On the bed surface, changing with incoming amount and flow intensity
Composition of moving sediment	A small part of the moving sediment	Main part of the moving sediment
Patterns of movement	Bed load and suspended load	Mainly suspended load
Transport rate	Determined from the flow intensity and less correlated with incoming sediment, but for heavily sediment laden rivers depending also on the incoming sediment	Determined mainly from the incoming sediment
Relationship between flow and sediment transport	Relationship established on basis of mechanics and may be estimated by sediment transport capacity formulas	Relations based on watershed characteristics, determined by measured or empirical data
Basic laws of movement	For alluvial rivers the laws of movement for bed material load and wash load are identical	
Significance	Transport rate determines bed stability	Transport rate determines rate of reservoir deposition
Criterion	Coarser than D_5 of the bed material	Finer than D_5 of the bed material

REFERENCES

[1] Jefffreys, H. "On the Transport of Sediment by Streams." *Proceeding Cambridge Philosophy Society*, Vol. 25, Pt. 3, 1929, pp. 272-276.

[2] Demenchiev, M.A. "Interference of Two Solid Bodies in Liquid Flow." Periodicals published and edited by All USSR's Scientific and Research Institute of Hydraulic, Vol. 15, 1935, pp. 28-48.

[3] Yegiazapov, I.V. "Generalized Formula for Sediment-Carrying Capacity in Alluvial Rivers and Threshold Velocity." *Collection of the 3rd National Hydrological Conference*, USSR, Vol. 5, Leningrad, 1960, pp. 117-132.

[4] Einstein, H.A., and E.A. El-Samni. "Hydrodynamic Forces on a Rough Wall." *Review of Modern Physics*, Vol. 21, No. 3, 1949, pp. 520-524.

[5] Cheng, D.H., and C.G. Glyde. "Instantaneous Hydrodynamic Lift and Drag Forces on Large Roughness Elements in Turbulent Open Channel Flow." *In Sedimentation, Symposium to Honour H.A. Einstein*, 1972, pp. 3-1 to 20.

[6] Chepil, W.S. "The Use of Evenly Spaced Hemi-Spheres to Evaluate Aerodynamic Forces on a Soil Surface." *Transition, American Geophysical Union*, Vol. 39, No. 3, 1958, pp. 397-404.

[7] Chepil, W.S. "The Use of Spheres to Measure Lift and Drag on Wind Eroded Soil Grains." *Proceeding of Soil Science Society of America*, Vol. 25, No. 5, 1961, pp. 343-345.

[8] Coleman, N.L. "The Drag Coefficient of a Stationary Sphere on a Boundary of Similar Spheres." *La Houille Blanche*, No. 1, 1972, pp. 17-21.

[9] Coleman, N.L. "A Theoretical and Experimental Study of Drag and Lift Forces Acting on a Sphere Resting on Hypothetical Stream Bed." *Proceeding of 12th Congress of International Association for Hydraulic Research*, Vol. 3, 1967.

[10] Garde, R.J., and S. Sethuraman. "Variation of the Drag Coefficient of a Sphere Rolling down an Inclined Boundary of Closely Packed Spheres." *La Houille Blanche*, No. 7, 1969, pp. 727-732.

[11] Watters, G.Z., and M.V.P. Rao. "Hydrodynamic Effects of Seepage on Bed Particles." *Journal of Hydraulic Division, Proceeding of American Society of Civil Engineers*, Vol. 97, No. HY3, 1971, pp. 321-439.

[12] River Study Group, Wuhan Institute of Hydraulic and Electric Engineering (WIHEE). "Review on the Einstein Theory of Bed Load Motion and Discussion on Bed Load Motion." *Bulletin of Wuhan Institute of Hydraulic and Electric Engineering*, No. 4, 1965, pp. 1-16. (in Chinese)

[13] Gessler, J. "The Beginning of Bed Load Movement of Mixtures Investigated as Natural Armoring in Channels." *Mitteilung Nr. 69*, Versuchsanstalt fur Wasserbau und Erdbau, Zurich, 1965.

[14] Christensen, B.A. "Erosion and Deposition of Cohesive Soils." *Journal of Hydraulic Division, Proceeding of American Society of Civil Engineers*, Vol. 91, No. HY5, 1965, pp. 301-308.

[15] Gessler, J. "Behavior of Sediment Mixtures in Rivers." *International Symposium of River Mechanics*, International Association for Hydraulic Research, Vol.1, 1973, pp.35-1 to 10.

[16] WIHEE. *River Dynamics*. China Industry Press, 1961, pp. 23-31.

[17] Dou, Guoren. "On the Threshold Velocity for Initiation of Sediment." *Journal of Hydraulic Engineering*, No. 4, 1960, pp. 44-60. (in Chinese)

[18] Derjaguin, B.V., and T.N. Voropayeva. "Surface Forces and the Stability of Colloid and Disperse System." *Journal of Colloid Science*, Vol. 19, No. 2, 1964, pp. 113-135.

[19] Tang, Cunben. "Laws of Sediment Initiation." *Journal of Hydraulic Engineering*,1963. (in Chinese)

[20] Dou, Guoren. "Theory of Threshold Motion of Sediment Particles." *Scientia Sinica*, Vol. 11, No. 7, 1960, pp. 999-1032. (in Russian)

[21] Rowe, P.N., and G.A. Henwood. "Drag Forces in a Hydraulic Model of a Fluidised Bed, part I." *Transition of Institute of Chemical Engineerings*, Vol.39, No.1, 1961, pp.41-54.

[22] Bagnold, R.A. "Experiments on a Gravity-Free Dispersion of Water Flow." *Proceeding of Royal Society*, London, Ser. A, Vol. 225, 1954.

[23] Martin, C.S., and M.A. Mostafa. "Seepage Force on Interfacial Bed Particles." *Journal of Hydraulic Division, Proceeding of American Society of Civil Engineers*, Vol. 97, No. HY7, 1971, pp. 1081-1100.

[24] Danel, P., R. Durand, and E. Condolier. "Introduction a letude de la Saltation." *La Houille Blanche*, No. 6, 1953, pp. 815-829.

[25] Kalinske, A.A. "Criteria for Determining Sand Transport by Surface Creep and Saltation." *Transition of American Geophysical Union*, Vol. 23, Pt. 2, 1942, pp. 639-643.

[26] Gilbert, G.K. "The Transportation of Debris by Running Water." *US Geological Survey*, Prof. Paper, No. 86, 1914, p. 259.

[27] Bagnold, R.A. "Some Flume Experiments on Large Grains but Little Denser than the Transporting Fluid and Their Implications." *Proceeding of Institute of Civil Engineerings*, 1955, pp. 174-205.

[28] Bagnold, R.A. "The Flow of Cohesionless Grains in Fluids." *Philosophy Transition*, Royal Society London, Ser. A., Vol. 249, 1956, pp. 235-297.

[29] Kresser, W. "Gedanken zur Geschieber und Schweb-stofführung der Gewässer." *Österreichische Wasser-wirtschaft*, Vol. 16, No. ½, 1964.

[30] Fredsøe, J. "The Distinction Between Bed Load and Suspended Load and its Implications in River Morphology and Sedimentation Problems." *Rep. No. 128*, the Danish Center for Applied Mathematics and Mechanics, Technical Univ. Denmark, Nov. 1977, p. 18.

[31] Einstein H.A., A.G. Anderson, and J. W. Johnson. "A Distinction between Bed Load and Suspended Load in Natural Streams." *Transition of American Geophysical Union*, Pt. 2, 1940, pp. 628-633.

[32] Einstein H.A., and Ning Chien. "Can the Rate of Wash Load be Predicted from the Bed Load Function." *Transition of American Geophysical Union*, Vol. 34, 1953, pp. 876-882.

[33] Chien, Ning. "Explanation of the Concepts of Bed Material Load and Wash Load." *Journal of Hydraulic Engineering*, No.3, 1957, pp.29-45. (in Chinese)

[34] Mai, Qiaowei, Ye-an Zhao, and Xiandi Pan. "Computing Method for Alluvial Process Downstream of Reservoirs on Heavily Sediment-Laden Rivers." *Yellow River Construction*, No. 3, 1965, pp. 28-34. (in Chinese)

[35] Chien, Ning, Ren Zhang, Jiufa Li, and Weide Hu. "A Preliminary Study on the Auto-Adjustment of Sediment Transport Capacity in the Lower Reach of the Yellow River." *Geographica Sinica*, Vol. 36, No. 2, 1981, pp. 143-156. (in Chinese)

[36] Partheniades, E. "Unified View of Wash Load and Bed Material Load." *Journal of Hydraulic Division, Proceeding of American Society of Civil Engineers*, Vol. 103, No. HY9, 1977, pp. 1037-1058.

[37] Bagnold R.A. "Autosuspension of Transported Sediment; Turbidity Current." *Proceeding of Royal Society London*, Ser. A, Vol. 265, No. 1322, 1962, pp. 314-319.

[38] Chien, Ning, and Qiaowei Mai. "Estimation of Sediment Load Downstream of Reservoirs in Heavy Sediment-Laden Rivers." *Journal of Hydraulic Engineering*, No. 4, 1962, pp. 9-20. (in Chinese)

CHAPTER 6

BED FORM MOVEMENT

The movement of sediments as bed load and suspended load usually refers to their movement as individual particles. Yet while sediment is being transported, the bed load particles move collectively in all sorts of ways along the river bed. Their motion in turn can cause changes in the configuration of the river bed in accordance with the variation of sediment transport rate. The collective movement of large quantities of sediment particles on the bed is called bed form movement.

6.1 DEVELOPMENT OF BED FORM

Fig. 6.1 shows the various phases of bed form development; they would occur in a progression if the flow rate were gradually increased.

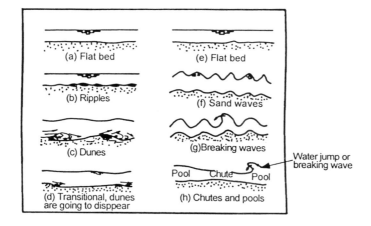

Fig. 6.1 Various phases of bedform development

6.1.1 Ripples

With low rates of flow over an initially flat stationary bed, as shown in Fig. 6.1a, no sediment moves; but once the flow velocity reaches a certain value, some particles are set in motion. Soon after that, a few particles may gather on the bed and form a small ridge; this ridge gradually moves downstream and tends to increase in length. Finally, the ridges connect and ripples with a regular shape form, as shown in Fig. 6.1b.

The longitudinal cross sections of ripples are usually not symmetrical. The upstream face is long and has a gentle slope, and the downstream face is short and steep. The former is generally between 2 and 4 times as long as the latter. The height of ripples is usually between 0.5 cm and 2 cm; the highest ripple is not more than 5 cm. The wave lengths normally do not exceed 30 cm, and they are usually within the range of 1 cm to 15 cm. Some ripples that form in deep-water regions are symmetrical, but this type occurs only rarely [1].

Ripples are the smallest of the bed configurations. They are related to the physical parameters near the river bed and have little correlation with the water depth. Their occurrence is the result of the unstable viscous layer near the boundary. They can form in both shallow and deep water. In plan, they either are parallel to each other or have a shape like fish scales. With an increase of the flow velocity, the plan form of the ripples gradually develops from straight lines to curves and then to a pattern like fish scales, symmetrical or unsymmetrical, as shown in Fig. 6.2 [2]. Fig. 6.3 is a plan view of tongue-shaped ripples in a natural river.

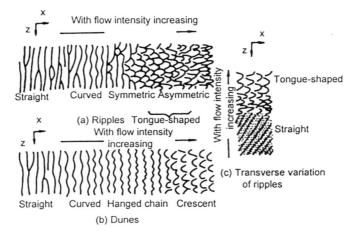

Fig. 6.2 Series of bedforms with increasing flow intensity (after Alan, J.R.L.)

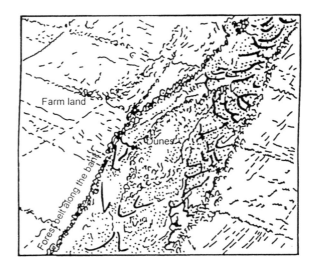

Fig. 6.3 Plan view of tongue-shaped ripples in a natural river

194

Fig. 6.4 is a comparison of the hydraulic parameters with ripple formation and for initiation of sediment motion, in which U_* is shear velocity, ω is particle fall velocity and $U_* D/\nu$ is particle Reynolds number [3]. This form of the Reynolds number is an indication of the ratio of the particle diameter to thickness of the viscous sub-layer. The figure shows that with the velocity increasing ripples and dunes form soon after sediment particles begin to move, and before intensive movement of sediment takes place. Between these two stages, the sediment concentration near the bed is low, and the difference in the density of this layer and the pure water above it is also small. Thus, the phenomena of ripple and dune formation cannot be the result of an unstable interface between two layers of fluid with different densities.

In his analysis [4], as shown in Fig. 6.4, Liu did not delineate the different phases of ripples and dunes. More recent research indicates that ripples are transformed into dunes if the grain Reynolds number, $U_* D/\nu$, exceeds a certain value, as indicated in Table 6.1 [5].

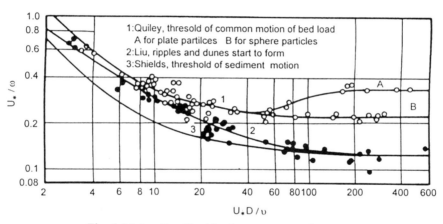

Fig. 6.4 Intensity of bed load movement at the moment of
formation of ripples and dunes (after Liu, H.K.)

Table 6.1 Critical grain Reynolds number for transformation from ripples to dunes
(after Engelund, F. and E. Hansen)

Particle size (mm)	0.19	0.27	0.28	0.45	0.93
$U_* D/\nu$	7.3	10.3	11.0	11.7	No ripple formation

Table 6.1 shows that the critical grain Reynolds number for the ripples to form dunes increases with particle size. If $U_* D/\nu$ is greater than 11.7, ripples do not form, and the flat bed is transformed directly into dunes. It should be noted that when $U_* D/\nu$ =11.6, the particle size equals the thickness of the viscous layer; hence, ripples can only form in the hydraulically smooth regime. Ripples usually do not form in open channel flow if the particles are larger than about 0.6 to 0.7 mm, or 0.5 to 0.9 mm according to some other's observation. Obviously, if fluids with a viscosity higher than that of water are used under laboratory conditions, ripples can form even for still larger

particles. In one example, in which the fluid medium was a mixture of kerosene and machine oil, ripples formed for a sand with a size of 1.68 mm. For this fluid, flows with finer sand particles did not readily produce ripples [6].

6.1.2 Dunes

With increasing flow velocity, ripples develop further and eventually become dunes (Fig. 6.1c). The size of a dune is closely related to the water depth. Fig. 6.5 shows that the heights and lengths of dunes vary significantly in different rivers. Fig. 6.6 shows the process of dune development in the Yellow River at Huayuanko. The relationship between the wave length and wave height is shown in Fig. 6.7.

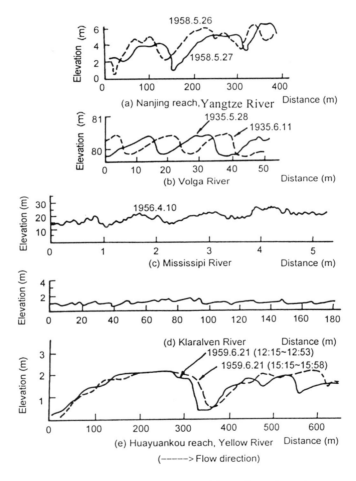

Fig. 6.5 Longitudinal profiles of dunes in various rivers

If the flow velocity is raised progressively, the dune pattern changes in plan from straight lines to curves; it then has a shape like a suspended cable or a moon crescent, as shown in Fig. 6.2 b. The dimension of the straight ridge in the transverse direction is larger than that along the streamwise direction. Fig. 6.8 shows contours of dune

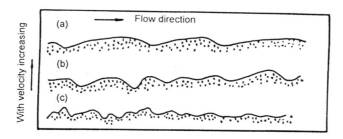

Fig. 6.6 Process of dune development in the lower Yellow River at Huayuanko

ridges in Klarälven River in Sweden. The dunes on both sides of the region of main flow curve forward at the edges and occur in clusters. Stripe dunes often occur on beaches on the convex bank of a river bend, and they stretch downstream towards the concave side of the river. The direction of the flow near the bed, which brings sediment to the convex side, is perpendicular to the ridge lines of the dunes; the latter form a larger angle with the direction of the surface flow, as shown in Fig. 6.9.

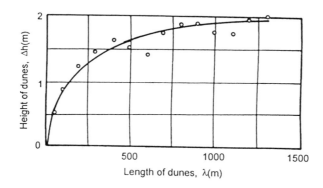

Fig. 6.7 Relationship between wave length and wave height of dunes at Huayuanko

If the amplitude of a sand wave is large compared to the water depth, the sand wave can affect the water surface. The surface is usually lower near the dune peak. Light reflected from the surface there is different from that from a flat water surface, and its color is dark red. Many waves of small amplitude form in this vicinity, and they occasionally make what looks from a distance like a stripe on the surface. It reflects, in fact, the location of the ridge line of the dune. People living near the lower Yellow River call this phenomenon "Lianzhishui." If the ratio of the dune height to the water depth is large enough, the vortex that originates in the separation region downstream of the dune peak is strong enough to carry a high concentration of sediment up to the water surface, it then has the appearence of water at its boiling point. Fig. 6.10 shows the regions of violent vortices and of high concentration of sediment at the water surface over dunes that form in the Brahmaputra River during the flood season [7]. Generally, the length of those regions is within the range from 15 to 50 m, but it can be as large as 250 m.

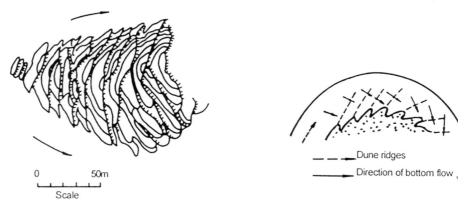

Dune ridges

Direction of bottom flow ,

Fig. 6.8 Contours of dune ridges in Klaralven
River in Sweden (after Sundborg, A.)

Fig. 6.9 Dune ridges on the
convex side of a river bank

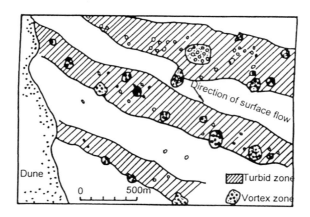

Fig. 6.10 Region of separation and vortex zones (which indicate approximately
the positions of the dune ridge) on the water surface during flood period in
Brahmaputra River (after Coleman, J.M.)

6.1.3 Flat bed

If the dune reaches a certain height and the flow velocity is then increased further, the dune decays; its wave length increases and its height gradually decreases to the form shown in Fig. 6.1d. With still further increase in velocity, the bed becomes flat again (Fig. 6.1e). Fig. 6.11 is a plot of the dune height against the depth of water in the Hankou section of the Yangtze River. The maximum height occurs if the water stage is 21.5 m. If the water stage is below this level, the dune height increases gradually with the discharge or with the water stage; if the stage is above it, the dune height decreases. If the water stage is 24.5 m or more, the river bed is flat again.[1]

---

[1] Hankou Survey Team, Yantze Valley Planning Office. "Basic Pattern of Sand Waves in the Yangtze River," April 1960, pp.29 .

6.1.4 Sand waves

The sediment transport rate is quite large in the second flat bed phase. If the velocity continues to increase, the flow approaches or becomes supercritical (Froude number of about unity or even larger), and the bed forms a sand wave, (Fig. 6.1f). A sand wave is a type of bed configuration that is in phase with the wave on the water surface, and these two waves interact strongly. The differences between a sand wave and a dune are as follows: the shape of a dune is non-symmetrical, and the streamlines of the flow separate at the dune peak; in contrast, a sand wave is symmetrical, more like a surface wave. The streamlines are almost parallel to the river bed and no separation occurs. If sand waves form, the water surface also

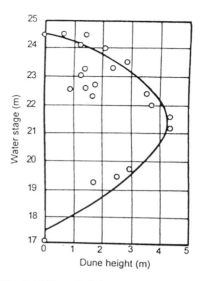

Fig. 6.11 Dune height vs. water stage in the Hangkou reach of the Yangtze River

undulates; this process is called "Gan" in the Yellow River, in order to distinguish it from wind-induced surface wave. With dunes, the bed configuration does not correspond with the surface wave; the surface wave and the bed configuration are out of phase, as already stated. Kondratev[8], using the relationship of specific energy and depth, showed that the former flow should be supercritical and the latter subcritical.

Sand waves can move either in the same direction as the flow, as do ripples and dunes, or in the opposite direction. The former is called a "downstreamward sand wave" and the latter is called a "upstreamward sand wave or antidune." Antidunes often form in shallow flows that are moving at high velocities. The amplitude of a sand wave is comparatively large. The flow must climb the upstream or rising side of the sand wave, and usually drops part of its sediment load there. On the downstream side of the peak, the flow possesses surplus energy and can entrain sediment from the bed. As a consequence, even though the movement and transport of every particle is in the direction of the flow, the sand wave as a whole profile, generally moves upstream.

The direction of individual sand wave movement should not be confused with the direction of movement of a series of sand waves. When the gravity waves move in still water, several of the waves at the front of a series of waves decay faster because the rate of energy transfer lags behind the wave celerity. As a result, new waves are produced at the rear of the wave series (details in Chapter 16). Similar phenomena also occur with the surface waves over sand waves. Although the individual surface waves and the corresponding sand waves appear to be moving against the flow, the whole series of waves is actually moving with the stream, because of some waves disappearing from the upstream end of the series and some others generating at the downstream edge.

In the development of antidunes, the amplitude of the surface wave may exceed that of the sand wave by a factor of 1.5 to 2. The trough of surface waves can even be below the crest of the sand waves (Fig. 6.12) [9]. In this instance, the waves on the water surface are unstable and break (Fig. 6.1g). The phenomenon is similar to waves breaking on a beach. Kennedy [10] found from flume experiments that the waves break if the ratio of wave height to wave length reaches a value of 0.142, which is similar to the theoretical limit for gravity waves. However, in a wave series, wave breaking usually does not occur at each wave simultaneously. Waves in the middle, developing faster than those at the front and back, break first.

At the moment a surface wave breaks, the velocity near the bed is almost zero, as shown in Fig. 6.13, and a certain volume of water is temporarily at rest in that section of the river. Once the sand wave disappears and the flow regains its normal state, this water begins to flow again, causing an increase in the flow rate downstream. The change in discharge can cause a sand wave located further downstream to disappear. The accumulation of these effects along the flow causes a surge wave to form. This kind of surge occurred periodically during movable bed experiments conducted in a 50 m flume and it repeated itself every few minutes. In natural creeks, periodic surge waves propagating in the downstream direction frequently occur. Generally, this type of surge wave forms in a river with a steep slope and rapid flow or in an unstable hyper-concentrated flow. The details of the mechanism described herein are not yet entirely clear.

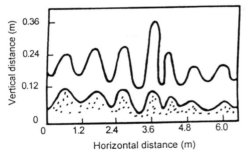

Fig. 6.12 Antidune on the verge of breaking
(after Simons, D.B., and E.V. Richardson)

Fig. 6.13 Schematic sketch of a breaking
surface wave (arrows showing the
direction of water particle movements)

The crest lines of sand waves in plan are not quite parallel to each other; in most cases, they are similar to short and wide sea waves in which the length and width are the same order of magnitude. Thus, they usually occupy only a part of the width of a river. The wandering reach of the lower Yellow River is quite wide, and sand waves that form along its course are generally either near the flood plain or at the confluences of channels. In transitional and wandering reaches of the Yellow River, especially in the straight sections downstream of bends, a special phenomenon, "Gan" occurs occasionally, and it is related to the formation of sand waves. For example, at the Tuchengzhi section of the lower Yellow River, whenever the water level is high, sand waves appear over the whole river cross section, (the width of the river is some 500 to 600 m), the height of the "Gan" corresponding to the sand wave is about 1 to 3 m, and its length is about 15 m. "Gan" generally occurs during the falling stage of a flood. Initially, the river may appear calm and be flowing along gently, then suddenly, a series of waves, usually numbering between 6 and 10, but sometimes up to 20, appear on the water surface. These surface waves develop and increase rapidly in size; then, after about 10 minutes, they gradually decay and disappear. Sometimes they make a sound like thunder as they break. The wave appears not to move, but with observations related to a reference point on the bank, one can see that it actually moves slowly upstream. Table 6.2 shows the flow conditions and the dimensions of the surface wave during the sand wave stage for some foreign rivers [10]. Because the formation of sand waves and their characteristics depend on the local velocity and depth, all of the data listed are not average values for an entire cross section. The published literature shows that sand waves usually occur during the rising stage of a flood.

Table 6.2 Flow conditions and the dimensions of surface wave (after Kennedy, J.F.)

River	Flow Depth (m)	Velocity (m/s)	Froude Number	Particle Size (mm)	Wave Length (m)
Little Colorado River	0.07	0.76	0.91	0.16	0.38
Fuji River	0.12	0.99	0.91	0.19	0.64
Pebble Gully	0.06	0.97	1.33	0.45	0.46
Pigeon Perch Creed	0.14	1.15	0.98	0.41	0.79
Pigeon Perch Creed	0.19	1.31	0.97	0.41	0.98
Dry Branch Gully	0.24	1.49	0.97	0.38	1.22
Dry Branch Gully	0.40	1.64	0.83	0.38	2.16
Dry Branch Gully	0.42	2.00	0.98	0.38	3.35
Kafawa Creek	0.91	2.43	0.81	0.46	3.04
Pigeon Perch Creed	0.95	1.98	0.65	0.41	4.57
Pigeon Perch Creed	1.21	2.34	0.68	0.41	4.27

6.1.5 Chutes and pools

If the velocity is higher than that for which sand waves form, the undulating bed resembles that of a mountain stream, with chutes and pools. The flow is supercritical at the chutes and subcritical in the pools. The transition from supercritical flow to subcritical flow is achieved through a hydraulic jump (Fig. 6.1h), and the entire bed form migrates slowly upstream. Severe erosion occurs at the chutes, and the sediment particles eroded from these regions are deposited in the pools. Deposition occurs even in the region of the hydraulic jump. In natural rivers on plains, the velocity is seldom high enough for this phenomenon to occur; the literature contains no mention of them in such locations.

From the foregoing discussion, the process of sand wave development vis-à-vis an increasing flow velocity can be divided into two distinct stages. Ripples and dunes form in the first stage and sand waves in the second. The transition between them is a flat bed stage or a bed with traces of dunes that are about to be washed out. In ordinary rivers, the most common bed features are ripples and dunes. Sand waves, chutes, and pools occur much less often.

In natural rivers, the process described above may not occur in a normal progression; various types of bed forms can exist at the same time, and the process of development may differ from one instance to another. Even in flume experiments, different bed forms can co-exist in different parts of a flume; showing that a strong relationship exists between the bed forms and the local turbulent structure. For certain combinations of flow and sediment, the initial flat bed can change directly into a dune or from a dune to an antidune without the usual transitional stages of ripples or a flat bed.

Another question that needs study is whether the process whereby bed forms develop is reversible. Fig. 6.1 shows the development of bed forms with a progressively increasing velocity. If, instead, the velocity gradually decreases, do the corresponding changes in bed form occur in reverse order--from the sand wave to flat, then dunes, then ripples, and finally back to the initial flat bed? Although few experimental data on this aspect have been presented, the bed form probably can develop in reverse order with decreasing velocity if enough time is available for each phase to develop. However, if the velocity decreases rapidly and bed forms have no

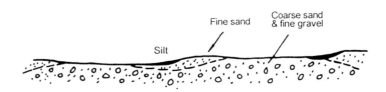

Fig. 6.14 Longitudinal profile of deposition on a dune

time to fully develop, the sequence of the stages shown in Fig. 6.1 can be disturbed. Fig. 6.14 shows the longitudinal profile of a dune in a stream after a flood. The top layer of the dune consists of fine sand, and the trough of the dune is covered by a layer of fine silt. Beneath the top layer of the dune is a layer of coarse sand and fine gravel with a symmetrical shape resembling that of a sand wave. The following situation can be imagined. The bed was composed of coarse sand and fine gravel, and sand waves formed during extreme flood events; particles smaller than fine sand were basically wash load and most of them would have been suspended and carried away in the water. After a flood, the velocity decreased and the coarse sand and fine gravel then no longer moved; if the flood velocity fell too rapidly for the corresponding bed form to develop, the original shape of the sand wave was maintained. At this time, some of the particles suspended in the water settled out, and as the sediment moved slowly on the surface of the "frozen" sand wave, dunes were formed. Again, the dune formation remained because the water level had decreased so rapidly. Finally, a layer of silt deposited in the trough of the dune.

In natural rivers, one often sees sand waves that formed during two different phases of development are superimposed, like the dune profile from the Mississippi River shown in Fig. 6.5. Furthermore, ripples on the upstream face of a dune can co-exist with the dune.

6.2 FLOW STRUCTURE AND SEDIMENT MOVEMENT ON THE SURFACE OF A DUNE

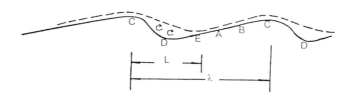

Fig. 6.15 Streamline near the surface of a dune

Fig. 6.15 shows a streamline near the surface of a dune. As the water flows over the dune, the streamline is not parallel to the dune surface. The water in the region near the bed flows along the adverse slope from A to the crest of the dune at C, where it separates. It meets the bed again at point E (called the point of reattachment). The distance L between the crest of the dune and point E is the length of the separation region. Experiments show that the ratio of this distance L to λ, the length of the dune, is a function of the Froude number, as shown in Fig. 6.16 [11]. For small Froude numbers, L/λ decreases sharply with increasing Fr, and that it approaches a constant value of about 0.32 if Fr is larger than about 0.2.

Fig. 6.17a shows the distributions of the pressure, velocity, and mean shear stress on the bed of a dune [12,13]. The flow separation affects the velocity profiles near the bed. The flow is accelerating on the front side and decelerating on the lee side. The velocity distribution at the crest, C, is almost uniform. The pressure attains its maximum value at the point of reattachment, E. The difference between the pressure on the front and lee side results in a force in the direction of flow. The reaction to this force is the form resistance of a dune (or dune resistance). On the one hand, the mean shear stress increases from zero (at point E) to a maximum at the crest (point C); on the other hand, the big difference in velocity between the upper and lower layers of the separation zone causes the flow there to be unstable and generates localized turbulent action. The intensity of turbulence is a maximum at point E; from there, it diffuses into the main flow region in such a way that the turbulence intensity near the bed becomes smaller and smaller with distance downstream.

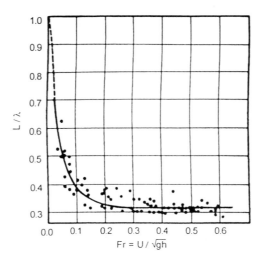

Fig. 6.16 Relationship between length of recirculating zone of a dune and flow conditions (after Karahan, M.E., and A.W. Peterson)

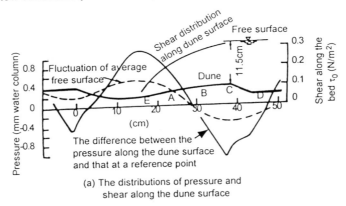

(a) The distributions of pressure and shear along the dune surface

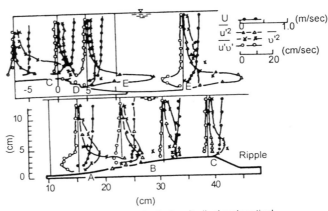

(b) Profiles of average velocity, longitudinal and vertical pulsing velocity and pulsing shear along the dune surface

Fig. 6.17 Flow pattern and dynamic condition on the bed of a dune (after Raudvivi, A.J.)

For the foregoing flow conditions, the characteristics of the sediment movement along the surface of a dune are as shown in Fig. 6.18. Under the combined effects of both the shear stress and the turbulence, sediment particles on the upstream side of a dune roll or jump as they move along the bed of the dune. Once these particles arrive

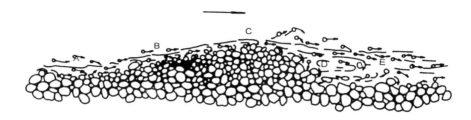

Fig. 6.18 Characteristics of sediment movement along the surface of a dune

at the crest of the dune, they fall into the eddy in the lee side and deposit there. Hence, the lee side grows downstream and causes the dune to move in the direction of the flow (Fig. 6.5). The upward fluctuating velocities caused by the separation near the crest are large enough to keep the finer particles in suspension, so that they do not deposit in the eddy of the lee side. Rather, they are transported to the next dune, where a portion of them deposit on the gentle front slope. Other particles move along the surface of the front side. The upward fluctuating velocities are not large enough to suspend the coarse particles moving near the bed. When they arrive at the crest, these particles are either captured by the eddy or roll down the slope, CD. In both cases, they deposit in the trough. As the dune moves forward, these coarse particles are covered by the dune and then exposed again after the dune has passed. A portion of the coarse particles on the exposed surface move up the dune, and the process is repeated. The movement of a coarse particle is clearly intermittent; also the "sorting" phenomenon described in Chapter 5 takes place so that the sizes of the particles in the upper layer differ significantly from those in the lower layer (Fig. 5.19). The slope of the lee side of the dune is slightly bigger than the angle of repose of the sediment as a result of the reverse flow in the eddy in the region of CE.

After a dune reaches a certain height, the size of the eddy behind the crest increases as the velocity increases. In this case, the sediment particles involved in the eddy no longer deposit on the lee side of the dune, but in a region some distance from the toe of the lee face. Jopling took photographs using a strong light to show the details of sediment movement in the eddy. He found that the distance between the central point of the sediment particles

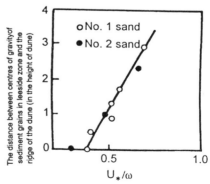

Fig. 6.19 Relationship between deposition site of sediment particles with flow and sediment characteristics (after Jopling, A.V.)

205

and the crest of the dune depends on a parameter, U_*/ω, as shown in Fig. 6.19 [14]. On the same movable bed, the bigger the friction velocity, the more is the quantity of sediment deposited beyond the toe of the lee slope and the less the amount of sediment that is deposited on the lee face. Furthermore, the increase of turbulence intensity in the separation zone (due to the increase in velocity) transports more of the sediment particles further downstream. These two factors cause the dune to disappear. In other words, while a dune grows, the factor causing it to disappear also develops. At a certain stage, the latter factor becomes dominant. The length of the dune gradually increases while its height gradually decreases; finally, the river bed becomes flat again.

6.3 MECHANISM OF FORMATION OF A SAND WAVE

A sand wave is a kind of bed configuration with a periodic and regular outline. It differs from a wind-induced surface wave, in that the latter can transfer mass. Both have their own characteristics. However, as an unstable phenomenon at the interface between the river bed and the flowing water, the sand wave has some characteristics that are similar to those of other natural waves at interfaces.

The occurrence and development of a sand wave as a bed configuration should be related to the flow conditions near the bed. The formation of a sand wave is probably associated with the stability of the viscous sublayer. Liu [4] first proposed this idea, but he did not explain the mechanism of the relationship in any detail. In fact, as mentioned in the foregoing, only the formation of ripples is associated with the stability of the viscous sublayer.

In addition, a sand wave has a sinusoidal profile. If it appears, the rate of sediment transport is already quite large. The sediment concentration near the bed is then much higher than that in the main flow region. Two layers with different densities form near the bed. The sand wave and the corresponding surface wave probably form because of unstable undulations at the interface between these layers. However, according to observations in the north-west region of China, the phenomenon of "Gan" (for which the water at the surface has a very high concentration) exists in some tributaries of the Yellow River during the flood reason. These streams are known all over the world for their high sediment concentrations. They can exceed a thousand kilograms per cubic meter after a torrential rain. The sediment concentration distribution is then fully uniform along both the vertical and the transverse directions. No two layers with different densities are present. Hence, it is not possible to explain the formation of sand waves as a mechanism of unstable undulation at the interface between density currents (treated in Chapter 14).

Another explanation that has been proposed is that the sand wave is caused by the surface wave in the following manner. A surface wave can cause velocity increases near the bed. The amplitude and the direction of these velocity increases are different at different positions near the sand wave. Near the trough, its direction is the same as the main flow, while at the wave peak, its direction is opposite to it. If the amplitude of the surface wave is large or the water depth small, the velocity caused by the surface wave is not negligible in comparison with that of the main flow, and the variable velocities near the bed in the direction of the flow thus created result in local scour and deposition. In the troughs, the bed will be scoured and sediment will be transported and deposited under the neighboring crest, thus forming a series of sand waves on the bed. Because bed forms formed in this way are synchronized with the water-surface waves, the two should be in phase, as already mentioned. This explanation would be far-fetched if applied to dune formation. In fact, dune formation can also be observed in closed duct flows that have no free surface. In addition, dunes often form in natural rivers at places where the water surface is as smooth as a mirror.

Typical concepts concerning the mechanism of sand wave formation are introduced in the following sections to illustrate various opinions about this phenomenon and their shortcomings.

6.3.1 Mechanism of ripple formation

Two types of explanation for the mechanism of ripple formation have been presented. One is based on the undulation of the viscous sublayer following a local disturbance, and the other on a balance of the forces acting on the bed. In addition, the formation of ripples induced by wind probably has still another mechanism, as discussed in detail in Chapter 15.

6.3.1.1 Undulation of the viscous sublayer induced by a local disturbance

Field and laboratory data have shown that ripples cannot form if the particles are quite large. The Laboratoire Nationae d'Hydraulique, Chatou, France [15] investigated the development of bedforms for grain Reynold numbers from 5 to 100. They found that ripples did not appear if this number was larger than 10. The experimental results agree with the data listed in Table 6.1. Thus, a typical ripple can be fully developed only in the hydraulically smooth or transitional region for which a viscous sublayer exists near the bed. Some connection may exist between the ripple formation and the stability of the viscous sublayer, but such a relationship has not yet been clearly demonstrated.

A bed absolutely flat is extremely difficult to maintain because an alluvial river bed consists of loose sediment particles of various sizes. Furthermore, turbulence causes the velocity to vary with time and place. For any location, if the turbulence intensity reaches a critical value, it can erode the bed and form a local scour hole. Just downstream of this location, the turbulence intensity may not be strong enough to carry away all the sediment coming from upstream, so that particles deposit. Therefore, for an alluvial river, if the flow intensity exceeds the threshold condition for sediment motion, a local undulation (i.e., a continuously varying bed) forms. The question now is how can a series of ripples developed from such a local hump.

If the river flow and its boundary are within the hydraulically smooth region, the thickness of the viscous sublayer along the bed is almost constant. A hump composed of sediment particles can form at some location on the bed because of an external action. The flow separates as it passes over the peak of the hump, and the turbulence intensity is higher within the separation zone. One consequence is that a more turbulent flow is transported downstream from the separation zone; it decreases the thickness of the viscous sublayer there and may even destroy it completely. This action increases the local shear stress and causes a local scour hole to form. At the downstream edge of this local scour hole, the bed slope forces the water to flow upwards. Consequently, the thickness of the viscous layer increases gradually and the bed shear decreases; then, sediment particles deposit and another hump forms. After the height of this hump reaches to a critical value, the flow separates again and interchanges between erosion and deposition continue as the process develops in the downstream direction. Finally, a series of local bed undulations have formed on the downstream bed through changes in the viscous sublayer thickness along the flow direction.

Exner [16] analyzed how undulations can develop into typical ripples with an asymmetrical geometry. He assumed the bed to be horizontal and the friction to be negligible. If ξ and η are vertical distances from a reference plane to the water surface and to the bed, respectively, the continuity equations of water and sediment can be written as:

$$Q = B\,(\xi - \eta)\,U = \text{constant} \tag{6.1}$$

$$\frac{\partial g_T}{\partial x} + \gamma_b \frac{\partial \eta}{\partial t} = 0 \tag{6.2}$$

in which

Q— discharge;

g_T — the rate of sediment transport per unit width by weight: if the deformation of the bed is only caused by the movement of sand wave, it will be the bed load transport rate;

γ_b — specific weight of sediment taking porosity into account;

B — average river width;

U— average velocity.

Exner also assumed that the rate of sediment transport is proportional to the velocity, and he combined the coefficient of proportionality and the constant γ_b in another coefficient α, and he rewrote Eq. (6.2) in the form

$$-\frac{\partial \eta}{\partial t} = \alpha \frac{\partial U}{\partial x} \tag{6.3}$$

From a combination of Eqs (6.1) and (6.3),

$$\frac{\partial \eta}{\partial t} = \frac{\alpha}{B} \frac{Q}{(\xi - \eta)^2} \frac{\partial(\xi - \eta)}{\partial x}$$

Generally, the undulation $\partial\xi/\partial x$ of the water surface is much smaller than that of the bed. If the undulation of the water surface is ignored, the above equation becomes

$$\frac{\partial \eta}{\partial t} = -\frac{\alpha Q}{B(\xi - \eta)^2} \frac{\partial \eta}{\partial x} = -\frac{M}{(\xi - \eta)^2} \frac{\partial \eta}{\partial x} \tag{6.4}$$

in which $M = \alpha\, Q/B$, and both M and ξ are constants. The solution of Eq. (6.4) can be expressed in the following form:

$$\xi - \eta = f\left[\frac{Mt}{(\xi - \eta)^2} - x\right] \tag{6.5}$$

in which f is an arbitrary function. The initial bed configuration at $t = 0$ can be approximated by a cosine curve, i.e.

$$\eta = A_0 + A_1 \cos kx \tag{6.6}$$

then, at any time t,

$$\eta = A_0 + A_1 \cos k[x - \frac{Mt}{(\xi - \eta)^2}] \tag{6.7}$$

Fig. 6.20 shows the bed configuration obtained from the above equation at various time intervals. It clearly describes how an undulation gradually develops into typical ripples as it moves downstream. Actually, the wave peaks cannot be maintained because the particles there slide down along the surface AB and deposit in the trough. Naturally, Exner's analysis is quite simplified. Neither the neglect of the friction force nor the assumption

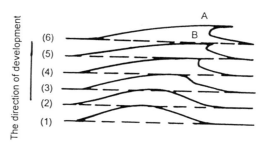

Fig. 6.20 Development of bed configuration into typical ripples (Exner's mathematical solution)

that the rate of sediment transport is directly proportional to the flow velocity are well founded. Modifications of Exner's work have been proposed [17], but these are not discussed herein.

6.3.1.2 Balance of forces acting on the bed

The presence of ripples on the bed increases the form drag. Because ripples are one of the stable bed patterns, their form drag should be included in the force balance equation. Bagnold [18] proposed an approach for ripple formation based on this concept.

The external force, on a unit area a distance y_o above the bed and in the flow direction, consists of two parts:

$$\gamma(h - y_o)J + (\gamma_s - \gamma)J \int_{y_o}^{h} S_v dy$$

The first term is the component of the weight of clear water in the flow direction, and the second is the component of the weight of the sediment, also in the flow direction. The reacting force on this unit area is also composed of two parts:

$$\tau' + (\gamma_s - \gamma) \tan\alpha \int_{y_o}^{h} S_{vb}\, dy$$

The first term is the shear stress in clear water, and the latter is the additional shear stress caused by the collision among the particles near the bed that is called the particle shear stress, as discussed in Chapter 5. In the two expressions:

h — water depth;

J — slope;

γ_s and γ — specific weight of sediment and water, respectively;

S_v and S_{vb} — total and bed load concentration at y, in volume;

$\tan\alpha$ — ratio of the particle shear stress and dispersive force at y_o. The last term is similar to the coefficient of dynamic friction and its value ranges between 0.32 and 0.75.

The extra force acting on a unit area that cannot be balanced by the particle shear stress is:

$$\tau_o' = \gamma h J + (\gamma_s - \gamma)J \int_0^h S_v dy - (\gamma_s - \gamma)\tan\alpha_o \int_0^h S_{vb} dy \tag{6.8}$$

in which the subscript "0" refers to the value at the bed. For a flat bed, the friction of the particles at rest on the bed should be equal to or greater than this extra force.

The friction among stationary particles on bed can be expressed as

$$D\, S_{vo}\, (\gamma_s - \gamma)\, f$$

in which

D — particle diameter;

S_{vo} — concentration of particles in the bed per unit volume;

f = coefficient of static friction.

Hence, the condition for the bed to remain flat is:

$$DS_{vo} f \geq \frac{\gamma}{\gamma_s - \gamma} h J + J \int_0^h S_v dy - \tan\alpha_o \int_0^h S_{vb} dy \tag{6.9}$$

If the above condition is not satisfied, then an unbalanced force acts on the flat bed, and the balance must be achieved by the form drag due to the ripples. Therefore, the condition for ripple formation is

$$DS_{vo}f \ < \ \frac{\gamma}{\gamma_s - \gamma}hJ + J\int_0^h S_v dy - \tan\alpha_o \int_0^h S_{vb}dy \qquad (6.10)$$

The form drag of ripples τ_0'' is given by

$$\tau_0'' = \gamma hJ + (\gamma_s - \gamma)J\int_0^h S_v dy - (\gamma_s - \gamma)\tan\alpha_o \int_0^h S_{vb}dy - (\gamma_s - \gamma)DS_{vo}f \quad (6.11)$$

From Eq. (6.9) or (6.10), the coarser the sediment particles, the easier it is for the bed to remain flat. This conclusion is supported by observations in the field. The mechanism of ripple formation can be likened to using an eraser to erase something on a piece of paper. If the paper is thin or the erasing force large, the paper folds and crumples easily. Conversely, if the paper is relatively thick or the force small, the paper more easily remains flat. In a similar way, if the bed material is coarser or the acting force of the flow is smaller, the bed of an alluvial river remains flat more easily; and on the contrary, if the bed is composed of fine sediment or the acting force of the flow is large, ripples form easily. Pratt [19] confirmed the concept proposed by Bagnold with flume experiments using uniform sands with a median diameter of 0.49 mm.

The mechanism of ripple formation is not as simple as the foregoing discussion would indicate. For example, local deposition on a bed may not necessarily develop into ripples. Sometimes it simply disappears as a result of the action of the flow. Also, one cannot readily explain why the above analysis can be used for the formation of ripples but not for that of dunes.

6.3.2 Mechanism of formation of dunes and sand waves

Two kinds of theories have also been proposed for the mechanism whereby dunes and sand waves form. In one of them, both dunes and sand waves are treated from the point of view of flow stability. In the other, dunes are presented as the result of disturbances due to turbulence and sand waves as the result of undulations of the water surface.

6.3.2.1 Stability analysis.

Many researchers have suggested that dune formation is a typical problem of stability, i.e. a flat bed is stable, whereas dunes or sand waves represent an instability of the flow. They hypothesized that undulations with small amplitudes form in the bed first, i.e. micro waves, and they then investigated the conditions that cause the micro waves to increase or decrease with time.

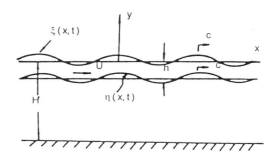

Fig. 6.21 Sketch used by Kennedy concerning sand wave formation using potential theory (after Kennedy, J.F.)

211

1. Kennedy's analysis [20,21] . Kennedy based an analysis on the assumptions that the flow is two-dimensional and non-viscous, the bed undulation has a sine curve configuration[1] and the bed wave amplitude is small enough compared with the wave length and the water depth that the analysis can be linearized. His analysis includes seven main steps:

(1) With reference to the coordinate system shown in Fig. 6.21, the undulation of the bed and the water surface can be expressed in the form:

$$\eta(x,t) = a(t)\sin k(x - ct) \tag{6.12}$$

$$\xi(x,t) = A(t)\sin k(x - ct) \tag{6.13}$$

in which a and A are the half amplitudes of the sand wave and the surface wave, respectively, c is the propagation speed of both waves, $k = 2\pi/\lambda$, and λ is the wave length.

(2) If the flow is a potential one, the water surface and the bed are streamlines. In this case, the flow is similar to open channel flow with velocity U over a horizontal boundary (the distance between the bed and water surface is H). An undulation, as described in Eq. (6.13), occurs at the free surface, and the bed is a streamline at a distance, h below the water surface, as expressed by Eq. (6.12). A potential function Φ is defined that follows the Laplace equation:

$$\frac{\partial^2 \Phi}{\partial x^2} + \frac{\partial^2 \Phi}{\partial y^2} = 0$$

The boundary conditions are: at the free surface, the velocity is parallel to the water surface and the pressure is constant, so that at the surface

$$\begin{cases} U\dfrac{\partial \xi}{\partial x} + \dfrac{\partial \xi}{\partial t} = \dfrac{\partial \Phi}{\partial y}, & \text{at } y = 0 \\[2ex] g\xi + U\dfrac{\partial \Phi}{\partial x} + \dfrac{\partial \Phi}{\partial t} = 0, & \text{at } y = 0 \end{cases}$$

and, at the bed, the velocity normal to the bed is zero. And

$$U\frac{\partial \eta}{\partial x} + \frac{\partial \eta}{\partial t} = \frac{\partial \Phi}{\partial y}, \qquad \text{at } y = -h$$

---

[1] Antidune can be indeed regarded as sine curve. For dunes, Kennedy suggested that the streamline near the bed can be roughly assumed to be a sine curve if the separation region on the downstream side of the dune peak is included

The potential function satisfying these conditions for $c << U$ can be expressed as

$$\Phi = Ux + UA(t)\frac{\cosh k(y+H)}{\sinh kH}\cos k(x-ct) \qquad (6.14)$$

the ratio of the amplitude of the sand wave to that of the water surface is,

$$\frac{a(t)}{A(t)} = \frac{\sinh k(H-h)}{\sinh kH} \qquad (6.15)$$

(3) The continuity equation for the sediment transport is given by Eq. (6.2).

(4) The sediment transport capacity is assumed to be proportional to the velocity near the bed at a distance δ upstream of the cross section raised to the power n, i.e.

$$g_T(x,t) = m[\frac{\partial \Phi(x-\delta,-h,t)}{\partial x}]^n \qquad (6.16)$$

in which the values of m, n and δ depend on the velocity, the flow depth and the characteristics of both the fluid and the solid particles.

In Eq. (6.16), Kennedy assumed that the change in the sediment transport capacity delays the changes in the bed velocity so that there is a time lag between them. The delay occurs because, once the bed velocity changes, all the other parameters, like the shear stress, velocity at various vertical distances, turbulence intensity, and parameters related to the sediment capacity, can only react to the consequences of this change after a certain distance of travel. Furthermore, the change in sediment concentration as the result of subsequent erosion or deposition also needs to move a certain distance before the sediment transport capacity can adjust. The distance δ plays an important role in this theory.

(5) If the velocity and sediment transport capacity obtained from Eqs. (6.14) and (6.16) are substituted into the sediment continuity equation, Eq. (6.2), the propagation speed and amplitude of the sand wave can be obtained:

$$c = -(n\overline{g_T}k / \gamma_b)\coth k(H-h)\cos k\delta \qquad (6.17)$$

$$a(t) = A(0)\frac{\sinh k(H-h)}{\sinh kH}\exp[t\frac{n\overline{g_T}k^2}{\gamma_b}\coth k(H-h)\sin k\delta] \qquad (6.18)$$

in which $\overline{g_T}$ = the average rate of sediment transport per unit width, and it can be expressed as mU^n.

(6) According to the general method for solving such a stability problem, the length of the wave that grows the fastest initially becomes that of the final wave. Thus,

$$\frac{d(da/dt)_{t=0}}{dk} = 0$$

and it can be expressed as a form of the Froude number

$$Fr^2 = \frac{U^2}{gh} = \frac{1 + kh\tanh kh + jkh\cot jkh}{(kh)^2 + (2 + jkh\cot jkh)kh\tanh kh} \qquad (6.19)$$

in which

$$j = \delta/h \qquad (6.20)$$

The corresponding wave propagation speed is

$$c = \frac{n\bar{g}_T k}{2\gamma_b}[\frac{\sinh 2kh + 2kh}{\sinh^2 kh - jkh\cot jkh - 1}]\cos jkh \qquad (6.21)$$

From Eqs. (6.19) and (6.21), one can show that the sand wave length depends primarily on the Froude number, the average water depth and the characteristic distance δ, the time lag between the sediment transport capacity, and the changes in the bed velocity. Also, the speed of propagation of the sand wave depends not only on h and δ, but also on the sediment discharge per unit width. If the water depth is much larger than δ, j approaches zero, and then, Eqs. (6.19) and (6.21) can be rewritten as

$$Fr^2 = \frac{2 + kh\tanh kh}{(kh)^2 + 3kh\tanh kh} \qquad (6.22)$$

$$c = \frac{n\bar{g}_T k}{2\gamma_b}[\frac{\sinh 2kh + 2kh}{\sinh^2 kh - 2}] \qquad (6.23)$$

(7) The criterion for judging the bedform depends on the following three conditions:

(i) Eq. (6.15) indicates whether the water surface and sand wave are in phase: If $h < H$, they are in phase and sand wave forms. If $h > H$ they are out of phase by 180° and dunes form.

(ii) Eq. (6.18) determines whether the amplitude of the sand wave increases, or decreases and becomes a flat bed;

(iii) Eq. (6.21) determines the direction the sand wave propagates.

The conditions for different sand wave patterns to form are presented in Table 6.3, and they are based on these three criteria. Fig 6.22 shows the water surface and the longitudinal distributions of erosion, deposition, and bed velocity for the four possible kinds of bedforms (antidunes, sand waves moving in flow direction, flat bed and dunes), at two successive times, t and $t + \Delta t$. The bed pattern in Fig. 6.22c is unstable. Because erosion at the peak of a sand wave and the deposition in the trough decrease the amplitude of the bed undulation, the bed finally becomes flat. The figure shows, for a given sand wave length, how the δ-value affects the pattern of the bedform.

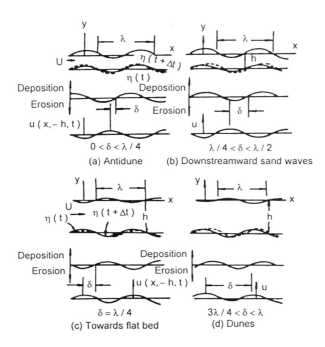

Fig. 6.22 Four types of bed patterns and their phase differences

For each value of j, the values of kh and Fr that correspond to the various bed patterns can be obtained from Eq. (6.19), and are as shown in Fig. 6.23. The relationship plotted in the figure

$$Fr^2 = \frac{\tanh(kh)}{kh}$$

is the line of demarcation between sand waves and dunes (above ABC is the sand waves region, the region below BCD is the dune region). In order to distinguish between the two sand wave patterns, it is easier to use the relationship between Fr and j shown in Fig. 6.24. The flat bed can be converted into an sand wave only if the Froude number is larger than 0.84. If the Froude number is larger than 1, a flat bed cannot exist.

Table 6.3 Conditions for formation of bed configurations

No	Sand wave and surface wave	H-h	$k\delta$	$\sin k\delta$	$\cos k\delta$	Direction of dune propagation	Sand wave pattern
1		+	$0< k\delta<\pi/2$	+	+	upstream	
2	in phase	+	$\pi/2$	+	0	standing wave	sand wave
3		+	$-\pi/2< k\delta<\pi$	+	-	downstream	
4a		-	$\pi< k\delta<1.5\pi$	-	+		
4b	flat	-	$0< k\delta<\pi$	+			flat
4c		+	$\pi< k\delta<2\pi$	-			
5	out of phase	-	$1.5\pi< k\delta<2\pi$	-	+	downstream	dune

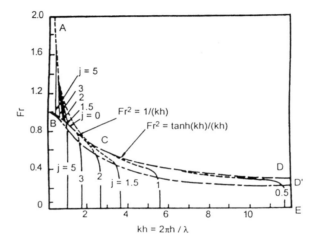

Fig. 6.23 Parameters characterizing bed patterns and sand wave lengths (Eq. 6.19)

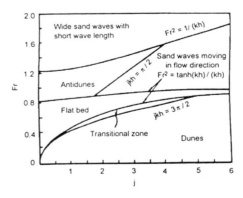

Fig. 6.24 Conditions for formation of bed configurations (theoretical solution)

216

From Airy's Equation

$$kU^2 / g = \tanh kH$$

Kennedy found that the minimum wave length corresponds to the limiting condition that H approaches infinity,

$$\lambda_{min} = 2\pi U^2 / g \qquad (6.24)$$

the critical condition for a 2-D sand wave is

$$Fr^2 = \frac{1}{kh} \qquad (6.25)$$

Only waves that are short and wide can form if the Froude number is larger than the value given by this critical condition. Reynolds [17] modified Kennedy's analysis and proposed a different critical condition:

$$Fr^2 = \frac{\coth kh}{kh} \qquad (6.26)$$

a relationship that can be obtained by substituting $j = \delta/h$ into Eq. (6.19). This result indicates that the 2-D sand wave reaches its limit if $\delta = \lambda$. Reynolds also studied 3-D sand waves. He found that both the 3-D dune and the 3-D sand wave can exist if the Froude number is larger than the critical value for a 2-D sand wave determined from potential theory. If h/λ is small, the results based on 2-D potential theory are close to those for 3-D flow; however, the difference between them increases if h/λ increases [17].

2. *Determination of the phase difference δ.* In Kennedy's analysis, the bed configurations mainly depend on the phase difference δ and the corresponding dimensionless parameter j, as shown in Eq. (6.20). Kennedy gave a physical meaning to j. By the introduction of the parameter j, a single curve Fr -kh is turned into a family of curves, mathematically such a treatment is in a sense arbitrary. If the parameter j, or δ, cannot be predicted, Kennedy's method cannot be used to determine the sand wave patterns for various flows and boundary conditions. Many investigations have been carried out in order to determine the phase difference δ.

In the Kennedy method, the phase difference δ ranges from 0 to λ, as shown in Fig. 6.22. The distance lag between the sediment transport rate and the change of the bed velocity can thus be about the same as the sand wave length. Physically, this result is difficult to accept. Hayashi modified the Kennedy method by introducing the effect of the bedform on the sediment transport rate. He used $n = 4$ for the exponent in Eq.

(6.16). He further assumed that, besides the local bed velocity, the local slope of the sand wave also influences the sediment transport capacity. That is,

$$g_T(x,t) = m[1 + \alpha \frac{\partial \eta (x - \delta, t)}{\partial x}][\frac{\partial \Phi (x - \delta, -h, t)}{\partial x}]^4 \qquad (6.27)$$

in which α is a coefficient. After this modification, the bed configuration ultimately depends on the Froude number and the dimensionless parameter Y,

$$Y = (\alpha / \delta)(U^2 / 2g) \qquad (6.28)$$

The range of Y is from 1.5 to 3, but generally, $Y = 2$ can be used.

Hayashi's modification makes Kennedy's method physically more realistic, but it still did not provide a method for predicting the phase difference. Yalin [23] noted that each constant j along the curve of OBD'E in Fig. 6.23 corresponds to a straight line parallel to the Fr-axis. This region accounts for 80% of the total dune region OBDE, within which $2\pi h/\lambda$ is a function only of j and is therefore independent of Fr; i.e.

$$\frac{2\pi h}{\lambda} = f_1(j) \qquad (6.29)$$

In contrast, in a dimensional analysis, Yalin found that the length of a dune is a function of the grain Reynolds number, Re_* and the dimensionless parameter $Z = h/D$,

$$\frac{\lambda}{D} = f_2(Re_*, Z) \qquad (6.30)$$

in which $Re_* = U_* D/v$. Eliminating λ from Eqs. (6.29) and (6.30), one obtains

$$f_1(j) = \frac{2\pi Z}{f_2(Re_*, Z)}$$

so that

$$j = f_3 (Re_*, Z) \qquad (6.31)$$

Fig. 6.25 is a plot of this relationship [23]. The dune length can be estimated from it and Fig. 6.23.

According to Parker [24], the reason that the sediment transport delays the changes in the flow conditions is the inertia of the sediment particles. The inertia effect can be expressed by a dimensionless parameter that he called the inertia parameter. For the sand wave, a relationship occurs between the inertia parameter and the phase difference. The details presented in his paper are not discussed herein.

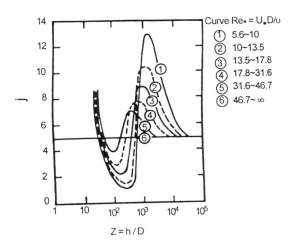

Fig. 6.25 Relationship for parameter j as function of Re_* and Z (after Yalin, M.S.)

3. Relationship between the modes of sediment transport and the phase difference. As mentioned in the foregoing, the local sediment discharge and the local bed velocity are not in phase; the difference is a distance lag or phase lag. The mechanism for this lag is different for bed load movement and for suspended load movement [25, 26].

If a bedform with a small amplitude appears, the flow contracts and accelerates locally in the region from trough to peak, and it decelerates and diffuses in the region from peak to trough. In terms of the local average value, the velocity is a maximum at the peak of the bedform.

As discussed in Chapter 5, the rate of bed-load transport depends on flow conditions near the bed. Fig. 6.17 shows that the combination of bed shear stress and turbulence intensity reaches a maximum upstream of the peak--not at the peak. Hence, the maximum of the bed-load transport rate occurs upstream of the point of maximum velocity. In contrast, because suspended material occurs not only near the bed but extends over the entire depth, the time for bed load to be picked up or for suspended load to deposit onto the bed is longer. Thus, the maximum concentration of suspended material generally occurs downstream of the point of maximum velocity.

At the beginning of dune formation, the rate of sediment transport is not large, and most of the sediment moves as bed-load. The phase lag between the bed pattern and shear stress combined with turbulence is important. In the stage of sand waves, the sediment transport rate is large and suspended load is dominant, so that bed load can be ignored; consequently, the phase difference between the sediment transport rate and

219

shear stress plays an important role. In the stage of a dune developing into a flat bed, both bed load and suspended load are important, and the corresponding phase lags have different signs. Thus, the transition from dune to flat bed depends strongly on the ratio of the transport rates of bed load and suspended load.

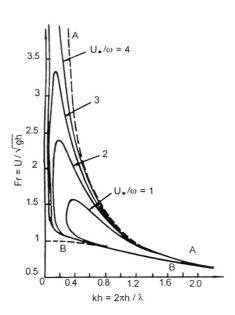

Fig.6.26 Bounds for the stable regions with dominant suspended load, $U/U_* = 17$ (after Engelund, F.)

From this physical picture, both the shear stress distribution along the sand wave and the action of turbulence must be included as parts of the problem of stability. However, it is difficult to include them in the traditional potential theory. Engelund [27] used the vortex transport equation to describe the real flow passing over the wavy boundary and described turbulent by using the viscous flow model. The shear stress between different layers of the flow depends on the product of the velocity gradient and eddy viscosity [as shown in Eq. (4.22)]. This coefficient was taken to be a constant in order to simplify the analysis. After a series of mathematical manipulations, Engelund found that the formation of sand waves for sediment moving mainly as suspended load depends on three parameters:

$$\left\{ \begin{array}{l} Fr = U / \sqrt{gh} \\ U_* / \omega \\ U / U_* \end{array} \right.$$

the effect of U/U_* is generally small. Fig. 6.26 shows the criterion for sand wave formation for $U/U_* = 17$. Within the region surrounded by a curve corresponding to a given value of U_*/ω, a sand wave can form, whereas the bed is stable outside this region. The equations for the upper and lower bounds, in Fig. 6.26 are as follows:

for AA,

$$Fr^2 = \frac{\coth(kh)}{kh} \tag{6.32}$$

and for BB,

$$Fr^2 = \frac{\tanh(kh)}{kh} \qquad (6.33)$$

These two equations can also be obtained from the potential flow theory [28]; Eq. (6.33) is also the line of demarcation between sand waves and dunes in Kennedy's work, as shown in Fig. 6.23; however, the physical meanings in the two cases are different.

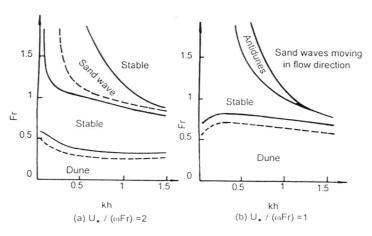

Fig. 6.27 Bounds for the stable region with both suspended and bed load, $U/U* = 21$

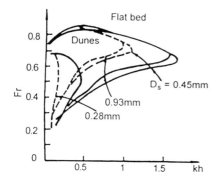

Fig. 6.28 Bounds for stable region in ($h = 0.2$ m, $T=20$ °C) (after Engelund, F.)

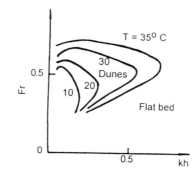

Fig 6.29 Bounds for stable dune which dune forms region at various temperatures ($h=0.01$ m, $D_s =0.2$ mm) (after Engelund, F)

If bed load movement is also taken into the consideration, another unstable region occurs for low Froude numbers; within it a dune will form [27]. Figs. 6.27a and b show the critical bounds of the stable regions, with $U/U_* = 21$, for two values of $U_*/(F_r\,\omega) = 2$ and 1, respectively. The dashed line represents the wave number for which the incremental rate is the greatest at a given Froude number. A comparison of Figs. 6.27a and b shows that the critical line between the stable and unstable regions is higher for coarser particles. The upper bound of the sand wave region that is shown in Fig. 6.27a is outside the area shown in Fig. 6.27b; the sand wave does not become stable and restore the flat bed unless the Froude number reaches 3.3.

In 1974, Engelund and Fredsфe [25] conducted another study of the condition for dune formation, and their results in Fig. 6.28 show various regions of the process for three sediment diameters under the conditions of $h = 0.2$ m, $T = 20^\circ$C. The figure shows that the finer the particles are, the greater the length of the dune wave and the smaller the critical Froude number at which the dune disappears. Fig. 6.29 shows the relationship between the dune formation and the temperature for $h = 0.1$ m and $D_s = 0.2$ mm. For a lower temperature, the viscosity of the water is higher and the fall velocity of the sediment lower. Also, the suspended load discharge and the ratio of the suspended load to bed load are larger. Finally, the transition from dune to flat bed is affected. Fig. 6.29 shows that a dune disappears as the dune length increases with decreasing temperature, and the critical Froude number for a dune to disappear and become a flat bed decreases. In some cases, even if the flow coming from upstream does not change, the dune can become flatter and the resistance to flow can decrease significantly with a decrease in temperature. Fig. 6.30 shows eight bed profiles for the Missouri River recorded between Sept. 8 and Nov. 8, 1966. The discharge, water depth and sediment size did not vary much, only the temperature changed; it decreased from 25°C to 5°C during this period. The corresponding change in the bed profile corresponds well with the theoretical results shown in Fig. 6.29. This phenomenon is discussed in more detail in Chapter 7.

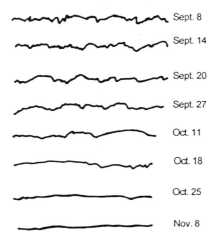

Fig. 6.30 Seasonal variation of bed configuration in Missouri River, Autumn ,1966 (after Yalin, M.S.)

6.3.2.2 Yalin's analysis

Yalin[29] stated that a clear distinction should be made between the mechanisms of formation for sand waves and dunes. A sand wave can form only if a free surface is present so that its formation should be related to the Froude number. In contrast, a dune can also form in a closed duct, and the Froude number should therefore not be a dominant factor. Thus, it is not productive to treat the mechanisms of the two processes together as one problem of stability.

1. Water surface waves are the causative factor in the formation of sand waves. Both laboratory and field data show that sand waves occur only in open channels. Since surface waves can exist for flow over a rigid bed, but the sand wave cannot form without a free surface, sand waves are the result from surface waves. Once sand waves are formed, they interact with the surface waves, too. Fig. 6.31 shows schematically how a surface wave induces a sand wave. As already mentioned, because of the non-uniformity of the sediment and the characteristics of turbulence, some undulations are always present on the bed of an alluvial river, and local discontinuities like a sand bump can form. In Fig. 6.31a, the interface between the rigid and movable bed is discontinuous. As the flow passes the discontinuity, the difference between the friction of the rigid bed upstream and that of the mobile bed produces a series of surface waves if the Froude number exceeds some critical value. The amplitude of these waves gradually decreases in the downstream direction. The water depth in the troughs, sections 1, 3, 5, ... is smaller than the average depth, and that under the crests, sections 2, 4, ... is larger than the average. Consequently, the average velocity, the shear stress, and the sediment transporting capacity at sections 1, 3, 5.., are larger than those at sections 2, 4,... Initially, the distribution of sediment transporting capacity, $g_T(0, x)$, and the variation of sediment transporting capacity along the flow direction, represented with $\partial g_T/\partial x \big|_{t=0}$, are as shown in Fig. 6.31a. From the sediment continuity equation, Eq. (6.2), the variation of sediment transport capacity causes local erosion and deposition, and the flat bed gradually deforms into an undulating bed, Fig. 6.31b. The wave length of the bedform is the same as that of the surface wave.

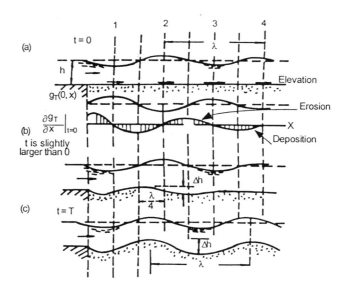

Fig. 6.31 Development of bed configuration induced by water surface (after Yalin, M.S.)

After sand waves form, their amplitude increases gradually with time, and they can affect the surface wave in such a way as to cause it to propagate downstream. In other words, the interaction between the sand wave and surface wave causes both to develop, in both the x and y directions. The development in the x direction has no limit, but that in the y direction is limited by the maximum slope of a sand wave. Initially, the phase difference between the surface wave and bed wave is $\lambda/4$. This phase difference approaches zero once the sand wave develops, as shown in Fig. 6.31c.

The foregoing mechanism of the formation of sand wave can be summarized in chart form:

discontinuous bed surface → wave in water surface → sand wave

A weakness of the above analysis is that experiments have shown that sand waves are still present even if the Froude number is as large as 2, whereas the surface wave is evident only if the Froude number is about 1. Although Yalin stated that surface waves can occur for larger Froude numbers, their amplitude is too small to be perceived. No explanation has shown how such small waves could induce the much larger sand waves that often occur.

2 Disturbances caused by large-scale turbulent structure produce dunes. As already discussed, the ratio of wave length to sediment diameter for a dune is a function of the grain Reynolds number ($Re_* = U_*D/v$) and the relative roughness for the bed ($Z = h/D$), as indicated by Eq. (6.30). This equation can be rewritten in the form

$$\lambda / h = f(\mathrm{Re_*}, Z) \qquad (6.34)$$

In flow over a hydraulically rough boundary, the water depth is much larger than the particle diameter, so that the effects of Re_* and Z are negligible; the wave length of a dune is then directly proportional to the water depth. The following discussion shows how this equation is related to the structure of large-scale turbulence in open channel flow.

Velikanov was the first to point out that many shape characteristics of alluvial rivers, including that of dune formation, are related to the structure of large-scale turbulence. The scale of the large eddies is the same order as the water depth. The flow characteristics fluctuate in time and space because of turbulence. For a flat bed, the stochastic procedures do not vary with either t or x; i.e., for x and t, the stochastic characteristics are the same.

If a flow characteristic like the average velocity is disturbed at $x = 0$, this disturbance will be effective within a distance ξ in the x direction because of the large-scale eddies. The ratio of ξ to h is approximately one. Because of variations in the size of eddies, the distance ξ also varies somewhat. In terms of the frequency f, defined as

$$f = 1/\xi$$

the disturbance in the average velocity has a certain frequency spectrum $S(f)$, like that shown in Fig. 6.32a, where the peak in the frequency spectrum corresponds to $f_0 = 1/h$. The autocorrelation function of the disturbance with such a frequency spectrum distribution can be written as

$$K(x) = e^{-ax} \cos(f_o x) \qquad (6.35)$$

or, approximately

$$K(x) \approx e^{-ax} \cos (x/h) \qquad (6.36)$$

Fig. 6.32b shows the variation of $K(x)$ with x. It indicates that the mean velocities have a positive correlation at all the even cross sections , with an interval distance of $\pi/f_o \approx \pi h$, and a negative one at the odd cross sections. Hence, if the mean velocity at $x = 0$ increases suddenly due to some action, the mean velocity at $x = 2\pi h$, $4\pi h$.. increases also, but the mean velocity at $x = \pi h$, $3\pi h$ ... decreases. The term e^{-ax} (a is positive) in Eqs. (6.35) and (6.36) cause a decrease with the distance x.

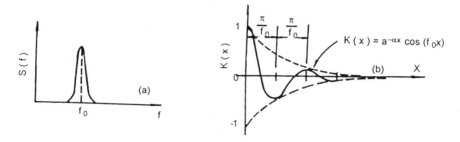

Fig. 6.32 Frequency spectrum of disturbed wave and effect of
large-scale eddies on flow downstream (after Yalin, M.S.)

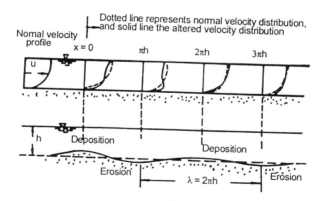

Fig. 6.33 Disturbed velocity distribution and its effect on flow
and bed configuration downstream (after Yalin, M.S.)

If the flow characteristic in question is the vertical velocity profile, instead of the average velocity, and if the bed is not continuous (if it has a local hump at $x = 0$), the velocity profile at $x = 0$ would change. The large-scale turbulent structures would then cause a sequence of changes in the vertical velocity profiles at all cross sections at an interval distance πh away, as shown in Fig. 6.33. The increase in the velocity gradient near the bed increases the shear stress and the sediment transport rate, and the decrease in velocity gradient near the bed decreases the shear stress and the transport rate; thus the sediment deposition at $x = 0$ leads to a local scour at $x \approx \pi h$, $3\pi h$... and local deposition at $x \approx 2\pi h$, $4\pi h$.... The flat bed then forms a series of undulations with the wave length

$$\lambda \approx 2\pi h \qquad\qquad (6.37)$$

Many of the observed data indicate that the dune wave length is indeed within the range of 5 to 7 times the water depth if the value of the grain Reynolds number and h/D are both large.

The above mode of dune formation can be summarized in a flow chart:

discontinuity of bed surface $\rightarrow$ disturbance of turbulent structure $\rightarrow$ dune

6.4 CRITERIA FOR PREDICTING BED CONFIGURATION

Prediction of the occurrence of the various possible bed configurations on the basis of the flow conditions and sediment characteristics is an important topic for research. In fact, such a study should lead to a better understanding of the mechanism of bed form development. Once the conditions for the formation of the various bed configurations are delineated, criteria for predicting the appearance of specific types of bed form can be established. Since no complete theoretical solution yet exists that enables one to describe the entire range of bed form phenomena, empirical relationships have frequently been used to delineate the conditions under which the various types of bed form appear. In the following discussion, the parameters employed in such empirical relations are the same as those used in the preceding discussion of the mechanism of bed form changes.

As stated in the preceding section, the appearance of ripples depends primarily on the grain Reynolds number

$$Re_* = U_* D / \nu$$

whereas the formation of sand waves is related to the Froude number

$$Fr = U / \sqrt{gh}$$

At the stage of dune formation, more parameters are involved; some of the ones commonly used by different researchers are

$$U_* D / \nu, \ h/D, \ Fr, \ U_* / \omega, \ U/U_*$$

The following sections examine the various types of empirical relationships that have been proposed concerning bed form phenomena.

6.4.1 Criteria for regions of flat bed, ripples, and dunes

Shields was the first to propose a demarcation of bed forms for regimes at low flows. Fig. 6.34, known as the Shields diagram for the initiation of sediment motion[30], indicates the various types of bed form on a plot of $\tau_o/(\gamma_s-\gamma)D$ versus U_*D/v. Fig. 6.35 [15] is a clearer demarcation of the bed forms based on the experimental data collected more recently at the Laboratoire Nationae d'Hydraulique, Chatou, France. It shows that for Re_* between 10 and 20, ripples can co-exist with dunes. In this case, the ripples can form on the upstream slope of dunes.

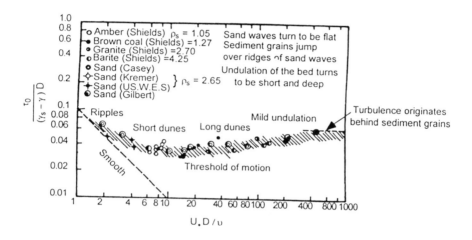

Fig. 6.34 Threshold of sediment transport and demarcation of bedforms (after Shields, A.)

As mentioned later, $\tau_o/(\gamma_s-\gamma)D$ is an important flow parameter, its magnitude is directly related to the intensity of bed load transport. It is usually represented by the symbol Θ,

$$\Theta = \frac{\tau_o}{(\gamma_s - \gamma)D} \qquad (6.38)$$

Another important quantity, $U_* = \sqrt{\tau_o/\rho}$, is known as the shear velocity, The relationship between the particle fall velocity ω and the grain diameter D is as follows (Chapter 4)

$$\omega^2 = \frac{4}{3C_D} \frac{(\gamma_s - \gamma)gD}{\gamma}$$

227

in which C_D — drag coefficient,

$$C_D = f_1(\frac{\omega D}{v}) = f_1(\frac{U_* D}{v} \frac{\omega}{U_*})$$

Consequently,

$$\frac{\tau_o}{(\gamma_s - \gamma)D} = \frac{4}{3C_D}(\frac{U_*}{\omega})^2 = f_2(\frac{U_* D}{v}, \frac{U_*}{\omega}) \qquad (6.39)$$

which is the relationship shown in Figs. 6.34 and 6.35. It also represents the relationship between U_*D/v and U_*/ω. In fact, these two parameters were used by Liu to study the conditions for ripple formation (Fig. 6.4). Albertson et al.[31] extended this method to judge other bed forms, and they proposed the criteria shown in Fig. 6.36.

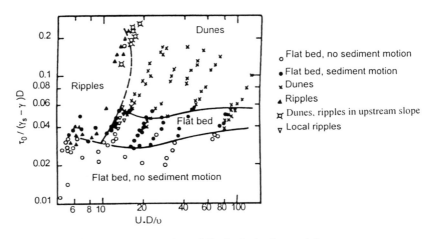

Fig. 6.35 Demarcation of flat bed, ripples, and dunes

An important factor is that the data used in Fig. 6.36 are primarily from laboratory flumes; the ranges for the parameters involved are accordingly limited. For the upper flow regime, for which the bed becomes flat again or sand waves form, the use of the grain Reynolds number as a characteristic parameter is difficult to justify on theoretical grounds.

The parameter, Θ, can be expressed in other forms, for example,

$$\Theta = \frac{\tau_o}{(\gamma_s - \gamma)D} = \frac{\gamma}{\gamma_s - \gamma}\frac{U_*^2}{gD} = \frac{\gamma}{\gamma_s - \gamma}\frac{(U_* D/v)^2}{gD^3/v^2} \qquad (6.40)$$

For natural sediment, one can distinguish the various bed forms using the grain Reynolds number and gD^3/v^2, as shown in Fig. 6.37[32]. The curve delimiting the formation of ripples intersects the Shields curve for the initiation for the value

$$gD^3/v^2 = 3,900$$

If gD^3/v^2 is greater than 3900, a flat bed will transform directly into a dune, bypassing the ripple phase. If $v = 0.01$ cm²/s, the corresponding critical grain diameter is 0.7 mm. This value agrees well with the experimental finding already discussed.

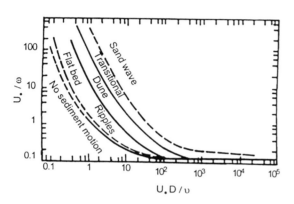

Fig. 6.36 Criteria for prediction of bed form   (after Albertson, M.L., D.B. Simons, and E.V. Richardson)

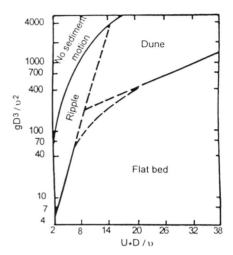

Fig. 6.37 Criteria for prediction of flat bed-ripples-dunes (after Hill et al.)

In summary, for the phases going from flat bed to ripples, and then to dunes, the main parameters for characterizing the bed forms are the grain Reynolds number and the flow intensity Θ. Other dimensionless parameters, although presented originally in various forms, can all be transformed into these two characteristic parameters.

6.4.2 Criteria for prediction of dunes, flat bed, and sand waves

For the upper flow regimes, the Froude number of the flow plays a significant role in determining the bed forms. In addition, the flow intensity Θ is also significant, as shown in Fig. 6.38 [33]. Some plots, following Yalin, use the equivalent of h/D as a representative parameter, such as R/D_{50} in Fig. 6.39 [34].

From the foregoing analysis, one sees that the different phases of bed form development depend on some of the following parameters:

$$U*D/v, \ \tau_0/(\gamma_s-\gamma)D \ (\text{or} \ U*/\omega), \ h/D, \ Fr$$

These parameters are, in fact, similar to those introduced in Section 6.3 in which the mechanism of bed-form formation was discussed. The results from Sections 6.3 and 6.4 are interrelated in the sense that they represent two different aspects of the same phenomenon.

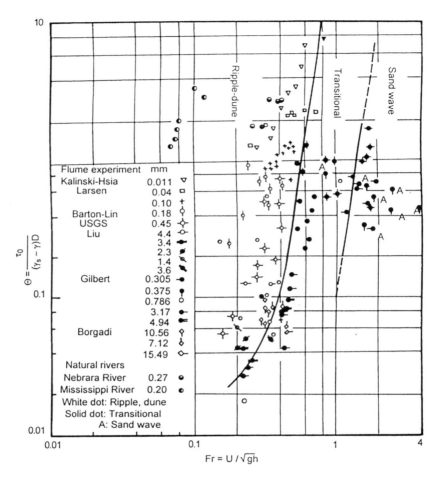

Fig. 6.38 Criteria for prediction of dune-flat bed-sand wave (after Garde, R.J., and M.L. Albertson)

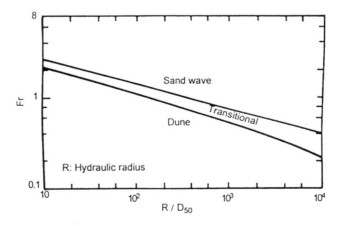

Fig. 6.39 Criteria for prediction of dunes-flat bed-sand wave (after Athaullah, M.)

Prior to the conclusion of this section, mention should be made of the point that the flow intensity Θ, used to evaluate bed load transport, is also one of the important parameters for determining the movement of sand waves. In current discussions of the sediment carrying capacity of flow, some researchers argue that the stream power supplied by the flow per unit time and per unit bed area, $\tau_o U$, is an important parameter. If so, a relationship should exist between $\tau_o U$ and the bed configuration. An illustration of this deduction is shown in Fig. 6.40 [2].

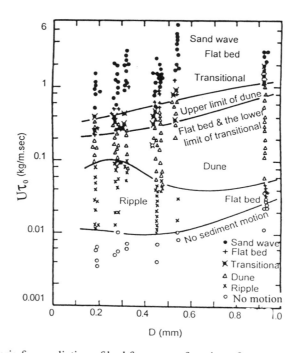

Fig. 6.40 Criteria for prediction of bed forms as a function of stream power

6.5 GEOMETRIC AND STATISTICAL CHARACTERISTICS OF BED FORMS

6.5.1 Statistical properties of bed forms [35,36]

As Fig. 6.5 shows, the geometric characteristics of bed forms are highly irregular. Serious limitations arise if only the average wave length and average wave amplitude are used to describe this complicated natural phenomenon. In fact, the geometrical characteristics of bed forms are stochastic in nature, and they should, therefore, be described statistically.

Fig. 6.41 is a longitudinal section of a bed form that has formed on a river bed. If the x-axis passes through the average level of a river bed, in the streamwise direction, and η is the height above the bed with mean value $\overline{\eta} = 0$, the longitudinal bed

231

configuration can be indicated by the function, $\eta = \eta\,(x,\,t)$. The function $\eta = \eta\,(t)$ can be obtained by recording the height of the bed level as it changes with time at a fixed location $x = x_o$. Similarly, one can record the height of the bed along the stream direction to establish the function $\eta = \eta\,(x)$ at any instant $t = t_o$. If the rate of flow and sediment discharge coming from upstream are constant, and the bed material does not change with time or distance, the function $\eta = \eta\,(x,\,t)$ does not vary with the location of the origin.

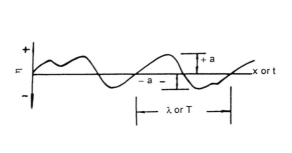

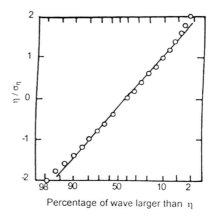

Percentage of wave larger than η

Fig. 6.41 Longitudinal profile of a bed form on a river bed (after Nordin, C.F., Jr)

Fig. 6.42 Cumulative frequency curve of bed level distribution (after Nordin, C.F., Jr)

Fig. 6.42 shows the cumulative frequency curve of the observed bed level η in a straight canal plotted on probability graph paper. Most of the data points can be approximated by a straight line, which indicates that η can be described by the normal distribution function. Of course, this model can only be an approximation. Because the water depth is finite, the range within which the bed level can vary is also finite. The slope of the upstream face of a dune is usually rather gentle, and that of the downstream face is steeper, close to the angle of repose of sand in water. As a result, the sand wave configuration is more regular than it would be if normally distributed.

If, nonetheless, the normal distribution function is used, the properties of such a stable process can be determined from the mean and standard deviation of the function. The standard deviation of $\eta(t)$ can be obtained from the product of $\eta(t)$ and η $(t+s)$,

$$C_\eta(s) = E\{\eta(t) \cdot \eta(t+s)\} = \lim_{T \to \infty} \frac{1}{T} \int_0^T \eta(t) \cdot \eta(t+s)dt \qquad (6.41)$$

in which s is a time interval. Alternatively, the standard deviation function can be replaced by a function of frequency spectra density,

$$G_\eta(\omega) = \frac{2}{\pi} \int_0^\infty C_\eta(s) \cos \omega s\, ds \qquad (6.42)$$

in which ω is the angular frequency. The relationship between ω and the linear frequency, f is

$$\omega = 2\pi f \qquad (6.43)$$

Actually, by means of the Fourier transform, $G_\eta(\omega)$ and $C_\eta(s)$ can be expressed as

$$C_\eta(s) = \int_0^\infty G_\eta(\omega) \cos \omega s\, d\omega \qquad (6.44)$$

The mean square deviation of $\eta(t)$ is

$$\sigma_\eta^2 = C_\eta(D) = \int_0^\infty G_\eta(\omega) d\omega \qquad (6.45)$$

and the average of the frequency spectra density is

$$\overline{\omega} = \frac{\int_0^\infty \omega G_\eta(\omega) d\omega}{\int_0^\infty G_\eta(\omega) d\omega} \qquad (6.46)$$

The above statistical parameters reflect the geometric characteristics of the bed configuration. For example, the standard deviation σ_η of the bed level is proportional to the value of 'a' in Fig. 6.41, which is equal to about half the height of the sand wave. The reciprocal of Eq. (6.46), which expresses the average period or wave length of the spectrum, is related to the average dune period and wave length in Fig. 6.41.

Fig. 6.43 shows the auto-correlation and frequency spectrum of the longitudinal bed configuration $\eta = \eta(x)$ in a straight channel. The auto-correlation can be used to assess the degree of correlation between the value of η at two points with a time interval s. Fig. 6.43 shows that in addition to the usual dunes, which have relatively small wave lengths, a much longer wave can also exist (in the frequency spectra, they are equivalent to a low frequency wave with a wave number of about 0.016 cycle/meter). The latter is actually the alternative bars in the straight channel and is not a property of the bed form discussed here.

Fig. 6.44 shows the relationship between the frequency spectra density and the wave characteristics (length and wave number) of ripples and dunes in flumes and channels of different widths. The figure shows that the dune development is affected markedly if the channel is quite narrow. The height and length of the bed form increase with the channel width. Although dunes and ripples are quite similar in shape, characteristic differences exist in the variation of the frequency spectra with wave number. For ripples, the frequency spectrum is rather uniformly distributed with the

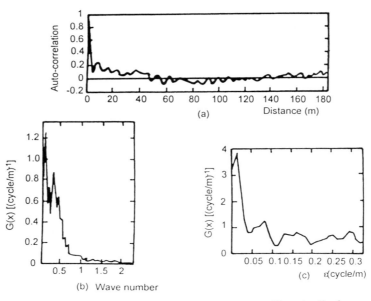

Fig. 6.43 Auto-correlation and frequency spectrum of longitudinal
bed profile in a straight channel (after Nordin, C.F., Jr)

wave number. A majority of the ripples have large wave numbers and are short. For dunes, most of the frequency spectrum are concentrated in the region of small wave numbers and long waves.

For a random variable, η, and if it is stable, the mean amplitude $\bar{\eta}$ and standard deviation σ_η^2 are constants. Hence, one can transform η into $(\eta - \bar{\eta}) / \sigma_\eta$, a quantity that has a mean value of zero and a standard deviation of 1. Thus standardized data from different flumes and channels can be unified. Hino [37] pointed out, from his theoretical analysis, that the function of frequency spectrum density of a bed form should be proportional to the square of the sand wave period and the cube of the wave length. Nordin [35] found this conclusion to be well supported by experimental data for the higher frequencies in the spectrum. Therefore, the speed of dune propagation ($c = \lambda/T$) should be proportional to the wave length raised to the power (-1/2); i.e., for a given flow rate, a small dune propagates faster than a large one.

For their effect on the resistance caused by a sand wave, lower frequency waves are more significant. Using dimensional analysis, Engelund [38] obtained two parameters to describe the frequency spectrum, in the lower and higher portions of the range,

$$\frac{8ghJ}{U^2}\frac{G(x)}{\sigma} \quad \text{and} \quad \sigma\varepsilon\frac{U^2}{8ghJ}$$

as shown in Fig. 6.45; in the latter parameter ε = wave number, $(8ghJ)/U^2$ = friction factor.

These statistical parameters of a bed form must be affected by both the flow and sediment characteristic. From the few available references on this aspect, the standard deviation of the bed level appears to vary with the average water depth and the discharge per unit width, as in Fig. 6.46.

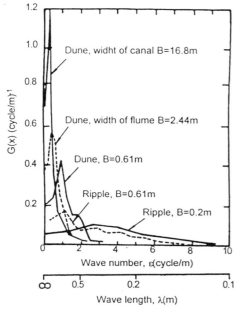

Fig. 6.44 Frequency spectrum density against wave length and wave number of ripples and dunes (after Nordin, C.F., Jr)

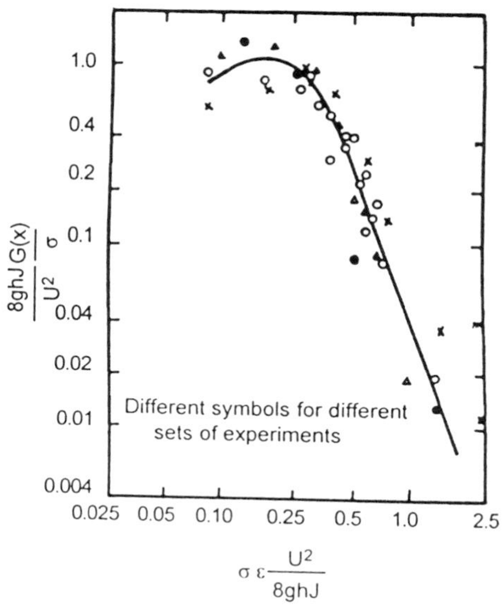

Fig. 6.45 Frequency spectrum of dunes against the wave number (after Engelund, F.)

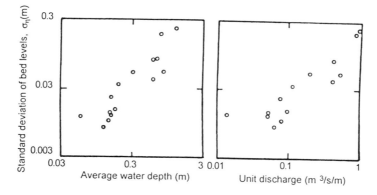

Fig. 6.46 Relationship between the standard deviation of bed levels and flow conditions

6.5.2 Correlation of length and height of bed forms with characteristics of flow and sediment

Statistical analysis of the bed form is potentially useful. However, the development of the process is still in its infancy. So far, studies on the correlation of the bed form and the characteristics of flow and sediment must be based on mean values of dune length and height. Certainly, all of the parameters of the flow and sediment that determine the properties of bed forms can play important roles in the way bed configurations form.

6.5.2.1 Ripple stage

In the ripple stage, the most important flow parameter is the grain Reynolds number. Fig. 6.47 shows the relationship between ripple length (as indicated by λ/D) and grain Reynolds number [39]. If $U_* D / \nu < 3.5$, a laminar sublayer forms near the bed, and the flow near the particles is also laminar. In this case, the length quantity that describes the flow pattern near the bed is only the thickness of the laminar sublayer

$$\delta = 11.6 \, \nu/U_* \tag{6.47}$$

Obviously, the ripple length must be a function of this length, as evidenced by those experimental data that are distributed on the 45^0 line in Fig. 6.47. The equation of this line is,

$$\lambda = 2,250\nu / U_* = 194 \, \delta \tag{6.48}$$

In the range $3.5 < U_* D/ \nu < 11.6$, although a laminar sublayer forms near the bed, the flow generates wakes behind the particles, so that both the grain Reynolds number and the particle diameter affect the ripple length. Fig. 6.47 shows the experimental data obtained by three different authors. The spread of the results may be due to the different sizes of the flumes used in these studies.

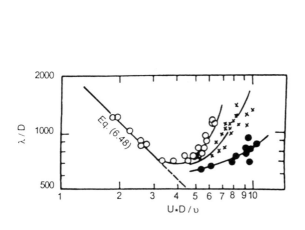

Fig. 6.47 Relationship between wave length of ripples and grain Reynolds number (after Yalin, M.S.)

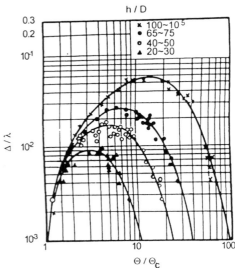

Fig. 6.48 Relationship between steepness of dune and flow conditions (after Yalin, M.S.)

6.5.2.2 Dune stage

The flow parameter Θ, Froude number, and relative roughness D/h play important roles in the dune stage. Fig. 6.48 shows the relationship between the ratio of dune height to dune length $(\varDelta/\lambda)$ and Θ/Θ_c, and the third parameter is h/D; the quantity Θ_c is the threshold condition for the initiation of sediment motion [39,40]. This diagram reflects all of the stages in the development of bed forms from their formation and growth to their elimination. The dune steepness increases as h/D increases, but it has a maximum value of 0.06. Shinohara and Tsubaki [41] related the dune size and the parameter

$$\frac{\gamma}{\gamma_s - \gamma} \frac{U_*^2}{gD} \frac{\tau_e}{\tau_o}$$

as shown in Fig. 6.49. They found that not all of the bed shear stress (τ_0) is related to the dune formation, rather only a part of it (τ_e). This question is discussed in detail in Chapter 7. A significant fact is that the formation of bed form follows different trends; for coarser sand, with diameters ranging between 0.62 to 1.46 mm, and for fine particles with a diameter of 0.21mm, they may have been represented in the two different stages of development shown in Fig. 6.48.

The factors included in Figs. 6.48 and 6.49 are not all the factors functioning in the phenomenon. Garde and Albertson [33] proposed that the bed form dimensions in

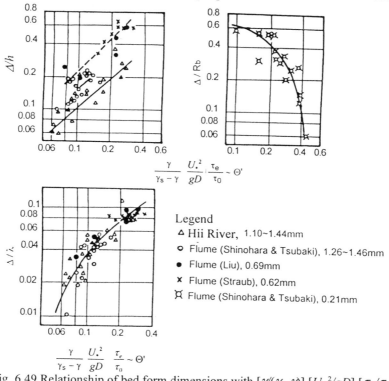

Legend
Δ Hii River, 1.10~1.44mm
○ Flume (Shinohara & Tsubaki), 1.26~1.46mm
● Flume (Liu), 0.69mm
✕ Flume (Straub), 0.62mm
✤ Flume (Shinohara & Tsubaki), 0.21mm

Fig. 6.49 Relationship of bed form dimensions with $[\gamma/(\gamma_s-\gamma)]\,[U_*^2/gD]\,[\tau_e/\tau_0]$
(after Shinohara, K., and T. Tsubaki)

the ripple stage are related primarily to the parameters $\dfrac{\tau_o}{(\gamma_s - \gamma)D}$ and $\dfrac{U_* D}{\upsilon}$; whereas

those for the dune stage are $\dfrac{\tau_o}{(\gamma_s - \gamma)D}$ and $\dfrac{U}{\sqrt{gh}}$, as shown in Fig. 6.50. The experimental data they used are not sufficient to define completely the relationships shown in the figure; but their attempt is undoubtedly a step in the right direction. From a limited set of data, Engelund and Hansen [42] obtained the following relationship:

$$\frac{\lambda}{h} J\Theta'^2 = 0.037 Fr^{5.4} \tag{6.49}$$

in which Θ' is the effective flow intensity and is similar to the parameter used by Shinohara and Tsubaki.

Ranga Raju and Soni [43] developed a correlation for bed form dimensions based

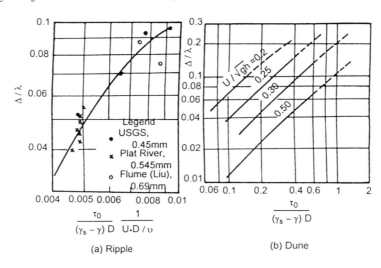

(a) Ripple

(b) Dune

Fig. 6.50 Relationship between bedform dimensions of ripples
and dunes against $\tau_o/(\gamma_s - \gamma)D$, $U/(gh)^{0.5}$ and $U*D/\nu$

on the relationship between bed form propagation and bed load transport. A relationship has been established between the rate of bed load transport per unit width and the bed form propagation speed c:

$$g_b = 0.5\,\gamma_b\,\Delta c \tag{6.50}$$

in which γ_b is the specific weight of bed material, and if e is its porosity; then

$$\gamma_b = (1-e)\,\gamma_s \tag{6.51}$$

In the deduction of Eq. (6.50), the ripples and dunes have been approximated to be triangular in shape.

One can write a dimensionless parameter Φ of the form,

$$\Phi = \frac{g_b}{\gamma_s}(\frac{\gamma}{\gamma_s - \gamma})^{1/2}(\frac{1}{gD^3})^{1/2} \qquad (6.52)$$

and combining Eqs. (6.50) and (6.51) into Eq. (6.52), one obtains

$$\Phi = (\frac{1-e}{2})\frac{\Delta}{D}\frac{c}{U}\frac{U}{\sqrt{gD}}\sqrt{\frac{\gamma}{\gamma_s - \gamma}} \qquad (6.53)$$

Kondap and Garde [44] proposed the relationships

$$\frac{c}{U} = 0.021 Fr^3 \qquad (6.54)$$

in which $Fr = \dfrac{U}{\sqrt{gR_b}}$, and R_b = hydraulic radius related to resistance of the bed. Eq. (6.53) can be re-written as

$$\Phi \sim (1-e)\frac{\Delta}{D}\sqrt{\frac{\gamma}{\gamma_s - \gamma}}\sqrt{\frac{R_b}{D}}Fr^4 \qquad (6.55)$$

In Chapter 9 it is shown that the bed load transport parameter Φ depends mainly on the flow parameter Θ', i.e.,

$$\Phi = f(\Theta')$$

and that the bed porosity is approximately constant. Therefore, one can derive the following relationship:

$$\frac{\Delta}{D}\sqrt{\frac{\gamma}{\gamma_s - \gamma}}\sqrt{\frac{R_b}{D}}Fr^4 = f(\Theta')$$

Fig. 6.51a shows the relationship based on laboratory data from eight flumes and field data from a river. Table 6.4 summarizes the range of the parameters. The data scatter, but their trend is a straight line on a log-log plot, with the equation

$$\frac{\Delta}{D}\sqrt{\frac{\gamma}{\gamma_s - \gamma}}\sqrt{\frac{R_b}{D}}Fr^4 = 6.5 \times 10^3 \Theta'^{3/8} \qquad (6.56)$$

Table 6.4 Summary of the range of the parameters

Researchers	D(mm)	R_b(m)	$J\times10^3$	U(m/s)	Δ (cm)	λ (m)
Guy et al.	0.19~0.93	0.09~0.33	0.15~6.50	0.21~1.05	0.15~19.8	0.09~5.40
Vanoni & Huang	0.21~0.23	0.06~0.27	0.46~2.90	0.17~0.56	1.10~1.70	0.12~0.23
Vanoni & Brooks	0.14	0.07~0.15	0.39~2.80	0.23~0.45	1.30~1.80	0.10~0.14
Williams	1.35	0.07~0.11	1.33~10.88	0.46~0.81	1.30~5.10	0.40~2.70
Larsen	0.10	0.05~0.24	0.43~1.86	0.33~1.02	1.90~3.40	0.14~0.17
Barton & Lin.	0.18	0.09~0.36	0.44~2.10	0.23~1.10	1.60~3.50	0.13~0.23
Tsubaki	1.26	0.16~0.47	1.61~1.73	0.58~0.76	2.10~8.20	1.10~1.60
Bart	0.60	0.04~0.13	1.00~7.00	0.26~0.81	0.27~2.00	0.15~0.59
Murtinec (River Luznice)	2.40	0.14~1.41	0.36~0.68	0.35~0.98	2.30~30.0	0.35~3.60

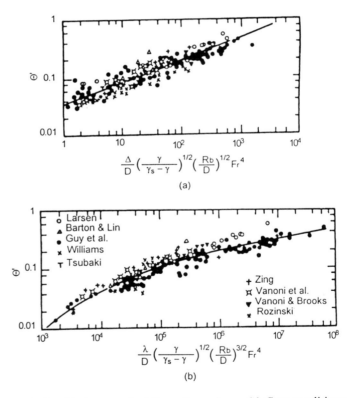

Fig. 6.51 Relationship between bed form dimensions with flow conditions
(after Raju, K.G. Ranga and J.P. Soni)

Using the concept that the height and length of a sand wave are governed by the same flow parameters, Raju and Soni also found a relationship between

$\dfrac{\lambda}{D}\sqrt{\dfrac{\gamma}{\gamma_s-\gamma}}\sqrt{\dfrac{R_b}{D}}Fr^4$ and Θ, but they found that the use of the power 3/2 on the factor R_b/D gave less scatter, Fig. 6.51b.

In another study, also based on laboratory and field data, researchers at Wuhan Institute of Hydraulic and Electric Engineering (WIHEE) derived the following relationship [45]:

$$\frac{\Delta}{h} = 0.086 Fr(\frac{h}{D})^{1/4} \tag{6.57}$$

Znaminskaya [46] conducted two sets of experiments using particles with median sizes of 0.18 and 0.8 mm, and obtained a relationship between the relative dune height and the parameters U/U_c and h/D, as shown in Fig. 6.52, in which U_c is the threshold velocity for the initiation of particle motion. Her parameter U/U_c is similar to Θ/Θ_c used by Yalin.

6.5.2.3 Antidune stage

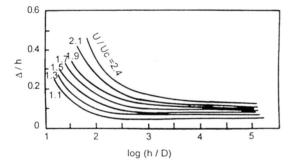

Fig. 6.52 Relative wave height of dunes

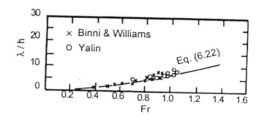

Fig. 6.53 λ/h versus Fr

In the antidune stage (zone ABC in Fig. 6.23), the various curves corresponding to different values of the relative phase difference j are so close to each other that it is difficult to separate them on the basis of the experimental data. From the practical point of view, j can be set equal to zero, so that the relationship between the antidune length and Froude number is given by Eq. (6.22). This equation corresponds well with the measured data. Fig. 6.53 shows the relationship between the wave length of a surface wave caused by the bed form and the Froude number. The measured data agree well with Eq. (6.22). These observations indicate that the surface waves were caused by antidune and that their wave lengths are equal.

6.5.3 Correlation of the migration speed of bed form and flow and sediment characteristics

Most of the existing formulas for the migration speed of bed form deal with that of dune movement. Much as for the dune dimensions, the parameters that characterize dune movement are as follows:

$$\frac{\tau_o}{(\gamma_s - \gamma)D} \, , \text{Fr} \, , h/D$$

Shinohara and Tsubaki [41] proposed the following relationship:

$$\frac{ch}{\sqrt{\frac{\gamma_s - \gamma}{\gamma} gD^3}} = a(\frac{\gamma}{\gamma_s - \gamma} \frac{U_*^2}{gD} \frac{\tau_e}{\tau_o})^m \tag{6.58}$$

in which c = the rate of dune propagation, and the values for a and m are as follows:

sediment size (mm)	a	m
0.1 ~ 0.21	76.1	2.5
0.69 ~ 1.46	48.6	1.5

The WIHEE [45] group and Knoroz [47] both found the Froude number to be the key variable. The former proposed the following equation:

$$c/U = 0.0144 \, U^2 / (gh) \tag{6.59}$$

A comparison of this equation with experimental data does not show good agreement. Knoroz [47] proposed the following equation:

$$\frac{c}{\sqrt{gh}} = 0.425(\frac{U - U_c}{\sqrt{gR}})^{2.5} \tag{6.60}$$

in which R is the hydraulic radius. This equation is valid only if the water depth is less than 0.5 m and the sediment is larger than 0.5 mm.

Using the dune data of the Yangtze River, the Hankou Survey Team from the Yangtze River Water Conservancy Office derived the following empirical equation:[1]

---

[1] Hankou Survey Team of YRCC. *Characteristics of Bed Form in the Yangtze River*, April 1960, p. 29

$$\frac{c}{U} = 0.012 \frac{U^2}{gh} - 0.043 \frac{gD}{U^2} \qquad (6.61)$$

in which gD/U^2 is the product of the parameters gh/U^2 and D/h.

6.6 SIGNIFICANCE OF BED FORM STUDIES

The study of the movement of bed form is important from the view points of both theory and engineering practice.

6.6.1 Bed forms as a primary contributor to resistance in alluvial rivers

Because the streamlines near the sand bed are not parallel to the bed surface once bed form appear, especially in the form of ripples and dunes, the flow separates at the crest of the bed form so that the pressures on the downstream and upstream sides differ (Fig. 6.17). The net force thus produced is the form resistance of the sand wave. As a major component of resistance, form resistance changes as the flow conditions change. Hence, the friction factor in an alluvial river is not just a constant, but varies with flow conditions.

Table 6.5 Bed forms and the corresponding flow resistance for various flow conditions (results from flume experiments)

Bed forms	0.28 mm Sand					0.45 mm Sand				
	Slope $\times 10^{-3}$	$\dfrac{U}{\sqrt{gh}}$	Sediment concentration $\times 10^{-4}$	n	f	Slope $\times 10^{-3}$	$U/\sqrt{gh}$	Sediment concentration $\times 10^{-4}$	n	f
Flat	0.11	0.17	0	0.016	0.0301	0.15	0.18	0	0.016	0.0359
Ripples	0.23~1.10	0.17~0.37	0.01~1.5	0.02~0.027	0.0635~0.1025	0.16~1.10	0.14~0.28	0.01~1.0	0.020~0.028	0.0521~0.133
Dunes	0.90~1..60	0.32~0.44	1.5~8.0	0.021~0.026	0.0612~0.0791	0.6~3.0	0.28~0.65	1~12	0.019~0.033	0.0489~0.149
Transition	1.30~1.70	0.55~0.67	10~24	0.014~0.017	0.0250~0.0344	3.7~4.9	0.61~0.92	14~20	0.016~0.022	0.0415~0.0798
Flat	1.50~1.80	0.71~0.92	15~31	0.013~0.014	0.0244~0.0262	—	—	—	—	—
Stand. wave	—	—	—	—	—	3.6~6.2	1.0~1.6	40~70	0.011~0.015	0.02~0.0406
Sand waves	3.3~10	1.0~1.3	50~420	0.014~0.022	0.0281~0.0672	6.6~10.0	1.4~1.7	60~150	0.012~0.014	0.0247~0.0292

Table 6.5 shows the results of laboratory experiments on bed forms and the corresponding energy losses under different flow conditions[9]. The energy loss is normally expressed in terms of either the Manning coefficient n or the friction factor, f:

$$\left. \begin{array}{l} n = \dfrac{1}{U} R^{2/3} J^{1/2} \\[2mm] f = \dfrac{8gRJ}{U^2} \end{array} \right\} \qquad (6.62)$$

During the transformation of a sand bed from a flat bed to ripples and then to dunes, the resistance the river flow encounters increases continuously. Then it decreases as the bed becomes plane again at still higher rates of flow. In the latter case, the resistance can even be less than it was before the ripples formed. This latter phenomenon may be due to the presence of suspended load and its effect on the turbulent structure of the flow. During the sand wave phase, the undulation of the sand bed is much more pronounced than it is in the ripple and dune phases, but the sand waves have a symmetrical shape with no flow separation at their crests. Therefore, their form resistance is smaller and the energy loss is less than that for ripples and dunes. The corresponding energy loss is only a little more than that for a plane bed, because the breaking of the sand waves generates a strong local turbulence that dissipates parts of the flow energy. Experience shows that the flow depth in a flume usually does not change appreciably with the discharges so that the different bed configurations are accompanied by changes in the slope of the water surface. In contrast, the depth in a natural river changes a lot as the rate of flow varies, whereas the surface slope, over a long distance, remains almost constant. Because the energy loss is a function of the relative roughness or the ratio of the amplitude of the sand wave to the flow depth, the energy loss due to the various bed configurations should be different in natural rivers and in flumes. A more detailed discussion of resistance in alluvial rivers is contained in the next chapter.

6.6.2 Unusual rating curve as a result of sand wave development

Different flow conditions lead to different bed form features. Similarly, the bed configuration changes if the flow conditions are altered. Because of the inertia inherent in the system, the response of a sand wave always lags behind the changes that alter the flow conditions. During the rising stage, the sand wave may not have enough time to develop fully, so that the flow could experience a smaller resistance for a given discharge. Conversely, during the falling stage, the response of the bed configuration to the decreasing discharge may also be incomplete. In this case, for a given discharge, the flow encounters a larger resistance at the lower discharge and thus requires a larger flow depth to

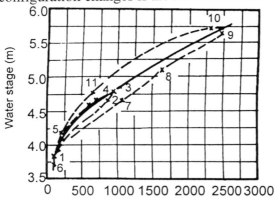

Fig. 6.54 Counter-clockwise loop of rating curve on Xingyi River

244

convey the same flow than it did during the rising stage. Therefore, the rating curve becomes a counter-clockwise loop instead of a single curve, as shown in Fig. 6.54 [48]. The foregoing description is for flow conditions in which the dune did not completely vanish. If the flow is large enough to change the dune into a flat bed or to form antidunes, the bed experiences tremendous changes; under certain conditions, it will have an unusual rating curve so that different discharges occur for the same flow depth. The causes and conditions for such an occurrence are also dealt with in Chapter 7.

6.6.3 Estimation of bed load transport rate from the characteristics and dimensions of bed forms

The bed load discharge in natural rivers is not easy to measure. A sand wave is a special type of collective bed load motion for which the size and velocity of the sand wave can be measured continuously over a large area; one can therefore quite naturally ponder over the possibility of estimating the bed load discharge from the characteristics of the bed form.

If the trough of the sand wave is used as the datum for the sand bed and the height of the sand wave at any point is denoted by η, then by comparison with Eq. (6.2), one obtains

$$\frac{\partial g_b}{\partial x} + \gamma_b \frac{\partial \eta}{\partial t} = 0$$

in which g_b = bed load discharge per unit width by weight, γ_b = specific weight of the bed material, including the soil porosity. If the shape of the sand wave remains unchanged during its motion, one obtains

$$\frac{\partial \eta}{\partial t} + c \frac{\partial \eta}{\partial x} = 0$$

in which c = celerity of sand wave movement.

From the two foregoing equations and with g_o = the bed load discharge per unit width for the condition $\eta = 0$, the average discharge between two adjacent crests of the sand waves can be expressed as

$$\overline{g_b} = \alpha \gamma_b \Delta c + g_o \tag{6.63}$$

The shape factor α for bed form would be 0.5 for a triangular one; g_o is the average bed load discharge on the bed surface between the lowest point at the less side of one bed form and the beginning of the next. Since the quantity g_o is much smaller than the first term of the right side of Eq. (6.63), it can be neglected. Then

$$\overline{g}_b = \alpha \gamma_b \Delta c \qquad (6.64)$$

From measurements of the shape of sand waves in rivers and flumes, Shinohara and Tsubaki obtained a value of 0.55 for α instead of 0.5, so that

$$\overline{g}_b = 0.55 \gamma_b \Delta c \qquad (6.65)$$

Fig 6.55 compares measured results with Eq. (6.65) and it shows that the measured values are slightly larger than the calculated ones, probably it is the result of neglecting g_o [41]. For most measurements in rivers, estimates of the bed load discharge can be made directly from the shape and the celerity of bed form. This method was used successfully for rivers in southern China[1]. However, the method is successful only if the sand waves are distributed uniformly on the bed surface and are essentially two-dimensional, as they usually are in small rivers. The distribution of sand waves on the bed surface in larger rivers is quite complex; therefore, the evaluation of the bed load transport rate by this method is not advisable in this instance.

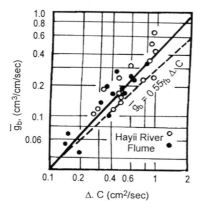

Fig. 6.55 Relationship between bed load transport rate and bed form characteristics.

Besides the foregoing three points, the deposition on the downstream side of a sand wave usually forms a clear diagonal bedding. With more deposition, such a bedding sometimes remains. The studies concerning the bedding of deposition are helpful in the interpretation of the historical paleo-geographical environments. The aggradation and degradation process in a river reflects to a certain extent the results of bed form movement. This aspect is not, however, included in this book.

---

[1] Bureau of Water Resources & Hydro-Power, Yiyang Prefecture, Hunan, Liushahe Gauging Station, Yiyang Prefecture, Hydrology Group of Geography Department, Zhongsha University. *Bed Form Survey on Liushahe River and the Calculation of Bed Load Transport Rate*, 1977, p.71.

REFERENCES

[1] Sundborg, A. "The River Klaralven - A Study of Fluvial Process." *Meddelanden Frau Uppsala* Univ. Geografiska Inst., Ser. A, No. 115, 1956, pp. 127-316.

[2] Allen, J. R. L. *Current Ripples: Their Relation to Pattern of Water and Sediment Motion.* Amsterdam, North Holland, 1968, p. 433.

[3] Chien, Ning. "The Present Status of Research on Sedement Transport." *Trans., Amer. Soc. Civil Engrs.*, Vol. 121, 1956, pp. 833-844.

[4] Liu, H.K. "Mechanics of Sediment Ripple Formation." *J. Hyd. Div., Proc. Amer. Soc. civil Engrs.*, Vol. 83, No. HY2, 1957, p. 23.

[5] Engelund, F. and E. Hansen. " A Monograph on Sediment Transport in Alluvial Streams." *Revised Ed., Teknisk Forlag*, Copenhagen, Denmark, 1972.

[6] Sahgal, P.B. "Effect of Characteristics of Bed Material and Fluid in Ripple Formation." *Intern. Symp. on River Mech.*, Vol. 1, Intern. Assoc. Hyd. Res., 1973, pp. 653-666.

[7] Coleman, J.M. "Brahmaputra River: Channel Processes and Sedimentation." *Sedim., Geol.*, Vol. 3, No. 2/3, 1969, pp. 129-239.

[8] Kondratev, N.E. "Kinematic Pattern of A Stream with the Dune Structure of the Bottom." *Soviet Hydrology*, Amer. Geophys. Union, Vol. 6, 1964.

[9] Simons, D.B., and E.V. Richardson. "Resistance of Flow in Alluvial Channels." *J. Hyd. Div, Proc., Amer. Soc. Civil Engrs.*, Vol. 86, No. HY5, 1960, pp. 73-99.

[10] Kennedy, J.F. "Stationary Waves and Antidunes in Alluvial Channels." *Rep. KH-R-2*, Keck Lab. Hyd. and Water Resources, Calif. Inst. Tech., 1961, p. 146.

[11] Karahan, M.E., and A.W. Peterson. "Visualization of Separation Over Sand Waves." *J. Hyd. Div., Proc., Amer. Soc. Civil Engrs.*, Vol. 106, No. Hy8, Aug. 1980, pp. 1345-1352.

[12] Raudkivi, A.J. *Loose Boundary Hydraulics.* 2nd ed. Pregamon Press, 1976, p. 397.

[13] Rifai, M.F., and K.V.H. Smith. "Flow Over Triangular Elements Simulating Dunes." *J.Hyd.Div., Proc., Amer. Soc. Civil Engrs.*, Vol. 97, No. Hy7, 1971, pp. 963-976.

[14] Jopling, A.V. "An Experimental Study on the Mechanics of Bedding." *Ph.D. Dissertation*, Harvard Univ., 1961.

[15] Chabert, J., and J.L. Chauvin. "Formation des Dunes et des Rides dans les Modeles Fluviaux." *Bull.* du Centre de Recherches et d'Essais de Chatou, No.4, 1963.

[16] Liliavsky, S. *An Introduction to Fluvial Hydraulics.* Constable and Co., London, 1955, pp. 24-26.

[17] Reynolds, A.J. "Waves on the Erodible Bed of an Open Channel." *J. Flu. Mech*, Vol. 22, Pt. 1, 1965, pp. 113-133.

[18] Bagnold, R.A. "The Flow of Cohesionless Grains in Fluids." *Phil. Trans.*, Royal Soc. London, Ser. A, Vol. 249, No. 964, 1956.

[19] Patt, C.J. "Bagnold Approach and Bed-Form Development." *J. Hyd. Div., Proc. Amer. Soc. Civil Engrs.*, Vol. 99, No. Hy1, 1973, pp. 121-138.

[20] Kennedy, J.F. "The Mechanics of Dunes and Antidunes in Erodible Bed Channels." *J. Flu. Mech.*, Vol. 16, Pt. 4, 1963.

[21] Kennedy, J.F. "The Formation of Sediment Ripples, Dunes, Antidunes." *Annual Rev. Flu. Mech.*, Vol. 1, 1969, pp. 147-168.

[22] Taizo, Hayashi. "Formation of Dunes and Antidunes in Open Channels." *J. Hyd .Div., Proc., Amer. Soc. Civil Engrs.*, Vol. 96, No. HY2, 1970, pp. 357-366.

[23] Yalin, M.S. "Determination of Kennedy's Parameter j for Dunes." *J. Hyd. Div., Proc. Amer. Soc. Civil Engrs.*, Vol., 99, No. HY8, 1973, pp. 1287-1290.

[24] Parker, G. "Sediment Inertia as Cause of River Antidunes." *J. Hyd. Div., Proc., Amer. Soc. Civil Engrs.*, Vol. 101, No. HY2, 1975, pp. 211-221.

[25] Engelund, F., and J. Fredsøe. "Transition From Dunes to Plane Bed in Alluvial Channels." *Ser. Paper 4*, Inst. Hydrodynamics and Hyd. Engin., Tech. Univ. Denmark, 1974, p. 46.

[26] Fredsøe, J. "The Distinction Between Bed Load and Suspended Load and Its Implication In River Morphology and Sedimentation Problems." *Rep. 128*, Danish Center for Applied Math. and Mech., Tech. Univ. Denmark, Nov. 1977, p. 18.

[27] Engelund, F. "Instability of Erodible Bed." *J. Flu. Mech.*, Vol. 42, Pt. 2, 1970, pp. 225-244.

[28] Engelund, F., and J. Fredsφe. "Three Dimensional Stability Analysis of Open Channel Flow Over An Erodible Bed." *Nordic Hydrology*, Vol. 11, No. 2, 1971, pp. 93-108.

[29] Yalin, M.S. *Mechanics of Sediment Transport*. Pergamon Press, 1972, p. 290.

[30] Shields, A. "Anwendung der Aechlichkeits-Mechanik und der Turbulenzforschung auf die Geschiebewegung." *Mitt. Preussische Versuchsanstalt fur Wasserbau und Schiffbau*, Berlin, 1936.

[31] Albertson, M.L., D.B. Simons, and E.V. Richardson. "Discussion-Mechanics of Sediment Ripple Formation." by H.K. Liu, *J. Hyd. Div, Proc., Amer. Soc. Civil Engrs.*, Vol. 84, No. HY1, 1958.

[32] Hill, H.M., A. J. Robinson, and V.S. Srinivassa. "On the Occurrence of Bed-Forms in Alluvial Channels." *Proc., 14th Cong. Intern. Assoc. Hyd. Res.*, Vol. 3, 1971, pp. 91-100.

[33] Grade, R.J., and M.L.Albertson. "Characteristics of Bed Forms and Regimes of Flow in Alluvian Channels." *Rep. CER 59 RJG 9*, Colorado State Univ., 1959, p. 18.

[34] Athaullah, M. "Prediction of Bed Forms in Erodible Channels." *Ph.D. Dissertation*, Colorado State Univ., 1968.

[35] Nordin, C.F. Jr. "Statistical Properties of Dune Profiles." *U. S. Geol. Survey*, Prof. Paper 562-F, 1971, p. 41.

[36] Nordin, C.F. Jr. "Alluvial Channel Bed Forms." In *Stochastic Approaches to Water Resources*, Vol. 2, By H.W. Shen(ed), 1976, p. 30.

[37] Hino Mikio. "Equilibrium Range Spectra of Sand Waves Formed by Flowing Water." *J. Flu. Mech.*, Vol. 34, Pt. 3, 1968, pp. 565-573.

[38] Engelund, F. "On the Possibility of Formulating A Universal Spectrum for Dunes." *Basic Res. Progress Rep.8*, Hyd. Lab., Tech. Univ. Denmark, 1969, pp. 1-4.

[39] Yalin, M.S. "On the Determination of Ripple Length." *J. Hyd. Div., Proc., Amer. Soc. Civil Engrs.*, Vol. 103, No. HY4, April 1977, pp. 439-442.

[40] Yalin, M.S., and E. Karahan. "Steepness of Sedimentary Dunes." *J. Hyd. Div., Proc., Amer. Soc. Civil Engrs.*, Vol. 105, No. HY4., April 1979, pp. 381-392.

[41] Shinohara, K., and T. Tsubaki. "On the Characteristics of Sand Waves Formed Upon the Beds of the Open Channels and Rivers." *Rep., Res. Inst. Applied Mech.*, Kyushu Univ. Japan, Vol. 7, No. 25, 1959, pp. 15-45.

[42] Engelund, F., and E. Hansen. "Investigation of Flow in Alluvial Streams." *Acta Polytechnics Scadinavica*, Ci. 35, 1966.

[43] Raju, K.G. Ranga, and J.P. Soni. "Geometry of Ripples and Dunes in Alluvial Channels." *J. Hyd. Res., Intern. Assoc. Hyd. Res.*, Vol. 14, No. 3, pp. 241-249.

[44] Kondap, D.M., and R.J. Garde. "Velocity of Bed Forms in Alluvial Channels." *Proc., 15th Cong., Intern. Assoc. Hyd. Res.*, 1973.

[45] Wuhan Institute of Hydraulic and Electric Engineering. *River Dynamics*. China Industrial Press, 1961, pp. 288.

[46] Znamenskaya, N.S. "Experimental Study on Bed Load Motion." *Proc. Soviet National Hydrology Institute*, Vol. 108, 1963, pp. 89-114. (in Russian)

[47] Levi, E.E., and V.S. Kronos. "New Development of Study on Bed Load Motion at the Soviet National Hydrology Institute." *Journal of Sediment Research*, Vol. 2, No. 2, 1957, pp. 13-17 (in Chinese)

[48] Carey, W.C., and M.D. Keller. "Systematic Changes in the Beds of Alluvial Rivers." *J. Hyd., Div., Proc., Amer. Soc. Civil Engrs.*, Vol. 83, No. HY4, 1957, p. 24.

CHAPTER 7

FLOW RESISTANCE IN ALLUVIAL STREAMS

The main purposes of studying flow and sediment movement are to know the discharge capacity and sediment transport capacity of alluvial streams. For a channel reach with a certain geometric shape and slope, what discharge can pass through it, how much can be carried in the main channel, how much spreads over the floodplains, and how large is the velocity? All these issues are, of course, related to the resistance of the stream channel to the flow. Furthermore, the resistance itself reflects the magnitude of the action of flow on the stream channel and indicates the intensity of sediment transport. From whatever point of view, the study of the resistance of alluvial streams is profoundly important.

7.1 TRANSFORMATION PROCESS OF FLOW ENERGY

Water flows in an open channel only if its surface slopes downward in the direction of flow. For uniform flow, the water surface slope and channel bed slope are the same and equal to the energy slope, which denotes the energy loss per unit weight of water in overcoming resistance as it flows a unit distance in the channel. All the lost energy is finally transformed into thermal energy. To study the resistance in alluvial streams is essentially to study the mechanism of the transfer of mechanical energy into thermal energy.

For uniform flow in an open channel, the mean velocity does not change from cross-section to cross-section. The flow energy comes totally from the potential energy. For the potential energy of each point in the water body, a small part transforms into thermal energy at any given location due to the viscosity of water, but most is transmitted to the boundary by the shear stress and transformed into turbulent kinetic energy. In the transformation, a portion of the energy is lost, and the rest is transformed into kinetic energy of eddies. The eddies leave the boundary, enter the main flow region, and break up into smaller eddies. Then the energy of the smaller eddies is dissipated locally into thermal energy due to the viscosity of water. These processes describe completely the way flow energy is transformed. In the following section, the details of the steps in the transformation are described one by one; for simplicity, only two-dimensional (2-D) flow is taken into consideration.

7.1.1 Energy provided by water flow

Fig.7.1 shows an element of water $abcd$ at a point y above the river bed in a 2-D open channel flow; it is dy in height, dx in length, and has a unit thickness. The energy loss for the water element in unit time is

$$\gamma Judxdy$$

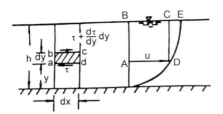

Fig.7.1 Forces acting on the water element
abcd in 2-D open channel flow

In the equation, γ--specific weight of water; J -- energy slope; u -- velocity at point y.

The energy taken in unit time from an element of water at y from the bed

$$w_b = \gamma J u \qquad (7.1)$$

From the balance of the forces acting on the element *abcd* in Fig.7.1, one can write

$$(\tau + \frac{d\tau}{dy} dy)dx - \tau dx + \gamma dx dy J = 0$$

In reduced form

$$\frac{d\tau}{dy} + \gamma J = 0$$

Substituting this relationship into Eq. (7.1), one gets

$$w_b = -u \frac{d\tau}{dy} \qquad (7.2)$$

7.1.2 Energy lost locally in overcoming resistance

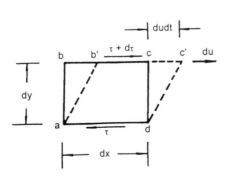

Fig.7.2 Deformation of a free water body due to forces in 2-D open channel flow

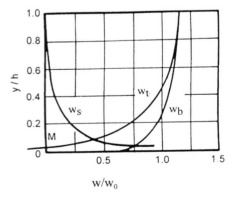

Fig. 7.3 Vertical distributions of the energy provided by flow, local energy loss, and energy transmitted to the boundary in a unit time step ($w_0 = \gamma J U$, U is the mean velocity over the vertical)

250

A part of the energy taken from the water body is lost in overcoming the resistance at a given location. Fig. 7.2 shows how the water element in Fig. 7.1 deforms due to the forces acting on it; after a time step dt, the water body $abcd$ has deformed to $ab'c'd$. The work done during the deformation is equal to the product of the shear stress τdx and displacement $dudt = \dfrac{du}{dy}dydt$. Hence in a unit time step, the energy loss for a unit water body at the point y above the bed in overcoming local resistance is

$$w_s = \tau \frac{du}{dy} \qquad (7.3)$$

The vertical distributions of w_b and w_s are not the same, as shown in Fig.7.3. The maximum of the mechanical energies provided in all flow layers for overcoming the resistance occurs at the water surface, and the minimum value of zero occurs at the stream bottom. In contrast, the energy loss due to overcoming the local resistance is zero at the water surface and has its maximum value at the stream bottom. Thus the energy of the flow is mostly in the main flow region, but the loss is concentrated near the boundary. Except at point M, the energy taken from each flow layer is not equal to the local energy loss. Above point M, $w_b > w_s$, which means that some surplus energy there can be transmitted to other regions; and below point M, $w_b < w_s$, so that the energy there is insufficient and energy must be supplied to it from other regions. To maintain an energy balance, some energy must be transmitted from the main flow region to the bottom.

7.1.3 Energy transmission

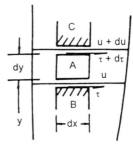

Fig.7.4   Mechanism   for transmission of flow energy to the boundary

The foregoing mechanism of energy transmission can be explained by the relationship of water body A to water bodies B and C, located below and above water body A, respectively, as shown in Fig.7.4. The relative movement of water bodies A and B makes the former do work τdxu on the latter; also in a unit time step, water body A receives the energy $(\tau + d\tau)dx(u + du)$ from water body C. In the balance, the energy transmitted from water body A to water body B is more than that received from water body C, i.e., some energy is transmitted toward the stream bed. The energy transmitted to the stream bed in a unit time step is $\tau dxu - (\tau + d\tau)dx(u + du)$ if the high-order term is neglected, the result becomes

$$-dx(\tau du + u d\tau) = -\frac{d}{dy}(\tau u)dydx$$

If the terms in the equation are divided by the volume of the water body $dxdy$, the energy transmitted from a unit water body at the point y above the river bed to the bed in a unit time step is obtained:

$$w_t = -\frac{d}{dy}(\tau u) \qquad (7.4)$$

The vertical distribution of w is shown in Fig. 7.3. Its value is positive above point M, and negative below point M.

7.1.4 Energy balance equation

In the main flow region, part of the energy taken from each layer is lost in overcoming the local resistance, and the surplus energy is transmitted to the boundary through the gradient of τu. However, the local energy near the boundary is not sufficient to overcome the local resistance, and must be supplemented from the main flow region. From the view point of the energy balance, it can be expressed as

$$w_b = w_s + w_t$$

If Eqs. (7.2), (7.3), and (7.4) are substituted into the above equation, the result is

$$-u\frac{d\tau}{dy} = \tau\frac{du}{dy} - \frac{d}{dy}(\tau u) \qquad (7.5)$$

Eq. (7.5) is the total different of τu, and the above derivation shows that all terms in the equation have a definite physical significance.

Summing up energy losses in all of the layers from the water surface to a point y above the river bed, one gets

$$\int_y^h (-u)\frac{d\tau}{dy}\,dy = \int_y^h \tau\frac{du}{dy}\,dy - \int_y^h \frac{d}{dy}(\tau u)\,dy$$

Because

$$\frac{d\tau}{dy} + \gamma J = 0$$

τ and u are the functions of y, and the above equation can be transformed into

$$\gamma J \int_y^h u\,dy = \int_{u(y)}^{u(h)} \tau\,du + \int_{\tau u(y)}^{\tau u(h)} d(-\tau u) \qquad (7.6)$$

$$(W_b) \qquad\qquad (W_s) \qquad\qquad (W_t)$$

For the entire flow region, all the mechanical energy provided by the flow to overcome resistance should equal the sum of all energy losses due to resistance, i.e.

$$\gamma J \int_0^h u\,dy = \int_0^h \tau\,du = W_0$$

$$\int_0^h d(-\tau u) = 0$$

In the foregoing equations, expressions for w_b and w_t are adopted for both laminar flow and turbulent flow; for the latter case Eq. (7.2) and (7.4) represent time averages. Because of the complexity of turbulent flow, w_s cannot be derived directly from Fig.7.2, but the total different in Eq. (7.5) is still tenable. That is, w_s expresses the energy loss of an unit water element due to the local turbulence in a unit time step.

7.1.5 Importance of near-bed flow region

Based on data measured in natural streams, Bakhmeteff and Allan plotted the vertical distributions of W_b, W_s, and W_t as shown in Fig. 7.5 [1]. The figure shows an

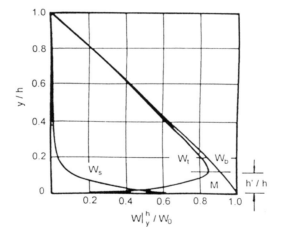

Fig.7.5 Vertical distributions of the total energy provided by flow, total local energy loss, and total energy transmitted to the boundary by the flow (after Bakhmeteff, B.A., and W. Allan)

inflection point M on the curve of W_t that divides the flow into two regions. The region above point M is called the main flow region; that below it is called the near-bed flow region. For natural streams, the near-bed flow region amounts to about 1/10 of the total water depth (h'/h= 0.11--0.12). Of the energy provided by flow, 92% comes from the main flow region, and 90% of it is transmitted to the near-bed flow region and lost to resistance there. The local energy loss in the main flow region accounts for only 8% of the total. Auborazuovski approximated the vertical velocity distribution by an exponential formula, i.e.,

$$u = u_0 (\frac{y}{h})^m \tag{7.7}$$

in which, u_0 is the velocity at the water surface.

From the equation one can derive [1]

$$h' = \frac{m}{1+m} h \tag{7.8}$$

$$\frac{W_{t\,max}}{W_0} = (\frac{m}{1+m})^m \tag{7.9}$$

Normally, the exponent m in the velocity distribution formula falls within the range 0.10 to 0.25, so h'/h falls between 0.09 and 0.20, and W_{tmax}/W_0 between 0.67 and 0.79. On average, the thickness of the near-bed flow region is somewhat less than 15% of the total water depth. But 73% of the total loss to resistance occurs in this region. Hence, most of the mechanical energy provided by flow is lost in the near-bed flow region as a result of the steep velocity gradient there.

The near-bed flow region is not only the region where most of the loss of flow energy is concentrated, but is also the place where the exchanges between the bed material and bed load, and between the bed load and suspended load, take place. If the vertical distribution of sediment concentration has a gradient, this gradient normally reaches a maximum in this region. The existence of a gradient for the sediment concentration affects the local flow characteristics. Thus, from the viewpoint of sediment movement, the near-bed flow region is the most important region, unfortunately, however, it is also the most difficult region in which to take measurements.

7.1.6 Dissipation of energy into thermal energy

For the flow, the mechanical energy provided to overcome resistance represents the energy that is lost as it is converted into thermal energy through the viscous deformation of the water. Because the process of transformation of mechanical energy to thermal energy is irreversible, the energy once lost can not be recovered by the flow. For turbulent flow, however, energy lost in overcoming resistance does not transform directly into thermal energy at that location, instead the transformation of a

---

[1] Teaching Group on Continental Hydrology of Hehai University. *Fluvial Processes*, 1961, pp 32-36.

part of the energy is accomplished by first transforming it into kinetic energy of turbulence.

As mentioned, most of the mechanical transfer of the energy of flow is concentrated in the near-bed flow region and transformed through the shear stress, this region is also the primary source of turbulence (Chapter 4). An eddy has a certain amount of kinetic energy as it departs from the boundary and rises into the main flow. Obviously, this kinetic energy can only be obtained from the mechanical energy of the flow that is concentrated in the region near the bed.

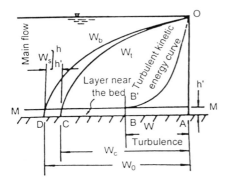

Fig.7.6 Energy transition in near-bed region

In Fig. 7.6, the energy concentrated in the region near the bed corresponds to portion AC. A small part of it dissipates directly into thermal energy due to resistance in the flow region near the bed, and another small part is lost in the process of eddy formation, only the portion AB transforms into turbulent kinetic energy in the form of eddies. These eddies, which have some size and circulation, rise into the main flow. However, they are not stable and continuously break up into smaller and smaller eddies until viscosity effects become dominant. Then, the turbulent kinetic energy finally dissipates into thermal energy through the viscous deformations in the flowing fluid. In Fig. 7.6, line $B'O$ represents the transformation of turbulent kinetic energy into thermal energy, i.e., the process of turbulence attenuation.

From the physical picture of the transformation of flow energy provided by Bakhmeteff and Allan, one reaches the following important conclusions:

1. The spatial distributions of the mechanical energy provided by the flow and the energy loss due to overcoming the local resistance are not the same, energy comes mainly from the main flow region, but the energy loss is concentrated in the flow region near the bed.

2. Conclusion (1) means that energy is continually transmitted from one flow layer to another. By means of this transmission, the energy lacking in the near-bed flow region is supplied from the main flow region.

3. The characteristics of turbulent flow are such that only a small part of the energy dissipates directly into thermal energy by viscous deformation in the main flow region, 80 to 85% of the total energy loss is transmitted and concentrated in the near-bed flow region by the shear stress, and is transformed there into turbulent kinetic energy. During the transformation, a part of the energy is lost, the remaining

part generates eddies which have some size and circulation that depart from the near-bed flow region with a circulation and rise into the main flow region. Just this part of the energy that generates the turbulent flow has a diffusion function and forms a special shear stress field. For the mechanical energy of the flow, the turbulent kinetic energy represents energy being lost, the flow can not recover any energy once it has reached this form.

4. The continuous break up of the eddies finally transforms the turbulent kinetic energy into thermal energy through viscous deformations in small eddies.

5. The complete process of energy transformation can be simply described as follows:

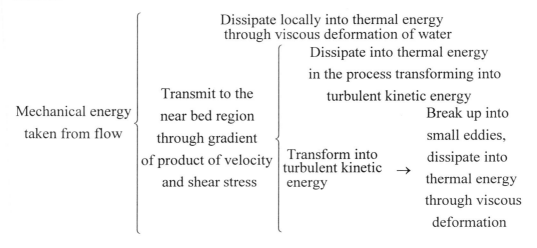

6. The foregoing discussion is for clear water. For flow carrying sediment, the energy transformation is much more complex, it is discussed in detail in Chapter 12.

7.2 COMPONENTS OF RESISTANCE

The resistance in alluvial streams, which differs from that in open channels, has a number of components, these are presented in the following paragraphs.

7.2.1 Grain friction

At any solid boundary, the sediment particles on an alluvial stream bed cause surface friction that is called the grain friction. Nikuradse glued uniform sediment particles to the inner surface of a pipe for his classical study of the resistance of flow through pipes. The particle diameter was taken as the size of the roughness in the pipe. For a natural stream with a non-uniform bed material, a representative particle size of the bed material may be taken as the roughness size that characterizes the grain friction. Research in recent years shows that the roughness size representing the grain friction may be larger even than the particles. Scientists have pointed out that even for a flat bed, the resistance of a movable stream bed may be quite different from that on a rigid bed.

7.2.2 Bed form resistance

In Chapter 6, the channel bed was shown to take various bed forms for various flow conditions. If ripples and dunes form, the flow separation at the peaks makes the pressure on the stoss face larger than that on the lee face, and in this way causes form resistance. With sand waves, although the stream lines near the bed are almost parallel to the bed and are without separation, the water surface wave corresponding to the sand wave can break and cause intensive local turbulence with an increase in the resistance loss. This kind of extra resistance is called bed form resistance. The relation between grain friction and bed form resistance can be illustrated as follows. For flow passing over a piece of flat sand paper, the flow encounters only grain friction; if, however, the sand paper has a wavy shape, the flow encounters not only the grain friction, but also the bed form resistance.

7.2.3 Bank and floodplain resistance

The material of natural river banks and floodplains is normally finer than the bed material. On the floodplains that are usually above the water level of normal floods, grasses and bushes often grow. The roughness due to them not only changes with their characteristics, density, stem height, and season, but it is also affected by the depth and velocity of the flow. In a shallow flow with a low velocity, the stems of grasses and bushes are erect, and they offer the maximum resistance to the flow. If the water depth and velocity are somewhat higher, the stems often bend down so that the flow area encountering resistance is less, and the bank and floodplain resistances are correspondingly less. At high flows, where the grasses and bushes tend to lie flat on the bed, the roughness is nearly constant and it changes very little with the discharge. If the bank is protected or lined, its roughness is unaffected by vegetation. For a mountain stream in a steep valley, bank resistance is especially important.

7.2.4 Channel form resistance

The friction loss during flow is also affected by the shape of a channel. For channels with sand bars, the flow can be braided or meandering; the flow width is then variable and the resistance to flow is high. The resistance is proportional to the square of velocity for flow in an open channel with a regular shape. But if a river meanders and the flow Froude number exceeds a critical value, the resistance can be much higher and would no longer follow the square law [2]. As "a low flow tends to be meandering, and a high flow tends to be straight," and the channel form resistance for low flows therefore is higher than that for high flow.

7.2.5 External resistance of artificial structures

Artificial structures built along a stream, like training and bank protection works, bridge, etc., tend to create local resistance. The magnitude depends on the shape, size, and orientation of the structure.

Due to the complexity and variability of components of the friction loss in alluvial streams, and to the changeable property of resistance for flow carrying sediment, a unified understanding of resistance in alluvial streams does not exist at present. Aside from the effect of sediment concentration on resistance discussed in detail in Chapter 12, the following articles underline the complexity of resistance in alluvial streams, and it treats the various roles played by components of resistance in sediment movement. In this way, a rational approach to determining the resistance in alluvial streams is developed.

7.3 RATIONAL APPROACH TO THE ISSUES OF RESISTANCE IN ALLUVIAL STREAMS

The magnitude of the resistance of a channel boundary to flow is related to the "roughness," and it is commonly expressed in terms of a "roughness coefficient." For an open channel flow with a rigid bed, the roughness can be treated as a constant. For turbulent flow over a rough surface, the Chezy or Manning equation is usually employed to express the flow resistance:

$$\text{Chezy equation} \qquad U = C\sqrt{RJ} \qquad (7.10)$$

$$\text{Manning equation} \qquad U = \frac{1}{n} R^{2/3} J^{1/2} \qquad (7.11)$$

in which C is the Chezy coefficient and n the Manning coefficient.

The two are interrelated through

$$C = \frac{1}{n} R^{1/6} \qquad (7.12)$$

in which R is the hydraulic radius. From measured data for canal flow, Pavlovski determined that the exponent on R in Eq. (7.12) is not a constant, and he proposed that it should be written as

$$C = \frac{1}{n} R^{y} \qquad (7.13)$$

in which

$$y = 2.5\sqrt{n} - 0.13 - 0.75\sqrt{R}(\sqrt{n} - 0.10) \qquad (7.14)$$

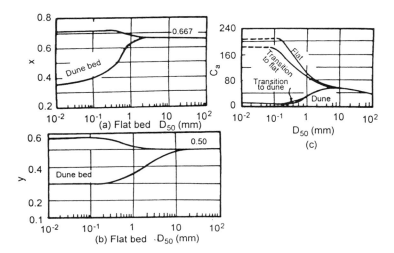

Fig. 7.7 Coefficients and exponents in generalized Manning equation as function of bed material and bed form (results of flume experiments after Liu, Hsin-Kuan, and S.Y. Hwang)

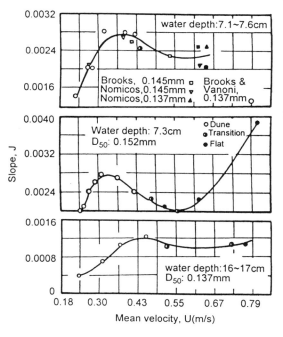

Fig. 7.8 For dunes of various sizes, different velocities occur for a given water depth and slope (results of flume experiments after Vanoni, V.A., and N.H. Brooks)

Eq. (7.14) is established within the ranges

$$0.1 \text{ m} < R < 3 \text{ m}$$

$$0.011 < n < 0.040$$

For sandy canal with the Manning roughness coefficient smaller than 0.03, the difference between the Manning equation and Pavlovski equation can be neglected [3]. A significant fact is that the Manning roughness coefficient in the two equations is considered to be a constant, one that does not vary with the flow conditions. Textbooks on hydraulics present tables of Manning roughness coefficients for various boundary materials under different conditions. Other tables of roughness (such as the tables for permanent streams by Podapov of Soviet Union) present different n values for different flow depths in stream channels having

259

same river characteristics [4]. Except for the special situation of vegetation growing in a stream channel, the *n* values do not change much with the depths. Barnes presented color photos of streams of various types with corresponding *n* values, and they are convenient to use.

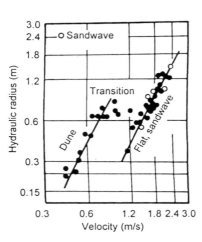

Fig. 7.9 Relationship between hydraulic radius and velocity for the Rio Grade River, USA (after Culbertson, J. K., and C. F. Nordin, Jr)

Fig. 7.10 for various dunes, the same discharge at different water depths for a constant slope (flume experiment after Kennedy, J. F.)

For alluvial streams, especially sandy streams, the pattern of roughness is much more complex. Among the factors contributing to the resistance, several depend on the flow condition, and not only on the boundary characteristics. In particular, the status of the bed configuration significantly affects the roughness. The roughness may increase by several hundred percent for streams with various bed configurations. In such cases, the roughness coefficient is surely not a constant. If one nonetheless takes the roughness coefficient to be a constant, the exponents of R and J in the Manning equation cannot remain at *2/3* and *1/2*, respectively. Lacey analyzed the data of canals in India (movable bed) and suggested the following equation as a substitute for the Manning equation:

$$U = 16 R^{2/3} J^{1/3} \tag{7.15}$$

Malhotra proposed the use of different exponents in the above equation [6]:

$$U = 18 R^{5/8} J^{1/3} \tag{7.16}$$

Clearly these two equations are only appropriate for certain conditions of flow and sediment, and one should not apply them widely. In fact, Liu analyzed a vast number of data from laboratory flumes and small natural canals, and found that for flow in moveable channels the Manning equation should be generalized in the form

$$U = C_a R_b^x J^y \qquad (7.17)$$

in which, R_b is the hydraulic radius for the bottom, and the coefficient C_a and exponents x and y vary over a wide range with the size of the bed material and the bed form, as shown in Fig. 7.7 [7]. However, until the formula suggested by Liu has been verified with more data from natural streams, it should not be used generally, but his study indicates both the complexity of the resistance of flow in channels with movable beds and the importance of the role played by the bed configuration.

The real situation is even more complex than that indicated so far. If a dune on the bed gradually diminishes in size, it can eventually disappear; the resistance experienced by flow then decreases greatly as the velocity increases. For the same slope and water depth, the bed may form dunes, or be flat without dunes, and the resistances for these two cases are quite different; correspondingly, the velocities are also different. As a result, different unit discharges can flow at the same slope and water depth. Fig. 7.8 shows the results of flume experiments by Vanoni and Brooks [8] in which four types of fine sediment were used. For each set of experiments, the water depth was kept constant, and the slope was adjusted. The results show that for a certain range of slopes, quite different velocities can occur, even for the same slope and water depth. The same situation can also occur in natural streams. Fig. 7.9 shows the relationship between the hydraulic radius and velocity measured for the Rio Grande in the United States [9]. For a hydraulic radius of about 0.7 m, the velocity ranged from 0.8 to 1.5m/s, and the unit discharge thus changed correspondingly. Furthermore, since different discharges can occur for given value of depth and slope, so also can different water depths occur for a given discharge and slope. Fig. 7.10 shows the results of flume experiments by Kennedy [10]. In effect, they confirmed the above possibility, although the phenomenon occurred only within a narrow range of flow conditions [10]. Obviously, these phenomena are difficult to comprehend because they cannot be explained in terms of any of the existing formula for flow resistance.

In this situation, only after the intrinsic relationship between the roughness and the flow has been determined, can one understand resistance in alluvial streams. The roughness of an alluvial stream has various components and each component has its own relationship to the flow. Hence, a rational approach should be first to clearly demonstrate how each component functions for various conditions of flow, and then to determine how these components combine and function to provide a comprehensive picture of roughness for an alluvial stream.

The overall resistance of an alluvial stream should reflect a suitable separation of the functions of each of the resistance components and their special significance in sediment movement. As pointed out in Section 1 of this chapter, the mechanical energy of flow is transmitted by shear stress, is concentrated near the bed boundary and produces turbulence there. Because each component of resistance is affected by different portions of the bed, and the turbulence created by each resistance component also occurs at different distances from the bed, so their contributions to sediment movement are not the same.

For grain friction, the eddies created by the corresponding flow potential energy from grains on the channel bed play a large role in the transportation of bed material. In contrast, the turbulence created by the floodplain resistance can have only a moderate effect on the sediment movement on the floodplain, and no direct effect on that in the main channel. The same conclusion can be reached for bed form resistance. Hence, Leopold et al. declared that the effective force for sediment movement can be decreased by the irregularity of the section geometry and the stream pattern to stable the channel bed [2].

The relationship between bed form resistance and sediment movement is not easy to establish. Although grain friction and bed form resistance both act on the bed surface, the ways in which they affect the movement of bed material are different. As already mentioned, the formation of bed form resistance is the result of the separation of flow at the peaks of sand waves and the unsymmetrical distribution of pressure on the stoss and lee faces. The turbulence created by bed form resistance occurs mainly in the separation on the lee face, and it occurs at some distance from the bed grains (Fig. 7.11). The role of the eddy created by bed form resistance on bed load movement is thus not as direct as that from the grain friction. The turbulence created by bed form resistance, of course, has an effect on the suspended load movement. The distribution of the concentration of suspended load in the vertical direction shows that its gradient is normally a maximum near the bed. There, the turbulence created by grain friction plays a decisive role. At some distance from the bed, the gradient is much less. At present, the difference between the combined turbulence, created by both grain friction and bed form resistance or that by grain friction only, can be neglected. Generally, the grain friction is dominant for bed load transportation. However, for the suspended load transportation, some scientists believe that both grain friction and bed form resistance play significant roles[11], and others think that the bed form resistance is much less important than the grain friction [12].

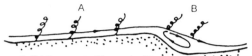

A:Eddies created by grain friction originate near the bed grains. **B**: Eddies created by bed form resistance originate in the separation zone on the lee face, some distance from the bed

Fig. 7.11 Sketch of turbulence created by grain friction and bed form resistance

Bagnold analyzed the results of flume experiments by Gilbert and the Waterway Experiment Station of USA. He found that as the shear stress τ_0 acting on the bed increases and ripples grow into dunes of large size, and the bed load movement becomes weak and discontinuous, as shown in Fig. 7.12 [13]. The figure indicates that, when dunes form on the bed, although the existence of bed form resistance increases the total resistance to the flow, this part of the resistance does not affect the sediment movement; on the contrary, the sediment concentration of bed load tends to be less.

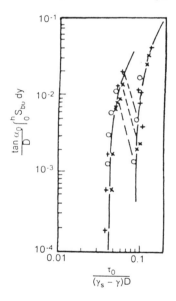

Fig. 7.12 Discontinuity of bed load movement in the process of dune evolution (S_{bv} is the bed load concentration by volume, τ_0 is the bed shear stress, α_0 is the ratio of the grain shear stress on the bed surface to the dispersion force among particles) (after Bagnold, R.A.)

Vanoni, Brooks, and Kennedy also found in flume experiment that the total shear stress on the bed, as a dependent variable to denote the intensity of sediment movement, is not an appropriate choice [8,10]. Fig. 7.13 shows the results of flume experiments by Nomicos, in which the water depth and temperature remained unchanged [8]. The figure shows that at the stage of dunes, the sediment concentration increases with the increase of shear stress, but the absolute concentration value is relatively small. And if the dunes attenuate and disappear, although the shear stress decreases, the concentration still increases. Once the bed becomes flat again, the sediment concentration increases with the increase of shear stress. The whole curve is S-shaped, i.e., for the same shear stress different sediment concentrations can occur. Similar results were observed in the experiment by Brooks, in which he confirmed the fact that the bed form resistance is almost useless in analysis of sediment transportation. Einstein and Chien assumed that only the grain friction is related to the sediment movement, they then reanalyzed the experimental data of Brooks, and obtained consistent results [14].

In summary, in order to understand correctly the mechanism of resistance in alluvial streams, one must treat the resistance components individually, i.e. study the relationship between them and the flow, and determine how the components combine together to produce the total roughness to flow in alluvial streams. Similarly, if the total roughness has been estimated from experimental data, one must also identify each component of the comprehensive resistance, especially the grain friction. Only then can one calculate the rate of sediment transport in alluvial streams through the mechanical relationship between the grain friction and sediment movement. The next steps are to treat the link between the comprehensive and component resistances, to introduce expressions for each component.

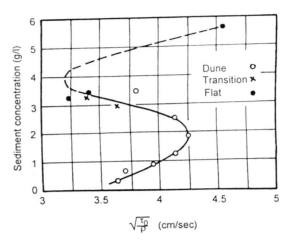

Fig. 7.13 Different sediment concentrations for the same bed shear stress (flume experiment, h=7.35cm, D_{90}=0.152 mm)

7.4 RELATIONSHIP BETWEEN THE COMPREHENSIVE RESISTANCE AND ITS COMPONENT PARTS

In treating the relationship between the comprehensive and component resistances, two situations should be dealt with separately. One is that the several component resistances act on the different elements of the boundary; for instance, one can readily identify the boundary elements on which floodplain resistance, bank resistance, and bed resistance act. Sometimes the roughnesses of different parts of the bed also differ, and on one part of the bed the water can be deeper, with sand waves, although another part of the bed is flat. In another situation, various component resistances act on the same boundary, for instance, grain friction and bed form resistance. The bed form resistance is a part of the bed resistance, but it is normally concentrated in one region; the relationship between it and grain friction and bed form resistance is similar to that between the local resistance and the boundary resistance along the course.

7.4.1 Resistance components acting on different boundaries

For convenience, a simple river channel with a rectangular cross section of flow area A and wetted perimeter $p(=B+2h)$ is considered initially. In a unit distance, due to the boundary friction to flow, the resistance to the flow is $p\tau_0$, in which, τ_0 is the

264

average shear stress on the boundary. Also, the flow is acted on by gravity, which has a component $A\gamma J$ along the direction of flow (γ is the unit weight of water, J is the gradient). For uniform flow, J is identical to the energy slope and the flow is not accelerating; hence

$$A\gamma J = p\tau_0$$

$$\tau_0 = \gamma \frac{A}{p} J = \gamma R J \qquad (7.18)$$

in which, R is the hydraulic radius.

This flow resistance actually has two parts, one is the bank resistance acting on the two bank walls, expressed by τ_w, the other is the bed resistance acting on the bed, expressed by τ_b. The wetted perimeter of the former is $p_w=2h$, and for the latter, $p_b=B$. Both parts can be expressed by a formula similar to Eq. (7.18) in which the variables should correspond to the component parts of the resistance.

7.4.1.1 Physical models to divide the bed resistance and bank resistance

In Eq. (7.18), the unit weight of water is the same for bed resistance and bank resistance; only the hydraulic radius R and energy slope J should be subdivided. If the hydraulic radius is taken as a geometric characteristic of the whole cross-section and a constant that is independent of the various resistance components, the energy slope must be divided into two parts, one J_w corresponds to the bank resistance, and the other J_b to the bank resistance, then

$$\left.\begin{array}{l} \tau_w = \gamma R J_w \\ \tau_b = \gamma R J_b \end{array}\right\} \qquad (7.19)$$

The energy slope actually indicates the energy loss of a unit weight of water in a unit length due to resistance. Except for the small part of the flow energy that dissipates directly into thermal energy, most of the energy loss is transmitted to the boundary through the shear stress field, and it is transformed into turbulent kinetic energy there. Division of the energy slope according to the resistance components implies that any unit water element (such as A in Fig.7.14) transmits energy to all resistance components separately, as shown in Fig.7.14a. Such a physical picture is hard to accept because that water element is close to the left side wall. The energy is thus more readily transmitted to the left side wall than to the remote right wall or channel bottom, as shown in Fig.7.14b.

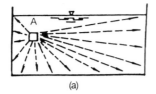
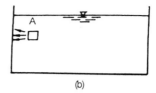

(a)   (b)

Fig. 7.14 Sketch of energy transmission for a unit water element
to the relevant boundary resistance

In another model, the energy slope is taken as a constant but the hydraulic radius is divided according to the resistance components. For the above example, the expressions for the bed resistance and bank resistance are

$$\left.\begin{array}{l} \tau_w = \gamma R_w J \\ \tau_b = \gamma R_b J \end{array}\right\} \quad (7.20)$$

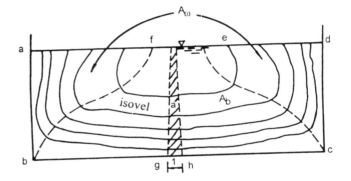

Fig. 7.15 Division of flow section into different energy regions

Because the energy slope is considered to be the same for both parts, the energy from a unit element of water transmits only in one direction: to the left wall, to the right wall, or to the channel bed. Thus from the view of energy the whole flow section can be divided into three parts, as shown in Fig. 7.15. The energy concentrated on the left wall ab (for a unit thickness in the flow direction) is from the potential energy of water body abf, and the turbulent kinetic energy created on ab finally dissipates into thermal energy in the region abf. Similarly, the flow region $bcef$ and the bottom bc combine together as one group in energy transmission, and the region cde and right wall cd as another. No energy is exchanged across the faces bf and ce.

The energy component is not limited in number. For the area gh on a unit boundary perimeter (Fig. 7.15), the energy transmitted is taken from the water body with a volume a. If the roughness is the same for all perimeter elements, the turbulence created from any area on a unit perimeter is also the same, and the volume of water to provide the turbulent energy is the same too. For a part of the boundary with wetted perimeter p, there are totally p areas with unit boundary perimeter. The energy transmitted to each area is from the water body with a volume a, so pa must equal to the whole volume of the flow, i.e.,

$$A = pa$$

266

or

$$a = \frac{A}{p} = R$$

Hence, the hydraulic radius itself does not fully represent the geometry of a flow section, but it does have a clear physical meaning. From the mechanism of the energy transformation of flow, the turbulence energy created on the area of a unit boundary element comes from the potential energy of a water volume R; the turbulent kinetic energy created there finally dissipates into thermal energy in the same volume of water. The hydraulic radius R represents the water volume possessing this part of the energy [15]. Division or summation of the hydraulic radius only implies the division or summation of the volume of the water body, an understandable process.

In common alluvial streams, the roughness coefficients of side walls and bed are different, and the turbulent intensities there are different, too. The energy taken from the flow by a unit area of the side wall is not equal to one on the bed, hence the hydraulic radius, R_w, corresponding to the bank resistance is not equal to R_b, corresponding to the bed form resistance, instead the two are

$$R_w = \frac{A_W}{2h}$$

$$R_b = \frac{A_b}{B}$$

Here A_w is the sum of areas *abf* and *cde* in Fig. 7.15, and A_b is the area *bcef*. Because the energy transmission must be accomplished through the shear stress field, energy is transformed wherever a velocity gradient exists; hence the boundary faces of energy components *bf* and *ce* are usually drawn perpendicular to the isovels, as shown in Fig. 7.15.

The above picture is only a simplified physical model of a complex problem. For a section with a regular geometry and relatively straight or symmetric isovels, the use of this physical model is convenient in obtaining a relationship between the total resistance and its parts. But if a section is irregular and has unsymmetrical isovels with large curvature, then dividing the energy components according to the method of Einstein would introduce appreciable errors [16].

7.4.1.2 Calculation of bed resistance and bank resistance

The method of calculating the bank and bed form resistances of a rectangular flow section according to the principle of division of energy into its component parts needs further discussion. Such an analysis involves the five known variables

$$A, U, J, p_w, p_b$$

The eight unknown variables are

$$A_w, A_b, U_w, U_b, R_w, R_b, n_w, n_b$$

Thus, eight equations are required, and these are as follows:

1. Continuity of geometry:

$$A = A_w + A_b \tag{7.21a}$$

2. Continuity of flow:

$$AU = A_w U_w + A_b U_b \tag{7.21b}$$

3. Definition:

$$R_w = A_w / p_w = \frac{A_w}{2h} \tag{7.21c}$$

$$R_b = A_b / p_b = A_b / B \tag{7.21d}$$

4. Manning equation:

$$U_b = \frac{1}{n_b} R_b^{2/3} J^{1/2} \tag{7.21e}$$

$$U_w = \frac{1}{n_w} R_w^{2/3} J^{1/2} \tag{7.21f}$$

5. Theoretical or empirical formulas:

$$n_b = f_1(R_b, U_b, v, \Delta_b, S_b) \tag{7.21g}$$

$$n_w = f_2(R_w, U_w, v, \Delta_w, S_w) \tag{7.21h}$$

S_b, S_w --shape coefficients of water body corresponding to the bed form and bank resistances, respectively; Δ_b, Δ_w --absolute roughnesses of the bed and side walls; v --kinematic viscosity of water.

Normally, the river bank is rigid, so that its roughness coefficient can be determined empirically or in other ways, but that for the river bed is not easy to determine in advance. Thus, of above eight equations, Eq. (7.21g) is, in fact, not established. Hence one more equation is needed to solve the problem.

At present, three different methods are used:

1. Jiang method [17]. From a summation of shear stress,

$$\tau_b p = \tau_0(p_b + p_w) = \tau_w p_w + \tau_b p_b$$

If the Manning equation is used for each resistance component in the above equation, the result is

$$\frac{m^2 U^2}{R^{1/3}} p = \frac{m_w^2 U_w^2}{R_w^{1/3}} p_w + \frac{m_b^2 U_b^2}{R_b^{1/3}} p_b$$

in which n is the effective roughness for the whole section, and

$$R = A / (p_b + p_w)$$

If

$$\frac{U^2}{R^{1/3}} = \frac{U_w^2}{R_W^{1/3}} = \frac{U_b^2}{R_b^{1/3}} \qquad (7.21i)$$

then

$$n^2 p = n_w^2 p_w + n_b^2 p_b \qquad (7.22)$$

The above equation implies that the resistances on the different parts of the boundary are distributed according to square of the Manning coefficients.

2. Einstein method [14]. Einstein assumed the additional condition that

$$U = U_w = U_b \qquad (7.21j)$$

then

$$R_w = (\frac{n_w U}{J^{1/2}})^{3/2} \qquad (7.23)$$

$$R_b = h(1 - 2 \frac{R_w}{B}) \qquad (7.24)$$

From Eq. (7.20), one can readily obtain the corresponding bank and bed form resistances.

If Eq. (7.21j) is used instead of Eq. (7.21i), one obtains

$$n^{3/2}p = n_w^{3/2}p_w + n_b^{3/2}p_b \qquad (7.25)$$

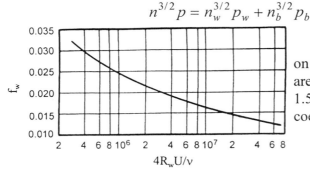

Fig. 7.16 Relationship between $4R_wU/\upsilon$ and f_w for a hydraulically smooth bank

In this case, the resistances on different boundary elements are distributed according to the 1.5 power of the Manning coefficient.

3. Lotter method [19].

Early in the 1930s, Lotter derived the expression

$$\frac{pR^{5/3}}{n} = \frac{p_wR_w^{5/3}}{n_w} + \frac{p_bR_b^{5/3}}{n_b} \qquad (7.26)$$

Knight and MacDonald analyzed the reliability of the additional condition in Eq. (7.21j) [20]. They carried out a series of experiments in a rectangular flume with rough bed and smooth side walls ($h/B = 0.1$-0.7); they found that U_w/U varied within the range of 0.927 to 1.10 and had an average value of 1.015. Yassin carried out an experiment in a flume with smooth side walls and a fixed bed composed of sand and gravel. The results were basically the same as those of Einstein [21]. In fact, according to Han's analysis, the additional condition Eq. (7.21j) possesses deeper physical meaning in that it follows the principle of minimum energy dissipation. [1] If the hydraulic factors (i.e. the comprehensive roughness n) and side wall roughness n_w of the section are known, and if the weighted distribution of the resistanceis accordance with Eq. (7.25), then the corresponding R_b is that for which the bed roughness n_b is a minimum. Nevertheless, the experiments of Motayed and Krishnamurthy indicated that among the above three methods, the Lotter method produced the smallest errors [22].

If the banks (side walls) are effectively hydraulically smooth, then one should adopt the Darcy-Weisbach resistance equation instead of the Manning equation. If f, f_w, f_b are the resistance coefficients of the whole section, bank and bed, respectively,

$$f = 8gRJ/U^2$$
$$f_w = 8gR_wJ/U_W^2$$
$$f_b = 8gR_bJ/U_b^2$$

---

[1] Yangtze River Hydraulic Research Institute. *Preliminary Analysis of Change in Roughnesses of the Three Gorges Reservoir before and after being Commissioned*, 1977, pp. 20.

If Eq. (7.21j) is tenable, then

$$R / f = R_w / f_w = R_b / f_b \tag{7.27}$$

If the bank is hydraulically smooth in effect, the standard relationship between f_w and Reynolds number $4R_w U / \upsilon$ exists, shown in Fig. 7.16. From this relationship and Eq. (7.27), the bank resistance coefficient f_w and bed resistance coefficient f_b can be obtained:

$$f_b = f + \frac{2h}{B}(f - f_w) \tag{7.28}$$

7.4.1.3 Separation of channel resistance and floodplain resistance

The channel resistance and floodplain resistance can also be divided on the foregoing principle, i.e., by using eight equations corresponding to Eq. (7.21). On floodplains, grasses and bushes normally grow, and they can be treated as fixed beds, so that the roughness can be estimated (details are given later); but the channel roughness is still unknown, so one additional condition is required. As the water on floodplains is much shallower than that in main channel and the roughness there is much larger than that on main channel, the assumption

$$U_{whole\ section} = U_{flood\ plain} = U_{channel}$$

is not suitable. Also the floodplains are often wide and a natural levee often forms adjacent to the main channel. The flow section can be divided into two parts by two vertical lines. One part is the river channel area ($A_{channel}$), where the flow energy is transformed into turbulence through the channel resistance. The other is the floodplain area ($A_{flood\ plain}$), where the flow energy is transformed into turbulence through resistance on the floodplain. Although the position of the dividing line may not be well-defined, the induced error is not likely to be large. From the flow areas of the channel and floodplains, the channel resistance and floodplain resistance can then be determined. In addition, the river channel often meanders, but the flow over the floodplains tends to be straight. Thus the slopes for the two flows are often different.

7.4.2 Treatment of resistance components acting on the same boundary

The grain friction and bed form resistance are used in an example to illustrate how to divide the two resistance components that combine to make up the channel resistance.

Although both grain friction and bed form resistance act on the bed surface, the turbulence created by the former originates directly from bed material particles near the bed, and that due to the latter originates in the separation zone near the downstream face of a sand wave peak. They are basically independent. For the energy, one can neglect any energy exchange between the flows in the two zones.

In the study of flow resistance, the skin friction (i.e., grain friction) and form resistance (i.e., bed form resistance) often exist simultaneously. For instance, the resistances borne by a ship running on a water surface consist of two parts: the skin friction on the surface and the form resistance of the ship. They are normally combined by using the simple sum of the component resistances to obtain the comprehensive resistance, i.e., the total resistance borne by the ship is equal to the sum of skin friction and form resistance. The next question is whether this principle of summation of resistance components can be applied to the bed resistance of alluvial streams.

To clarify this problem, Einstein and Banks conducted two sets of experiments [23, 24]. The first set was conducted on a rigid bed. The resistance components were formed by concrete blocks placed on a flat surface; the blocks were alternately high and low, and bolts of various shapes were inserted in the blocks in different patterns and densities. As shown in Fig. 7.17, the resistance includes the skin friction of the blocks and also the form resistance created by the blocks and the bolts.

A series of experiments was conducted with water flow over different bed forms: (1) concrete blocks were on flat surface to obtain the surface resistance of blocks τ_b; (2) then the bed consisted of flat blocks and bolts, the form resistance of bolts τ_p could be obtained by subtracting τ_b from the total resistance borne by flow; (3) next, water flowed the bed with alternately high and low blocks so that the form resistance τ_s could be obtained by subtracting τ_b from the measured resistance; (4) last, the bed consisted of alternately high and low blocks and bolts, the measured resistance was compared with the sum of τ_b, τ_p, and τ_s to test the principle of summation of resistance components.

An analysis of the experimental results gave

$$\frac{\tau_b}{\frac{1}{2}\rho U^2} = 0.0104$$

$$\frac{\tau_p}{\frac{1}{2}\rho U^2} = 0.00036N$$

$$\frac{\tau_s}{\frac{1}{2}\rho U^2} = 0.01456$$

in which N is the number of

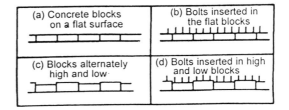

Fig. 7.17 Disposition of the experiments for testing the principle of summation of resistance components (the first set)(after Einstein, H.A., and R.B. Banks)

272

bolts inserted on a unit area (m^2). Thus, if all three resistances were involved, the principle of summation of resistance gave the total resistance for the flow

$$\frac{\tau}{\frac{1}{2}\rho U^2} = 0.0104 + 0.00036N + 0.01456 \tag{7.29}$$

Fig. 7.18 shows the comparison of the experimental results with Eq. (7.29). The data fall near the line and on both sides of it, so the principle of summation of resistance in the experiment range is tenable.

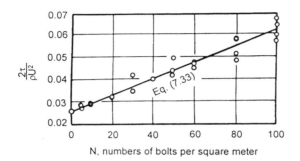

Fig. 7.18 Test of the principle of summation of resistance (first set)

The second set of experiments by Einstein and Banks was conducted on a movable bed in which regularly space baffles were inserted, as shown in Fig. 7.19. Under this condition, the flow resistance consisted of grain friction, bed form resistance, and form resistance caused by the baffles. The experimental procedures were the same as in the first set; the three components of the resistance and the total resistance were determined and are shown in Fig. 7.20. The results again support the principle that the resistances can be added together to obtain the bed resistance, even for alluvial streams. In the experiments, lightweight sediment (specific weight 1.052) was used so the results were more nearly comparable to those of natural sediment in rivers.

The principle of summation of resistance is tenable if the faces on which the resistance components act are fully separated, and if they partly overlap but have different heights, so the resistance components do not affect each other. If sand waves form on the bed, the separation zone on the lee side of a sand wave will reduce the bed area in direct contact with the flow, and correspondingly reduce the skin friction due to sediment particles. Also, where the flow returns to the stoss side of the next sand wave, the turbulence intensity is greater, so the local resistance loss is greater. In balance, the grain friction and sand wave resistance can be taken as independent in their effect contributions to the total resistance. For these conditions, the principle of summation of resistance is applicable. However, this principle is only a tool used to deal with the resistance problem; it is tenable only within certain ranges and cannot be employed for others.

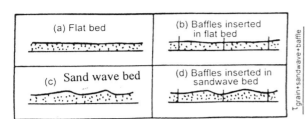

Fig. 7.19 Disposition of the experiments for testing the principle of summation of resistance in alluvial streams (second set)

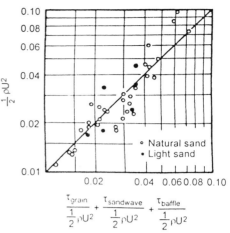

Fig. 7.20 Calibration of the principle of summation of resistance (second set)

Two different approaches are used in calculations based on the principle of summation of resistance. One is that under a certain velocity and flow depth without bed form on the stream bed, the bed slope can be smaller; however, if bed form also exists, the flow requires larger slope. Thus, one can divide the energy slope into two parts corresponding to the grain friction and bed form, respectively [25]. Because the grain friction and bed form resistance occur on the same boundary, it is reasonable that a unit weight water transmits energy simultaneously to the two resistance components. Also in dealing with the grain friction and bed form resistance, Einstein divided the hydraulic radius in the same way as for the bed resistance and bank resistance. Consequently, these two approaches are currently used for dividing the total resistance of alluvial streams into corresponding components in the following ways:

$$
\text{total resistance} (\gamma R J) \begin{cases} \text{bank resistance}(gR_w J) \\ \text{bed resistance}(gR_b J) \begin{cases} \text{grain friction} \dashrightarrow \begin{array}{l} (1)\ gR_b' J \\ (2)\ \gamma R_b J' \end{array} \\ \text{sand wave resistance} \rightarrow \begin{array}{l} (1)\ gR_b'' J \\ (2)\ \gamma R_b J'' \end{array} \end{cases} \end{cases}
$$

A similar approach can be used to combine bed form resistance and the resistance induced by artificial structures and other localized resistances. The floodplain resistance is discussed in preceding sections. In the Three Gorges region of the Yangtze River, narrow and wide gorges alternate so that the cross-section geometry changes significantly in the longitudinal direction; the form resistance induced by these changes accounts for some 5 to 60% of the total resistance. In dealing with this issue, Hui and Chen, using the approach of Einstein, calculated the bed resistance and the bank resistance, and then divided form resistance induced by

the sudden enlargement or contraction of the cross-sections from the total resistance.
[1)]

7.5 COMPONENTS OF RESISTANCE

Expressions for the grain friction, bed form resistance, bank resistance, and floodplain resistance combine to make up the total. Grasses and bushes normally grow on floodplains, so the problem of floodplain resistance is actually one of determining the effect of vegetation on resistance.

7.5.1 Grain friction

Grain friction is usually determined from the resistance acting in two-dimensional flow (without the influence of bank walls) for a flat channel bed. The evolution of bed form described in Chapter 6 indicates that the bed in alluvial streams is flat for two stages of flow. The first stage is for sediment that is either not moving or moving only at a low rate of transport. At this stage, the sediment has no appreciable effect on the flow. The resistances on rigid beds and movable beds do not differ significantly. In the second stage, dunes disappear, but antidunes have not formed even though sediment transport intensity is high. At this stage, sediment movement has some effect on the flow pattern. the movable bed resistance may differ from that of rigid bed.

Flow resistance is usually represented by the resistance coefficient of the boundary, using the Chezy coefficient C, Manning roughness coefficient n, or Darcy-Weisbach friction coefficient f, and they are interrelated with each other:

$$\frac{C}{\sqrt{g}} = \frac{R^{1/6}}{\sqrt{g}}\frac{1}{n} = \sqrt{\frac{8}{f}} = \frac{U}{U_*}$$ (7.30)

Because the resistance coefficients can also be expressed in terms of the ratio of the mean velocity of flow to the friction velocity, which can be obtained from the vertical velocity distribution, the resistance is directly related to the velocity distribution.

7.5.1.1 Grain friction on rigid bed

Nikuradse, in the thirties, was the first one to stick uniform sediment particles on the inner pipe wall to study grain friction on rigid surface [26]. For this situation, the

---

[1)] Hui, Yujia & Chen, Zhicong. *"Preliminary Analysis of the Roughness on the Three Gorges Reaches of the Yangtze River,"* *Research Report of Dept. Of Hydraulic Engineering*, Tsinghua University, 1981.

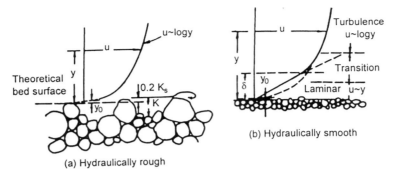

(a) Hydraulically rough

(b) Hydraulically smooth

Fig. 7.21 Flow on rough bed and flow on smooth bed

bed roughness can be expressed as K_S. The Nikuradse study has had a far-reaching influence on later research. However, questions concerning his experimental results were raised in the late 1960s.

1. Hydraulically smooth or rough bed. Turbulence is not fully developed near the bed where the flow is dominated by viscosity, as discussed in Chapter 4; a near-bed laminar sublayer forms with an effective thickness *of* δ. If the size of the roughness is much larger than the thickness of the laminar sublayer, then the sediment particles on the bed protrude into the turbulent region so that the flow is directly affected by roughness, and the flow can even separate from the tops of the particles [as shown in Fig. 7.21a], thus more eddies occur and promote the instability of the flow. This kind of channel bed is called hydraulically rough. At the other extreme, if the roughness is much smaller than the thickness of the laminar sublayer, then the sediment particles are submerged within the laminar sublayer most of the time, as shown in Fig. 7.21b. Only when eddies coming from outside of the laminar sublayer sweep the bed surface do the sediment particles on the bed make direct contact with the turbulent flow. Because the thin laminar sublayer that is dominated by viscosity comes between the bed and main flow, the resistance to the flow is rather small and has no direct relationship to the shape and size of particles. This kind of bed is called hydraulically smooth. Whether the bed is hydraulically smooth or rough does not depend on the absolute size of the bed material but on the ratio of the particle size to the thickness of the laminar sublayer. Generally, if

$$K_s / \delta > 10 , \quad \text{the bed is hydraulically rough}$$

$$K_s / \delta < 0.25 , \quad \text{the bed is hydraulically smooth}$$

$$0.25 < K_s / \delta < 10 , \quad \text{the bed is in the transitional region}$$

2. The logarithmic velocity distribution formula. From the principle of kinetic momentum exchange and the hypothesis of turbulence similarity, pointed out in

Chapter 4, the following relationship exists between the shear stress field and the velocity field

$$\tau = \rho k^2 \frac{(du/dy)^4}{(d^2u/dy^2)^2} \tag{7.31}$$

For 2-D flow, the shear stress is distributed linearly in the vertical direction,

$$\tau = \tau_0(1 - y/h) \tag{7.32}$$

in which: τ_0 -- the shear stress at a point y from the bed;

τ -- the shear stress on the bed $(=\gamma h J)$.

Introducing the above equation into Eq. (7.31), one obtains

$$\frac{u_{max} - u}{U_*} = -\frac{1}{k}\left\{(1 - \frac{y}{h})^{1/2} + 1n[1 - (1 - \frac{y}{h})^{1/2}]\right\} \tag{7.33}$$

in which: u is the velocity at a point y from the bed, u_{max} the surface velocity ($y = h$), U_* the friction velocity $(= \sqrt{ghJ})$.

If the friction velocity is a constant, the difference between the maximum velocity and the local velocity is a function only of the position within flow, and it is therefore not related to the discharge or the roughness.

The shear stress does not change significantly near the bed, and it can be taken as a constant and equal to the shear stress on the bed τ_0. Therefore,

$$\frac{d^2u/dy^2}{(du/dy)^2} = -k/U_*$$

Solving, one obtains

$$\frac{u}{U_*} = \frac{1}{k}1n(\frac{y}{y_0}) \tag{7.34}$$

in which y_0 denotes the distance from the point of zero velocity to the bed.

Since the turbulent mixing length l is equal to:

$$l = k\frac{du/dy}{d^2u/dy^2}$$

one can show that setting $\tau = \tau_0$ (a constant) is equivalent to the approximation

$$l = ky \tag{7.35}$$

The turbulent mixing length l is a kind of measure of the eddy size. In the region near the bed, the eddy size should be proportional to the distance to the bed (Fig. 4.32).

Although Eq. (7.34) is limited to the bed region, experimental results show that the error induced by extending its range of application into the main flow region is not significant. Since Eq. (7.34) is much simpler than Eq. (7.33), that formula is generally employed to describe the vertical velocity distribution in 2-D open channel flow.

In essence, y_0 in Eq. (7.34) is clearly one of the characteristic lengths of turbulence near the bed, and it varies with the friction velocity, roughness, and other factors. For a hydraulically smooth bed, the factors affecting the velocity distribution are primarily the shear stress on the bed and the density and viscosity of water. From dimensional analysis, one can readily infer

$$y_0 \sim v / U_*$$

Introducing this expression into Eq. (7.34), and determining the constant from the results of Nikuradse's pipe experiments, one obtains

$$\frac{u}{U_*} = 5.5 + 5.75 \log(\frac{yU_*}{v}) \tag{7.36}$$

The above equation was derived from pipe flow experiments. In turbulent flows, the equation for the velocity distribution derived from a watercourse with a certain geometry can be applied to the flow section with another geometry for the reason that the velocity at any point near the boundary is mainly dependent on the shear stress on the boundary near that point. Thus it has little relationship to the shear stresses on other parts of the boundary. In the region far from the boundary, the velocity gradient is relatively small. Keulegan showed that Eq. (7.36) can also be used for 2-D open channel flow over a hydraulically smooth bed [27].

For a hydraulically rough bed, the form resistance induced by boundary roughness is much greater than the shear due to viscosity, in this case

$$y_0 \sim K_s$$

The coefficient can be determined from Nikuradse's results and the open channel experiments of Bazin, and it can be written as

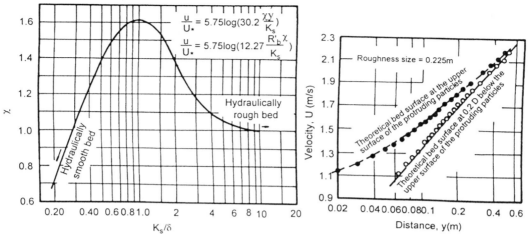

Fig. 7.22 Relationship between χ and K_s/δ in the logarithmic velocity distribution formula for hydraulically smooth and rough beds (after Einstein, HA.)

Fig. 7.23 Velocity distribution measured in flow over a hydraulically rough bed (after Einstein, H.A., and A. El-Samni)

$$\frac{u}{U_*} = 8.5 + 5.75\log(\frac{y}{K_s}) \tag{7.37}$$

Eqs. (7.36) and (7.37) can be combined into one equation:

$$\frac{u}{U_*} = 5.75\log(30.2\frac{yx}{K_s}) \tag{7.38}$$

in which χ is a function of K_s/δ. If $K_s/\delta > 10$, i.e., the bed is hydraulically rough, $x=1$; if $K_s/\delta < 0.25$, i.e. the bed is hydraulically smooth, $\chi = 0.3K_sU_*/v$; if $K_s/\delta = 0.25\sim10$, i.e., the bed is in a transitional region, χ is also shown in Fig.7.22 [12]. Recent experimental results indicate that the curve is for a roughness composed of uniform sediment particles, as in Nikuradse's experiments. If the bed material is not uniform, the curve is parallel but somewhat lower [28].

From Eq. (7.38), the velocity at the stream bed ($y=0$) would be negatively infinite, a condition that is, of course, physically unreal; hence, the logarithmic velocity distribution formula has been questioned by some scientists. In essence, the logarithmic formula has no meaning in the region near the boundary. For a hydraulically rough bed, the theoretical mean bed surface is not at the upper surface of the protruding particles, but a distance mK_s below it, in which values of m from the experiment results of different researchers vary in the range of 0.15-0.35 [29]. For instance, Einstein and El-Samni used hemispherical particles as the bed material; if they took the plane at the top of the hemispheres as the theoretical bed surface, the data for the velocity distribution formed a curve on semi-logarithmic paper; if they took the plane 0.2 D (D is the diameter of the hemisphere) below the tops as the nominal bed surface, then the data fell on a straight line, as shown in Fig. 7.23 [30].

Thus, if y is close to zero, the flow region extends to levels within the zone of bed roughness. Because of the separation of flow from the top of protruding particles, the flow usually contains eddies, and the velocity distribution, of course, cannot be expressed by Eq. (7.38). Similarly, for a hydraulically smooth bed, a laminar sublayer forms near the bed. The flow there is dominated by viscosity, and the velocity distribution is naturally different from that of turbulent flow. In the laminar sublayer near the bed,

$$\tau_0 = \mu \frac{du}{dy}$$

If the shear stress is a constant, then

$$\frac{u}{U_*} = \frac{U_* y}{v} \tag{7.39}$$

Eq. (7.36) intersects Eq. (7.39) at $y=11.6v/U_*$, as shown in Fig. 7.21b. And the corresponding velocity is $11.6U_*$. Generally the flow region below the intersection is taken as the laminar sublayer, and its thickness is

$$\delta = 11.6v / U_* \tag{7.40}$$

Actually, the transition from laminar flow to turbulent flow is more gradual and occurs over a region; the effect of viscosity of water is significant up to about $y=30v/U_*$

Recently, Dou combined the velocity distribution formulae for the main flow region and near-bed laminar sublayer into one expression for flow over hydraulically smooth bed [31]:

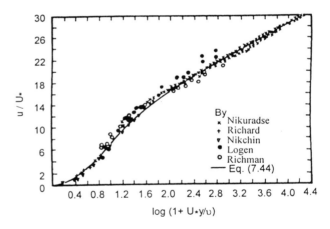

Fig. 7.24 Comparison of Eq. (7.41) with experimental results for hydraulically smooth bed

$$\frac{u}{U_*} = 2.5\ln(1 + \frac{U_* y}{5v}) + 7.05 \left[\frac{\dfrac{U_* y}{5v}}{1 + \dfrac{U_* y}{5v}} \right]^2 + 2.5 \left[\frac{\dfrac{U_* y}{5v}}{1 + \dfrac{U_* y}{5v}} \right] \tag{7.41}$$

If $\dfrac{U_* y}{5v}$ is large, this equation is equivalent to Eq. (7.36), but if $\dfrac{U_* y}{5v}$ is small, it is consistent with Eq. (7.39). Fig. 7.24 compares Eq. (7.41) with the experimental results.

If one knows the velocity distribution, one can determine the mean velocity for a flow section by integration. For the observed data of Bazin, and for 3-D flow over a hydraulically smooth bed, Keulegan obtained

$$\frac{U}{U_*} = 3.25 + 5.75\log(\frac{RU_*}{v}) \tag{7.42}$$

For a hydraulically rough bed,

$$\frac{U}{U_*} = 6.25 + 5.75\log(\frac{R}{K_s}) \tag{7.43}$$

in which

$$U_* = \sqrt{gRJ}$$

With the unique correction value χ from Eq. (7.38), Eqs. (7.41) and (7.42) may be combined into one

$$\frac{U}{U_*} = 5.75\log(12.27\frac{R\chi}{K_s}) \tag{7.44}$$

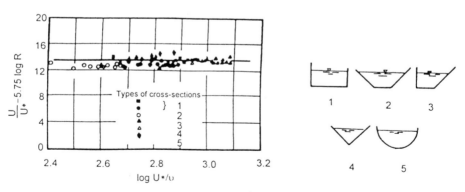

Fig. 7.25 Relationship between $U/U_* - 5.75logR$ and U_* / v for flow sections with the same roughness and different geometry (after Keulegan, G.H.)

Keulegan showed that $\dfrac{U}{U_*} - 5.75\log R$ is basically a constant for flow sections with the same roughness but different geometries; hence, the slope, roughness, and hydraulic radius are the same, the mean velocity does not change with the geometry, as shown in Fig. 7 25 [27].

The remaining question is how to determine the roughness size K_S in Eq. (7.38) or Eq. (7.44). In Nikuradse's experiments with uniform sand particles, the K_S value was taken to be the diameter of the glued

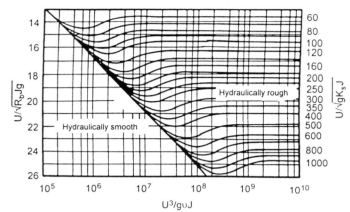

Fig.7.26 Graph for deriving the hydraulic radius of grain friction based on mean velocity

particles, D. For nonuniform sediment, experimenters usually select a relative coarse representative diameter for the roughness size like D_{65} [30], D_{75} [32], or D_{90} [25]. A review of Nikuradse's work showed that he stuck sediment particles on the inner surface and then applied a layer of varnish to ensure that the particles were fixed in place, this process reduced the bed roughness somewhat, so that the effective roughness smaller than the actual particle size. Engelund [33] and Bayazit [29] proposed the use of a value 2.5 times the diameter as the roughness size for grain friction (For nonuniform particles, the particle size should be one whose fall velocity equals to the average fall velocity). Kamphuis recommended $K_S=2D_{90}$ [34]. For streams with gravel beds, because protruding coarse particles have a rather large effect on the roughness, the roughness value should be larger. Bray suggested a value of $3.5D_{84}$ [35] and Charlton one of $3.5D_{90}$ [36].

Finally, the above derivation is only applicable for open channel flow without sand waves on the bed; for mean velocity of cross-section, the roughnesses of two side walls and bed are the same. If sand waves exist or if the roughness of two side walls is not equal to that of bed, the grain friction is related to sediment movement, formulas are still available according to the Einstein system. However, the hydraulic radius in the formulae should be replaced by the hydraulic radius that corresponds to the grain friction, R_b', and the friction velocity, U_* should be replaced by the friction velocity that corresponds to the grain friction, $U_*'(= \sqrt{R_b'gJ})$. Thus, if Eq. (7.44) is used to derive R_b', a trial calculation is required. Use of graphic method of Fig. 7.26 simplifies the process [8].

3. Manning formula. The Manning formula is widely used

$$U = \frac{1}{n} R^{2/3} J^{1/2} = \frac{0.32}{n} R^{1/6} \sqrt{RgJ}$$

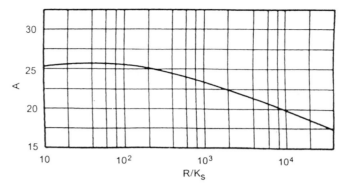

Fig. 7.27 Relationship between coefficient A and relative roughness K_s/R

i.e.,

$$\frac{U}{U_*} = 0.32\frac{R^{1/6}}{n} \tag{7.45}$$

From dimensional analysis, n should be proportional to the one-sixth power of a characteristic length; if no sand waves exist, the characteristic length should be the roughness size K_s. If one lets

$$n = \frac{1}{A}K_s^{1/6} \tag{7.46}$$

in which K_s, in meters, is substituted into Eq. (7.45), the result is

$$\frac{U}{U_*} = 0.32A(\frac{R}{K_s})^{1/6} \tag{7.47}$$

Eq. (7.47) is equivalent to Eq. (7.43),

$$0.32A(\frac{R}{K_s})^{1/6} = 6.25 + 5.75\log(\frac{R}{K_s})$$

From this, a relationship can be derived between the coefficient A and relative roughness K_s/R, as shown in Fig. 7.27. Strickler obtained $A=24$ from an analysis of data from some streams with gravel beds in Europe [37]. For the data of the Lower Yellow River, $A=19$ [38]. Fig.7.27 shows that these results are special cases for specific relative roughnesses. At $R/K_s=30$, A reaches the maximum value of about 26. Generally, values of A for large rivers are smaller than those for small rivers.

Introducing the Strickler coefficient into Eq. (7.47), one obtains the Manning-Strickler formula[33]:

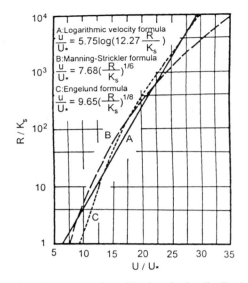

$$\frac{U}{U_*} = 7.68(\frac{R}{K_s})^{1/6}$$

(7.48)

In a somewhat different approach, Engelund proposed an exponent of 1/8 instead of 1/6 commonly used in the Manning formula [33];

$$\frac{U}{U_*} = 9.45(\frac{R}{K_s})^{1/8}$$

(7.49)

Fig. 7.28 Comparison between the logarithmic velocity distribution formula and those of Manning-Strickler and Engelund

Fig. 7.28 is a comparison of equations Eqs. (7.43), (7.48) and (7.49) in the range

$$2<R/K_s<1,500$$

the difference between the Manning-Strickler formula and the logarithmic velocity distribution formula is not significant. And for the range

$$13<R/K_s<15,000$$

the difference between Engelund formula and the logarithmic velocity distribution is less than 5%.

If the Manning formula is combined with the logarithm formula, Eq. (7.43), the result is

$$n = \frac{R^{1/6}}{19.56 + 18\log(R/K_s)}$$

For data from natural streams, Limerinos adjusted the coefficients in the above formula and obtained the following equation in terms of the depth h [39]:

$$n = \frac{h^{1/6}}{10.27 + 17.7\log(h/D_{84})}$$

(7.50)

The data from some streams with gravel beds in Canada correspond with Eq. (7.50) [35, 40].

7.5.1.2 Grain friction with high rates of sediment transport

For high rate flow, dunes do not occur, the bed is flat, and the intensity of sediment movement is quite high; not only does bed load movement exist, but also a large amount of sediment is carried in suspension. So much sediment in the flow affects the flow structure, of course, and it also affects the loss due to flow resistance. This process is discussed in detail in Chapter 12.

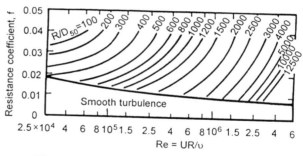

Fig. 7.29 Grain friction to flow on movable bed
(after Lovera, F., and J.F. Kennedy)

Lovera and Kennedy collected data for flow over flat beds in flumes and natural streams (for particle sizes in the range of 0.09--0.79 *mm* and velocities of 0.5--2.4 *m/s*), and they plotted the relationship between resistance coefficient and Reynolds number to obtain the result shown in Fig. 7.29 [41]. They found that the resistance on a rigid bed is quite different from that on a movable bed. For open channel flow on a rigid bed, the resistance coefficient

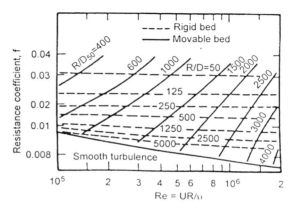

Fig.7.30 Comparison of grain friction on rigid bed and on movable bed

is not related to the Reynolds number if the stream bed is in the hydraulically rough region; but for that on a movable bed, the resistance coefficient not only varies with the relative roughness, but it also increase with increasing Reynolds number. The relationship shown in Fig. 7.30 [41] shows that, for the same relative roughness, the resistance on a movable bed is usually larger than that on a rigid bed. Hence, a question arises as to whether all of the data selected by Lovera and Kennedy were for the condition of a flat bed. That is, aside from grain friction, was any other kind of resistance acting? In their analysis, especially for the data from natural streams, one cannot know the real condition at the stream bed; hence, this question has not been fully resolved.

In addition, Vanoni and Nomicos studied the flow over rigid and movable beds for the same discharge, water depth, sediment concentration, and cross-sectional geometry. The results are shown in Table 7.1. The results indicate that the resistance for a movable bed was not significantly larger than that for a rigid bed.

Table 7.1 Comparison of resistance to flow on rigid and movable bed
(after Vanoni, V.A., and G.N. Nomicos)

Mean diameter (mm)	Bed configuration	Bed	Concentration (g/l)	f_b
0.091	small dunes	movable	4.60	0.0211
		rigid	3.63	0.0211
0.091	flat bed	movable	6.92	0.0170
		rigid	6.82	0.0164
0.148	flat bed	movable	3.61	0.0227
		rigid	3.27	0.0222

7.5.2 Bed form resistance

7.5.2.1 Bed form resistance and loss from sudden enlargement of cross-section

Suppose a series of dunes or ripples exists with wave length of λ, height of Δ, and lee side angle of $\varnothing$. The flow over the peak of a sand wave causes a sudden enlargement in cross-section and a loss in flow energy. The mode of the resistance loss at such bed forms is much like the losses due to a local enlargement in open channel flow.

The head loss due to a local enlargement h_L can be estimated by means of the Kalnold formula:

$$h_L = \alpha \frac{(U_1 - U_2)^2}{2g} \qquad (7.51)$$

in which, U_1 is the velocity at the wave peak, U_2 is the velocity downstream of the wave peak and α is a constant. If q is the unit discharge, then

$$U_1 = \frac{q}{h - \dfrac{\Delta}{2}}$$

$$U_2 = \frac{q}{h + \dfrac{\Delta}{2}}$$

Introducing these values into Eq. (7.51), one obtains

286

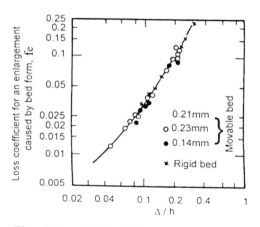

Loss coefficient for an enlargement caused by bed form, f_e

Movable bed
0.21mm
o 0.23mm
• 0.14mm

× Rigid bed

Δ / h

$$h_L = \frac{\alpha q^2}{2g}\left[\frac{1}{h-\frac{1}{2}\Delta} - \frac{1}{h+\frac{1}{2}\Delta}\right]^2 \approx \frac{\alpha U^2}{2g}(\frac{\Delta}{h})^2$$

If f_e is the loss coefficient for an enlargement caused by bed form,

$$h_L = f_e \frac{U^2}{2g}$$

Fig. 7.31 Relationship between the loss coefficient for a local enlargement caused by the bed form and relative height of sand waves (after Chang, F.M.)

and

$$f_e = \alpha(\frac{\Delta}{h})^2 \qquad (7.52)$$

Chang conducted experiments to determine f_e, and the results are shown in Fig. 7.31 [43]. For $\Delta/h>0.1$,

$$f_e = 1.9(\frac{\Delta}{h})^{1.8} \qquad (7.53)$$

This result is close enough to that of Eq. (7.52) to indicate that the resistance induced by bed form is like that at an enlargement.

If f_b'' is the coefficient of bed form resistance,

$$f_b'' = \frac{8gh}{U^2}\frac{h_L}{\lambda} = \frac{4h}{\lambda}f_e \qquad (7.54)$$

Another parameter

$$e = \frac{\Delta \cot \phi}{\lambda} \qquad (7.55)$$

is used that actually reflects the distribution density of ripples or dunes. If Eqs. (7.53) and (7.55) are substituted into Eq. (7.54), the final result is

$$f_b'' = 7.6 \frac{e}{\cot \phi}(\frac{\Delta}{h})^{0.8} \qquad (7.56)$$

Thus, the bed form resistance is primarily dependent on the relative height and distribution density of sand waves.

Vanoni et al. proposed the following formula based on data obtained in a laboratory channel [44].

$$\frac{1}{\sqrt{f_b''}} = 3.5\log\frac{R_b}{\Delta \cdot e} - 2.3 \qquad (7.57)$$

This equation is structurally different from Eq. (7.56), but the loss coefficients from them are about the same.

7.5.2.2 Hydraulic factors determining bed form resistance

In the preceding subsection, the bed form resistance was shown to be like that due to local enlargements, and it is thus closely related to the geometry of the sand wave. However, the equation cannot be used to predict the value of bed form resistance under a given flow condition. Still, the analysis in Chapter 6 makes it clear that, as the bed configuration is in the process of development from ripples to dunes, and then to antidunes, the hydraulic factor playing the determinant roles in the geometry of bed forms progresses from particle Reynolds number ($Re*$) to the parameter of flow intensity (Θ or ψ), and then to the Froude number (Fr). The relative roughness (D/R) also affects the geometry, as does the Froude number. Evidently, these parameters should determine the bed form resistance. Several sets of representative results demonstrate this progression. Differences between flume results and data obtained from natural streams may be important.

The primary differences are:

1. For low rates of sediment transport, ripples often form in flumes, whereas in natural streams, the primary bed forms are dunes.

2. In flume experiments, the velocity can be increased simply by adjusting the flume slope, so that one can easily reproduce sand waves and corresponding flow phenomena. But in natural streams, even if sand waves form, they normally appear on only a part of the river bed.

3. In natural streams, other localized resistance components are hard to separate from the bed form resistance. For example, the flow in the main channel in dry seasons often meanders. Then the plan form resistance induced by the continuous change of flow direction can affect the bed form resistance. This situation is not so common in flumes where the flow is restricted by the side walls.

4. In flumes, the bed forms are limited laterally by the width of the flume and may form in parallel belts with the entire flow passing over the stoss side of bed forms, a process that restrains the development of bed forms. But on a wide natural stream, the bed forms normally are separated and scattered. They do not combine.

Thus, part of the flow may go around a sand wave, and the bed forms may have more chance to develop freely.

The overall result is that bed form resistance in flumes is normally smaller than it is in natural streams under similar flow conditions.

Various expressions for bed form resistance have been presented.

1. Einstein and Barbarossa method [45]. Based on data from 10 rivers, US, Einstein and Barbarossa established a relationship for bed form resistance. As the data were measured mainly for resistance caused by dunes, a special flow parameter

$$\Psi' = \frac{\gamma_s - \gamma}{\gamma} \frac{D_{35}}{R_b'J} \qquad (7.58)$$

was introduced for dune resistance, as shown in Fig. 7.32; D_{35} -- the particle size of sediment of which 35% by weight is finer; R_h' -- the hydraulic radius related to gain friction; $U_*'' = \sqrt{gR_b''J}$, R_b'' -- the hydraulic radius related to bed form resistance.

With an increase of flow intensity, i.e., a decrease of ψ', dunes tend to diminish, and the dune resistance decreases correspondingly.

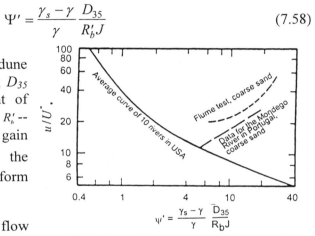

Fig 7.32 Relationship between bed form resistance and flow parameter ψ' (after Einstein, H.A., and N.L. Barbarossa)

Among the 10 rivers analyzed by Einstein and Barbarossa, 8 had values of D_{35} smaller than 0.5 mm, and the other two D_{35} had values of 0.7 and 1.0 mm. In later experiments with coarser bed material, the result departed from the mean curve of the 10 U.S. rivers, as shown in Fig.7.32. The bed form resistance for coarse sand was shown to be smaller than that for medium and fine sand. In 1967, Cunha provided data for the Mondego River in Portugal (D_{90}=6.1 mm, D_{35} =1.7 mm). The trend of data points is the same as for the flume results, but its position is a little lower [46].

2. Shen's correction to the Einstein-Barbarossa curve [47]. Based on the flume results using uniform sediment (including natural sediment and lightweight sediment), Shen showed that in the expression of bed form resistance in alluvial streams, in addition to the flow parameter ψ', another parameter $\omega D/\upsilon$ is required, in which ω is the sediment fall velocity. His formula for bed form resistance is

$$\frac{U_*''}{U} = A_1 + B_1 \log \frac{\Psi'}{\alpha} \tag{7.59}$$

for which, the coefficients A_1 and B_1 and the parameter α are listed in Table 7.2.

Table 7.2 shows that for $\omega D/\upsilon > 100$ (for normal water temperature, this value corresponds to a natural sediment size greater than 1 mm), ripples do not form on a bed, and the bed configuration is mainly that of dunes. In this case, the flow parameter Ψ' plays a determinant role. For $\omega D/\upsilon < 100$, because Shen's analysis is based mainly on a large amount of flume experiment data, the bed configuration is mostly in the form of ripples. In this case, in addition to the parameter Ψ', the Reynolds number $\omega D/\upsilon$ must be included. Chapter 6 shows that the two key parameters are the flow parameter Ψ' and the grain Reynolds number $U_* D/\upsilon$. If the bed configuration is in the ripple-dune phase, the effects of the flow parameter ψ' and the grain Reynolds number should both be used in determining the bed form resistance.

Table 7.2 Values of coefficients and parameter used in Eq. (7.59) (after Shen, H.W.)

$\omega D/\upsilon$	ψ'	A_1	B_1	α
>100	<10	0.03	+0.01	7.12
	>10	0.06	-0.09	7.12
1—100	—	0.03	+0.11	$=(\omega D/\upsilon)^{1/2}$

3. Engelund method [33]. Engelund used the flow parameters suggested by Einstein. His method is considered to be reliable, particularly in Europe, and is widely used there. Hence, its derivation should be presented and compared with Einstein method.

First, following Mayer-Peter's work, Engelund divided the bed resistance τ_b in 2-D alluvial streams into grain friction τ_b' and bed form resistance τ_b'' according to the energy slope,

$$\tau_b = \gamma h J = \tau_b' + \tau_b'' = \gamma h J' + \gamma h J'' \tag{7.60}$$

For grain friction, Engelund suggested

$$\tau_b' = \gamma h J' = \gamma h' J$$

For bed form resistance, the resistance loss can be written much as in Eq. (7.52). If each resistance component is replaced by the corresponding Darcy-Weisbach coefficient of resistance, then Eq. (7.60) can be written as

$$f_b = f_b' + \frac{4\alpha\Delta^2}{h\lambda} \tag{7.61}$$

Furthermore, Engelund introduced a parameter for the flow intensity

$$\Theta = \frac{hJ}{\dfrac{\gamma_s - \gamma}{\gamma} D} \tag{7.62}$$

This parameter is simply the inverse of the Einstein flow parameter, ψ'. Correspondingly,

$$\Theta' = \frac{h'J}{\dfrac{\gamma_s - \gamma}{\gamma} D} \tag{7.63}$$

and,

$$\Theta'' = \frac{1}{2} F_r^2 \frac{\alpha\Delta^2}{\dfrac{\gamma_s - \gamma}{\gamma} D\lambda} \tag{7.64}$$

in which,

$$Fr = \frac{U}{\sqrt{gh}}$$

Now Eq. (7.60) can be written as

$$\Theta = \Theta' + \Theta''$$

Engelund next presented his similarity hypothesis; namely, if two rivers (denoted by subscript 1 and 2) are dynamically similar, they must meet the following two criteria:

First, the effective shear stresses on these beds must be equal ($\Theta_1' = \Theta_2'$);

Second, the ratios of local enlargement losses caused by bed configuration to total energy loss must be equal.

From the second criterion,

$$f_{b1} / f_{b2} = f'_{b1} / f'_{b2} \tag{7.65}$$

Since

$$f'_b / f_b = \Theta' / \Theta$$

So

$$\Theta_1 / \Theta_2 = \Theta'_1 / \Theta'_2 \tag{7.66}$$

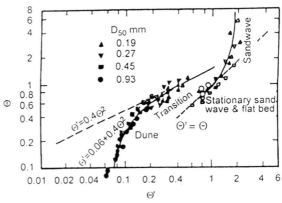

Fig. 7.33 Relationship between grain friction and total bed resistance (after Engelund, F., and E. Hansen.)

From the first criterion

$$\Theta'_1 = \Theta'_2, \Theta_1 = \Theta_2$$

Hence, if Θ is a function of only Θ', Eq. (7.66) is tenable,

$$\Theta = f(\Theta') \tag{7.67}$$

From the foregoing analysis, Engelund plotted the $\Theta \sim \Theta'$ curve using data from flume experiments, as shown in Fig.7.33. In the dune phase,

$$\Theta' = 0.06 + 0.4\Theta^2 \tag{7.68}$$

As Θ decreases, Θ' gradually approaches the constant value of 0.06, which corresponds to the condition of incipient motion. If $\Theta > 0.4$,

$$\Theta' = 0.4\Theta^2 \tag{7.69}$$

In contrast, for high transport rates of sediment and with sand waves forming on the bed, the data fall near the other curve. For flat bed or for stationary sandwaves without local enlargement loss,

$$\Theta' = \Theta$$

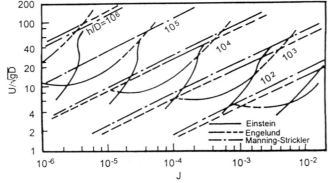

Fig. 7.34 Comparison of resistance formulas
(after Chollet, J.P., and J.A. Cunge)

But in the sand wave phase, as a result of the additional energy loss caused by the breakage of the water surface, Θ' is smaller than Θ. Engelund was able to express the resistance losses for all phases of bed configuration, except for the ripple phase, in a single figure.

Because Engelund and Einstein used the same flow parameter to express the bed form resistance, the two methods can be readily compared. After some mathematical transformation, Chollet and Cunge expressed the equations of Einstein, Engelund, and Manning-Strickler as relationships between $U/\sqrt{gD}$ and J with h/D as a third parameter, as shown in Fig.7.34 [48]. The figure shows that

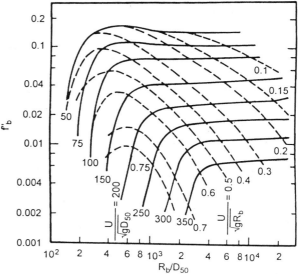

Fig. 7.35 Relationship between bed form resistance, Froude number and relative roughness (after Alam, A.M.Z., and J.F. Kennedy)

(1) The Manning-Strickler formula expresses only grain friction, and should not be used for flows with bed forms;

(2) If the bed is flat, the trend of the three formulas is essentially the same: a straight line with a slope of 1/2. The different formulas used to express the mean velocity did not affect the results significantly;

(3) Because Engelund included results both high and low sediment transport rates, the formation of a flat bed was possible; hence, the two ends of his curve fall on the straight line with a slope of 1/2. But the Einstein formula includes only dune resistance; if the sediment transport rate is high and dunes tend to disappear, the curve intersects with the straight line with a slope of 1/2. For a low sediment transport rate, the curve does not automatically transfer into the resistance formula for a flat bed;

(4) In the dune phase, if the bed is fairly steep and the relative roughness (D/h) is large, the Einstein and Engelund formulas differ very little; but for streams in an alluvial plain with small relative roughness, the difference between them is quite large.

4. Alam and Kennedy method [49]. Among the formulas for bed form resistance, some include the Froude number and relative roughness as parameters. The work of Alam and Kennedy is representative of this type.

They used the curves in Fig. 7.29 to obtain the grain friction, and then the principle of summation of component resistances to obtain the corresponding bed form resistance. They plotted the curve for the bed form resistance coefficient against $U/\sqrt{gR_b}$, using R_b/D_{50} as a third parameter, as shown in Fig. 7.35. In the same figure, they plotted curves for $U/\sqrt{gD_{50}}$ as a third parameter. For large rivers, the plots of $U/\sqrt{gD_{50}}$ are straight lines parallel to the abscissa, which indicates that bed form resistance is not related to the relative roughness but depends mainly on a Froude number of the type of $U/\sqrt{gD_{50}}$. In small rivers and flumes, both $U/\sqrt{gD_{50}}$ and R_b/D_{50} affect the bed form resistance significantly.

7.5.3 Bank resistance

The bank resistance can usually be estimated using the Manning equation:

$$U = \frac{1}{n_w} R_w^{2/3} J^{1/2}$$

in which the bank roughness n_w depends on the material in the bank wall and can be obtained from roughness tables like Table 7.3:

Table 7.3 Bank roughness coefficient

Situation of banks	n_w	
	range	common value
Cement pavement	0.011—0.015	0.013
Stone blocks paved with cement	0.017—0.030	0.025
Dry rock paving	0.025—0.035	0.030
Smooth ground banks	0.017—0.025	0.0225
Ground banks with weed	0.027—0.035	0.030
Sand and gravel banks	0.020—0.030	0.027
Gravel banks	0.025—0.030	0.030

In irrigation canals, artificial methods are sometimes used to promote the deposition of fine sediment in a certain stretch to prevent the banks from scouring. In such cases, the material of the banks is mostly fine sediment, and the bank surface is smooth enough that it can be treated as hydraulically smooth as already discussed. However, in mountain rivers, especially in gorges reaches with high and precipitous banks and a low ratio of width to depth, the bank roughness is often quite large. For instance, in the river reaches of Wuxia Gorge and Qutangxia Gorge of the Yangtze River, the Manning roughness coefficient for the bank can exceed 0.10.

7.5.4 Floodplain resistance

The vegetation on floodplains includes grasses, brush, trees, and even forest. When the flood reaches the floodplains, the resistance to flow is mainly either ordinary friction or form resistance due to vegetation. In some of the older textbooks on hydraulics, the vegetation density was used to determine the roughness. Now, however, the use of only this one parameter to indicate the effect of vegetation on resistance is known to be far from sufficient. More studies of this aspects have been made, but the results do not provide a final solution.

The study of the effect of vegetation on resistance needs to reflect two quite different situations. Depending on the flow velocities and the stiffness of the vegetation, grasses and brush can either stand up or bend over. In the first situation, the roughness is comparable to that of large-sized particles protruding out from bed, and the resistance is related to the projected area and the vegetation density. For the second, the bed can have an undulating shape that may fall within the hydraulically smooth region for turbulent flow. In this latter case, the resistance can be affected somewhat by the Reynolds number. But, if the flow depth is less than the vegetation height, the result is more like the first case.

7.5.4.1 Prone vegetation

The first scientists to study systematically the flow capacity of canals with vegetation were Ree and Palmer [50]. They found that as the flow velocity increases, vegetation progressively bends over and the resistance continuously decreases. Using the concept of relative roughness, they established a correlation between the canal roughness and the product of mean velocity U and hydraulic radius R, as shown in Fig.7.36. Various types of vegetation with different heights fall on different curves. Later, Kouwen and Unny conducted flume experiments using styrene strips 10 cm long, 0.5 cm wide, and 0.3 mm thick to simulate vegetation, and they identified the two primary strip positions as vertical and horizontal [51]. If the styrene strips were fully bent over, the bed was more or less hydraulically smooth and the resistance coefficient varied with the Reynolds number. Their data, in Fig. 7.36, coincided closely with the results of Ree and Palmer. As the viscosity of water varies little for normal conditions, the parameter UR can be used in place of the Reynolds number for the abscissa in the figure. That is, if the vegetation is quite pliable, it bends over with the flow, and the bed becomes either hydraulically smooth or within the transition region from hydraulically smooth to rough. Morris called this region quasi-smooth and suggested that it is like a continuous bed with individual, scattered resistance components [52].

If water flows over floodplains covered by vegetation, the velocity is often rather low so that the flow can be laminar. For this case, the relationship between resistance coefficient and Reynolds number is that shown in Fig. 7.37[53]. Although the resistance coefficient is then inversely proportional to the Reynolds number, the constant of proportionality is much larger than that for a relatively stable bed formed of glued sand particles, and it increases with the ground slope J_0,

$$f = \frac{510000 J_0^{0.662}}{R_e} \qquad (7.70)$$

The coefficient of proportionality in the relationship between the resistance coefficient and the Reynolds number is not equal to 24 for low flow. This change with bed roughness has been reported in several studies of flow over a bed covered by vegetation, but the reason for the coefficient to be so large and to vary with ground slope is still unknown. The data of Ree and Palmer, Kouwen and Unny are also shown in Fig.7.37. The trend of the data indicates that their experimental results fall within the transitional region between laminar and turbulent flow. For the same Reynolds number and ground slope, the resistance of a trapezoidal cross-section appears to be larger than that for a rectangular cross-section.

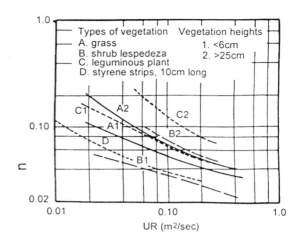

Fig. 7.36 Relationship between the roughness, vegetation and flow condition in channels with vegetation (after Ree,W.O., and V.J. Palmer)

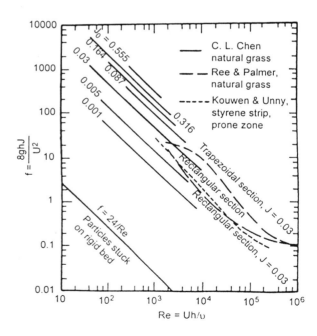

Fig. 7.37 Relationship between resistance coefficient and Reynolds number on a bed covered with vegetation (in laminar and transitional regions) (after Chen, C.L.)

296

7.5.4.2 Standing vegetation

For vegetation standing up straight, the vertical velocity distribution is that shown in Fig. 7.38a. The vegetation height is k, and the velocity at that level is u_k. For the experimental data with styrene strips, u_k is a function of the friction velocity U_*, the data for the vegetation are concentrated on separate belts. In the main flow above the vegetation, the velocity distribution follows the usual logarithmic law, and forms a straight line on semi-logarithmic paper (Fig. 7.38c).

Kouwen and Unny found that the flow over standing vegetation was similar to that over a rough surface. The resistance loss does not vary with the Reynolds number and thus depends only on the relative roughness k/h. The measured data are shown in Fig. 7.39. The range of the Reynolds numbers in the experiment was not large enough to define the curve fully, especially if the vegetation extends nearly to the surface. In fact, the phenomenon is more complex than that shown in Fig. 7.39. The resistance loss caused by standing vegetation, somewhat like that for large protruding roughness on a bed, is mainly dependent on the projected area of vegetation normal to flow and the vegetation density. For scattered vegetation, each tree or blade of grass has its individual resistance. With a greater density of vegetation, the wakes caused by separation on the lee side tend to overlap. Once the vegetation covers the plane surface, the flow contacts only the tops of the vegetation; it is therefore still hydraulically rough, but the roughness size is certainly less than the bended height k. Petryk and Bosmajian III studied the resistance for flow depths smaller than the vegetation height [54], and derived the expression

$$n = n_b \sqrt{1 + \frac{C_D}{2g} e'(\frac{1}{n_b})^2 R^{4/3}} \qquad (7.71)$$

in which, n_b is bed surface roughness excluding the effect of vegetation, C_D the resistance coefficient due to vegetation, e' the vegetation density

$$e' = NA' / A \qquad (7.72)$$

N is number of trees on bed area A, A' the projected area of trees normal to the flow.

If the resistance is mainly induced by vegetation, Eq. (7.72) can be simplified to

$$n = R^{2/3} \sqrt{\frac{C_D}{2g} e'} \qquad (7.73)$$

Eqs. (7.71) and (7.72) should be used only for scattered vegetation without mutual disturbance.

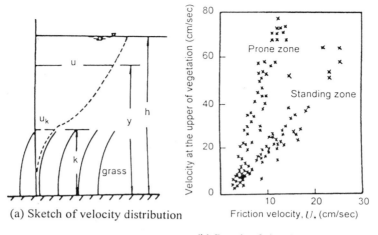

(a) Sketch of velocity distribution

(b) Result of simulation experiments

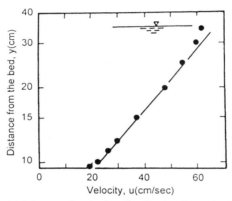

(c) Measured velocity distribution in main flow region

Fig.7.38 Vertical velocity distribution for a bed covered by vegetation
(after Kouwen, M., and Y.E. Unny)

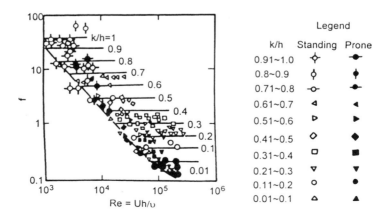

Fig. 7.39 Resistance due to vegetation (after Kouwen, M., and Y.E. Unny)

298

The above analysis shows that the variation of roughness with flow depth is closely related to the distribution of projected area of vegetation on a vertical plane. If the vegetation resistance is mainly due to tree stems, and the stem diameter does not vary significantly over the depth, or if the decrease in stem diameter with increasing depth tends to be affected by the increase in branches and leaves, the roughness coefficient is proportional to the 2/3 power of R. In contrast, if the area of the vegetation decreases with the increasing flow depth, so that it is inversely proportional to the 3/4 power of R, then the roughness coefficient may not change much with water level and is approximately constant. The field data show that both of these situations occur as well as a transition between them, from the type of vegetation and the season, one can determine the vegetation density and the resistance it presents.

7.6 COMPREHENSIVE RESISTANCE FORMULAE

The method for determining a composite resistance from resistance elements presented in the foregoing section is quite complex. Sometimes, however, one does not need to know grain friction and its corresponding hydraulic radius, but wants to know the discharge that passes through a cross-section with a certain slope. A relatively simple and comprehensive resistance formula can be used for the latter purpose.

7.6.1 Chien-Mai comprehensive resistance formula [38]

Chien and Mai used the Manning-Strickler formula in the form

$$U = \frac{A}{K_s^{1/6}} R^{2/3} J^{1/2}$$

If $K_S = D_{65}$, and if the cross-section is wide and shallow so that the bank resistance is negligible, the formula can be written in the form

$$U = \frac{A}{D_{65}^{1/6}} h^{2/3} J^{1/2} \tag{7.74}$$

They did not treat A as constant but related it to the factors that dominate the evolution of sand waves. As a first approximation, they expressed A as a function of the flow parameter Ψ' [Eq. (7.58)]; Fig. 7.40 presents the results of the data for the Lower Yellow River in this form. If the velocity is low, the sediment transport rate not high, and the corresponding ψ' value quite large, the parameter A is small because of the action of bed form resistance or other channel form resistance. If ψ'

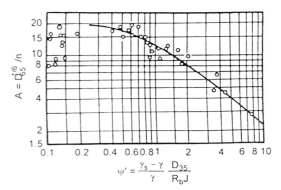

Fig. 7.40 Relationship between comprehensive roughness and flow parameter on the Lower Yellow River

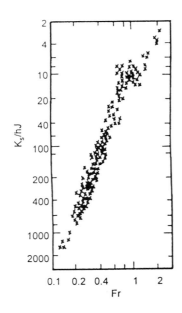

Fig. 7.41 Relationship between comprehensive roughness and Froude number (after Hideo Kikkawa, Hiroyoshi Shi-Igai, and Shoji Fukuoka)

decreases, as the flow increases, A increases accordingly. The bed form obtained for the Huayuankou reach of the Lower Yellow River shows that if ψ' is about 0.4-0.5, the bed forms tend to disappear. Fig. 7.40 also shows that if ψ' is in this range, A is approximately 19 and hence not a function of ψ'. Thus, the only resistance is grain friction, and the Manning coefficient is related only to the size of bed material size,

$$n = D_{65}^{1/6} / 19$$

The coefficient in the foregoing formula is for the Lower Yellow River. If the relative roughness of another river is quite different from that of the Lower Yellow River, the coefficient may have some other constant value (Fig. 7.27). If the parameter ψ' is still lower, sand waves may form, and A then decreases and varies considerably because of the instability of the antidunes.

7.6.2 Method of Kikkawa, Shi-Igai and Fukuoka [55]

From the formula for the logarithmic velocity distribution Eq. (7.43), Kikkawa et al. found that the roughness size K_s of the characteristic resistance is a function not only of the size of the sediment on the bed, but even more of the height and geometry of sand waves. They found that the Froude number dominates the bed form geometry. They plotted K_s/hJ versus F_r using the flume data with the result shown in Fig. 7.41. The data in the figure follow a well defined trend. For subcritical flow, K_s/hJ is approximately inversely proportional to the 2nd power of F_r.

7.6.3 Resistance formula of Li and Liu [56]

Kikkawa stated that the sediment size did not affect the roughness size, a finding that is hard to understand physically. Using the logarithmic velocity distribution, Li and Liu suggested

$$K_s = \alpha D_{50}$$

in which the ratio α is related to the bed form geometry, and is thus related to by the flow parameter

$$U/U_a$$

in which, U_c is the velocity for incipient motion of sediment. In fact, this parameter is another form for the parameter Ψ

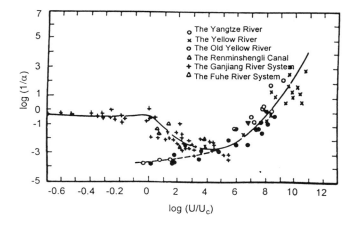

Fig. 7.42 Relationship between comprehensive resistance and flow parameter (proposed by Li and Liu)

used by Einstein and the Θ of Engelund. Fig. 7.42 shows the relationship between α and U/U_c for Chinese rivers. If U/U_c is less than one, then α=2; i.e., the roughness size of grain friction is two times the median size of the bed material. Once sediment begins to move, sand waves gradually form, and the equivalent roughness size increases. If $log(U/U_c)$=0.4, i.e., U/U_c =2.5, dunes have their maximum size, and the roughness also reaches the maximum. After this, dunes will gradually disappear. From $log(U/U_c)$=0.8, i.e., U/U_c =6.5, the resistance is again no different from that for a rigid bed. The increase of the intensity of the flow further reduces the resistance. However, the small values of U/U_c from the Yangtze and old Yellow Rivers are not consistent with the above trend; they follow the dashed line in the figure.

7.7 DISCUSSION OF SPECIAL ISSUES AFFECTING RESISTANCE

The method of calculating the resistance for mountain streams with coarse bed materials and the impacts of water temperature and bed seepage on resistance are discussed in this section.

7.7.1 Method for calculating the resistance due to large scale roughness

In some mountain streams, the bed material is usually quite coarse, and sand waves do not normally form. Also, as the roughness size and water depth often have the same order of magnitude, the roughness characteristics are quite different from that of grain friction presented in Section 5.

Two situations arise. One is that large protruding rocks and boulders can occur on the channel bed. The other is that the bed material is very coarse, but rather uniform. In the former situation, the interaction of the flow with protruding rocks and boulders is a special occurrence and it involves the distribution density of the protruding rocks. In the latter situation, the flow is more nearly similar to grain friction, but the value of K_s/h is very large and the whole flow area is directly affected by the bed material.

7.7.1.1 Individual and protruding large rocks and boulders on bed

If separate large rocks and boulders protrude from a river bed, the resistance depends mainly on their protruding height, arrangement, and distribution density. In early studies of this issue, scientists distributed regular objects (like cubes, balls, or sand particles) in various geometrical arrangements over the flume bed to simulate the natural situation [57,58]. Later, Rouse analyzed these data and proposed a relationship between the effective roughness and the geometric pattern of the protruding materials, as shown in Fig. 7.43 [59]. In the figure, Δ is the height of the protruding materials; e' is their distribution density from Eq. (7.72). The density differs little from that used for sand waves in Section 5 [Eq. (7.55)]. The former uses the projecting area of rocks on a plane perpendicular to flow, whereas the latter uses the projecting area of the lee side of sand waves, also normal to the direction of flow.

If the number of protruding rocks is small, their effect on the total resistance is

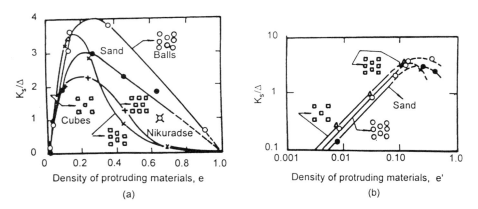

Fig. 7.43 Relationship between the effective roughness and the shape, disposition, and density of protruding materials (after Rouse, H.)

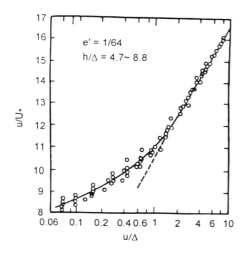

e' = 1/64
h/Δ = 4.7~ 8.8

Fig.7.44 Vertical velocity distribution for flow over large protruding rocks (after O'Laughlin, E.M., and V.S. Annambhotla)

also small, of course. But, if the rocks are tightly spaced, their wakes overlap, so that a part of the form resistance is eliminated; in the latter case, the bed roughness felt by the flow is obviously much smaller than the rock size. Thus some density between these extremes should exist for which the roughness is a maximum. Fig.7.43a shows that this critical density is in the range of 15 to 25%, and it depends somewhat on the shape and disposition of the protruding materials, as shown in Fig.7.43b. For uniform natural sediment, the resistance reaches maximum if the density is 21.5%. As a reference value Nikuradse obtained

$$K_s / \Delta = K_s / D = 1$$

in his experiments, which corresponds to a particle density of 64%.

O'Loughlin et al. measured the vertical velocity distribution for various rock densities in flume experiments. The results for $e'=1/64$ are shown in Fig. 7.44 [60]. It shows that if $y/\Delta h >2$, the velocity distribution follows the logarithmic law. But below that level, the flow is affected by the wakes behind the protruding rocks, and the velocity distribution departs from the logarithmic law.

Bathurst, using data from mountain streams in the UK, proposed the following formula for the resistance due to large scale roughness [61]:

$$(\frac{8}{f})^{1/2} = (\frac{R}{0.365D_{84}})^{2.34}(\frac{B}{h})^{7(e'-0.08)} \tag{7.75}$$

$$e' = 0.139\log(1.91\frac{D_{84}}{R}) \tag{7.76}$$

in which, f is the Darcy-Weisbach coefficient, D_{84} the Grain size for which 84% by weight of the bed material is finer.

From experiments, Simons et al. found that if gravel and sand carried by floods fill the spaces among the large rocks and boulders on mountain streams, the roughness and the Manning coefficient is only a half to two-thirds compared to normal situation [62]. Experience shows that for the channel bed where debris flow often occurs, the roughness is also small as the range of particle sizes is large.

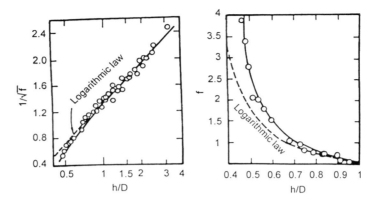

Fig. 7.45 Resistance for uniform large roughness (after Bayazit, M.)

7.7.1.2 Uniformly distributed large scale roughness

Bayazit used a dense arrangement of hemispheres (diameter $D=2.3$ cm) in a flume to simulate large-scale roughness [29]. He found that the resistance follows the logarithmic law if $h/D>1.0$; but if the water depth and roughness size are about the same, the resistance coefficient is larger than that from the logarithmic law, as much as 50% larger, as shown in Fig. 7.45. In the experiments, the nominal bed surface, from which the water surface is measured, was taken to be $0.35D$ below the hemispheres.

7.7.2 Effect of water temperature on resistance

In Chapter 6, the point was made that water temperature can significantly affect the bed configuration. The effect of water temperature is primarily the result of changes in the water viscosity. On the one hand, the variation of water temperature changes the thickness of the laminar sublayer, and thus changes the flow near the particles on the bed. On the other, it affects the sediment fall velocity, and thus directly affects the sediment suspension. Both factors affect grain friction and bed form resistance, so that temperature can strongly affect the resistance in alluvial streams. In most situations, the effect on grain friction is negligible [63]; however, its effect on bed form resistance is not.

In natural streams, the resistance tends to decrease with a decrease in the water temperature, and the water level for a given discharge drops accordingly. For example, on the Rio Grand in the United States, a straight reach of 60 m wide and 500 m long near Wenden City was selected for research. For a discharge in the range of 23.9 to 26.5 m³/s, the bed form changed from dunes to ripples and then to a flat bed as the water temperature dropped from $25°C$ to $4°C$; the bed form resistance also decreased, and the velocity increased from 0.64 m/s to 0.94 m/s [64]. In the middle Loup River, the Manning coefficient varied with water temperature as shown in Fig. 7.46 [65]. During the measurements, the discharge was in the range of 9 to 15 m³/s . As the water temperature fell from $25°C$ to $0°C$, n decreased from 0.033 to 0.021. The most complete data on this effect

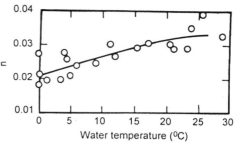

Fig. 7.46 Effect of water temperature on resistance (in the Middle Loup River) (after Colby, B.R., and C.H. Scott)

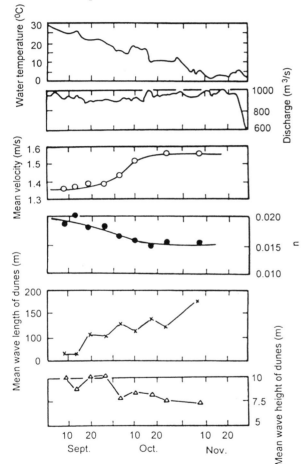

Fig. 7.47 Variation in hydraulic parameters and bed form size due to decreasing water temperature in the Missouri River (after U.S. Army Corps of Engineers)

are the measured results for the Missouri River cited in Chapter 6; the details are shown in Fig. 7.47 [66]. Even though the discharge changed by only a small amount as the water temperature fell from $28°C$ to around $0°C$, the mean wave length of dunes increased from 70 m to 180 m, and the mean wave height declined from 10 m to 7.5 m. As a consequence, the Manning coefficient decreased from 0.020 to 0.015, and the mean velocity increased from 1.36 m/s to 1.56 m/s.

The data from some rivers indicate that, within a specific flow range, the effect of water temperature on bed form is sizable. Outside this flow range, the effect is not as large. For example, in the Elkhorn River, only for discharges of 70 to 140 m³/s did water temperature play an appreciable role in the resistance; in the Mississippi River, only if the discharge exceeded 11,000 m³/s was the effect of water temperature on resistance noticeable.

In contrast, some flume experiments showed a trend in the opposite direction to that for the rivers. For example, Franco, using natural sediment with a median diameter of 0.23 mm in his flume experiment, found that as the water temperature fell, the resistance increased; but when he used powdered coal with a median diameter of 2.2 mm and a specific weight of 1.30 in his experiment, the results showed the same trends as for natural rivers [67].

These phenomena and the contradiction can be explained on the basis of the following physical picture [68].

Fig. 7.48a illustrates how the resistance coefficient changes with the increase of friction velocity (or mean velocity) over the entire range. The resistance coefficient has a maximum value. To the left of this value, the decrease in water temperature causes an increase of water viscosity; for some flow conditions, the shear stress τ_0 increases correspondingly. Since

$$U_* = \sqrt{\tau_0 / \rho}$$

the friction velocity U_* increases with a decrease in water temperature. Thus, if the initial bed form corresponds to point A in the figure, as the water temperature falls and U_* increases, the resistance increases. In contrast, for point B in the figure, a drop in the water temperature causes U_* to increase, and the resistance decreases. Also, for flow is as at point B, the effect of water temperature on resistance is significant. But if the flow condition is that for point C, the resistance does not change greatly with water temperature.

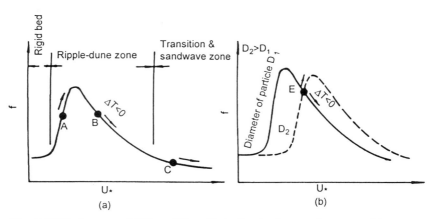

Fig. 7.48 Various possibilities of the effect of water temperature on resistance

The f-- U_* curves for two different sizes of sediment that are plotted in Fig. 7.48b intersect at point E. If the conditions for incipient motion of two types of sediment are located near point E, then a decrease of water temperature would cause the resistance of a channel consisting of fine sediment to decrease; however, the resistance of channel consisting of a coarse sediment would increase.

7.7.3 Influence of bed seepage on resistance

Although the effect of bed seepage has rarely been considered by river and sedimentation scientists, the experimental results have shown that the effect of bed seepage on resistance may not be negligible [69].

If an exchange occurs between the flows in the river and in the ground, the

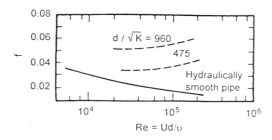

Fig. 7.49 resistance to flow in permeable pipes (after Ippen, A.T.)

Karman constant in the logarithmic velocity distribution formula can decrease from 0.40 to 0.26. The results of a pipe experiment with sand particles glued on the pipe wall showed that the resistance loss will increase continuously with an increase in the Reynolds number, that is, it does not approach some constant value corresponding to its relative roughness, as shown in Fig. 7.49. The situation is somewhat similar to the grain friction curve (Fig. 7.29) obtained by Lovera and Kennedy. For a given Reynolds number, the resistance coefficient increased with the increase of $d/K^{0.5}$, d is the pipe diameter and K is the seepage coefficient of the pipe wall [70].

REFERENCES

[1] Bakhmeteff, B. A., and W. Allan. "The Mechanism of Energy Loss in Fluid Friction." *Trans., Amer. Soc. Civil Engrs.*, Vol. 111, 1946, pp. 1043-1102.

[2] Leopold, L.B.; R.A. Bagnold; M. G. Wolman; and L. M. Brush, Jr. "Flow Resistance in Sinuous or Irregular Channels." *U. S. Geol. Survey, Prof. Paper 282-D*, 1960, pp. 111-134.

[3] Wuhan Institute of Hydraulic and Electric Engineering. *River Dynamics*. China Gongye Press, 1961, p. 269-288. (in Chinese)

[4] Wuhan Institute of Hydraulic and Electric Engineering, Hydraulics, Hydraulic and Electric Press, 1960, pp. 452. (in Chinese)

[5] Barnes, H.H.Jr. "Roughness Characteristics of Natural Channels." *U. S. Geol. Survey, Water Supply Paper 1849*, 1967, p. 213.

[6] Schnackenberg, E.C. "Slope Discharge Formulae for Alluvial Streams and Rivers." *Proc., New Zealand Inst. Civil Engrs.*, Vol. 37, 1951, pp. 340-449.

[7] Liu, Hsin-Kuan, and S.Y. Hwang. "Discharge Formula for Straight Alluvial Channels." *J. Hyd. Div., Proc., Amer. Soc. Civil Engrs.*, Vol. 85, No. Hy11, 1959, pp. 65-97.

[8] Vanoni, V.A., and N.H. Brooks. "Laboratory Studies of the Roughness and Suspended Load of Alluvial Streams." *Report E-68*, Sedimentation Lab., Calif. Inst. Tech., 1957, p. 121.

[9] Culbertson, J.K., and C.F. Nordin, Jr. "Discussion of the Paper--Discharge Formula for Straight Alluvial Channels." *J. Hyd. Div., Proc. Amer. Soc. Civil Engrs.*, Vol. 86, No. Hy6., Pt. 1, June 1960, pp. 98-102.

[10] Kennedy, J.F. "Further Laboratory Studies of the Roughness and Suspended Load of Alluvial Streams." *Report KH-R-3*, V. M. Keck Lab. Hyd. and Water Res., Calif. Inst. Tech., 1961, p. 36.

[11] Ackers, P., and W.R. White. "Sediment Transport: New Approach and Analysis." *J. Hyd. Div., Proc., Amer. Soc. Civil Engrs.*, Vol. 99, No. HY11, 1973, pp. 2041-2060.

[12] Einstein, H.A. "The Bed Load Function of Sediment Transportation in Open Channel Flows." *U. S. Dept. Agri., Tech. Bull. 1026*, 1950, p. 71.

[13] Bagnold, R.A. "The Flow of Cohesionless Grains in Fluids." *Phil. Trans.*, Royal Soc. London, Ser. A, Vol. 249, No. 964, Dec., 1956, pp. 235-297.

[14] Einstein, H.A., and Ning Chien. "Discussion of the Paper - Mechanics of Streams with Movable Beds of Fine Sand." *Trans., Amer. Soc. Civil Engrs.*, Vol. 123, 1958, pp. 553-562.

[15] Einstein, H.A. "Der Hydraulische Oder Profil Radius." *Scherizerisch Bauzeitung*, Bd. 103, No. 8, 1934.

[16] Taylor, R.H.Jr. "Exploratory Studies of Open Channel Flow Over Boundaries of Laterally Varying Roughness." *Rep. KH-R-4*, W. M. Keck Lab. Hyd. & Water Res., Calif. Inst. Tech., July 1961, p. 65.

[17] Jiang, Guogan. "The Effect of the Flume Walls on Critical Shear Stress." *Study Report B-1*, Central Hydraulic Lab, 1948. (in Chinese)

[18] Einstein, H.A. "Method of Calculating the Hydraulic Radius in A Cross Section with Different Roughness." *Appen. II of the paper Formulas for the Transportation of Bed Load*, Trans., Amer. Soc. Civil Engrs., Vol. 107, 1942, pp. 575-577.

[19] Lotter, G.K. "Considerations on Hydraulic-Design of Channels with Different Roughness of Walls." *Trans., All-Union Sci. Res. Inst. Hyd. Engin.*, Leningrad, U.S.S.R., Vol. 9, 1933.

[20] Knight, D.W., and J.A. MacDonald. "Open Channels with Varying Bed Roughness." *J. Hyd. Div.*, Proc., Amer. Soc. Civil Engrs., Vol. 105, No. HY9, 1979, pp. 1167-1183.

[21] Yassin, A.M. "Mean Roughness Coefficient in Open Channels with Different Roughness of Bed and Side-Walls." *Mitt. aus der Versuchsanstalt fur Wasserbau und Erdbau*, Zurich, Nr. 27, 1953, pp. 90.

[22] Motayed, A.K., and M. Krishnamurthy. "Composite Roughness of Natural Channels." *J. Hyd. Div.*, Proc., Amer. Soc. Civil Engrs., Vol. 106, No. HY6., 1980 pp. 1111-1116.

[23] Einstein, H.A., and R.B. Banks. "Fluid Resistance of Composite Roughness." *Trans., Amer. Geophys. Union*, Vol. 31, No. 4., Aug. 1950, pp. 603-610.

[24] Einstein, H.A., and R.B. Banks. "Linearity of Friction in Open Channels." *Extract du Tome III*; Assoc. Intern. d'Hydrologie Scientifique, Assemblee Generale de Bruxelles, 1951, pp. 488-498.

[25] Meyer-Peter, E., and R. Muller. "Formulas for Bed Load Transport." *Trans., Intern. Assoc. Hyd. Res.*, 2nd. Meeting, Stockholm, 1948, pp. 39-65.

[26] Nikuradse, J. "Stromungsgesetze In Rohren." *Ver. Deut. Ing.*, Forschungsheft 361, 1933.

[27] Keulegan, G.H. "Laws of Turbulent Flow in Open Channels." *J. Res.*, Nat. Bureau of Standards, Vol. 21, No. 6, Dec. 1938, pp. 707-741.

[28] Nikitin, I.K. *Turbulent Open Channel Flow and Processes Near the Boundary.* Academy of Sci., Kiev, USSR, 1963. (in Russian)

[29] Bayazit, M. "Free Surface Flow in A Channel of Large Relative Roughness." *J. Hyd. Res., Intern. Assoc. Hyd. Res.*, Vol. 14, No. 2. 1976, pp. 115-125.

[30] Einstein, H.A., and A. El-Samni. "Hydrodynamic Forces on A Rough Wall." *Rev. Modern Phys.*, Vol. 21, No. 3. 1949.

[31] Dou, Guoren. "Turbulent Structure in Open Channel and Pipe Flow." *J. China Science*, No. 11, 1980, pp. 1115-1124. (in Chinese)

[32] Lane, E.W., and E.J. Carlson. "Some Factors Affecting the Stability of Canals Constructed in Coarse Granular Materials." *Proc., Minnesota Intern. Hyd. Conf.*, 1953, pp. 37-48.

[33] Engelund, F., and E. Hansen. "A Monograph on Sediment Transport in Alluvial Streams." *Teknisk Forlag*, Copenhagen, 1972, p. 62.

[34] Kamphuis, J.W. "Determination of Sand Roughness for Fixed Beds." *J. Hyd. Res., Intern. Assoc. Hyd. Res.*, Vol. 12, No. 2, 1974, pp. 193-203.

[35] Bray, D.I. "Estimating Average Velocity in Gravel Bed Rivers." *J. Hyd. Div., Proc., Amer. Soc. Civil Engrs.*, Vol. 105, No. HY9, 1979, pp. 1103-1127.

[36] Charlton, F.; P.M. Brown; and R. W. Benson. "The Hydraulic Geometry of Some Gravel Rivers in Britain." *Rep. IT 180*, Hyd. Res. Sta., Wallingford, England, 1978.

[37] Strickler, K. "Beitrage Zur Frage der Geschwindigkeitsformel und der Raukigeitszahlen Fur Strom Kanale und Geschlossene Leitungen." *Mitt. No. 16 des Eidgenossische Amtes fur Wasser Wirtschaft*, Bern, Switzerland.

[38] Chien, Ning; Qiaowei Mai; Rujia Hong; and Cifen Bi. "The Roughness of the Lower Yellow River." *J. Sediment Research*, Vol. 4, No. 1, 1959, pp. 1-15. (in Chinese)

[39] Limerions, J.T. "Determination of Manning Coefficient from Measured Bed Roughness in Natural Channels." *Water Supply Paper 1898-B*, U. S. Geol. Survey, 1970.

[40] Lewis, C.P., and B.C. McDonald. "Rivers of the Yukon North Slope." *Proc. Hydrology Symp. on Fluvial Processes and Sedimentation*, Univ. Alberta, 1973, pp. 251-271.

[41] Lovera, F., and J.F. Kennedy. "Friction Factors for Flat-Bed Flows in Sand Channels." *J. Hyd. Div., Proc., Amer. Soc. Civil Engrs.*, Vol. 95, No. HY4, 1969, pp. 1227-1234.

[42] Vanoni, V.A., and G.N. Nomicos. "Resistance Properties of Sediment-Laden Streams." *Trans., Amer. Soc. Civil Engrs.*, Vol. 125, 1960, pp. 1140-1175.

[43] Chang, F.M. "Ripple Concentration and Friction Factor." *J. Hyd. Res., Proc., Amer. Soc. Civil Engrs.*, Vol. 96, No. HY2, 1970, pp. 417-430.

[44] Vanoni, V.A., and L.S. Hwang. "Relation Between Bed Forms and Friction in Streams." *J. Hyd, Div., Proc., Amer. Soc. Civil Engrs.*, Vol. 93, No. HY3, 1967, pp. 121-144.

[45] Einstein, H.A., and N.L. Barbarossa. "River Channel Roughness." *Trans., Amer. Soc. Civil Engrs.*, Vol. 117, 1952, pp. 1121-1146.

[46]Cunha, L.V.Da. "About the Roughness in Alluvial Channels with Comparatively Coarse Bed Materials." *Proc., 12th Cong., Intern. Assoc. Hyd. Res.*, Vol. 1, 1967, pp. 76-84.

[47] Shen, H.W. "Development of Bed Roughness in Alluvial Channels." *J. Hyd. Div., Proc., Amer. Soc. Civil Engrs.*, Vol. 88, No. HY3, 1962, pp. 45-58.

[48] Chollet, J.P., and J.A. Cunge. "New Interpretation of Some Head Loss-Flow Velocity Relationships for Deformable Movable Beds." *J. Hyd. Res., Intern. Assoc. Hyd. Res.*, Vol. 17, No. 1, 1979, pp. 1-14.

[49] Alam, A.M.Z., and J.F. Kennedy. "Friction Factors for Flow in Sand-Bed Channels." *J. Hyd. Div., Proc., Amer. Soc. Civil Engrs.*, Vol. 95, No. HY6., 1969, pp. 1973-1992.

[50] Ree, W.O., and V.J. Palmer. "Flow of Water in Channels Protected by Vegetative Linings." *US Dept. Agri., Tech. Bull. No. 967*, 1949, p. 115.

[51] Kouwen, N., and T.E. Unny. "Flexible Roughness in Open Channels." *J. Hyd. Div., Proc., Amer. Soc. Civil Engrs.*, Vol. 99, No. HY5, 1973, pp. 713-728.

[52] Morris, H.M. *Applied Hydraulics in Engineering*. Ronald Press Co., N. Y., 1963, pp. 51-65.

[53] Chen, C.L. "Flow Resistance in Broad Shallow Grassed Channels." *J. Hyd. Div., Proc., Amer. Soc. Civil Engrs.*, Vol. 102, No. HY3, 1976, pp. 307-322.

[54] Petryk, S., and G. Bosmajian III. "Analysis of Flow Through Vegetation." *J. Hyd. Div., Proc., Amer. Soc. Civil Engrs.*, Vol. 101. No. HY7, 1975, pp. 871-884.

[55] Hideo Kikkawa; Hiroyoshi Shi-Igai; and Shoji Fukuoka. "On the Effects of Suspended Sediments to the Bed Roughness." *Proc., 12th Cong. Intern. Assoc. Hyd. Res.*, Vol.1, 1967, pp. 49-56.

[56] Li, Changhua, and Jianmin Liu. "The Resistance of Alluvial Streams." *Proc. of Nanjing Hydraulic Research Institute*, 1963. (in Chinese)

[57] Schlichting, H. "Experimentelle Untersuchungen zum Rauhigkeitsproblem." *Ing. Arch.*, Vol. 7, 1936.

[58] O'Laughlin, E.M., and E.G. MacDonald. "Some Roughness-Concentration Effects on Boundary Resistance." *La Houille Blanche*, No. 7, 1964.

[59] Rouse, H. "Critical Analysis of Open Channel Resistance." *J. Hyd. Div., Proc., Amer. Soc. Civil Engrs.*, Vol. 91. No. HY4, 1965.

[60] O'Laughlin, E.M., and V.S. Annambhotla. "Flow Phenomena Near Rough Boundaries." *J. Hyd. Res., Intern. Assoc. Hyd. Res*, Vol. 7, No. 2, 1969, pp. 231-250.

[61] Bathurst, J.C. "Flow Resistance of Large Scale Roughness." *J. Hyd. Div., Proc., Amer. Soc. Civil Engrs.*, Vol. 104, No. HY12, 1978, pp. 1587-1603.

[62] Simons, D.B.; K.S. Al-Shalkh-Ali; and Ruh-Ming Li. "Flow Resistance in Cobble and Boulder Riverbeds." *J. Hyd. Div., Proc., Amer. Soc. Civil Engrs.*, Vol. 105, No. HY5, 1979, pp. 477-488.

[63] Chien, Ning. "The Effect of Temperature on Sediment Movement." *J. Sediment Research*, Vol.3, No. 1, 1958, pp. 15-28. (in Chinese)

[64] Harms, J.C., and R.K. Fabnestock. "Stratification, Bed Forms, and Flow Phenomena (With An Example from the Rio Grande)" In Primary Sedimentary Structures and Their Hydrodynamic Interpretation, Edited by G. V. Middleton, *Sp, Pub. No. 12*, Soc. Econ. Paleontologists and Mineralogists, 1965, pp. 84-115.

[65] Colby, B.R., and C.H. Scott. "Effects of Water Temperature on the Discharge of Bed Material." *US Geol. Survey, Prof. Paper 462-G*, 1965.

[66] U.S. Army Corps of Engineers, "Missouri River Channel Regime Studies, Omaha District." *MRD Sediment Series No. 13B*, 1969.

[67] Franco, J.J. "Effect of Water Temperature on Bed Load Movement." *J. Waterways and Harbor Div., Proc., Amer. Soc. Civil Engrs.*, Vol. 90, No. WW3, 1968, pp. 342-352.

[68] Cunha, L.V.Da. "The Influence of Water Temperature on the Roughness of Alluvial Flows." *Memoria No. 437*, Laboratorio Nacional de Engenharia Civil, Lisboa, 1974, p. 13.

[69] Ippen, A.T. "Boundary Stresses and Resistance Coefficients in Free Surface Streams." *Proc., Seminar on Hydraulics of Alluvial Streams*, New Delhi, India, 1973, pp. 1-19.

[70] Chu, Y.H., and L.W. Gelhar. "Turbulent Pipe Flow with Granular Boundaries." *Tech. Rep. 148*, R. M. Parsons Lab. of Water Res. and Hydrodynamics, Massachusettes Institute of Tech., 1972.

CHAPTER 8

INCIPIENT MOTION OF SEDIMENT

Incipient motion is an important critical condition, for which sediment starts to move under the action of flow. In 1753, Brahms suggested that the velocity for incipient motion is proportional to the grain weight raised to the one-sixth power. This concept agreed quite well with the knowledge of incipient motion of sediment at that time. At the end of the nineteenth century, people began to study the problem from the concept of a balance of forces acting on the grains. In 1914, Forchheimer systematically summarized and evaluated the knowledge that had accumulated by that time, and he discussed the influence of sediment gradation, sorting and armoring on the incipient motion of sediment. In 1936, Shields applied the methods of dimensional analysis, prevalent at that time to sediment motion and developed the well-known Shields diagram [1], one that is still used widely. In the 1950s, Lane employed the concept of drag force in canal design and proposed a design of regime canals that is more soundly based on theoretical grounds [2]. More recently, investigations of the incipient motion of the sediment have focused on conditions for non-uniform and cohesive sediments. For the former, the armoring process at the bed should be included as part of the process of erosion, and for the latter, the sediment motion is related to the physico-chemical properties at the surface of fine grains. Both of these cases are quite complex. After long experience, people gradually began to conceive of incipient motion of sediment as a stochastic phenomenon. Its study must use an approach that combines the theories of probability and fluid mechanics if one is to understand the physical essence of the incipient motion of sediment.

This chapter treats first the stochastic nature of the condition for incipient motion and then proceeding from the simple to the complicated, the conditions for non-cohesive uniform sediment, for non-cohesive non-uniform sediment, and for cohesive sediment.

8.1 STOCHASTIC PROPERTY OF THE PHENOMENON OF THE INCIPIENT MOTION OF SEDIMENT

8.1.1 Description of the phenomenon of incipient motion

If the intensity of flow exceeds a certain value, sediment particles begin to move. The flow condition that corresponds to this critical limit is called incipience.

Although the flow condition for which the sediment grains on the bed start to move is a well-defined physical concept, many difficulties are encountered in determining the actual threshold condition for specific cases. A typical bed surface is composed of innumerable sediment grains of various combinations of sizes, shapes, specific gravity, orientations, and locations. Besides, water flow also has fluctuating characteristics. Therefore, the forces exerted on sediment grains at different places are different at any one time, and they are also different at various times for grains located at a given place. Thus, even for uniform sediment, the grains do not start to move or come to rest all together. For non-uniform sediment, conditions are much more complicated. Even for given flow conditions, one cannot define a specific grain size such that larger particles remain at rest and smaller particles are all in motion. Also, the spatial distribution of sediment movement at a certain instant is such that grains move at some places and remain at rest at others. And at certain location of the bed, sediment moves during one time interval and fails to move at another. The incipient motion of sediment is clearly a stochastic phenomenon.

In fact, in alluvial rivers, a certain amount of sediment moves whenever water flows; only the opportunity for movement varies for various grain sizes. Helland-Hansen et al. carried out a long series of experiments in a flume with gravel having a median diameter of 2.5 cm [3]. Although the hydraulic variables used in the experiment combine to form a parametric value that is much less than the critical one in the

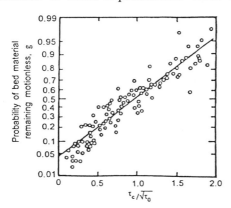

Fig. 8.1 Probability for grains to remain motionless under action of various surface drags (after Gessler, J.)

Shields diagram, the gravel did not remain absolutely motionless. Sometimes, only one or two particles of gravel moved during an hour; still, a definite functional relationship between sediment movement and the flow conditions existed. In the statistical sense, a certain intensity of sediment movement takes place for any flow condition. What is called the critical incipient condition for incipient motion is really an intensity of sediment movement, or a probability of specific sediment being in motion (or motionless). As for the Shields diagram that is used so comprehensively, someone suggested that it may represent the flow condition at which the probability of the sediment on a bed moving is equal to that of it remaining at rest. Experiments have shown that the critical flow condition in the Shields diagram is equivalent to

$$g_{b\upsilon} / (U_*D) = 10^{-2}$$

in which, $g_{b\upsilon}$ is the volumetric transport rate of bed load per unit width [4]. From experiments with uniform sediment, the relationship between the shear ratio (critical shear stress τ_c divided by the average shear stress on bed $\overline{\tau_0}$) and the probability of sediment remaining motionless is shown in Fig 8.1 [5]. In it, if $\overline{\tau_0}$ is less than τ_c, some sediment is in motion, but the probability of a particle moving is small. With the same reasoning, if $\overline{\tau_0}$ is greater than τ_c, not all sediment particles are in motion, part of them remain motionless.

Since the incipient motion of sediment is a stochastic phenomenon, some liken it to the concept of the most probable flood in river — a concept that lacks any real meaning. In his system, Einstein did not include the concept of incipient motion. He proposed another approach: in a statistical sense, and for a specific flow condition, which sediment and how much of it can be moved are the things to be determined.

8.1.2 Criterion of incipient motion of sediment

The criterion of a specific condition for the incipient motion of sediment must involve arbitrary concepts and definitions. Kramer recommended that the movement of bed load be divided into four stages [6]:

1. No sediment motion: all sediment on a bed is motionless.

2. Sparse sediment motion: only a few fine grains on a bed move, here and there.

3. Mean sediment motion: sediment grains smaller than the median diameter move everywhere on a bed, and the movement is too intense for one to count the number of grains that are moving in a given area.

4. Strong sediment motion: all sizes of sediment move and the bed configuration changes progressively.

Obviously, if the criterion for incipient motion were determined according any one of the conditions 2, 3, or 4, the results would be quite different. The incipient condition for coarse sediment is shown in Fig. 8.2; for the simplest condition, the critical drag force is a function only of the grain diameter. The curves in Fig. 8.2 cover a broad band because of the various decisions as to a standard for incipience that were made by the various investigators. Most of the curves follow an equation of the type

$$\tau_c = \Theta_c(\gamma_s - \gamma)D \tag{8.1}$$

in which, Θ_c has values within the range 0.017 to 0.076.

In order to achieve a quantitative standard for defining this condition, Yalin suggested the following parameter as the criterion of incipient motion [7],

$$\varepsilon = \frac{m}{At}\sqrt{\frac{\rho D^5}{\gamma_s - \gamma}} \tag{8.2}$$

in it, m is the number of grains moving from a given area A in a given time interval t. For one grain size, ε should be a constant. If one repeats an experiment for two groups of sediment with the same specific gravity but sizes differing by a factor of 10, the value of m/At for the coarser group should be less than that for the finer group by the

313

factor of $10^{5/2}$. Thus, one should adopt a different point of view and take a much longer time for any experiment.

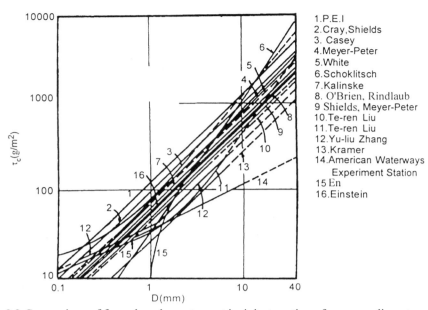

Fig. 8.2 Comparison of formulas: shear stress at incipient motion of coarse sediment

Dou also studied incipient motion and used the velocity near the bed as the hydraulic parameter that determines the incipient motion of sediment[8]. He included the fluctuation of the flow, but not the frequency distribution for the incipient velocity. According to his analysis, the three probabilities for incipient motion that correspond to Kramer's criteria for bed load movement are as follows:

1. Occasional individual motion,

$$p_{c1} = p[u_0 > u_c = \bar{u}_c + 3\sigma_{u_0} = 2.11\bar{u}_c] = 0.00135$$

2. Sparse motion,

$$p_{c2} = p[u_0 > u_c = \bar{u}_c + 2\sigma_{u_0} = 1.74\bar{u}_c] = 0.0227$$

3. Strong motion,

$$p_{c3} = p[u_0 > u_c = \bar{u}_c + \sigma_{u_0} = 1.37\bar{u}_c] = 0.159$$

in which, $\bar{u}_c$ is the time average of the critical velocity near the bed, σ_{u_0} is the standard deviation of the velocity fluctuation near the bed. One can choose any of the three probabilities as a general standard for defining incipient movement.

In fact, the incipient motion of sediment depends on the mutual action of two variables, each with its characteristic frequency distribution.

314

1. Every grain on a bed will move if a certain critical shear stress acts on it. Even for uniform sediment, the critical shear stresses on various particles are different because of the different positions of the grains. For example, in Fig. 8.3, grains numbered 1, 3, 4, 6, 7, 8, 10, and 11 are easy to move, and those numbered 0, 2, 5, 9, and 12 are hard to move, in these present locations. If effects of size, shape, gradation and orientation of the grains are included, the value of τ_c for grains on a bed at various location are variable in the statistical sense. Its distribution reflects the characteristics of the sediment on bed.

Fig. 8.3 Grains in different positions on bed

2. The turbulence near the bed causes the shear stress that is exerted on bed to fluctuate. Thus, the distribution of this stress is also stochastic in a way that reflects the random characteristics of the flow.

In the simplest case, the flow near the bed is laminar, without any turbulence, and the shear stress exerted on the bed is designated as τ_o. Along the direction of the flow, the distribution of τ_c could be like the blocks shown in Fig. 8.4. For various values of τ_o (as for the lines A-A, B-B and C-C in Fig. 8.4), grains at various locations will move, and the grains projecting farthest into the flow will move first. If τ_o increases, more and more grains will move, and the intensity of the sediment movement will thus increase continuously with the flow.

However, the condition of no turbulence near the bed is rare. Even if the bed is smooth and a laminar sub-layer exists, eddies with large moment will frequently enter the boundary layer from the region of the main flow and cause fluctuations there (as discussed in Chapter 4). In this case, the relationship between τ_o and τ_c is like that shown in Fig. 8.4b. A determination of the location at which sediment will move first

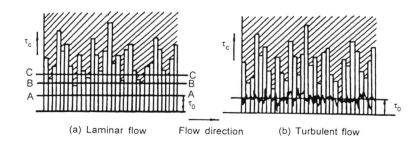

Fig. 8.4 Distribution of τ_c and τ_o along the bed

315

is again stochastic. The grain projecting farthest into the flow may not necessarily be set in motion first even though it is the easiest to move.

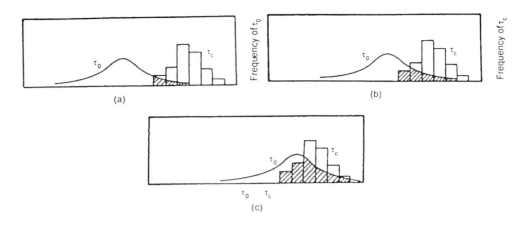

Fig. 8.5 Frequency distribution of flow and shear stress for incipient motion
of bed grains and their interrelationship (after Grass , A.J.)

The curves of frequency distributions of τ_c and τ_0 are shown in Fig. 8.5 [9]. Since $\overline{\tau_0} < \overline{\tau_c}$, the shaded area for which the two curves overlap reflects the amount of the sediment moved and its intensity. Fig. 8.5a shows that the drag force with a large intensity coincides with the part of sediment that is easy to entrain, but the frequency of the condition is quite low; hence, sediment particles that can move are scarce. In Fig. 8.5b, the larger shear stress can set more of the grains in motion, but the amount of sediment in motion is still limited. In Fig. 8.5c, the intensity of the flow is so large in comparison with the grain sizes that the drag is big enough to entrain sediment nearly everywhere, and a large part of the sediment on the bed then moves. From such an analysis, one can select a size of shaded area as the criterion for identifying incipient motion.

The treatment of incipient motion on the basis of Figure 8.5 is logical, but how to determine the frequency distributions of the drag force at bed and of the critical drag force of the sediment in a useful way remains a key question. Also, the fluctuation of the forces on the particles or of the velocity at bed may or may not comply with the normal error law, as mentioned in Chapter 5. With more detailed experimental information, one could unify the understanding of this process. The determination of the frequency distribution of the critical drag force for a grain on the bed is also difficult. No systematic experimental results are available for such a determination. In such circumstances, one must use a simplified and average conditions for the bed material, an approach that neglects important stochastic properties.

8.2 CONDITION OF INCIPIENT MOTION FOR NON-COHESIVE UNIFORM SEDIMENT

8.2.1 Various ways to express the condition for incipient motion

The hydraulic parameters for this condition can be expressed by using shear stress (drag force), average velocity, or flow power.

8.2.1.1 Shear stress for incipient motion

The definition can be approached in the following three ways:

1. Formula for shear stress in non-uniform channel flow

First, in uniform flow, the component of the weight of water acting on the bed surface along the direction of flow is proportional to $\gamma h J$. This force component must be equal to the drag force exerting on bed by the flow.

$$\tau_0 = \gamma h J \qquad (8.3)$$

For non-uniform flow, the above formula is still valid, but the slope must be the energy gradient for the flow. From the viewpoint of energy, one can prove this point in the following way:

The process of the transfer of flow energy has been discussed. The energy that is dissipated per unit volume of water per unit time is

$$\omega_s = \tau \frac{du}{dy}$$

The total energy dissipated per unit time integrated over the depth h is

$$W_0 = \int_0^h \tau \frac{du}{dy} dy \qquad (8.4)$$

The vertical distribution of velocity can be expressed as

$$u = U f(y)$$

in which the average velocity is

$$U = \frac{1}{h} \int_0^h U f(y) dy$$

Also,

317

$$\frac{1}{h}\int_0^h f(y)dy = 1$$

In two-dimensional flow, the distribution of shear stress along a vertical is

$$\tau = \tau_0(1 - y/h)$$

If these relationship are combined with Eq. (8.4),

$$W_0 = \int_0^h \tau_0(1-y/h)U\frac{df(y)}{dy}dy = \tau_0 U\left[\int_0^h \frac{df(y)}{dy}dy - \frac{1}{h}\int_0^h y\frac{df(y)}{dy}dy\right] = \tau_0 U \quad (8.5)$$

In addition, the energy slope J_e is physically equal to the energy per unit weight of water dissipated in a unit distance, and it can be expressed as

$$W_0 = \gamma h U J_e \tag{8.6}$$

From Eqs. (8.5) and (8.6), one can obtain Eq. (8.3), in which J_e is the energy slope of the flow.

2. Shields diagram and its modification

The Shields equation for incipient drag force for non-cohesive uniform sediment can be used. For simplification, the sediment are approximated by spherical grains with a submerged weight of

$$W' = (\gamma_s - \gamma)\frac{\pi D^3}{6}$$

The forces acting on grains on the bed, as discussed in Chapter 5, are mainly the drag force

$$F_D = C_D \frac{\pi D^2}{4}\frac{\rho u_0^2}{2} \tag{8.7}$$

and the uplift

$$F_L = C_L \frac{\pi D^2}{4}\frac{\rho u_0^2}{2} \tag{8.8}$$

in which, C_D and C_L are the drag and uplift coefficients, respectively, and u_0 effective velocity acting on grains at the bed.

As the bed is composed of uniform grains, the vertical distribution of velocity from Eq. (7.38) is

$$\frac{u}{U_*} = 5.75 \log 30.2 \frac{\chi y}{\alpha_1 D} \qquad (8.9)$$

in which α_1 is a constant of about 2, and χ is a function of the grain Reynolds number

$$\chi = f_1(\frac{U_* D}{\upsilon}) \qquad (8.10)$$

The velocity u_o in Eqs. (8.7) and (8.8) can be taken as the velocity at $\alpha_2 D$, in which α_2 is a constant of about 1. As a result,

$$u_0 = 5.75 U_* \log 30.2 \frac{\alpha_2}{\alpha_1} \chi = U_* f_2(\frac{U_* D}{\upsilon})$$

The condition for a grain on the bed to begin to slide is, as in Eq. (5.13),

$$F_D = f(W' - F_L)$$

in which, f is the friction coefficient among grains. With the corresponding forces substituted into the above formula,

$$\frac{\tau_c}{(\gamma_s - \gamma)D} = \frac{4}{3} \frac{f}{(C_D + fC_L)\left[f_2(\frac{U_* D}{\upsilon})\right]^2} \qquad (8.11)$$

The drag coefficient C_D depends on the shape of a grain and the Reynolds number of the flow. If grains are approximately spherical, it is mainly a function of Reynolds number.

$$C_D = f_3(\frac{U_0 D}{\upsilon}) = f_3\left[\frac{U_* D}{\upsilon} \frac{u_0}{U_*}\right] = f_3\left[\frac{U_* D}{\upsilon} f_2(\frac{U_* D}{\upsilon})\right] = f_4(\frac{U_* D}{\upsilon})$$

A similar result can be obtained for the uplift coefficient C_L. Thus, the quantity on the right side of Eq. (8.11) depends only on the grain Reynolds number.

$$\frac{\tau_c}{(\gamma_s - \gamma)D} = f(\frac{U_* D}{\upsilon}) \qquad (8.12)$$

This is the formula Shields used for the drag force. In his original derivation he neglected the effect of uplift. Actually, the basic structure of the formula is not changed whether uplift is included or not. Hence, the formula shows that when grains start to move, the ratio of the drag force acting on a grain to its weight is a function of the grain Reynolds number.

The form of the function f in Eq. (8.12) must be determined by experiment. The results of experiments conducted by Shields for sediment with four different specific

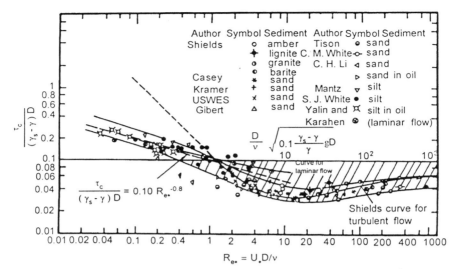

Fig. 8.6 Condition for incipient motion for non-cohesive sediment
(Shields curve and its modification)

gravities are shown in Fig. 8.6. From these data and data from other investigators, an average curve was obtained. This curve has the following characteristics:

(1) It has a saddle shape. A minimum value of $\tau_c/(\gamma_s-\gamma)D$ occurs for a grain Reynolds number of about 10. For that value, the thickness of the laminar sub-layer is approximately equal to the grain diameter of the sediment that is most easily entrained.

(2) If the bed is smooth, a sediment particle is shielded by the laminar sub-layer so that a larger drag force is needed to move it. If $U_* D/v < 2$, that is, if the thickness of sub-layer is more than six times the grain size, the proposed function is a straight line with a slope of 45^0 and incipient motion does not depend on the grain size.

(3) If $U_*D/v > 10$, the stability of the grain is greater because the incipient drag force increases as the grain weight increases. If the grain Reynolds number is greater than 1,000, $\tau_c/(\gamma_s-\gamma)D$ has a constant value of about 0.06.

After the work of Shields, a number of other researchers studied the incipient motion of sediment; these include White [10], Mantz [11], Tison [12], and Li [13]. Their results are included in Fig. 8.6. A belt for the incipient drag force can be drawn to represent the data. This belt differs from the original Shields curve in two important ways:

(1) Shields concluded that if the grain Reynolds number is less than 2, the curve would be a straight line with a 45° slope. However, he had no experimental data in that region. He may have relied on an analogy with the relationship between the drag coefficient and the Reynolds number, and simply decided that $\tau_c/(\gamma_s-\gamma)D$ should be inversely proportional to the grain Reynolds number if the latter is sufficiently small. The later experiments, shown in Fig. 8.6, indicate that this concept does not agree with

reality. In this range of Reynolds number, $\tau_c/(\gamma_s - \gamma)D$ is proportional to Re. with an exponent of -0.3.

(2) If Re. is quite large, Shields proposed that $\tau_c/(\gamma_s - \gamma)D$ would have the value 0.06. The results now available show that this value should be taken as an upper limit; the lower limit should be about 0.04. In Fig. 8.6, most of the results fall within this range. From the experimental data of Paintal [14], Miller et al. recommended that the ratio approaches the value 0.045, not 0.06, if $Re.$ is quite large [15].

The experimental work done by Shields and other investigators was carried out for turbulent flow. More recently, Yalin and Karahan studied the incipient motion of sediment for laminar flow [16]. Their results are also shown in Fig. 8.6 and they show that incipient motion for laminar flow is different from that for turbulent flow. The drag force required with laminar flow is generally larger than that for turbulent flow. For turbulent flow with a smooth boundary, the flow pattern near the boundary is similar to that for laminar flow because of the laminar sub-layer. For this condition, the data for the two regimes fall together, and they have an asymptote that is given by the following formula:

$$\frac{\tau_0}{(\gamma_s - \gamma)D} = 0.1 R_{e*}^{-0.3} \tag{8.13}$$

In the Shields diagram, the parameter $U_*(\tau_0 = \rho U_*^2)$ occurs in both the abscissa and the ordinate. Hence, in determining the incipient drag force. one must proceed by trial and error. To simplify the process, a set of lines for constant values of the parameter of $\frac{D}{\upsilon}\sqrt{0.1\frac{\gamma_s - \gamma}{\gamma}gD}$, with a slope of 2, has been drawn in Fig. 8.6. The intersections of these lines with the Shields curve are the corresponding drag forces for incipient motion.

3. Effect of fluctuations of the flow on the drag force at incipient motion

Although the Shields curve included the fundamental factors that determine the shear stress for incipient motion, his physical interpretation of it is incomplete because it does not include the effects of flow fluctuations. Chepil derived the following formula in his study of incipient motion of sediment under the action of wind [17]

$$\overline{\tau}_c = \frac{0.66\eta \tan\phi(\gamma_s - \gamma)D}{(1 + 0.85\tan\phi)T} \tag{8.14}$$

in which η is a constant reflecting the extent that a sediment grain extends above the surrounding grains (with the other grains remained at a given level), φ is the angle of repose for the sediment, and T is the ratio of the maximum drag and uplift forces exerted on the bed to their average values; in this way, it reflects the intensity of air flow fluctuations. The value of T in Eq. (8.14) can be determined from the following equation [18],

$$T = \frac{3\sigma_p + \bar{p}}{\bar{p}}$$

in which $\bar{p}$ is the average pressure acting on bed, and σ_p is the standard deviation of the pressure fluctuation.

Experimental data show that T varies from 2.1 to 3.0, with an average of 2.5. If the values $\eta = 0.5$-0.75, $\phi = 24°$, $T = 2.5$ are substituted into Eq. (8.14), the result is

$$\bar{\tau}_c = (0.04\sim0.06)(\gamma_s - \gamma)D$$

so that it has the form of the Shields result for shear stress on coarse sediment.

8.2.1.2 Velocity for incipient motion

The relationship between the velocity field and shear stress field, as discussed in Chapters 4 and 7, is the logarithmic velocity formula

$$U = 5.75U_* \log 12.27 \frac{\chi R}{K_s} = 5.75\sqrt{\frac{\tau_0}{\rho}} \log 12.27 \frac{\chi R}{K_s}$$

If this formula is substituted into Eq. (8.12), the formula for incipient drag force can be transformed into the following relationship:

$$\frac{U_c}{\sqrt{\frac{\gamma_s - \gamma}{\gamma} gD}} = 5.75\sqrt{f(\frac{U_* D}{\upsilon})} \log 12.27 \frac{\chi R}{K_s}$$

For the belt zone in Fig. 8.6 with Re_* larger than 60, the $f(\frac{U_* D}{\upsilon})$ has a value within the range of 0.03 to 0.06.

Hence,

$$\frac{U_c}{\sqrt{\frac{\gamma_s - \gamma}{\gamma} gD}} = (1 \sim 1.4) \log 12.27 \frac{\chi R}{K_s}$$

For natural sediment, $\dfrac{\gamma_s - \gamma}{\gamma}$ can be taken as 1.65, and the formula is then

$$\frac{U_c}{\sqrt{gD}} = (1.28 \sim 1.79)\log 12.27\frac{\chi R}{K_s} \tag{8.15}$$

Many formulas for the critical velocity have this form. The best known are those of Goncharov and Levy. The Goncharov formula [19] is

$$\frac{U_c}{\sqrt{\dfrac{\gamma_s - \gamma}{\gamma}gD}} = 1.06\log\frac{8.8h}{D_{95}} \tag{8.16}$$

and the Levy formulas [20] are

for $R/D_{90} > 60$,

$$\frac{U_c}{\sqrt{gD}} = 1.4\log\frac{12R}{D_{90}} \tag{8.17}$$

for $R/D_{90} = 10\text{-}40$,

$$\frac{U_c}{\sqrt{gD}} = 1.04 + 0.87\log\frac{10R}{D_{90}} \tag{8.18}$$

Both of these scholars used the coarsest portion of the sediment as the representative friction grain size. The constants in the logarithmic term in these two formulas are different because the structure and coefficients of the formulas they used are somewhat different.

Some scholars prefer to use an exponential velocity formula instead of a logarithmic one. Their formulas involve the water depth or hydraulic radius raised to some power. A representative one is the Shamov formula [21].

$$\frac{U_c}{\sqrt{gD}} = 1.47\left(\frac{h}{D}\right)^{1/6} \tag{8.19}$$

Orlov suggested a similar formula but proposed a coefficient of 1.56 instead of 1.47.

Shields studied the critical shear stress for incipient motion as a function of the Reynolds number. Other investigators, including Bao-ru Li [22], Zheng and Wang [23] and Knoros[24] sought to establish a relationship between the Reynolds number and a

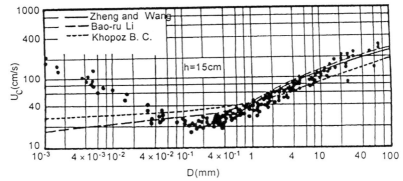

Fig. 8.7 Relationship between critical velocity and grain size

critical velocity for incipient motion. Table 8.1 contains the formulas proposed by various investigators for various ranges of roughness. In Fig. 8.7, curves are drawn for the various formulas for a depth of 15 cm and the kinematic viscosity of $0.01 \text{cm}^2/\text{sec}$.

For comparison, the experimental results obtained by a score of investigators are plotted in same figure. For coarse sediment, the trends of the various formulas and experimental data agree quite well with only small differences among them. For fine sediment, the measured data show that the incipient velocity increases as the grain size decreases, and none of the formulas correctly reflect this trend. Obviously, the cohesion among grains can not be neglected for fine sediment. The inclusion of the effect of the laminar sub-layer alone does not give a complete picture of the occurrence. This point is treated later in Paragraph 8.4.

Before concluding the discussion of incipient velocity, a brief introduction of Yang's work may be helpful [25]. From a consideration of the drag force and uplift acting on bed material, Yang decided that the relationship between drag coefficient and u_oD/v corresponds to the relationship between the drag coefficient C_D and $\omega D/v$ for a particle settling in quiescent water (see Coleman's result in Table 5.1 and Fig. 5.3). In addition, he assumed that the coefficients of drag and uplift have about the same magnitude. Then, from the usual equation for the force balance during the settling of sediment,

$$C_D \frac{\pi D^2}{4} \frac{\rho \omega^2}{2} = (\gamma_s - \gamma) \frac{\pi D^3}{6}$$

the combined coefficient can be replaced by the fall velocity ω. A series of transformations like those in the derivation of the Shields formula [(Eq. (8.12)] leads to the following relationship

$$\frac{U_c}{U_*} = f(\frac{U_* D}{\upsilon}, \frac{D}{h})$$

Table 8.1 Incipient velocity formulas in consideration of effect of laminar sub-layer

Range of Drag	B.R. Li formula		Z.Z. Zheng formula		Knoros formula	
	Range	Form of Formula	Range	Form of Formula[1]	Range	Form of Formula
Smooth Zone	$\dfrac{U_cD}{v}\left(\dfrac{D}{R}\right)^{1/6} < 50$	$U_c = 55.8D^{0.1}v^{0.158}R^{1/6}$	$\dfrac{U_cD}{v} < 60$	$U_c = 4.88v^{0.313}R^{0.187}N^{0.381}$	$\dfrac{U_*D}{v} < 3.5$	$U_c = \dfrac{90D^{0.05}h^{0.125}}{\sqrt{h^{0.27}}+7.5}$
Transition Zone	$\dfrac{U_cD}{v}\left(\dfrac{D}{R}\right)^{1/6} = 50\sim300$	$U_c = 57.3D^{0.333}R^{1/6}$	$\dfrac{U_cD}{v} = 60\sim120$	$U_c = 3.04v^{0.095}R^{0.169}N^{0.452}D^{0.188}$	$\dfrac{U_*D}{v} = 3.5\sim25$	$U_c = 25D^{0.3}\log\dfrac{3.6h}{D^{1.6}}$
	$\dfrac{U_cD}{v}\left(\dfrac{D}{R}\right)^{1/6} = 300\sim1200$	$U_c = 36.8D^{0.77}v^{-0.294}R^{1/6}$	$\dfrac{U_cD}{v} = 120\sim1700$	$U_c = 0.96v^{-0.193}R^{0.179}N^{0.595}D^{0.607}$		
Rough Zone	$\dfrac{U_cD}{v}\left(\dfrac{D}{R}\right)^{1/6} > 1200$	$U_c = 79.75D^{1/3}R^{1/6}$	$\dfrac{U_cD}{v} > 1700$	$U_c = 3.16R^{0.167}N^{0.5}D^{0.333}$	$\dfrac{U_*D}{v} > 25$	$U_c = 1.2\sqrt{gD}\log\dfrac{14.7h}{D^{0.75}}$

1) $N = \dfrac{2}{3}\left(\dfrac{\gamma_s - \gamma}{\gamma}\right)g\mu_s$, in which, μ_s is the consolidation of bed material in the calculation of the curve in Fig. 8.7, $\mu_s = 0.5$

In Fig. 8.8, Yang plotted such a function including the results of 153 sets of experimental data (including part of the data in Fig. 8.6). He obtained formulas.

$$1.2 < \frac{U_* D}{\upsilon} < 70 \qquad \frac{U_c}{\omega} = \frac{2.5}{\log(\dfrac{U_* D}{\upsilon}) - 0.06} + 0.66 \qquad (8.20)$$

$$\frac{U_* D}{\upsilon} > 70 \qquad \frac{U_c}{\omega} = 2.05 \qquad (8.21)$$

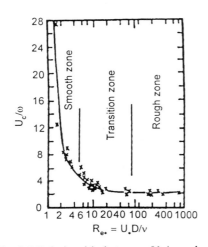

Fig. 8.8 Relationship between U_c/ω and grain Re(from experiment in wooden flume)(after Yang Chih-Ted)

For coarse sediment, the direct link between U_c and ω, showing that the critical velocity is roughly equal to twice the fall velocity, is quite useful in both theory and application. But the significance of relationship in Fig. 8.8 is limited because the data for it were all obtained from flume studies. The range of water depth in the experiments was not large enough to reflect the full effect of relative water depth on the critical velocity. Therefore, Eqs. (8.20) and (8.21) should not be used for natural rivers.

8.2.1.3 Incipient flow power

Because sediment grains move under the action of flowing water, the flow must do work and expend energy to keep the grains moving. To link the rate of bed load transport with the flow energy expended per unit time is an important way to study bed load movement (Chapter 9). Instead of using a critical shear stress acting on grains to express the condition for incipient sediment motion, one can also adopt a critical value for the flow power required to maintain the grains in motion.

Chapter 5 points out that the energy to keep bed load in motion comes from the potential energy of the flow, and that the potential energy of water per unit area lost in a unit time is

$$W_0 = \gamma h J U = \gamma q J$$

in which q is the discharge per unit width. Hence, the following relationship must exist.

$$\frac{\gamma_s - \gamma}{\gamma_s} g_b = f(W_0) \qquad (8.22)$$

in which g_b is the transport rate of bed load per unit width (dry weight).

From the experimental results for No. 1 Sand (median diameter 0.59 mm) conducted at the Waterways Experiment Station (USA), Bagnold plotted the functional relationship of Eq. (8.22) with the result shown in Fig. 8.9 [26]. The data tend to be distributed along a straight line but with some scatter. The intersection of this line and the abscissa is the flow power that is needed for the incipient motion of sediment. The data points for small transport rates fall somewhat above the line because part of the coarser sediment remains at rest for such a flow condition. For larger flows, all of the grains move, and the data points fall rather evenly on both sides of the line. Therefore, Eq. (8.22) can be expressed as

$$\frac{\gamma_s - \gamma}{\gamma_s} g_b = K(W_0 - W_c) \tag{8.23}$$

in which W_c is the flow power for incipient motion. Bagnold found it to be proportional to the grain diameter raised to the power 3/2; so that,

$$W_0 = \gamma h J U = \tau_0 U \sim \tau_0 U_* \sim \tau_0^{3/2}$$

For coarser sediment, the shear stress for incipient motion is proportional to the grain diameter.

So the incipient motion condition for coarser sediment can be expressed as:

$$\frac{W_c}{\frac{\gamma}{g}\left(\frac{\lambda_s - \gamma}{\gamma} gD\right)^{3/2}} = C \tag{8.24}$$

More work is needed to determine how much the constant C is.

For a specified slope, Eq. (8.24) can be written as

$$q_c = const. \frac{(\frac{\gamma_s - \gamma}{\gamma} gD)^{3/2}}{gJ} \tag{8.25}$$

For natural sediment,

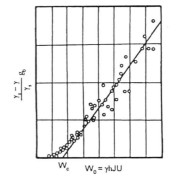

Fig. 8.9 Relationship between power used to maintain bed load motion and power supplied by flow (after Bagnold, R.A.)

$$q_c = K \frac{D^{3/2}}{J} \tag{8.26}$$

in which q_c is the discharge per unit width for the incipient motion. From experimental data, Schoklitsch obtained the following condition for incipient motion [27]:

$$q_c = 0.0194 \frac{D}{J^{4/3}} \tag{8.27}$$

in which q_c is in m^3/s.m and D is in m. The structures of Eq. (8.27) and Eq. (8.26) have some similarity.

8.2.2 Comparison of the three ways of expressing the condition for incipient motion

Although the drag force, velocity, and flow power for incipient motion provide three different expressions for the same phenomenon and can be mutually transformed from one to another, they still represent three different study approaches based on three different concepts. Each has advantages and disadvantages. Because the concept of critical flow power is a relatively new one and its theoretical expression and formulation are still being perfected, the following discussion is primarily a comparison between the use of the critical values for shear stress and velocity.

8.2.2.1 Consideration on the cause of initiation of sediment motion

The incipient motion of sediment is a dynamic process. The force that causes sediment to start to move, in the final analysis, is the drag force exerted by the flow on the particles. Thus, the use of a critical value of force as a criterion of incipient condition may well be the most straightforward procedure. To measure this condition for sediment motion by some type of velocity is less direct because the functional relationship between them involves two relationships: one is that between the acting force and the velocity near bed, and the other is that of the velocity near the bed to the average velocity.

In seeking to determine whether drag force or velocity is the more direct approach to describe the condition for sediment as it starts to move, a set of special experiments could be designed: the drag force of the flow would be the same everywhere but the average velocities at each section would vary. Thus, if flow force were the main factor for the incipient motion of sediment, all grains would begin to move at the same time as the flow intensity gradually increases up to the critical condition. In the other, if the velocity is the main factor for incipient motion, the phenomenon of incipient motion of sediment must be observed at a certain cross section. The latter experiment is the one White conducted in 1940 [28].

White considered two cases in his experiments. In the first, the sediment was submerged in the laminar sub-layer so that the bed was effectively smooth. According to the boundary layer theory, the relationship between drag force τ_0, average velocity U, and distance from entry x is as follows:

$$\tau_0 = C\rho\sqrt{\frac{\upsilon U^3}{x}} \qquad (8.28)$$

in which, C is a constant of proportionality. Thus, the drag force at every point along the boundary remains constant if U^3 is proportional to x. The special flume designed by White for the first set of experiments is shown in Fig. 8.10a, for which

$$\frac{b}{l} = 0.1\left(\frac{l}{x}\right)^{1/3} \tag{8.29}$$

In the second set of experiments, the bed was rough. According to the theory for a turbulent boundary layer,

$$1.5\frac{\chi}{K_s} = e^{\theta}(\theta - 1) - \frac{\theta^2}{2} + 1 \tag{8.30}$$

in which

$$\theta = \frac{U}{2.5}\sqrt{\frac{\rho}{\tau_0}} \tag{8.31}$$

From these equations, one readily obtains a relationship between U and x that gives a constant value of τ_0 for various values of x. The shape of such a flume is shown in Fig. 8.10b. The results of these two sets of experiments proved that motion of sediment started at almost the same time over the entire test area. One can conclude that the drag force of the flow is thus directly involved in the incipient motion of sediment.

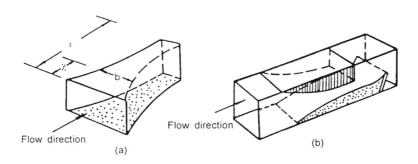

Fig. 8.10 Special flumes designed by White for studying the effect of drag force and velocity on incipient motion of sediment (after White, C.M.)

8.2.2.2 Effect of turbulence near the bed

As already mentioned, the use of the average velocity as a hydraulic criterion for the incipient motion of sediment involves a relationship between the average velocity and some velocity near the bed. Therefore, some investigators suggested using the velocity near the bed directly in order to study the critical condition. However, the relationship shows that the velocity near the bed does not reflect the drag force acting on bed well unless the bed roughness remains constant.

Lyles and Woodruff conducted experiments with four different types of bed roughness [29].

bed surface	type of roughness	
S1		smooth bed
S2		
S3	spherical particles glued on bed with the diameter of	$\begin{cases} 0.61 \ \text{cm} \\ 1.64 \ \text{cm} \\ 2.45 \ \text{cm} \end{cases}$
S4		

If the standard deviation σ_u' of the velocity fluctuations is taken as the parameter to express turbulence intensity, the vertical distribution of the ratio of σ_u' to U_* remains constant within the boundary layer even though the roughness of the bed varies as shown in Fig. 8.11a. But the vertical distribution of the ratio of σ_u' to u_y follows various curves, the rougher the bed, the larger the value of σ_u'/u_y is at a given elevation as shown in Fig. 8.11b for the different value of bed roughness. Thus, if one conducts experiments on the incipient condition with various erodible materials , like those shown in Table 8.2, the incipient velocity near the bed is smaller for a bed that is rougher. In this experiment, Lyles took the velocity at a point 1.22 cm above the bed as the velocity at the bed. The results show that the bed velocity for incipient motion is not a good criterion if the roughness of bed is a variable. In fact, if sand dunes form on the bed, the critical bed velocity would then be smaller than that on a flat bed [30]. In addition, if the friction velocity U_* is selected as a hydraulic parameter, the incipient U_* remains constant for a specific sediment with various bed roughness. Thus, in considerations of the effect of turbulence on the incipient motion of sediment, the drag force is a better critical index than the bed velocity.

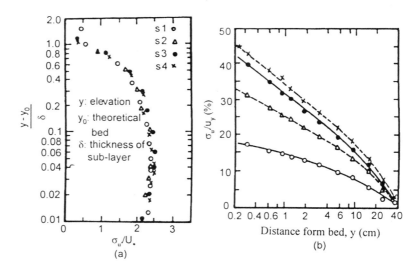

Fig. 8.11 Vertical distribution of the ratio of velocity fluctuation to longitudinal velocity and friction velocity for various bed roughness (after Lyles, L., and N.P. Woodruff)

Table 8.2 Incipient velocity near bed and drag force for various bed roughness

erodible	grain size	incipient velocity near bed (cm/s)			incipient friction velocity (cm/s)		
material	(mm)	S_2	S_3	S_4	S_2	S_3	S_4
sand	0.18-0.30	415	363	335	41.9	44.3	46.0
sand	0.42-0.59	617	498	448	62.3	60.7	61.2
sand	0.59-0.84	699	568	539	70.7	69.4	72.9
soil	0.42-0.59	499	419	385	50.4	51.2	52.1

8.2.2.3 Consideration of practical applications

In applications, both of these expressions are convenient, but both also present problems.

An important advantage of the formula for the critical drag force is that it can be taken as a constant for a particular flow condition and for a specific grain size even though it is a function of the grain Reynolds number. In contrast, the corresponding critical velocity varies with the grain Reynolds number, and it depends strongly on the water depth. As a result, calculated values diverge for the various formulas that are used, and the method is inconvenient to use. In addition, natural flows are not two-dimensional, and the shear stresses along the boundary are not uniformly distributed. Therefore, one must consider the location and value of the maximum shear stress and endeavor to control it within a specific limit. In the design of a stable slope for a canal bank, the component of gravitational force along the slope should be used. All such factors are easier to treat if the concept of drag force is used. For the movement of wind-blown sediment, the concepts of discharge and average velocity lose significance because the flow has no well-defined upper boundary. Only with the use of the drag force on the bed, or the corresponding frictional velocity, can the critical condition for wind blown sediment be analyzed.

The most serious disadvantage of the drag force concept is that the slope is included in the formula. Because the measurement of slope in rivers requires high precision, the results obtained are less reliable than those based on the average velocity; the latter is measured regularly at hydrological stations. Furthermore, the concept of velocity and water depth is easier for people to visualize. Thus, the concept of critical velocity also has convenient features.

8.2.3 Incipient motion for sediment on a sloping surface

So far, the component of gravitational force of grains in the direction of flow has been neglected because the slopes are usually quite small. For rivers or canals with steep gradients, this omission can introduce appreciable errors. Besides, the component of the gravitational force along the bank slope should be considered. This force makes sediment easier to move and many studies have done so [2,31,32].

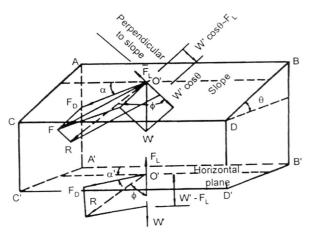

Fig. 8.12 Forces acting on grains at O on a slope

In the general case of a plane ABCD with a dip angle of θ as shown in Fig. 8.12, the water flows along the direction with an angle α measured from the horizontal axis of the channel. The forces acting on grains at point O on the slope are,

drag force F_D, in the flow direction, Eq. (8.7),

uplift F_L, perpendicular to the slope, Eq. (8.8),

weight W', downward and equal to $\frac{\pi}{6}D^3(\gamma_s - \gamma)$.

The resultant of the drag force and the component of the gravitational force along the slope, $W'\sin\theta$, is

$$F = \sqrt{(F_D \sin\alpha + W'\sin\theta)^2 + F_D{}^2 \cos^2\alpha}$$

the condition for incipient motion of sediment on a slope in still water is

$$\frac{F}{W'\cos\theta - F_L} = \tan\phi \qquad (8.32)$$

in which, ϕ is the angle of repose for submerged sediment.

If τ_c represents the incipient shear stress on a flat bed and τ_c' that on a sloping one [Eq. (5.13) and (8.32)], the following formula can be derived for the incipient motion of a grain on a flat bed and on a sloping one.

$$\frac{\tau_c'}{\tau_c} = \left\{ (1 + \frac{C_l}{C_D}\tan\phi)(\cos^2\theta - \sin^2\theta\cot^2\phi)\cot\phi \right\} \cdot \left\{ \left(\frac{F_L}{F_D}\cos\theta + \sin\theta\sin\alpha\cot^2\phi \right) + \right.$$

$$\left. \cot\phi \sqrt{\cos^2\theta - \sin^2\theta\cot^2\phi + (\frac{F_L}{F_D})^2\sin^2\theta + 2\frac{F_L}{F_D}\sin\theta\cos\theta\sin\alpha + \sin^2\theta\sin^2\alpha\cot^2\phi} \right\}^{-1} \qquad (8.33)$$

Two special and prevalent conditions are the following:

1. If $\alpha = 0$, the direction of the flow is parallel to the banks. The critical condition of the incipient motion of sediment on canal slope is

$$\frac{\tau_c{}'}{\tau_c} = \frac{(\cot^2 \theta - \cot^2 \phi)\left(\cot \phi + \dfrac{C_L}{C_D}\right)\sin \theta}{\dfrac{C_L}{C_D}\cot \theta + \cot \phi \sqrt{\cot^2 \theta - \cot^2 \phi + \left(\dfrac{C_L}{C_D}\right)^2}} \qquad (8.34)$$

If uplift is neglected,

$$\frac{\tau_c{}'}{\tau_c} = \cos\theta \sqrt{1 - \frac{\tan^2 \theta}{\tan^2 \phi}} \qquad (8.35)$$

This result coincides with that given by Lane and Carter.

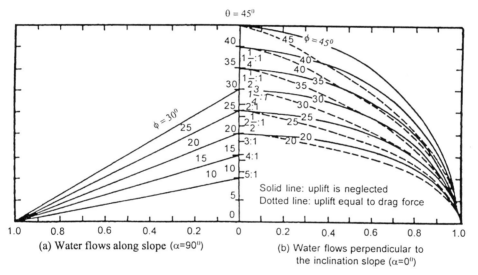

Fig. 8.13 Relationship between $\tau_c{}'/\tau_c$ and the dip angle θ and sediment repose angle ϕ (after Stevens, M.A., and D.B. Simons)

2. If $\alpha = 90°$, water flows at right angle to the banks.

$$\frac{\tau_c{}'}{\tau_c} = \cos\theta - \frac{\sin\theta}{\tan\phi} \qquad (8.36)$$

The lift force does not appear in Eq. (8.36). It reveals that the result does not depend on the lift force.

Fig. 8.13 is a graphical representation of these two cases. The solid line represents the result if uplift is neglected, and the dotted line if it is included and equal to the drag force ($F_L/F_D = C_L/C_D = 1$). For determinations of the stability of a canal slope, neglect of the uplift leads to values of $\tau_c{}'/\tau_c$ that are too large.

333

8.2.4 Distribution of shear stress on the channel boundary

In the preceding treatment, the flow was taken to be two-dimensional, so that shear stress acting on the bed can be simply expressed as $\gamma h J$, from Eq. (8.3). However, in natural rivers, the shear stress over the boundary is not uniformly distributed. The shear stress can be quite large at certain location. In order to maintain the stability of a canal against erosion, the design of the canal conveying clear water must meet the requirement that the drag force due to the flow should not exceed the drag force for incipient motion of the material anywhere on the bed. Therefore, the distribution of the shear stress over the boundary is a key factor in determining the condition for incipient motion of sediment.

8.2.4.1 Distribution of shear stress on simple cross-sections

Fig. 8.14 shows two distributions of shear stress for a rectangular cross-section[33], one with a smooth boundary and the other a rough one. The maximum shear stress occurs near the water surface, on both sides, and at the middle on the bottom. With a rough boundary, the location of the maximum shear stress is further down the side on the bank and spreads more on the bottom toward the sides. The maximum value is usually less than 1.4 times the average shear stress, as shown by line 3 in Fig. 8.15 [34]. For a trapezoidal cross-section, the distribution of shear stress is more irregular as a result of secondary currents (Fig. 8.16). The shear stress at the sides near the water surface is zero, and the location of the maximum value occurs at points nearer the corners where the side intersects the bottom. The flatter the slope of the trapezoidal sides, the closer the point of maximum shear stress is to the corner.

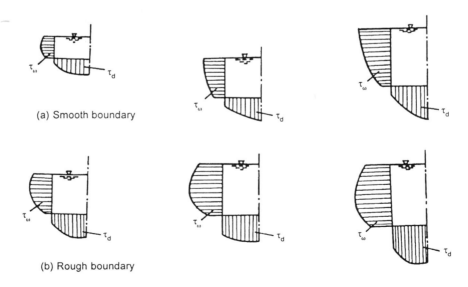

(a) Smooth boundary

(b) Rough boundary

Fig.8.14 Distribution of shear along the boundary for a rectangular cross section
(after Ghosh, S.N., and N. Roy)

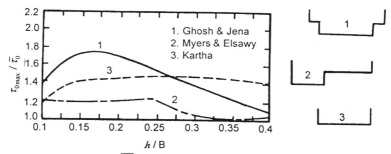

Fig. 8.15 $\tau_{0\,max} / \overline{\tau_0}$ for various cross-section with various water depths
(after Kartha, V.C., and HJ. Leutheusser)

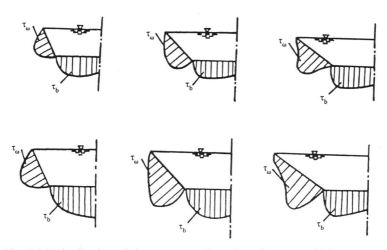

Fig. 8.16 Distribution of shear stress on boundary for trapezoidal cross-sections (rough
boundary) (after Kartha, V.C., and HJ. Leutheusser)

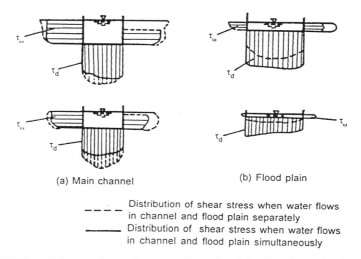

(a) Main channel   (b) Flood plain

- - - -   Distribution of shear stress when water flows
          in channel and flood plain separately
———   Distribution of shear stress when water flows
          in channel and flood plain simultaneously

Fig. 8.17 Effect of flow exchange between channel and flood plain on the distribution of
boundary shear stress (after Myers, R.C., and E.M. Elsawy)

Rajaratnam and Muralidhar studied the distribution of shear stress on a rectangular cross-section for supercritical flow (Fr=1.1-2.4) [35]. They found that the maximum shear stress always occurred at the middle of the bottom if the boundary was smooth. The value of maximum shear stress was nearly equal to that for two-dimensional flow (rhJ) if the width to depth ratio exceeded 15.

8.2.4.2 Distribution of shear stress on complex cross-sections

To learn more about the exchange of flows between the main channel and the flood plain, Myers and Elsawy compared the distribution of shear stress for the two parts, with water flowing in them simultaneously and separately [36]. Fig. 8.17 shows the results for the two cases. If the water flows in both parts simultaneously, the shear stress on the main channel is less because of the exchange of momentum between the channel and the flood plain. Within the range of the experiment, the maximum reduction in the shear stress was 22% for the channel, and the maximum increase for the flood plain was 26%. If the water flowed separately in the two parts, the distribution of shear stresses in both the channel and on the flood plain were symmetrical about their center lines. If water flows in them simultaneously, however, the locations of the maximum shear stress for both channel and flood plain were closer to their junction. Fig. 8.15 shows that the ratio of maximum to average shear stress varies with the shape of the cross-section. For such complex cross sections, the exchange of momentum between channel and flood plain is so significant that the maximum shear stress on the channel bed is only about 20% larger than the average value; the reason for this result is the cross-sections selected by Myers and Elsawy had a comparatively broad flood plain. For the symmetrical complex cross-section with narrow flood plains studied by Ghosh and Jena, the maximum shear stress is nearly 80% larger than the average value [37].

8.2.4.3 Distribution of shear stress at a bend

The distribution of shear stress at a bend is naturally quite complicated. Fig. 8.18 shows the results obtained by Ippen and Drinker [38]. The cross-section they used in their experiment was a trapezoid with a side slope of 1:2. The central angle of the bend was 60°. The ratio of channel width to the radius of curvature (B/R) varied from 0.29 to 0.80. Contour lines for the ratio of the shear stress at any point to the average value at the entry ($\tau_0 / \bar{\tau}_0$) are shown in Fig. 8.18 for two channels. The shaded area mark the region within which $\tau_0 / \bar{\tau}_0$ is larger than 1.4.

The experiments showed that the distributions of shear stress and of its relative value depend mainly on the geometry of the river channel; it was nearly independent of both the water depth and the velocity at the point of entry. As the curvature increased, both the magnitude of shear stress and the area of large shear stress increased also. For a larger curvature, the maximum shear stress occurred at the convex bank within the bend and at the concave bank downstream of the bend (Fig. 8.18a). For a smaller curvature, the larger shear stress occurred at the concave bank along the lower part of the bend (Fig. 8.18b). In both cases, the region of large shear stress extends well into the straight reach downstream of the bend.

A similar experiment was conducted by Nouh and Townsend [39]. They found that due to the effect of circulation induced by the bend, large shear stresses occurred at the convex bank and small shear stresses at the concave bank near the bend entry; the

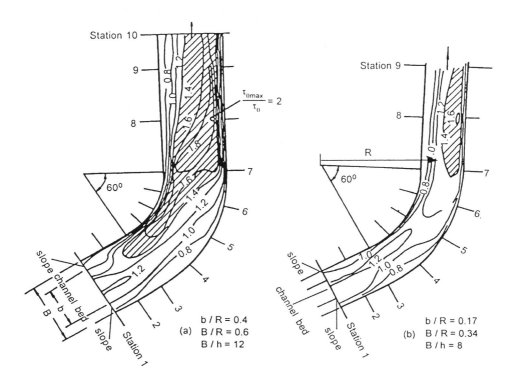

Fig. 8.18 Contour lines for boundary shear stress at bends (after Ippen, A.T., and P.A. Drinker)

values at the bend exit were just the reverse. As is generally known, bank caving usually occurs at the convex bank at entry and concave bank at exit, and these are naturally the regions of large shear stress. Where the flow in the bend enters the straight reach, the shear stress is less at the concave bank and larger at the convex bank. Over some distance, the circulation gradually diminishes and the distribution of shear stress becomes more nearly uniform.

The foregoing results are all for experiments in flumes. Nowadays, with modern instruments, some observers have been able to measure the distribution of shear stress along the cross sections of natural rivers and at the river mouth area and have obtained useful results [40,41].

8.3. INCIPIENT MOTION OF NON-COHESIVE NON-UNIFORM SEDIMENT

8.3.1 Physical meaning of condition for incipient motion for non-uniform sediment

Much confusion has arisen as to what is meant by the criterion of incipient motion for non-uniform sediment. Actually, two different meanings are involved.

1. In strict terms, no single critical grain diameter can be determined to distinguish which part of the sediment moves and which does not for any specific conditions. As already stated, more of the fine sediment moves, but part of it also remains at rest. In addition, even if less of the coarse sediment is in motion, not all of it is at rest. Lane and Carlson reached this conclusion after comprehensive investigations of canals in the western part of the United States [42]. Even for uniform sediment, the problem of an absolute distinction between motion and motionless does not exist. Any condition for incipience implies really a certain intensity of sediment movement or a probability of motion. Therefore, for non-uniform sediment, a critical probability of incipient motion can also be assigned; for it, the grain size corresponding to that probability is the critical one, and it characterizes the portion of the sediment that is entrained. To be able to determine such a critical grain size is significant in any application. For instance, after a large project has been completed on a river, the bed material in the reach downstream should be studied to determine which part will be scoured and which part will remain to form an armoring blanket.

2. For uniform sediment, the boundary condition at the bed is not changed substantially by sediment motion. Therefore, incipience for uniform sediment is not changed if part of the sediment is removed; thus the occurrence can be treated as a steady phenomenon. For non-uniform sediment, the situation is different. Because the fine sediment is easier to move, incipient motion of bed material is also the beginning of the armoring process of the bed surface, and the phenomenon is an unsteady one. In the latter case, one must study the incipient motion of non-uniform sediment together with the armoring process of the bed material once erosion takes place.

8.3.2 Critical grain size for given conditions of flow and sediment gradation

Although incipient motion of sediment and movement of bed load are treated in separate chapters, the principle of incipient motion and of the movement of bed load are closely related. The former is really a special case of the latter in which the transport rate of the bed load is equal to zero, or more correctly, the intensity of movement is extremely small. This limiting condition can be derived from formulas for the rate of bed load transport. The Meyer-Peter formula, which is widely used in Europe, can be written in the form

$$\Phi = \left(\frac{4}{\psi} - 0.188\right)^{3/2} \tag{8.37}$$

in which

$$\Phi = \frac{g_b}{\gamma_s}\left(\frac{\gamma}{\gamma_s - \gamma}\right)^{1/2}\left(\frac{1}{gD^3}\right)^{1/2} \qquad (8.38)$$

and

$$\psi = \frac{\gamma_s - \gamma}{\gamma}\frac{D}{hJ} \qquad (8.39)$$

The parameter Φ represents the transport intensity of bed load and the parameter ψ is the hydraulics of the flow and g_b is the transport rate of bed load in dry weight.

From Eq. (8.37), if the rate of bed load transport is zero,

$$\psi_c = \frac{4}{0.188} = 21.3 \qquad (8.40)$$

the sediment is then just on the verge of moving.

Another bed load formula that is widely used is the Einstein formula,

$$1 - \frac{1}{\sqrt{\pi}}\int_{-0.143\psi-2}^{0.143\psi-2} e^{-t^2}\,dt = \frac{43.5\Phi}{1 + 43.5\Phi} \qquad (8.41)$$

the relation between Φ and ψ can be plotted as in Fig. 9.8. If

$$\psi_c = 27 \qquad (8.42)$$

Φ=0.0001, the transport rate of bed load is so low that it can be used as the criterion for incipient motion.

The parameter ψ in Eq. (8.39) is the reciprocal of the parameter used for the ordinate in the Shields diagram,

$$\frac{\tau_0}{(\gamma_s - \gamma)D} = \frac{1}{\psi} = \Theta \qquad (8.43)$$

the critical value of Θ_c for incipient motion, as derived from the formulas of Meyer-Peter and Einstein, are 0.047 and 0.037, respectively.

For non-uniform sediment, the combination of grains of various sizes and their mutual influence must be considered; these include the sorting of grains and the shielding of fine grains by coarse ones (Chapter 9). With the inclusion of these factors, Eq. (8.39) can be changed as follows

$$\psi_* = \frac{\xi Y(\beta/\beta_x)^2}{\theta}\psi \qquad (8.44)$$

in which, ξ, Y, $(\beta/\beta_x)^2$ and θ are used to modify the equation so as to include the effect of the non-uniformity of bed material. For sediment with grains of various sizes, the value of ψ_* can be calculated. All of the grains for which $\psi_* < 27$ will start to move for this flow condition. By using this relationship, Harrison calculated the extreme grain size of the armoring layer and good results were achieved.

8.3.3 Critical shear stress for incipient motion of a mixture with armoring

In his study of incipient motion [44~46], Gessler considered only the effect of flow fluctuations and thus neglected the stochastic distribution of the corresponding shear stress for grains on the bed. In turbulent flow,

$$\tau_0 = \overline{\tau}_0 + \tau_0'$$

in which, τ_o, $\overline{\tau}_0$ and τ_o' are the instant, temporal average and fluctuating value of the drag force exerted by the flow on the bed, respectively. If

$$\tau_0 < \tau_c$$

and also

$$\frac{\tau_0'}{\overline{\tau}_0} < \frac{\tau_c}{\overline{\tau}_0} - 1$$

sediment cannot be entrained.

From numerous flume experiments, Gessler concluded that the fluctuation of drag force follows a normal error distribution. Thus, the probability of sediment staying on the bed is (Fig. 8.1),

$$q = \frac{1}{\sigma\sqrt{2\pi}} \int_{-\infty}^{\frac{\tau_c}{\overline{\tau}_0}-1} \exp\left(\frac{x^2}{-2\sigma^2}\right) dx \qquad (8.45)$$

in which, σ is the standard deviation of $\tau_o'/\overline{\tau}_0$. For coarse sediment, σ is about 0.57.

Because the probability of incipient motion of the sediment with grain diameter D is $(1-q)$, the size distribution curves of the bed materials washed away and remaining on the bed can be derived. If the maximum and minimum grain diameters of the original bed material are known and the weight percentage of grain with diameter of D is $p_0 (D)$, the accumulated percentage of the sediment with a diameter less than D is

$$\int_{D_{min}}^{D} p_0(D) dD$$

340

For the armoring layer of the bed after scouring, the frequency of grains with diameter D is

$$p_a(D) = C_1 q p_0 dD$$

in which, the coefficient C_1 can be determined by means of the following equation,

$$\int_{D_{min}}^{D_{max}} p_a(D)dD = 1$$

Thus, the sediment size distribution of the armoring layer is

$$p_a(D) = \frac{\int_{D_{min}}^{D} q p_0(D)dD}{\int_{D_{min}}^{D_{max}} q p_0(D)dD} \tag{8.46}$$

and the size distribution of the sediment washed out is

$$p_l(D) = \frac{\int_{D_{min}}^{D} (1-q) p_0(D)dD}{\int_{D_{min}}^{D_{max}} (1-q) p_0(D)dD} \tag{8.47}$$

Gessler gave numerical example to demonstrate the way the size distribution of sediment changes during the scouring process. If the original size distribution curve of the bed material is a straight line on semi-logarithmic paper, the condition for incipient motion of the grains with various sizes can be determined from the Shields diagram. The size distributions of the sediment washed away and that in the armoring layer are as shown in Fig. 8.19 under the action of various temporal averages of the shear stresses. From the figure, if the shear stress exerted by the flow on the bed is small, only fine sediment is washed out and the coarse particles remain on the bed. Hence, the size distribution of the sediment washed out and that of the sediment staying on the bed have a large difference with that of the original bed material. As the shear stress increases, more and more coarse sediment moves, and the size distribution of the sediment washed out and that of the sediment staying on the bed approach the size distribution of the original bed material. That is , under the action of shear stress, there is an extreme value of size distribution for the armoring layer, for which the effective diameter of the armoring layer reaches a maximum.

The maximum shear stress that a mixture can withstand is the sum of the shear stresses that grains with different sizes, weighted with their frequencies, can withstand. For a grain with diameter D, the maximum shear stress that can be resisted is

$$\tau_c = \Theta_c (\gamma_s - \gamma) D$$

in which, Θ_c is a function of the grain Reynolds number. Hence, the total shear stress that the armoring layer can withstand is

$$\tau = (\gamma_s - \gamma) \int_{D_{min}}^{D_{max}} \Theta_c D p_a(D) dD \qquad (8.48)$$

If the shear stress reaches a maximum, the bed is in its most stable configuration. Gessler took this value as the critical shear stress for mixed sediment. If it is exceeded, the stability of bed surface will be less.

For coarse sediment, the coefficient Θ_c is approximately constant and the extreme condition of Eq. (8.48) is then

$$\int_{D_{min}}^{D_{max}} D p_a(D) dD = Max. \qquad (8.49)$$

the left side of the above formula is the effective grain size of the mixture. Therefore, the critical shear stress for the mixture proposed by Gessler is also the shear stress that occurs if the effective grain size of the armoring layer has its maximum value. Fig. 8.20 shows the relationship between shear stress and effective grain size for the example shown in Fig. 8.19. From the figure, the effective grain size reaches maximum if $\tau = 3 kg/m^2$, and the shear stress can then be taken as the critical shear stress of the mixture.

As an acceptable extension of the Gessler theory, the average of the probability for sediment remaining in the armoring layer can be used to symbolize the stability of the armoring layer,

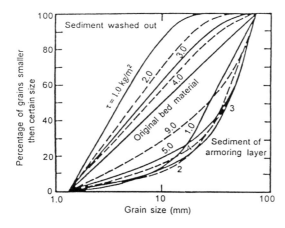

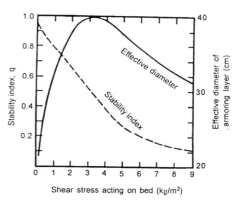

Fig. 8.19 Comparison between size distributions of armoring layer, sediment washed out and original bed material (after Gessler, J.)

Fig. 8.20 Relationship between stability coefficient, effective diameter of armoring layer, and drag force on bed (after Gessler, J.)

$$\bar{q} = \frac{\int_{D_{\min}}^{D_{\max}} q^2 p_0(D) dD}{\int_{D_{\min}}^{D_{\max}} q p_0(D) dD} \qquad (8.50)$$

Gessler called this value the stability coefficient. If the shear stress is zero, then $\bar{q}=1$; if the shear stress approaches infinity, $\bar{q}=0.04$. For the foregoing example, the relationship between shear stress and the stability coefficient is as shown in Fig. 8.20. As the shear stress reaches the critical value for the mixture, the stability coefficient is about 0.5. In an application to canal design, a safety factor of 1.3 could be used, and the stability coefficient would then be 0.65. In this context, the study by Lane and Carlson shows that the stability coefficient in a natural regime canal varies within the range 0.55 to 0.74 and has an average of 0.65.

8.4. CONDITION FOR INCIPIENT MOTION OF COHESIVE SEDIMENT

A critical diameter can be found in the usual way by a study of incipient motion of sediment with various sizes. For sediment coarser than the critical size, more intense flow is needed to induce motion because of the larger weight. Also, for sediment finer than the critical size, it becomes harder and harder to move as they become smaller and smaller. The cause is not only the shielding effect of the laminar sub-layer but also the effect of cohesion between the finer grains of sediment.

In the study of critical conditions for the motion of cohesive sediment, two different cases arise. One case is that of unconsolidated sediment newly deposited during the natural process of siltation. This kind of sediment can be treated as individual grains but with some cohesion. Another case is the cohesive sediment formed during a long-term process of deposition that has undergone physico-chemical action. The incipient motion of this kind of sediment is not that of single particles but rather the movement of clusters. Although a distinction should be made between the two, it is not easy to do so in some application.

8.4.1 Incipient conditions for newly deposited cohesive sediment

The cohesion of fine sediment is discussed in detail in Chapter 5. Cohesion acts like a downward force on grains and thus increases their stability. The condition for incipient motion of the sediment can be derived from an equation for the balance of forces acting on a particle; these include the weight, drag, and uplift.

An example is the following equation suggested by Tang for incipient motion [47]:

$$\tau_c = \frac{1}{77.5}\left[3.2(\gamma_s - \gamma)D + \left(\frac{\gamma_b}{\gamma_{b_0}}\right)^{10}\frac{k}{D}\right] \qquad (8.51)$$

in which

343

γ_b — the unit weight of sediment on bed;

γ_{b_0} — the unit weight of consolidated sediment ($=1.6 \text{g/cm}^3$);

k — a constant equal to 2.9×10^{-4} g/cm.

The distinguishing feature of the formula is that the relative consolidation of the sediment on the bed is included in the term for cohesion. Undoubtedly, such a relationship is qualitatively correct. The unit weight of sediment he used is not the sediment concentration usually adopted to reflect the degree of consolidation. Therefore, the exponent on the ratio of unit weight has the high value of 10 in the formula; it results from the relatively small variation of the unit weight of sediment. A shortcoming of the formula is that the viscosity of water was not included in its derivation. The relationship between the shear stress for incipient motion and the grain size of the fully consolidated sediment ($\gamma_b = \gamma_{b_0}$) is shown in Fig. 8.21. The shear for incipient motion has a minimum value for a grain size of about 0.08 mm.

For the velocity at incipient motion for cohesive sediment that has been consolidated recently, Dou [48], Wuhan Institute of Hydraulic and Electric Engineering [49], Tang [47], and Sha [50] obtained the formulas for incipient motion of various sizes of sediment given in Table 8.3. Fig. 8.22 is a comparison of these formulas and the experimental data for a water depth of 15 cm and the specific gravity of sediment of 2.65g/cm^3. For fine sediment, these formulas represent a large improvement over the formulas shown in Fig. 8.7. But for coarse sediment ($D>5$ mm), the data tent to scatter. One cannot easily decide which formula is the most reliable. For the larger particles, experimental data are relatively scarce and their expressions for cohesion differ

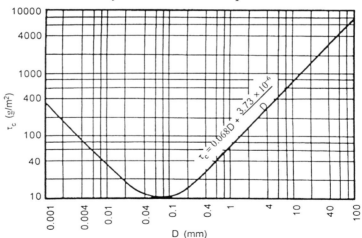

Fig. 8.21 Relationship between critical shear stress, including effect of cohesion among grains and grain size (Cunben, Tang's formula)

considerably. Consequently, one cannot readily see why the formulas for the relevant velocity agree with the experimental data so well. However, a closer analysis shows that all of these formulas include two coefficients that must be determined experimentally. Possibly, the adjustment of these coefficients helps the corresponding curves to coincide with the data. These formulas properly belong to the class of semi-theoretical and semi-empirical formulas. However, because of the complications inherent in sediment laden flow, one must often use them in the study of practical cases since nothing better is available.

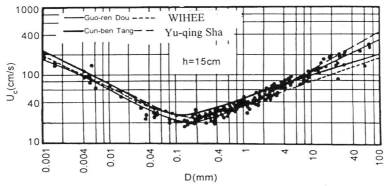

Fig. 8.22 Relationship between critical velocity for incipient motion and grain size

Fine sediment that has been deposited takes a long time to consolidate. For various degrees of consolidation, the conditions of incipient motion are different. Detailed studies on the incipient motion of clayey mud were conducted by Migniot [51,52]. The relationship between the concentration of the deposit and the friction velocity for incipient motion is shown in Fig. 8.23. In the figure, as the degree of consolidation and the concentration of deposit increase, the corresponding friction velocity (and also the shear stress) increases sharply. Nevertheless, within the range

Table 8.3 Incipient velocity formulas in consideration of the cohesion among grains

Author	Form of formula	Remarks
G.R. Dou	$$\frac{U_c^2}{gd} = \frac{\gamma_s - \gamma}{\gamma}(6.25 + 41.6\frac{h}{h_a}) + (111 + 740\frac{h}{h_a})\frac{h_a\delta}{D^2}$$	atmospheric pressure measured by height of water column, δ: thickness of water molecules $(=3\times10^{-8}\,\mathrm{cm})$
Y.Q. Sha	$$U_c = R^{1/5}\sqrt{(1520\frac{D^{5/3}}{\omega^{4/3}} + 194D)(f\cos\theta - \sin\theta)}$$	all length in m, ω:fall velocity of grains, f: fiction coefficient $=1.42D^{1/8}$, θ:inclination of river bed
WIHEE	$$U_c = \left(\frac{h}{D}\right)^{0.14}\left(17.6\frac{\gamma_s - \gamma}{\gamma}D + 0.000000605\frac{10+h}{D^{0.72}}\right)^{1/2}$$	—
C.B. Tang	$$U_c = \frac{h^{\frac{1}{m}}}{D}\frac{m}{m+1}\left[3.2\frac{\gamma_s - \gamma}{\gamma}gD + \left(\frac{\gamma_b}{\gamma_{bo}}\right)^{10}\frac{C}{\rho D}\right]^{1/2}$$	$m=\begin{cases}4.7(h/D)^{0.06} & Flume\ data\\ 6 & Natural\ river\end{cases}$ $C=2.9\times10^{-4}\,g/cm$

345

of the experiments, the rate of increase should not cause the function to have an exponent as large as the value 10 in Eq. (8.51). The samples taken from various locations form various curves in the figure because of their different mineral compositions. This kind of clayey mud belongs to the category of a Bingham fluid. The incipient friction velocity is closely related to the Bingham shear stress as shown in Fig. 8.24. If τ_B is less than 15 dyne/cm^2, the clayey mud is in a plastic state,

$$U_{*c} = 0.95\tau_B^{1/4} \tag{8.52}$$

For τ_B greater than 15 dyne/cm^2, the clayey mud has become consolidated,

$$U_{*c} = 0.50\tau_B^{1/2} \tag{8.53}$$

the friction velocity U_{*c} is in cm/sec and the Bingham shear stress is in dyne/cm^2. As pointed out in the discussion of the properties of fine sediment in Chapter 3, the deposit has the property of a Bingham fluid because it contains a certain amount of sediment finer than 0.01 mm. Actually, if pure sand is placed in clay suspension with a concentration of 470 g/l, the incipient friction velocity will not change if the ratio of sand does not exceed 30%.

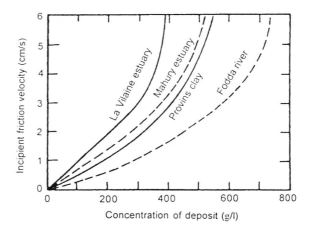

Fig. 8.23 Relationship between incipient friction velocity of clay mud and consolidation of sediment (after Migniot, C.)

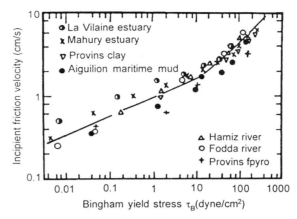

Fig. 8.24 Relationship between incipient friction velocity of clay
mud and Bingham yield stress (after Migniot, C.)

8.4.2 Incipient motion of consolidated cohesive sediment

8.4.2.1 Experimental result

In the 1950s, investigators followed the methods used for non-cohesive sediment in an effort to establish an experimental relationship between the drag force or velocity for incipient motion and the properties of the sediment [53,54]. The sediment properties generally used were:

1. moisture content — the ratio of the weight of water in a sediment sample to the weight of solid grains

2. shearing strength — the shearing strength of the deposited sediment using the experiment methods of soil mechanics

3. consistency — the degree of hardness at a specific moisture content in the terms of its limiting value. According to the experimental method suggested by Atterberg, soil can be classified in four groups according to their moisture contents: solid state, semi-solid state, plastic state and liquid state. The value of moisture content that separate these four states are shrinkage limit (W_s), plasticity limit (W_p) and liquidity limit (W_l). The difference between the liquidity limit and the plasticity limit is called the plasticity index.

4. dispersity,

$$\text{dispersity } (\%) = a/b \times 100 \tag{8.54}$$

in which, a is the amount of sediment content with $D<0.001$ mm in micro-aggregate analysis (light dispersive sample), b is the sediment content with $D< 0.001$ mm in mechanical analysis (high dispersive sample). The larger the dispersity, the more unstable the micro-structure of the soil is.

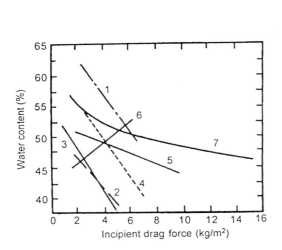

Fig. 8.25 Relationship between incipient drag force of cohesive sediment and its water content

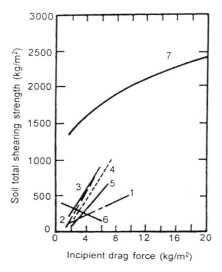

Fig. 8.26 Relationship between incipient drag force and total shearing strength of cohesive sediment

Smerdon and Beasley concluded that a relationship exists between the shear stress for incipient motion of cohesive sediment τ_c and both the plasticity index (PI) and the dispersity (Dr) [55].

$$\tau_c = 0.017(PI)^{0.84} \tag{8.55}$$

$$\tau_c = 1.04(D_r)^{-0.63} \tag{8.56}$$

in which τ_c is in kg/m^2 ; many investigators use the moisture content in sediment or the shearing strength as the index of the scouring-resistant strength of the soil. Fig. 8.25 and Fig. 8.26 are plots of the experimental results for various samples of soil. The differences between them are quite large; they even reflect different qualitative trends. Even for non-cohesive sediment, results for the incipient condition given by different investigators vary considerably. For cohesive sediment, the differences in the incipient condition are much larger, even up to 200 times for different soil properties. The condition for incipient motion of consolidated cohesive sediment depends mainly on the magnitude and characteristics of the forces among the grains. The soil properties already mentioned are simply the manifestations of the system of cohesion. If these soil properties are used as the main criterion for the incipient motion of the cohesive sediment, the results may have only limited value.

348

The cohesion among clayey grains is quite complicated. On the one hand, it depends on the mineral composition of the soil, the property and concentration of the ions of the liquid, the temperature and acidity of the liquid, and other such factors. On the other hand, it is also related to the framework formed by grains of clay. The first group affects directly the property of the film water at the grain surface and the attractive force among grains, and the second factor is related to the overall stability of the system. Grissinger studied the effect of soil property on scouring and reached the following conclusions [56]:

1. The arrangement and combination among the clayey grains has a strong effect on scouring.
2. The increase of clay content in soil generally increases its resistance to scouring.
3. The moisture content in a soil sample is closely related to its rate of scouring.
4. The rate of scouring doubles if the water temperature rises from $20°$ to $35°$.

These statements describe the phenomena, but do not reveal their essence. Recently, investigators have began to study the influence of the electro-chemical effects among clayey grains. A few results have been published but they are not systematic enough for one to make a comprehensive summary [57,58].

The foregoing discussion is limited to the case of cohesive sediment with preliminary consolidation. Partheniades compared the conditions for incipient motion of cohesive sediment that was newly deposited, so that it had a loose structure, with those for a consolidated cohesive sediment [59]. According to the normal results of soil mechanics, the total scouring-resistant strength of these two samples would differ by a factor of 100. The rates of scouring for various drag forces due to the flow are shown in Fig. 8.27. Although the rate of scouring of the consolidated clay is smaller than that of the newly deposited sediment, the incipient shear stress of these two samples are nearly same. Thus, the incipient motion of cohesive sediment induced by flow is entirely different from the mechanism of the shearing destruction for soil under the action of an external force. For many samples of soil, the total shearing strength can be as much as 1000 kg/m^2 , but still soil particles can be entrained by a drag force due to the flow that is less than 1 kg/m^2. The trend shown in Fig. 8.27 for the influence of concentration of deposit on incipient motion is in contradiction with that mentioned previously. In the experiment conducted by Partheniades, the consolidated soil samples were tested for the same conditions as for Fig. 2.25b and Fig. 2.25c. The distance between clayey grains in flocs of these two samples as well as the cohesion are the same. Therefore, the incipient drag force for these two samples should be same as long as the shear stress induced by flow in sediment does not exceed its shearing strength. The value should have no relationship with the total shearing strength of the sediment. From another aspect, the cohesive force between clayey grains has a determinant effect on the condition for the incipient motion of cohesive sediment.

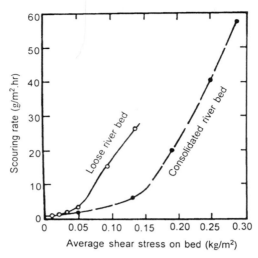

Fig. 8.27 Scouring rate of cohesive sediment under action
of various shear stress (after Partheniades, E.)

8.4.2.2 Analysis of field data

From the operation of canals and rivers, waterways can be classified according to their stability and to the extent that scouring occurs. Then, further studies can be conducted to analyze the effect of the soil properties on the shearing strength for the material used in these waterways. Results of such studies show strong regional characteristics. Hence, no general index can be provided. Schroeder established a relationship between soil property and the limiting drag force due to flow from data collected for 121 rivers (and canals) in middle-western America, with and without plantings; the results are as shown in Fig. 8.28 [54]. In the analysis, the shear stress on the bed caused by the average annual peak discharge is taken as the shear stress due to the flow, and the grain size of the bed material for non-cohesive sediment and the plasticity index for cohesive sediment are used as the corresponding soil properties. Flaxman analyzed the data of 13 rivers in western America and obtained a relationship between the incipient work and the shearing strength of soil as shown in Fig. 8.29 [60].

From the data of 46 rivers and canals in western America, Gibbs used the plasticity index and liquidity limit as two parameters to divide the waterways according to their shearing strength, as shown in Fig. 8.30 [61]. The figure shows which soils are easily eroded but it does not reveal whether a specific soil can be eroded for a given flow condition.

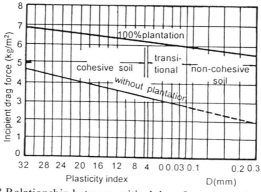

Fig. 8.28 Relationship between critical drag force for incipient motion and soil properties for waterways in middle-western America (after Task committee on erosion of cohesive sediments, Amer. Soc. Civil Engrs.)

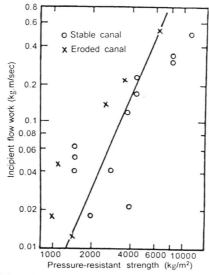

Fig. 8.29 Relationship between incipient flow work and soil properties for 13 rivers in western America (after Flaxman, E.M.)

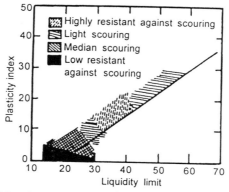

Fig. 8.30 Relationship between resistance to scouring and soil properties of waterways in western America (after Graf, W.H.)

351

REFERENCES

[1] Shields, A. "Anwendung der Aechlichkeitsmechanik und der Turbulenzforschung auf die Geschiebewegung." *Mitt. Preussische Versuchsanstalt fur Wasserbau und Schiffbau*, Berlin, 1936.

[2] Lane, E.W. "Progress Report of Studies on the Design of Stable Channels by the Bureau of Reclamation." *Proc., Amer. Soc. Civil Eengrs.*, Vol.79, No. 280, 1953.

[3] Helland-Hansen, F., R.T. Milhous, and P.C. Klingman. "Sediment Transport at Low Shields-Parameter Values." *J. Hyd. Div., Proc. Amer. Soc. Civil Engrs.*, Vol.100, No.HY1, 1974, pp. 261-265.

[4] Taylor, B.D. "Temperature Effects in Alluvial Streams." *Report KH-R27*, W.M. Keck Lab. of Hyd. & Water Resources, Calif. Inst. Tech., 1971.

[5] Gessler, J. "Beginning and Ceasing of Sediment Motion." In *River Mechanics*. Edited by H.W. Shen, Vol. 1, 1971, pp. 22.

[6] Karmer, H. "Sand Mixtures and Sand Movement in Fluvial Models." *Trans., Amer. Soc. Civil Engrs.*, Vol.100, 1935, pp.798-838.

[7] Yalin, M.S. *Mechanics of Sediment Transport*. Pergamon Press, 1972, p.74-110.

[8] Dou, Guoren. "Theory on Threshold Motion of Sediment Particles." *Scientia Sinica*, Vol. 11, No.7, 1962, pp.999-1032 (in Russian)

[9] Grass, A.J. "Initial Instability of Fine Sand Bed." *J. Hyd. Div., Proc., Amer. Soc. Civil Engrs.*, Vol. 96, No. Hy3, 1970, pp. 619-632.

[10] White, S.J. "Plain Bed Thresholds for Fine Grained Sediments." *Nature*, Vol. 228, No. 5267, 1970, pp. 152-153.

[11] Mantz, P.A. "Incipient Transport of Fine Grains and Flakes by Fluids-Extended Shields Diagram." *J. Hyd. Div., Proc., Amer. Soc. Civil Engrs.*, Vol. 103, No. HY6, 1977, pp. 601-616.

[12] Tixon, L.J. "Etude des Conditions dans Lesquelles les Particules Solides sont Transportees dans les Courants a lit Mobiles." *Proc., Assoc. Intern. Sci. Hydro.*, Assemblee Ggnerale d'Oslo, Tome 1, 1948, pp. 293-310.

[13] Li, Changhua, and Meixiu Sun. "Criteria for Threshold Shear Stress and Ripple Formation." *Proc. Nanjing Hydraulic Research Institute (River & Habour Division)*, 1964. (in Chinese)

[14] Paintal, A.S. "A Stochastic Model for Bed Load Transport." *J. Hyd. Div., Proc., Amer. Soc. Civil Engrs.*, Vol. 97, 1971, pp. 527-553.

[15] Miller, M.C., I.N. McCave, and P.D. Komar. "Threshold of Sediment Motion Under Unid
erectional Currents." *Sedimentology*, Vol. 24, No. 4, 1977, pp. 507-527.

[16] Yalin, M.S., and E.K Arahan. "Inception of Sediment Transport." *J. Hyd. Div., Proc., Amer. Soc. Civil Engrs.*, Vol. 105, No. HY11, 1979, pp.1433-1443.

[17] Chepil, W.S. "Equilibrium of Soil Grains at the Threshold of Movement by Wind." *Proc., Soil Sci. Soc. Amer.*, Vol.23, No. 6, 1959, pp. 422-428.

[18] Chepil, W.S., and N.P. Woodruff. "The Physics of Wind Erosion and Its Control." *Advances in Agron.*, Vol. 15, 1963, pp. 211-302.

[19] Goncharov, V.N. *Basic River Dynamics*. Hydro-Meteorological Press, Leningrad, 1962. (in Russian)

[20] Levy, E.E., *River Mechanics*. National Energy Press, Moscow, 1956. (in Russian)

[21] Shamov, G.E. "Formulas for Determining Near-Bed Velocity and Bed Load Discharge." *Proc. Soviet National Hydrology Institute*, Vol. 36, 1952. (in Russian)

[22] Li, Baoru. "Calculation of Threshold Velocity of Sediment Particles." *Journal of Sediment Research*, Vol. 4, No. 1, 1959, pp. 71-77. (in Chinese)

[23] Zeng, Zhaozen, and Shangyi Wang. "Study on Threshold Criteria for Granular Sediment." *Journal of Tianjin University*, 1963, pp. 19-40. (in Chinese)

[24] Knoros, V.S. "Non-scouring Velocity for Non-cohesive Soil and Its Application." *Proc. Soviet National Hydrology Institute*, Vol. 59, 1958. (in Russian)

[25] Yang, Chih-Ted. "Incipient Motion and Sediment Transport." *J. Hyd. Div., Proc., Amer. Soc. Civil Engrs.*, Vol. 99, No. HY10, 1973, pp. 1679-1704.

[26] Bagnold, R.A. "Sediment Discharge and Stream Power." *U. S. Geol. Survey Circular 421*, 1960, p. 23.

[27] Schoklitsch, A. "Der Geschiebetrieb und die Geschie befracht." *Wasserkraft und Wasserwirtschaft*, Vol. 29, No. 4, 1934, pp. 37-43.

[28] White, C.M. "Equilibrium of Grains on Bed of Stream." *Proc., Royal Soc. London*, Ser. A, Vol. 174, 1940, pp. 322-334.

[29] Lyles, L., and N.P. Woodruff. "Boundary-Layer Flow Structure: Effects on Detachment of Noncohesive Particles." In *Sedimentation*, Edited by H.W. Shen, 1972, p. 16.

[30] Rathbun, R.E., and H.P. Guy. "Measurement of Hydraulic and Sediment Variables in A Small Recirculating Flume." *Water Resources Res.*, Vol. 3, No. 1, 1967, pp. 107-122.

[31] Stevens, M.A., and D.B. Simons. "Stability Analysis for Coarse Granular Material on Slopes." In *River Mechanics*, Edited by H.W. Shen, Vol. 1, 1971.

[32] Christensen, B.A. "Incipient Motion on Cohesionless Channel Banks." In *Sedimentation*, Edited by H.W.Shen, 1972, p. 22.

[33] Ghosh, S.N., and N. Roy. "Boundary Shear Distribution in Open Channel Flow." *J. Hyd. Div. Proc., Amer. Soc. Civil Engrs.*, Vol. 96, No. HY4, 1970, pp. 967-994.

[34] Kartha, V.C., and H.J. Leutheusser. "Distribution of Tractive Force in Open Channel Flow." *J. Hyd. Div., Proc., Amer. Soc. Civil Engrs.*, Vol. 96, No. HY4, 1970, pp. 967-994.

[35] Rajaratnam, N., and D. Muralidhar. "Boundary Shear Distribution In Rectangular Open Channels." *La Houille Blanche*, Vol. 24, No. 6, 1969, pp. 933-946.

[36] Myers, R.C., and E.M. Elsawy. "Boundary Shear in Channel with Flood Plain." *J. Hyd. Div., Proc., Amer. Soc. Civil Engrs.*, Vol. 101, No. Hy7, 1975, pp. 933-946.

[37] Ghosh, S., and S.B. Jena. "Boundary Shear Distribution in Open Channel Compound." *Proc., Inst. Civil Engrs.*, Vol.49, Aug.1971, pp.417-430.

[38] Ippen, A.T., and P.A. Drinker. "Boundary Shear Stresses in Curved Trapezoidal Channels." *J. Hyd. Div., Proc., Amer. Soc. Civil Engrs.*, Vol. 88, No. HY5, 1962, pp. 143-179.

[39] Nouh, M.A., and R.D. Townsend. "Shear Stress Distribution in Stable Channel Bends." *J. Hyd. Div., Proc., Amer. Soc. Civil Engrs.*, Vol. 105, No. HY10, pp. 1233-1245.

[40] Bathurst, J.C., C.R. Thorne, and R.D. Hey. "Secondary Flow and Shear Stress at River Bends." *J. Hyd. Div., Proc., Amer. Soc. Civil Engrs.*, Vol. 105, No. HY10, 1979, pp. 1277-1295.

[41] Nece, R.E., and J.D. Smith. "Boundary Shear Stress in Rivers and Estuaries." *J. Waterways and Harbor Div., Proc., Amer. Soc. Civil Engrs.*, Vol. 96, No. WW2, 1970, pp. 1277-1295.

[42] Lane, E.W., and E.J. Cartson. "Some Factors Affecting the Stability of Canals Constructed in Coarse Granular Material." *Proc. 5th Cong.*, Intern. Assoc. Hyd. Res., 1953.

[43] Harrison, A.S. "Bed Sediment Segregation in A Degraded Bed." *Tech. Rep.33-1*, Inst. Engin. Res., Univ. Calif., 1950.

[44] Gessler, J. "Self Stablizing Tendencies of Sediment Mixtures with Large Range of Grain Sizes." *J. Waterways and Harbor Div., Proc., Amer. Soc. Civil Engrs.*, Vol. 96, No. WW2, 1970.

[45] Gessler, J. "Critical Shear Stress for Sediment Mixtures." *Proc., 14th Cong.*, Intern. Assoc. Hyd. Res., Vol. 3, 1971, pp. 1-8.

[46] Gessler, J. "Stochastic Aspects of Incipient Motion on Riverbeds." In *Stochastic Approach to Water Resources*, Vol. 2, Edited by H.W. Shen, 1976, p. 26.

[47] Tang, Cunben. "Law of Sediment Threshold." *Journal of Hydraulic Engineering*, 1964. (in Chinese)

[48] Dou, Guoren. "On Threshold Velocity of Sediment Particles." *Journal of Hydraulic Engineering*, April 1960, pp. 22-31. (in Chinese)

[49] WIHEE. *River Mechanics*. Industry Press, April, 1960, pp. 44-60 (in Chinese).

[50] Sha, Yuqing. *Elementary Sediment Mechanics*. Industry Press, 1965, p. 302 (in Chinese).

[51] Migniot, C. "Study on Physical Properties of Fine Sediment (silt) and Their Properties under Flow Dynamics." *La Houille Blanche*, Vol.7, 1968

[52] Migniot, C. "Effects of Flow, Wave and Wind on Sediment." *La Houille Blanche*, No. 1, 1977

[53] Vanoni, V.A. *Sedimentation Engineering*. Amer. Soc. Civil Engrs., 1975, pp. 107-113.

[54] Task Committee on Erosion of Cohesive Sediments, Amer. Soc. Civil Engrs. "Erosion of Cohesive Sediments." *J. Hyd. Div., Proc. Amer. Soc. Civil Engrs.*, Vol. 94, No. HY4, 1968.

[55] Smerdon, E.T., and R.P. Beasley. "Tractive Force Theory Applied to Stability of Open Channels in Cohesive Soils." *Res. Bull.* No. 715, Agri. Exp. Sta., Univ. Missouri, Oct. 1959.

[56] Grissinger, E.H. "Resistance of Selected Clay Systems to Erosion by Water." *Water Resources Res.*, Vol. 2, No. 1, 1966, pp. 131-138.

[57] Sargunam, A. et al. "Physico-Chemical Factors in Erosion of Cohesive Soils." *J. Hyd. Div., Proc., Amer. Soc. Civil Engrs.*, Vol. 99, No. HY3, 1973, pp. 555-558.

[58] Arulanandan, K. "Fundamental Aspects of Erosion of Cohesive Soils." *J. Hyd. Div., Proc., Amer. Soc. Civil Engrs.*, Vol. 101, No. HY5, 1975, pp. 635-638.

[59] Partheniades, E. "Erosion and Deposition of Cohesive Materials." In *River Mechanics*, Edited by H.W. Shen, Vol. 2, 1971, p. 91.

[60] Flaxman, E.M. "Channel Stability in Undisturbed Cohesive Soils." *J. Hyd. Div., Proc., Amer. Soc. Civil Engrs.*, Vol. 89, No. HY2, 1963, pp. 87-96.

[61] Graf, W.H. *Hydraulics of Sediment Transprt*, McGraw-Hill Book Co., 1971, p. 513.

CHAPTER 9

BED LOAD MOTION

Following on the discussions of patterns of sediment motion and its initiation in Chapters 5 and 8, this chapter introduces various theories and formulas of bed load motion. The first section treats theories and formulas for uniform bed load, and the second and the third sections discuss the effects of non-uniformity of particle size on the transport process.

9.1 LAWS OF MOTION OF UNIFORM BED LOAD

In the last part of the 19th century, the French scientist DuBoys advanced a theory for bed load motion based on shear stress [1]. Since then, many scientists have studied the phenomenon and they have proposed a number of formulas for bed load transport. These formulas are based on different modes of motion and employ different parameters, including shear stress, flow velocity, and power. Though the formulas have a variety of structures and forms, they can be classified into the following groups according to the approaches used:

1. Empirical formulas--based on huge amounts of experimental data, represented by the Meyer-Peter Formula;

2. Semi-theoretical formulas--based on physical concepts and through mechanical analysis, represented by the Bagnold Formula;

3. Probability theory-based formulas--approached through a combination of probability theory and mechanics, represented by the Einstein Bed Load Function;

4. Dimensional analysis-based formulas--deduced from the concepts of Einstein or Bagnold by using dimensional analysis and calibrations with measured data, represented by Engelund, Yalin, and Ackers-White Formulas.

In the following treatment, the theoretical backgrounds and deductions of six representative formulas are discussed in detail; then, some formulas with the average velocity as the main parameter are introduced briefly.

9.1.1 Meyer-Peter formula [2]

First, Meyer-Peter developed a simple empirical bed load formula by using a similarity law and data from his preliminary experiments. The formula involved only a few simple parameters. Then he applied the formula to more complex cases involving the variation of additional parameters and found a systematic difference between the measured data and the formula. He analyzed the difference and determined its cause. He studied the effect of each new parameter by separating it from the others in additional experiments and then incorporated the parameter into his formula. In this way, he studied the effects of density, size composition of the sediment and bed form step by step and eventually obtained a comprehensive bed load formula. This approach requires a long time, but it is effective for studying problems involving numerous parameters. The steps he took in the development of his formula were the following:

9.1.1.1 Preliminary form of the formula

In 1934, Meyer-Peter conducted an experiment with uniform sediment with a density of 2.68 g/cm^3 and obtained the following empirical formula[3]:

$$\frac{(\gamma q)^{2/3} J}{D} = a_1 + b_1 \frac{g_b'^{2/3}}{D} \qquad (9.1)$$

in which q is the discharge per unit width, g_b' is the rate of bed load transport per unit width in terms of submerged weight, a_1 and b_1 are constants. In kg-m-s system, $a_1 = 17$ and $b_1 = 0.547$. If the effect of the walls is not negligible, q is given by

$$q = UR_b$$

in which R_b is the hydraulic radius as affected by the bed resistance and defined by Eq. (7.21).

If the two sides of Eq. (9.1) are divided by $\gamma^{2/3} g^{1/3}$ their dimensions are $U/(gL)^{1/2}$, as in the Froude number, in which L is a characteristic length. Eq. (9.1) also implies the existence of a critical condition for the initiation of bed load motion:

$$\left. \frac{(\gamma q)^{2/3} J}{D} \right]_c = a_1 \qquad (9.2)$$

9.1.1.2 Energy slope correction

Eq. (9.1) was based on a limited amount of experimental data; the energy slope especially varied within a narrow range in the experiments. Later Meyer-Peter conducted more experiments with a wider range of energy slopes and found that the constant a_1, characterizing the initiation condition, varies with the energy slope. He then rearranged the formula into

$$\left. \frac{(\gamma q)^{2/3} J}{D} \right]_c = f J^{1/3} \qquad (9.3)$$

356

in which f is a constant. With $q = UR_b$ and using the Manning-Strickler formula for resistance,

$$U = \frac{A}{D^{1/6}} R_b^{2/3} J^{1/2}$$

Combining this with Eq. (9.3), he obtained the following formula:

$$\left. \frac{\gamma R_b J}{D} \right]_c = \frac{\gamma^{1/3} f}{A^{2/3}} \left(\frac{D}{R_b} \right)^{1/9} = K \left(\frac{D}{R_b} \right)^{1/9}$$

the variation of D/R_b on the right side was neglected because the exponent on it was small, so that

$$\left. \frac{\gamma R_b J}{D} \right]_c \approx const$$

Thus, to accommodate the expanded scope of the experiments, the relationship becomes

$$\frac{\gamma R_b J}{D} = a_2 + b_2 \frac{g_b'^{2/3}}{D} \tag{9.4}$$

in place of Eq. (9.1).

9.1.1.3 Correction for bed configuration

The bed configuration and the movement of sand waves are bound to affect bed load motion. Meyer-Peter found that if a moving sand wave formed on the bed surface, the rate of bed load transport calculated from Eq. (9.4) was higher than the measured value. This result implies that not all the drag stress is effective in causing bed load motion.

To eliminate the effect of the walls, Meyer-Peter used the Manning-Strickler formula in the form:

$$U = K_b R_b^{2/3} J^{1/2} \tag{9.5}$$

to obtain the velocity; in which K_b is a coefficient for bed resistance. Furthermore, Meyer-Peter divided the energy slope into two parts, one part, used to overcome the grain resistance, is J', and the other part is the resistance caused by sand waves, J''. Since only the first part is effective for bed load motion, the energy slope J' is substituted into Eq. (9.5),

357

$$U = K'_b R_b^{2/3} J'^{1/2} \tag{9.6}$$

in which K'_b is the roughness coefficient due to grain resistance.

From Eqs. (9.5) and (9.6),

$$J' = \left(\frac{K_b}{K'_b}\right)^2 J \tag{9.7}$$

If J in Eq. (9.4) is replaced by J', the bed load formula becomes:

$$\frac{\gamma R_b J}{D}\left(\frac{K_b}{K'_b}\right)^2 = a_2 + b_2 \frac{g_b'^{2/3}}{D}$$

and it includes the effect of sand waves.

Further analysis indicated that if the exponent on the term (K_b / K'_b) is changed from 2 to 3/2, the calculated results conform better to the data,

$$\frac{\gamma R_b J}{D}\left(\frac{K_b}{K'_b}\right)^{3/2} = a_3 + b_3 \frac{g_b'^{2/3}}{D} \tag{9.8}$$

To separate the wall resistance, Meyer-Peter used a technique similar to that shown in Fig.7.15. He divided the wetted perimeter into two parts, one part is affected by the wall and the other by the bed. Therefore, the part of the discharge pertaining to the bed is

$$Q_b = B R_b U$$

The total discharge through the cross section is

$$Q = B h U$$

Hence,

$$R_b = \frac{Q_b}{Q} h \tag{9.9}$$

Substituting this expression into Eq. (9.8) results in

$$\gamma \frac{Q_b}{Q}\left(\frac{K_b}{K'_b}\right)^{3/2} \frac{hJ}{D} = a_3 + b_3 \frac{g_b'^{2/3}}{D} \tag{9.10}$$

9.1.1.4 Comprehensive formula valid for different sediments

Meyer-Peter conducted experiments with a heavy sediment (γ_s =4.2 g/cm³) and with lighter brown coal cinders (γ_s=1.25 g/cm³). The measured results showed parallel relationships for different densities of sediment in a plot based on the parameters in Eq. (9.10); hence, the constant a_3 in Eq. (9.10) is in fact a function of the density of sediment. By taking this into account and using g_b, the rate of bed load transport per unit width by dry weight he obtained, to replace g'_b, a revised form of Eq. (9.10),

$$\gamma \frac{Q_b}{Q}\left(\frac{K_b}{K'_b}\right)^{3/2} hJ = a_4(\gamma_s - \gamma)D + b_4\left(\frac{\gamma}{g}\right)^{1/3}\left(\frac{\gamma_s - \gamma}{\gamma_s}\right)^{2/3} g_b^{2/3} \quad (9.11)$$

and this is known as the Meyer-Peter Formula. The formula was calibrated against measured data, as shown in Fig. 9.1, in order to determine the two constants:

$$\begin{cases} a_4 = 0.047 \\ b_4 = 0.25 \end{cases}$$

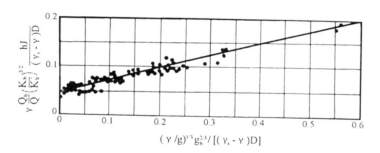

Fig. 9.1 Comparison of the Meyer-Peter formula with measured data

The Meyer-Peter Formula is based on a large quantity of experimental data. The main variables in his experiments varied within the following ranges:

Width of flume: 0.15~2 m

Flow depth: 0.01~1.2 m

Energy slope: 0.04~2 %

Density of sediment: 1.25~4 g/cm³

Diameter of sediment: 0.40~30 mm

These data have quite a large scope, especially notable are those for gravel with a size of 30 mm. Therefore, the formula is more reliable than some of the others for rivers carrying coarse sand and gravel. The Meyer-Peter formula has been widely used in Europe, and the results obtained from it are generally satisfactory.

A point to note is that the flow velocities in Meyer-Peter's experiments were relatively high so that almost all of the sediment could be carried by the flows. For mountain rivers in China, however, many of the particles on the bed cannot be moved except in extreme events. The Meyer-Peter Formula predicts a larger bed load transport rate than the observed one in such case [4].

9.1.2 Bagnold formula

9.1.2.1 Basic physical picture of bed load motion

Bagnold believed that the movement of bed load, like other physical phenomena, obeys some basic laws of motion; the following are the principal ones [5]:

1. Sediment-laden flow is a result of shear, including the shear between sediment layers and that between the sediment and surrounding liquid. To sustain such a shear, a tractive force must act in the flow direction. For an open channel flow, the force is the component of the gravitational force in the direction of motion.

2. Sediment particles have a larger density than water and, hence, tend to settle to the bottom. To maintain them in continuous motion requires a force to balance their submerged weight. This force is created as part of the process of shearing.

3. The force supporting particles is transferred from the channel bed to the particles dispersed in the flow in two different ways. For bed load, the force is generated from collisions between particles and the corresponding component of momentum exchange during the collisions. It is the dispersive force discussed in Chapter 5. For suspended load, however, the force is caused by the eddying motion of turbulence and the momentum exchanges between eddies in the vertical direction.

4. As shown in Fig. 9.2, a shear stress in a sediment-laden flow is composed of two parts:

$$\tau = T + \tau'$$ (9.12)

in which T stands for the dispersive shear stress generated from collisions between particles, and τ' is the shear stress created from the deformation of the liquid as it flows around a particle. The former involves solid contacts, but the latter is related only to the liquid phase. Chapter 7 treats the transformation of energy and indicates that the potential energy of flow is transmitted to the bed zone through the shear stress τ' and results in turbulence there. A part of the potential energy, which is transmitted to the bed zone through T, is transformed into heat through friction as particles strike the bed with some velocity. No turbulence is generated in the process.

Thus, if the bed load transport rate increases, T increases and τ' becomes smaller. Consequently, the intensity of turbulence decreases. An experiment showed that if the volume concentration of particles was over 50%, the shear stress between the particles was more than 100 times the shear stress in the liquid, even though no cohesive particles were involved [6]. In this extreme case, turbulence was suppressed and the flow was laminar.

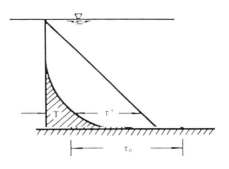

Fig. 9.2 Shear stress composition and distribution

5. The relative movement of two solids in contact with each other induces a pressure p; it also requires a shear stress T in the tangential direction that is given by

$$T / p = \tan \alpha \qquad (9.13)$$

in which α is the angle of friction; its value depends on the properties of the contacting surfaces, not on the relative velocity. For surfaces without cohesion, $\alpha = 33°$, and $\tan \alpha = 0.63$ [7].

Similarly, relative movement between two layers of particles also requires a shear stress T and a dispersive pressure p that are generated as a result of collisions between the moving particles. Bagnold measured T and p with an instrument involving rotating and fixed cylinders [6]. In this case, $\tan \alpha$ is the ratio of the component in the flow direction to that in the vertical direction of the exchange of momentum during the collisions. Fig. 9.3 presents the measured values of $\tan \alpha$ as a function of the Shields shear parameter Θ [5]; in it $\tan \alpha$ varies in the range of 0.37 to 0.75. Bagnold suggested in his later publications that $\tan \alpha$ for normal sand is equal to 0.63.

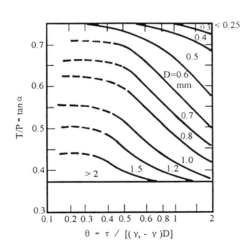

Fig. 9.3 Kinematic friction factor as a function of flow intensity and sediment diameter (after Bagnold, R.A.)

6. Continuous motion consumes energy because of the resistance

361

encountered. The speed of the motion is determined finally by the rate of energy supply, whatever the mechanism of energy transfer. If flowing water is viewed as a machine for transporting solid particles, the law of conservation of energy can be expressed as follows:

Energy consumed for transporting particles in unit time

= Rate of energy supply of the flowing water multiplied by Efficiency

The energy comes from the potential energy of flowing water; therefore, the expression can be rewritten as:

Energy consumed for transporting particles in unit time

= Rate of potential energy loss of the flowing water multiplied by Efficiency

This expression is essentially equivalent to the following well-known formula:

Electrical energy generated per unit time

= Rate of potential energy loss of the flowing water multiplied by Efficiency

The rate of energy supply from flowing water per unit area of channel bed is as follows:

$$pE = \frac{Q\,g\,J\,[\rho + S_v(\rho_s - \rho)]}{B} = \tau_0 u \qquad (9.14)$$

in which S_v is the volume concentration averaged over the depth, U the average velocity, and τ_0 the shear stress, exerted on the bed surface by the flow.

If W_b' denotes the submerged weight of the bed load over a unit area of channel bed and u_b the average velocity of the bed load motion, the rate of bed load transport per unit width is given in terms of submerged weight by

$$g_b' = W'_b\,\bar{u}_b \qquad (9.15)$$

g_b' has the dimension of power [Force . Length/Time], but it is not equal to the power for transporting sediment because W_b' is in a direction different from the direction of bed load motion. Hence, one must multiply $W_b'\,u_b$ in Eq. (9.15) by $\tan\alpha$ which can be expressed as

$$\frac{Shear\ \text{stress for maintaining sediment motion}}{Submerged\ \text{weight of bed load over unit area}} = \frac{T}{p} = \tan\alpha$$

362

and obtain the power needed for transporting bed load: $W_b' \bar{u}_b \tan \alpha$. In this way, the following bed load transport formula was obtained :

$$W_b' \bar{u}_b \tan \alpha = \tau_0 U e_b \qquad (9.16)$$

in which e_b stands for efficiency of the flowing water in the process of transporting bed load.

From his experimental data, Bagnold deduced a relationship between e_b and the mean velocity and sediment diameter, as shown in Fig. 9.4. The figure shows that the efficiency is in the range of 0.11~0.15 for flow velocities of 0.3~3 m/s and sediment diameters of 0.03~ 1.0 mm [5].

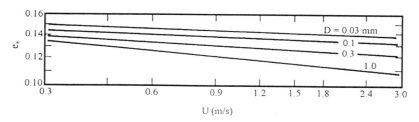

Fig. 9.4 Relationship of bed load transporting efficiency with velocity and diameter (after Bagnold, R.A.)

9.1.2.2 Mechanical analysis of bed load motion [7]

Saltation is the primary mode of bed load motion. It bears an analogy with throwing a flat stone on a water surface so that the stone skips along the water surface. As indicated in Fig. 9.5, a particle falls down to the bed; its momentum component in the flow direction is $m'u_1$, in which m' is the submerged mass, the mass of the particle minus the mass of water having the same volume, and u_1 is the velocity component of the particle in the flow direction when it hits the bed. The particle is acted on by a force, or a momentum, with a component opposite to the flow direction by the bed. The component of the momentum is then $-m'u'$, in which $-u'$ is the reduction in the particle velocity in the direction of flow because of its collision with the bed. Because of the quantity $-m'u'$, the saltating particle slows down and finally stops moving. To sustain the saltation of the particle, the flowing water must act on the particle so as to provide a momentum component $m'u'$ within the time interval Δt between successive collisions of the particle with the bed. In other words, flowing water has to exert a force on the particle with a component in the flow direction of

$$\bar{F}_x = \frac{m'u'}{\Delta t} = \frac{W'u'}{g\Delta t} \qquad (9.17)$$

in which W' is the submerged weight of the particle in water.

363

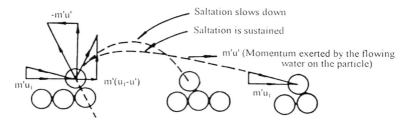

Fig. 9.5 Change of a particle's momentum during saltation

If u_b denotes the average velocity of the particle, then

$$\textit{Work done by the flowing water on the particle } = \ \overline{F_x u_b}$$

Also,

$$\textit{Energy consumed in unit time by the flow } = \ W' \overline{u_b} \tan \alpha$$

A combination of these two expressions yields the following equation:

$$\frac{\overline{F_x}}{W'} = \tan \alpha = \frac{u'}{g \, \Delta t} \qquad (9.18)$$

In reality, the force exerted on the particle varies with the distance of the particle from the bed. If the distance y_n is the location at which the particle is acted upon by a force equal to $\overline{F_x}$ and it is accelerated from $u_1 - u'$ to u', a difference must exist between the particle velocity $\overline{u}_b$ and velocity of flowing water u_n at that elevation that is given by

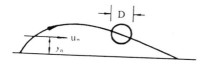

$$\overline{u}_r = u_n - \overline{u}_b \qquad (9.19)$$

as indicated in Fig.9.6 .

If many particles move along a bed, then

$$T\overline{u}_b = W_b' \, \overline{u}_b \tan \alpha = g_b' \tan \alpha$$

so that

$$g'_b \ = \ \frac{T}{\tan \alpha}(u_n - \overline{u}_r) \qquad (9.20)$$

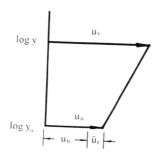

Fig. 9.6 Definitions of relative velocity
and velocity profile near bed

in which T is the grain shear stress at $y = y_n$; from Eq. (9.12), it is given as

$$T = a\,\tau_0 \tag{9.21}$$

in which a is a coefficient yet to be determined. If the flow velocity follows the logarithmic formula in the zone $y > y_n$, and the velocity at 0.4 of the depth from the bed is taken to be the average velocity, then

$$u_n = U - 5.75U_* \log\left(\frac{0.4h}{y_n}\right) \tag{9.22}$$

A combination of Eqs. (9.21) and (9.22) with Eq. (9.20) yields

$$g_b' = \frac{a\,\tau_0}{\tan\alpha}\left[U - 5.75U_*\log\left(\frac{0.4h}{y_n}\right) - \bar{u}_r\right] \tag{9.23}$$

in which a, y_n, and $\bar{u}_r$ must still to be determined.

1. Determination of a-value. According to the nature of shear induced by flow, a equals 0 at the beginning of bed load motion and 1 for an extremely high intensity of bed load motion. Accordingly, Bagnold assumed:

$$a = \frac{U_* - U_{*c}}{U_*} \tag{9.24}$$

in which U_{*c} is the critical shear velocity for initiation of motion of bed sediment.

2. Determination of $\bar{u}_r$. The force exerting on a particle can be expressed as follows:

$$\overline{F}_x = \frac{1}{2}C_{Dx}\frac{\pi D^2}{4}\rho\bar{u}_r^2 = W'\tan\alpha \tag{9.25}$$

in which C_{Dx} is the resistance coefficient in the flow direction. For a particle falling in still water, a force $\overline{F}_y$ acts on the particle. If the submerged weight of the particle is balanced by this force, the particle falls at a constant velocity ω. Therefore,

$$\overline{F}_y = \frac{1}{2}C_{Dy}\frac{\pi D^2}{4}\rho\omega^2 = W' \tag{9.26}$$

365

in which C_{Dy} is the drag coefficient for a particle settling in the vertical direction. Combining Eq. (9.25) and Eq. (9.26), one obtains

$$\bar{u}_r = \omega \left(\frac{C_{Dy} \tan \alpha}{C_{Dx}} \right)^{1/2} \tag{9.27}$$

Measured data show that [8]

$$C_{Dy} \approx C_{Dx}$$

and

$$(\tan \alpha)^{1/2} \approx 1$$

therefore,

$$\bar{u}_r \approx \omega \tag{9.28}$$

That is, the relative velocity between the flow and the bed load particle is approximately equal to the fall velocity of the particle.

3. Determination of y_n. If no sand dunes have formed on the bed, the average elevation of the saltating particles is proportional to their diameter

$$y_n = mD \tag{9.29}$$

in which m depends on the flow intensity. From experiment,

$$m = K \left(\frac{U_*}{U_{*c}} \right)^{0.6} \tag{9.30}$$

Bagnold analyzed the trajectories of particles in saltation measured by Francis [9], and found that $K=1.4$. Williams conducted an experiment with uniform sand, $D =1.1$ mm and also concluded that $K=1.4$ [10]. In rivers, bed material is usually composed of fine and coarse sand with a wide range of sizes. A particle of fine sand can bounce quite high after striking a large particle, and the K value can be as much as 2.8, in such a case [11]. In streams with gravel beds, the bed material is still coarser, and the height reached by saltating particles is even greater. The value of K then varies in the range 7.3~9.1.

If Eqs. (9.24), (9.28) and (9.29) are combined with Eq. (9.23), the result is :

$$g_b' = \frac{U_* - U_{*c}}{U_*} \frac{\tau_0 U}{\tan \alpha} \left[1 - \frac{5.75 U_* \log\left(\dfrac{0.4h}{mD}\right) + \omega}{U} \right] \qquad (9.31)$$

The energy efficiency for transporting sediment is given by

$$e_b = \frac{g_b' \tan \alpha}{\tau_0 U} = \frac{U_* - U_{*c}}{U_*} \left[1 - \frac{5.75 U_* \log\left(\dfrac{0.4h}{mD}\right) + \omega}{U} \right] \qquad (9.32)$$

Eq. (9.31) is the Bagnold formula, and it agrees well with the data of Williams.

Eq.(9.31) shows that the larger the depth, the smaller the rate of bed load transport. This change results from the use of average velocity as a key parameter, even though the bed load motion depends mainly on the flow near the bed. Neither the flow near the bed nor the intensity of bed load motion increases much with an increase in depth, but the average velocity does increase. Therefore, flow depth is introduced into the formula. Consequently, if the average velocity is employed as the parameter in a flume study, the results cannot be applied in river flows without a correction for the depth.

9.1.3 Einstein bed load theory

Einstein evolved the stochastic nature of bed load motion from his long term observations in flume experiments. He noted also the continual exchanges that take place between moving particles and sediment in the bed, and he introduced a theory of bed load motion founded on both this physical picture and modern fluid mechanics. The principles of his theory had already been formulated in the 1940s [12]. He obtained a relationship between the intensity of bed load transport and intensity of flow from a plot of data from flume experiments. No theoretical derivation or physical explanation of the relationship was given at that time. Applying probability theory, Einstein derived a mathematical expression for the relationship and expanded it to include non-uniform sediment in 1950 [13]. Then he combined the theory of bed load motion and the diffusion theory of suspended load motion based on the concept of exchanges between bed load and suspended load, and finally proposed an approach for the calculation of the sediment transport capacity of the flow. His approach is the most comprehensive method that has been developed for calculating sediment transport; however, it is complicated and some parts could still be improved. It is presented in this and the next two chapters; some of its shortcomings and possible improvements are discussed in Chapter 11.

9.1.3.1 Exchange between bed material and moving bed load

The concept of continual exchange between sediment in the bed and that moving as bed load is introduced in Chapter 5. Equilibrium of this exchange is important, as it implies that the amount of sediment that settles down on to the bed equals the amount of sediment eroded from it. The Einstein bed load theory is based on this type of equilibrium.

Einstein conducted two sets of experiments with colored sand, also presented in Chapter 5. The experiments showed how the exchanges between the bed material and moving bed load take place. Bed load particles move intermittently rather than continuously. Once lifted from the bed, a particle is usually carried downstream some distance by the flow, and it then settles to the bed and stops, so that it becomes part of the bed again. After a period of time, a second cycle begins. If the intensity of bed load motion is low, the particle spends more time at rest on the bed than it does in motion. Fig. 9.7 shows the process of movement of gravel measured at the Goumen Village Station of the Taohe River. The median diameter of the gravel was 32.2 mm, the water depth was 3.6 m and the average velocity was 2.03 m/s. In an observation period of a little more than a hour, 3,954 seconds, the gravel was at rest on the bed for 3,807 seconds. That is, the gravel remained stationary 96.3% of the time. The mean velocity of the gravel only while moving was measured at 3.16 cm/s. If the time spent at rest was included in the calculation of the velocity, however, the average velocity would be only 0.118 cm/s. Therefore, the rate of bed load transport is indeed dependent on the time the particles are at rest. The longer the time is, the smaller the rate of transport.

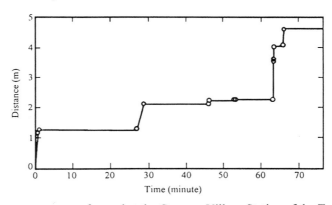

Fig. 9.7 Movement process of gravel at the Goumen Village Station of the Taohe River

9.1.3.2 Statistical analysis of sediment exchange

In his experiment with colored sand, Einstein found that if the sand was fed into a channel section that was in a state of equilibrium, a part of the colored sand deposited near the entrance of the section and became part of the bed. However, it did not stay there long; it gradually moved downstream in a series of steps. The distribution of the number of colored particles varied regularly, as shown in Fig. 9.8.

Bed load theory deals with a large number of sediment particles that are, more or less, exposed to the same flow. Probability theory is a useful tool in the study of their intermittent motion. It is well known that an increase in the number of tests or observations increases the reliability of knowledge about the probability of such an event. Consequently, a statistical analysis

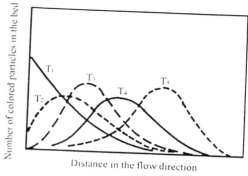

Fig. 9.8 Variation of the distribution of colored particles

becomes more feasible. Statistical analysis does not treat the behavior of a single event but rather, it predicts the events that are likely to occur for a large number of particles in a given flow situation.

Whether a particle moves or remains stationary on the bed, the possibility or probability depends only on the diameter and the shape of the particle and the flow conditions; it is independent of the history of its movement and of the existence of other particles. The nature of bed load motion in flowing water differs from the movement of blown sand. A wind-blown sand particle moves mainly in saltation and

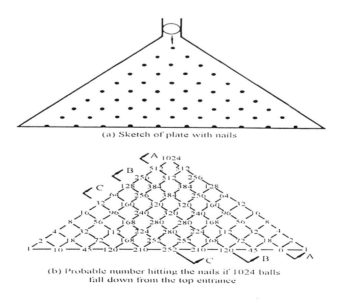

(a) Sketch of plate with nails

(b) Probable number hitting the nails if 1024 balls fall down from the top entrance

Fig. 9.9 Model for bed load motion by steps

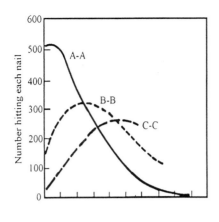

Fig. 9.10 Distribution of the balls
hitting the nails

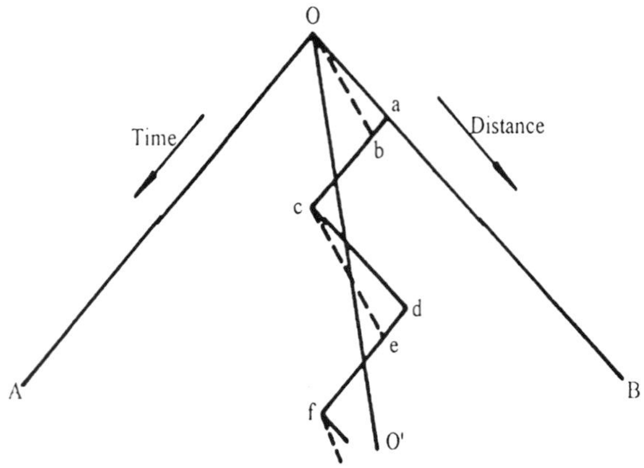

Fig. 9.11 A coordinate system for recording
particle movement

has a large momentum. It hits the bed particles when it falls to the sand surface and induces some of them to jump up into the wind. Therefore, the probability of bed particles being carried away depends on: (1) how many particles are hit by falling particles; and (2) the impact force, which in turn is related to the height the saltating particles reach before they fall.

The essential nature of bed load motion can be illustrated by using the concept of "random walk." Fig. 9.9(a) shows a triangular-shaped vertical plate. There are many nails fixed on the plate with a uniform spacing. Balls fall down from the apex of the triangle. Whenever a ball hits a nail, it must change direction, moving to the right or the left. The probability of going either way is 0.5. The track of one ball is to this extent completely random. If many balls are released at the entrance, however, the way they are distributed follows a well defined pattern. Fig. 9.9b presents the distribution of 1024 balls falling from the top. Fig. 9.10 shows the distributions of those balls hitting along the *A-A, B-B*, and *C-C* lines. Apparently, the distributions have the same form as those in Fig. 9.8.

By analogy with the model, a coordinate system, as shown in Fig. 9.11, can be used to describe the movement of a bed load particle; *OA* is the time axis and *OB* the distance axis. A line segment parallel to the time axis implies that a particle does not move but time elapses. A line segment parallel to the distance axis means that the particle moves foreword one step at an infinite velocity, in zero time. Were the movement of the particle continuous, the track of the particle in the coordinate system would be the line *OO'* and the slope of the line would represent the velocity. Instead, bed load particles move intermittently as they alternate between moving and standing still. If a particle starts from point *O*, it travels a distance *Oa* in a time *ab* and stops there for the time period *bc*, then it travels a distance cd in a time period of de and stops again for the time period *ef*, and on to the third cycle.... Thus, *Obcef* describes the particle's movement, and it is similar to the measured process of movement of bed load shown in Fig. 9.7. Because the times of motion (*ad*, *de* ...) are negligible compared with the times at rest (*bc*, *ef*,...), the movement can be described approximately by the broken line *Oacdf*... Two particles can follow completely different paths. For a great number of particles with all of them starting at the point *O*, however, the distributions of the number of particles along the channel bed, with various period of time, are similar to the curves shown in Fig. 9.8 and Fig. 9.9.

Many useful concepts can be obtained from the foregoing model, even though it is greatly simplified. In the random-walk model, a ball has one choice, to go right or left, at each step. The probabilities are equal, and are independent of the position of the nails. Correspondingly, a bed load particle makes a choice at each step, to stop and become an element of the bed sediment or to continue moving. These probabilities are also independent of the position of the particles. Nevertheless, the probabilities of the two choices depend on the flow intensity so that they can be quite different from 0.5.

After reviewing a series of preliminary experiments and using probability theory in this model, Einstein reached the following conclusions [14]:

1. Sediment on the bed surface and the bed load in motion form an inseparable entity. Consecutive exchanges take place between them. The behavior of a particle can be described as "move - stop - move again - ...." The rate of bed load transport is then determined by the time the particles are resting on the bed.

2. Movement of bed load should be studied from a statistical point of view, and it should be based on observations of the nature of movement of a large number of particles rather than one or only a few.

3. The probability that a bed particle will be lifted from the bed depends on the properties of the particle and the flow in the vicinity of the bed, but it is independent of the history of the particle. The lift force is the primary cause for the particle to move. A particle enters into motion whenever the lift force is greater than the submerged weight of the particle.

4. After traveling some distance, a particle will choose again, whether to stop and rest on the bed or to continue moving. The distance, called the "step length," depends on the diameter and shape of the particle, but is independent of the flow, bed sediment composition and rate of bed load transport. For a spherical particle, the average step length is about 100 times the diameter.

5. After traveling a step length, a particle stops or deposits on the bed if the local flow is not strong enough to maintain the particle in motion. The probability for the particle to stop moving is identical over the entire bed surface.

371

9.1.3.3 Einstein bed load function

Einstein derived a bed load function to be used for calculating the rate of bed load transport under the conditions that the exchanges between bed sediment and particles in motion are in equilibrium, i.e. the amount of sediment removed from the bed is equal to the amount of sediment depositing on the bed in a given period of time.

1. Rate of deposition

The average traveling distance L_0 is the distance a particle travels from its starting point until it is deposited on the bed. The step length of a particle of diameter D can be expressed as λD and for spherical particles, $\lambda \approx 100$. If after a particle travels a step length, it falls on the bed at a point where the local lift force is greater than the submerged weight of the particle, the particle does not stop moving but travels a second step length. If p is the probability of the lift force being greater than the submerged weight, $N(1-p)$ particles will deposit on the bed after traveling a step length, in which N is the number of particles in motion. Thus, only Np particles continue moving. After traveling the second step length, $Np(1-p)$ more particles stop moving and only Np^2 particles remain in motion. In this way, all N particles will stop on the bed after some time has elapsed. The average traveling distance can be expressed as:

$$L_0 = \sum_{n=0}^{\infty}(1-p)p^n(n+1)\lambda D = \frac{\lambda D}{1-p} \qquad (9.33)$$

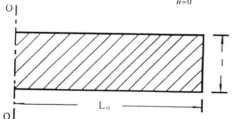

Fig. 9.12 Deposition area for particles passing the O-O cross section per unit time

The line *O-O* designates the cross section of a channel at which motion begins, as shown in Fig. 9.12; the particles passing that section per unit time will deposit within a length of the channel that is equal to L_0, no matter from where they started moving. If g_b represents the rate of bed load transport in dry weight, then

$$rate\ of\ deposition\ on\ unit\ area\ = \frac{g_b}{1 \times L_o} = \frac{g_b(1 - p)}{\lambda D} \qquad (9.34)$$

2. Rate of erosion

The rate at which particles are removed from the bed depends on how many particles are exposed to the flow and their probability of entering into motion. The number of particles per unit area can be estimated as $1/A_1 D^2$, and their total weight

372

is $(A_2 \gamma_s D^3) / A_1 D^2$. If p is the probability for a particle to begin to move, sediment with a total weight of $(A_2 \gamma_s / A_1)pD$ is eroded from the bed per unit time, in which A_1 and A_2 are coefficients related to the shape of the particles.

The next question is within what length of time a particle will be removed from the bed. This time period is a function of the properties of the particle and is independent of the flow intensity, even though the probability of removal is determined by the flow intensity. The time for a particle to be removed is assumed proportional to the time for the particle to fall a length of one diameter in still water. Thus

$$t \sim \frac{D}{\omega} = A_3 \sqrt{\frac{D}{g} \frac{\gamma}{\gamma_s - \gamma}} \qquad (9.35)$$

in which ω is the fall velocity of the particle, A_3 is a constant. Therefore, the rate of erosion per unit area of bed surface is

$$\frac{\dfrac{A_2 \gamma_s}{A_1} pD}{A_3 \sqrt{\dfrac{D}{g} \dfrac{\gamma}{\gamma_s - \gamma}}} = \frac{A_2}{A_1 A_3} p \gamma_s g^{1/2} D^{1/2} \left(\frac{\gamma_s - \gamma}{\gamma}\right)^{1/2} \qquad (9.36)$$

What is implied by the probability for a particle to be removed? The probability for an event to occur is usually expressed as follows:

$$p = \frac{number\ of\ cases\ in\ which\ the\ event\ occurs}{total\ number\ of\ cases}$$

The probability for a particle to be removed from the bed is equal to the probability of the local lift force being greater than the submerged weight of the particle. For any position of the bed surface, the probability represents the ratio of the time during which particles can be removed to the entire time. Correspondingly, for a selected instant, the probability stands for the fraction of the area for which the lift force is greater than the submerged weight of the particles.

For instance, the value $p=0.5$ can be interpreted in three ways: (1) the instantaneous lift force is greater than the submerged weight of particles with a probability of 50%; (2) for any position on the bed, sediment there is moving 50% of the time and is resting the other 50%; and (3) for any instant of time, on 50% of the bed surface, sediment is being eroded and on the rest 50% of the surface sediment is depositing.

3. Equilibrium of sediment transport

Sediment transport is in equilibrium if the amount of sediment eroded from the bed is equal to the amount of sediment deposited on the bed. A combination of Eq. (9.34) and Eq. (9.36) yields

$$\frac{g_b(1-p)}{\lambda D} = \frac{A_2}{A_1 A_3} p \gamma_s g^{1/2} D^{1/2} \left(\frac{\gamma_s - \gamma}{\gamma}\right)^{1/2}$$

which can be rewritten as

$$\frac{p}{1-p} = \frac{A_1 A_3}{\lambda A_2} \frac{g_b}{\gamma_s g^{1/2}} \left(\frac{\gamma}{\gamma_s - \gamma}\right)^{1/2} \frac{1}{D^{3/2}} = A_* \Phi \qquad (9.37)$$

in which

$$A_* = \frac{A_1 A_3}{\lambda A_2}$$

and

$$\Phi = \frac{g_b}{\gamma_s} \left(\frac{\gamma}{\gamma_s - \gamma}\right)^{1/2} \left(\frac{1}{gD^3}\right)^{1/2} \qquad (9.38)$$

The quantity Φ is called bed load transport intensity, and the probability is given by

$$p = \frac{A_* \Phi}{1 + A_* \Phi} \qquad (9.39)$$

The submerged weight of the particles is

$$W' = (\gamma_s - \gamma) A_2 D^3$$

and the lift force is given by

$$F_L = C_L A_1 D^2 \frac{1}{2} \rho u^2 \qquad (9.40)$$

Einstein and El-Samni found that for uniform sediment, if the velocity at elevation $y=0.35D$ is taken as the effective velocity in Eq. (9.40), the distribution of fluctuating lift force follows the normal distribution with a standard deviation equal to half the mean value and $C_L = 0.178$ [15]. Hence, the lift force can be expressed as follows:

$$F_L = 0.178 A_1 D^2 \frac{1}{2} \rho \cdot 5.75^2 R_b' Jg \log^2(10.6)(1+\eta) \qquad (9.41)$$

in which $\eta = f(t)$ represents the fluctuating component of the lift force. If one employs the standard deviation of the lift force η_0 instead of η,

$$\eta = \eta_0 \eta_* \qquad (9.42)$$

in which η_* is a dimensionless number that represents the fluctuation of the lift force, this force can be expressed in the following form:

$$F_L = \frac{0.178 A_1 575^2}{2} \rho D^2 R_b' Jg \log^2(10.6) \cdot (1 + \eta_* \eta_0) \qquad (9.43)$$

The term p stands for the probability that W'/F_L is smaller than 1, i.e.

$$1 > \frac{W'}{F_L} = \left[\frac{1}{1 + \eta_* \eta_0} \right] \left\{ \frac{\gamma_s - \gamma}{\gamma} \frac{D}{R_b' J} \right\} \left(\frac{2 A_2}{0.178 A_1 5.75^2} \right) \frac{1}{\log^2(10.6)} \qquad (9.44)$$

If the following terms are defined

$$\psi = \frac{\gamma_s - \gamma}{\gamma} \frac{D}{R_b' J} \qquad (9.45)$$

$$B = \frac{2 A_2}{0.178 A_1 5.75^2}$$

$$\beta = \log(10.6)$$

$$B' = B / \beta^2$$

Eq. (9.44) can be rewritten as

$$1 > [\frac{1}{1 + \eta_* \eta_o}] B' \psi$$

The fluctuation of the lift force is caused by the fluctuation of the velocity. Whether the fluctuation of velocity is positive or negative, the lift force is always positive. Thus, the inequality for the lift force can be written as

$$|1 + \eta_* \eta_o| > B' \psi$$

or

$$\left|\frac{1}{\eta_o} + \eta_*\right| > \frac{B'}{\eta_o}\psi = B_*\psi$$

in which

$$B_* = B' / \eta_0$$

The critical condition for particles to be removed from the bed is

$$\eta_* = \pm B_*\psi - \frac{1}{\eta_o} \qquad (9.46)$$

Between the two values, no bed load motion occurs.

As stated, the lift force follows a normal distribution, therefore, η_* also follows a normal distribution, i.e.

$$1 - p = \frac{1}{\sqrt{\pi}} \int_{-B_*\psi - 1/\eta_0}^{B_*\psi - 1/\eta_0} e^{-t^2} dt$$

or

$$p = 1 - \frac{1}{\sqrt{\pi}} \int_{-B_*\psi - 1/\eta_0}^{B_*\psi - 1/\eta_0} e^{-t^2} dt \qquad (9.47)$$

Fig. 9.13 represents Eq. (9.47). In the shaded area, the lift force is greater than the submerged weight of a particle and bed load motion occurs.

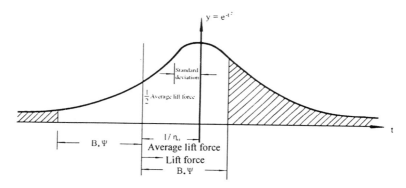

Fig. 9.13 Probability density distribution of lift force and probability of bed load motion
(Shaded area=Probability of bed load motion)

If Eq. (9.39) is combined with Eq. (9.47), the result is

$$1 - \frac{1}{\sqrt{\pi}} \int_{-B_*\psi - 1/\eta_0}^{B_*\psi - 1/\eta_0} e^{-t^2}\, dt = \frac{A_* \Phi}{1 + A_* \Phi} \qquad (9.48)$$

in which the constants, determined through experiments, are as follows:

$$1 / \eta_0 = 2.0$$

$$A_* = \frac{1}{0.023}$$

$$B_* = \frac{1}{7}$$

Eq. (9.48) is the Einstein bed load function. Fig. 9.14 is a comparison of the function with measured data, and it shows that the function represents the data quite well.

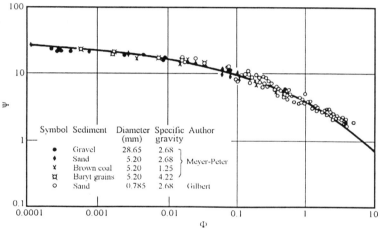

Fig. 9.14 Comparison of Einstein bed load function with measured data
(uniform sediment)(after Einstein, H.A.)

Vellikanov and Dou developed the Einstein theory further, but they do not agree on the definition of *(1-p)* for the probability of bed load to stop moving [16,17]. They used the probability of the vertical component of fluctuating velocity being greater than the fall velocity of particles $p(v_0' < \omega)$ instead. $p(v_0' < \omega)$ is the probability for suspended load to deposit rather than the probability for bed load to stop moving. Dou acknowledged that his bed load formula involves a part of the suspended load.

9.1.4 Yalin bed load formula [18]

Like Bagnold, Yalin also started his derivation from Eq. (9.15)

$$g_b' = W_b' \overline{u}_b$$

If u_{bx} and u_{by} represent the velocity components of a sediment particle in the flow direction and the vertical direction, and F_x and F_y are the force components of flow acting on the particle in the flow direction and vertical direction, respectively, then the equation of motion for a particle is

$$\left. \begin{aligned} F_x &= m' \frac{du_{bx}}{dt} \\ -F_y - W' &= m' \frac{du_{by}}{dt} \end{aligned} \right\} \tag{9.49}$$

in which

$$\left. \begin{aligned} F_x &= \frac{\pi}{8} C_{Dx} \rho D^2 (u - u_{bx})^2 \\ F_y &= \frac{\pi}{8} C_{Dy} \rho D^2 u_{by}^2 \end{aligned} \right\} \tag{9.50}$$

in which u is velocity of the water, m' is the submerged mass and W' is the weight of the particle.

A particle jumps up from the bed under the action of a lift force F_L. The lift force then decreases with distance from the bed and is equal to W' at an elevation where the particle reaches its maximum vertical velocity component. The maximum vertical velocity component can be obtained from the following differential equation:

$$-F_y - W' + F_L = m' \frac{du_y}{dt} \tag{9.51}$$

This equation represents the initial condition of the two equations, (9.49).

To help solve these equations, Yalin made the following assumptions:

(1) $F_L / W' \sim e^{-(y/D)}$ (9.52)

(2) C_{Dx} and C_{Dy} can be treated as constants;

(3) u / U_* can be treated as a constant at the vicinity of the bed;

In this way, he obtained an expression for u_{bx}, and then he determined its average value over the time it is in motion; it is given by

$$\bar{u}_b = U_* C_1 \left[1 - \frac{1}{as} \ln(1 + as) \right] \tag{9.53}$$

in which

$$s = \frac{\Theta - \Theta_c}{\Theta_c} \tag{9.54}$$

$$a = 2.45\sqrt{\Theta_c}\left(\frac{\gamma}{\gamma_s}\right)^{0.4} \tag{9.55}$$

and C_I is a constant yet to be determined. The term Θ is defined in Eq. (6.38), and it is indeed the reciprocal of the parameter Ψ; Θ_c is the critical value for initiation of motion of sediment particles.

Yalin determined the submerged weight of bed load discharge from dimensional analysis. He linked bed load motion with sand wave motion. As discussed in Chapter 6, a sand wave is a massive movement of bed load. If the intensity of the motion is not strong, the movement of a sand wave depends mainly on the following two parameters:

$$\Theta = \frac{\gamma}{\gamma_s - \gamma} \frac{R_b J}{D}$$

$$Re_* = \frac{U_* D}{\nu}$$

It follows that

$$\frac{W_b'}{(\gamma_s - \gamma)D} = f_1(\Theta, Re_*) \tag{9.56}$$

The grain Reynolds number can be expressed as

$$Re_* = \sqrt{\frac{(\gamma_s - \gamma)D^3}{\rho \nu^2}\Theta}$$

Therefore, Eq. (9.56) can be rewritten as

$$\frac{W_b'}{(\gamma_s - \gamma)D} = f_2\left[\Theta, \frac{(\gamma_s - \gamma)D^3}{\rho \nu^2}\right] \tag{9.57}$$

At the initiation of the bed load motion, $W_b'=0$, and

$$f_2\left[\Theta_c, \frac{(\gamma_s - \gamma)D^3}{\rho v^2}\right] = 0 \tag{9.58}$$

A combination of Eqs. (9.57) and (9.58) yields

$$\frac{W_b'}{(\gamma_s - \gamma)D} = f_3(\Theta, \Theta_c) \tag{9.59}$$

Yalin assumed

$$\frac{W_b'}{(\gamma_s - \gamma)D} = C_2 \frac{\Theta - \Theta_c}{\Theta_c} \tag{9.60}$$

in which C_2 is also a constant to be determined.

Substituting Eqs. (9.53) and (9.60) into (9.49) and determining the constants from measured data, Yalin obtained the following bed load formula, and it is known by his name,

$$\frac{g_b'}{(\gamma_s - \gamma)DU_*} = 0.635s\left[1 - \frac{1}{as}\ln(1 + as)\right] \tag{9.61}$$

9.1.5 Engelund formula [19]

Engelund treated sediment particles as spheres of diameter D, so that there are approximately $1/D^2$ spherical particles in a unit area of the bed surface. For a certain flow intensity, the proportion of the particles on the bed surface that are moving is p. Here, p is indeed the probability of motion suggested by Einstein. Hence, the rate of bed load transport is given by

$$g_b = \frac{\pi}{6}D^3\gamma_s \frac{p}{D^2}\overline{u_b} \tag{9.62}$$

in which $\overline{u_b}$ is the mean velocity of the bed load particles. The main concern is the way he determined p and $\overline{u_b}$.

The tractive force acting on a bed load particle is

$$C_D \frac{1}{2}\rho[\alpha U_* - \overline{u_b}]^2 \frac{\pi}{4}D^2$$

in which αU_* represents the velocity of flow at the level of the bed particle. If the particle is at a distance of one to two particle diameter above the bed, $\alpha = 6\sim10$. The frictional force on the moving particle is

$$(\gamma_s - \gamma)\frac{\pi D^3}{6}\beta$$

in which β is a kinetic frictional coefficient. For equilibrium, the tractive force and the frictional force are equal, and the following equation can be derived:

$$\frac{\bar{u}_b}{U_*} = \alpha\left[1 - \sqrt{\frac{\Theta_0}{\Theta}}\right] \qquad (9.63)$$

in which Θ is defined by Eq. (6.38), and Θ_0 is given by

$$\Theta_0 = \frac{4\beta}{3\alpha^2 C_D}$$

Eq. (9.63) implies that Θ_0 is the value of Θ as a particle stops moving so that $\bar{u}_b = 0$. The value is comparable to the threshold value for the initiation of sediment motion Θ_0, the value used for the Shields curve in Chapter 8. In fact, Θ_c is the critical value for the initiation of motion of a particle in a compactly arranged bed, and Θ_0 is the critical value for a particle protruding from the bed surface. Of course, Θ_0 is smaller than Θ_c. Measured data showed that Θ_0 is equal to $0.5\Theta_c$. Thus, Eq. (9.63) can be rewritten as

$$\frac{\bar{u}_b}{U_*} = \alpha\left[1 - 0.7\sqrt{\frac{\Theta_c}{\Theta}}\right] \qquad (9.64)$$

For a sandy river bed, $\alpha = 9.3$. Eq. (9.64) was verified by a comparison with many measured results; however, the coefficient α is not always the same because it varies somewhat with the shape of the sediment particles [20].

According to Bagnold, the shear stress of flow is composed of grain shear stress τ and fluid shear stress τ', as indicated by Eq. 9.12. Furthermore, he suggested that the fluid shear stress τ' equals the critical shear stress for initiation of motion of bed particles. Hence,

$$\tau_0 = \tau_c + T = \tau_c + n\bar{F}_x \qquad (9.65)$$

in which n is the number of moving particles per unit area of bed surface, and F_x is the drag force acting on the particles. Engelund assumed that

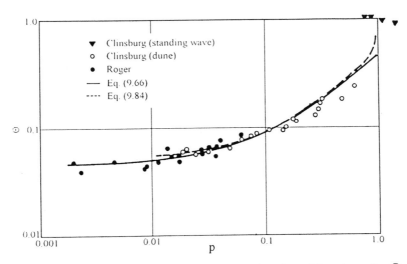

Fig. 9.15 Probability of motion of particles as a function of flow intensity Θ
(after Engelund, F. and J. Fredsøe)

$$\overline{F}_x = (\gamma_s - \gamma)\frac{\pi D^3}{6}\beta$$

and that $p = nD^2$, and he substituted these values into Eq. (9.65) to obtain

$$\Theta = \Theta_c + \frac{\pi}{6}\beta p$$

or

$$p = \frac{6}{\pi\beta}(\Theta - \Theta_c) \tag{9.66}$$

Fig. 9.15 compares Eq. (9.66) with the probability of motion of particles, p, as a function of Θ, and he obtained the values $\beta=0.8$ and $\Theta_c = 0.046$; Eq. (9.66) agrees well with the measured data.

Combining Eqs. (9.64) and (9.66) with Eq. (9.62), Engelund obtained his bed load formula:

$$g_b = \frac{9.3}{\beta}\frac{D\gamma_s}{\sqrt{\Theta}}U_*(\Theta - \Theta_c)(\sqrt{\Theta} - 0.7\sqrt{\Theta_c}) \tag{9.67}$$

9.1.6 Ackers-White formula [21]

Ackers and White collected 1,000 sets of experimental data from previous researchers. Following Bagnold's approach, they derived a functional relationship

382

between dimensionless parameters. Then they conducted a regression analysis with the data to determine a functional relationship. Their formula includes both bed load and suspended load; its derivation is given in the next chapter. Nevertheless, the formula was simplified into a bed load formula for natural sand coarser than 2.5 mm, in the following form:

$$Y = 0.025\left(\frac{M}{0.17} - 1\right)^{1.5}$$

(9.68)

in which

$$M = \frac{U}{\sqrt{g\frac{\gamma_s - \gamma}{\gamma}D}} \frac{1}{\sqrt{32}\frac{10h}{D}}$$

(9.69)

$$Y = \frac{\gamma S_{Wb}h}{\gamma_s D}$$

(9.70)

in which S_{Wb} represents the average concentration of sediment load in weight per unit volume.

9.1.7 Bed load formulas with velocity as the main parameter

The preceding six bed load formulas use either shear stress or power as the main parameter. In the former USSR, scientists employed the average velocity instead. Derivations of these formulas are similar to those of Bagnold and Yalin.

First, Eq. (9.15) is rewritten in a more general form as follows:

$$g_b = \gamma_s h_b S_{vb} \bar{u}_b$$

(9.71)

in which h_b stands for the thickness of bed load layer (equivalent to Bagnold's y_n) and S_{vb} for the volume concentration of bed load in the layer.

Second, the researchers made various assumptions about h_b, S_{vb} and u_b, and thus they obtained different formulas. Table 9.1 presents three representative ones [22,23,24]. The three, especially the Sharmov formula and Levy formula, share much in their approach, although they differ in details. The Gongcharov formula is more complicated because it includes suspended load. Still, the formula is discussed in this chapter, because it is more appropriate for flow in which bed load is dominant.

Sharmov and Levy assumed

$$\bar{u}_b \sim (u - u_c)$$

in which u is the velocity acting on the bed particle and u_c is the velocity for the initiation of motion of a particle on the bed. The assumption is similar to Bagnold's:

$$\bar{u}_b \sim (u - \omega)$$

Table 9.1 Bed load formulas with velocity as the main parameter

Author	$\bar{u}_b$	h_b	s_{vb}	g_b (Kg/m/s)	Valid range	Note
Sharmov	$\left(U - \dfrac{U_c}{1.2}\right)\left(\dfrac{D}{h}\right)^{1/4}$	$K'D$	$K''\left[\dfrac{\frac{U}{U_c}}{1.2}\right]^3$	$0.95\sqrt{D}\left[\dfrac{\frac{U}{U_c}}{1.2}\right]^3$ $\cdot\left(U - \dfrac{U_c}{1.2}\right)\left(\dfrac{D}{h}\right)^{1/4}$	$0.2<D<0.73$ mm, and $13<D<65$ mm; $1.02<h<3.94$ m, $0.18<h<2.16$ m $0.4<U<1.02$ m/s $0.8<U<2.95$ m/s	K' and K" are function of D. Therefore the formula includes the square root of D instead of D.
Levy	$\alpha'(U - U_c)$	$\alpha''D$	$\alpha'''\left(\dfrac{U}{\sqrt{gD}}\right)^3$ $\cdot\left(\dfrac{D}{h}\right)^{1/4}$	$2D\left(\dfrac{U}{\sqrt{gD}}\right)^3$ $\cdot(U - U_c)\left(\dfrac{D}{h}\right)^{1/4}$	$0.25<D<23$ mm $5<h/D<500$ $1<U/U_c<3.5$	——
Gongcharov	$\alpha_k U\left[1 - \dfrac{\left(\frac{U_c}{1.4}\right)^3}{U^3}\right]$	$(\alpha_1 + \alpha\zeta)D\dfrac{U - \frac{U_c}{1.4}}{\frac{U_c}{1.4}}$	$\alpha_4\dfrac{1 + \alpha_6\frac{\frac{U_c}{1.4}}{U}}{1 + \frac{\alpha_3\frac{U_c}{1.4}}{\zeta U}}$ $\cdot\dfrac{U^2}{\left(\frac{U_c}{1.4}\right)^2}$	$(3.0\sim5.3)(1+\zeta)D$ $\left[\dfrac{U^3}{\left(\frac{U_c}{1.4}\right)^3} - 1\right]$ $\cdot\left(U - \dfrac{U_c}{1.4}\right)$	$0.08<D<10$ mm $10<h/D<1550$ $0.72<U/Uc<$ 13.1	The coefficient 3 is suitable for river flow and 5 for flumes; ς is a coefficient related to turbulence

Levy further assumed that u is proportional to the average velocity over the vertical profile. Sharmov employed the velocity distribution and introduced the term $(D/h)^{1/4}$. Gongcharov included a part of the suspended load and added a correction term $(U_c/U)^2$ because suspended load moves faster than bed load, i.e.,

$$\overline{u}_b \sim U - \frac{U_c}{1.4}\left[\frac{U_c}{1.4U}\right]^2$$

As the flow velocity increases, more and more sediment becomes suspended; the second term in the preceding formula then approaches zero and the average velocity of the sediment approaches the average velocity of flow.

Sharmov and Levy assumed the thickness of the bed load layer to be proportional to the diameter of the sediment. Einstein did the same, as discussed in the next chapter. Gongcharov believed the thickness should be related to D and to the parameter $[1.4U/U_c - 1]$. His way is similar to that of Bagnold, who related the thickness to D and to the parameter $[U_*/U_{*c}]^{0.6}$.

Sharmov and Levy assumed that the concentration of sediment in the bed load layer is proportional to $[U/U_c]^3$, but Levy used the additional term of $(D/h)^{1/4}$. In the approach by Gongcharov, the concentration was taken to be proportional to $[U/U_c]^2$ because concentration of suspended load is lower than that in the bed load layer.

9.2 COMPARISON OF BED LOAD FORMULAS [25]

The bed load formulas presented in the preceding section have common properties and give similar results under certain conditions, even though they have different forms. In this section, the Meyer-Peter formula, Bagnold formula, and Yalin formula are transformed into formulas using the bed load transport intensity parameter Φ and the flow intensity parameter Ψ. Their similarities and differences between the formulas are indicated.

9.2.1 Transformation of the formulas

9.2.1.1 Meyer-Peter formula

As discussed in Chapter 7, the resistance consists of grain resistance and sand wave resistance, if sand waves form on the river bed. Einstein and Meyer-Peter separated the sand wave resistance from the grain resistance in different ways. Einstein split the hydraulic radius into a hydraulic radius for the bed configuration and another for grain friction; Meyer-Peter divided the energy slope into sand wave slope and grain slope. The results are as follows:

	Einstein	*Meyer-Peter*
Bed resistance	$U = K_b R_b^{2/3} J^{1/2}$	$U = K_b R_b^{2/3} J^{1/2}$
Grain resistance	$U = K_b' R_b'^{2/3} J^{1/2}$	$U = K_b' R_b'^{2/3} J'^{1/2}$

Therefore

$$\frac{K_b}{K_b'} = \left(\frac{J'}{J}\right)^{1/2} = \left(\frac{R_b'}{R_b}\right)^{2/3} \tag{9.72}$$

If the Einstein method is employed, the Meyer-Peter formula is transformed into

$$\gamma R_b' J = a_4 (\gamma_s - \gamma) D + b_4 \left(\frac{\gamma}{g}\right)^{1/3} \left(\frac{\gamma_s - \gamma}{\gamma_s}\right)^{2/3} g_b^{2/3}$$

If both sides are divided by $(\gamma_s - \gamma)D$, it becomes

$$\frac{\gamma}{\gamma_s - \gamma} \frac{R_b' J}{D} = a_4 + b_4 \left[\frac{g_b}{\gamma_s}\left(\frac{\gamma}{\gamma_s - \gamma}\right)^{1/2}\left(\frac{1}{gD^3}\right)^{1/2}\right]^{2/3} \tag{9.73}$$

Substituting Eqs. (9.38) and (9.45) into Eq. (9.73), one obtains

$$\frac{1}{\psi} = a_4 + b_4 \Phi^{2/3}$$

After some manipulation, the Meyer-Peter formula can be written in the following form [26]:

$$\Phi = \left(\frac{4}{\psi} - 0.188\right)^{3/2} \tag{9.74}$$

For initiation of bed load motion, $\Phi = 0$,

$$\left(\frac{1}{\psi}\right)_c = \Theta_c = 0.047$$

and for large Θ

$$\Phi = 8\Theta^{3/2} \tag{9.75}$$

9.2.1.2 Bagnold formula

If, in the derivation of the Bagnold formula, u_n is determined directly from the logarithmic formula instead of from Eq. (9.22) and the relationships

$$\tau_0 = \rho U_*^2$$

$$g_b = \frac{\gamma_s}{\gamma_s - \gamma} g_b{}'$$

are used, Eq. (9.31) can be rewritten as

$$g_b = \frac{\gamma_s}{\gamma_s - \gamma} \frac{U_* - U_{*c}}{\tan \alpha} \rho U_*^2 \left[5.75 \log 30.2 \frac{mD}{K_s} - \frac{\omega}{U_*} \right]$$

The equation can be further transformed into

$$\Phi = \frac{1}{\psi} \left(\frac{1}{\sqrt{\psi}} - \frac{1}{\sqrt{\psi_c}} \right) \left[\frac{1}{\tan \alpha} \left(5.75 \log 30.2 \frac{mD}{K_s} - \frac{\omega}{U_*} \right) \right] \qquad (9.76)$$

The term in the second parenthesis on the right hand side of the equation is a function of flow and sediment, but it varies within a rather small range; hence, the rough assumptions that the velocity distribution is logarithmic and that the relative velocity between particle and water can be represented by the fall velocity of particle have little effect on calculation of the transport rate.

9.2.1.3 Yalin formula

If Eq. (9.61) is multiplied by

$$\frac{1}{\sqrt{\psi}} = \sqrt{\frac{\rho U_*^2}{(\gamma_s - \gamma)D}}$$

the Yalin formula is then transformed into

$$\Phi = 0.635 \frac{s}{\sqrt{\psi}} \left[1 - \frac{1}{as} \ln(1 + as) \right] \qquad (9.77)$$

At the initiation stage of bed load motion, $\Theta - \Theta_c$ is small and as is also small. Hence

$$\frac{1}{as} \ln(1 + as) \approx 1 - \frac{as}{2}$$

$$\Phi = 0.32 \frac{as^2}{\sqrt{\psi}} = 0.78 \left(\frac{\gamma_s}{\gamma} \right)^{0.4} \left(\frac{1}{\psi} \right)^{1/2} \psi_c^{3/2} \left(\frac{1}{\psi} - \frac{1}{\psi_c} \right)^2 \qquad (9.78)$$

387

For a high intensity of bed load motion, Θ is large, $as \to \infty$, and $\ln(1 + as)/as \to 0$; then

$$\Phi = 0.635 \frac{s}{\sqrt{\Psi}} = \frac{0.635}{\Psi^{3/2}}(\Psi_c - \Psi) \tag{9.79}$$

9.2.1.4 Engelund formula

Dividing Eq. (9.67) by

$$\gamma_s \sqrt{\frac{(\gamma_s - \gamma)}{\gamma} gD^3}$$

and setting $\beta = 0.8$, one obtains

$$\Phi = 11.6(\Theta - \Theta_c)(\sqrt{\Theta} - 0.7\sqrt{\Theta}) \tag{9.80}$$

If $\Theta_c << \Theta$,

$$\Phi = 11.6\Theta^{3/2} \tag{9.81}$$

9.2.1.5 Ackers-White formula

Because

$$g_b = S_{wb}\,\gamma\,Uh$$

and, for a flat river bed,

$$\frac{h}{D} = \frac{1.65}{J}\Theta$$

Eq. (9.68) can be transformed into

$$\Phi = 2.02\log\left(\frac{16.5}{J}\Theta\right)\sqrt{\Theta}(\sqrt{\Theta} - 0.17)^{1.5} \tag{9.82}$$

The Ackers-White formula expresses the intensity of bed load transport as a function of flow intensity and bed slope. For the initiation of bed load motion,

$$\Theta_c = 0.029$$

and for a high intensity of bed load motion,

$$\Phi \sim \Theta^{5/4} \log\left(\frac{16.5}{J}\Theta\right) \qquad (9.83)$$

9.2.2 Comparison of the formulas of Meyer-Peter, Einstein, Bagnold, Yalin, Engelund, and Ackers-White

The typical bed load formulas have been transformed into functional relationships between Φ and Θ (or $1/\Psi$), so that they can be compared more directly.

The comparison is based on the following conditions:

First, the channel bed is flat and the characteristic roughness is the sediment diameter.

Second, except for the Ackers-White formula, the threshold condition of the initiation of bed load motion is taken as $\Theta_c = 0.047$.

Fig. 9.16 shows is a comparison of the Meyer-Peter, Einstein, Bagnold, and Yalin formulas. For the Bagnold formula, his value of 0.63 is used for $tan\,\alpha$. The $\Phi \sim \Psi$ relationships for particle diameter of 0.2 mm and 2 mm are given and they give similar results. The curve for the Bagnold formula in Fig. 9.16 is the average of the two cases. The figure shows that for $\Psi > 2$, the Bagnold, Einstein, and Meyer-Peter formulas are close together, but the Yalin formula yields smaller values for the bed load transport rate. The Meyer-Peter formula follows the experimental data better than does the Einstein bed load function for low intensities of bed load motion, but

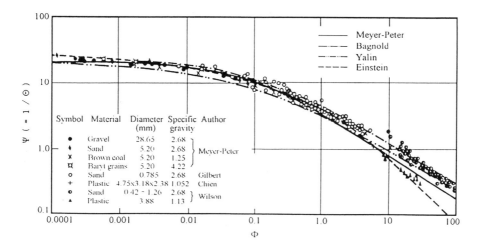

Fig. 9.16 Comparison of the formulas of Meyer-Peter, Einstein, Bagnold, Yalin

the opposite is true for moderate intensities. The Bagnold formula gives a larger transport rate than the measured data in this range. The main differences among the formulas arise for the higher intensities of bed load motion ($\Psi <2$).

Fig. 9.17 is a comparison between the Engelund and Meyer-Peter formulas. The two formulas present almost the same results for $\Psi >2$. They differ only for high intensities of bed load motion. The Engelund formula gives a Φ-value 45% larger than that of Meyer-Peter formula for $\Psi <2$. The Engelund formula involves the determination of p, the probability that a bed particle will move; clearly p should not be larger than 1 from its intrinsic meaning. However, Eq. (9.66) gives p values that are larger than 1 for $\Theta >0.5$ or $\Phi < 2$, as shown in Fig. 9.15, even though the formula conforms to the data for $\Theta < 0.5$. The Engelund formula, expressed as Eq. (9.80), is not valid for $\Theta >0.5$. Engelund suggested a change in the $p \sim \Theta$ relationship with the following form:

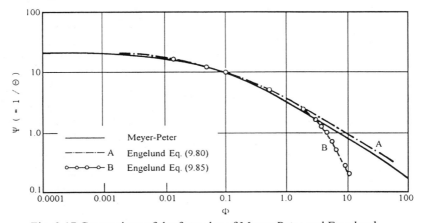

Fig. 9.17 Comparison of the formulas of Meyer-Peter and Engelund

$$p = \left[1 + \left(\frac{\pi \beta}{6(\Theta - \Theta_c)} \right)^4 \right]^{-1/4} \tag{9.84}$$

Fig. 9.15 shows that this formula differs little from Eq. (9.66) for $\Theta < 0.4$ and approaches $p=1$ for $\Theta > 0.5$. If Eq. (9.84) is substituted into Eq. (9.62), the Engelund formula is then

$$\Phi = 4.87 \left[1 + \left(\frac{\frac{\pi}{6} \beta}{\Theta - \Theta_c} \right)^4 \right]^{-1/4} \left(\sqrt{\Theta} - 0.7\sqrt{\Theta_c} \right) \tag{9.85}$$

and it is shown as curve *B* in Fig. 9.17. For $\Theta < 1$, $\Phi \sim \Psi^{1/2}$; thus it differs from the other bed load formulas and does not fit the data well.

Fig. 9.18 is a comparison of the Meyer-Peter and Ackers-White formulas. For the Ackers-White formula, two curves with energy slopes of *J*=0.01 and 0.001 are plotted. The Ackers-White formula gives larger bed load transport rates for low intensities of bed load motion because the formula employed a smaller threshold value for the initiation of bed load motion than did the Meyer-Peter formula. If the threshold value used in the Meyer-Peter formula is employed, the Ackers-White formula takes the form

$$\Phi = 2.02 \log \left(\frac{16.5}{J} \Theta \right) \sqrt{\Theta} \left(\sqrt{\Theta} - 0.217 \right)^{1.5} \tag{9.86}$$

Curves of Eqs. (9.86) for *J*=0.01 and 0.001 are also shown in Fig. 9.18, and they are quite close to the Meyer-Peter curve.

In the following sections two important problems are discussed:

Structure of bed load formulas at low intensity of bed load transport

For low intensity of bed load transport, the $\Phi \sim \Psi$ curves slope gently because all of the formulas, except the Einstein formula, have a term of $(\Theta - \Theta_c)^m$ or $(\sqrt{\Theta} - \sqrt{\Theta_c})^m$ and the exponent *m* is larger than *1*. If $\Theta \ll 1$, a slight variation in Θ responds to great change in Φ. This trend is more apparent if Θ is close to Θ_c. In other words, bed load transport is quite sensitive to the flow if the transport intensity is low. Some researchers employed the average velocity in their bed load formulas. For this type of bed load formula, the exponent on the average velocity is large for low intensities of bed load motion. For instance, students in the Hydrology Division, Department of Geography of Zhongshan University, measured the bed load transport rate in the Beijiang River in Guangdong Province and the Liusha River in Hunan

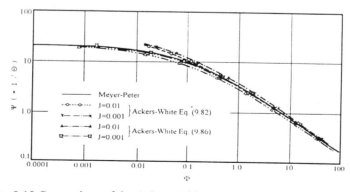

Fig. 9.18 Comparison of the Ackers-White and the Meyer-Peter formulas

province. They found that the transport rate is proportional to the velocity raised to a power of 5.7~7.5. [1] Furthermore from flume experiments at low intensities of bed load transport, Mantz obtained the relationship [20]:

$$\frac{g_b}{(\gamma_s - \gamma)DU_*} = 2.6 \times 10^{21}\Theta^{17} \tag{9.87}$$

For flake-shaped bed particles with a structure like fish scales, the coefficient in the formula is 10^{13}. In fact, for bed load formulas that use some power of U or Θ to represent the flow intensity, the exponents on the flow intensity terms are not constant but vary with the rate of bed load transport. The lower the intensity of bed load transport, the larger the exponents.

For low intensities of bed load transport, the curves of the Meyer-Peter, Einstein, Bagnold, Engelund, and the modified Ackers-White formulas give similar results, as shown in Fig. 9.16, 9.17, and 9.18. Nevertheless, the maximum difference between the formulas represents a factor of some 2-3. In fact, none of the formulas can provide accurate results for $\Psi > 15$ or $\Theta < 0.067$ because of errors in the measurements of flow intensity. For river training, however, hydraulic engineers are often required to answer questions such as what will be the cross section and equilibrium slope of a river bed for a given bed load transport rate, how to design a canal for sediment-laden flow and how to determine the slope of the upper section of a reservoir delta. For this question, the formulas can give satisfactory results [25].

Structure of bed load formulas at high intensity of bed load transport

The bed load formulas diverge for $\Psi < 2$. The $\Phi \sim 1/\Psi$ curves in this range approach straight lines on a log-log plot. The Meyer-Peter, Bagnold, Yalin, and Engelund formulas approach lines indicating an exponent of 1.5 on the ($1/\Psi$) term. In contrast, the Einstein bed load function approaches the line

$$\Phi = \frac{7.9}{\Psi}$$

and the exponent is therefore 1. The exponent for the Ackers-White formula is 1.35 ~ 1.45, a value that falls between the other two values. The data for high

---

[1] Department of Geography of Zhongshan University. *Preliminary Report on the Measurement and Calculation of Bed Load Transport at the Shijiao Section of the lower reaches of the Beijiang River*. 1976, p.37. and *Report on the Measurement and Calculation of Movement of Sand Wave and Bed Load Transport in the Liusha River in Hunan Province*. 1977, p.71.

intensities of bed load motion is insufficient. Wilson conducted experiments with a high intensity of bed load transport in a closed rectangular conduit [27]. He used both a natural sand of diameter 0.42~1.26 mm and plastic material of diameter 3.88 mm and specific weight of 1.38 g/cm³. The measured data with the natural sand are close to the Bagnold and Yalin formulas but the data with the plastic material are nearer to the Einstein curve. So far, one can not conclude which formula is the best one to use. A serious difficulty arises from the suspension of the material with high intensity bed load transport; in this case, one can not readily separate suspended load from bed load. Engelund believed that all measured data should follow the line with a slope of 1.5. The cause of the dispersion of Wilson's data along two lines, in Fig. 9.16, is the difference in density of the two kinds of sand. This can be explained in terms of Bagnold's dispersive force [28].

In Chapter 5, the sediment transport in streams is shown to be similar to wind-blown sand if the rate of bed load transport is high. For blown sand, Bagnold presented the following formula for the transport rate of saltation load and contact load [29]:

$$g_b = C_1 \left(\frac{D}{D_1}\right)^{1/2} \rho \frac{U_*^3}{g}$$ (9.88)

in which C_1 is a constant, D_1 is what he called the standard diameter and is equal to 0.25 mm. The formula can be rewritten in the form

$$\Phi = C_1 \left(\frac{D}{D_1}\right)^{1/2} \frac{\gamma_s - \gamma}{\gamma_s} \frac{\Psi^{-3/2}}{g}$$ (9.89)

For wind-blown sand in a desert, the diameter range is small in that all particle diameters differ little from D_1 because of the long term sorting action of wind. Because the density of air is much smaller than that of sand, $(\gamma_s-\gamma)/\gamma_s$ is close to 1. Therefore, the transport intensity is proportional to Ψ raised to the -3/2 power. This result is consistent with the formulas of Bagnold, Meyer-Peter, Engelund, and Yalin.

Finally, the Meyer-Peter formula can be expressed in another way. For a flat channel bed, the left hand side of Eq. (9.73) is

$$\frac{\gamma}{\gamma_s - \gamma} \frac{R_b J}{D} = \frac{\tau_0}{(\gamma_s - \gamma)D}$$

For initiation of bed load motion, i.e. $g_b = 0$, one can deduce

$$a_4 = \frac{\tau_c}{(\gamma_s - \gamma)D}$$

Substituting the two expressions into Eq. (9.73), one obtains the expression

$$g_b = \frac{8}{\rho^{1/2}}\left(\frac{\gamma_s}{\gamma_s - \gamma}\right)(\tau_0 - \tau_c)^{3/2} \qquad (9.90)$$

Using the data of Gilbert, O'Brien, and Rindlaub developed a bed load formula in 1934 [30]

$$g_b = k(\tau_0 - \tau_c)^m \qquad (9.91)$$

in which the exponent m is in the range of 1.3~1.4. Experiments with sediment of median diameters of 0.21-4.08 mm, conducted in the US Water Experiment Station in 1935, revealed that the coefficient k is inversely proportional to the Manning roughness coefficient and m is within the range 1.5~1.8 [31]. These early results are close to the Meyer-Peter formula.

9.2.3 Comparison of Sharmov, Levy, and Gongcharov formulas

The formulas based on velocity have similar structures so that they can be compared directly. Fig. 9.19 presents a comparison of the three formulas for h=15 cm, γ_s =2.65 t/m³, T=20°C and for mean velocities of 0.5, 1.0 and 1.5 m/s. The Sharmov and Levy formulas are rather close to each other, but the Gongcharov formula gives a higher transport rate than do the others because it includes some suspended load.

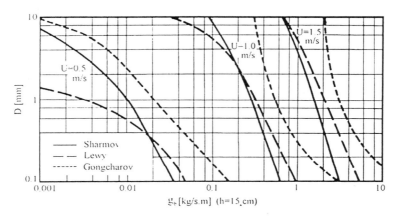

Fig. 9.19 Comparison of Sharmov, Levy, and Gongcharov formulas

9.3 LAWS OF NON-UNIFORM BED LOAD MOTION

The foregoing are only for uniform sediment. However, sediment in natural rivers is always non-uniform and the laws of bed load motion needed to describe its motion are much more complicated because of the consequences of having particles of various sizes.

Two techniques are used to deal with non-uniform bed load motion. If only the total bed load transport rate is required, the bed load formulas introduced in the Section 9.1 can be used directly, but a representative diameter must be determined. If instead, the transport rates of various diameters are required, the mutual effects of the various particle sizes must be studied.

9.3.1 Determination of representative diameter

The use of the uniform bed load formulas to estimate the transport rate of non-uniform bed load requires the determination of a representative diameter. Einstein found from data measured in both small streams and flume experiments that D_{35} can be used as the diameter in the bed load formulas. D_{35} stands for the diameter for which 35% of the bed material is finer [32]. Meyer-Peter suggested another form of representative diameter :

$$D_m = \frac{\sum D_i \Delta p_i}{100}$$

(9.92)

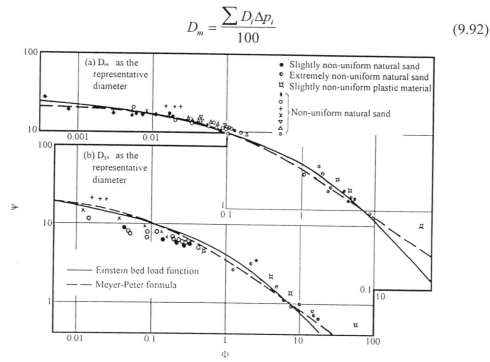

Fig. 9.20 Comparison of measured bed load transport rates for non-uniform sediment with results calculated using different representative diameters

in which Δp_i stands for the percentage of particles of diameter D_i in the bed material [2]. The senior author of the book examined the two representative diameters and the results are presented in Fig. 9.20 [26]. The results show that D_m is preferable to D_{35} for low intensities of bed load motion, but he found no difference between the two for high intensities. Essentially, if the flow intensity is much higher than the threshold value, the choice of a diameter to use in the formulas makes little difference. This result is clearly shown by the Meyer-Peter formula. For high flow intensities, the term $a_4 (\gamma_s - \gamma) D$ in Eq. (9.11) is negligible, and the other two terms do not contain the sediment diameter. Therefore, the calculated bed load transport rates are the same for different diameters. Of course, the bed load transport rate depends in an essential way on the composition of bed material because the latter affects the flow pattern near the bed. Still, this dependency is indirect for high intensities of transport.

The representative diameter of non-uniform sediment should be selected according to the problem that is to be studied. For a study of river bed roughness, a larger representative diameter should be used because coarse particles constitute the main component of bed roughness. But for a study of sediment movement, a smaller representative diameter should be selected because the moving particles are usually finer than the bed material, and they are more likely to be transported by the flowing water.

9.3.2 Bed load transport rates of various grain sizes

Many engineering situations require not only the calculation of total bed load transport rate but also estimates of transport rates of the various grain sizes of a non-uniform sediment. Major hydraulic projects can destroy the equilibrium between sediment transport and the channel bed and induce changes in the composition of the river bed. For example, bed composition usually undergoes a fining process in the upstream reaches of a dam, whereas an armoring layer develops in downstream reaches. The mechanism of the river responses can not be well understood without studying the adjustment of bed composition according to the flow conditions, and the fining or coarsening processes along the river bed can be evaluated only if the movement of the various sizes is adequately known.

Few researchers have studied the movement of various sizes of non-uniform sediment because the mutual interactions between the various sizes are complex. Some preliminary results of Einstein and Chien are presented in this section.

9.3.2.1 Laws of movement for slightly non-uniform sand

Sediment uniformity is usually indicated by the following sorting coefficient

$$S_0 = \sqrt{\dfrac{D_{75}}{D_{25}}} \qquad (9.93)$$

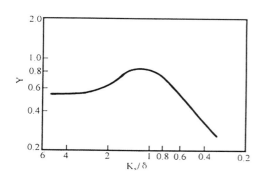

Fig. 9.21 Modification factor of lift force for non-uniform sediment as a function of K_s/d

The bed material is uniform if S_0 is equal to one. The larger the coefficient is, the more non-uniform the material. For slightly non-uniform sediment with S_0 only a little more than one, Einstein found that his bed load function can be used if the following modifications are made:

1. The weight of sediment particles with diameter D falling on a unit bed area per unit time is proportional to the percentage composition i_b in bed load. That is, the settling rate of particles with the diameter D per unit bed area is

$$Settling\,rate = \dfrac{i_b g_b (1-p)}{\lambda\,D} \qquad (9.33a)$$

2. The weight of sediment particles of diameter D scoured away from a unit bed area per unit time depends on the probability that the sediment is lifted by the flow (not the same for different grain sizes), and also on the number of particles of that diameter on the bed; the latter number can be taken to be equal to the fraction of the particles in the composition of bed material. If i_b is the percent of particles with diameter D in bed material, then the scouring rate of such particles from unit area is

$$Scouring\,rate = \dfrac{A_2}{A_1 A_3} i_0 p \gamma_s g^{1/2} D^{1/2} \left(\dfrac{\gamma_s - \gamma}{\gamma} \right)^{1/2} \qquad (9.36a)$$

3. In the expression for the lift force Eq. (9.40), the flow velocity at $0.35X$ [definition of X is given by Eqs. (9.94 and 95)] should be taken as u, the lift coefficient is $0.178/Y$, in which Y is a modification factor and a function of K_S/δ as shown in Fig. 9.21, and the fluctuation of the lift force follows the normal distribution and its coefficient of standard deviation remains unchanged. Therefore, the lift force is written as

$$F_L = \dfrac{0.178}{Y} A_1 D^2 \dfrac{1}{2} \rho 5.75^2 R_b' J g \log^2 \left(10.6 \dfrac{X}{\Delta} \right)(1 + \eta) \qquad (9.41a)$$

4. Non-uniform bed material is viewed as an entity in evaluating the lift force. However, the particles of different sizes are shielded differently. The differences are presented for two cases. The first case is for a smooth bed so that the fine particles,

hiding within the laminar sublayer near the bed, are not often affected by the turbulence of the flow. The second case is for a rough boundary for which the fine particles in the wakes behind the coarse particles that protrude from the bed surface do not contact the main flow. X is the maximum diameter subjected to the shielding effects in the bed composition. Then for the second case,

$$X = 0.77\Delta \quad (\Delta/\delta > 1.8) \tag{9.94}$$

and for the first case

$$X = 1.39\delta \quad (\Delta/\delta < 1.8) \tag{9.95}$$

in which Δ is equal to K_S/X in the formula for the logarithmic velocity distribution (7.38).

For the particles finer than X, the lift force given by Eq. (9.41a) should be divided by a shielding factor ξ, which is a function of D/X shown by curve A in Fig. 9.22. The two shielding effects, the laminar sublayer and eddies behind coarse particles, can not be separated at present and the corrections due to them are represented by a single curve. The correction factor ξ is equal to one for D/X larger than one. If D/X is smaller than one, the curve is a line with a slope of -2 on a log-log plot except for a smooth transition region near $D/X=1$. With this correction factor, the lift force is written as

$$F_L = \frac{0.178}{Y\xi} A_1 D^2 \frac{1}{2}\rho 5.75^2 R_b' Jg \log^2\left(10.6\frac{X}{\Delta}\right)(1+\eta) \tag{9.41b}$$

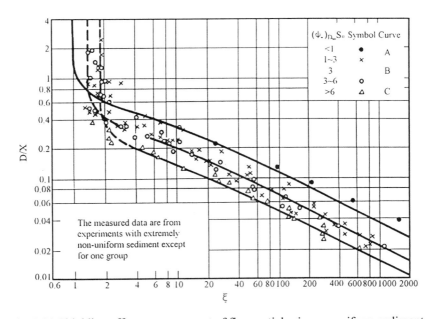

Fig. 9.22 Shielding effect on movement of fine particles in non-uniform sediment

5. If Eqs. (9.36) and (9.41) replaced by Eqs. (9.36a) and (9.41b) and modified as in the derivation of the Einstein bed load function, Section 9.1, the following formula is obtained :

$$1 - \frac{1}{\sqrt{\pi}} \int_{-B*\psi*-1/\eta_0}^{B*\psi*-1/\eta_0} e^{-t^2} \, dt = \frac{A_* \Phi_*}{1 + A_* \Phi_*} \tag{9.96}$$

and it is suitable for various group of grain sizes: in it

$$\left. \begin{aligned} \Phi_* &= \frac{i_b}{i_0} \Phi \\ \Psi_* &= \xi Y \left(\frac{\beta^2}{\beta_X^2} \right) \Psi \end{aligned} \right\} \tag{9.97}$$

$$\beta_x = \log \left(10.6 \frac{X}{\Delta} \right) \tag{9.98}$$

9.3.2.2 Transportation of extremely non-uniform sediment [33]

The particles used in the flume experiments by Einstein were nearly uniform; their sorting coefficients fell mostly within the range from 1.13 to 1.35; a few were larger, with the largest being 1.67. Such sediment moved more or less like uniform sediment and no sorting took place. The composition of the bed surface did not differ from the main bed material. In 1950 to 1952, the first author of the book carried out experiments with eight different kinds of non-uniform sediments with sorting coefficient of 1.18 for one kind and 1.61 to 2.45 for the other seven kinds. One of the non-uniform sediments was a mixture of particle sizes ranging from 0.005 mm to 4mm. Experiments with the sediments revealed that the bed material was sorted by the flow, with large and small particles gathering at different places. Thus, the effective quantities of various size-groups of bed material changed. Because the experiment involved very fine particles, the movement mechanism of sediment with grain Reynolds number less than 3.5 could be studied for the first time.

The sorting of bed material refers to the separation of coarse sediment particles with low mobility from fine particles with high mobility during flow, and it causes different sizes to gather in different layers. Fig. 9.23 shows the size distributions of the moving sediment and the sediment in different layers beneath the bed surface in such an experiment. Curve A is the original size distribution of the sediment put on the bed before the experiment. Fine sediment was washed away from the bed by the flow during the experiment and its size distribution is shown as curve B. Then, the bed material was composed of two layers; the covering layer was fine with the size distribution shown as curve C, and the second layer was coarse and is shown as curve D.

If no sorting occurs, coarse particles protrude from the bed with fine particles around them. Even if a laminar sublayer develops on a smooth bed, some coarse particles may extend clear through it. These particles form shelter zones in which fine sediment could not be easily removed by the flow. Eq. (9.94) represents the sheltering effect due to coarse particles. Even for a smooth boundary with a laminar sublayer, coarse sediment can play a role in the shielding effects, which is included in Eq. (9.94). If sorting occurs, coarse sediment is covered by a layer of fine sediment, so that the shielding zones due to coarse sand are much fewer and the sheltering effect on the movement of fine particles is also less.

One sees that the shielding effects on the movement of fine particles in non-uniform sediment depend on the extent of sorting. A factor associated with the degree of sorting of the bed material should be introduced, as shown in Fig. 9.22. This parameter should include the degree of scatter of bed material sizes, and can also reflect the relatively low intensities of motion for coarse particles. The factor that was chosen is the following:

$$\left(\Psi_*\right)_{D_{90}} S_0 = \frac{\gamma_s - \gamma}{\gamma} Y \frac{D_{90}}{JR_b'} \sqrt{\frac{D_{75}}{D_{25}}} \tag{9.99}$$

In Fig. 9.22, curve A is essentially for values of $\left(\Psi_*\right)_{D_{90}} S_0$ less than one, as initially selected by Einstein. For larger value of $\left(\Psi_*\right)_{D_{90}} S_0$, the sorting of bed material is more pronounced and the curve for $D/X \sim \xi$ is lower. The distance below curve A is larger if the value of $\left(\Psi_*\right)_{D_{90}} S_0$ is higher. However, the sorting of bed material reaches a limit and the curve of $D/X\text{-}\xi$ does not change if $\left(\Psi_*\right)_{D_{90}} S_0$ is greater than six.

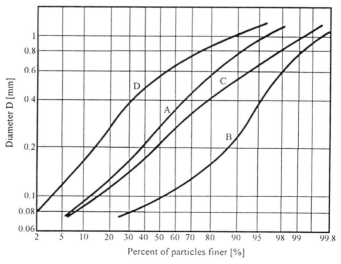

Fig. 9.23 Sorting phenomenon of bed sediment

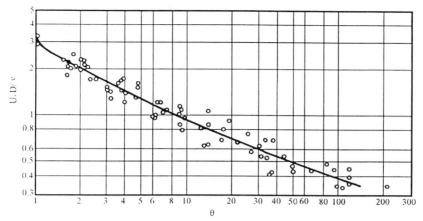

Fig. 9.24 Correction coefficient for lift as a function of grain Reynolds number

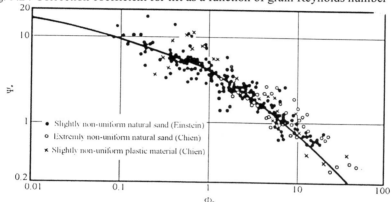

Fig. 9.25 Comparison of non-uniform bed load formula
with data measured by Einstein and Chien

The finer the bed material, the stronger is the shielding effect on its movement. But if the grain Reynolds number $(=U_*D/v)$ is less than 3.5, the trend is quite different because the fine sediment is completely within laminar flow region. The effective velocity acting on the particles is small and the coefficient of lift varies. In such cases, Eq. (9.41b) underestimates the lift and should be modified by a coefficient θ, therefore

$$F_L = \frac{0.178\theta}{Y\xi} A_1 D^2 \frac{1}{2}\rho 5.75^2 R_b' Jg \log^2\left(10.6\frac{X}{\Delta}\right)(1+\eta) \qquad (9.41c)$$

The coefficient θ is a function of the grain Reynolds number, as shown in Fig. 9.24. It equals 1 if U_*D/v exceeds 3.5 and increases steeply with decreasing grain Reynolds number, following a line with a slope of -2 in the diagram, if the Reynolds number is smaller than 3.5.

The bed load function Eq. (9.96) can be used for extremely non-uniform sediment, but only after some modifications; the flow parameter Ψ_* is redefined as

401

$$\Psi_* = \frac{\xi Y (\beta / \beta_x)^2}{\theta} \Psi \qquad (9.97a)$$

Fig. 9.25 is a comparison of measured transport rates for various sizes of non-uniform sediment with this formula; experimental results of plastic material with a specific gravity of 1.05 are included. The conformity between the formula and the data is satisfactory although the data points scatter because of the complexity of the phenomenon and the various difficulties encountered in the measurements. Light material is often used in physical model tests to simulate natural sand; therefore, a sediment transport equation is useful in designs based on model studies only if it is valid for both natural sand and light material. The foregoing method meets this requirement.

Suspended load motion may occur even under medium flow intensity if the bed material contains many fine particles. It can be difficult to distinguish between bed load and suspended load. Because exchanges occur between bed load and suspended load, Einstein took the average concentration of the bed load layer as a boundary condition for suspended load motion. He combined bed load motion and suspended load motion and derived a unified formula to calculate the total sediment transport rate (detailed discussion in Chapter 11). With his approach, the bed load transport rate can be calculated from the measured total transport rate. The data points in Fig. 9.25 were obtained in this way rather than by direct measurement of bed load transport. Recently, some researchers measured the single step length, thickness of bed load layer and migration velocity of bed material by means of high speed photography. The results run counter to some of the deductions of Einstein and Bagnold [34]. It is still too early to interprete the results because no systematic data are available.

REFERENCES

[1] DuBoys, P. "Etudes du Regime du Rhone et l'Action Exercee par des Eaux sur un lit a Fond de Graviers Indefiniment Affouillable." *Annales des Ponts et Chausses, Ser. A*, Vol. 18, 1879, pp. 141-195.

[2] Meyer-Peter, E., and R. Muller. "Formula for Bed Load Transport." *Proc., 2nd Meeting, Intern. Assoc. Hyd. Res.*, Vol. 6, 1948.

[3] Meyer-Peter, E., H. Favre, and H.A. Einstein. "Neuere Versuchsresultate uber den Geschiebetrieb." *Schweiz Bauzeitung*,Vol. 103, No. 12, 1934, pp. 147-150.

[4] Du, Guohan, Yunze Peng, and Deyi Wu. "Reconstruction of the Dujiangyan Project and Its Gravel Bed Load Problem." *J. of Sediment Research*, 1980, pp. 12-22. (in Chinese)

[5] Bagnold, R.A. "An Approach to the Sediment Transport Problem from General Physics." *U.S. Geol. Survey*, Prof. Paper No. 442-I, 1966, p. 37.

[6] Bagnold, R.A. "Experiments in a Gravity-free Dispersion of Water Flow." *Proc. Royal Soc. London*, Ser. A, Vol. 225, 1954.

[7] Bagnold, R.A. "The Nature of Saltation and of Bed Load Transport in Water." *Proc. Royal Soc., London*, Ser. A, Vol. 332, 1973, pp. 473-504.

[8] Coleman, N.L. "The Drag Coefficient of a Stationary Sphere on a Boundary of Similar Spheres." *La Houille Blanche*, No. 1, 1972, pp. 17-21.

[9] Francis, J.R.D. "Experiments on the Motions of Solitary Grain along the Bed of a Water Stream." *Proc. Royal Soc., London*, Ser. A, Vol. 332, 1973, pp. 443-471.

[10] William, G.P. "Flume Width and Water Depth Effects in Sediment Transport Experiments." *USGS Prof. Paper No. 562-H*, 1970.

[11] Bagnold, R.A. "Bed Load Transport by Natural Rivers." *Water Resources Research*, Vol. 13, No. 2, April 1977, pp. 303-312.

[12] Einstein, H.A. "Formula for the Transportation of Bed Load." *Trans. Amer. Soc. Civil Engrs.*, Vol. 107, 1942, pp. 561-597.

[13] Einstein, H.A. "The Bed Load Function for Sediment Transport in Open Channel Flows." *U.S. Dept. Agri., Tech. Bull. 1026*, 1950, p. 71.

[14] Einstein, H.A. "Der Geschiebetrieb als Wahrscheinlichkeits Problem." *Mitl. Versuchanst. fuer Wasserbau, au die Eidg.* Tech. Hochschule in Zurich, Verlag Rascher & Co. Zurich, 1937.

[15] Einstein, H.A., and A. El-Samni. "Hydrodynamic Forces on a Rough Wall." *Rev. Modern Physics*, Vol. 21, 1949, pp. 520-524.

[16] Velikanov, M.A. *Fluvial Processes of Rivers*. Physics Mathematics Literature Press, Moscow, 1958. (in Russian)

[17] Dou, Guoren. "Laws of Bed Load Motion." *Report Series (River and Port)*, Nanjing Institute of Hydraulic Research, 1964 (in Chinese).

[18] Yalin, M.S. *Mechanics of Sediment Transport*. Pergamon Press, 1972, p.290.

[19] Englund, F., and J. Fredsφe. "A Sediment Transport Model for Straight Alluvial Channels." *Nordic Hydrology*, Vol. 7, 1976, pp. 293-306.

[20] Mantz, P.A.. " Low Sediment Transport Rate over Flat Beds." *J. of Hyd. Div., Proc., Amer. Soc. Civil Engineers*, Vol. 106, No. HY7, 1980, pp. 1173-1190.

[21] Ackers, P., and W.R. White. "Sediment Transport: New Approach and Analysis." *J. of Hyd. Div., Proc.* ASCE, Vol. 99, No. HY11, 1973, pp. 2014-2060.

[22] Sharmov, G.I. *River Sedimentation*. Hydrology and Metrology Press, Leningrad, 1959, pp. 84-93.

[23] Levy, I.I. *River Dynamics*. National Energy Resources Press, Moscow, 1957, pp.127-140.

[24] Gongcharov, B.N. *River Dynamics*. Hydrology and Metrology Press, Leningrad, 1962, pp. 226-236 and p. 252. (in Russian)

[25] Chien, Ning. "Comparison of Bed Load Formulas." *Chinese J. of Hydraulic Engineering*, 1980, No. 4, pp. 1-11. (in Chinese)

[26] Chien, Ning. "Meyer-Peter Formula for Bed Load Transport and Einstein Bed Load Function." *Missouri River Div. Sediment Series No. 7*, Missouri River Div., Corps Engrs, 1954, p. 23.

[27] Wilson, K.C. "Bed Load Transport at High Shear Stress." *J. of Hyd. Div.*, ASCE, Vol.92, No. HY6, 1966, pp.49-59.

[28] Engelund, F. "Transport of Bed Load at High Shear Stress." *Progress Report No.53,* Institute of Hydrodynamics and Hydraulic Engineering, Tech. Univ. of Denmark, 1981, pp. 31-35.

[29] Bagnold, R.A. *The Physics of Blown Sand and Desert Dunes*. Methuen and Co., London, 1941.

[30] O'Brien, M.P., and R.D. Rindlaub. "The Transportation of Bed Load by Streams." *Trans. American Geophas. Union*, Vol. 15, Pt. 2, 1934, pp. 593-603.

[31] U.S. Waterway Experiment Station. "Studies of River Bed Materials and Their Movement, with Special Reference to the Lower Mississippi River." *Paper 17*, U.S. Waterway Experiment Station, 1935, p. 161.

[32] Einstein, H.A. "Bed Load Transportation in Mountain Creek." *Tech. Paper, No. 55*, Soil Conservation Service, U.S. Dept. of Agriculture, 1944, p. 54.

[33] Einstein, H.A., and Ning Chien. "Transport of Sediment Mixture with Large Ranges of Grain Sizes." *Missouri River Div. Sediment Series No.2*, Missouri River Div., U.S. Corps Engrs, 1953, p. 49.

[34] Luque, R.F., and R. van Beek. "Erosion and Transport of Bed Load Sediment." *J. of Hyd. Div.*, Intern. Assoc. Hyd. Res., Vol.14, No. 2, 1976.

CHAPTER 10

MOTION OF SUSPENDED SEDIMENT

Diffusion of turbulent flow results in exchanges of both momentum and sediment particles between layers of the flow. If the fall velocity of a sediment particle is small enough, sediment can move in suspension. In this chapter, an extension of the treatment in Chapter 4, the process of sediment suspension is presented on the basis of an analysis of the burst behavior of turbulence and the trajectories of sediment particles. Then, the diffusion equation of sediment suspension is derived according to the classical theory of turbulence. In addition, the vertical distribution and streamwise variation of suspended sediment concentration are discussed in terms of a simplified diffusion equation for two special cases. The sediment discharge per unit width can be readily computed if the vertical profiles of the sediment concentration and the flow velocity are known. In recent years, more and more attention is being paid to environmental protection. Actually, the diffusion of pollutants in water bodies is essentially the same as that of suspended sediment, except that in the former case, three dimensional effects are more often included because of the three dimensionality of the flows involved. Finally, the dispersion of pollutants, an important addition to their diffusion, is also treated.

10.1 BURSTS OF TURBULENCE AND THEIR ROLE IN SEDIMENT SUSPENSION

In Chapter 4, the burst phenomenon was characterized as an alternation of the lifting of bands of low-velocity fluid from the bed, followed by the downward sweep of a band of flow with high-velocity fluid from the main flow to the bed. The effect of turbulence bursting on sediment suspension has been demonstrated experimentally.

10.1.1 Experimental conditions

In 1978, Sumer and Oguz observed both the turbulence structure and the motion of sediment near the bed [1]. The following year, Sumer and Deigaard extended these observations and obtained additional results [2].

The latter experiment was conducted in a flume using both smooth and rough beds. In the latter case, particles of gravel with a size of 3.6 mm were glued to the bed, a roughness characterized by $KU_* / v = 81$, in which K stands for the mean distance from the top of the gravel to the bed. In the experiments, a single plastic particle that had been specially made was placed on the bed, and its trajectory was recorded by means of a camera mounted above the flume that moved at a speed equal to the mean flow velocity. Photographs were taken with stroboscopic lighting using an inclined mirror to also give the side view. From an analysis of the photos, the trajectory of the particle and the instantaneous velocity components of the particle in the longitudinal and vertical directions were obtained.

Table 10.1 Property of trace particles used by Sumer and Deigaard
(after Sumer, B.M., and R. Deigaard)

bed surface	Property of plastic trace particle			
	type	diameter(mm)	specific weight	settling velocity (cm/s)
smooth	B	3.0	1.0075	1.33
	C	3.1	1.0258	3.06
rough	D_1	3.0	1.0090	1.49

Since the trace particles were to be carried in suspension, yet should be large enough to protrude from the layer near the bed, large plastic spheres were used. The diameter, specific weight, and settling velocity of the particles are listed in Table 10.1, and they were used for flows for both smooth and rough boundaries. In the experiment, the flow depth was kept nearly constant and varied only from 68 cm to 70 cm, and the mean velocity ranged between 19.1 cm/s and 29.8 cm/s. The initial position of the plastic particle on the bed surface was the same for tests with both smooth and rough beds.

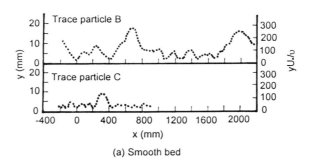

(a) Smooth bed

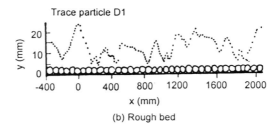

(b) Rough bed

Fig. 10.1 Trajectories of sediment particles in turbulence bursts
(after Sumer, B.M., and B. Oguz)

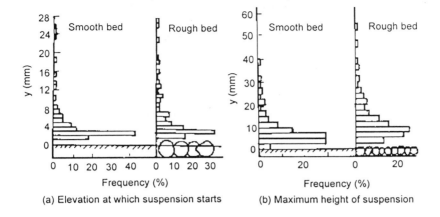

(a) Elevation at which suspension starts

(b) Maximum height of suspension

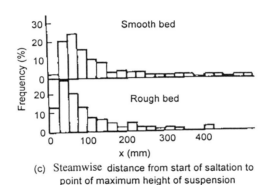

(c) Steamwise distance from start of saltation to point of maximum height of suspension

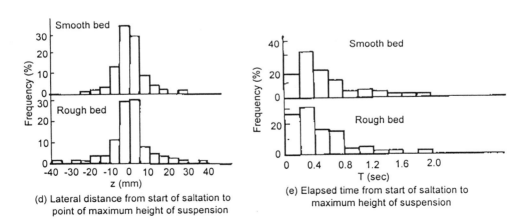

(d) Lateral distance from start of saltation to point of maximum height of suspension

(e) Elapsed time from start of saltation to maximum height of suspension

Fig. 10.2 The probability density distribution of the characteristic quantities for the trajectories of particles during uplift (after Sumer, B.M., and B. Oguz)

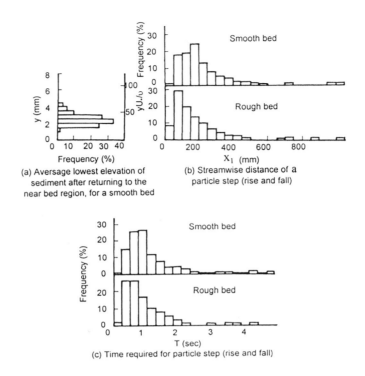

Fig. 10.3 Probability density distribution of characteristics of trajectories of falling sediment particles (after Sumer, B.M., and B. Oguz)

10.1.2 Trajectory of sediment particle

Fig. 10.1 shows the trajectory of three different trace particles, each point representing the position of the trace particle at intervals of 0.086 seconds. The vertical scale is greatly enlarged in the figure. The figure shows that the trace particle C moved in saltation, with the saltation height being no larger than twice the particle diameter. The B and D_l particles moved in suspension, with their trajectories showing both upward and downward movements. The downward motion can cause them to reach the bed region and even to contact the bed surface (for a smooth bed, Sumer defined the near bed region as $5 < yU_*/v < 70$).

10.1.3 Process from initiation of sediment suspension until particle reaches highest level

Figs.10.2 and 10.3 show the behavior of suspended particle moving in turbulent flow due to burst action, and Table 10.2 lists some of characteristic values. From the figures and table, the following features become clear.

1. A sediment particle moves into suspension quite near the bed. For a smooth bed, the average elevation from which suspension begins corresponds to $yU_*/v = 59$; that is, at an elevation within the near bed layer. For a rough bed, the region from which suspension begins is larger, but it is less than 15% of the flow depth for the test conditions.

2. For a smooth bed, the average maximum height the sediment reached corresponded to $\bar{y}U_* / v = 12.5$. For a rough bed, particles could go higher because the lift acting on a saltating particle was greater. Occasionally, particles came close to the water surface.

Table 10.2 Characteristics of suspended particles during burst action in turbulent flow (after Sumer, B.M., and B. Oguz)

(1) Lifting process

bed type	elevation at which suspension starts(mm)		maximum height of suspension (mm)		statistics for lifting process from the start of saltation to highest position:						
					streamwise distance (mm)		lateral distance (mm)		required time (sec)		
	$\bar{y}$	σ_y	$\bar{y}$	σ_y	$\bar{x}$	σ_x	$\bar{z}$	σ_z	$\bar{T}$	σ_T	$\bar{T}_{u\,max} / h$
smooth bed	3.9	3.8	8.1	6.1	106.1	85.2	-0.3	6.6	0.508	0.356	2.4
rough bed	5.3	4.5	11.6	7.7	84.2	74.2	-0.3	9.1	0.436	0.324	2.0

(2) Falling process, after reaching the point of maximum height

bed type	lowest elevation after returning to the near bed (mm)			statistics for the process from the start of saltation to the return to the bed					
				streamwise distance (mm)			duration (sec)		
	$\bar{y}$	σ_y	$\bar{y}U_*/v$	$\bar{x}_1$	σ_{x_1}	$\bar{x}_1 / h$	$\bar{T}_1$	σ_{T_1}	$\bar{T}_{1u\,max}$
smooth bed	2.5	0.6	38	211.6	170.8	3.1	0.972	0.663	4.6
rough bed	--	--	--	176.3	157.2	2.5	0.853	0.577	4.0

Note: σ_y is the standard deviation of y, with similar usage for other quantities

3. Sediment particles advance with the flow, and as they do so, they also move laterally. The streamwise distance from the beginning of saltation to the maximum height of suspension is shorter for a rough bed than for a smooth one, and so is the time required, T. For a smooth bed, $\overline{T}_{u\max}/h=2.4$ (u_{max} represents the maximum flow velocity over the depth). This result is consistent with Jackson experimental finding that the time duration of a turbulence burst, expressed in dimensional form, is 2.3 for a smooth bed, and 1.3 for a bed with small-scale ripples or dunes [3].

4. For a smooth bed, 77% of the suspended particle fell directly back to the near bed region after reaching their initial maximum level; the other 23% entered the eddies above the near bed region and were lifted up for a second cycle of suspension. The average lowest elevation of sediment particles after returning to the near bed region corresponds to a value of 38 for yU/v. The average streamwise distance of a single step for a suspended particle, from start to return is 3.1 times the flow depth; the time duration corresponds to a value of 4.6 for the ratio $\overline{T}_{1u\max}/h$, a value of the same order as the average cycle of bursting, which is 5.0. For a rough bed, the suspended particle also returns to the bed region, but both the length of a single step and the time required are shorter.

10.1.4 Velocity of sediment motion

Fig.10.4 shows the streamwise and vertical velocities of sediment particles

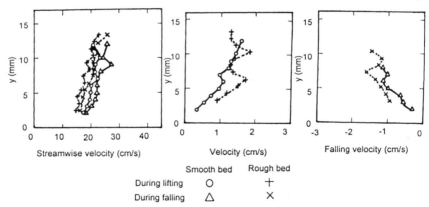

Fig. 10.4 Velocity of sediment particles in the lifting and falling process

during the processes of lifting and falling. The streamwise velocity is about the same as the flow velocity; the velocity near a rough bed is smaller than that near a smooth one, and the resistance of the former is larger. An ascending particle, which is affected by the high-speed streak of flow due to the bursting, has a lower velocity than a descending particle; the latter is affected by the higher speed in the region of flow. The vertical velocity of a particle ascending from a rough bed is larger than that from a smooth bed because the bursting intensity of the high-speed streak of flow is greater. For a descending particle, the results are reversed.

10.1.5 Mechanism of sediment moving in suspension

The observations of the trajectories of sediment particles and velocity distributions presented in Chapter 4 and the measurements of the burst phenomena of turbulent flow allow one to summarize the mechanism of sediment moving in suspension.

Table 10.3 Trajectory and mechanism of sediment moving
in suspension over smooth bed

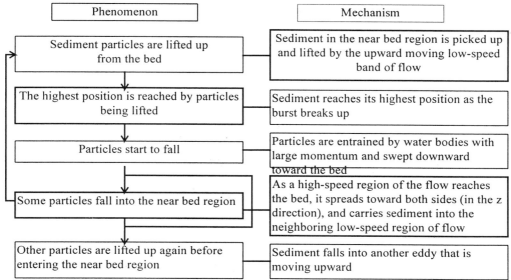

Table 10.3 shows the observed phenomena and the related mechanism of particle suspension for a smooth bed. If the low-speed streak of flow due to a burst of turbulence results in a lifting motion near the bed, the sediment there is carried upward. If the fall velocity of a particle is large, the particle will quickly fall back through the low-speed streak of the flow, and quickly return to the bed. Such particles are part of the saltation load. If, in contrast, the fall velocity is small, the sediment can be carried upward along with the low-speed water element until the latter breaks up; at that moment the sediment has reached its highest position and begins to settle back down. As the particles fall, some of them, caught in the downward moving part of the high-speed streak of flow, will return to the near bed region, while others, caught in an upward moving eddy, are lifted again. The higher the turbulence intensity and the smaller the particle size, the greater is the portion of the particles that are lifted up. The high-speed streak of flow sweeping the bed spreads out to both sides, and a new upward low-speed zone forms between the two high-speed zones (Fig. 4.13). A particle falling down in the high-speed streak of flow can also be pushed toward one side or the other before reaching the bed. There, it enters another low-speed zone, and subsequently starts a new cycle of suspension. In

this way, some sediment is kept in suspension; however, in the process, a continuous exchange occurs between the suspended sediment and the sediment in the near-bed region.

The trajectories and mechanisms of particles moving in suspension over a coarse bed are essentially the same as those over a smooth one, except that the region where suspension starts is broader, the lifting velocity is stronger, and the maximum height the particles attain is greater. The average elevation of the highest position of the particles is related to the nature of the flow near the bed; it is dependent on the ratio v/U^* for a smooth bed, and to the height of the roughness K for a rough bed.

The experiments of Sumer et al. were conducted under quite simple conditions. A fixed bed was used to study the suspension of single particles, the particles used were much larger than the thickness of near bed layer. For a movable bed, the way turbulence bursting affects sediment suspension is not yet fully understood. One can speculate that, in the process of the downward sweeping of the zone of high-speed flow followed by the lifting of the low-speed flow, most of the sediment would come from particles moving as bed load; however, a certain amount of the sediment could come directly from the bed material lying at the top of the bed surface. That is, the major exchange would occur between suspended load and bed load rather than the exchange between suspended load and bed material.

10.2 DIFFUSION EQUATION OF SEDIMENT MOTION

As a simple demonstration of diffusion, some dyed water is injected into a two-dimensional flow at the instant t_1, as shown in Fig. 10.5. As the dyed element is carried forward by the mean flow, it spreads both vertically and laterally due to the diffusion of turbulent flow. At a later time t_3, the centroid of the dyed water element reaches cross section AB. For a water element of length Δx, height Δy, and unit thickness, the inflow and outflow volumes of the dyed element in the time interval Δt can be expressed as shown in Fig. 10.5b. If the dye solution has a specific weight larger than that of water, they will move downward with some velocity ω. From the law of mass conservation, the concentration of the dyed water in the element varies with time if the inflow and outflow rates are not equal as follows

$$\left[-\frac{\partial}{\partial x}(uS_v) - \frac{\partial}{\partial y}(vS_v) + \omega \frac{\partial S_v}{\partial y} \right] \Delta x \cdot \Delta y \cdot \Delta t = \frac{\partial S_v}{\partial t} \Delta x \cdot \Delta y \cdot \Delta t \qquad (10.1)$$

in which u and v are the x- and y-components of the instantaneous velocity, respectively, and S_v is the instantaneous concentration of the dyed water element by volume.

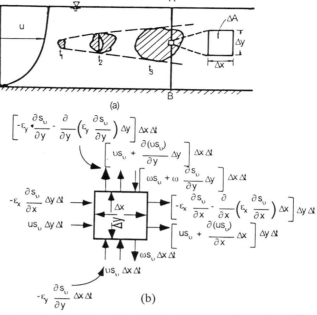

(a)

$$\left[-\varepsilon_y \cdot \frac{\partial s_\upsilon}{\partial y} - \frac{\partial}{\partial y}\left(\varepsilon_y \frac{\partial s_\upsilon}{\partial y}\right)\Delta y \right]\Delta x\,\Delta t$$

$$\left[\upsilon s_\upsilon + \frac{\partial(\upsilon s_\upsilon)}{\partial y}\Delta y \right]\Delta x\,\Delta t$$

$$\left[\omega s_\upsilon + \omega \frac{\partial s_\upsilon}{\partial y}\Delta y \right]\Delta x\,\Delta t$$

$$-\varepsilon_x \frac{\partial s_\upsilon}{\partial x}\Delta y\,\Delta t$$

$$\upsilon s_\upsilon\,\Delta y\,\Delta t$$

$$\left[-\varepsilon_x \frac{\partial s_\upsilon}{\partial x} - \frac{\partial}{\partial x}\left(\varepsilon_x \frac{\partial s_\upsilon}{\partial x}\right)\Delta x \right]\Delta y\,\Delta t$$

$$\left[\upsilon s_\upsilon + \frac{\partial(\upsilon s_\upsilon)}{\partial x}\Delta x \right]\Delta y\,\Delta t$$

$$\omega s_\upsilon\,\Delta x\,\Delta t$$

$$\upsilon s_\upsilon\,\Delta x\,\Delta t$$

$$-\varepsilon_y \frac{\partial s_\upsilon}{\partial y}\Delta x\,\Delta t \qquad \text{(b)}$$

Fig.10.5 Diffusion of dyed liquid injected in two-dimensional flow

For turbulent flow, the velocity and concentration fluctuate with time. Fig. 10.6a shows the vertical distribution of u and S_v at section AB, and Fig. 10.6b their variation with time. The following expressions are useful:

$$u = \bar{u} + u'$$

$$S_v = \bar{S}_v + S_v'$$

$$v = \bar{v} + v'$$

The bar indicates a time-average value, and the prime a fluctuating component. The three expressions can be substituted into Eq. (10.1), and their average taken

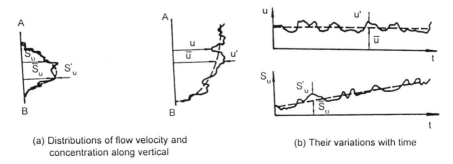

(a) Distributions of flow velocity and concentration along vertical

(b) Their variations with time

Fig. 10.6 Distribution of flow velocity and concentration along vertical (a) and for their variations with time (b) for section AB

413

over a long period of time. Because the time average of the fluctuating quantities is equal to zero[1], the flow is two-dimensional

$$\frac{\partial \bar{S}_v}{\partial t} = -\bar{u} \frac{\partial \bar{S}_v}{\partial x} - \bar{S}_v \frac{\partial \bar{u}}{\partial x} - \frac{\overline{\partial(u'S_v')}}{\partial x} - \frac{\overline{\partial(v'S_v')}}{\partial y} + \omega \frac{\partial \bar{S}_v'}{\partial y}$$

For uniform flow,

$$\frac{\partial \bar{u}}{\partial x} = 0$$

So that, the equation becomes

$$\frac{\partial \bar{S}_v}{\partial t} = -\bar{u} \frac{\partial \bar{S}_v}{\partial x} - \frac{\overline{\partial(u'S_v')}}{\partial x} - \frac{\overline{\partial(v'S_v')}}{\partial y} + \omega \frac{\partial \bar{S}_v}{\partial y} \qquad (10.2)$$

In Chapter 4, the Prandtl mixing-length theory was introduced. One can assume that in turbulent flow several fluid elements move randomly. The motion of each element starts with a flow behavior characteristic of its initial position. As the fluid body moves, the characteristics of the body remain comparatively unchanged over a distance l, one mixing length, then the fluid mixes with the local fluid, loses its initial characteristics, and behaves like any other element of the local flow. The change in the value of any property of the fluid element between its initial and final position is a temporal fluctuation. According to this concept,

$$S_v' = -l_y \frac{\partial \bar{S}_v}{\partial y}$$

From the definition of the turbulent diffusion coefficient (4.50)

$$\varepsilon_y = \sqrt{\overline{v'^2}}\, l_y$$

Hence,

$$\overline{v'S_v'} = -\varepsilon_y \frac{\partial \bar{S}_v}{\partial y} \qquad (10.3)$$

And similarly,

---

[1] Actually, $\bar{v} = 0$ only if the particle size and the sediment concentration are small; this aspect is treated in section 7.

$$\overline{u'S_v'} = -\varepsilon_x \frac{\partial \overline{S}_v}{\partial x} \tag{10.4}$$

in which, ε_x and ε_y stand for the diffusion coefficients of sediment motion in the streamwise and vertical directions, respectively.

Substitution of Eqs. (10.3) and (10.4) into Eq. (10.2) yields

$$\frac{\partial \overline{S}_v}{\partial t} = -\overline{u}\frac{\partial \overline{S}_v}{\partial x} + \varepsilon_x \frac{\partial^2 \overline{S}_v}{\partial x^2} + \frac{\partial \varepsilon_x}{\partial x}\frac{\partial \overline{S}_v}{\partial x} + \varepsilon_y \frac{\partial^2 \overline{S}_v}{\partial y^2} + \frac{\partial \varepsilon_y}{\partial y}\frac{\partial \overline{S}_v}{\partial y} + \omega \frac{\partial \overline{S}_v}{\partial y}$$

In the following treatment, only the time-average of the various quantities are included, and for simplicity, the bar representing the time-average in the foregoing equation is dropped, thus the equation can then be rewritten in the simpler form

$$\frac{\partial S_v}{\partial t} = -u\frac{\partial S_v}{\partial x} + \varepsilon_x \frac{\partial^2 S_v}{\partial x^2} + \frac{\partial \varepsilon_x}{\partial x}\frac{\partial S_v}{\partial x} + \varepsilon_y \frac{\partial^2 S_v}{\partial y^2} + \frac{\partial \varepsilon_y}{\partial y}\frac{\partial S_v}{\partial y} + \omega \frac{\partial S_v}{\partial y} \tag{10.5}$$

If sediment particles take the place of the dye injection, then Eq. (10.5) is the diffusion equation for sediment in two-dimensional flow. The first term on the right-hand side of this equation represents advection, the second, third and fourth terms are diffusion terms; and the last term is the settlement of the particles.

The foregoing procedure can also be used to derive the diffusion equation for three-dimensional flow

$$\frac{\partial S_v}{\partial t} = -u\frac{\partial S_v}{\partial x} + \varepsilon_x \frac{\partial^2 S_v}{\partial x^2} + \frac{\partial \varepsilon_x}{\partial x}\frac{\partial S_v}{\partial x} + \varepsilon_y \frac{\partial^2 S_v}{\partial y^2}$$
$$+ \frac{\partial \varepsilon_y}{\partial y}\frac{\partial S_v}{\partial y} + \varepsilon_x \frac{\partial^2 S_v}{\partial z^2} + \frac{\partial \varepsilon_z}{\partial z}\frac{\partial S_v}{\partial z} + \omega \frac{\partial S_v}{\partial y} \tag{10.6}$$

in which ε_z is the diffusion coefficient of sediment in the z-direction, the transverse direction.

10.3 VERTICAL DISTRIBUTION OF CONCENTRATION OF SUSPENDED LOAD

In turbulent flow the movement of water elements, as they change positions between water layers, causes sediment exchanges between the layers also. At the same time, sediment particles, because of their greater specific weight, tend to settle out and move toward the bed. As a result, the sediment concentration is greater near the bed than it is at a point some distance above the bed. This tendency is illustrated by a typical concentration distribution of suspended load in the vertical direction of a natural river channel like those shown in Fig. 10.7. Because of this variation in

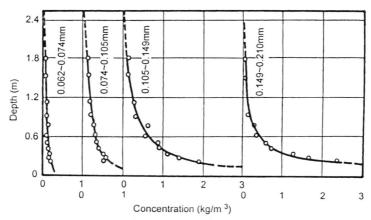

Fig. 10.7 Vertical concentration profiles of different sizes
of suspended sediment particles in a natural river

concentration, water elements moving upward carry a greater amount of sediment than do water bodies moving downward; thus, the exchange between the upward and downward water elements of the same volume results in a net transport of sediment in the upward direction. The amount of the upward sediment flux is proportional to the concentration gradient. As a consequence, when clear water flows over the sandy bed of a canal, the vertical concentration profile of suspended load near the canal intake is like curve A in Fig. 10.8. At this cross-section, the suspended sediment is concentrated near the bed, and the concentration gradient is large. Further downstream, more sediment has diffused upwards toward the water surface. The gradient of the concentration is less there, and the vertical concentration profile is as shown by curve B. Still further downstream, the amount of sediment carried upward by turbulence balances the amount settling down due to gravitation, and the vertical concentration profile as shown by curve C is then in equilibrium. In this paragraph the vertical concentration profile of suspended load in a state of equilibrium is studied.

Two different theories have been proposed for the vertical concentration distribution of suspended load: the diffusion theory and the gravitation theory. Both are presented in the following sections.

10.3.1 Diffusion theory

10.3.1.1 The first approximation to the solution of the diffusion theory

1. Derivation of equation

If the vertical concentration profile of suspended load is in equilibrium, the sediment diffusion process is steady and uniform; hence, all derivatives with respect

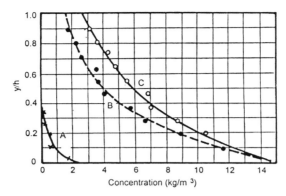

Fig.10.8 Diffusion of suspended sediment upward from the bed concentration (kg/m³)

to distance x and time t in Eq. (10. 5) are equal to zero. In this special but common case, the equation takes a more simplified form

$$\varepsilon_y \frac{d^2 S_v}{dy^2} + \frac{d\varepsilon_y}{dy}\frac{dS_v}{dy} + \omega\frac{dS_v}{dy} = 0$$

and it is the derivative with respect to y of the following equation:

$$\varepsilon_y \frac{dS_v}{dy} + S_v\omega = 0 \tag{10.7}$$

In 1925, Schmidt derived this equation on the basis of the diffusive characteristics of turbulent flow and used it to describe the distribution of dust in the atmosphere. In the 1930s, O'Brien (in America) and Makaviev (in Russia) applied this equation to the distribution of suspended load in flowing water.

2. Solution of differential equation

In order to solve the differential equation Eq. (10.7), one must determine the vertical distribution of ε_y. The simplest procedure is to assume that it is a constant. The solution is then

$$\frac{S_v}{S_{va}} = e^{-\omega(y-a)/\varepsilon_y} \tag{10.8}$$

in which S_{va} is a reference concentration of the suspension a distance "a" above the bed. In 1929, Hurst took measurements of sediment concentration at different depths in a cylinder in which water and uniform sediments of various particle sizes (median diameters 0.9, 0.4, and 0.2 mm) were uniformly mixed by a series of propellers [7]. The experimental results agree well with formula Eq. (10.8). Later on,

417

Rouse obtained the results shown in Fig. 10.9 by a similar method; he used a series of grids moving in simple harmonic motion in a cylinder [8]. The figure shows that the experimental results essentially follow the theoretical curve for all except the coarsest particles. Lane and Kalinske made analyses of data from a natural river, and found that Eq. (10.8) also gave satisfactory results for this practical case. They suggested that ε_y could be expressed as follows.

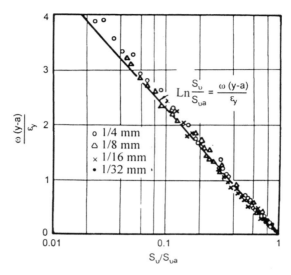

Fig. 10.9 Vertical distribution of sediment concentration for various particle sizes in a sediment mixture (Rouse experiments with uniform stirring) (after Rouse, H.)

$$\varepsilon_y = \frac{\kappa U_* h}{6} \tag{10.9}$$

in which κ is the Karman constant in the logarithmic formula for the velocity distribution. If the usual value of $\kappa = 0.4$ is taken then

$$\varepsilon_y = 0.067 U_* h \tag{10.10}$$

The sediment exchange coefficient is not nearly a constant but is a function of position in space. With current experimental techniques, however, determination of the vertical distribution of ε_y is difficult. From the theory of turbulent flow, the diffusion coefficient is equivalent to the momentum exchange coefficient ε_m, and it is related to the velocity gradient in the following way

$$\varepsilon_m = \frac{\tau}{\rho \dfrac{du}{dy}} \tag{10.11}$$

418

For simplicity one assumes that

$$\varepsilon_y = \varepsilon_m$$

For two-dimensional flow, the shear stress is linearly distributed along the depth, so that

$$\tau = \tau_0 \left(1 - \frac{y}{h} \right) \tag{10.12}$$

in which τ_0 stands for the shear stress at the bed. For a logarithm velocity profile

$$\frac{u}{U_*} = \frac{1}{\kappa} \ln\left(\frac{y}{y_0} \right)$$

Differentiating, one obtains

$$\frac{du}{dy} = \frac{U_*}{\kappa} \frac{1}{y} \tag{10.13}$$

Then the substitution of Eqs. (10.12) and (10.13) into Eq. (10.11) yields

$$\varepsilon_y = \varepsilon_m = \kappa U_* y \frac{h-y}{h} \tag{10.14}$$

By integrating this expression and taking the average, one obtains Eq. (10.9).

The substitution of Eq. (10.14) into Eq. (10.7) yields

$$\kappa U_* y \frac{h-y}{h} \frac{dS_v}{dy} + S_v \omega = 0 \tag{10.15}$$

which, after integration, gives the vertical concentration profile of suspended load for the case of low concentrations

$$\frac{S_v}{S_{va}} = \left(\frac{h-y}{y} \frac{a}{h-a} \right)^z \tag{10.16}$$

in which

$$z = \frac{\omega}{\kappa U_*} \tag{10.17}$$

For a dune-covered bed, and in the absence of more experimental data for U_*. Einstein suggested that U_* may be replaced by the shear velocity relevant to grain friction $U'_* = (R'_b g J)^{0.5}$.

Expression (10.16) shows that S_v approaches infinity for $y = 0$. This illogical result is the consequence of the form of the logarithmic velocity profile. Velikanov suggested another form for the logarithmic law,

$$\frac{u}{U_*} = \frac{1}{\kappa} \ln\left(1 + \frac{y}{\Delta}\right) \tag{10.18}$$

in which Δ is a parameter that depends on the bed roughness; and this relationship leads to the following form for the vertical concentration distribution of suspended load:

$$\frac{S_v}{S_{va}} = \left(\frac{h - y}{\Delta + y} \frac{\Delta + a}{h - a}\right)^z \tag{10.19}$$

This equation gives a finite value for S_v at $y = 0$ and avoids the illogical aspect of Eq. (10.16). However, such a modification is no more than a formality. Neither Eq. (10.16) nor Eq. (10.19) is valid for sediment motion near the bed because the sediment there does not move in suspension. This point is discussed further in the following chapters.

The exponent z in the expression for suspended load affects the distribution of the sediment concentration. Fig. 10.10 compares the relative vertical distributions of suspended load concentration obtained from Eq. (10.16). The figure shows that a smaller value of z results in a more uniform distribution. Thus, the height of the suspension is also a function of z. In the case of $z = 5$, the amount of sediment carried in suspension is very small; the discharge ratio of suspended load to bed load is then *1:4*, according to an estimation based on the Einstein method. From the practical point of view,

$$\frac{\omega}{\kappa U_*} = 5 \tag{10.20}$$

can be taken as the threshold value for sediment suspension. However, various researchers have used other threshold values. Bagnold [10] used the value *3* and Engelund [11] the value *2*; these values yield ratios of suspended load to bed load, of 2:1 and 0.9:1, respectively.

3. Verification of diffusion theory for suspended load

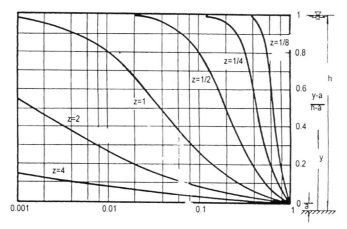

Fig. 10.10 Relative distribution of suspended load
obtained from diffusion theory (after Rouse, H.)

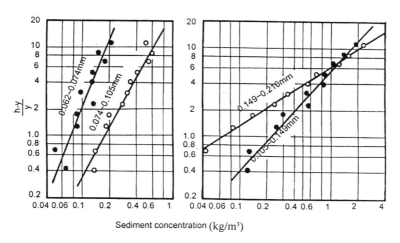

Sediment concentration (kg/m³)

Fig. 10.11 Vertical distribution of suspended load obtained by replotting data
in Fig. 10.7 in accordance with diffusion theory

Since the form of expression Eq. (10.16) was derived analytically in the 1930s, a number of studies have been conducted to test the diffusion theory against field observations and laboratory data. The verification has two aspects: whether the formula structure is correct, and whether the analytical expression for the exponent z is valid.

If a logarithmic scale is used to plot the relationship between *(h-y)/y* and S_v, the Eq. (10. 16) should appear as a straight line with a slope equal to the value of z. The data for suspended load distribution shown in Fig. 10.7 from natural rivers is replotted in Fig. 10.11 using this method. For each particle size the points are distributed normally along the corresponding straight lines. The data observed at the

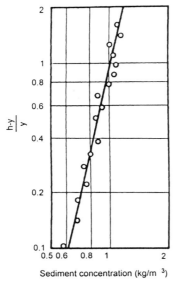

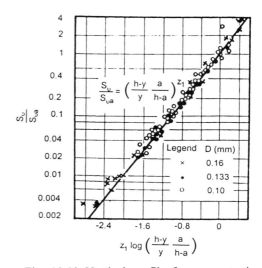

$$\frac{S_v}{S_{va}} = \left(\frac{h-y}{y}\cdot\frac{a}{h-a}\right)^{z_1}$$

Legend	D (mm)
×	0.16
•	0.133
o	0.10

$$z_1 \log\left(\frac{h-y}{y}\cdot\frac{a}{h-a}\right)$$

Fig. 10.12 Vertical concentration profile of suspended load measured at Nanjing Station on the Yangtze River

Fig. 10.13 Vertical profile for concentration of suspended load for laboratory data (after Vanoni, V.A., and N.H. Brooks)

Nanjing station of the Yangtze River in Fig. 10.12 are comparable. Vanoni and Ismail, using laboratory data, verified the concentration as given by diffusion theory. The experimental results of Vanoni for three different sizes of sediment (0.10, 0.133 and 0.16 mm) are plotted in Fig. 10.13.

Despite the foregoing correspondence, the measured value of the exponent z_1 differs from the theoretical value of z; still, the form of the Eq. (10.17) is confirmed. Vanoni was the first to find in a laboratory study that z_1 is in general smaller than z, and that the measured vertical concentration profile of suspended load is more uniform than the theoretical one. Ismail found that the ratio of z to z_1 is approximately constant, and its value is *1.5* for sediment particles of 0.1 mm, and 1.3 for particles of 0.16 mm. The range of the data used by Ismail was quite narrow. For a wide range of particle sizes, the relation between z_1 and z can be obtained from the plot in Fig. 10.14. If the value of z is small, the case of fine particles and strong turbulence intensity, the discrepancy between the theoretical and measured values is small, but it increases with z; in the limit, z_1 approaches a constant.

The discrepancy between the theoretical and measured values of the exponent in the expression for the distribution of suspended sediment is generally attributed to the assumption that the sediment exchange coefficient equals the momentum exchange coefficient. In fact, the two coefficients differ significantly. Eddies that are nearly the same size as the sediment particles do not contribute to sediment exchange in the way they do to momentum exchange. Because of inertial effects, sediment particles do not follow the motion of the water elements during fluctuations at higher frequencies. Vanoni determined the vertical distributions of the two exchange coefficients from the measured profiles of sediment concentration and velocity as shown in Fig. 10.15. Although the two curves appear to be similar, their absolute values are not the same. Generally, the sediment exchange coefficient is somewhat larger than the momentum exchange coefficient. Hence, Vanoni proposed that

$$\varepsilon_y = \beta \varepsilon_m \qquad (10.21)$$

in which β is a proportionality factor that is greater than unity and depends on the properties of the sediment in suspension. Thus, Eq. (10. 17) can be rewritten as

$$z = \frac{\omega}{\beta \kappa U_*} \qquad (10.22)$$

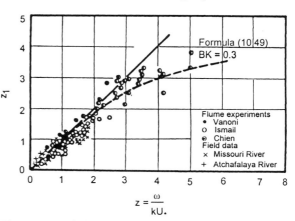

Fig. 10.14 Relationship between the observed exponent (z_1) and the theoretical one (z) for vertical concentration profiles of suspended load in natural rivers and flumes

In contrast, other experimenters have obtained results showing a trend opposite to that of Vanoni. Brush studied the relationship between ε_y and ε_m by observing directly the diffusion of sediment in a submerged jet [16]. His experimental setup is shown in Fig. 10.16. Since the jet is directed vertically downward, the diffusive motion of the sediment is normal to the direction of settlement. For simplicity, Brush neglected the effect of settlement. His experimental results for the distributions of velocity and sediment concentration at different cross sections are shown in Fig. 10.17; the solid curve represents the transverse distribution of velocity. The figure shows that the velocity distribution is

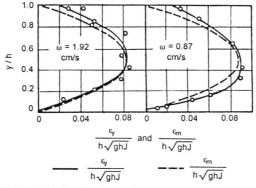

Fig.10.15 Vertical distributions of exchange coefficients of sediment and momentum (after Vanoni, V.A., N.H. Brooks)

the same as for the concentration distribution for glass spheres with diameter of 0.19 mm, but that it is more uniform than those for spheres with larger diameters, 0.32 mm, and 0.55 mm. This result demonstrates that the diffusion process of fine sediment is roughly the same as that of flow momentum, but that the movements of coarse sediment lag behind due to the greater inertia. The measured data yielded the following values for the ratio of the sediment exchange coefficient to the momentum exchange coefficient:

diameter (mm)	$\beta = \varepsilon_y/\varepsilon_m$
0.55	0.15
0.32	0.50
0.19	1.00

The trend of these experimental results is just opposite to that observed by Vanoni.

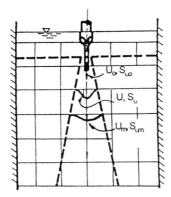

Fig. 10.16 Experimental setup used by Brush to study relationship between sediment diffusion coefficient and momentum exchange coefficient (U_m and S_{vm} are the velocity and sediment concentration along the center line) (after Brush, L.M., Jr)

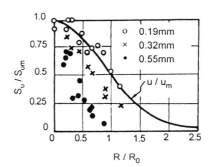

Fig. 10.17 Distributions of velocity and sediment concentration in a submerged downward jet (R_0 is distance from center line to the position of $U = 0.5U_m$) (after Brush, L.M., Jr)

To explain the conflict between the aforementioned results, Jobson and Sayre pointed out the existence of two different aspects of sediment exchange. On one hand, as flow is accelerating or decelerating in its fluctuation, the greater inertia of the heavier sediment causes it to lag behind the water. From this point of view, the sediment exchange coefficient should be smaller than the momentum exchange coefficient. On the other hand, a greater sediment exchange coefficient would result if the effective mixing and diffusion were intensified due to the greater centrifugal force of sediments particle within the rotating eddies; the centrifugal force acting on

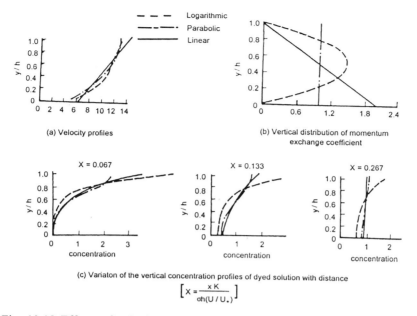

(a) Velocity profiles

(b) Vertical distribution of momentum exchange coefficient

(c) Variaton of the vertical concentration profiles of dyed solution with distance

$$\left[X = \frac{xK}{\sigma h(U/U_*)} \right]$$

Fig. 10.18 Effects of velocity profile on the vertical distributions of momentum exchange coefficient and concentration profile of a dyed solution (after Jobson, H.E, and W.W. Sayre)

the heavier sediment particles is larger than that acting on a water element. Thus the particles could move further away from the eddy center than does the surrounding water. It follows that the relative magnitudes of ε_y and ε_m depend on the balance between these two effects.

In fact, the vertical distribution of sediment concentration does not depend significantly on the form of the vertical distribution of the sediment diffusion coefficient[17]. Fig. 10.18a shows three arbitrary forms chosen for the velocity profile: logarithmic, parabolic, and linear. The corresponding vertical distributions of momentum exchange coefficient, expressed in terms of the dimensionless parameter $6\varepsilon_m/khU_*$, are shown in Fig. 10.18b. Obviously, these distributions differ considerably because of the different velocity profiles. If the sediment exchange coefficient is taken to be equal to the momentum exchange coefficient, however, these three quite different types of vertical distribution for the sediment exchange coefficient do not cause much difference in the vertical distribution of the sediment concentration, as shown in Fig. 10.18[1]. The above result indicates that the discrepancy between the measured and theoretical values of the exponent in the first

---

[1] Fig.10.18c shows the results calculated for the diffusion of a dyed solution with the same specific weight as water. At the inlet, when x is zero, the greatest concentration of dyed solution occurred at the free surface where the dyed solution was injected.

approximation for the diffusion theory cannot be completely attributed to this questionable illogical assumption Eq. (10.14); rather it is probably due to some intrinsic property that is discussed in the presentation of the second approximation to the diffusion theory.

4. Sediment concentration at free surface

According to Eq. (10.16), the sediment concentration at the free surface should be zero, a result that does not agree with observations of natural rivers. In Fig. 10.11, the momentum exchange coefficient ε_m is zero at free surface, but sediment exchange coefficient ε_y has a finite value there. This phenomenon can be explained in the following way. For momentum exchange, the relationship

$$\tau = -\rho \overline{u'v'} \tag{10.23}$$

is valid, and it shows that the shear stress τ is generated by momentum exchange only at the position where u' and v' have a certain degree of correlation(from the definition of the correlation coefficient Eq. (4.36)). In contrast, sediment suspension depends primarily on the vertical fluctuation component v', and much less on u'. Hence, even if u' and v' are not correlated, sediment can still be transported. Thus, even if the momentum exchange coefficient is zero at the free surface, sediment exchange can take place and produce some sediment at that level.

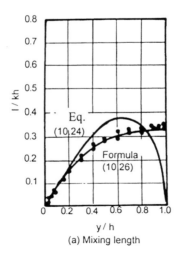
(a) Mixing length

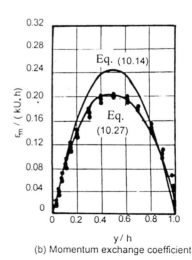
(b) Momentum exchange coefficient

Fig. 10.19 Comparison of vertical distributions of mixing length and momentum exchange coefficient with measured data

Further analysis shows that the vertical distribution of the momentum exchange coefficient, expressed in the form of Eq. (10.14), corresponds to the following vertical distribution of mixing length,

$$l = \kappa y \sqrt{\frac{h-y}{y}} \tag{10.24}$$

Fig. 10.19 shows that neither Eq. (10.24) nor Eq. (10.14) agrees well with the measured data. If

$$\frac{u_{max} - u}{U_*} = \frac{2}{\kappa} \tanh^{-1} \left(\frac{h-y}{h} \right)^{3/2} \tag{10.25}$$

is adopted instead of the logarithmic distribution for the velocity profile, the mixing length and momentum exchange coefficient are then as follows:

$$l = \frac{\kappa}{3} h \left[1 - \left(\frac{h-y}{h} \right)^3 \right] \tag{10.26}$$

$$\varepsilon_m = \frac{\kappa}{3} U_* h \sqrt{\frac{h-y}{h}} \left[1 - \left(\frac{h-y}{h} \right)^3 \right] \tag{10.27}$$

and these agree well with the measured data as shown in Fig.10.19. If Eq. (10.21) is also used for the relationship between ε_y and ε_m, then the differential equation for the vertical concentration profile of suspended load can be obtained in the form

$$S_v \omega + \beta \frac{\kappa}{3} U_* h \sqrt{\frac{h-y}{h}} \left[1 - \left(\frac{h-y}{h} \right)^3 \right] \frac{dS_v}{dy} = 0 \tag{10.28}$$

and the solution takes the form [18]

$$\frac{S_v}{S_{va}} = e^{z\varnothing} \tag{10.29}$$

in which z can be expressed by

$$\varnothing = \frac{1}{2} \ln \frac{\left[\left(\frac{h-y}{h} \right)^{3/2} + 1 \right] \left[\left(\frac{h-y}{h} \right)^{1/2} - 1 \right]^3}{\left[\left(\frac{h-y}{h} \right)^{3/2} - 1 \right] \left[\left(\frac{h-y}{h} \right)^{1/2} + 1 \right]^3} + \sqrt{3} \tan^{-1} \left\{ \frac{\sqrt{3} \sqrt{\frac{h-y}{h}}}{\left[\frac{h-y}{h} - 1 \right]} \right\}_{y=a}^{y} \tag{10.30}$$

Fig. 10.20 is a comparison between the relative vertical concentration profile of Eq.(10.29) and the first approximate solution to the diffusion theory from Eq. (10.16). The figure shows that the difference between these two distributions is not great over most of the flow region. However, the modification does allow one to avoid the zero value for the sediment concentration at the free surface. Zhang, using the Wang velocity profile, obtained an expression for the vertical concentration profile that does not have a zero value at free surface [19]. Since his expression is rather complicated, it is not introduced herein. From the practical point of view, Eq. (10.16) has already provided a satisfactory interpretation, so that further computations of a more tedious and elaborate procedure seem unnecessary.

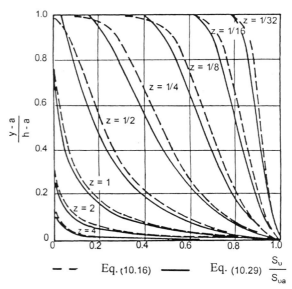

Fig. 10.20 Comparison of two expressions for vertical concentration profiles

10.3.1.2 The second approximation to the solution of the diffusion theory

After a detailed analysis of the conceptual model on which the diffusion theory is based, Einstein and Chien pointed out that the derivation of the differential equation for the vertical concentration profile was based on a series of assumptions concerning turbulence structure and sediment exchange. As a result, the solution is bound to be an approximate one. If these assumptions are suitably modified, then an improved theoretical relation between z_1 and z, one that better reflects reality, can be obtained. This solution is referred to as the second approximation to the diffusion theory [20].

The differential equation for the vertical concentration profile should be examined from another point of view in order to reveal the physical background of the assumptions and the conceptual model on which the differential equation is based. They can then be suitably modified.

Turbulent motion causes upward and downward flows to pass through a representative horizontal plane located a distance y above the bed. Some flow moves upward and some downward, each occupying about half of a unit flow area as shown in Fig. 10.21. If the mean value of the mixing length at elevation y is l, the water moving upward comes from the elevation $y-(l/2)$ and that moving downward comes from $y + (l/2)$. Since sediment particles settle downward with a velocity ω, the rate of upward sediment flux per unit volume is

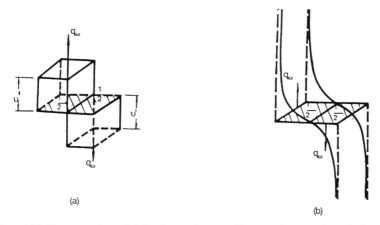

Fig. 10.21 Conceptual model for the exchange of water elements in turbulent flow

$$S_{v(y-l/2)} \frac{v' - \omega}{2}$$

and that for a downward sediment flux is

$$S_{v(y+l/2)} \frac{v' + \omega}{2}$$

For steady flow

$$S_{v(y-l/2)} \frac{v' - \omega}{2} - S_{v(y+l/2)} \frac{v' + \omega}{2} = 0 \qquad (10.31)$$

The expression for the sediment concentration at various elevation can be expanded in a Taylor series,

$$S_{v(y-l/2)} = S_{vy} - \frac{1}{2}l \frac{ds_{vy}}{dy} + \frac{1}{8}l^2 \frac{d^2 S_{vy}}{dy^2} - \frac{1}{48}l^3 \frac{d^3 S_{vy}}{dy^3} + \cdots$$
$$S_{v(y+l/2)} = S_{vy} + \frac{1}{2}l \frac{ds_{vy}}{dy} + \frac{1}{8}l^2 \frac{d^2 S_{vy}}{dy^2} + \frac{1}{48}l^3 \frac{d^3 S_{vy}}{dy^3} + \cdots$$
$$(10.32)$$

and if higher-order terms are dropped,

$$S_{v(y-l/2)} = S_{vy} - \frac{1}{2}l \frac{dS_{vy}}{dy}$$

$$S_{v(y+l/2)} = S_{vy} + \frac{1}{2}l \frac{dS_{vy}}{dy}$$
$$(10.33)$$

429

Substitution of Eq. (10.33) into Eq. (10.31) yields

$$\frac{1}{2}lv'\frac{dS_v}{dy} + S_v\omega = 0 \qquad (10.34)$$

If the same conceptual model is used to describe the momentum exchange induced by the upward and downward flow, one obtains the following result

$$\frac{1}{2}lv' = \kappa U_* y \frac{h-y}{h}$$

Substitution of the above expression into Eq. (10.34) yields once again the differential equation presented as Eq. (10.7).

In the foregoing, the Prandtl concept of mixing length for turbulent flow structure has been used, and the sediment exchange is presumed to be essentially the same as the momentum exchange. As pointed out in Chapter 4, the Prandtl concept does not describe a continuous process, i.e., the sediment carried by a water element does not mix with the local flow until it has moved a distance of one mixing length. In fact, the moving water element is continually mixing with the local flow, and the mixing intensity is the highest at the starting position and decreases with time. In other words, the mixing length should be a quantity that follows some distribution, rather than a constant with one definite value. The mixing length can have various values, and the probability of occurrence for each value is different. For the fluctuating velocity of flow, the situation is the same. At various elevations the time average velocity (expressed by its r.m.s.) has a certain value, but the instantaneous value of the pulsation at different times could be great or small, also the probability distribution in the region outside the near-bed layer follows a normal error distribution. This probability distribution and that of the intensity have not been properly incorporated into the current theory for the diffusion of suspended sediment.

In addition, the higher order derivatives of S_v with respect to y were neglected so as to simplify Eq. (10.32) into Eq. (10.33). For fine particles or high turbulence intensities, i.e., if the exponent z of the expression for vertical concentration profile is small, the distribution is relatively uniform, so that to keep only the first two terms of the Taylor series are enough. However, if z is large so that the distribution is not uniform, the neglect of the higher-order derivatives results in some error because the concentration can vary significantly within the distance of a mixing length.

In the following derivation, the important assumptions are modified. First, a Gaussian distribution is introduced for the variation of the velocity

$$p = \frac{1}{\sigma\sqrt{2\pi}}e^{-v'^2/2\sigma^2} \qquad (10.35)$$

in which p is the probability of occurrence of the velocity v', and σ is its standard deviation. The physical process of the exchanges of water elements in turbulent flow can be conceptualized in two ways. One is to use the conventional concept shown in Fig. 10.21a. In it, the velocity of fluid mass at a certain moment is uniformly distributed over a plane with the absolute value v' varying with time. The other is to assume a steady situation in which the velocity fluctuation is not uniformly distributed over the plane but varies from $-\infty$ to $+\infty$, with the v' value taking an area of pdv', as shown by Fig. 10.21b. In the latter concept, the flow exchange rate over the plane is

$$q_w = \int v'dA = \int_0^\infty v'pdv'$$

Substitution of the above expression into Eq.(10.35) gives

$$q_\omega = \frac{\sigma}{\sqrt{2\pi}} \tag{10.36}$$

For sediment exchange, if $v'<\omega$, the water element moving upward cannot carry any sediment upwards. That is, if S_{vu} and S_{vd} stand for the mean sediment concentrations of the water bodies moving upward and downward, respectively, then the sediment flux passing across the plane are

$$upward \quad \int_\omega^\infty S_{vu}(v' - \omega)pdv'$$

$$downward \quad \int_{-\omega}^\infty S_{vu}(v' + \omega)pdv'$$

For a state of equilibrium

$$\int_\omega^\infty S_{vu}(v' - \omega)pdv' - \int_{-\omega}^\infty S_{vd}(v' + \omega)pdv' = 0 \tag{10.37}$$

A better formulation of the continuous exchange of water element in motion is that the water element progressively loses its identity as it mixes with the surrounding flow. The distribution of the volume lost as it moves one mixing length is shown in Fig. 10.22a, in which the shaded area is equal to the volume of the water element. A second conceptual model can be used to describe such a phenomenon. According to the classical assumption, the mixing process is discontinuous, i.e., water element in motion remains unchanged until it has traveled a distance of mixing length, and then it mixes fully with the local flow and losses its original character. The difference between the new model and the classical one is that at any elevation the mixing length varies with a certain probability of occurrence rather than being a constant.

The distribution of the mean mixing length along a vertical is as follows

$$l = B\kappa y\left(1 - \frac{y}{h}\right)^{1/2}$$ (10.38)

in which B is a coefficient. For simplicity, the flow depth is assumed to be infinite, and the above expression can then be simplified into

$$l = B\kappa y$$ (10.39)

which indicates that the water element starts to mix with the local flow when it reaches the elevation $(1+B\kappa)y$.

For the sake of mathematical manipulation, the y-coordinate system is transformed into Y-coordinate system in accordance with

$$Y = \ln y$$ (10.40)

In the Y-coordinate system, a water element begins to mix after moving from Y to $Y+\ln(1+B\kappa)$, That is, in the Y-coordinate system the mixing length is a constant

$$L = \ln(1 + B\kappa)$$ (10.41)

as shown in Fig. 10.22b. In the above systems, for a water element starting from a certain position, the mixing length might be different, with the mean value given by Eq. (10.41). It is difficult to predict the value of the mixing length for a given water element at a given instant, but it is possible to express the probability of a given mixing length.

In the Y-coordinate system, Y varies from $-\infty$ to ∞. According to Eq. (10.41) , the mixing length is the same for any Y, or the field of probability of motion exchange is uniform. Einstein showed that in a uniform field, the probability of

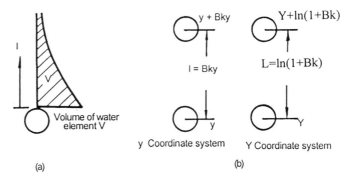

(a) (b)

Fig. 10.22 Two conceptual models for the mixing length of turbulent flow

occurrence of mixing length $|Y_2 - Y_1|$ is

$$e^{-\frac{|Y_1 - Y_2|}{L}}$$

in which Y_1 and Y_2 are the initial and final elevations for an exchange. This probability increases as the mixing length decreases. If $|Y_2 - Y_1|$ is large, the probability of occurrence of the corresponding mixing length is quite small, approaching zero as a limit. Physically, this model is the same as the one shown in Fig.10.22. The sum of all the probabilities is

$$\int_{Y_1}^{\infty} e^{\frac{Y_1 - Y_2}{L}} \frac{dY_2}{L} = 1$$

and the mean mixing length is

$$\int_{Y_1}^{\infty} (Y_2 - Y_1) e^{\frac{Y_1 - Y_2}{L}} \frac{dY_2}{L} = L$$

For a section of unit area in a horizontal plane at elevation Y, both upward and downward flows occur each in one-half of the plane. Starting at Y_1 $(Y_1 < Y)$ a water element moves upward with a mixing length of $Y_2 - Y_1$, and the flow rate is q. Then this take various value with different probability. From the statistical point of view, only the portion of the water element that has a mixing length greater than $Y - Y_1$ is able to reach the plane at Y and cross it. So the volume of a water element that crosses a unit area of a plane at Y, moving upward, can be expressed mathematically as

$$\frac{1}{2} \int_{Y}^{\infty} q e^{\frac{Y_1 - Y_2}{L}} \frac{dY_2}{L}$$

Not all of the water flowing upward across the plane comes from Y_1, any water element between $-\infty$ and Y can transport some water cross the reference plane at Y. Hence, the total discharge moving upward across the plane at Y is

$$q_w = \frac{1}{2} \int_{-\infty}^{Y} \int_{Y}^{\infty} q e^{\frac{Y_1 - Y_2}{L}} \frac{dY_2}{L} dY_1$$

If the flow depth is infinity, q does not vary with the position of Y_1. From the law of continuity, the above expression also represents the total discharge of water moving downward between Y and ∞ across the reference plane at Y.

The volume of sediment flux moving upward from the region between $-\infty$ and Y and crossing the reference plane at Y is

$$\frac{1}{2}\int_{-\infty}^{Y}\int_{Y}^{\infty}S_{vY1}qe^{\frac{Y_1-Y_2}{L}}\frac{dY_2}{L}dY_1$$

The volume of sediment flux moving downward between Y and ∞ across the plane at Y is

$$\frac{1}{2}\int_{\infty}^{Y}\int_{Y}^{-\infty}S_{vY1}qe^{\frac{Y_1-Y_2}{L}}\frac{dY_2}{L}dY_1$$

in which S_{vyl} stands for the sediment concentration at elevation Y_1. Since S_{vyl} is the real concentration, it contains all the high order terms in Eq. (10.32).

With the upward and downward discharges of water and sediment, one can calculate the mean sediment concentration in the form

$$\left.\begin{array}{l} S_{vu} = \dfrac{\int_{-\infty}^{Y}\int_{Y}^{\infty}s_{vY1}qe^{\frac{Y_1-Y_2}{L}}\dfrac{dY_2}{L}dY_1}{\int_{-\infty}^{Y}\int_{Y}^{\infty}qe^{\frac{Y_1-Y_2}{L}}\dfrac{dY_2}{L}dY_1} \\[20pt] S_{vd} = \dfrac{\int_{\infty}^{Y}\int_{Y}^{-\infty}s_{vY1}qe^{\frac{Y_1-Y_2}{L}}\dfrac{dY_2}{L}dY_1}{\int_{\infty}^{Y}\int_{Y}^{-\infty}qe^{\frac{Y_1-Y_2}{L}}\dfrac{dY_2}{L}dY_1} \end{array}\right\} \tag{10.42}$$

One finds that the first approximate solution of the vertical concentration profile, in the form of Eq. (10. 16) is still applicable, only the exponent is somewhat different from the theoretical value given by Eq. (10.17). Hence, the following relationship is obtained

$$\frac{S_{vy}}{S_{va}} = \left(\frac{h-y}{y}\frac{a}{h-a}\right)^{z_1} \tag{10.43}$$

The relation between the exponent z_1 in the above expression and the exponent z from Eq. (10.17) is obtained from the second approximate solution. If h is infinity, the above expression can be converted into

$$\frac{S_{vy}}{S_{va}} = \left(\frac{a}{y}\right)^{z_1} \tag{10.44}$$

If y is transformed into Y-coordinate system the expression becomes

$$S_{vy} = Ee^{-Yz_1} \tag{10.45}$$

in which E is a constant.

Substitution of Eq. (10.45) into Eq. (10.42) gives

$$S_{vu} = \frac{Ee^{-Yz_1}}{1 - Lz_1}$$

$$S_{vd} = \frac{Ee^{-Yz_1}}{1 + Lz_1}$$

(10.46)

The same approach can be used to analyze the momentum exchange. For the latter, the sediment concentration should be replaced by ρu_{y1}, in which ρ is the density of water, u_{y1} the flow velocity at Y_1 in the Y-coordinate system. From the equation for momentum conservation

$$q_w = \frac{\kappa U_*}{2L}$$

(10.47)

Hence

$$z = \frac{\omega}{\kappa U_*} = \frac{\omega}{2Lq_w}$$

Substituting Eq. (10.36) into the above expression yields

$$z = \frac{\omega\sqrt{2\pi}}{2\sigma L}$$

(10.48)

Substitution of Eqs. (10.35), (10.46) and (10.48) into Eq. (10.37) gives the solution for z_1

$$z_1 = \frac{z}{e^{-L^2 z^2/\pi} + zL\frac{2}{\sqrt{2\pi}}\int_0^{\sqrt{\frac{2}{\pi}}Lz} e^{-x^2/2}dx}$$

(10.49)

which is the theoretical relationship between z_1 and z. In Eq. (10.49), L may be determined from Eq. (10.41), the value of integration term in Eq. (10.49) is available in tabulations of the probability function . If z is very large

$$\frac{2}{\sqrt{2\pi}}\int_0^{\sqrt{\frac{2}{\pi}}Lz} e^{-x^2/2}dx \rightarrow 1$$

so that

$$z_1 \to \frac{1}{L}$$

or z_1 approaches a constant.
If z is very small

$$\frac{2}{\sqrt{2\pi}} \int_0^{\sqrt{\frac{2}{\pi}} Lz} e^{-x^2/2} dx \to 0$$

which gives

$$z_1 \to z$$

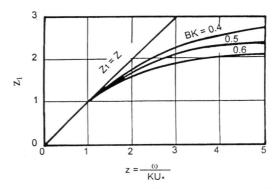

$$z = \frac{\omega}{KU_*}$$

Fig. 10.23 Theoretical relationship between z & z_1

Fig. 10.23 shows the relationship between z_1 and z from Eq. (10.49) for various values of $B\kappa$. The result is comparable to that shown in Fig. 10.14.

Turbulence exchange is a complicated physical process, and further study is needed to obtain practically significant results. The elementary considerations presented here are no more than an outline because of the several assumptions that are still involved and the uncertainty of the parameter B. The condition that the depth is infinite indicates that attention should be paid to the near-bed layer, in which the sediment concentration and its vertical gradient are comparatively large. If a theoretical approach can be developed to determine the vertical concentration profile in the near-bed layer, then the application of the theory to the region of low concentration and small concentration gradient will no longer involve significant errors.

10.3.2 Gravitational Theory

Velikanov proposed what he called the Gravitational Theory for the vertical profile of suspended sediment, and he developed it as an application of the principle of energy conservation [22]. However, some of the assumptions made in his derivation are questionable.

He considered a unit volume of a mixture of water and sediment that flows from a higher position to a lower position in a unit time interval. E_1 and E_2 stand for the amounts of energy supplied by the water and sediment phases, respectively; E_3 and E_4 stand for the energy lost in the water and sediment phases to overcome frictional resistance, respectively; E_5 stands for the amount of energy needed to maintain the suspension of sediment. Then Velikanov wrote the following energy balance equations: for the water-phase,

$$E_1 = E_3 + E_5$$

and for the sediment phase,

$$E_2 = E_4$$

For two-dimensional uniform flow,

$$E_1 = g\rho(1 - \bar{S}_v)\bar{u}J$$

$$E_2 = g\rho_s\bar{S}_v\bar{u}J$$

$$E_3 = -\bar{u}\frac{d\tau}{dy} = \bar{u}\frac{d}{dy}\left[(1 - \bar{S}_v)\overline{\rho u'v'}\right] = \rho\bar{u}\frac{d}{dy}\left[(1 - \bar{S}_v)\overline{u'v'}\right]$$

$$E_4 = \rho_s\bar{u}\frac{d}{dy}(\overline{\bar{S}_v u'v'})$$

$$E_5 = g(\rho_s - \rho)\bar{S}_v\omega(1 - \bar{S}_v)$$

In the above expressions, $\bar{u}$ is the time average of the velocity in the flow direction, and ρ_S density of the sediment particles. In the derivation, Velikanov took the fall velocity of a sediment particle to be equal to the difference

$$\bar{v} - \omega$$

in which $\bar{v}$ is the time average velocity component perpendicular to flow direction and ω the fall velocity of a single particle in still water of infinite extent. The continuity equation for sediment passing through a unit area located at a distance y from the bed is

$$\overline{S_v(v - \omega)} = 0$$

and the continuity equation for water flow is

$$\overline{v(1 - S_v)} = 0$$

If the instantaneous value is expressed as the sum of the time averaged value and the fluctuation value, the result is as follows

$$\bar{v}\bar{S}_v - \omega\bar{S}_v + \overline{v'S_v'} = 0$$

$$\bar{v} - \bar{v}\bar{S}_v - \overline{v'S_v'} = 0$$

Adding the preceding two formulas yields

$$\bar{v} = \bar{S}_v\omega \tag{10.50}$$

which indicates that the time average of the vertical velocity component becomes zero only for low concentration of fine particles. Hence, $(1-S_v)\omega$ is used rather than ω in the expression for E_5.

Substitution of the related energy terms into the energy balance equation yields

$$g(1-\overline{S}_v)\overline{u}J = \overline{u}\frac{d}{dy}\left[(1-\overline{S}_v)\overline{u'v'}\right] + \frac{\rho_s - \rho}{\rho}g\omega\overline{S}_v(1-\overline{S}_v) \tag{10.51}$$

$$g\overline{S}_vJ = \frac{d}{dy}(\overline{S}_v\overline{u'v'}) \tag{10.52}$$

In these two equations above, $\overline{S}_v$, $\overline{u}$ and $\overline{u'v'}$ are unknowns, so that an additional condition is needed to solve them. One can be derived from the formula for the velocity profile.

Velikanov used Eq. (10.18) for the velocity profile. The expression can be rewritten in terms of the variables $\eta = y/h$ and $\alpha = \Delta/h$

$$\overline{u} = \frac{\sqrt{ghJ}}{\kappa}\ln\left(1 + \frac{\eta}{\alpha}\right) \tag{10.53}$$

The solution for the vertical concentration profile is obtained by combining these three equations and integrating.

Dividing Eq. (10.51) by $\overline{u}$ and adding it to Eq. (10.52), one obtains

$$gJ = \frac{d}{dy}\overline{u'v'} + \frac{\rho_s - \rho}{\rho}\frac{g\omega\overline{S}_v(1-\overline{S}_v)}{\overline{u}}$$

which, in integral form, becomes

$$\int_y^h gJdy = \int_y^h \frac{d}{dy}\overline{u'v'}dy + \frac{\rho_s - \rho}{\rho}g\omega\int_y^h \frac{\overline{S}_v(1-\overline{S}_v)}{\overline{u}}dy$$

Thus, the following expression is obtained

$$-gJ(h-y) = \overline{u'v'} + \frac{\rho_s - \rho}{\rho}g\omega\int_y^h \frac{\overline{S}_v(1-\overline{S}_v)}{\overline{u}}dy$$

The second term on the right-hand side is much smaller than the first term and can be neglected. The equation then is simplified to

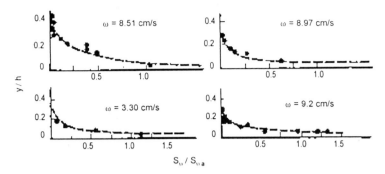

Fig. 10.24 Comparison of experimental results with profiles derived from Gravitational theory for vertical concentration profile

$$\overline{u'v'} = -gJ(h-y)$$

$$\frac{d\overline{u'v'}}{dy} = gJ \tag{10.54}$$

For small concentrations, $1-S_v \approx 1$, and the substitution of Eqs. (10.53) and (10.54) into Eq. (10.51) yields the differential form for the vertical concentration profile

$$\frac{dS_v}{S_v} = \beta \frac{d\eta}{(1-\eta)\ln[1+(\eta/a)]} \tag{10.55}$$

in which

$$\beta = \frac{\gamma_s - \gamma}{\gamma} \frac{\kappa\omega}{J\sqrt{ghJ}} \tag{10.56}$$

If

$$\xi(\eta,\alpha) = \int_{\eta_a}^{\eta} \frac{d\eta}{(1-\eta)\ln[1+(\eta/\alpha)]}$$

is used, then the vertical concentration profile can be expressed as

$$\frac{S_v}{S_{va}} = e^{-\beta\zeta} \tag{10.57}$$

Velikanov obtained the variation of ζ with η and α by numerical integration. The results are presented both in figures and tables for convenience in application. Fig. 10.24 is a comparison of Eq. (10.57) with experimental data, and the correspondence is satisfactory.

The most serious shortcoming of the gravitational theory is that the energy balance equation is not scientifically sound. In reality, the energy for suspension E_5 comes from the energy of turbulence that functions as the energy loss of the flow in order to overcome resistance. Hence, in the energy balance equation, that part of the dissipated energy should not be taken into account two times. Detailed comments on the gravitational theory have been presented by Zhang [23] and are not included herein.

10.4 TRANSPORT RATE OF SUSPENDED LOAD

10.4.1 Einstein formula for suspended load transport

If the vertical profiles of both the concentration S_{vy} and the velocity u_y are known, the discharge of suspended load passing through a cross section of unit area at y per unit time is $u_y S_{vy}$; integration of $u_y S_{vy}$ over the depth yields the discharge of suspended sediment per unit width.

In practical applications, one difficulty remains because neither the diffusion theory nor the gravitational theory gives more than a relative quantity. From Eq. (10.16) or Eq. (10.57), the concentration at any position remains unknown unless S_a, the concentration at the reference position a distance "a" above from the bed, is known. In fact, S_a cannot be determined theoretically. This point is explained in more detail in the next chapter.

Another difficulty is that the upper and lower limits for the integration need to be determined. The simplest way is to integrate from the bed to the free surface to get the total sediment discharge. But since Eqs. (7.38) and (10.16) both approach infinity for the velocity and concentration at $y = 0$, some researchers have adopted Eqs. (10.18) and (10.19) and integrated them instead [24]. However, even though the mathematical difficulty is thus eliminated, the physical meaning of the result thus obtained remains questionable because the validity of the suspended load theory near the bed is questionable.

Essential condition for the theory of suspended load are that the particle size be much smaller than the sediment-carrying water element and that the horizontal components of particle velocity and flow velocity are the same. In the vertical direction, sediment particle can move relative to the surrounding water. The size of a turbulent eddy in turbulent flow is known to be about the same as the mixing length. In the region near the bed, the mixing length is proportional to the distance from the bed, and sizes of the particles and the eddies are about the same. Furthermore, due to the large velocity gradient near the bed, the velocities at the top and bottom of a particle can be quite different. The flows at the two points can even be in opposite directions, so that the particle velocity differs greatly from the local flow. Besides, near the bed a particle can be lifted up into suspension by the water at one instant, and then its downward motion can carry it into another eddy or down to the bed at another. Often sediment in this region moves through the various eddies and its weight is supported by the bed rather than by the turbulence. In Chapter 5 this region is called the bed layer. In it, bed load moves by sliding, rolling, or saltating. And as pointed in Chapter 9, the law of bed load motion is completely different from that for suspended load. Since the bed load motion is dominant in the bed layer, (the layer below the suspension region and above the bed), the extension of the concentration distribution for the suspended load to the region near the bed is not theoretically feasible. As pointed out in Section 10.1, some sediment in the near bed region can be lifted up from the bed by turbulence bursting, but a major portion of the sediment that is lifted there would probably move in the form of bed load rather than suspension load.

If a in Eqs. (10.16) and (10.57) denotes the thickness of the bed layer, then the suspended sediment discharge per unit width can be expressed as

$$g_s = \gamma_s \int_a^h S_{vy} u_y dy \qquad (10.58)$$

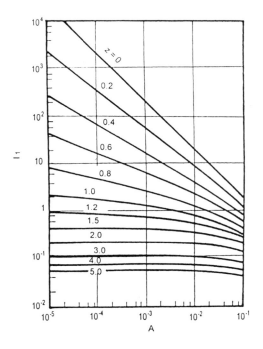

Fig. 10.25 Relationship of I_1 and A for suspended sediment discharge with z as a parameter (after Einstein, H.A.)

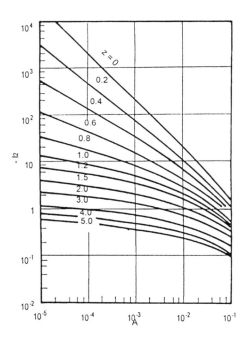

Fig. 10.26 Relationship of I_2 and A for suspended sediment discharge with z as a parameter (after Einstein, H.A.)

$$A = \frac{a}{h} \tag{10.59}$$

is used, then after substituting Eqs. (7.38) and (10.16) into Eq. (10.58) and simplifying, one obtains

$$g_s = 11.6\gamma_s U_* S_{va} a \left\{ 2.303 \log\left(\frac{30.2h}{\Delta}\right) \cdot I_1 + I_2 \right\} \tag{10.60}$$

in which

$$I_1 = 0.216 \frac{A^{z-1}}{(1-A)^z} \int_A^1 \left(\frac{1-y}{y}\right)^z dy$$

$$I_2 = 0.216 \frac{A^{z-1}}{(1-A)^z} \int_A^1 \left(\frac{1-y}{y}\right)^z \ln y \, dy \tag{10.61}$$

Clearly, I_1 and I_2 are functions of A and z, and their values can be obtained by numerical integration with the results shown in Figs. 10.25 and 10.26.

10.4.2 Velikanov formula for suspended load transport

An expression for the suspended sediment discharge can also be derived using the gravitational theory. Velikanov integrated the differential equation Eq. (10.51) rather than that of Eq. (10.57) and took the bed as the lower limit of integration. If the sediment concentration is small, $1 - \bar{S}_v \approx 1$, and one can then integrate Eq. (10.51) over the entire flow depth in the form

$$\int_0^h g\bar{u} J dy = \int_0^h \bar{u} \frac{d}{dy} \overline{u'v'} dy + \int_0^h \frac{\rho_s - \rho}{\rho} g\omega \bar{S}_{vy} dy \tag{10.62}$$

in which $d(\overline{u'v'})/dy$ represents τ/ρ. Since τ is proportional to the square of the velocity,

$$\int_0^h \bar{u} \frac{d}{dy} \overline{u'v'} dy = bU^3$$

can be used, in which b is a coefficient and U the velocity averaged over the depth. Eq. (10.62) can now be written in the form

$$gJUh = bU^3 + \frac{\rho_s - \rho}{\rho} g\omega S_{vm} h$$

in which S_{vm} is the average concentration over the depth. If

$$\lambda = \frac{ghJ}{U^2}$$

is introduced, this expression can be simplified as follows:

$$1 = \frac{bU^2}{ghJ} + \frac{\rho_s - \rho}{\rho}\frac{\omega}{JU}S_{vm} = \frac{b}{\lambda} + \frac{\rho_s - \rho}{\rho}\frac{\omega}{JU}S_{vm} \qquad (10.63)$$

For clear water, $S_{vm}=0$, so

$$b = \frac{ghJ}{U^2}\bigg|_{S_{vm}=0} = \lambda_0$$

Under certain flow conditions, the suspended load may reach the state of saturation (the maximum sediment carrying capacity of the flow). In this case,

$$\lambda = \lambda_\kappa$$

In the following, λ_0 / λ_k is approximately taken as a constant.

Substitution of the foregoing conditions into Eq. (10.63) yields

$$\frac{\rho_s - \rho}{\rho}\frac{\omega}{JU}S_{vm} = 1 - \frac{\lambda_0}{\lambda_k} \qquad (10.64)$$

in which S_{vm} is the saturated average concentration over the depth.

The average flow velocity can be determined from the following

$$U = \frac{1}{h}\int_0^h u_y\, dy = \frac{1}{h}\int_0^h \frac{\sqrt{ghJ}}{k}\ln\left(1+\frac{y}{\Delta}\right)dy = c\frac{\sqrt{ghJ}}{k} \qquad (10.65)$$

in which

$$c = (1+\alpha)\{\ln(1+\alpha)-1\} = f(\alpha)$$
$$\alpha = \Delta / h$$

Eq. (10.65) gives the relationship between the mean velocity and the friction velocity. The substitution of Eq. (10.65) into Eq. (10.64) yields

$$\frac{\rho_s - \rho}{\rho}\frac{\omega\kappa}{cJ\sqrt{ghJ}}S_{vm} = 1 - \frac{\lambda_0}{\lambda_k}$$

or

$$\frac{\beta}{c} S_{vm} = 1 - \frac{\lambda_0}{\lambda_k} = \text{const}$$

Hence,

$$S_{vm} \sim \frac{c}{\beta} = \frac{\rho}{\rho_s - \rho} \frac{\kappa^2}{f^2(\alpha)} \frac{U^3}{gh\omega}$$

The above expression can be rewritten in a more general form

$$S_{vm} = K \frac{U^3}{gh\omega} \qquad (10.66)$$

in which K is a constant that must be determined. Eq. (10.66) is the formula for the depth-averaged concentration of suspended load.

Researchers at WIHEE (Wuhan Institute of Hydraulic and Electric Engineering) made an extensive analysis of field data collected from rivers and canals including Yangtze River, Yellow River, Yongding River, People's Victory Canal, and Qingtong Xia Irrigation System; they concluded that Eq. (10. 66) should be modified to the following [26]

$$S_{vm} = k \left(\frac{U^3}{gh\omega} \right)^m \qquad (10.67)$$

in which the coefficients k and m are functions of ($U^3 / gh\omega$) as shown in Fig. 10.27.

10.4.3 Bagnold formula for suspended load transport [10]

Bagnold investigated the suspended load discharge using the same method he used for the bed load discharge.

In water flow, sediment particles are known to settle with a fall velocity ω, but the centroid of the entire suspended load can still be maintained at a certain elevation. Thus the water flow must continuously lift sediment with an upward velocity equal to ω. In the water column above a unit bed area, the work done by turbulence to suspend sediment is equal to $W_s' \omega$, where W_s' is the total submerged weight of suspended sediment in the column. The suspended sediment discharge can be expressed as

$$g_s' = W_s' \overline{u}_s \qquad (10.68)$$

in which $\overline{u}_s$ is the velocity of suspended load averaged over the depth

The turbulence energy required to maintain the suspension of sediment actually comes from the potential energy of the water flow. Hence, a relationship should exit between the energy required for sediment suspension and the potential energy loss. In a manner similar to that for Eq. (9.16), which represents the amount of flow potential energy used to sustain the bed load motion, the work done for sediment suspension can be expressed as

$$W_s'\omega = \tau_0 U(1-e_b)e_s \tag{10.69}$$

in which e_b and e_t are the efficiencies for the motions of bed load and suspended load, respectively.

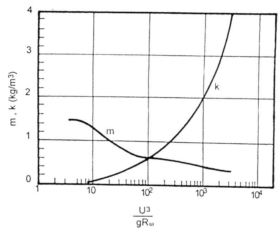

Fig.10.27 Variation of k and m with $U^3/gh\omega$, Eq.(10. 27)

By combining Eq. (10.68) with Eq. (10.69), one can readily obtain the expression

$$g_s' = \tau_0 U \frac{e_s \overline{u}_s}{\omega}(1-e_b) \tag{10.70}$$

Since suspended particles move with the same velocity as the local flow, one can write

$$\overline{u}_s = \frac{1}{h-\alpha}\int_\alpha^h S_v u dy$$

in which "a" is the distance from the lower boundary of the suspension region to the bed, and S_v the volume concentration of suspended load at a distance y above bed. Furthermore,

$$U = \frac{1}{h} \int_0^h u \, dy$$

Since the velocity increases and the sediment concentration decreases with y, the average velocity of what is generally smaller than the depth-averaged velocity U. If α stands for the ratio of these two velocities ($\alpha = \overline{u_s}/U < 1$), then Eq. (10.70) can be written

$$g'_s = \tau_0 U \frac{U}{\omega} e_s (1 - e_b) \alpha \tag{10.71}$$

Bagnold reviewed the laboratory data and obtained

$$e_s (1 - e_b) \alpha = 0.01$$

Thus, the Bagnold formula for suspended load discharge becomes

$$g'_s = 0.01 \tau_0 U \frac{U}{\omega} \tag{10.72}$$

In Chapter 9, the transport efficiency of bed load transport e_b was shown to vary generally within the range of 0.11 to 0.15. As an example, if $e_b = 0.13$ and $\alpha = 0.25$, then the transport efficiency of suspended load transport is 0.046; if $\alpha = 0.5$, the efficiency is 0.023. These values indicate that only a small portion of the potential energy of the flow is consumed by turbulence in suspending the sediment.

One can show that the Velikanov formula for suspended sediment discharge can be derived from the concept proposed by Bagnold in the following way. First, Eq. (10.69) is rewritten in the form

$$E_5 = e_s (1 - e_b) E_1$$

If the sediment concentration, is small, so that $1-S_V = 1$, the following relationships hold:

$$E_1 = \gamma U J$$

$$E_5 = (\gamma_s - \gamma) S_{vm} \omega$$

Hence

$$S_{vm} = e_s (1 - e_b) \frac{\gamma}{\gamma_s - \gamma} \frac{UJ}{\omega}$$

With the Chezy resistance formula, the above expression becomes

$$S_{vm} = e_s \left(1 - e_b\right) \frac{\gamma}{\gamma_s - \gamma} \frac{g}{C^2} \frac{U^3}{gh\omega} \tag{10.73}$$

in which C is the Chezy coefficient. A comparison of Eqs. (10.66) and (10. 73) indicates that

$$K = e_s \left(1 - e_b\right) \frac{\gamma}{\gamma_s - \gamma} \frac{g}{C^2} \tag{10.74}$$

10.5 NON-EQUILIBRIUM SEDIMENT TRANSPORT

Section 10.3 treats the vertical concentration distribution of suspended load for steady uniform flow of both water and sediment. This section treats the special case of non-equilibrium sediment transport in which the distribution of concentration varies in the streamwise direction even though the flow of water is steady and uniform. Typical examples of such a transport are the degradation process induced by clear water erosion downstream of a newly built dam and the aggradation process in a sediment silting canal.

As shown in section 10.1, the diffusion equation of sediment motion in *2-D* flow is Eq. (10.5). For simplicity, the following approximations are introduced.

1. Sediment motion is steady

$$\partial S_v / \partial t = 0$$

2. The streamwise variation of the sediment exchange coefficient is negligible

$$\partial \varepsilon_x / \partial x = 0$$

3. The second derivative of sediment concentration with respect to x is negligible compared to that in the y direction.

$$\partial^2 S_v / \partial x^2 << \partial^2 S_v / \partial y^2$$

For these conditions, the diffusion equation of sediment transport becomes

$$u \frac{\partial S_v}{\partial x} = \varepsilon_y \frac{\partial^2 S_v}{\partial y^2} + \frac{\partial \varepsilon_y}{\partial y} \frac{\partial S_v}{\partial y} + \omega \frac{\partial S_v}{\partial y} \tag{10.75}$$

For a uniform sediment, the equation of non-equilibrium sediment transport is the solution to this differential equation with suitable boundary conditions.

10.5.1 Recovery of sediment concentration along the flow direction by scouring

10.5.1.1 For the case of ε_y independent of y

If the variation of sediment exchange coefficient with elevation can be neglected and its depth-averaged value (Eq. (10.10)) is used, then Eq. (10.75) can be further simplified,

$$u\frac{\partial S_v}{\partial x} = \varepsilon_y \frac{\partial^2 S_v}{\partial y^2} + \omega \frac{\partial S_v}{\partial y} \tag{10.76}$$

Several studies have been based on the integration of this equation, by Hou[*], Mai [27], Apmann [28], and Zhang [29].

Hou and his colleagues studied the condition that the velocity profile at inflow is uniform, and they defined the boundary conditions as follows:

1. Free surface condition. At the free surface, $y = h$, the upward transport by turbulent diffusion is the same as that due to sediment settling, so that no sediment crosses the free surface.

$$\varepsilon_y \frac{\partial S_v}{\partial y} + \omega S_v = 0$$

2. Channel bed condition. The sediment concentration at the bed approaches the saturation value S_{v0} within a relative short distance. Thus, at $y = 0$,

$$S_v = S_{vo}$$

3. At the entrance to the section, $x = 0$,

$$S_v = S_{vo} f(y)$$

If the inflowing water is clear, then $f(y) = 0$.

The boundary conditions and the process of recovery of sediment concentration in the direction of flow are shown in Fig.10.28. The objective is to determine the sediment concentration distribution $S_v(x,y)$ throughout the flow field.

For these conditions, the solution to the differential equation has the form

---

[*] Hou et al. Theoretical Analysis of Streamwise Recovery of Sediment Concentration along River Course, *Research Report*, Institute of Water Conservancy and Hydroelectric Power Research, Oct. 1964.

$$S_v(x, y) = S_{v0} \exp\left(-\frac{\omega}{2\varepsilon_y} y\right)\left[\exp\left(-\frac{\omega y}{2\varepsilon_y}\right) - \exp\left(-\frac{\omega^2 x}{4\varepsilon_y U}\right)\sum_{n=1}^{\infty} A_n \exp\left(-\frac{\varepsilon_y \beta_n^2 x}{U}\right)\sin \beta_n y\right]$$

$$(10.77)$$

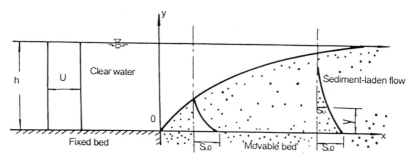

Fig. 10.28 The variation of sediment concentration in a channel with movable bed starting with clear water at the point of inflow

in which

$$A_n = \cfrac{4}{\cfrac{1}{2}\beta_n h\left[\left(\cfrac{\omega}{\varepsilon_y \beta_n}\right)^2 + 4\right] + \cfrac{\omega}{\varepsilon_y \beta_n}}$$

$$\cdot \cfrac{1}{\cfrac{h}{2} - \cfrac{\varepsilon_y \omega}{\omega^2 + 4\varepsilon_y^2 \beta_n^2}} \int_0^h f(y) \cdot \exp\left(\frac{\omega y}{2\varepsilon_y}\right)\sin \beta_n y\, dy$$

$$(10.78)$$

and the coefficient β_n can be calculated from

$$\tan \beta_n h = -\frac{2\varepsilon_y \beta_n}{\omega}$$

$$(10.79)$$

The depth-averaged concentration can be obtained from the integral of Eq. (10.77) with respect to y over the depth h.

An example of the recovery of sediment concentration resulting from clear water erosion as calculated from Eq. (10.77) is shown for flow with a slope of $J = 0.0001$, depth $h = 2.4$ m, mean velocity $U = 1.9$ m/s, and particle sizes of 0.04 mm and 0.1 mm. The computed results, shown in Fig. 10.29, indicate that the distance required for the recovery of concentration from clear water to the saturation state is in general not long if the sediment is uniform and the streamwise variation of sediment size gradation caused by clear water erosion is negligible. In the example shown in Fig.10.29, the concentration recovers 89% of the saturated value within a distance of 800 m.

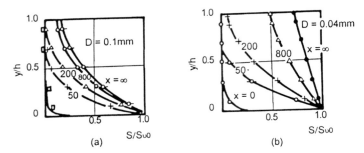

Fig. 10.29 Recovery of sediment concentration by clear water erosion along a channel.

10.5.1.2 Case in which ε_y varies with y

Hjelmfelt and Lenau studied the effect of the variation of the sediment exchange coefficient with y [30]. The nature of the function is expressed as follows

$$\varepsilon_y = \beta \kappa y U_* \left(1 - \frac{y}{h}\right)$$ (10.80)

in which β is the ratio of the sediment exchange coefficient to the momentum exchange coefficient.

In the study by Hou et al., sediment was considered to be suspended over the entire flow region; however, the real situation is somewhat different. Hjelmfelt and Lenau made a distinction between the suspension region and the bed layer, and they allocated the sediment motion in the bed layer to bed load. In their derivation for non-equilibrium sediment transport, however, they included only the suspended load.

To make the differential equation Eq. (10.75) dimensionless, Hjelmfelt and Lenau used the quantities

$$\eta = \beta \kappa \frac{U_*}{U}$$

$$X = \eta \frac{x}{h}$$

$$Y = \frac{y}{h}$$ (10.81)

$$C = \frac{S_v}{S_{va}}$$

$$z = \frac{\omega}{\beta \kappa U_*}$$

450

in which S_{va} is the concentration at the lower limit of the suspension region, where $y = a$. Then they rewrote Eq. (10.75) in the form

$$\frac{\partial C}{\partial X} = z \frac{\partial C}{\partial Y} + \frac{\partial}{\partial Y}\left[Y(1-Y)\frac{\partial C}{\partial Y}\right]$$

(10.82)

The boundary conditions are as follows:

1. Free surface condition. At $y = h$ $(Y=1)$, the sediment concentration is zero, or

$$C = 0$$

2. Lower limit of the suspension region. At $y = a$ $(Y=a/h=A)$,

$$S_y = S_{ya} \quad and \quad C = 1$$

3. Initial condition. At $x = 0$ $(X=0)$, $a < y \le h$ $(A<Y \le 1)$,

$$C=0$$

For these boundary conditions, the solution is

$$C(X,Y) = \left(\frac{A}{1-A}\right)^z \left(\frac{1-Y}{Y}\right)^z$$

$$+2\sum_{m=1}^{\infty} \frac{\alpha_m P(Y;\alpha_m)}{\left(\alpha_m^2 - \frac{1}{4}\right) \dfrac{\partial P(A;\alpha_m)}{\partial x}} \exp\cdot\left[-X\left(\alpha_m^2 - \frac{1}{4}\right)\right]$$

(10.83)

the depth-averaged concentration for the suspended load is

$$C_m(X) = \frac{A^z}{(1-A)^{1+z}} \int_A^1 \left(\frac{1-Y}{Y}\right)^z dy$$

$$+2A\sum_{m=1}^{\infty} \frac{\alpha_m \dfrac{\partial P(A;\alpha_m)}{\partial Y}}{\left(\alpha_m^2 - \frac{1}{4}\right) \dfrac{\partial P(A;\alpha_m)}{\partial x}} \cdot \exp\left[-X\left(\alpha_m^2 - \frac{1}{4}\right)\right]$$

(10.84)

in which

$$P(Y;\alpha_m) = (1-Y)^z F\left(\frac{1}{2}+z+\alpha_m, \frac{1}{2}+z-\alpha_m, 1+z, 1-Y\right)$$

In the above expression, F is a hypergeometric function that can be expressed as

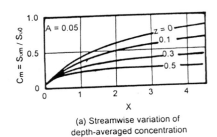

(a) Streamwise variation of
depth-averaged concentration

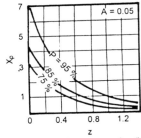

(b) Distance for the recovery of sediment
concentration expressed in term of percent P

Fig. 10.30 Recovery of sediment concentration due to clear water erosion with inclusion of variation of ε_y with respect to y (after Hjelmfelt, A.T. and C.W. Lenau)

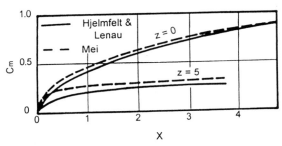

Fig. 10.31 Comparison of the recovery of sediment concentration due to clear water erosion with and without the vertical variation of ε_y

$$F(d,b,c,Y) = \sum_{n=0}^{\infty} \frac{d_n + b_n}{c_n} \frac{Y^n}{n!}, \quad |Y| < 1$$
$$\begin{cases} d_n = d(d+1)(d+2)\cdots(d+n-1); n \geq 1 \\ d_0 = 1 \end{cases}$$

a_m (m=1,2,3...) stands for the roots of the following equation

$$F\left(\frac{1}{2} + z + \alpha, \ \frac{1}{2} + z - \alpha, \ 1 + z, \ 1 - A\right) = 0$$

α is a constant.

Fig. 10.30 shows graphically the solution to Eq. (10.84) for $A = 0.05$. The curves in Fig. 10.30a show the relationship of C_m to X for various values of z; X_p in Fig. 10.30b stands for the distance (expressed as the dimensionless parameter X) required for the recovery of the concentration. The results are expressed in terms of P, the percent of the saturation concentration. The figures show that, the distance required to reach the saturation condition increases with as z decreases, and for a smaller value of z, more sediment diffuses toward the upper regions of the flow.

Fig. 10.31 is a comparison of the analytical results of Hjelmfelt and Lenau with those of Mei. Mei did neglect the vertical variation of the sediment exchange coefficient, as in Hou's analysis. The figure shows that the effect of the variation of ε_y with respect to y is small. A detailed comparison shows that if the vertical variation of ε_1 is not considered, the recovery is faster near the entrance, erosion in the downstream region takes place more slowly, and the distance to reach the equilibrium is shorter.

10.5.1.3 Armoring phenomenon in degradation of channel bed

In the foregoing discussion, the sediment was taken to be uniform. Hence, the bed material does not change during degradation; i.e., the sediment carrying capacity of the flow does not vary along the river course unless the cross section of the flow changes. For this condition, the recovery distance is the distance over which the river bed is scoured. The studies conducted by various authors confirm that this distance is usually not long. In nature, however, the bed material is composed of sediment that is a mixture of particle sizes. Because the flow can carry fine particles more readily than coarse ones, most of the fine sediment is carried away while the coarse sediment stays in place. The result is called the armoring of bed, and it causes a decrease in the sediment-carrying capacity. This phenomenon starts upstream and progresses downstream. For this reason, the distance for the sediment concentration to recover differs from that for erosion. Although the former is rather short, the latter distance is quite long [31].

To resolve the more complex situation of armoring, one must use an equation for sediment transport that is applicable to sediment of various sizes. In addition, one must determine the thickness of the bed region in which the bed material exchanges with the material carried by the flow during degradation. On a study of the degradation and armoring in the Yellow River caused by the clear water released from Sanmanxia reservoir, the authors took the thickness of the bed region, in which the bed material exchanges with the bed load, to be two times the amplitude of bed level variation in a year. The Einstein formula for sediment carrying capacity was applied for various sub-reaches and time intervals; the agreement achieved between the computed result and the real situation was fairly good (Chapter 11).

10.5.2 Decrease of sediment concentration along the flow due to deposition

The analysis of deposition is quite similar to that for the recovery of sediment concentration; only the boundary and initial conditions are different.

From Zhang's analysis [29], the depth-averaged value of the sediment exchange coefficient ε_y was used, and the effect of its variation in the vertical direction was neglected. The diffusion equation Eq. (10.76) is applicable. Two conditions for the channel bed are introduced. In the first, no re-suspension takes place because the velocity is too low, as in a desilting basin or in the deeper part of a reservoir. In other words, once sediment settled out and is in contact with the bed, it can not be lifted up or re-suspended. In the second, re-suspension is possible. The velocity is large enough that sediment that reaches the bed can be re-suspended, as in the shallow region of a reservoir.

The boundary conditions are as follows:

1. At the free surface, $y = h$,

$$\varepsilon_y \frac{\partial S_v}{\partial y} + S_v \omega = 0$$

2. At the bed, $y = 0$, for no re-suspension,

$$\frac{\partial S_v}{\partial y} = 0$$

re-suspension is possible

$$\varepsilon_y \frac{\partial S_v}{\partial y} + \omega S_{v0} = 0$$

in which S_{v0} is the saturation concentration at the bottom.

3. At the entrance, $x = 0$,

$$S_v = S_{vi} \exp\left(-\frac{\omega}{\varepsilon_y} y\right)$$

in which S_{vi} is the concentration at the bed surface ($S_{vi} > S_{v0}$).

The solutions to the diffusion equation for these conditions are as follows:

First, For the case of no re-suspension,

$$\frac{S_v(X,Y)}{S_{vi}} = 2K_1 \exp\left(-\frac{K_1}{2} Y\right) \cdot \sum_{n=1}^{\infty} \frac{\alpha_n^2}{\left(\alpha_n^2 + \frac{K_1^2}{4}\right)\left(\alpha_n^2 + \frac{K_1^2}{4} + K_1\right)} \cdot \exp\left[-\left(\alpha_n^2 + \frac{K_1^2}{4}\right)K_2 X\right]$$

(10.85)

$$\frac{S_{vm}(X)}{S_{vi}} = 2K_1^2 \sum_{n=1}^{\infty} \frac{\alpha_n^2}{\left(\alpha_n^2 + \frac{K_1^2}{4}\right)^2 \left(\alpha_n^2 + \frac{K_1^2}{4} + K_1\right)} \cdot \exp\left[-\left(\alpha_n^2 + \frac{K_1^2}{4}\right)K_2 X\right]$$ (10.86)

where

$$X = \frac{x}{h}$$

$$Y = \frac{y}{h}$$

$$K_1 = \frac{\omega h}{\varepsilon_y}$$

(10.87)

$$K_2 = \frac{\varepsilon_y}{Uh}$$

and α_n is the solution to the following equation

$$2 \cot \alpha_n = \frac{\alpha_n}{K_1/2} - \frac{K_1/2}{\alpha_n}$$

(10.88)

Second, For the case that re-suspension can occur,

$$\frac{S_v(X,Y)}{S_{vo}} = \exp(-K_1 Y) + \frac{S_{vi} - S_{vo}}{S_{vo}} \exp\left(-\frac{K_1}{2} Y\right)$$

$$\cdot \sum_{n=1}^{\infty} \frac{2\alpha_n^2 \left(\cos\alpha_n Y + \frac{K_1}{2\alpha_n}\sin\alpha_n Y\right)}{\left(\frac{\alpha_N^2}{K_1} + \frac{K_1}{4}\right)\left(\alpha_n^2 + \frac{K_1^2}{4} + K_1\right)} \cdot \exp\left[-\left(\alpha_n^2 + \frac{K_1^2}{4}\right)K_2 X\right]$$

(10.89)

$$\frac{S_{vm}(X) - S_{v*}}{S_{vi} - S_{v0}} = \frac{K_1}{1 - e^{-K_1}} \sum_{n=1}^{\infty} \frac{2\alpha_n^2}{\left(\frac{\alpha_n^2}{K_1} + \frac{K_1}{4}\right)^2 \left(\alpha_n^2 + \frac{K_1^2}{4} + K_1\right)} \cdot \exp\left[-\left(\alpha_n^2 + \frac{K_1^2}{4}\right)K_2 X\right]$$

(10.90)

in which S_{v*} is the depth-averaged value of the saturation concentration and can be expressed as

$$S_{v*} = \frac{S_{v0}}{K_1}\left[1 - \exp(-K_1)\right]$$

(10.91)

Since the expressions on the right-hand sides of Eq. (10.86) and Eq. (10.90) are the same, the relationship between $S_{vm}(X)/S_{vi}$ and $K_2 X$ is the same as that between $[S_{vm}(X) - S_{v*}]/[S_{vi} - S_{v0}]$ and $K_2 X$, as shown in Fig.10.32. The figure shows that finer sediment requires a greater distance to deposit. If re-suspension occurs, the distance for sediment to settle out is much longer. Because the bed conditions given by Zhang are not really satisfactory, his results are only approximations to the real situations.

If the sediment being deposited is non-uniform, 1. the original top layer of bed material is getting finer due to deposition; and 2. a size gradation due to sorting during deposition causes the top layer to become finer and finer with increasing distance. The effect of the streamwise variation of bed material affects the sediment-carrying capacity of the flow in a way that is just opposite to that of the flow depth. The former tends to increase the sediment-carrying capacity in the streamwise direction, whereas the latter causes a streamwise decrease. Hence, for a given condition of flow, the distance for the sediment mixture to deposit is greater than that for uniform sediment. In addition, the situation in the reservoir, where flow is not uniform, differs from that in a desilting canal. Han [33] made a detailed study of non-equilibrium sediment transport in a reservoir, he took into account the effects of non-uniformity of suspended load.

10.6 SPREADING OF CONTAMINANT IN A WATER ELEMENT [34]

In recent years, environmental pollution has become progressively more of a public concern. Waste disposal into rivers and the consequent pollution are highly important concerns for environmental protection. Contaminants diffuse rapidly because of molecular motion and turbulence once they are released into a river, lake or coastal area during the disposal of industrial and agricultural waste water. Velocity gradients in both vertical and horizontal directions, cause particles to move at different speeds in water bodies, and this process also contributes to the spreading of contaminants that may have been confined initially within a relatively small region. Diffusion and advection carry the contaminants downstream with the water flow, and these processes also help to spread them to all parts of a water element.

The major part of the substances that come to a body of water from industrial and agricultural waste water move as suspended load; only a small part moves as bed load [35]. Hence the following section of this chapter deals with the spreading of suspended load. The differences between the contaminants and the suspended sediment lie in the facts that most of the contaminants are dissolvable or colloidal and the difference in the specific weights of these materials and water is small. Such materials are often called

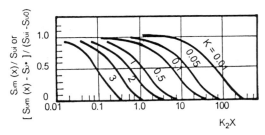

Fig. 10.32 Sediment deposition in a desilting canal for two cases: sediment that falls to the bottom can or cannot be re-suspended.

neutrally suspended material or just neutral material. The fall velocity of neutral material is negligible, so they are less complicated to deal with than is sediment. Nevertheless, difficulties arise from the three-dimensionality of contaminant diffusion. The waste water enters a river, lake or coastal area from a point source or a line source. Because of the many factors involved, contaminant spreading is then much more complex than the one-dimensional diffusion discussed in section 10.3.

Spreading of contaminants in a water element is an important part of environmental science, and much research has been published on related topics. This chapter is a brief introduction to selected problems, and it explains how the mechanics of sediment motion is a part of a fundamental science that is used in various other scientific fields.

10.6.1 Diffusion and dispersion of neutral material

Relevant results have been obtained from field measurements on the Mississippi

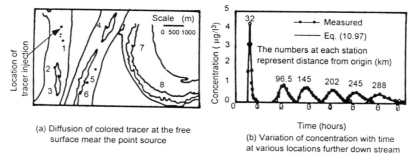

(a) Diffusion of colored tracer at the free surface mear the point source

(b) Variation of concentration with time at various locations further down stream

Fig. 10.33 Diffusion and dispersion of dyed tracer in Mississippi River
(after McQuivey, R.S. and T.K. Keefer)

River, U.S.[36]. A dyed tracer solution with the same specific weight as water was injected into the river, the river discharge at that time was 22,430 m^3/s. The concentration of the tracer was then observed at several downstream stations. As tracer moved downstream with the flow, it diffused rapidly in all directions. Fig.10.33a shows the location of the dyed solution at the free surface, from observations at several locations 13 km downstream of the point of injection. The diffusion of the dyed solution in both the streamwise direction and the transverse direction is shown in the figure. The measured data also indicate that, at a point 15 km downstream of the source, the concentrations at the top and bottom, with a depth of 15 m, are nearly the same, indicating complete diffusion in the vertical direction. The diffusion in the horizontal direction is completed only at a distance of some 80 km from the source. And further downstream from this location, the dyed tracer continues to spread in the streamwise direction, as shown by both the decrease in the peak concentration, and the elongation of the time the dyed tracer is detected at various locations, shown in Fig. 10.33b. A distinction is made between the spreading of dyed tracer between the source and a point some 80 km downstream, which is called diffusion, and the spreading further downstream, which is called dispersion.

The diffusion phase described above is equivalent to that shown in Fig. 10.5. The random motion that is characteristic of turbulent flow is the major cause of the spreading of the dyed tracer. The diffusion equation for the tracer can be obtained, for a unit volume of water at a certain position in the flow by using the mass conservation equation Eq. (10.6). Because the settling velocity of the material is negligible, the three-dimensional diffusion equation can be simplified into the form

$$
\frac{\partial S_v}{\partial t} = -u\frac{\partial S_v}{\partial x} - v\frac{\partial S_v}{\partial y} - \omega\frac{\partial S_v}{\partial z} + \varepsilon_x\frac{\partial^2 S_v}{\partial x^2} + \frac{\partial \varepsilon_x}{\partial x}\frac{\partial S_v}{\partial x}
$$
$$
+ \varepsilon_y\frac{\partial^2 S_v}{\partial y^2} + \frac{\partial \varepsilon_y}{\partial y}\frac{\partial S_v}{\partial y} + \varepsilon_z\frac{\partial^2 S_v}{\partial z^2} + \frac{\partial \varepsilon_z}{\partial z}\frac{\partial S_v}{\partial z}
$$

(10.92)

The change to the dispersion phase of the dyed tracer is illustrated in Fig. 10.34. At some instant t_4, the dyed tracer has spread over the entire cross-section. Subsequently, at t_5, after the dyed tracer has moved still further downstream the way the dyed tracer concentration can vary in the flow direction is limited; the three-dimensional process then reduces to a one-dimensional one. If

$$
\bar{u} = U + u''
$$
$$
\bar{S}_v = S_v + S_v''
$$

in which $\bar{u}$ and $\bar{S}_v$ are the time-averaged velocity and the time averaged concentration at a point, U and S_v are the mean velocity and concentration over a cross section, and u'' and S_v'' are the differences between the time-averaged and the mean values.

If the velocity and concentration in Eq. (10.92) represent the time-averaged values, then one can substitute the above expressions into Eq. (10.92) and take the averages over the cross-section to obtain

$$
\frac{\partial S_v}{\partial t} + U\frac{\partial S_v}{\partial x} = \varepsilon_x\frac{\partial^2 S_v}{\partial x^2} + \frac{\partial(\overline{-u''S_v''})}{\partial x}
$$

(10.93)

The double bar denotes the additional averaging over the cross-section. With the Taylor simplification that the advection term due to u'' is also proportional to the streamwise concentration gradient as in Eq. (10.3) and Eq. (10.4), then by adding the term $\left(\varepsilon_y\frac{\partial S_v}{\partial x}\right)$, i.e., by assuming that [37]

$$E_x \frac{\partial \bar{S_v}}{\partial x} = -\overline{u''S_v''} + \varepsilon_x \frac{\partial \bar{S_v}}{\partial x} \quad S_v'' \langle\langle S_v \tag{10.94}$$

in which E_x is the longitudinal dispersion coefficient. The substitution of Eq. (10.93) into Eq. (10.94) yields

$$\frac{\partial \bar{S_v}}{\partial t} + U \frac{\partial \bar{S_v}}{\partial x} = E_x \frac{\partial^2 \bar{S_v}}{\partial x^2} \tag{10.95}$$

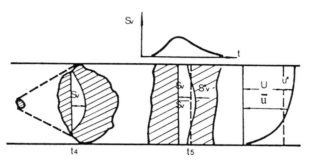

Fig.10.34 Dispersion of dyed material in two-dimensional flow
(after McQuivey, R.S., and T.K. Keefer)

and this is the equation for the dispersion of a dyed tracer. From Eq. (10.94) the longitudinal dispersion obviously depends on both the velocity profile and on diffusion in the transverse direction, because u'' depends on the former and S_v'' on the latter.

Diffusion and dispersion are different processes. Unfortunately, they are often used confusedly in some papers. To summarize the difference, diffusion refers to the mixing due to molecular motion (in a laminar flow field) or to turbulent fluctuations, thus it is used to describe the mixing at a given point of the flow field. In contrast, dispersion refers to the streamwise variation of mean concentration averaged over the cross section, and the variation is caused by the non-uniform velocity distribution over the cross section [38]. Of course, dispersion process includes some diffusion, but that part is quite small in comparison with the convection due to the non-uniformity of velocity distribution.

10.6.2 Determination of diffusion coefficients

For given boundary and initial conditions, one can readily solve Eq. (10.92) for the diffusion of the dyed tracer. As usual, one must make certain simplifications in accordance with the specific conditions in order to surmount the mathematical difficulties. A crucial step is to determine the diffusion coefficient for each of the three principal directions.

1. The vertical diffusion coefficient

As pointed out in section 10.2, the vertical distribution of diffusion coefficient ε_y is as shown in Fig.10.15, and the ratio of the depth-averaged value of $\varepsilon_y/U_* h$ is a constant equal to 0.067 as in Eq. (10. 10).

2. The transverse diffusion coefficient

Table 10.4 contains the results obtained for the transverse diffusion coefficient from laboratory tests and field measurements [39]. The tracers used in the laboratory tests included both soluble and floating material. Fig.10.35 shows the vertical distribution of the transverse diffusion coefficient ε_z , and it is quite different from the vertical distribution of the vertical diffusion coefficient (Fig.10.15).

From the data of flume studies using floating materials shown in Table 10.4, the mean value at the free surface is approximated

$$\frac{\varepsilon_z}{hU_*} \approx 0.20$$

and the depth-averaged value as

$$\frac{\varepsilon_z}{hU_*} \approx 0.15$$

Lau and Krishnappen conducted tests in a rough flume for various ratios of width to depth, and they found that $\varepsilon_z/(hU_*)$ appears to vary with both the friction coefficient f and the ratio B/h, as shown in Fig.10.36[40]. With greater frictional resistance, both the turbulence intensity and the ratio of the transverse diffusion

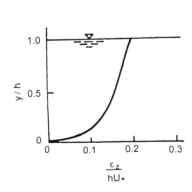

Fig.10.35 Vertical distribution of transverse diffusion coefficient (after Okoye, J.K.)

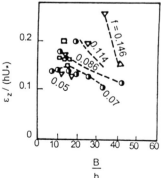

Fig.10.36 Variation of transverse diffusion coefficient with both resistance coefficient and ratio of width to depth (after Lau, Y.Lam and B.G. Krishnappen)

Table 10.4 Results of both laboratory and field measurements for the transverse diffusion coefficient. (after Okoye, J.K.)

(1) flume tests using tracers of colored soluble liquids

Author	Location	Channel bed	Flume width (m)	Mean depth (cm)	Mean velocity (cm/s)	Frictional velocity (cm/s)	Width-depth ratio	Frictional coefficient f	ε_z (cm²/s)	$\dfrac{\varepsilon}{U.h}$
Elder	-	smooth	0.36	1.2	21.6	1.59	0.033	0.043	-	0.16
Engmann	-	metal-bar roughness	0.12	4.0~6.5	19.7~26.4	-	0.032~0.053	0.051~0.068	0.86~1.55	0.152~0.180
Glover	-	smooth	2.42	14.1	66.3	7.16	0.058	-	-	0.07
Holly	-	concrete	1.20	9.7	11.2	5.19	0.081	0.023	0.93	0.16
Kalinshi and Pien	-	smooth	0.69	15.8	94.5	5.19	0.229	0.024	6.56	0.08
Lau et al	-	smooth and rough	0.3~0.6	1.27~4.96	15.5~33.7	0.93~2.80	0.0234~0.112	0.0127~0.145	0.34~1.41	0.108~0.259
Miller et al	-	hexagonal-bar rough	0.59	12.47~13.11	30.5~81.4	-	0.21	0.076~0.368	3.72~36.24	0.098~0.180
Okoyo	-	smooth and stone block rough	0.85~1.1	1.52~17.3	27.1~50.4	-	0.15~0.204	0.017~0.156	0.64~7.48	0.094~2.35
Sayre and Zhang	-	wood-block rough	2.39	14.8~37.1	23.5~47.6	-	0.0062~0.0155	0.098~0.211	9.57~35.76	0.179~0.192
Sullivan	-	smooth	0.80	7.32~10.20	19.3~22.9	-	0.092~0.128	0.022~0.023	0.91~1.18	0.108~0.133

Table 10.4 Continued

Prych	-	smooth and metal-bar rough	1.10	3.9~11.0	35.4~46.0	-	0.0355~0.101	0.015~0.080	1.06~3.52	0.136~0.179
(2) flume tests using tracers of floating materials										
Englund	-	sand material	2.20	5.5~17.3	30~34.4	1.6~3.6	0.025~0.079	-	4.0~6.5	0.204~0.234
Orlob	-	wide metal mesh	1.22	1.62~12.8	30.8	3.68	0.044	-	-	0.16
Sayre and Chamberlain	-	sand material	2.44	17.7	48.7	3.51	0.073	-	15.0	0.24
Sayre and Zhang	-	wood pier	2.38	14.7~37.1	23.5~47.5	3.9~6.04	0.062	-	13.3~59.2	0.20~0.26
Prych	-	smooth and rough	1.10	3.9~11.1	34~44.8	1.96~4.23	0.036~1.0	-	1.33~5.73	mean0.20
(3) natural watercourses										
Fisher	irrigation channel	sand material	18.3	67.4	64.6	6.25	0.037	-	-	0.24
Glover	Columbia River	sand material	305	306	135	8.76	0.010	-	-	0.73
Patterson and Gloyna	Colorado River	sand material	-	-	57.3~80.8	-	0.025	-	492	-
Yotsukura et al	Missouri River	sand material	226	270	175	7.40	0.012	-	-	0.60

462

coefficient to (U_*h) are greater. Their data also indicate that for a given depth, the secondary currents become weaker as the flume becomes wider, which corresponds to a decrease in the value of $\varepsilon_z/(hU_*)$. This result is in contradiction to that of Okoye [39].

Fischer used the Rozovskii formula for the transverse distribution of the flow velocity at a river bend, and he derived the following expression for the transverse diffusion coefficient for flow in a river bend

$$\frac{\varepsilon_z}{hU_*} \approx 0.25 \left(\frac{U}{U_*}\right)^2 \left(\frac{h}{R_c}\right)^2 \frac{1}{\kappa^5} \qquad (10.96)$$

in which R_C is the bend radius, and κ the Karman coefficient. The secondary currents at a river bend enhance the transverse diffusion, and a smaller radius results in more transverse diffusion.

Irregularity of cross section and secondary currents in bends, along with the many other types of secondary currents in a natural river, cause the transverse diffusion to be much greater there than it is in a flume, in which is nearly two-dimensional flow. In a deep river like the Missouri, $\varepsilon_z/(hU_*)$ can be as large as 0.60. In a more regular river, the depth limits the scale of turbulence. The transverse diffusion can be strong in a tributary or bay area where both the river channel and the length scales of the turbulence are large. Measurements in the southern part of San Francisco Bay, USA, yielded a value of $\varepsilon_z/(hU_*)=1$ [42].

3. Longitudinal diffusion coefficient

Experimental data for the longitudinal diffusion coefficient is quite limited. What is known indicates that the value of ε_x can be three times as large as ε_z [43,44].

In summary, the relative magnitude of diffusion coefficients in the three directions display the following trend:

$$\varepsilon_y < \varepsilon_z < \varepsilon_x$$

From Fig. 10.33a, diffusion in the longitudinal direction is much more rapid than is that in the transverse direction.

10.6.3 Solution of dispersion equation and determination of dispersion coefficients

10.6.3.1 Solution of dispersion equation

The standard form of the dispersion equation Eq. (10.95) can be rewritten in the form

$$S_v(x,t) = \frac{W}{2\gamma A\sqrt{\pi E_x t}}\exp\frac{-(x-Ut)^2}{4E_x t} \qquad (10.97)$$

in which W is the weight of colored tracer injected, γ the specific weight of both the water and colored tracer, and A the cross-sectional area of the channel.

Eq. (10.97) indicates that the time distribution of the colored tracer concentration at any location follows the normal curve, and that the peak value decreases as the distance of travel increases. For the data from the Mississippi River in Fig.10.33b, Eq. (10. 97) cannot be used to describe the longitudinal dispersion of the colored tracer until the colored tracer has traveled the distance L_x. Near the region of tracer injection, the concentration distribution is not symmetrical. This phenomenon is easily understood from the following example. The colored tracer is injected uniformly over an initial cross section, as shown in Fig.10.37a, and then the streamwise variation of the cross-sectional mean concentration is measured at some point downstream. Initially, since the effect of turbulent diffusion is not significant, the colored tracer spreads only as a result of the vertical velocity profile. As a result, the tracer near the free surface moves faster, and the tracer near the bed moves more slowly, as shown in Fig. 10.37b; also, neither the streamwise concentration distribution at a fixed instant nor the time distribution of concentration for a fixed location is symmetrical, as shown in Fig. 10. 37(c). This example indicates that in the advection region, where the non-uniformity of the velocity distribution is effective, the dispersion of the colored tracer can not be described by Eq. (10.97). Eq. (10.97) does not give reliable results unless the tracer has traveled far enough for the turbulence diffusion to make the cross-sectional distribution of tracer concentration nearly uniform. The factors of possible importance are the skewness and kurtosis of the concentration distribution before the tracer has traveled this distance.

These physical concepts and the use of dimensional analysis lead one to the following approximation relationship

$$L_x \sim \frac{l^2}{\varepsilon}U$$

in which l is the characteristic length scale, and ε is the turbulence diffusion coefficient. The magnitude of l can be taken as half the width of a symmetric cross section for a man-made channel, or the distance between the centerline of the main channel to the farthest bank for a natural river. As for ε, the transverse diffusion coefficient ε_z should be used because its action is the dominant one. From the fact

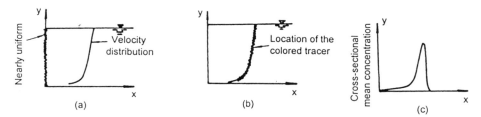

Fig.10.37 Spreading of colored tracer near point of tracer injection(after Fischer, H.B.)

that ε_z is proportional *to* $hU*$ and also from experimental results, the foregoing expression can be rewritten as [47]

$$L_x = 1.8 \frac{l^2}{R} \cdot \frac{U}{U_*}$$ (10.98)

in which the hydraulic radius R is used instead of the depth h.

Methods have been developed for the computation of dispersion in which advection is dominant, and details are given in relevant literature [48].

10.6.3.2 Determination of dispersion coefficient

1. Dispersion coefficient for two-dimensional open channel flow

Taylor was the first to present the dispersion equation in the form of Eq. (10.95), and he also derived the following analytical solution for the longitudinal dispersion coefficient in pipe flow

$$E_x = 10.1 r U_*$$ (10.99)

in which r is the radius of a circular pipe. Elder extended the method to two-dimensional open-channel flow and obtained [49]

$$E_x = 5.9 h U_*$$ (10.100)

The many subsequent experimental studies show that the order of magnitude of Eq. (10.100) is correct, but that the coefficient varies over the range from *5.9* to *13.0*. That is, Elder's value appears to be the minimum for this simple condition. The dispersion coefficient would be larger

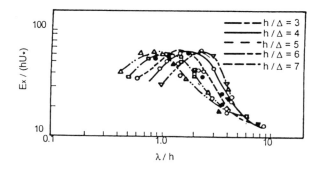

Fig. 10.38 Effect of roughness on the dispersion coefficient (after El-Hadi, N.D.Abd., and K.S. Davar)

465

than the value given by Eq. (*10.100*) if any transverse gradient of the velocity varies significantly.

The dispersion coefficient, like the diffusion coefficient, varies with the bed surface. For rough beds, $E_X/(hU_*)$ can vary with the bed resistance. El-Hadi and Davar conducted experiments, using a rough bed formed of wooden bars of height A and spacing *L*, and obtained the results shown in Fig.10.38 [50]. The figure shows that, if *L*< 6, the ratio $E_X/(hU_*)$ varies considerably with the relative height *A/h* of the wood bars and with the relative spacing *L/h*. If the relative roughness is large, the measured longitudinal dispersion coefficient is much greater than the theoretical value proposed by Elder. For the various combinations of the wood bar parameters in their experiments, the maximum value of $E_X/(hU_*)$ was 60.

Richardson and Miller showed that the dispersion coefficient is not greatly dependent on flow resistance if the velocity is small, and that the longitudinal dispersion coefficient increases with flow resistance if the velocity exceeds *0.5* m/s [51]. They also showed experimentally that the longitudinal diffusion coefficient ε_X is not more than *4%* of E_x. The result indicates the dominant effect of dispersion in the longitudinal spreading of contaminants.

2. Dispersion coefficient for three-dimensional open-channel flow

Table 10.5 gives some measured results for the dispersion coefficient in three-dimensional open-channel flow [45]. Fischer's results were obtained from tests in a flume with roughness elements added to the side walls that produced a transverse gradient of velocity like that for the three-dimensional flow in a natural river. From the table, the longitudinal dispersion coefficient for three-dimensional flow is much larger than that for two-dimensional flow. Observations for the Missouri River yielded values of $\varepsilon_X/(hU_*)$ ranging from 900 to 7,500.

The three factors that cause the dispersion coefficient to be so much greater for three-dimensional flow than it is for two-dimensional flow are the following.

1. In a natural river, the irregular cross-sections, sand dunes, and tributaries cause the velocity profile to be quite non-uniform so that the colored tracer spreads much more by advection.

2. The secondary currents in the bends of a natural river affect dispersion in two different ways. One is that the secondary currents enhance the transverse mixing of the contaminant, which will considerably reduce the longitudinal dispersion. The other is that the secondary currents cause a region of high velocity near the inner bank and one of low-velocity near the outer bank. The difference between them results in a large transverse gradient of velocity and enhances the longitudinal dispersion of the colored tracer. Usually, the latter effect exceeds the former.

Table 10.5 Results for longitudinal dispersion coefficient in three-dimensional flow, both laboratory and field measurements (after Fischer, H.B.)

Author	Channel	Depth (m)	Width (m)	U_* (cm/s)	E_X (m²/s)	E_x / hU_*
Fischer	trapezoidal channel with rough side walls	0.021~0.047	0.19~0.43	2.02~3.88	0.123~0.415	150~392
Fukuoka	sinusoidal channel with smooth side walls	0.023~0.070	0.13~0.25	1.1~2.7		5.8~35
Thomas	Chicago Canal	8.07	48.8	1.91	3.0	2.0
Schuster	Jumamasa Canal	3.45		3.45	0.76	8.6
Godfrey and Frederick	Kochila Canal	1.56	24	4.3	9.6	140
Godfrey and Frederick	Cobal Greek	0.49~0.85	16~18	8.0~10.0	9.5~21.0	245~500
Godfrey and Frederick	Clink River (Tennessee, US)	0.85	47	6.7	14	235
Godfrey and Frederick	Clink River (Tennessee, US)	2.1	53~60	10.4~10.7	47~54	210~245
Godfrey and Frederick	Bowale River	0.85	34	5.5	9.5	200
Godfrey and Frederick	Clink River (Virginia, US)	0.58	36	4.9	8.1	280
California state	Sacramento River (California, US)	4.0		5.1	15	74
Owens et al	Dalwant River	0.25		14	4.6	131
Glover	South Purart River	0.46		6.9	16.2	510
Fischer	Green-Tuwamisi River	1.10	20	4.9	6.5~8.5	120~160
Yotsukura et al	Missouri River	2.7	200	7.4	1,500	7,500
McQuivey & Keefer	Mississippi River				210	900

3. In a natural river, regions of reverse flow and dead water are common. Once the tracer enters these regions, it can stay in them for considerable periods of time before gradually moving on. These delays further decrease the peak concentration of the colored tracer in the streamwise direction, and retard the decrease of concentration after the peak passes. This action also increases somewhat the

dispersion coefficient. This phenomenon has been confirmed in both experimental and theoretical studies[52-54].

REFERENCES

[1] Sumer, B.M., and B. Oguz, "Particle Motions Near the Bottom in Turbulent Flow in an Open Channel." *Journal of Fluid Mechanics*, Vol. 86, Pt. 1, 1978, pp. 109-127.

[2] Sumer, B.M., and R. Deigaard. "Experimental Investigation of Motions of Suspended Heavy Particles and the Bursting Process." *Series Paper 23*, Inst. Hydrodynamics and Hydraulic Engineering, Technical University of Denmark, 1979, p. 106.

[3] Jackson, R.G. "Sedimentological and Fluid Dynamic Implications of the Turbulent Bursting Phenomenon in Geophysical Flows." *Journal of Fluid Mechanics*, Vol. 77, 1976, p. 531.

[4] Schmidt, W. "Der Massenaustausch in Freier Luft und Verwandte Erscheinungen." *Probleme der Kosmischen Physik*, Vol. 7, Hamburg, 1925.

[5] O'Brien, M.P. "Review of the Theory of Turbulent Flow and Its Relation to Sediment Transportation." *Trans.*, Amer. Geophys. Union, Vol. 14, 1933, pp. 487-491.

[6] Makaweif, V.M. "Theory of Turbulence and Suspended Sediment." *Proc. Soviet National Hydrology Institute*, No. 32, 1931.(in Russian)

[7] Hurst, H.E. "The Suspension of Sand in Water." *Proc., Royal Soc. London*, Ser. A, Vol. 124, 1929, pp. 196-201.

[8] Rouse, H. "Experiments on the Mechanics of Sediment Suspension." *Proc., 5th. Intern. Cong. for Applied Mech.*, 1938, pp. 550-554.

[9] Lane, E.W., and A.A. Kalinske. "Engineering Calculations of Suspended Sediment." *Trans.*, Amer. Geophys. Union, 1941, pp. 603-607.

[10] Bagnold, R.A. "An Approach to the Sediment Transport Problem from General Physics." *U.S. Geol. Survery, Prof. Paper 422-*, 1966, p. 37.

[11] Engelund, F. "A Criterion for the Occurrence of Suspended Load." *La Houille Blanche*, No. 6, 1965, p. 607.

[12] Vanoni, V.A. "Transportation of Suspended Sediment by Running Water." *Trans., Amer. Soc. Civil Engrs.*, Vol. 111, 1946, pp. 67-133.

[13] Vanoni, V.A., N. H. Brooks, and J. F. Kennedy. "Lecture Notes on Sediment Transportation and Channel Stability." *Rep. KH-R-1*, W. M. Keck Lab. of Hyd. & Water Resources, Calif. Inst. Tech., 1960.

[14] Ismail, H.M. "Turbulent Transfer Mechanism and Suspended Sediment in Closed Channels." *Trans., Amer. Soc. Civil Engrs.*, Vol. 117, 1952, pp. 409-446.

[15] Brush, L.M., Jr. "Exploratory Study of Sediment Diffusion." *J. Geophys. Res.*, Vol. 67, No. 4, 1962, pp.1427-1433.

[16] Jobson, H.E., and W.W. Sayre. "Vertical Transfer in Open Channel Flow." *J. Hyd. Div., Proc., Amer. Soc. Civil Engrs.*, Vol. 96, No. HY3, 1970, pp. 703-725.

[17] Jobson, H.E., and W.W. Sayre. "Predicting Concentration Profiles in Open Channels." *J. Hyd. Div., Proc., Amer. Soc. Civil Engrs.*, Vol. 96, No. HY10, 1970, pp.1983-1996.

[18] Zagustin, K. "Sediment Distribution in Turbulent Flow." *J. Hyd. Res.*, Intern. Assoc. Hyd. Res., Vol. 6, No. 2, 1968, pp.163-171.

[19] Wuhan Institute of Hydraulic and Electric Engineering. *River Sedimentation Engineering*. Vol. I, Hydraulic and Electric Press, 1981, pp119-123. (in Chinese)

[20] Einstein, H.A., and Ning Chien. "Second Approximation to the Solution of the Suspended Load Theory." *M.R.D. Sediment Series No. 3*, Missouri River Div., U. S. Corps Engrs., 1954, p.30.

[21] Einstein, H.A. "Der Geschiebetrieb als Wahrscheimlichkeits Problem." *Mitt. Versuchsanst. fuer Wasserbau, an der Eidg. Techn. Hochschule in Zurich*, Verlag Rascher & Co., Zurich, 1937.

[22] Velikanov, M.A. *Fluvial Process of Rivers*. Physics Mathematics Literature Press, Moscow, 1958, pp. 241-245.

[23] Zhang, Ruijin. "Discussion of Gravitation Theory and Mechanism of Sediment Suspension." *Chinese J. of Hydraulic Research*, Vol. 3, 1963, pp. 11- 23. (in Chinese)

[24] Yan, Jinghai et al. "A Preliminary Study of Sediment-Carrying Capacity of Open-Channel Flow." *Construction of Yellow River*, Vol. 12, 1957, pp. 30-34. (in Chinese)

[25] Einstein, H.A. "The Bed Load Function for Sediment Transportation in Open Channel Flows." *U. S. Dept. Agri., Tech. Bull. 1026*, 1950, p. 71.

[26] Wuhan Institute of Hydraulic and Electric Engineering. *Mechanics of River Engineering*. China Industry Press, 1961, pp. 60-63. (in Chinese)

[27] Mei, C.C. "Non-Uniform Diffusion of Suspended Sediment." *J. Hyd. Div., Proc., Amer. Soc. Civil Engrs.*, Vol. 96, No. HY1, 1969, pp. 581-584.

[28] Apmann, R.P., and R.R. Rumer. "Diffusion of Sediment in Developing Flow." *J. Hyd. Div., Proc., Amer. Soc. Civil Engrs.*, Vol. 96, No. HY1, 1970, pp. 109-123.

[29] Zhang, Qisun "Study of Sediment-Diffusion Mechanism for Open-Channel Flow and Its Application." *Chinese J. of Sediment Research*, Vol. 1, 1981, pp. 37- 52. (in Chinese)

[30] Hjelmfelt, A.T., and C.W. Lenau. "Non-Equilibrium Transport of Suspended Sediment." *J. Hyd. Div., Proc., Amer. Soc. Civil Engrs.*, Vol. 96, No. HY7, 1970, pp. 1567-1586.

[31] Chien, Ning, Ren Zhang, and Zhide, Zhou. *Fluvial Processes*. Science Press, 1986. (in Chinese)

[32] Chien, Ning. "A Study of Bed Armoring in the Downstream Reach of the Yellow River." *Chinese J. of Sediment Research*, Vol. 4, No. 1, 1959, pp. 16-23. (in Chinese)

[33] Han., Qiwei. "A Study on the Non-Equilibrium Transport of Non-Uniformly Graded Suspended Sediment." *Chinese Science Communication*, Vol. 17, 1979, pp. 808-808. (in Chinese)

[34] Zhou, Zide. "On Turbulence Diffusion and Longitudinal Dispersion of Open-Channel Flow." *J. of Nanjing Hydraulic Research Institute*, China, Vol. 3, 1980, pp. 94-110. (in Chinese)

[35] Shen, H.W., and H.F. Cheong, "Dispersion of Contaminated Sediment Bed Load." *J. Hyd. Div., Proc., Amer. Soc. Civil Engrs.*, Vol. 99, No. HY11, 1973, pp. 1947-1966.

[36] McQuivey, R.S., and T.N. Keefer. "Dispersion - Mississippi River Below Baton Rouge, La.." *J. Hyd. Div., Proc., Amer. Soc. Civil Engrs.*, Vol. 102, No. HY10, 1976, pp. 1425-1437.

[37] Taylor, G.J. "The Dispersion of Diffusion of Matter in Turbulent Flow Through A Pipe." *Proc., Royal Soc. London*, Ser. A, Vol. 223, 1954, pp. 446-468.

[38] Holley, E.R. "Unified View of Diffusion and Dispersion." *J. Hyd. Div., Proc., Amer. Soc. Civil Engrs.*, Vol. 95, No. HY2, 1969, pp. 621-631.

[39] Okoye, J.K. "Characteristics of the Transverse Mixing in Open Channel Flows." *Report KH-R-23*, W. M. Keck Lab. of Hydraulics and Water Resources, Calif. Inst. Tech., 1970, p. 269.

[40] Lau, Y.L., and B.G. Krishnappen. "Transverse Dispersion in Rectangular Channels." *J. Hyd. Div., Proc., Amer. Soc. Civil Engrs.*, Vol. 103, No. HY10, 1977, pp. 1173-1190.

[41] Fischer, H.B. "The Effect of Bends on Dispersion in Streams." *Water Resources Res.*, Vol. 5, No. 2, 1969, pp. 496-506.

[42] Ward, P.R.B., and H.B. Fischer. "Some Limitations on Use of the One-Dimensional Dispersion Equation, with Comments on Two Papers by R. W. Paulson." *Water Resources Res.*, Vol. 7, 1971, pp. 215-220.

[43] Keefer, T. "The Relation of Turbulence to Diffusion in Open Channel Flows." *Ph.D. Dissertation*, Colorado State Univ., 1971.

[44] Sayre, W.W. "Dispersion of Mass in Open Channel Flow." *Hydraulics Paper No.3*, Colorado State Univ., 1968, p. 73.

[45] Fischer, H.B. "Longitudinal Dispersion and Turbulent Mixing in Open Channel Flow." *Annual Rev. of Fluid Mech.*, Vol. 5, 1973, pp. 59-78.

[46] Chatwin, P.C. "Presentation of Longitudinal Dispersion Data." *J. Hyd. Div., Proc., Amer. Soc. Civil Engrs.*, Vol. 106, No. HY1, 1980, pp. 71-84.

[47] Fischer, H.B.,"Longitudinal Dispersion in Laboratory and Natural Streams." *Ph. D. Dissertation*, Calif. Inst. Tech., 1966.

[48] Mc Quivey, R.S., and T.N. Keefer. "Convective Model of Longitudinal Dispersion." *J. Hyd. Div., Proc., Amer. Soc. Civil Engrs.*, Vol. 102, No. HY10, 1976, pp. 1409-1424.

[49] Elder, J.W. "The Dispersion of Marked Fluid in A Turbulent Shear Flow." *J. Fluid Mech.*, Vol. 5, NO. 4, 1959, pp. 544-560.

[50] El-Hadi, N.D. Abd., and K.S. Davar. "Longitudinal Dispersion for Flow over Rough Beds." *J. Hyd. Div., Proc., Amer. Soc. Civil Engrs.*, Vol. 102, No. HY4, 1976, pp. 483-498.

[51] Richardson, E.V., and A.C. Miller. "Diffusion and Dispersion in Open Channel Flow." *J. Hyd. Div., Proc., Amer. Soc. Civil Engrs.*, Vol. 100, No. HY1, 1974, pp. 159-172.

[52] Valentine, E.M., and I.R. Wood. "Longitudinal Dispersion With Dead Zones." *J. Hyd. Div., Proc., Amer. Soc. Civil Engrs.*, Vol. 103, No. HY9, 1977, pp. 975-990.

[53] Sabol, G.V., and C.F. Nordin, Jr. "Dispersion in Rivers as Related to Storage Zones." *J. Hyd. Div., Proc., Amer. Soc. Civil Engrs.*, Vol. 104, No. HY5, 1978, pp. 695-708.

[54] Valentine, E.M., and I.R. Wood. "Experiments in Longitudinal Dispersion with Dead Zones." *J. Hyd. Div., Proc., Amer. Soc. Civil Engrs.*, Vol. 105, No. HY8, 1979, pp. 999-1018.

CHAPTER 11

SEDIMENT TRANSPORT CAPACITY OF THE FLOW

The amount of sediment that can pass through a given river reach for given conditions of the flow and boundary is termed the sediment transport capacity of the flow. This capacity should include both bed load and suspended load. In view of the difference in the processes for the transport of bed material load and wash load (Chapter 5), formulas expressing these two parts of sediment transport capacity and the relevant computation methods have different forms. Therefore, they should be dealt with separately.

11.1 FORMULA FOR SEDIMENT TRANSPORT CAPACITY OF BED MATERIAL LOAD

In principle, one should be able to establish sediment transport capacity through the basic relationships of mechanics. However, the present level of knowledge is not sufficient, so that some links still have to be treated by semi-empirical or even purely empirical methods and relationships.

11.1.1 Theoretically based formula

The mechanism of bed load movement differs from that of suspended load movement. Chapter 9 treats relationships and characteristics of bed load, and Chapter 10 discusses those for the suspended load movement. The sum of the amounts of bed load and suspended load is the total bed material load that can be transported for a given flow and for given boundary conditions. The studies conducted by Einstein and Bagnold deal with these two processes.

11.1.1.1 Einstein's work

1. Bed load function

Einstein's bed load function [1] provides a method for computing the bed material load, and it takes into consideration bed material, bed load, and suspended load in combination. Bed load moves in the forms of rolling, sliding, and saltation, and exchanges between the bed material in the surface layer of river bed occur continuously during the process of movement. The suspended load moves in the main current at the same velocity as the flow, and it also exchanges material continuously with the bed load. The transition from bed load to suspended load takes place gradually and smoothly in nature. However, for convenience, one can assume that the transition occurs entirely at one elevation; i.e., below some elevation bed load movement prevails, and above this elevation, suspension prevails. Results of flume experiments reveal that unless the movement of sediment is quite intense, this critical elevation is about two grain diameters above the river bed.

In Chapters 9 and 10, the Einstein formulas for the sediment carried as bed load and that as suspended are given as follows:

For bed load

$$1 - \frac{1}{\pi} \int_{-B_*\psi_*-1/\eta_0}^{B_*\psi_*-1/\eta_0} e^{-t^2}\, dt = \frac{A_*\phi_*}{1 + A_*\phi_*} \tag{11.1}$$

in which

$$\phi_* = \frac{i_b}{i_0}\frac{g_b}{\gamma_s}\left(\frac{\gamma}{\gamma_s - \gamma}\right)^{1/2}\left(\frac{1}{gD^3}\right)^{1/2} \tag{11.2}$$

$$\psi_* = \xi Y \frac{(\beta/\beta_X)^2}{\theta}\frac{\gamma_s - \gamma}{\gamma}\frac{D}{R_b' J} \tag{11.3}$$

For suspended load

$$i_s g_s = 11.6 U_* S_{ua} a(PI_1 + I_2)\gamma_s \tag{11.4}$$

in which

$$P = \frac{1}{0.434}\log(30.2\frac{h}{K_S/X}) \tag{11.5}$$

$$I_1 = 0.216\frac{A^{z-1}}{(1-A)^z}\int_A^1 (\frac{1-y}{y})^z\, dy \tag{11.6}$$

$$I_2 = 0.216\frac{A^{z-1}}{(1-A)^z}\int_A^1 (\frac{1-y}{y})^z \ln y\, dy \tag{11.7}$$

$$A = \frac{a}{h} \tag{11.8}$$

$$z = \frac{\omega}{\kappa U_*} \tag{11.9}$$

The quantities i_0, i_b, and i_s are the portions of sediment with median diameter D in the bed material, bed load, and suspended load respectively; g_b refers to the sediment discharge of bed load and g_s to the suspended load, and by weight per unit width.

The next question is how to determine the sediment concentration at the interface between the two (at the elevation $a=2D$); it is to be used as the specific reference concentration S_{va} in Eq. (11.4). The mean sediment concentration (expressed in percent by volume) in the bed surface layer is

$$\frac{i_b g_b}{2 D \bar{u}_b \gamma_s}$$

in which U_b denotes the mean velocity of bed load movement.

If the sediment concentration at the top of the bed surface layer is proportional to the mean value of sediment concentration in the layer and if $\overline{u_b}$ is proportional to the friction velocity, then

$$S_{va} = \zeta \frac{i_b g_b}{2 D U_* \gamma_s}.$$

The coefficient ζ has been shown in artificial flume experiments to be the reciprocal of the well known constant 11.6. Thus, the above expression can be rewritten as

$$i_b g_b = 11.6 S_{va} U_* a \gamma_s$$

Substituting it into (11.4) one obtains

$$i_s g_s = i_b g_b (PI_1 + I_2) \tag{11.10}$$

If g_T denotes the total discharge of bed material expressed by weight per unit width, and including both bed load and suspended load, and i_T denote the portion of sediment with diameter D in bed material load, then

$$i_T g_T = i_b g_b (1 + PI_1 + I_2) \tag{11.11}$$

This is the Einstein formula for the sediment transport capacity as bed material load, in which the term $i_b g_b$ can be deduced from the Einstein bed load function. If sand waves exist on the bed surface, the term U_* should be replaced by U_*' ($= (R_b' g J)^{0.5}$). The computation method and procedure in details can be found in the original works of Einstein. The method is rather elaborate, but, the whole procedure has been programmed and is easy to use with computers[1].

---

1) Chien, Ning, et al. "Preliminary Study on Mechanism of Self-adjustment of Sediment Transport Capacity of Sediment-Laden Rivers." *Report 80-2*, Dept. of Hydraulic Engineering, Tsinghua University, 1980.

2. Evaluation of the Einstein theory

Einstein was one of the outstanding leaders in the field of sediment studies. Although his theory was developed mainly in the late 1940s and early 1950s and he himself passed away so many years ago, his many insights, voluminous works and outstanding achievements still exert a profound influence. The preceding chapters present key elements of his theory in considerable detail. Here an attempt is made to give an overall evaluation of them; i.e., to confirm the proven contributions and to point out certain inadequacies. This step may also serve as a guide to support further research on the mechanics of sediment motion.

Einstein's contributions can be summarized in four principal aspects [2]:

1. He was the first to put forward clearly the idea that the sediment load should be divided into bed material load and wash load according to its origin, and that it should be based on the characteristics of the rating curves for measured flow discharge against measured sediment discharge obtained at hydrometric stations. The relationship of sediment transport capacity obtained from flume experiments refers only to bed material load, and the amount of wash load must be evaluated by a different approach.

2. Instead of taking the resistance to the flow as being represented solely by the roughness of the perimeter of the channel bed, Einstein considered the resistance to the flow through a process of the conversion of potential energy into kinetic energy of turbulence and heat energy; thus, he linked the resistance with sediment movement. From the concept that turbulence originates from the channel bed, he expressed the need to subdivide the total resistance into two parts so as to distinguish the resistance of bed configuration from that of grain resistance on the bed surface that is directly related to the bed load movement. This concept has now been widely accepted. He was the first to develop the computational method for determining resistance of bed configurations by using field data from natural rivers.

3. Einstein initiated the approach to the studying of bed load movement that incorporates the stochastic process with a dynamic analysis. Sediment particles have a dual nature in their movement. On the one hand, each particle moves in various patterns and obeys the laws of mechanics under the actions of external forces; on the other hand, their motion has stochastic properties because of fluctuations of the flow and the probabilities of occurrence for large number of sediment particles. Early in the 1940s, he opened this new approach for the study of sediment movement, but his initiative did not draw the response from other researchers that it merited. Only in recent decades, as researchers are challenged by environmental pollution, has this method of stochastic-mechanics analysis attracted wide attention. Over time, a great deal of research related to incipient motion, settling, diffusion of sediment, and other problems has been conducted, and it includes a variety of significant advances.

4. Einstein noticed the continuous exchanges that take place among bed material, bed load, and suspended load, and he pointed out that one has to be aware of the different rules that govern bed load movement and suspended load movement as well as the active links among the three types of sediment movement. Starting from this concept, Einstein established the sediment transport capacity for bed material load that includes simultaneously both the bed load and the suspended load. He further introduced the effect of flow on different-sized sediment particles and the interactions of these particles, thus he extended his computational method to include sediment mixtures. Up to now, this method is the only effective way to calculate the transport rate of sediment that is a mixture of different grain sizes.

The scope of Einstein's research is broad. His achievements are far from limited to these four aspects. For example, early in the 1950s, Einstein and Li clearly pointed out the instability of the laminar sublayer through a simple experiment. Even on a smooth boundary, the influence of flow turbulence can be felt directly. Recently, as a result of improvements in measuring techniques, the nature of bursting of turbulence has been developed. It has greatly promoted the study of turbulent shear flow, and it reveals that confirmation and a deeper understanding of the bursting phenomenon may cause developments concerning the theory of sediment movement.

Still, as knowledge advances unceasingly, new findings inevitably expose additional inadequacies in Einstein's theory. To bring to light the weak points will undoubtedly benefit all of us in planning future theoretical and experimental research.

First, Einstein's work included only the case of flat bed configuration [3]. Under such conditions, sediment particles involved in bed load movement can be characterized by a definite step length and a definite period of rest. For a river bed covered by dunes, Einstein only differentiated the form resistance of bed configuration from the grain resistance, and all terms involving the flow depth that occurs in the expression of friction velocity were replaced by the hydraulic radius relevant to sediment grain resistance. Except for these considerations, the river bed was still taken as being flat. In so doing, he did not take into account more of the details induced by dunes. For example, as pointed out in Chapter 6, sediment particles moving near the bed fall into the eddies in the separation zone behind a dune and stay there until the whole sand dune advances for a distance of wave length; then the particles may be exposed again and come into direct contact with the flow current so as to continue their travel. Evidently if a bed configuration exits, the mean value of step length and rest time of bed load particle should be related to the nature of sand dunes. Based on flume experiment of single sediment particle, Grigg found that, with bed configuration, the mean step length and mean rest time are functions of wave length and celerity of sand dunes, respectively, see Fig. 11.1 [4]. These figures reveal that the whole picture of bed load movement is more complicated once sand dunes have formed, if details of bed load movement are taken into account.

Second, as mentioned above, Einstein's theory of bed load movement is based on the continual exchange between bed material and bed load. The entrainment of particles from the bed surface into motion is the effect of lift forces acting on particles resting on bed. If the intensity of sediment movement is not too high, this physical picture is undoubtedly correct. But the channel bed of an alluvial river is composed of loose sediment particles. The drag of the flow can act not only on the sediment particles on the bed surface, it can also be transmitted to the underlying strata. If the flow intensity is high enough, sediment particles some depth below the bed surface can be brought into motion. However, confined in space, these particles take the form of laminated load, as mentioned in Chapter 5. Under such circumstances the picture

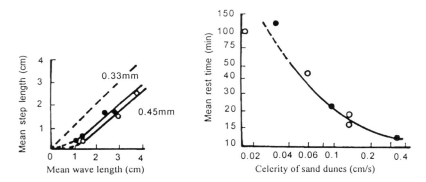

Fig. 11.1 Relationship of mean step length to wave length, and mean rest time to celerity of sand dunes (after Grigg, N.S.)

proposed by Einstein is no longer valid. For this reason and some other reasons given in the following paragraphs, Einstein's formula for sediment discharge may underestimate the sediment load if the intensity of sediment movement is high.

Third, in spite of his outstanding work, Einstein was limited by the accuracy and the scope of experimental data available at that time, so that he oversimplified some of the links in the deduction and determined only roughly some of the parameters. From material now available, at least the following points should be given for further consideration.

1. Length of single step

Einstein thought that the distance of a single step length in bed load movement L was a function only of sediment size, i.e.,

$$L = \lambda D \qquad (11.12).$$

in which the coefficient λ is about 100 for sediment particles that are nearly spherical. In fact, one should expect that the higher the flow intensity, the longer a step should be. Fig. 11.2 shows the experimental results obtained by Yalin [5], using light material with a specific weight of 1.145 to 1.301 and a diameter of 3.4 to 3.7 mm; and he found that λ is proportional to Θ if Θ is not small.

2. Time of uplift from the bed

Einstein thought that whether a sediment particle could be lifted from the bed depended only upon the flow conditions. Also during the process of being lifted, he believed that the time required is unrelated to the flow conditions, but is a function of its inherent properties. Hence, he assumed

$$t \sim D / \omega \qquad (9.35)$$

in which ω is the fall velocity. The validity of this assumption has been questioned by many persons. From common sense, the higher the flow intensity is, the shorter the time for lifting from the bed should be.

Yalin deduced the rate of erosion per unit area from a different viewpoint. If t_* denotes the cyclic appearance of eddies appearing in the region from y to D above the bed, then from dimensional concerns

$$t_* \sim l / \sqrt{\overline{v'^2}}$$

where l is the size of the eddy. According to Prandtl,

$$l = ky$$

near the bed, the order of magnitude for l and D are the same; v' is the vertical component of the fluctuating velocity. Based on results of measurement presented in Chapter 4,

$$v' \sim U_*$$

hence,

$$t_* = A_3 \frac{D}{U_*}$$

$$(11.13)$$

in which A_3 is a coefficient of proportionality. Yalin thought that once an eddy forms, a sediment particle has a chance of being lifted from the bed, and the probability of its happening is denoted as p. Within a time

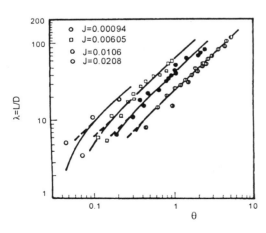

Fig. 11.2 Relationship of single step length vs. flow intensity (after Yalin, M.S.)

period of T, the total weight of sediment particles lifted up per unit area of the bed is

$$\frac{A_2 \gamma_s}{A_1} pD \frac{T}{t_*}$$

The rate of erosion per unit area is

$$\frac{A_2 \gamma_s}{A_1} pD \frac{1}{t_*}$$

A comparison of this result with Eq. (9.36) shows that t_* corresponds to the term t in Eq. (9.36) proposed by Einstein. The only difference is that t is unrelated to the flow condition, whereas t_* is inversely proportional to U_*.

3. Coefficients A_* and B_*

In the Einstein bed load formula Eq.(9.48), the two main coefficients A_* and B_* are taken to be constants, to be based on experimental data. According to their definition, the coefficients can be expressed as:

$$A_* = \frac{A_1 A_3}{\lambda A_2} \qquad B_* \sim \frac{A_2}{A_1 C_L \eta_0}$$

Among the parameters involved, A_1 and A_2 can be considered as constants because they are related only to sediment shape;

$$\lambda \sim \Theta = \frac{(\gamma_s - \gamma)D}{\rho U_*^2}$$

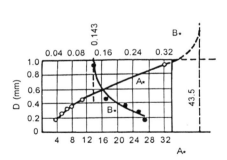

Fig. 11.3 A_* vs. D and B_* vs. D (after Bishop, A.A.; D.B. Simons; and E.V. Richardson)

Coefficient A_3, C_L and η_0 for the uplift force and its standard deviation, should be related to the grain Reynolds number and can be taken as a constant only for turbulent flow passing a rough boundary. Therefore, in more rigorous terms, A_* and B_* should be functions of both the flow conditions and the sediment properties. From the experimental data of Bishop, A_* and B_* are function of sediment size, as shown in Fig. 11.3[6]. In those experiments, natural sand of D=0.2 to 0.9 mm was used. The experiments did not include effects of variables γ_s, ρ and v.

4. Velocity of bed load movement $\bar{u}_b$

Einstein assumed that $\bar{u}_b$ is proportional to U_*, a point that he deduced indirectly from transport data of total load without direct experimental support. Chapter 9 points out that Engelund found the following relationship based on experimental data for the movement of a single particle [7]

$$\bar{u}_b = \alpha U_* \left[1 - 0.7 \sqrt{\frac{\Theta_c}{\Theta}} \right] \tag{9-64}$$

Only for the case of $\Theta \gg \Theta_c$, is the mean velocity of bed load movement proportional to the shear velocity U_*.

5. Thickness of bed surface layer

Einstein assumed the thickness of the bed surface layer (i.e., the zone of bed load movement) to be about twice the sediment diameter. This assumption applies only if the intensity of sediment movement is not strong. One can expect, however, that this thickness increases with the increasing of flow intensity. In an extreme case, as mentioned in Chapter 5, all of the sediment particles in the whole flow region can move as bed load. The interpretation adopted by Bagnold is different from that of Einstein. For Eq. (9.30), Bagnold assumed the validity of

$$a \sim D\left(\frac{U_*}{U_{*c}}\right)^{0.6} \tag{11.14}$$

Only for the case of low intensity of sediment movement (i.e., if U_* approaches U_{*c}), is the thickness of the bed surface layer proportional to the sediment diameter. As the flow intensity increases, the thickness increases rapidly. Both Eq. (9.30) and Eq. (11.14) are deduced indirectly rather than being based on direct measurements of the thickness of the zone of bed load movement.

6. Effective drag force

One of the major contributions of Einstein was to differentiate the drag force (or resistance) into the two forms of grain resistance of the bed surface and form resistance of bed configuration, so that only the grain resistance is effective for bed load movement. Later on, in dealing with suspended load, he assumed that the shear velocity should also be replaced by a different shear velocity related to grain resistance for flow involving sand waves. This method, of course, is a simplification. Vittal et al. conducted research on this aspect [8]. From a large volume of both flume data and field data from natural rivers moving with a flat bed, they developed the diagram shown in Fig. 11.4, in which Φ_T is the transport intensity of all bed material load including bed load and suspended load.

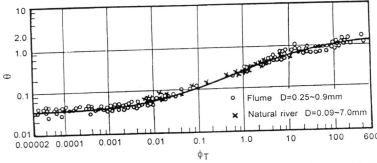

Fig. 11.4 Relationship between total load transport intensity Φ_T and flow intensity Θ for a flat bed
(after Vittal, N., K.G. Ranga Raju, and R.J. Garde)

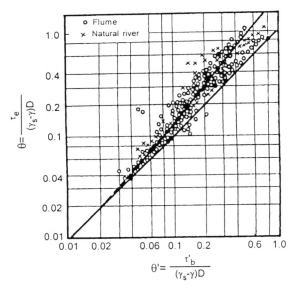

Fig. 11.5 Relationship between effective drag force and grain resistance on the bed surface

$$\Phi_T = \frac{g_T}{\gamma_s}\left(\frac{\gamma}{\gamma_s - \gamma}\right)^{1/2}\left(\frac{1}{gD^3}\right)^{1/2} \qquad (11.15)$$

Without sand waves, the resistance involved in the parameter for flow intensity Θ is the grain resistance. It is expressed as the effective drag force so that

$$\Theta = \frac{\tau_e}{(\gamma_s - \gamma)D} \qquad (11.16)$$

Next, Vittal et al. analyzed flume data for flow with sand waves. If the relationship shown in Fig. 11.4 still holds true, then the effective drag force τ_e with sand waves can be computed according to the measured transport rate of bed material load. A comparison between τ_e, the total drag force τ_0, and the grain resistance τ_b' computed from the Manning-Strickler formula reveals that τ_e is close to τ_b'; however,

480

with increasing intensity of both flow and suspended load, τ_e and τ_b' differ more and more, with τ_e becoming greater than τ_b'. Within the scope of the experimental data, the effective drag force could be twice as large as τ_b'. Thus, after the appearance of sand dunes, the local turbulence occurring at the separation of dune crest also exerts an effect on suspended load movement.

Recently Ranga Raju, Garde, et al. analyzed 54 sets of experimental and field data from rivers and canals obtained by various researchers. They found that

$$\tau_e = \tau_b'(\frac{\tau_b'}{\tau_0})^m \tag{11.17}$$

in which

$$
\begin{aligned}
m &= 0 & &\text{if} & \frac{U_*}{\omega} &\leq 0.5^{1)} \\
m &= 0.2\frac{U_*}{\omega} - 0.1 & &\text{if} & \frac{U_*}{\omega} &\geq 0.5
\end{aligned}
\left.\rule{0pt}{32pt}\right\} \tag{11.18}
$$

Fig. 11.4 can be used to determine the total transport rate if ripples or dune bed configurations exist [9].

Furthermore, for the stochastic process of bed load movement, Einstein assumed that both the step length and the rest period follow exponential distributions. Studies of these movements show, however, that only the rest period follows an exponential distribution; that for the step length is more like a gamma distribution [10]. This point will not be discussed in detail.

11.1.1.2. Bagnold's work [11]

Bagnold's formulas for sediment transport capacity for both bed load and suspended load, as introduced in Chapters 9 and Chapter 10, are as follows:

$$g_b' = \tau_0 U \frac{e_b}{\tan \alpha} \tag{9.16}$$

$$g_s' = 0.01\tau_0 U \frac{U}{\omega} \tag{10.72}$$

in which the efficiency of bed load movement e_b is as expressed in Eq. (9.32). Then the transport rate of total bed material load by submerged weight is

---

1) If Karman constant is taken as 0.4, U_*/ω corresponds to the condition, under which sediment moves as bed load, that is, $z \geq 5$.

$$g_T = \tau_0 U \left(\frac{e_b}{\tan \alpha} + 0.01 \frac{U}{\omega} \right) \tag{11.19}$$

He verified Eq. (11.19) by various flume data for which D was within the range from 0.11 to 5 mm with satisfactory results.

11.1.2. Empirical or semi-empirical formulas

On the basis of presently available theory, prediction of the sediment transport capacity of natural rivers by Einstein's and Bagnold's formulas show a rather large divergence. (details discussed in the following section). In order to solve problems arising in practice, one has to rely on empirical formulas. Some of these have some theoretical basis, but the derivations are oversimplified; hence, they belong to the categories of semi-empirical or even purely empirical formulas.

11.1.2.1 Basic expressions for sediment transport capacity

1. Factors affecting sediment transport capacity

From preceding discussions, factors that affect sediment transport capacity of a flow include the following four:

(1) Flow conditions: including velocity U, flow depth h, slope J, gravitational acceleration g.

(2) Physical properties of water: including specific weight and kinematic viscosity.

(3) Physical properties of sediment: including specific weight γ_s, sometimes replaced by effective specific weight $(\gamma_s - \gamma)$, fall velocity ω , sediment diameter D; since ω and D are correlated, only one of the two needs to be included.

(4) Boundary conditions: including bed material composition and channel width.

For a river bed composed of cohesionless sediment, a representative diameter can be used for the bed material. If only bed material load is involved, only one of the two variables, diameter of the bed material and the fall velocity of moving sediment particles needs to be considered. The channel width needs to be taken into account only if wall effects are significant.

Thus the sediment transport capacity of flow, represented by the average sediment concentration at a cross-section, can be written in the form

$$S_m = f(U, h, g, J, \gamma, \gamma_S - \gamma, v, \omega, D, B) \tag{11.20}$$

In addition, the sediment-laden flow has to obey the resultant equation

$$U = \frac{1}{n} h^{2/3} J^{1/2}$$

in which the Manning coefficient n is a function of the bed material D and the velocity U. In other words, the variables U, h, J, and D are interrelated, so that only three of them can be used as independent variables in Eq.(11.20). If U, h and D are selected, then the equation can be rewritten as

$$S_m = f(U, h, g, \gamma, \gamma_s - \gamma, v, \omega, D, B) \tag{11.21}$$

If h, J, and D are chosen, and sediment transport rate per unit width g_T is used to express the sediment transport capacity, then Eq (11.20) becomes

$$g_T = f(J, h, g, \gamma, \gamma_s - \gamma, v, \omega, D, B)$$

In current practice, instead of the slope J, use of the variable U_* (with the dimension of velocity) is usually preferred, then the above expression takes the form of

$$g_T = f(U_*, h, g, \gamma, \gamma_s - \gamma, v, \omega, D, B) \tag{11.22}$$

Thus the task is to establish formulas for sediment transport capacity in the form of Eq. (11.21) or Eq. (11.22) that are based on the available data by means of regression analyses or empirical curve fitting.

2. Basic functional relationship

Both Eq. (11.21) and Eq. (11.20) involve a number of variables that have dimensions, and these must be converted into dimensionless parameters by means of dimensional analysis. The results thus obtained are affected by which basic variables are selected. In the following, two cases are presented:

(1) If U, h, and γ are the basic variables, the dimensionless form of Eq. (11.21) is

$$S_m = f(\frac{U^2}{gh}, \frac{\gamma_s - \gamma}{\gamma}, \frac{Uh}{v}, \frac{U}{\omega}, \frac{D}{h}, \frac{B}{h}) \tag{11.23}$$

For natural rivers, $\frac{\gamma_s - \gamma}{\gamma}$ is a constant, and both the viscosity and wall-effect can be neglected, and the expression can be further simplified,

$$S_m = f(\frac{U^2}{gh}, \frac{U}{\omega}, \frac{D}{h}) \tag{11.24}$$

(2) If U_*, D and $\gamma_s - \gamma$ are the basic variables, the dimensionless form of Eq. (11.22) is

$$\frac{g_T}{U_* D(\gamma_s - \gamma)} = f(\frac{U_*^2}{gD}, \frac{\gamma_s - \gamma}{\gamma}, \frac{U_* D}{\nu}, \frac{U_*}{\omega}, \frac{D}{h}, \frac{B}{D})$$ (11.25)

Eqs. (11.23) and (11.25) are the two basic expressions for the sediment transport capacity of a flow. A great number of existing formulas can be reduced to one of these two types of formulas. For example, in spite of the complexity of the Einstein bed load function, the variables involved do not go beyond this scope. For the simple case of two dimensional flow, uniform sediment and flat bed, the relevant formulas from his bed load function and the variables involved are as follows:

Formula	Dimensionless parameters
(11.3)	$\dfrac{\gamma_s - \gamma}{\gamma}$, $\dfrac{D}{hJ}(= \dfrac{gD}{U_*^2})$
	$Y[= f(\dfrac{U_* D}{\nu})]$
(11.5)	$\dfrac{h}{K_S}(= \dfrac{h}{D})$, $X[= f(\dfrac{U_* D}{\nu})]$
(11.6, 11.7)	$A(\sim \dfrac{D}{h})$, $z(\sim \dfrac{\omega}{U_*})$

For three dimensional flow, Einstein developed a method for excluding the wall-effect from the total resistance: the details are presented in Chapter 7. The method is based on the width-depth ratio $B / h(= \dfrac{B}{D} \cdot \dfrac{D}{h})$ In his analysis, the variables are the six contained in Eq. (11.25).

11.1.2.2 Some of representative examples of empirical and semi-empirical formulae [12]

1. Formula of Wuhan Institute of Hydraulic and Electric Engineering (WIHEE)

For rivers flowing over alluvial plains, suspended load predominates so that the bed load is generally negligible. In such cases, the suspended sediment transport capacity is approximately equal to the total transport capacity. Among such formulas, the one most widely used in China is that developed at WIHEE:

$$S_{vm} = k(\frac{U^3}{gh\omega})^m$$ (10.67)

A similar one to that formula is the Velikanov formula

$$S_{vm} = K(\frac{U^3}{gh\omega})$$ (10.66)

The principle parameter in these formulas is the product of U^2/gh and U/ω. In fact, Eq. (10.14) shows that the coefficient K in the Velikanov formula is a function of $\frac{\gamma_s - \gamma}{\gamma}$ and the Chezy coefficient C, and that the latter is related to D/h if the river bed is not covered by sand dunes. Hence, such expressions for sediment transport capacity of the flow belong to the category that is based on Eq. (11.21). In Fig. 11.6, a comparison of Eq. (10.67) with field data and in the laboratories, displays some scatter.

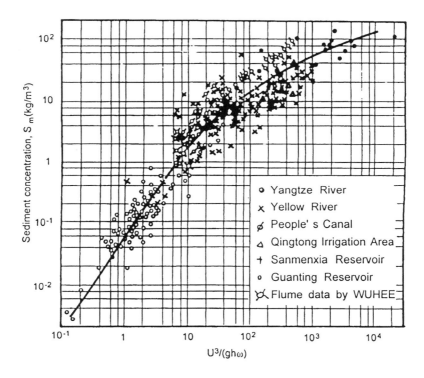

Fig. 11.6 Comparison of Eq. (10.67) with observed data

A few formulas of sediment transport capacity depend on parameters other than U, h, and ω, but they can still be derived from these parameters, in effect. For example, the term $U-U_c$ can be derived from the mean velocity U, in which U_c refers to the critical velocity for incipient motion of bed movement or the velocity for the incipient motion required to suspend particles. The flow depth can be replaced by the hydraulic

485

radius R, and the fall velocity by the mean diameter of the suspended sediment. Table 11.1 contains formulas of this type along with the parameters used and the exponents on them [13].

Table 11.1 Formulas for sediment transport capacity of flow deduced from the transport of suspended load

Name	Formula	Exponent on corresponding factor			
		U	h(or R)	ω	D
Zamarin	$S_m = 0.022 \left(\dfrac{U}{\omega}\right)^{3/2} \sqrt{RJ}$, suitable for $\omega > 0.002$m/s	2.5	—	-1.5	—
	$S_m = 11U \sqrt{\dfrac{RJU}{\omega}}$, suitable for $\omega < 0.002$m/s	2.5	—	-0.5	—
Lapating	$S_m = \dfrac{4h^{0.5}J}{n^2\omega} = \dfrac{4UJ^{0.5}}{n\omega h^{0.17}}$	2.0	-0.67	-1.0	—
Horst	$S_m = k \dfrac{RJU}{\omega}$	3.0	—	-1.0	—
Gostysky	$S_m = 3200 \dfrac{J^{3/2}h^{1/2}}{\omega}$	3.0	-1.0	-1.0	—
Abalyanze	$S_m = 0.02 \dfrac{U^3}{h\omega}$	3.0	-1.0	-1.0	—
Fan, Jiahua	$S_m = 0.0234 \dfrac{U^4}{\omega R^2}$	4.0	-2.0	-1.0	—
Irrigation canal by the Yellow River	$S_m = 6600 \dfrac{U^2}{gR} \dfrac{U - U_c}{\omega} \dfrac{D}{R}$	-3.0	-2.0	-1.0	1.0
Stem and tributaries of the Yellow River	$S_m = 0.031 \dfrac{U^{2.25}}{R^{0.74}\omega^{0.77}}$	2.25	-0.74	-0.77	—

Note: (1) Units used in the formula are S_m in kg/m^3, u and ω in m/s, h, R, and D in m;

(2) In the first four formulas, either $U = C\sqrt{RJ}$ or $U = C\sqrt{hJ}$ is used to eliminate the slope J.

In Table 11.1, the common factors are U, h, and ω, except for the next to the last formula, in which the ratio of D/R is included. However, the exponents on these parameters show great differences. The exponent on the velocity term varies from 2 to 4, and is mostly between 2.5 and 3.5; and the exponent on flow depth or hydraulic radius varies from 2/3 to 2. Also, in some cases the relative roughness is taken as an independent parameter, while in other cases, it- is included in the final overall coefficient. Hence, the large variation for the exponent on the flow depth is not surprising. The exponent of fall velocity varies between 0.5 and 1.5, but usual has the value of 1. The listed formulas are all empirical, and the data used to establish these formulae are sometimes not comparable. For example, the sediment in the Yellow River is much finer and sediment concentrations are much higher than the values for rivers or canals in the former USSR. Hence the divergence is hardly surprising. In fact, from Eq. (10.67) and Fig. 10.27, one can conclude that for variables of such widely different ranges the relevant exponential values could hardly be the same.

2. Sha Yuqing formula [14]

Some researchers used data for both suspended load and bed load, because they thought that there was no substantial difference between them, and the statistical analyses could be made without differentiation. Sha's formula is the representative of such an approach.

Sha collected flume data from Mayer Peter, Gilbert, Waterways Experiment Station, U.S. and others, and he also used field data from the Jinhe and Weihe irrigation systems of the Yellow River, canals in Central Asia of the former USSR, the Yellow River, and the Yongding River and others, with a total of about 1,000 data sets in his analysis. The scope of the data is rather wide, as shown in Table 11.2 Most flume data are for bed load, whereas most field data are for suspended load.

Table 11.2 Scope of data used by Sha, Yuqin

Variable	Max.	Min.	Geometric mean
Concentration S_m (Kg/m^3)	1000	0.0025	3.47
Slope J ($\times 10^{-4}$)	6300	1.6	128.5
Mean velocity U (m/s)	5	0.2	0.89
Effective velocity $U-U_c$ (m/s)	5	0.03	0.51
Hydraulic radius R (m)	10	0.01	0.29
Wetted perimeter p (m)	10,000	0.16	6.98
Sediment diameter D (mm)	40	0.004	0.12
Fall velocity ω (cm/s)	6.3	0.004	0.40

Table 11.3 Correlation between sediment concentration and various independent variables

Variable	Correlation	Degree of correlation
Slope J	-0.002	nil
Mean velocity U	+0.451	medium
Effective velocity $U-U_c$	+0.674	strong
Hydraulic radium R	+0.243	weak
Perimeter P	+0.062	nil
Diameter D	-0.539	strong
Fall velocity ω	-0.615	strong

In the analysis, Sha took the sediment concentration as the dependent variable and the others as independent variables to determine the correlation coefficients between sediment concentration and each of the independent variables (Table 11.3).

Table 11.3 shows that sediment concentration depends mainly upon an effective flow velocity, and the fall velocity (or sediment diameter), and it depends to a lesser degree upon flow velocity and hydraulic radius; in contrast, slope and wetted perimeter have little effect. Since the effective velocity already reflects to some extent the effect of velocity, only four independent variables are used in the final regression analysis: U-U_c, ω, D and R. The final expression then takes the form

$$S_m = 5.14 \frac{(U - U_c)^{1.58}}{R^{0.64} D^{0.16} \omega^{0.33}}$$

It can be rewritten as

$$S_m = \frac{5.14}{\omega^{1/3}} (\frac{U - U_C}{\sqrt{R}})^{5/3} (\frac{R}{D})^{1/6} \tag{11.26}$$

in which

S_m — sediment concentration in kg/m³;

U — mean velocity in m/s;

U_c — critical velocity for incipient motion of sediment particle m/s;

R — hydraulic radius in m;

D — sediment diameter in mm;

ω — fall velocity in mm/s.

The regression equation was obtained by means of a statistical analysis of the data, and it can only reflect a general trend for average conditions. In order to obtain a better fit for various specific conditions, the author rewrote Eq. (11.26) in a more general form

$$S_m = \frac{A_m}{\omega^{1/3}} (\frac{U - U_{C1} R^y}{\sqrt{R}})^m (\frac{R}{D})^{1/6} \tag{11.27}$$

and calibrated it against additional flume and field data to determine values for the coefficient A_m and the exponents m and y. The term U_{C1} in Eq. (11.27) refers to the critical threshold velocity for h (or R) = 1 m, and is called the specific velocity for sediment transport capacity.

The analysis indicated that $y = 0.2$ and m is either 2 or 3

for $J<0.0025$ or $Fr<0.8$, $m = 2$

for $J>0.0025$ or $Fr>0.8$, $m = 3$

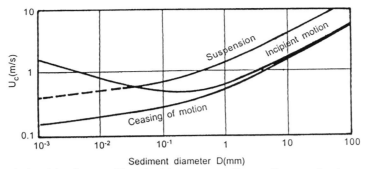

Fig. 11.7 Relationships for specific velocities versus sediment diameter for the cases of suspension, incipient motion and cessation of motion

In which Fr is the Froude number for sediment transport capacity with the status of sediment movement: if all of the sediment particles move as suspended load, the specific velocity for sediment transport capacity should be taken as that for suspension for h (or R) = 1m ; if the flow erodes the river bed and picks up sediment particles, the specific velocity should refer to the critical velocity of incipient motion; if sediment is in a transitional state from motion to rest, the velocity should take the critical velocity at which sediment particles stop moving for a flow depth of 1 m. Fig. 11.7 shows relationships between specific velocities for suspension, incipient motion, or the ceasing of motion. The coefficient A_m varies with m. For $m = 2$, $A_m = A_2$; for $m = 3$, $A_m = A_3$. Plots of both A_2 vs. D and A_3 vs. D are shown in Fig. 11.8.

Sha's formula also has a functional relationship with the basic form of Eq. (11.24), but it is not dimensionally homogeneous. Also, after analysing Brook's flume data, Sha found that for the same sediment diameter, the coefficient of sediment transport capacity for a flat bed differs from that for a bed covered with dunes. That is, Sha's original formula did not take into account the effect of bed configuration on the sediment movement. Furthermore, instead of the drag force, even if the velocity is involved as a hydraulic index, the effects of bed configuration should also be considered.

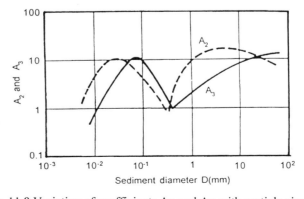

Fig. 11.8 Variation of coefficients A_2 and A_3 with particle size D

489

3. Engelund-Hansen formula [15]

The Engelund-Hansen formula is broadly recognized as one of the most reliable. The authors of it concluded that a dune-covered bed is the principal type of bed configuration in natural rivers. If λ denotes the wave length and Δ the wave height of a sand dune, the energy required for the sediment load of g_T to move from the trough to the crest of dune in unit time and per unit width is

$$\frac{\gamma_s - \gamma}{\gamma_s} g_T \Delta$$

The potential energy thus gained must equal the work done by shear stress on the moving sediment particles. Since suspended load has no motion relative to the flow, the suspended load is not subjected to shear force during its motion. The sediment in motion referred to by them is in fact the movement of bed load particles. Following Bagnold's concept, Engelund and Hansen considered the shear force acting on moving sediment particles to be $\tau_0 - \tau_c$ (Eq. 9.65). If a bed is covered by dunes, the grain resistance τ_b' should be used instead of τ_0. They did not use Eq. (9.64) for the mean velocity of sediment movement but included only the first term; i.e., they assumed the mean velocity of sediment movement to be proportional to the friction velocity U_*, then

$$\frac{\gamma_s - \gamma}{\gamma_s} g_T \Delta = \alpha(\tau_b' - \tau_c) \lambda U_*$$

in which α is a constant of proportionality. The preceding expression can be rewritten as

$$f \frac{g_r}{\gamma_s} \left(\frac{\Delta}{f\lambda} \right) = \alpha \frac{\tau_b' - \tau_0}{(\gamma_s - \gamma)D} U_* D \tag{11.28}$$

in which f is the Darcy-Weisbach resistance coefficient as in Eq. (7.30). From flume data, Engelund and Hansen took $\Delta/f\lambda$ to be effectively a constant. By using the parameters of flow intensity Θ of Eq. (6.40) and of sediment transport intensity ϕ_T of Eq. (11.15), they converted Eq. (11.28) into

$$f\Phi_T \sim (\Theta' - \Theta_c)\sqrt{\Theta} \tag{11.29}$$

As mentioned in Chapter 7, Engelund in an analysis of dune resistance, found for dune-covered beds

$$\Theta' = 0.06 + 0.4\Theta^2 \tag{7.68}$$

The constant 0.06 actually refers to Θ_c. Hence, Eq. (11.29) can be simplified as

$$f\Phi_T \sim \Theta^{5/2}$$

Fig. 11.9 compares the above formula against flume data, and reveals that the Engelund-Hansen formula fits well not only for the dune-covered bed configuration but also for one with antidunes. The coefficient of proportionality is 0.4, i.e.,

$$f\Phi_T = 0.4\Theta^{5/2} \tag{11.30}$$

If the mean velocity of sediment movement is taken to be proportional not to the friction velocity U_*, but to the friction velocity U_*' relevant to grain resistance on the bed surface, then the final expression for the Engelund-Hansen sediment transport capacity formula takes the form

$$f\Phi_T = 0.3\Theta^2 \sqrt{\Theta^2 + 0.15} \tag{11.31}$$

and it is shown by the dotted line in Fig. 11.9.

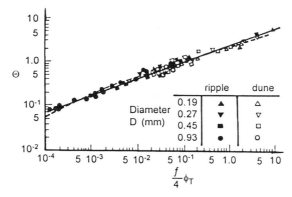

Fig .11.9 Comparison of the Engelund-Hansen formula against flume data
(after Engelund, F., and E. Hansen)

if Θ is small, $f\Phi_T \sim \Theta^2$

if Θ is large, $f\Phi_T \sim \Theta^3$

Starting from the bed load movement, the exponent of Θ increases from 3/2 to 3 as the flow intensity increases, and the suspended load movement increase correspondingly.

The Engelund-Hansen formula for sediment transport capacity is simple in structure and fits well with the measured data. However, some links in the process of deduction seem to be farfetched. Thus, the formula is considered to be an empirical or semi-empirical formula.

4. Ackers-White formula [16]

The Ackers-White formula adapted for coarse sediment is discussed in Chapter 9. Here, the formula is reintroduced briefly in a more generalized form. They used a dimensionless parameter X to divide all sediment into three groups: coarse, fine, and medium

$$X = \left[\frac{g \dfrac{\gamma_s - \gamma}{\gamma}}{v^2} \right]^{1/3} D \qquad (11.32)$$

If $X > 60$, the sediment is coarse, and the value corresponds to $D > 2.5$ mm for natural sediment; if $X < 1$, the sediment is fine, and it corresponds to $D < 0.04$ mm for natural sediment; if X is in between, $1 \le X \le 60$, the sediment is in the transitional region between the two for natural sediment.

Coarse sediment particles move in the form of bed load. If a river bed is covered by sand waves, the effective drag force responsible for bed load movement is only the part that corresponds to grain resistance. Ackers and White set this part of the drag force equal to the drag force on a flat bed induced by the same flow velocity, i.e.,

$$\sqrt{\frac{\tau_b'}{\rho}} = \frac{U}{\sqrt{32} \log(10 \dfrac{h}{D})} \qquad (11.33)$$

The constants used differ somewhat from those in the formula for a logarithmic velocity profile, introduced in Chapter 7. For fine sediment moving in the form of suspended load, they considered that the turbulence intensity is related to the total drag force, and hence, that the drag force responsible for the sediment movement should be

$$\sqrt{\frac{\tau_0}{\rho}} = \sqrt{ghJ}$$

For the condition that the intensity of sediment movement is related to the ratio of the effective drag force to the submerged weight of sediment particles at the bed surface layer, Ackers and White called their ratio a parameter of sediment mobility. After simplification, the parameter can be expressed as follows:

$$M = \frac{U_*^n}{\sqrt{gD \dfrac{\gamma_s - \gamma}{\gamma}}} \left[\frac{U}{\sqrt{32} \log(\dfrac{10h}{D})} \right]^{1-n} \cdot \qquad (11.34)$$

in which

$$\begin{cases} n = 0 & \text{for coarse sediment} \\ n = 1 & \text{for fine sediment} \\ n = f(X) & \text{for sediment in the transition region} \end{cases}$$

The parameter of sediment mobility M is no different from the parameter of flow intensity Θ, which is referred to frequently in preceding sections and chapters.

In establishing their formula for sediment transport capacity, Ackers and White also adopted the Bagnold concept that the intensity of sediment transport is related to the power provided by the flow; and they assumed the efficiency of sediment transport to be proportional to M. By combining the efficiency in M, they attained the following parameter for sediment transport

$$Y = \frac{S_{wT}h}{\frac{\gamma_s}{\gamma}D}(\frac{U_*}{U})^n \tag{11.35}$$

in which S_{wT} is the sediment concentration in percentage by weight for a water column above a unit element of the bed surface. From a large amount of flume data, they found this parameter of sediment transport to be a function of M and X. By analysing 1000 sets of flume data, they obtained the final expression as

$$Y = c(\frac{M}{A} - 1)^m \tag{11.36}$$

The condition of incipient motion of sediment is

$$M = A \tag{11.37}$$

For coarse sediment

$$\left.\begin{array}{l} n = 0 \\ A = 0.17 \\ c = 0.025 \\ m = 1.5 \end{array}\right\} \tag{11.38}$$

For sediment in transition region

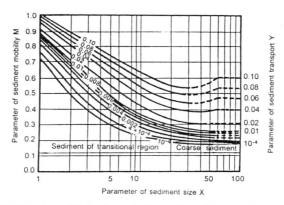

Fig. 11.10 Ackers-White formula for sediment transport capacity
(after Ackers, P., and W.R. White)

$$
\left.
\begin{aligned}
n &= 1 - 0.56 \log X \\
A &= \frac{0.23}{\sqrt{X}} + 0.14 \\
\log c &= 2.86 \log X - (\log X)^2 - 3.53 \\
m &= \frac{9.66}{X} + 1.34
\end{aligned}
\right\}
\tag{11.39}
$$

If fine sediment exhibits any effect of cohesion among the particles, the above-mentioned formulas are not applicable. Fig. 11.10 is a graphical representation of Eq. 11.36.

Both the Engelund-Hansen and the Ackers-White formulas for sediment transport capacity represent functional relationships in the basic form of Eq. (11.25). However, the parameters used in these formula are not as complete as those in the Einstein bed load function; they include only the first two sets of dimensionless parameters of Eq. 11.25. Besides, in the Ackers-White formula, the effect of D/h has been taken into account.

5   *Colby method* [17]

Many formulas of the afore mentioned empirical type have been proposed, mainly on the basis of flume experiments. In view of the great difference in the scales of flumes and natural rivers, the conversion of the results obtained in flume's flows to natural rivers raises difficult questions. In addition, other empirical formulas for calculating sediment transport are guided by modern theories of sediment movement and were established using field data observed at hydrometric stations. Among these formulas, the Colby method, the modified Einstein procedure [18], and the Toffaletti formula [19] have been widely used in western countries.

The Colby method is suitable for rivers with beds of medium to fine sand. Eq. (11.24) shows that the sediment transport capacity of a river depends mainly on three factors: the velocity, flow depth, and sediment diameter (or fall velocity). Instead of using regression analysis or empirical curve fitting to express the effects of these factors on sediment transport capacity, Colby developed a set of graphs, and these are shown in Fig. 11.11. Altogether 24 curves are included, and they correspond to values of h varying by factors of 1,000 and to various values of median diameter. The curves in Fig. 11.11 are for a temperature of 60°F, D_{50} = 0.2-0.3 mm, and for flows with a negligible amount of fine silt and clay. If conditions differ from these, then the sediment transport found on the chart should be multiplied by a correction factor

$$1 + (k_1 k_2 - 1)0.01k_3$$

in which k_1, k_2 and k_3 are correction coefficients for temperature, content of fine silt and clay and median diameter of bed material, respectively, shown in Fig. 11.12. Although a large portion of the curves involves extrapolations, comparisons show that 75% of the data have a discrepancy ratio within the range

$$0.5 \le \frac{g_{Tc}}{g_{Tm}} \le 2$$

in which g_{TC} is the computed sediment transport rate and g_{Tm} the measured one.

11.1.3 Comparison of various formulas and examination of scope of data

11.1.3.1 Reliability of various formulas

Up to now, no one has made an overall examination and comparison of the various existing formulas for sediment transport capacity. Fig. 11.13 is such a comparison and it shows a plot of the Einstein bed load function, Engelund-Hansen formula, and Colby methods against the field data for the Niobrara River [18]. The slope of this river is 0.00129; the median diameter of the bed material is 0.277 mm, and the flow depth was in the range of 20 to 40 cm when the field measurements were taken. Fig. 11.13 shows that the Colby method and the Engelund-Hansen formula are closer to the measured data; the trend of the curve computed by the Einstein formula is also similar to the measured one, but it generally underestimates the amount of sediment transport. In another comparison, White et al. examined the Ackers-White formula, Engelund-Hansen formula, Einstein formula, and Bagnold formula. They are plotted with 1020 sets of flume data and 260 sets of field data (excluding those with Froude numbers larger than 0.8) in Fig. 11.14 [20][21]. One can conclude that, limited by the present level of theoretical understanding, the theoretically based formulas are less reliable than the empirical ones. The Bagnold formula, within the scope of the data used for self-verification, is valid for such applications, but beyond this scope, the formula appears to be unreliable because of the oversimplified way it deals with the transport of suspended sediment.

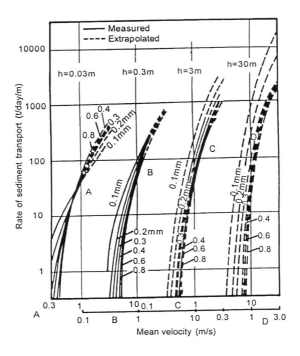

Fig. 11.11 Work chart for the Colby relationship for sediment transport capacity in terms of mean velocity (m/s) for six median diameters, four flow depths and a temperature of 60°F. (after Colby, B.R.)

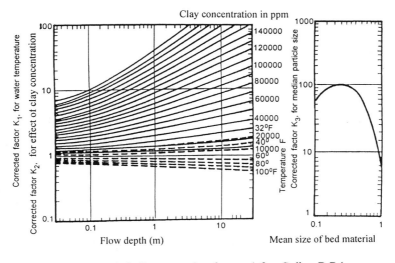

Fig. 11.12 Colby correction factors (after Colby, B.R.)

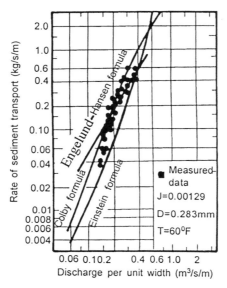

Fig. 11.13 Comparison of formulas of sediment transport capacity against field data of the Niobrara River (after Colby, B.R., and C.H. Hembree)

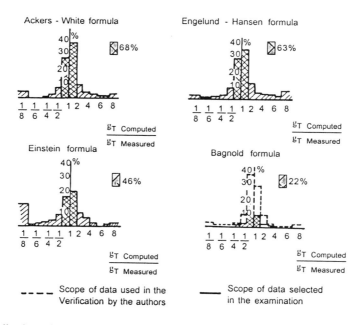

Fig. 11.14 Distribution of the discrepancy ratio for sediment transport capacity as computed by means of various formulas in comparison with the measured data (after Ackers, P., and W.R. White)

497

11.1.3.2 Scope of the existing flume data

Research on sediment transport capacity of a flow depends greatly upon data from flume experiments. Up to now, incomplete statistics show that more than 4000 sets of experimental data on sediment transport from flume tests have been obtained at various institutions. In 1972, Cooper et al. summarized 3,709 sets of flume data observed in United States, Europe, and India [22]. They used 125 kinds of sediment. The scope of the variation of the principal items in these flume experiments is given in Table 11.4. A comparison of the scope of the flume data with that of field data from natural rivers is shown in Fig. 11.15; it shows that only 2% of the flume data have values of h/D_{50} larger than 1,000, whereas in natural rivers h/D_{50} can be as high as 100,000. For the same h/D_{50}, the range for the median diameter of bed material used in flume experiments also seems too narrow. Faced with such a deficiency, especially for use in northern China where the sediment is rather fine, one should supplement the results of flume experiments with data for large ratios of h/D_{50}.

Table 11.4 Scope of principal variables in flume experiments
(after Cooper, R.H., A.W. Peterson, and T. Blench)

Item	Scope of variation	Percentage
Discharge $Q(m^3/s)$	<0.6	98
Flow depth $h(m)$	<0.34	98
Width of flow $B(m)$	<2.44	98
B/h	1-80	98
Slope J(percent)	0.012-2.2	98
Median diameter	>1	30
D_{50} (mm)	>8	2
h/D_{50}	>1000	2

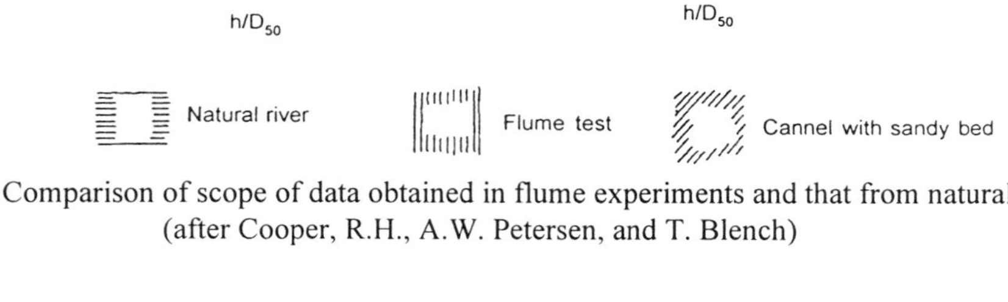

Fig. 11.15 Comparison of scope of data obtained in flume experiments and that from natural rivers
(after Cooper, R.H., A.W. Petersen, and T. Blench)

11.2 ESTIMATION OF THE TOTAL SEDIMENT LOAD INCLUDING THE WASH LOAD

Formulas of sediment transport capacity presented in the preceding sections that were established on the basis of mechanics should be used to compute only the sediment discharge in the form of bed material load. For the wash load, the relationship between sediment transport rate and flow rate is based on factors related to the common background of the watershed, as already pointed out in Chapter 5. Such a relationship can be established only from data observed in the field, including: (1) data of sediment load measurement at hydrometric stations; (2) information of sediment yield in drainage basins; (3) measurements of sediment deposits in reservoirs [23]. The following section contains a discussion of the nature of such data and methods for the treatment of the three kinds of data.

11.2.1 Annual sediment load evaluated from the relationship of flow discharge to sediment transport rate as measured at hydrometric stations

Regular measurements of flow discharge and sediment sampling are the routine work of hydrometric stations. In most cases, data are obtained for sediment transport rates related to various rates of flow. By means of the relationship between flow discharge and sediment transport rate and the frequency curve for flow, one can evaluate the total sediment load at a given hydrometric station. However, this approach is affected by three procedural difficulties: (1) The observed field data usually do not include the measurement of bed load; also, the portion of suspended load near the bed surface is not easy to measure so that the measured data do not fully reflect the total sediment load carried by the flow. (2) In some rivers, the measured data points scatter widely; thus it is difficult to set up a relationship of flow discharge versus sediment transport rate by conventional methods of curve fitting. (3) Fewer data for sediment transport rate are available than for flow discharge, and data series may be too short to be representative of average conditions. Efforts made to find rational solutions to these difficulties are discussed in the following sections.

11.2.1.1 Evaluation of total sediment load based on measurement of suspended load

Because of its size, a suspended load sampler cannot make measurements in the vicinity of the bed. Also, the bed load is well outside the scope of suspended load sampling. For the wash load, which is composed mainly of fine sediment that is uniformly distributed along the vertical, the mean sediment concentration obtained by the conventional sampling of methods should represent satisfactorily the true mean value. But for the bed material load, especially particles coarser than fine sand, a considerable part is concentrated near the river bed, and it is not included in the results of suspended sediment sampling. How to estimate the unmeasured sediment load is a major concern. Samples are taken both at points and by depth-integration devices, and the necessary corrections to be made for these two methods are different .

1. Correction for point samples [24]

In China, the instantaneous horizontal trap sampler has been widely used to measure the sediment concentrations at various heights; simultaneously, the velocity u_y is measured at the same point by current meter. The product of S_{vy} and u_y represents the sediment transport rate per unit area normal to the flow direction at that point. The weighted arithmetic mean of this product (usually taken at three to five points) multiplied by the flow depth h is taken to be the sediment transport rate per unit width for that vertical. The ratio of the computed sediment transport rate to the real sediment transport rate can be expressed as

$$\theta_1 = \frac{h\sum_{i=1}^{N} W_i S_{vyi} u_{yi}}{i_T g_T} \tag{11.40}$$

in which θ_1 is the correction factor for unit sediment transport rate for sampling using the point method, W_i is the weight of $S_{vyi}u_{yi}$ at the i-th point, and N is the number of measuring points. In addition, one has the relationships

$$\frac{S_{vy}}{S_{v2D}} = (\frac{h-y}{y} \frac{2D}{h-2D})^z$$

$$u_y = 5.75 U_*^' \log (30.2 \frac{y}{\Delta})$$

$$S_{u2D} = \frac{i_b g_b}{23.2 D U_*^'}$$

$$i_T g_T = i_b g_b (1 + PI_1 + I_2)$$

If one substitutes these four equations into Eq. (11.40) and uses

$$y_i = x_i h$$

one obtains

$$\theta_1 = 0.215 \frac{A^{z-1}}{(1-A)^z} \frac{1}{1+PI_1 + I_2} \sum_{i=1}^{N} W_i (\frac{1-x_i}{x_i})^z \times (P + \frac{1}{0.434} \log x_i) \tag{11.41}$$

If the number of measurement points and the appropriate corresponding weights are known, Eq. (11.41) indicates that θ_1 is a function of A, z, and P. The parameter P varies within a narrow range. For the hydrometric stations on the main stem and tributaries of the Yellow River, P_{max} is 14.5, P_{min} 10.2, and most values of P fall between 12.5 and 14.0 with a mean value of 13.

At present, the three-point method is generally used at hydrometric stations on the Yellow River; measurements are taken at relative depths of 0.2 (x_1=0.8), 0.6 (x_2=0.4)

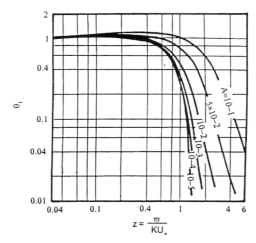

Fig. 11.16 Correction factor to be used for sampling by point method

and 0.8 (x_3=0.2), and the corresponding weights for these points are 1:2:1. If the average value 13 is used for P, then Eq. (11.40) becomes

$$\theta_1 = (0.61 \times 4^z + 1.30 \times 1.5^z + 0.69 \times 0.25^z) \times \frac{A^{Z-1}}{(1-A)^z} \frac{1}{1+13I_1 + I_2} \quad (11.42)$$

Fig. 11.16 is a graphical solution of Eq. (11.42). For small values of z, the curve is approximately horizontal and the value of θ_1 is quite close to 1.0. If an error of ±15 % is acceptable, then the three-point method gives satisfactory results for values of z less than 0.5. If z is greater than 0.5, Eq. (11.42) or Fig. 11.16 can be used to correct the measured data. But if the z-value is very large and A is rather small, the curves for the correction factor are steep. In such cases, most sediment particles are concentrated near the river bed, and only a small portion enters the sampler. The measured value can then hardly give accurate results. In such cases, small measured values of θ_1 must be multiplied by factors of 10 or even several 10s, and the probable error becomes unacceptably large to tolerate. In this range the evaluation of the total sediment transport rate from the measured suspended load is meaningless. In general, the smaller the distance between the measured point and the river bed, the more effective sampling by the point method is. However, the size of the sampler always limits how close the sampler can be to river bed.

2. Correction for depth-integrating method [25]

Sampling by the depth-integrating method is a process that samples continuously and automatically as the sampler is lowered at a given rate from water surface to river bed. The sample thus obtained represents the mean sediment concentration along the vertical. Multiplied by the discharge per unit width, it yields a value for the sediment transport rate per unit width at that vertical. Even if the sampler is lowered to the river bed, its nozzle is still some distance α above the bed surface, i.e., the sediment sample actually provides the mean sediment concentration from the water surface down to a point that is α above the bed. If g_{sa} denotes the suspended sediment transport rate in this range, and i_{sa} denotes the percentage of sediment with diameter D in the sediment sample taken by depth-integrating method, then

$$i_{sa}g_{sa} = \int_{\alpha}^{h} S_{v2D}(\frac{h-y}{y}\frac{2D}{h-2D})^{z}$$

$$\times 5.75 U_{*}' \log(30.2\frac{y}{\Delta})dy$$

Taking a procedure similar to the deduction of Eq. (11.10), one obtains

$$i_{sa}g_{sa} = i_{b}g_{b}(\frac{2D}{\alpha})^{z-1}(\frac{h-\alpha}{h-2D})^{z}(PI_{1\alpha} + I_{2\alpha}) \qquad (11.43)$$

in which

$$\left.\begin{array}{l} I_{1\alpha} = 0.216\dfrac{A_{\alpha}^{z-1}}{(1-A_{\alpha})^{z}}\int_{A_{\alpha}}^{1}(\dfrac{1-y}{y})^{z}dy \\[4mm] I_{2\alpha} = 0.216\dfrac{A_{\alpha}^{z-1}}{(1-A_{\alpha})^{z}}\int_{A_{\alpha}}^{1}(\dfrac{1-y}{y})^{z}\ln y dy \end{array}\right\} \qquad (11.44)$$

$$A_{\alpha} = \frac{\alpha}{h} \qquad (11.45)$$

If A in Eq. (10.61) is replaced by A_{α}, then Eq. (11.44) is obtained. The graphical solutions in Fig. 10.24 and Fig. 10.25 can still be used, but only if the value A in the figures is replaced by A_{α}.

$$\frac{i_{sa}g_{sa}}{i_{T}g_{T}} = (\frac{2D}{\alpha})^{z-1}(\frac{h-\alpha}{h-2D})^{z}\frac{PI_{1\alpha} + I_{2\alpha}}{1 + PI_{1} + I_{2}} \qquad (11.46)$$

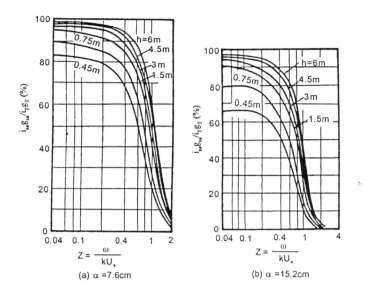

Fig. 11.17 Variation of $i_{sa}g_{sa}/i_Tq_T$ in percentage vs. h and z.

If P is taken to be a constant (for example, $P=10$), then Eq. (11.46) can be represented approximately by the family of curves shown in Fig. 11.17, in which the α values in parts (a) and (b) are taken as 7.6 cm and 15.2 cm, respectively. For an α value between these two, a solution can be obtained by interpolation. If P has values within the range of 9 to 12, a comparison of the computed results by means of Eq. (11.46) with the graphical solution given in Fig. 11.17 shows a discrepancy of less than 5%.

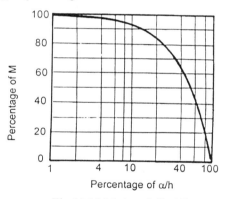

Fig.11.18 M vs. $\alpha/h(P=10)$

In applications of the depth-integrating method, the product of the mean sediment concentration obtained by the sampler multiplied by the unit discharge is taken as the sediment transport rate along the vertical. If M denotes the ratio of the unit discharge within the scope of sampling and the real unit discharge along the vertical, then the correction factor for the depth-integrating method is

$$\theta_2 = \frac{i_{sa}g_{sa}}{i_Tg_T}\frac{1}{M} \tag{11.47}$$

The relationship of M vs. α/h $(P=10)$ is shown in Fig. 11.18.

As was the case for the point method, if $z > 0.8$, errors in the computed total sediment transport rate, which are likely to be obtained by applying a correction from Eq. (11.46) to the result obtained by depth-integrating method, may still be unacceptably large. In such cases, bed load measurements are indispensable.

11.2.1.2. Method of establishing a relationship for discharge-sediment transport rate from scattered data points

As mentioned in Chapter 2, if the wash load in the drainage area is large and both regional factors (like vegetation cover, topography, soils, etc.) and the rainfall distribution are strongly non-uniform, the data points for measured sediment transport rate plotted against measured discharge usually display a wide band of scatter. If the relationship was established by following through the trend of the data, considerable error would result in the computation of the annual sediment load by using that relationship and the corresponding frequency curve for river discharge.

The wide scatter of the data points could result from two circumstances. First, due to the large spatial differences, runoff formed in different areas may bring about quite different sediment concentrations, sometimes high, and other times low. Second, the scatter could be due to temporal differences in runoff. For example, in early spring, runoff is large because of the melting ice and snow; in summer and autumn, the heavy rainstorms cause floods. The sediment concentrations for these two cases differ greatly. In some drainage areas, both conditions occur and the situation is then even more complicated. In addition, heavily sediment-laden rivers, because of self-regulation of the channel, are characterized by the phenomenon of "more sediment may be released if more sediment is supplied." Such a situation can enhance the extreme scatter of data points in a plot of sediment transport rate against water discharge.

In analyses of hydrological data, one can sometimes determine the concrete causes for the scatter of data points. One can then work out a set of relationships for discharge vs sediment transport rate and the corresponding discharge frequency curve that is in accord with the specific events of runoff originating in different source areas or occurring in different seasons. The total sediment transport rates for given time periods are then evaluated separately.

The data treatment techniques applied to the Yellow River serve as an example. The main source of sediment for the Yellow River lies in the middle reaches in Shanxi and Shaanxi Provinces; the soil loss in the upper reaches is much less. Thus, sediment concentration of the runoff originating from these different areas can be quite different.

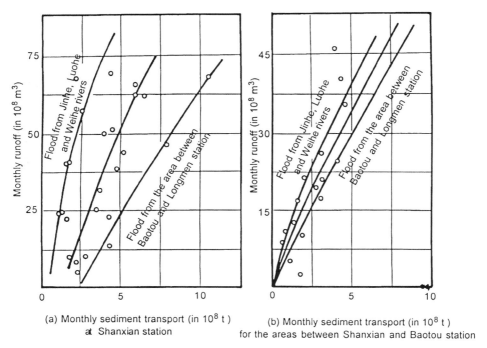

(a) Monthly sediment transport (in 10^8 t)
at Shanxian station

(b) Monthly sediment transport (in 10^8 t)
for the areas between Shanxian and Baotou station

Fig. 11.19 Example taken from the Yellow River for the demonstration of rating curves
with different source areas of sediment yield

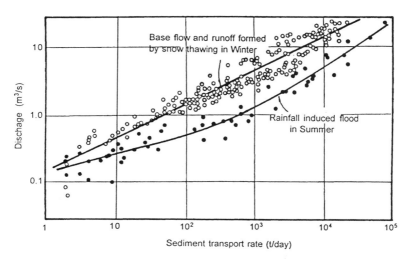

Fig. 11.20 Typical example of relationships for discharge vs. sediment transport rate
with different types of runoff

Fig. 11.19a shows the monthly relationship between runoff and sediment load at
the Shanxian hydrometric station. Despite the use of monthly averages, data points
scatter rather widely. Detailed investigation reveals that the data points group naturally

505

into belts according to the source of the runoff. Those for runoff originating from the region between Baotou and Longmen are located to the right side of the diagram, whereas those originating from the drainage area of Jinhe, Luohe, and Weihe Rivers are to the left. For an identical discharge, the sediment load coming from the former areas is considerably greater than that coming from the latter ones. The total runoff measured at Shanxian hydrometric station includes also that portion of the flow that originates in the upper reaches of the main stem. In order to eliminate its effect, the amounts of runoff and sediment load at Baotou hydrometric station are subtracted from those at Shanxian hydrometric station. Then the regional runoff and sediment load relationship between Baotou and Shaanxian stations is as shown in Fig. 11.19b. Obviously these data points are more concentrated.

For rivers with small drainage areas, the spatial differences are generally not great. Yet data points may still distribute themselves into belts according to seasons. Fig. 11.20 shows the ratio of discharge to sediment transport rate for the San Rafael River in the United States. In it, all data points of rainstorm-induced floods in summer form one belt, and those of the base flow in other seasons and those due to thawing snow in winter fall in another.

Similarly, the discharge frequency curve for runoff originating from different source areas, or occurring in different seasons and of different types may be established. With the aid of such graphs together with curves like those given in Fig. 11.19b and Fig. 11.20, one can compute more accurately the total amount of sediment load under various conditions.

11.2.1.3 Method of dealing with short series of data

Discharge data on most rivers have longer periods of record than do data for sediment transport rate. Whether the short series of data can be used to represent the real situation is a serious concern. Miller analyzed 19 years of field data for the San Juan River. He plotted the ratio of the annual discharge Q to Q_m , the average for the 19 years, as the ordinate, against the annual sediment load G_T to G_{Tm}, also the 19-year average as the abscissa. The result is shown in Fig. 11.21. The figure shows that a data series in which Q is greater than 90% of Q_m as a representative series will yield the errors of the estimated sediment transport rate of less than 20%. In contrast, if the data series for a low flow year is taken to be representative, the error could be rather large. In other words, if the data record is short, one should select a relationship for the discharge vs. sediment transport rate for a rather high flow year to deduce the perennial averaged sediment load of a longer time period. The data of low flow years should be eliminated.

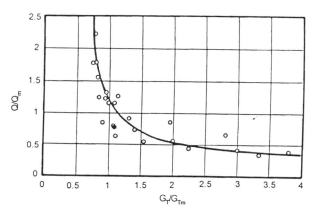

Fig. 11.21 The validity of the use of one year data record to deduce the average sediment transport rate for long time series (after Miller, C.R.)

11.2.2 Estimation of sediment load originating from the drainage area and conveyed into the river based on various factors within the river basin

Since soil erosion is the source of sediment, the amount of sediment conveyed in the river system is naturally related to the various factors that affect soil erosion in the watershed. If such relationships can be shown graphically or expressed by mathematical equations, the amount of sediment originating from the watershed and conveyed into the river can be deduced from the characteristics of the river basin factors; such a process can be useful if field data are lacking.

In practical applications, two approaches are available. One is to establish a direct relationship for the sediment load conveyed into the river expressed in terms of the characteristics of the given watershed, and based on measured data from the hydrometric network. The other is to estimate first the amount of soil eroded from the ground surface and then to estimate how much of that material can be carried into the river.

11.2.2.1 Direct estimation of sediment load in a river in terms of watershed factors

Anderson analyzed measured data for 29 watersheds in the state of Oregon, U.S. (watershed areas ranging from 145 to 18,850 km^2) to establish a relationship between suspended load and various regional characteristics of those watersheds that had hydrometric stations (he assumed that the bed load was negligible) [28]. The factors he included comprise a set that appears to be quite complete. The standards for measurement, units and physical interpretation of these factors are given in Table 11.5.

By multiple regression, Anderson established the following empirical relationship between the annual suspended sediment yield per unit area SS_1 and the regional factors of the watershed:

$$\log SS_1 = (\text{constant}) + 0.116 \log A + 1.673 \log FQ_p + 1.244 \log MA_q$$
$$+ 0.401 \log J + 0.0486 SC + 0.482/B + 0.0280 BC \qquad (11.48)$$
$$- 0.0036 OC + 0.942 R + 0.0086 RC$$

in which SS_1, in units of $t/km^2/a$, varies from 3.48 to 292 and has a mean value of 62.6.

Table 11.5 Principal factors affecting sediment yield in western Oregon, U.S.
(after Anderson, H.W.)

Factors in the watershed		Symbol	Unit	Average value	Range	Physical meaning
Flow factors	average annual runoff	MA_q	m^3/km^2	0.0325	0.0114~0.0817	Magnitude of rainfall
	steepness of flood discharge	FQ_p	–	3.56	1.98~4.30	Intensity of runfall
Soil factors	the content of silt and clay in the surface soil layer (15 cm)	SC	%	23.0	19.1~32.0	Representing the source of suspended load (fine sediment is easy to be suspended and carried away by the flow
	aggregate ratio[1)]	B	$\dfrac{\%}{cm^2/g}$	1.37	0.56~3.84	It reflects the permeability and the ability of withstanding erosion of the soil)
Geography factors	area of watershed	A	km^2	2,000	145~18,850	-
	river gradient	J	m/km	172	40~286	Average gradient of surface soil in the watershed
Factors of land utilization	road	R	%	0.3	0.05~0.6	Road construction induces water soil erosion
	forest cut within latest ten years	RC	% in ten years	6.0	0~30.4	Cutting forest destroys the protection effect by the forest
	cultivated land with thin cover	BC	%	4.0	0~22	Soil erosion will be greatly reduced if the
	cultivated land different from BC	OC	%	12	0~48	land surfaces is covered by plant The same as above
	cultivated land =(BC+OC)×A	C	km^2	20.7	0~173.5	-
	eroded bank	EB	m	5,180	0~62,500	Soil due to bank erosion directly enter the river

Note:1) Definition of soil aggregate ratio B and the technique for measuring the term B are given in Ref [28].

According to different origins of sediment, Anderson classified the source of sediment into three categories: sediment from forest areas, from cultivated areas and from collapses of river banks,

$$SS_2 = -1485 + SS_3 + 84C + 15.25EB \qquad (11.49)$$

in which SS_2 is the annual yield of suspended sediment in the whole watershed in t/a; the average value is 130,700 t/a and it ranges from 1,814 to 1,775,000 t/a; SS_3 is the annual mean sediment yield of suspended load originating from the forest areas in the watershed; the average value is 61,600 t/a, and it ranges from 27,200 to 420,000 t/a.

In Eq. (11.48), a variable having a larger coefficient has more effect on the sediment yield in the watershed. Hence one can deduce that:

1. Rainfall is the primary cause of soil erosion, and flood flows play a much larger role than does the mean runoff.

2. If the main features of the watershed are included one by one, then the sediment yield of the watershed is not affected by the drainage area.

3. The function of the soil aggregate ratio B and that of the slope are equally important.

4. Because it destroys a natural balance, road construction probably results in accelerated erosion. In some small watersheds the collapse of river banks is the major source of sediment.

The results of Anderson's research indicate not only the effect of various features of the drainage area on soil erosion, but they also point out the potential for soil erosion in other districts according to their distribution of rainfall, topography, soil use and land use. Thus watershed management and planning of soil conservation practice in various regions can be made on a scientific basis in accordance with need and priority.

Jiang and Song developed the following empirical formula to predict the sediment yield from small watersheds in the loess region from a study of field data for small watersheds in the gullied and hilly loess region of the middle Yellow River [29]:

$$M_s = 0.37 M^{1.15} JKP \qquad (11.50)$$

in which

M_s— modulus of sediment yield in one storm event (t/km^2);
M — modulus of flood volume in one storm event (m^3/km^2) that can be evaluated from the rainfall, watershed area, the length and slope of the principal gully, and other factors;

J— average slope of the watershed;

K—erosivity of soil, expressed by the ratio of sand and silt content per unit volume;

P— coefficient representing vegetation cover.

11.2.2.2 Estimation of sediment load from soil loss in a watershed

In an effort to investigate soil erosion from the watersheds in terms of regional factors, a large number of experimental stations were established for the observation of runoff in watersheds with naturally differing geographical conditions. Many graphs and formula have been developed from the resulting of observation. In the process of transport and convergence toward the main stem of the river system, a part of the sediment load settles out in depressed areas and regions with gentle slopes. The ratio of the sediment entering the river to that eroded from the watershed is called the delivery ratio. If one can determine the amount of soil loss and the delivery ratio, the product of the two is an estimate of the amount of sediment that enters the river [30].

1. Amount of soil loss

Numerous formula have been presented for the prediction for the soil erosion in terms of various features of a watershed. Most of them are strongly affected by regional factors and should be applied only to those places in which the natural conditions are similar. For example, the Xifeng experimental station for soil conservation [29] used data for cultivated land in the loess region during the season without vegetation cover; the relationship between soil loss per unit area M_e (kg/m^2) and the variables kinetic energy of rainfall E (kg/m^2), surface slope J (degrees), rainfall intensity for 30 min I (mm/min) can be expressed as

$$M_e = 3.27 \times 10^{-5}(EI)^{1.57} J^{1.06} \tag{11.51}$$

The relationship between slope length and soil erosion is rather complicated. For a steep slope, soil erosion increases with the slope length; but for a gentle slope and low rainfall intensity, soil erosion decreases with the slope length.

In the United States, the Universal Soil Loss Equation (USLE) has been widely applied for computing soil loss:

$$M_e = RKLJCP \tag{11.52}$$

in which R, K, L, J, C, and P are, respectively, parameters for rainfall, soil erodibility, slope length, land gradient, crop management, and soil conservation practice. All these factors should be determined in accordance with the specific condition for the local area. USLE relationship is applicable only for sheet erosion. If gully erosion is important, then the amount of sediment originating from it should also be assessed and added.

2. Delivery ratio

Various factors affect the delivery ratio; these include the pattern of sediment yield (sheet erosion or gully erosion), distance from the source of the sediment, specific features of the network (density and capability of keeping sediment in transportation), features of depressed areas, composition of eroded material and characteristics of the drainage basin. Among them, drainage area appears to be dominant. Fig. 11.22 is a plot of delivery ratio vs. drainage area for five different regions in United States [32]. Although the data points scatter considerably, a clear tendency is still revealed for the delivery ratio to vary in proportion to the several of the drainage area raised to a power of about 0.2. This curve is frequently applied in the United States for rough estimates of delivery ratio for drainage basins of different sizes, but it surely depends upon the local regional conditions. For example, the gullied and hilly loess region of the Yellow River Basin in China has no flat depressed areas between the tops of mounds and the ridges of the branch gullies or creeks. All of the main stem and the branches, including small gullies, function as conveyance channels for sediment transport because no obvious locations exist for deposition or erosion. Hence the delivery ratio is close to one [33]. In other words, a reduction of soil erosion in the watershed by one ton means a reduction of the sediment load entering the Yellow River of one ton also.

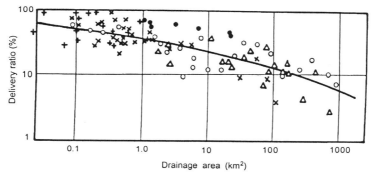

Fig. 11.22 Sediment delivery ratio vs. drainage area (after Boyce, R.C.)

11.2.3 Estimation of sediment yield of a watershed from reservoir deposition

If a large reservoir is constructed in a river, all of the sediment load from the upstream areas will be intercepted by the reservoir. Thus measurement of the amount of deposition in the reservoir is a reliable way to assess the sediment yield of the drainage area.

If, on the contrary, the storage capacity of the reservoir is not large relative to the volume of runoff, then part of the sediment load may be carried on downstream. Thus, the sediment yield based on deposition in the reservoir must include the efficiency of the reservoir in trapping sediment. Fig. 11.23 shows the relationship of the ratio of sediment outflow to inflow during flood events to the characteristics of the reservoir, and with sediment size and concentration as additional parameters [34]; in it, V is the storage volume of the reservoir, Q_i the inflow, and Q_0 the outflow. The abscissa of the diagram is VQ_i/Q_0^2 has the dimension of time; it reflects the time of flood detention in the reservoir. In addition, the efficiency of sediment release is also related to sediment size and sediment concentration. Fine sediment can be released much more easily than coarse sediment. If the sediment concentration exceeds 50 kg/m^3, the fall velocity of the sediment is less, and more sediment is released.

In addition to the above three approaches, physical and mathematical models have been used recently in the study of the formation and confluence of runoff, including the concept of sediment yield.

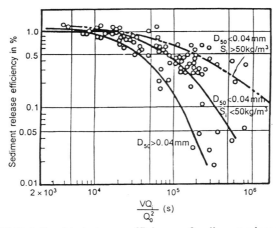

Fig. 11.23 Relationship between efficiency of sediment release and
characteristics of reservoir and sediment load

11.3 DISCUSSION OF SEDIMENT TRANSPORT CAPACITY

The following sections introduce to methods for computing the sediment transport capacity of a flow, and discuss some aspects of the sediment transport capacity.

11.3.1 The dual-value (or multi-value) for sediment transport capacity

The data measured at hydrometric stations show that the sediment load carried by the flow can vary to some extent even if the flow conditions do not differ; i.e., sometimes a given flow carries more sediment load, and at others it carries less. Naturally, people associate this anomaly with the possibility of multiple values of sediment transport capacity. Although the capacity of a flow has an upper and a lower limit, it can vary significantly between them. In other words, if the oncoming sediment load from the upstream reach exceeds the upper limit of the sediment transport capacity, deposition will occur; if the oncoming sediment load does not reach the lower limit of this transport capacity, erosion will take place; if the oncoming sediment load lies in between the two limits, all of the sediment load can pass the reach without deposition or scouring. From the viewpoint of just the two limits on sediment transport capacity, the sediment transport capacity of the flow may appear to be as a dual-valued. Still, since the flow can also transport sediment in equilibrium between these limits, one can view the process as a multi-valued one.

In order to clarify the physical basis of the dual or multi-valued relationship of sediment transport capacity, the author conducted a series of flume experiments [35, 36]. Since the work of Gilbert, who initiated the study of sediment movement in flumes, a large number of researchers have conducted such studies. An examination of the results of these studies indicates that despite great differences in flow conditions and sediment properties, and a consequent divergence in the conclusions reached, the methodology adopted by various researches is much the same. The experiments usually begin with a well-formed bed layer of sediment set at a given slope. Then a flow of clear water is begun and sediment is fed at a given rate at the entrance of the flume; simultaneously, the sediment falling into a desilting basin at the outlet of the flume is measured. The process continues until the sediment discharge equals the sediment supply so that the channel bed is in balance. In some experiments using fine sediment, the water-sediment mixture is circulated until the sediment concentration no longer changes. At that time, the movable bed is in equilibrium. In both cases, the sediment load carried by the flow comes from the erodible bed in the flume. Therefore if a dual-valued relationship of the sediment transport capacity really exists, the results of such flume experiments should correspond to the lower limit of the sediment transport capacity of the flow.

If the flow possesses an upper limit to sediment transport capacity that is different from that obtained in the afore-mentioned experiments, the difference between the two should be revealed through one of the following approaches to flume experiments.

1. After an equilibrium under erosion is reached for a process that was eroding, the rate of sediment supply is increased. Then, if the sediment transport capacity of the flow is a single-valued function, the additional sediment supply exceeding the balanced rate of sediment transport should deposit on the bed until the changes in the bed composition have adjusted to the increase. Also the sediment transport capacity of the flow equals the higher rate of sediment supply and a new equilibrium state of channel bed is reached. If on the contrary, the sediment transport capacity of the flow is a dual-valued function, unless the increased rate of sediment supply exceeds the upper limit, all the increased sediment supply should pass through the flume and be released without changing either the bed material composition or the flow conditions.

2. After the equilibrium state corresponding to the lower limit of sediment transport capacity is reached, all deposits of bed material are stirred up, so that each deposited particle has a chance to be carried by the flow. Those parts of sediment load that exceed the sediment transport capacity then deposit again on bed and reform it; the amount of sediment load that the flow is capable of carrying should then be the maximum sediment transport capacity of the flow for the new bed material.

3. Instead of providing a layer of sediment on the bed at the beginning, the experiment is conducted by adding sediment at a given rate at the entrance of the flume. A part of the sediment thus added is carried by the flow, while the excess deposits on the bed; in this way an alluvial bed gradually forms. The sediment transport capacity in equilibrium state should then correspond to the upper limit of the transport capacity of the flow.

The author conducted flume experiments following these three different approaches for two different uniform sediment and eight different mixtures of non-uniform sediment to compare them with results obtained for equilibrium conditions. The experiments showed that no matter what experimental approach was adopted, if the flow conditions were basically unchanged, the rate of sediment transport obtained for equilibrium of deposition was always larger than that for equilibrium of scouring; sometimes the divergence was quite large. In Table 11.6, two sets of experimental results are listed. In set B, the sediment concentration increased continuously with the increase of sediment supply. Sediment concentration in B-5 is about 1.45 times higher than that in B-1. In set C, the difference between the two equilibrium sediment concentrations, for deposition and that for scouring, is quite striking. The sediment concentration of C-3 was about 12 times that of C-5, although the hydraulic flow (discharge of flow, depth, velocity, energy gradient, etc.) remained basically unchanged, these results and other similar observations appear to confirm that a multi-valued function for the sediment transport capacity of the flow does exist.

The question that remains is how should one rationalize this result in terms of the sediment transport capacity of a flow. If the sediment transport capacity refers to the relationship between sediment concentration and hydraulic parameters, then the above conclusion is undoubtedly correct. On the contrary, as pointed out in Chapter 5, if the physically intrinsic nature of the increasing exchange between bed material and sediment load in motion, the sediment transport capacity of the flow should be interpreted as a function of three variables: the flow condition, the bed material composition and the amount of sediment load which can be transported to the downstream reaches, then the conclusion will be different. Table 11.6 shows that if the rate of sediment supply from upstream is larger, a part of the fine sediment deposits on the bed and makes the top layer finer; if the rate of sediment supply from upstream is less, the fine sediment in the top layer is eroded initially, and so a coarsening of the surface of the bed takes place. Such an increase or decrease of fine sediment in the bed material composition is not large enough to cause deposition or scouring of the bed on a relatively large scale. Still, the change is enough to alter the effective bed material composition, and this, in turn, changes the sediment concentration through the relationship of flow conditions and bed material composition to the rate of sediment transport, as shown in Table 11.6. If the regularity of sediment movement under the equilibrium state of scouring and deposition is examined on the basis of the foregoing logic, the two cases can be unified. Fig. 5.20 illustrates the essence of the argument. If the sediment transport capacity is the capacity of the flow to carry sediment load for a given bed material composition, then the relationship should be a single-valued function. For a function of three variables, if one variable is neglected and only the relationship between the other two variables is taken into account, the result is certain to be a multi-valued relationship. After the change in sediment supply, the more rapidly the bed material composition adjusts itself, the more marked will be the response, and the larger will be the amplitude of changes in the sediment concentration for a given water discharge.

Table 11.6 Comparison of sediment transport capacity of a flow for the equilibrium states of deposition and scour (results of flume experiments)

No.	Set of experiment	Number of test runs	Condition of equilibrium	Flow condition					Conc. (kg/m³)	Bed material (mm)		
				Q (l/s)	Flow depth	Velocity (m/s)	Energy slope J (%)	Water temperature (°F)		D_{90}	D_{65}	D_{35}
B	The bed was eroded first,	1	Scour	63.2	18.7	1.09	39.5	55	12.9	0.595	0.316	0.157
	after an equilibrium has been	2	Dep.	63.2	20.0	1.02	41.9	58	13.5	0.540	0.245	0.132
	approached, more sediment	3	Dep.	63.0	19.9	1.03	33.3	48	16.4	0.420	0.195	0.117
	was added	4	Dep.	63.2	19.8	1.04	35.8	52	17.2	0.440	0.225	0.112
		5	Dep.	63.2	19.9	1.03	36.6	51	18.7	0.440	0.174	0.104
C	Sediment was added firstly, in the process of deposition	1	Dep.	63.2	19.2	1.06	30.5	61	80.0	0.470	0.225	0.104
	equilibrium sediment transport rates at different stages were	2	Dep.	62.5	20.2	1.00	39.3	69	118.0	0.490	0.208	0.097
	measured. Then the sediment concentration was reduced, and	3	Dep.	62.5	19.5	1.04	36.8	58	149.0	0.475	0.190	0.086
	the bed was eroded correspondingly. In the process of	4	Scour	62.5	19.5	1.04	34.6	60	35.2	0.540	0.260	0.134
	erosion, equilibrium sediment transport rates at different stages were measured	5	Scour	62.5	19.3	1.05	39.4	57	12.5	0.615	0.305	0.167

Note: In our experiments the velocity was high and no dune existed on bed. As a result, the variation of size composition of bed material only caused the change of grain resistance, and did not effect the bed form resistance. As the change of grain resistance was negligible, the flow condition in the experiments could be considered as unchanged with the size composition of bed material. In natural rivers coarsening or getting finer quite often causes dramatic change of form resistance and a consequent change of flow condition.

11.3.2 Effect of water temperature on sediment transport capacity

From the analysis of field data observed at Tailor's Ferry on the Colorado River, in 1949, Lane found that the rate of sediment discharge in winter is much larger than that in summer for a given flow discharge [37]. His paper received a lot of attention. Fig. 11.24 shows the hydrographs of the measured sediment transport rate, sediment concentration, water temperature, and flow discharge in the period 1943 to 1951 at the Tailor's Ferry station. As a consequence of the regulation of a reservoir, the discharge at Tailor's Ferry was nearly uniform throughout the year. Sediment load came mainly from channel scouring downstream of the reservoir and it was bed material load with a median diameter of 0.32 mm. The contrast between the graph of water temperature and that of the sediment discharge is quite striking, given that the flow conditions are the same. The drop in temperature clearly increases the sediment transport capacity. The sediment discharge tends to reduce year after year as the armouring of the channel bed progresses. Taking into account this factor, Lane pointed out that sediment discharge at the lowest temperature (11.5°C) was two and one-half times that at the highest temperature (28°C).

Water temperature affects the sediment transport capacity in several ways. Chapter 7 indicates that a variation in temperature can lead to a considerable variation in bed configuration and thus directly affect the flow velocity. The viscosity of water is a function of temperature. An increase or decrease in temperature, on the one hand, changes bed load movement by causing a variation of the thickness of laminar sublayer, and on the other hand, it increases or decreases the fall velocity and thus affects the distribution of suspended sediment with depth. With the variation in

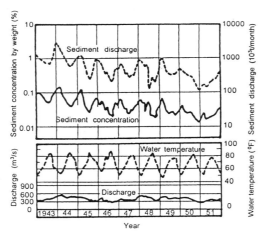

Fig. 11.24 Hydrographs of sediment discharge, sediment concentration, water temperature and flow discharge (1943-1951) at Tailor's Ferry, Colorado River (after Lane, E.W., E.J. Carlson, and O.S. Hanson)

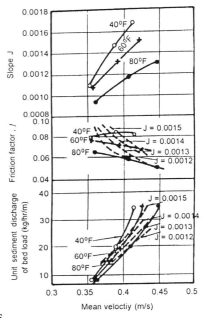

Fig. 11.25 Effect of water temperature on bed load movement (D_{50} =0.23 mm, q=5.76 $l/s/m$) (after Franco, J.J.)

temperature, adjustment of these factors takes place in a complicated way, and they can affect the sediment transport capacity in the opposite directions. Therefore, the overall response to variation in water temperature is quite complicated. A drop in temperature can either increase the sediment transport capacity or decreased it. Even for a given non-uniform sediment, the effect of variation in water temperature is different for the different particle sizes. Besides, as the conditions for flume experiment and those in natural rivers are different, so may the experimental results differ to some extent.

11.3.2.1 Effect of water temperature on bed load movement

Franco carried out flume experiments with fine sand (median diameter of 0.23 mm) to study the temperature effects on bed load movement. The results are shown in Fig. 11.25 [38]. The figure shows that for a constant velocity, a drop in temperature causes an increase of sediment transport rate, and also causes increases in the friction factor and the energy slope. For a constant slope, the drop in temperature resulted in decreases of both flow velocity and sediment transport rate and an increase in the friction factor.

The formulas for bed load movement given in Chapter 9 indicate that the movement is related to $\tau_0 - \tau_c$ in various ways; if the flow intensity is low, or if τ_0 is close to τ_c, any change in τ_c may lead to large changes in the bed load transport rate. Thus, the effects of water temperature on bed load movement can be interpreted as an effect on the threshold drag force. In the Shields diagram for the threshold condition, three zones can be classified according to the grain Reynolds number R_{e*} (= U_*D/ν). The temperature effects on bed load movement are different for the following three different zones:

zone	grain Reynolds number	variation in water temperature	variation in threshold drag force τ_c	variation in bed load transport rate g_b
I	<10	decrease	increase	reduce
II	10 to 600	decrease	decrease	increase
III	>600	decrease	unchanged	unchanged

Tailor and Vanoni studied the effects of water temperature on bed load movement with a flat bed configuration [39]. Seven kinds of natural sediment were used with sizes ranging from 0.14 to 3.95 mm, and clay balls with D=18.5 mm and a specific gravity of 1.37 were also used. Their results essentially confirm the above conclusions; only the demarcation grain Reynolds numbers differ somewhat from the Shields curve. For the demarcation between zone I and zone II, they found Re_*=13, instead of 10, and for that between zone II and zone III, Re_*=200, instead of 600.

The author tackled the problem of temperature effects on sediment movement from a conceptual theoretical approach following the Einstein bed load function [40]. For bed load movement on a flat bed, a variation in water temperature mainly affects the thickness of the laminar sublayers, and it does so in three different ways: (1) changes the relative roughness of the bed; (2) changes the external forces exerting on sediment particles on the bed; (3) changes the sheltering effect for particles lying on the bed. The trends of these three effects may differ, so they may intensify or weaken the bed load movement. Hence, the change in water temperature may not influence the bed load movement at all. The three zones can be identified by the use of K_s/δ and D/δ as governing parameters, as shown in Fig. 11.26. The parameter K_s/δ represents the relative roughness of the bed, and it refers to the river bed as a whole; the parameter D/δ represents the effect of the laminar sublayer on different sediment sizes within the bed material. Here K_s denotes the representative roughness of sediment on the bed surface, δ is the thickness of the laminar sublayer, and D is the diameter of the bed material. Fig. 11.26 shows the effects on the bed load movement for a drop in the water temperature from 30°C to 10 to 20°C, which is equivalent to an increase of viscosity in the range of 25 to 63%.

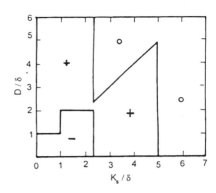

Note: + increase of sediment transport rate
     - decrease of sediment transport rate
     0 no change

Fig. 11.26 Effect of a drop in water temperature on bed load movement

11.3.2.2 Effect of water temperature on suspended load movement

A change in water temperature changes the viscosity of the flow and hence the fall velocity of the sediment particles. Fig. 11.27 shows the changes in fall velocity for various sediment sizes and the

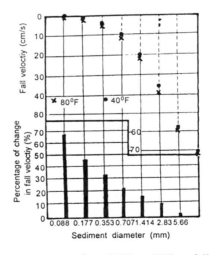

Fig. 11.27 Effect of a drop in water temperature from 80°F to 40°F on fall velocity of various sizes of sediment (after Colby, B.R., and C.H. Scott)

corresponding percentage changes that result from a drop in water temperature from 80°F to 40°F(or 26.7°C to 4.5°C) [41]. The effect of temperature change on the fall velocities of fine sediment is much larger than that on coarse sediment. In the afore mentioned field data obtained from Colorado River, the effect of the water temperature drop on sediment concentrations (Fig. 11.24) of bed material load resulted mainly from its effects on sediment finer than 0.295 mm; the effect on coarse sediment were much smaller.

Reduction of fall velocity due to a drop in water temperature makes the vertical concentration profile more nearly uniform. Fig. 11.28 shows how the concentration profiles for various sizes differ for water temperatures of 27.7°C and 5°C [42]. In this example, not only does the sediment concentration profile become more uniform, but also the concentration itself also increases at all the depths. This result is not unique; it is just one of the three possible cases that may occur. As the concentration at the lowest point of the distribution profile depends upon the intensity of the bed load movement, the absolute value of sediment concentration for the suspended load in the near-bed region depends to a large degree on the transport rate of the bed load. Only in the region well above the bed does the suspension index z play a dominant role [40]. Therefore, with regard to the absolute value of the suspended sediment concentration profile resulting from the water temperature drop, any one of three cases may occur: (1) an increase in sediment concentration for all the points along the vertical; (2) a decrease of sediment concentration for all points along the vertical; (3) a decrease of sediment concentration in the near-bed zone and an increase of sediment concentration in the zone some distance above the bed: The results of flume experiments with natural sand of 0.14 mm and 0.23 mm conducted by Tailor and Vanoni support this inference [43].

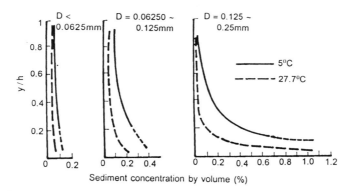

Fig. 11.28 Effect of water temperature on the concentration profiles of suspended load (after Straub, L.G.)

11.3.2.3 Disparity between results from flume experiments and natural rivers

In natural rivers, a channel slope once formed can hardly change later on. Instead, a variation in friction factor can cause the flow depth and velocity to adjust themselves for a given flow discharge. Thus the water temperature drop may cause changes in bed configuration that increase the flow velocity and consequently increase in the sediment transport capacity of the flow. In flume experiments the total amount of water in the flume is usually fixed. For a constant discharge, both the flow depth and the velocity remain unchanged; only the slope can adjust to a variation in water temperature. Thus the effects of water temperature variations on the sediment transport capacity through the adjustment of bed configuration are much weaker than those occurring in natural rivers. Adjustments performed in different ways bring about different results. A suggestion of this fact can be detected in Fig. 11.25.

In addition, flow depths in natural rivers are much greater than those in flumes. The significance of the near-bed zone in comparison with the total depth is thus much smaller for natural rivers than for flume experiments. Consequently, for flow in natural rivers, sediment concentrations at all points of the concentration profile are primarily affected by the suspension index z. A drop of water temperature causes the z-value to reduce also; the results are a more uniform concentration profile and an increase in the absolute values of sediment concentration.

In summary, although flume experiment may display various combinations of the possible conditions, natural rivers generally show an increase in sediment transport capacity of the flow with a drop in water temperature, as shown in Fig. 11.24.

REFERENCES

[1] Einstein, H.A. "The Bed Load Function for Sediment Transportation in Open Channel Flows." *U. S. Dept. Agri.*, Tech. Bull.1026, 1950, p. 71.

[2] Shen, H.W. "Hans A. Einstein's Contributions in Sedimentation." *J. Hyd. Div., Proc., Amer. Soc. Civil Engris.*, Vol. 101, No. HY5, 1975, pp. 469-488.

[3] Research Group of River Sedimentation, Wuhan Institute of Hydraulic and Electric Engineering (WIHEE). "Review on the Einstein theory of bed load movement and discussion of the process of bed load movement." *Journal of WHIEE*, 1965, No. 4, pp. 1-16 (in Chinese).

[4] Grigg, N.S. "Motion of Single Particles in Alluvial Channels." *J. Hyd. Div., Proc., Amer. Soc. Civil Engrs.*, Vol. 96, No. HY12, 1970, pp. 2501-2518.

[5] Yalin, M.S. *Mechanics of Sediment Transport*. Pergamon Press, N.Y., 1972, pp. 134-141.

[6] Bishop, A.A., D.B. Simons, and E.V. Richardson. "Total Bed Material Transport." *J. Hyd. Div., Proc., Amer. Soc. Civil Engrs.*, Vol. 91, No. HY2, 1965, pp. 175-191.

[7] Meland, N., and J.D. Normann. "Transport, Velocities of Single Particles in Bed Load Motion." *Geografiska Annaler*, Vol. 48, A, 1966.

[8] Vittal, N., K.G. Ranga Raju, and R.J. Garde. "Sediment Transport Relations Using Concept of Effective Shear Stress." *Proc., Intern. Symp. on River Mech.*, Vol. 1, Intern. Assoc. Hyd. Res., 1973, pp. 489~499.

[9] Ranga Raju, K.G., R.J. Garde, and R.C. Bhardwaj. "Total Load Transport in Alluvial Channels." *J. Hyd. Div., Proc., Amer. Soc. Civil Engrs.*, Vol. 107, No. HY2, 1981, pp. 179-192.

[10] Yang, C.T., and W.W. Sayre. "Stochastic Model for Sand Dispersion." *J. Hyd. Div., Proc., Amer. Soc. Civil Engrs.*, Vol. 97, No. HY2, 1971, pp. 265-288.

[11] Bagnold, R.A. "An Approach to the Sediment Problem from General Physics." *U.S. Geol. Survey, Prof. Paper 422-I*, 1966, p. 37.

[12] Wuhan Institute of Hydraulic and Electric Engineering (WIHEE). *River Dynamics*. China Industrial Press, 1961. pp. 60-63. (in Chinese)

[13] Mai, Qiaowei, and Suli Zhao. "Preliminary Study on Sediment Carrying Capacity of the Yellow River." *Journal of Sediment Research*, Vol. 3, No. 2, 1958, pp.1-39. (in Chinese)

[14] Sha, Yuqin. *Introduction to Sediment Movement*. China Industrial Press, 1965, pp. 302. (in Chinese)

[15] Engelund, F., and E. Hansen. "A Monograph on Sediment Transport in Alluvial Streams." *Teknisk Forlag*, Copenhagen, 1972, p. 62.

[16] Ackers, P., and W.R. White. "Sediment Transport: New Approach and Analysis." *J. Hyd. Div., Proc., Amer. Soc. Civil Engrs.*, Vol. 99, No. HY11, 1973, pp. 2041-2060.

[17] Colby, B.R. "Discharge of Sands and Mean-Velocity Relationships in Sand Bed Streams." *U.S. Geol. Survey, Prof. Paper 462-A*, 1964, p. 47.

[18] Colby, B.R., and C.H. Hembree. "Computation of Total Sediment Discharge, Niobrara River Near Cody, Nebraska." *U.S. Geol. Survey, Water Supply Paper 1357*, 1955.

[19] Toffaletti, F.B. "Definitive Computations of Sand Discharge in Rivers." *J. Hyd. Div., Proc., Amer. Soc. Civil Engrs.*, Vol. 95, No. HY1, 1969, pp. 225-246.

[20] White, W.R., H. Milli, and A.D. Crabbe. "Sediment Transport Theories: A Review." *Proc., Inst. Civil Engrs.*, Vol. 59, Pt. 2, 1975, pp. 265-292.

[21] Ackers, P., and W.R. White. "Bed Material Transport: A Theory of Total Load and Its Verification." *Proc. Intern. Symp. on River Sedimentation*, Beijing, China, Vol. 1, 1980, pp. 249-271.

[22] Cooper, R.H., A.W. Petersen, and T. Blench. "Critical Review of Sediment Transport Experiments." *J. Hyd. Div., Proc, Amer. Soc. Civil Engrs.*, Vol. 98, No. HY5, 1972, pp. 827-843.

[23] Agricultural Research Service, U.S. Dept. Agri. "Present and Prospective Technology for Predicting Sediment Yields and Sources." *Proc. Sediment Yield Workshop*, U.S. Dept. Agri. Sedimentation Lab., 1975, p. 285.

[24] Chien, Ning, Zhaohui Wan. "Possible Error Induced by the Sampling Point Method for Suspended Load in Determination of Sediment Transport Rates." *Journal of Sediment Reseasch*, Vol. 1, No. 2, 1956, pp. 55-64. (in Chinese)

[25] Chien, Ning. "The Efficiency of Depth-Integrating Suspended-Sediment Sampling." *Trans., Amer. Geophys. Union*, Vol. 33, No. 5, 1952, pp. 693-698; Vol.34, No.5, 1953, pp. 796-797.

[26] Chien, Ning, Ren Zhan, Jiufa Li, and Weide Hu. "Preliminary Research on the Mechanism of Self-Adjustment of Sediment Carrying Capacity in the Lower Yellow River." *Acta Geographia Sinica*, Vol. 36, No. 2, 1981, pp. 143-156. (in Chinese)

[27] Miller, C.R. "Analysis of Flow Duration, Sediment-Rating Curve Method of Computing Sediment Yield." *U.S. Bureau of Reclamation, Hydrology Branch*, 1951, p.55.

[28] Anderson, H.W. "Physical Characteristics of Soils Related to Erosion." *J. Soil and Water Conservation*, 1951, pp. 129-133.

[29] Jiang, Zhongshan, and Wenjin Song. "Sediment Yield in Small Watersheds in the Gullied-Hilly Loess Areas along the Middle Reaches of the Yellow River." *Proc. of Int. Symp. of River Sedimentation*, 1980, Vol. 1, pp. 63-72. (in Chinese)

[30] Renfro, G.W. "Use of Erosion Equations and Sediment-Delivery Ratios for Predicting Sediment Yield." In *Present and Prospective Technology for Predicting Sediment Yields and Sources*, by Agri. Res. Service, U.S. Dept. Agri., 1975, pp. 33-45.

[31] Foster, G.R.(ed.). *Soil Erosion: Prediction and Control.* Sp. Pub. No. 21, Soil Conservation Society of America, 1977, p. 393.

[32] Boyce, R.C. "Sediment Routing with Sediment-Delivery Ratios." In *Present and Prospective Technology for Predicting Sediment Yields and Sources*, by Agri. Res. Service, U.S. Dept. Agri., 1975, pp. 61-73.

[33] Gong, Shiyang, and Guishu Xiong. "The Origin and Transport of Sediment of the Yellow River." *Proc. of Int. Symp.on River Sedimentation*, 1980, Vol. 1, pp. 43-52. (in Chinese)

[34] Xia, Zhenhuan, Qiwei Han, and Enze Jiao. "The Long-term Capacity of a Reservoir." *Proc. Int. Symp. on River Sedimentation*, 1980. Vol. 2, pp. 753-762. (in Chinese)

[35] Chien, Ning. "Investigation of the Maximum Equilibrium Rate of Bed-Load Movement." *Ph. D. Dissertation*, Univ. Calif., 1951, p. 96.

[36] Einstein, H.A., and Ning Chien. "Transport of Sediment Mixtures with Large Ranges of Grain Sizes." *M.R.D. Sediment Series No.2*, Missouri River Div., U.S. Corps. of Engrs., 1953, p. 49.

[37] Lane, E.W., E.J. Carlson, and O.S. Hanson. "Low Temperature Increases Sediment Transportation in Colorado River." *Civil Engin.*, Sept. 1949, pp. 45-46.

[38] Franco, J.J. "Effect of Water Temperature on Bed Load Movement." *J. Waterways & Harbor Div., Proc., Amer. Soc. Civil Engrs.*, Vol. 94, No.WW3, 1965, pp. 343-352.

[39] Taylor, B.D., and V.A. Vanoni. "Temperature Effects in Low-Transport, Flat Bed Flows." *J. Hyd. Div., Proc., Soc. Civil Engrs.*, Vol. 98, No. HY8, 1972, pp.1427-1445.

[40] Chien, Ning. "Effects of Water Temperature on Sediment Movement." *Journal of Sediment Research*, Vol. 3, No. 1, pp.15-28. (in Chinese)

[41] Colby, B.R., and C.H. Scott. "Effects of Water Temperature on the Discharge of Bed Material." *U.S. Geol. Survey, Prof. Paper 462-G*, 1965, p. 25.

[42] Straub, L.G. "Effect of Water Temperature on Suspended Load in an Alluvial River." *Proc. 6th General Meeting*, Intern. Assoc. Hyd. Res., Vol. 4, 1955, p. 5.

[43] Taylor, B.D., and V.A. Vanoni. "Temperature Effects in High Transport, Flat Bed Flows." *J. Hyd. Div., Proc., Amer. Soc. Civil Engrs.*, Vol. 98, No. HY12, 1972, pp. 2191-2206.

CHAPTER 12

INFLUENCE OF THE EXISTENCE OF SEDIMENT ON FLOW

Chapter 1 mentions that the complexity of sediment-laden flow on a movable bed is due to the existence of two feedback systems. One is that the flow can alter the boundary, and the boundary in turn affects the flow. The other is that the flow carries sediment, and the existence of sediment changes the physical properties and turbulence structure of the flow; consequently, it influences the energy dissipation, velocity profile and concentration profile of the flow. The first point was discussed in Chapter 6 on bed configuration and in Chapter 7 on resistance, but the second has not been treated. The foregoing chapters treat only the question of how a clear water flow carries and transports sediment; they do not treat the influence of the existence of sediment on the flow. Such a simplified treatment is acceptable if the sediment concentration is low. But if the concentration is high, such a treatment can deviate significantly from the real situation.

The effect of the sediment on the flow is treated in this chapter. The available data that show this effect are limited, so the study of this topic is still provisional. Furthermore, the limited data reveal significant deviations and even contradictions among them. Interpretations of the relevant processes lead to even more divergence and debate. Disagreements are focused on the following issues: whether the turbulence is intensified or weakened, and whether the energy dissipation is increased of decreased in a sediment-laden flow. At the present level of knowledge it is difficult to judge the various opinions or to give satisfactory explanations of conflicting results. Possibly, different consequence have different causes, and critical conditions for different situations are not fully understood. This chapter is a comprehensive introduction to these data and illustrate the authors' provisional points of view. A complete solution to these problems must await more accurate and more reliable data and these must await an improvement in instruments.

12.1 EFFECT OF SEDIMENT PARTICLES ON THE STRUCTURE OF TURBULENCE

This section introduces the pertinent data and the differences in turbulence structure between sediment-laden flow and clear water flow. The introduction is followed by a discussion.

12.1.1 Measurements of turbulence intensity

Because of restrictions on the available instrumentation, measured data on the intensity of turbulence for sediment-laden flows are rather limited. Furthermore, serious contradictions and disagreements exist among those that do exist.

12.1.1.1 Data that show a decrease of intensity in a sediment-laden flow

1. Flume experiments at Wuhan Institute of Hydraulic and Electric Engineering (WIHEE)

Turbulence characteristics for clear water flow and sediment-laden flow at average concentration of 0.13 kg/m³ were measured by means of a strain resistance sensor at WIHEE, and the results are shown in Fig. 12.1[1]. The turbulence in a sediment-laden flow was less intense than in a clear water flow, particularly near the bed.

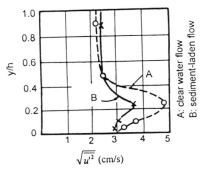

Fig. 12.1 Comparison of turbulence intensity in clear water and sediment-laden flow

Fig. 12.2 Photograph showing the water surface is as smooth as a mirror that reflects clearly the image of trees on the bank of canal (after Vanoni, V. A., and G. N. Nomicos)

2. Bagnold's flume experiments withlight weight material

Bagnold conducted flume experiments with paraffin-stearate spheres with a diameter of 1.36 mm and a specific weight of 1.004, a value quite close to that of water. The water depth was kept constant at 6 cm, and the turbulence intensities were measured for gradually increasing concentrations of sediment [2]. The experiments showed that the potential energy transferred by shear, caused by collisions among particles, constituted a higher and higher percentage of the total, so that the energy transferred by shear in the liquid was lower and lower. The tests revealed that the turbulence level gradually decreased as the concentration increased. Once the mean volumetric concentration reached 25%, random turbulence appeared to be replaced by an orderly secondary circulation, that is, by a secondary flow that ascended near the side walls and descended in the middle of the flume. At a volumetric concentration of about 30%, the turbulence appeared to be markedly weaker. At this level, the vertical concentration became uniform. At a volumetric concentration of about 35%, both turbulence and secondary flow disappeared entirely, and the uniformly distributed sediment particles moved forward in a state of laminar flow. For a volumetric concentration of about 60%, the spaces between particles were so small that particles could no longer move, and a "freezing" phenomenon occurred; sediment particles then blocked the whole channel.

3. Field observations of hyperconcentrated flow

Observations of hyperconcentrated flows in a river or a canal on the loess plateau of Northwest China, also show that turbulence is greatly damped; in particular, small-scale eddies almost disappear. Sometimes, even if the flow velocity is rather high, the water surface is smooth like a mirror, and one can see quite clearly the reflection of surrounding mountains. Fig. 12.2 is a picture of a hyperconcentrated flow taken at Luohui Irrigation Canal, Shaanxi Province, China. No waves appear on the water surface, and trees and people along the banks are clearly reflected.

4. Experimental data on hydrotransport in pipelines

Bruhl and Kazanskij measured turbulence intensities of flow with an inductance velometer in a 100-mm pipe, before and after adding fine sediment [3]. In experiments the discharge increased step by step. At first, experiments were conducted using 1.0 mm uniform sand at a volumetric concentration of 20%; then fine sediment with a median diameter of 0.01 mm at a concentration 6.5% was added to the flow. Measurements were carried out along two levels, as shown in Fig. 12.3, and the root mean squares for turbulence velocities at the two levels were measured, denoted by u_1 and u_2. The turbulence was obviously damped by the fine sediment, and instead of decreasing with further increases in the average velocity, it approached limiting values for $U=3.25$.

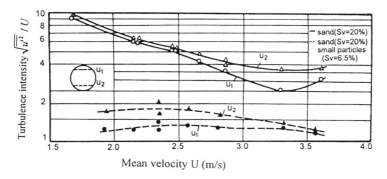

Fig. 12.3 Comparison of turbulence intensity of flows in pipe
with or without adding fine particles (after Bruhl, H., and I. Kazanskij)

12.1.1.2 Data that show an increase of turbulence for sediment-laden flow

1. Flume experiments of Elata and Ippen [4]

Elata and Ippen conducted experiments in a flume with a smooth-bed. In experiments the flow carried plastic beads with diameters in the range of 0.10 to 0.155 mm and specific weight of 1.05. Details of the experiments are contained in section 12.2.1. The measured turbulence intensities increased with an increase of concentration, as shown in Fig. 12.4

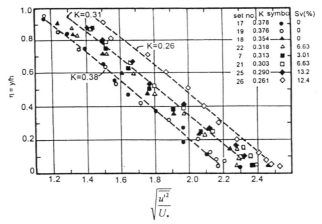

Fig. 12.4 Comparison of turbulence intensity in sediment-laden and clear water flow
(after Elata, C., and I. T. Ippen)

2. Muller's flume experiments [5]

Muller measured and compared turbulence intensities in flows of clear water and sediment-laden water using a laser velometer. The conditions for his two series of experiments were essentially the same except for the concentration, as shown in Table 12.1.

Table 12.1 Conditions of experiments conducted by Muller

	Discharge (cm³/s)	Sediment discharge (g/s)	Slope	Mean depth (cm)	Mean velocity (cm/s)	Fr	τ_0 (dyne/cm²)
Sediment-laden flow	2650	23.6	0.0227	2.57	68.8	1.37	42.7
Clear water	2600	0	0.0227	2.44	71.0	1.44	42.1

The turbulence intensity for sediment-laden flow, shown in Fig. 12.5, is definitely higher than that for clear water flow.

3. Bohlen's flume experiments [6]

Using silicone oil, instead of water, Bohlen carried out experiments in a flume 20 cm wide. Turbulence intensities were measured for flows with and without suspended load at the

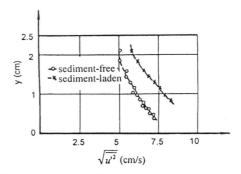

Fig. 12.5 Comparison of turbulence intensity in sediment-laden and clear water flow (after Muller, A.)

depth 1.75 cm and the velocity 28.15 cm/s. The diameter of the sediment used was 0.595 mm, and it had a fall velocity in silicone oil 0.37 cm/s.

The experiments revealed that for concentrations S_v lower than 0.8%, turbulence intensities in the sediment-laden flow were lower than that for flow without sediment in the region near the bed, but turbulence intensities in the sediment-laden flow increased with increasing concentration. In the main flow, turbulence intensities with suspended load were consistently higher for the various concentrations, and they, too, increased with increasing concentration. For the maximum concentration used in the tests of S_v=3.4%, the turbulence intensities of the sediment-laden flow were higher than those of clear water flow over the entire depth.

Results of these two groups of measurements are clearly contradictory. Particularly, for the experiments of Bagnold and of Elata and Ippen that were conducted with suspended particles of nearly neutral density; but even their results show opposite trends. Reasons for this divergence cannot yet be explained.

12.1.2 Changes of turbulence characteristics caused by sediment particles

Direct measurement of turbulence intensities is extremely difficult, particularly for high concentrations of sediment. Therefore, another method is used: to measure velocity profiles and to deduce parameters reflecting turbulence characteristics from variations in the velocity profiles. Variations of the Karman constant are discussed in the following section. Here changes of mixing length l, momentum exchange coefficient ε_m, the turbulence structure and vertical profiles of various energy indexes are discussed.

Einstein and Chien conducted special experiments for studying the effect of suspended sediment on the flow [7], and the details of the experiments are described in the following section. Fig. 12.6 is a comparison of vertical profiles of the main parameters for clear water flow and for sediment-laden flow [8].

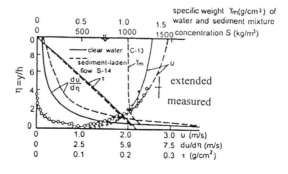

Fig. 12.6 Vertical distribution of major factors in sediment-laden and clear water flows

From the principle of the mixing length theory, the following equation can be obtained:

$$\tau = \rho \ell^2 \left(\frac{du}{dy}\right)^2 = \tau_o(1-\eta) \tag{12.1}$$

$$\frac{\ell}{h} = \frac{\sqrt{1-\eta}}{\dfrac{du_r}{d\eta}} \tag{12.2}$$

in which τ is the shear stress at a distance y from the bed, τ_o shear stress at the bed, h water depth, η relative depth y/h, u_r relative velocity u/U_* and l mixing length; the ratio l/h represents the relative scale of an eddy or vortex. The vertical profile of the relative vortex size l/h over the depth can be deduced from the velocity profile by the use of Eq. (12.2), with the result shown in Fig. 12.7. The profiles measured by Nikuradse in clear water flow are also plotted in the figure.

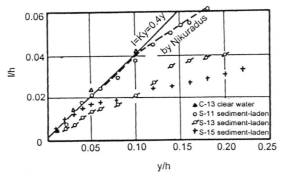
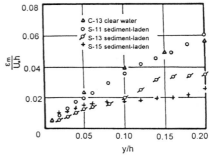

Fig. 12.7 Variation of relative mixing length in a vertical of sediment-laden and clear water flows (refer to Table 12.5 for condition of experiments)

Fig. 12.8 Variation of turbulence exchange coefficient in a vertical of sediment-laden and clear water flows (refer to Table 12.5 for condition of experiments)

The intensity of turbulence exchange can also be represented by the momentum exchange coefficient ε_m:

$$\varepsilon_m = \frac{\tau}{\rho \dfrac{du}{dy}} \tag{12.3}$$

Since $\tau_o = \rho U_*^2$, the preceding equation can be transformed into the following dimensionless form:

$$\frac{\varepsilon_m}{U_* h} = \frac{1-\eta}{\dfrac{du_r}{d\eta}} \tag{12.4}$$

In a similar way, the vertical profile of the dimensionless momentum exchange coefficient $\varepsilon_m/U_* h$ can be computed from the velocity profile data; it is shown in Fig. 12.8.

Fig. 12.7 and Fig. 12.8 show that both l/h and $\varepsilon_m/U_* h$ in sediment-laden flow are smaller than the corresponding values in clear water flow, and that the higher the concentration, the more the reduction is. That is, the existence of sediment reduces vortex sizes and weakens turbulent exchanges.

Changes of turbulence characteristics and temporal-mean velocity profiles caused by suspended particles induce corresponding changes in the spatial distribution of energy dissipation. In Fig. 12.6 the vertical profiles of velocity, velocity gradient, and shear stress of a clear water flow are contrasted with those of a sediment-laden flow for nearly similar conditions. In the figure, the profile of shear stress in sediment-laden flow is close to that in clear water flow, and the velocity gradient of a sediment-laden flow is larger than that of a clear water flow. In the region $y/h>0.19$, the velocity of a sediment-laden flow is larger than that in clear water flow; in the region $y/h<0.19$, the velocity in clear water flow is the larger. Profiles of some key parameters reflecting energy dissipation can be deduced from profiles of shear stress and velocity. Fig. 12.9 contains profiles of the supplied energy ω_b, the local dissipation energy ω_s and the energy ω_t transferred to the boundary per unit volume of flow in unit time, in which

$$\omega_b = -u\frac{d\tau}{dy} \tag{7.2}$$

$$\omega_s = \tau\frac{du}{dy} \tag{7.3}$$

$$\omega_t = \omega_b - \omega_s = -\frac{d}{dy}(\tau u) \tag{7.4}$$

Fig. 12.9 shows the following: (1) In the upper part of the main flow region, a sediment-laden flow supplies more energy than does a clear water flow; in the lower part, the situation is the reverse. (2) In most of the flow, the locally dissipated energy of a sediment-laden flow is more than that of a clear water flow because the vocity gradient of the former is larger; in the region close to the bed, a clear water flow dissipates more energy than does a sediment-laden flow. (3) The energy transferred from the main flow to the layer near the bed is not as concentrated in a sediment-laden flow as it is in a clear water flow.

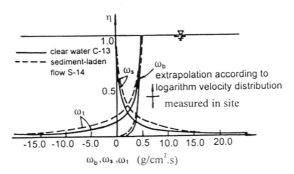

Fig. 12.9 Variation of $\omega_b, \omega_s, \omega_t$ in a vertical of sediment-laden and clear water flows

12.1.3 Discussion of the effect of sediment on turbulence structure

12.1.3.1 Relative importance of changes in the intensity and scale of turbulence

The common belief is that suspended sediment damps turbulence; Yalin analyzed the question: in a sediment-laden flow, is turbulence intensity weakened or are turbulence sizes reduced? [9] He compared a clear water flow with a sediment-laden flow with given slope J and same water depth h. He denoted the shear stress at y in a clear water flow and that in a sediment-laden flow by τ and τ_m respectively, in the expressions

$$\tau_m = \overline{\rho_m} gJh\left(1 - \frac{y}{h}\right)$$

$$\tau = \rho gJh\left(1 - \frac{y}{h}\right)$$

He obtained the following equation:

$$\frac{\tau_m}{\tau} = \frac{\overline{\rho_m}}{\rho} \tag{12.5}$$

in which $\overline{\rho_m}$ is the density of a sediment-laden flow averaged over the range between the water surface and elevation y:

$$\overline{\rho_m} = \frac{1}{h-y} \int_y^h \rho_m dy \tag{12.6}$$

and ρ_m is the density of a sediment-laden flow at elevation y.

For normal concentration profiles, the density is higher in the lower part and lower in the upper part,

$$\rho < \overline{\rho_m} < \rho_m$$

Therefore

$$1 < \frac{\tau_m}{\tau} < \frac{\rho_m}{\rho} \tag{12.7}$$

In regions outside the laminar sublayer near the bed, the total shear stress can be set equal to the turbulent shear stress; therefore

$$\frac{\tau_m}{\tau} = \frac{\dfrac{\rho_m}{g} \overline{u'_m v'_m}}{\dfrac{\rho}{g} \overline{u'v'}} \tag{12.8}$$

From Eq. (12.7) and Eq. (12.8),

$$\frac{\rho}{\rho_m} < \frac{\overline{u'_m v'_m}}{\overline{u'v'}} < 1 \tag{12.9}$$

in which u', v' are the horizontal and vertical turbulence velocities for clear water, and u_m', v_m' are the corresponding values for a sediment-laden flow.

For low concentrations, ρ_m is not much larger than ρ, and the reduction of turbulence velocity caused by the existence of sediment is correspondingly small.

From the mixing length theory, the following expressions can be deduced:

$$\tau = \rho l^2 \left(\frac{du}{dy}\right)^2$$

$$\tau_m = \rho_m l_m^2 \left(\frac{du_m}{dy}\right)^2$$

in which the subscript m again denotes the values for sediment-laden flow, and these two equations yield the expression

$$\frac{\left(\dfrac{du_m}{dy}\right)^2}{\left(\dfrac{du}{dy}\right)^2}\left(\frac{l_m}{l}\right)^2 = \frac{\tau_m}{\tau} \cdot \frac{\rho}{\rho_m} \qquad (12.10)$$

For low concentrations and a nearly uniform vertical distribution of the concentration, the difference between ρ_m and $\overline{\rho_m}$ is not large, hence

$$\frac{\left(\dfrac{du_m}{dy}\right)^2}{\left(\dfrac{du}{dy}\right)^2}\left(\frac{l_m}{l}\right)^2 \approx 1$$

therefore

$$\frac{\dfrac{du_m}{dy}}{\dfrac{du}{dy}} \approx \frac{l}{l_m} \qquad (12.11)$$

The preceding equation shows that the increase in the velocity gradient caused by the existence of suspended sediment is nearly in the same proportion to the reduction of mixing length.

To sum up, Yalin suggested that the damping of turbulence by suspended load is primarily displayed as a reduction in the turbulence scale. Hino reached the same conclusion theoretically–that the existence of sediment reduces the average eddy size [10]

12.1.3.2 Change of turbulence scale

1. Change caused by suspended particles and wakes behind settling particles

As discussed in Chapter 10, the existence of a certain concentration distribution for suspended sediment is the comprehensive result of two opposing effects; the continuous tendency of the particles to settle is counteracted by a net upward transport due to turbulence. That is, particles moving upward come from regions of relatively higher concentration and vice verse. Naturally, the eddies must be much larger than the sediment particles in order to lift them effectively; i.e., the lifting of particles requires kinetic energy from the turbulent eddies that are much larger than the particles themselves. Also, particles have a motion relative to the surrounding water in the vertical direction as they settle. Part of the converted potential energy is directly transformed into heat and it is dissipated by friction. Another part of the potential energy turns into kinetic energy of small eddies in the wakes behind sediment particles. The sizes of these eddies are of the same order as sediment particles themselves. Thus, in the processes of lifting and settling of sediment particles part of the turbulence energy is dissipated. At the same time, part of energy contained in large scale eddies is transferred to small eddies. The result is a change in the energy spectrum of turbulence. Monin obtained the results shown in Fig. 12.10 [11], in which the abscissa of the figure, K' is the wave number per unit length; and it is proportional to the reciprocal of the eddy size. The ordinate is the turbulence kinetic energy contained in eddies with sizes corresponding to K'. As shown in the figure, the peak of curves corresponding to sediment-laden flow ($d\rho/dy<0$) is to the left of that for clear water flow ($d\rho/dy=0$); i.e., part of the energy is transferred from larger eddies (smaller K') to smaller eddies (larger K'). In addition, the total area below the curve is less so that the total kinetic energy of turbulence is less; i.e., the total kinetic energy of turbulence is lower. One can deduce that coarse particles have more effect on the change of turbulence energy spectrum than fine particles do.

2. Disappearance of the smallest eddies because of the higher viscosity of sediment-laden flow

In Chapter 2, the existence of sediment particles was shown to increase the effective viscosity of a sediment-laden flow. The discussion on turbulence structure in Chapter 4 indicated that, taking turbulence kinetic energy from bulk flow, large-scale eddies transfer energy to eddies of next order and so on; finally, the smallest eddies transform mechanical energy to heat through the effect of viscosity. The size of the smallest eddies follows the criterion that their eddy Reynolds number is equal to one. The higher viscosity of sediment-laden flow affects the processes of energy transfer. In particular, the smallest eddies in it are significantly larger than those in clear water. In other words, the smallest eddies in clear water flow do not exist in sediment-laden flow. As fine sediment particles have a more pronounced effect on the increase of viscosity, they also are more effective in causing small-scale eddies to disappear. The extremely calm water surface observed for hyperconcentrated flow shown in Fig. 12.2 reflects this fact.

3. Change of eddy characteristics caused by the rheological properties of sediment-laden flow

A large increase in the content of fine particles causes sediment-laden fluid to

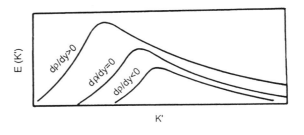

Fig. 12.10 Effect of density gradient on energy spectrum

be non-Newtonian. For a Newtonian fluid, the viscosity is a constant reflecting the physical properties of the fluid, and it is unaffected by the dynamics of the flow. Specifically, viscosity does not change with the shear rate du/dy. Non-Newtonian fluids are different. If the following form is adopted for an apparent viscosity μ_a

$$\mu_a = \frac{\tau}{du/dy} \tag{12.12}$$

the apparent viscosity for a Bingham fluid is

$$\mu_a = \frac{\tau_B}{du/dy} + \eta \tag{12.13}$$

For a pseudo-plastic fluid, it is

$$\mu_a = K \left(\frac{du}{dy} \right)^{m-1} \quad (m<1) \tag{12.14}$$

in which τ_B and η are yield stress and rigidity of Bingham fluid, and K and m are stickiness and plastic indexes for a pseudo-plastic one.

From the two preceding formulas, the apparent viscosity of either a Bingham fluid or a pseudo-plastic one tends to increase with a decrease in the shear rate. Since shear rates at different points in the flow are different, the apparent viscosities at these points are also different, so that viscosity gradients exist in non-Newtonian fluids. Obviously, the existence of viscosity gradients greatly affects the origin and structure of turbulence.

As already mentioned, turbulent flow is full of eddies of different sizes that originate in the region near the bed and diffuse from there into the main flow. At the outer edges of eddies, velocity gradients are large and the apparent viscosities are small, but in the central part of eddies, the velocity gradient is close to zero and the apparent viscosity is large. Hence, an eddy experiences intensive damping as it rotates, and small-scale rotations can hardly be maintained. The result is an eddy pattern quite different from that in clear water flow. Shaver and Merrill injected dye into a pseudo-plastic flow. They found that the dye solution did not mix with the surrounding flow and gradually disappear, as it does in a turbulent flow of clear water; instead, it maintained its original form except that it was distorted and expanded. The effect of eddies appeared to be a continuous squeezing of the dye filament that made them deform rather than diffuse [12]. This difference explains why the rotation of small-scale eddies cannot be maintained. Such a phenomenon has been observed in some rivers in Northwest China where hyperconcentrated flows sometimes occur. The water surface is smooth, and no ripples occur. Although distorted and deformed along their course, cloud-like patterns on the water surface maintain their identity over a rather long distances; clearly they do not mix in the usual way. Fig .12.11 is a picture taken on the Luohui Irrigation Canal of the passing of a hyperconcentrated flow, in which these distorted, deformed patterns are visible.

12.1.3.3 Variation of turbulence intensity

1. Causes of less turbulence intensity in a sediment-laden flow

(1) The energy for producing turbulence is less in a sediment-laden flow as a result of the transfer of part of the energy by the bed load. According to Bagnold's experiments, bed load has the effect of transferring shear stress. If sediment is carried as bed load, particularly if it moves intensively, not all the potential energy of the flow is transferred to the boundary through shear stress from the liquid phase; a

Fig. 12.11 Distorted pattern on the water surface of the hyperconcentrated flow in the Luohui Canal

part of the energy is transferred through shear stress due to collisions among particles. The part of the potential energy transferred through granular shear stress does not contribute directly to the origin of turbulence. Therefore, with bed load motion, turbulence is damped. If the bed load motion is intensive, the effect is, of course, more pronounced.

(2) Turbulence intensities are weakened as eddies pass through a layer with a density gradient. Near the bed, the concentration is a maximum, and correspondingly, the density gradient (density decreasing upwards) is larger too. The near bed region is also the source of turbulence. In a flow field in which the density gradually decreases upwards, an upward eddy will move into a fluid with less density. Consequently, it will experience an effective downward force due to gravity that reduces its upward velocity. Also, a downward eddy will move into a fluid with larger density, and it experiences buoyancy that reduces the downward velocity. Thus, density gradient damps the turbulence.

(3) The variation of the physical properties of a sediment-laden flow also has consequency. As discussed in the preceding paragraph, the viscosity of a sediment-laden flow is larger than that of clear water and its rheological properties vary. These properties change the turbulence scales and weaken turbulence, particularly the small scale eddies.

2. Causes of more intense turbulence

Hino made a thorough analysis of this process [10] and deduced the following expression for the ratio between the root-mean-square of the turbulence velocity of sediment-laden flow and that of clear water flow:

$$\frac{\sqrt{\overline{u_m'^2}}}{\sqrt{\overline{u'^2}}} = \left(\frac{1}{1-\alpha^3 \overline{s_v}}\right)^{\frac{1}{3}}\left(\frac{\overline{\rho_m}}{\rho}\right)^{\frac{1}{3}}\left[1-\frac{2\kappa_o \sigma \xi}{\left(1+2\overline{s_v}\right)\left\{1+\left[1+4\kappa_o B(1+2\overline{s_v})\xi\right]^{1/2}\right\}}\right]^{1/3} \quad (12.15)$$

in which

$$\xi = \frac{g\left(\frac{\gamma_s-\gamma}{\gamma}\right)\omega}{U_*^3 \ln{h}/\delta}\overline{s_v}(h-\delta) \quad (12.16)$$

$$\sigma = \frac{\ln{h}/\delta}{\ln{h}/\delta - 1} \quad (12.17)$$

$$\alpha^3 = 1.64\left[1+\int_\delta^h \frac{s_v - \overline{s_v}}{\overline{s_v}}dy\right] \quad (12.18)$$

In these equations, variables without subscript or with subscript 0 are for clear water flow, and variables with subscript m are for sediment-laden flow, B is a constant, κ is the Karman constant, δ is the thickness of laminar sublayer (for a smooth bed) or the height of roughness protrusions (for a rough bed), ω is the fall velocity of a sediment particle. From Eq. (12.15), the ratio between the turbulence intensity of clear water flow and that of sediment-laden flow depends on the concentration, as well as on the specific weight of sediment particles and other factors. For suspended particles with a neutral density ($\gamma_s=\gamma$), the second term in the last parentheses in Eq. (12.15) equals zero, and the turbulence intensity increases with an increase in concentration. If the density of particles is larger, the turbulence intensity may be less.

In addition, Elata and Ippen pointed out that the existence of sediment particles induces an instability in the zone near the boundary that intensifies the turbulence [4]. Photographs of sediment particles in a wind tunnel, which are shown in Chapter 15, show that particles rotate rapidly while in motion. Kazanskij and Bruhl suggested that this rotation of particles in the flow enhances large scale turbulence. They substantiated their point of view by data obtained from a pipeline experiment [13].

Before concluding the discussion in this section, we wish to emphasize that the energy for suspending sediment comes from the kinetic energy of turbulence, which in turn comes from the effective potential energy. In Fig. 12.16, which is discussed in section 12.2.2, the ordinate E is the ratio between suspension energy and effective potential energy. The figure shows that suspension energy generally accounts for not more than 4 to 5% of the effective potential energy. Even with hyper-concentration, this ratio is no more than 10% because of the significant reduction of sediment fall velocity caused by the increase of viscosity. The suspension efficiency proposed by Bagnold (Chapter 10), which is a few percent in general, possesses a similar significance, i.e. the percentage of the energy for suspension of sediment with respect to the whole effective energy, the usual effect of suspended load on turbulence can hardly be detected. If a sophisticated apparatus for measuring this tiny value is not available, no definitive conclusion can be reached. This dilemma is one of the reasons for the great divergence of opinion on this topic.

12.2. EFFECT OF SEDIMENT PARTICLES ON VELOCITY PROFILE

12.2.1. Experimental data

Many researchers have conducted experiments specially planned to study this problem. The main experiments among them are the following.

12.2.1.1 Experiments of Einstein and Chien[7]

Einstein and Chien conducted experiments in a tilting, circulating steel flume. The flume had a width of 30 cm, a depth of 36 cm, and a length of 12 m. Sediment of the same diameter as the suspended material was glued on the flume bottom. A steep slope was used to produce high velocities (Froude numbers larger than 1). Since sediment particles did not settle to the bed, the boundary condition remained unchanged. Experiments were run using both clear water and water carrying various concentrations of sediment. Velocity profiles and concentration profiles were measured. Three kinds of sediment were used: coarse sand, median sand, and fine sand. The fine sand had a median diameter of 0.274 mm. In the experiments, the maximum average concentration in the vertical was 65 kg/m^3, and the maximum concentration at a point was 625 kg/m^3.

The measured velocity profiles for some runs with fine sand are shown in Fig. 12.12, and the corresponding concentration profiles are shown in Fig. 12.13. In the figure the abscissa u/U_* shifts a certain distance for each run and the values of u/U_* are marked. The figure shows that velocity profiles in clear water flow follow the logarithmic formula. In sediment-laden flow, the velocity profile in the main flow region also follows the logarithmic law but with a reduced value of κ. Thus the velocity profile is not as uniform as that in clear water flow. Near the bed, the velocity profile deviates from the logarithmic formula. The higher the concentration and the smaller the value of κ, the more the velocity profiles deviate from the logarithmic formula and the wider is the range in which the velocity profiles deviate from the logarithmic formula.

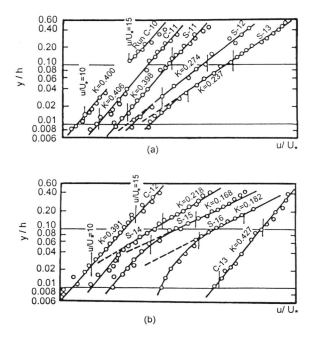

(a)

(b)

Fig. 12.12 Effect of the existence of sediment on the velocity profile
(refer to Table 12.5 for condition of experiments)

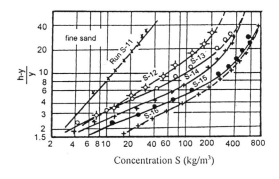

Fig. 12.13 Vertical distribution of sediment concentration

12.2.1.2 Vanoni and Nomicos experiments [14]

Vanoni and Nomicos conducted experiments in a flume that was 26.7 cm wide, 25.4 cm deep and 12 m long. Two kinds of sediment, with the properties shown in Table 12.2, were used in the experiments.

The initial experiments were run with a movable bed. After the bed had approached a stable configuration, a layer of chemical synthetic paint was sprayed on its surface to fix the sediment particles in place. Then the other experiments were carried out, using the same discharge and water depth, for both clear-water flow and sediment-laden flows of different concentrations. In this way, the boundary conditions were the same for all experiments.

Table 12.2 Properties of sediment used in Vanoni and Nomicos's experiment

No. sediment.	No. experiment	D₅₀ (mm)	Standard deviation	Average diameter (mm)	Average fall velocity at 20°C (cm/s)
6	I, II, III	0.091	1.16	0.105	0.945
7	IV	0.148	1.16	0.161	1.89

Measured velocity profiles are shown in Fig. 12.14. The velocity profile for clear water follows a logarithmic formula; the velocity profiles of sediment-laden flows basically follow a logarithmic formula also, but they deviate from it somewhat near the bed. The higher the concentration was, the smaller the value of Karman constant.

12.2.1.3   Elata   and   Ippen experiments

The purpose of the Elata and Ippen experiments was to determine which of the following factors affects the turbulence characteristics: sediment concentration, difference in density between sediment particles and water, or some other. For this purpose, they used plastic spheres with a specific

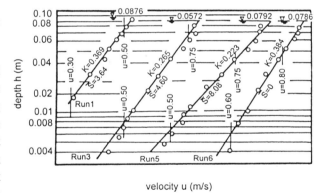

Fig. 12.14 Effect of the existence of sediment on the velocity profile (after Vanoni, V. A., and G. N. Nomicos)

gravity of 1.05: because they were only slightly heavier than water, they were uniformly distributed throughout the flowing fluid. Most of the plastic spheres (85% of them) had diameters within the range of 0.100 to 0.155 mm and a fall velocity of about 0.1 cm/s. Experiments were carried out in a flume that was 26.7 cm wide, 28.1 cm deep, and 12 m long, and its bed and walls were smooth. Before the experiments, the flume was thoroughly cleaned. During the experiments, a high velocity was maintained in order to avoid sediment deposition. Measured velocity profiles are presented in Fig. 12.15 in the form of velocity differences. In the figure, u_{max} is the velocity at the water surface, which was the highest velocity along the vertical, and U_* is the shear velocity. The figure shows that the velocity profile follows the logarithmic formula for clear water flow. In flow with uniformly distributed suspended particles, the velocity profiles are quite similar to those for flows with large concentration gradients shown in Fig. 12.12; that is, in the main flow region they follow the logarithmic formula, and near the bed they deviate from it. The higher the concentration was, the smaller the value of κ and the larger the range in which the velocity profiles deviate from the logarithmic formula.

12.2.1.4 Kalinske and Hsia experiments

Kalinske and Hsia conducted similar experiments [15] using very fine sediment. Even though the concentration reached a value of 11% by weight, the measured velocity profiles still followed the logarithmic formula and the Karman constant κ was about 0.40.

12.2.2 Velocity profile in main flow region

A summary of the preceding experimental results obtained by various researchers leads to the following results: the velocity profile for a sediment-laden flow follows the logarithmic formula in the main flow region, but the Karman constant κ no longer has the value 0.40. Thus, the study of the velocity profile in sediment-laden flows turns out to be a study of the variation of the κ-value.

12.2.2.1 Studies of the variation of κ from the point view of energy

Einstein and Chien first analyzed the effect of suspended load on the Karman constant [7]. They established a relationship between it and the parameter E as shown in Fig. 12.16,

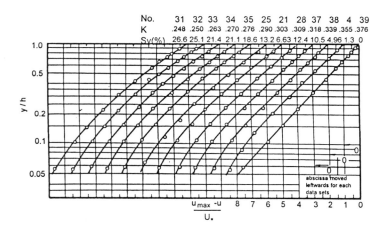

Fig. 12.15 Effect of the existence of sediment on the velocity profile

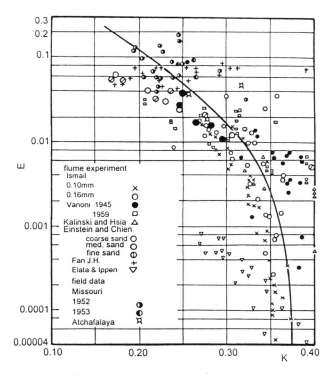

Fig. 12.16 Effect of suspended sediment on the Karman constant

$$E = \frac{\rho_s - \rho}{\rho_s} \frac{\sum \overline{s_w} \omega}{U J_e} \qquad (12.19)$$

in which ρ_s and ρ are the densities of sediment and water, respectively; U vertically averaged velocity; J_e energy slope; S_w vertically averaged concentration of sediment particles (fall velocity ω) by weight. The parameter E is the ratio of the energy for suspending sediment to the total dissipated potential energy (per unit weight of water in unit time). The value of κ gradually decreases with the increase of the parameter E, that is, with the increase of the energy required for suspending sediment. The figure displays the data of all the researchers mentioned in section 12.2.1. In the Kalinske and Hsia experiments, the concentration was high, but the sediment was fine and the corresponding fall velocity ω small; therefore, the E values were not large, and the κ values were still close to 0.40.

Second, Nomicos stated that high concentrations occurred near the bed, the region where turbulent vortexes are formed. To emphasize the effect of this layer, they took the average concentration $\overline{s_v}$ (by volume) in the layer between $0.001h$ and $0.01h$ as the concentration to use in the parameter E. Based on their experimental data, they established a relationship between κ and a parameter expressed as $\dfrac{\rho_s - \rho}{\rho} \dfrac{\overline{s_v}\omega}{UJ_e} \dfrac{0.01h - 0.001h}{h}$, shown in Fig. 12.17 [14]. The physical interpretation of this parameter is the ratio between the energy needed for suspension and the total potential energy dissipated in near the bed. In the figure, the experimental data follow the curve rather well.

12.2.2.2 Variation of κ from viewpoint of density gradient

First, Tsubaki wrote down a balance equation for turbulent kinetic energy. He equated the energy taken by turbulence from the mean motion to the sum of the energy required to overcome the density gradient and that transformed into heat through viscous deformation, and deduced the following relationship [16]:

$$\frac{\kappa}{\kappa_o} = f(\phi,\zeta) \tag{12.20}$$

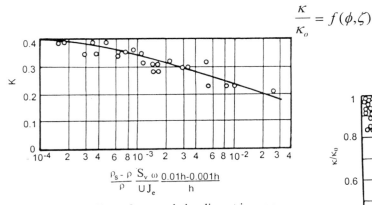

$$\frac{\rho_s - \rho}{\rho} \frac{S_v\,\omega}{UJ_e} \frac{0.01h - 0.001h}{h}$$

Fig. 12.17 Effect of suspended sediment in near bed zone on the Karman Constant (after Vanoni, V. A., and G. N. Nomicos)

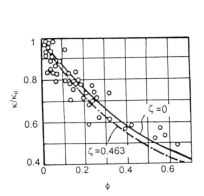

Fig. 12.18 Relationship of Karman constant and energy consumed in overcoming the density gradient

in which κ, κ_o are Karman *constants* for sediment-laden flow and clear water flow, respectively; ζ is a parameter related to the heat dissipated through viscous deformation;

$$\phi = \frac{\overline{\omega s_v}}{U_* J_e} \frac{\rho_s - \rho}{\rho_m} \left[\frac{\left(\dfrac{1 - \dfrac{y}{h}}{\dfrac{y}{h}}\right)^{z-1}}{\displaystyle\int_A^1 \left(\dfrac{1 - \dfrac{y}{h}}{\dfrac{y}{h}}\right)^{z} d\left(\dfrac{y}{h}\right)} \right]_{average} \tag{12.21}$$

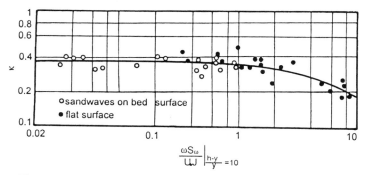

Fig. 12.19 Relationship of Karman constant and Richardson number
(after Barton, J.R., and P. N. Lin)

in which ρ_m is the density of the mixture of water and sediment; z, the exponent in the formula of vertical distribution of suspended load in diffusion theory; A, the ratio of the distance from the bed to the lowest limitation of suspension region to water depth. For the Vanoni and Nomicos data, he got the relationship between κ/κ_o and ϕ shown in Fig. 12.18. In the figure, the solid line is for $\zeta=0$, and the dotted line for $\zeta=0.463$.

Second, Barton and Lin conducted a similar study, and they pointed out that the Karman constant should be related to the Richardson number [17]. The Richardson number varies with location. They used the Richardson number at elevation *(h-y)/y=10* as the characteristic value and omitted *(ρ_s-ρ)/ρ_s* because it was essentially a constant; in this way, they established the relationship between κ and $\omega S_w/U_*J$ at *(h-y)/y=10* shown in Fig. 12.19.

12.2.2.3 Unifying parameters of different researchers from the viewpoint of Richardson number

The authors have discussed the parameters used by various researchers, unified them and pointed out their meanings [8]. The Richardson number is a significant parameter for flows with a density gradient. It represents the percentage of the total energy supplied that is represented by the energy required to overcome a vertical density gradient. In Chapter 7, the rate of energy supplied from unit volume of water is represented as

$$\omega_s = \tau \frac{du}{dy} \qquad (7.3)$$

Except close to the boundary, the shear stress is simply the Reynolds stress, and the preceding equation can therefore be written as

$$\omega_s = \rho_m \varepsilon_m \left(\frac{du}{dy} \right)^2 \qquad (12.22)$$

The rate of energy consumption per unit volume in overcome the density gradient is

$$\omega_d = -\varepsilon_m \frac{d\gamma_m}{dy} = -\varepsilon_m g \frac{d\rho_m}{dy} \qquad (12.23)$$

in which ε_m is the momentum exchange coefficient and γ_m specific weight of sediment-laden flow.

Therefore, the Richardson number can be written as:

$$R_i = \frac{\left(-\dfrac{g}{\rho_m} \dfrac{d\rho_m}{dy} \right)}{\left(\dfrac{du}{dy} \right)^2} \qquad (12.24)$$

The density gradient of a sediment-laden flow is caused by the concentration gradient, so that Eq. (12.23) can be written as:

$$\omega_d = -\varepsilon_s \frac{\rho_s - \rho}{\rho_s} \frac{dS}{dy} \qquad (12.25)$$

in which ε_s is sediment exchange coefficient, S the concentration expressed in sediment weight per unit volume. By using the diffusion equation

$$\varepsilon_s \frac{dS}{dy} + S\omega = 0 \qquad (10.7)$$

one can transform Eq. (12.25) into

$$\omega_d = \frac{\rho_s - \rho}{\rho_s} S\omega \qquad (12.26)$$

Actually the product $S\omega$ represents the energy consumed in suspending sediment, and it also reflects the effect of the concentration gradient (density gradient) on turbulence.

From Eq. (12.26), the Richardson number of a sediment-laden flow can be written as:

$$R_i = \frac{\dfrac{\rho_s - \rho}{\rho_s} S\omega}{\varepsilon_m \rho_m \left(\dfrac{du}{dy}\right)^2} \tag{12.27}$$

The Richardson number can be written in another form. Due to the non-uniform distribution of concentration in a sediment-laden flow, the vertical distribution of shear stress is not strictly linear, as it is for in a clear water flow. But in practice it does not deviate far from a straight linear except near the bed (Fig. 12.6). If the distribution of shear stress is taken to be linear, Eq. (7.3) can be written as

$$\omega_s = \tau_o \frac{h-y}{h} \frac{du}{dy} \tag{12.28}$$

If also, the velocity profile follows the logarithmic formula,

$$\frac{du}{dy} = \frac{U_*}{\kappa y}$$

then the Richardson number can be transformed into the following:

$$R_i = \frac{\dfrac{\rho_s - \rho}{\rho_s} S\omega}{\tau_o \dfrac{h-y}{h} \dfrac{U_*}{\kappa y}} = \frac{\omega S}{U_* J} \frac{\rho_s - \rho}{\gamma_m \rho_s} \kappa \frac{y/h}{1-y/h}$$

If the concentration is not high, $S_w \approx \dfrac{S}{\gamma} \approx \dfrac{S}{\gamma_m}$; then the preceding formula can be simplified to

$$R_i = \frac{\omega S_w}{U_* J} \frac{\rho_s - \rho}{\rho_s} \kappa \frac{y/h}{1-y/h} \tag{12.29}$$

From Eq. (12.27) and Eq. (12.29), the Richardson number clearly varies along a vertical, whereas the Karman constant is an overall parameter. Hence, to establish a relationship between them, one has the problem of how to choose a representative Richardson number. The following discussion shows that the different ways of making a selection lead to the several different Richardson numbers in the preceding paragraphs.

Barton and Lin obtained their parameter $\omega S_w/U_* J$ by taking the Richardson number for the point $(h-y)/y=10$ and by excluding some values that were constant or nearly so. Transforming S_w in Eq. (12.29) into S_v, introducing a diffusion formula for vertical concentration distribution in Eq. (10.16), integrating along the vertical and taking the average, one can obtain the parameter ϕ , adopted by Tsubaki [Eq. (12.21)].

If one uses the average value of the Richardson number over a vertical, instead of that arbitrarily selected of the elevation, Eq. (12.27), and then take the average values of the numerator and that of the denominator, one obtains:

$$R_i = \frac{\dfrac{1}{h}\dfrac{\rho_s - \rho}{\rho_s}\int_0^h S\omega dy}{\dfrac{1}{h}\int_0^h \tau \dfrac{du}{dy} dy}$$

For the condition that the average unit weight of sediment-laden water does not differ much from that for clear water, then, as already mentioned in Chapter 7:

$$\gamma J \int_0^h u dy = \int_0^h \tau du$$

Hence, the preceding formula can be written as:

$$R_i = \frac{\dfrac{1}{h}\dfrac{\rho_s - \rho}{\rho_s}\int_0^h \omega S dy}{\gamma U J} = \frac{\dfrac{\rho_s - \rho}{\rho}\left[\dfrac{1}{h}\int_0^h \omega S_w dy\right]}{UJ} \qquad (12.30)$$

The expression $\left[\dfrac{1}{h}\int_0^h \omega S_w dy\right]$ is nothing other than $\overline{S_w \omega}$ first suggested by Einstein and Chien. Suspended sediment is often non-uniform; in practical application, one can let S_w represent the part of sediment that corresponds to the fall velocity ω. Then the Richardson number in the form of Eq. (12.30) becomes the parameter E in the form of Eq. (12.19).

The use of E as a basic parameter in a relationship with κ may avoid the difficulty caused by the variation of Ri with y, but it can introduce other difficulties. Coarse sediment particles concentrate near the bed as they move, and a large density gradient forms there, but in the rest of the flow region, above the near bed zone, either the concentration gradient is small or it contains no sediment at all; in such cases, the energy consumed in supporting suspended particles (or for overcoming the density gradient) is concentrated in a narrow region near the bed. If one then averages the energy consumed over a vertical, the effect of concentrated energy dissipation is diluted. Furthermore, such an effect is different for different conditions of flow and sediment transport. The significance of adopting E as a primary parameter is reduced. To some extent, the scatter of the data points in Fig. 12.16 reflects just this fact.

Vanoni and Nomicos took the energy consumed in suspending sediment in near the bed as the numerator, and established a relationship between the parameter thus formed and κ, the data points then show less scatter.

12.2.2.4 Mechanism of the effect of neutral suspended particles on value of κ

The preceding analyses are based on the following considerations that sediment is heavier than water and that its presence (or the density gradient caused by its presence) consumes a part of the energy of the turbulence in the flow. Because Ippen and Elata used light material with a specific weight close to that of water, the suspension did not consume much energy. Also, the sediment particles were uniformly distributed in flow so that no density gradient existed. Even so, the velocity profile in a flow carrying neutral sediment particles is substantially different from that for a clear water flow. Their experimental results are included in Fig. 12.16, and the data points clearly deviate from those obtained in experiments with natural sediment. Hence, the pattern of variation for the velocity profile in a sediment-laden flow can not be explained from the viewpoint of energy alone.

Ippen studied the variation of κ from a completely different point of view [18]. He considered the particles in a common sediment-laden flow to be mostly concentrated near the bed, where the concentration was so high that the effective viscosity was affected. This effect, a local concentration of higher viscosity, could be described by Einstein formula [Eq. (2.14)]. In this way, Ippen deduced that the following ratio exists between the Karman constant in sediment-laden flow and that in clear water flow:

$$\frac{\kappa}{\kappa_o} = \frac{1 + \overline{S_v}\left(\frac{\rho_s}{\rho} - 1\right)}{1 + 2.5 S_v} \tag{12.31}$$

in which and $\overline{S_v}$ and S_v are average concentrations over the entire depth and within the laminar sublayer, respectively. Eq. (12.31) also indicates that κ decreases continuously as the concentration increases; Ippen also verified Eq. (12.31) with the experimental data of other researchers. However, due to some uncertainties in his deduction, the reliability of Eq. (12.31) is not fully established.

In another study involving two-phase flow in a pipeline, Oda et al. pointed out that if there are solid particles in flow, not only does the suspension of these particles take a part of energy from turbulence; but also, if the content of solid particles is so large that the spacing of the particles is smaller than the characteristic length of turbulence, the existence of the solid particles inevitably affects the origins of turbulence [19]. Such an effect would be more pronounce for the small-scale turbulence. In the region not far from the bed, the mixing length of turbulence is proportional to the distance to the bed, and the ratio is κ. Hence, a reduction of the turbulence would reflect a reduction of the Karman constant. This deduction is not affected by the specific weight of solid material, so that it also applies to neutral suspended material. This explanation appears to be rational, but it needs more study before it can be expressed analytically.

12.2.3 Velocity profile in region near the bed

As already mentioned, sediment-laden flows usually have a pronounced density gradient due to the concentration gradient near the bed. In this region, also the velocity profile clearly deviates from the logarithmic formula. One can analyze this problem by adopting results obtained in meteorology [8].

Because the temperature at the ground surface is different from that at high altitude, a density gradient also occurs in air near the ground; thus, the wind velocity profile can be described by:

$$\frac{du}{dy} = \alpha y^{-\beta} \tag{12.32}$$

which can be transformed into the following in dimensionless form:

$$\frac{du_r}{d\eta} = \alpha' \eta^{-\beta} \tag{12.33}$$

in which $u_r = u/U_*$, $\eta = y/h$, α is a coefficient of proportionality and $\alpha' = \alpha h^{1-\beta} / U_*$; the exponent β is a function of the Richardson number and has the following characteristics, as shown in Fig. 12.20:

if $d\rho/dy > 0$, $R_i < 0$ $\beta > 1$

         $d\rho/dy = 0$, $R_i = 0$ $\beta = 1$

i.e., the velocity profile follows logarithmic law.

         $d\rho/dy < 0$, $R_i > 0$ $\beta < 1$

A sediment-laden flow corresponds to the situation $d\rho/dy < 0$, and the velocity profile in Fig. 12.12 is similar to the curve for $d\rho/dy < 0$ in Fig.12.20. By using data from Fig. 12.12, one can plot the relationship between $du_r/d\eta \sim \eta$ on log-log paper, as shown in Fig. 12.21. In the main flow region, where the density gradient is not large, the experimental points fall on a straight line with an inclination of 45 degrees, and the velocity profile follows the logarithmic law; near the bed, the points are distributed in accordance with Eq. (12.33) with exponent $\beta < 1$.

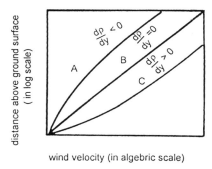

Fig. 12.20 Sketch of wind velocity profile near ground surface for different temperature gradients

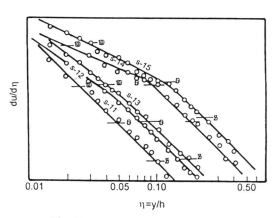

Fig. 12.21 Velocity profile using data sets plotted according to Eq. (12.33)

One can deduce that the parameters α' and β should be related to E, a parameter reflecting the effect of density gradient. Such a relationship does exist and is shown in Fig. 12.22. Eq. (12.33) can be used to predict the velocity gradient. The boundary value, required to determine the actual velocity distribution is obtained from the intercept η' of the straight line of velocity profile in the main flow region with the ordinate in a semi-logarithmic plot. The relationship between η' and E is plotted in Fig.12.22; the data points fit the curve well enough to indicate that such a relationship does exist.

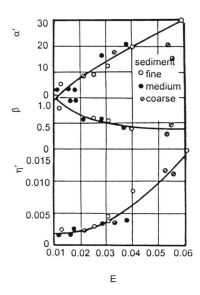

Fig. 12.22 Relationship of E and parameters α', β, and η in the expression of velocity profile of a sediment-laden flow

A division of the velocity profile along the water depth into two regions fits with the physical picture of the occurrence. Because of the large density gradient near the bed, the turbulence characteristics and velocity profile should not be the same as for clear water in main flow with no density gradient. Although eddies originating near the bed are affected by the sediment in that region and possess less turbulent kinetic energy than do those in clear water flow, these eddies still follow the same diffusion law as they diffuse into the main flow. As a result, the velocity profile there follows the logarithmic law. The only differences are that the value of κ is less and the velocity distribution tends to be more uneven because turbulence mixing is weaker.

12.2.4 Velocity profile over the whole flow region, including region near the bed

The above treatment, based on dividing the flow into the main flow region and the region near the bed, gives a good picture, but the relationship describing velocity profile is rather complex. Hence, the authors have attempted to obtain a unified formula that represents the velocity distribution over the whole depth.

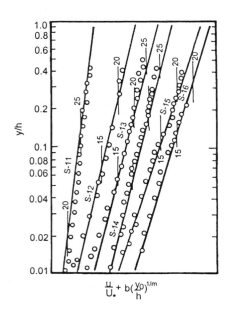

Fig. 12.23 Velocity profile plotted according to Eq. (12.34) (fine sediment set)

Integrating Eq. (12.33) and letting $\eta_o = K_s/30.2h$ represent the point at which the velocity is zero, they obtained

$$u_r = b\left(\eta^{1/a} - \eta_0^{1/a}\right)$$

(12.34)

The relationship, plotted together with the measured data form Fig. 12.12, is shown in Fig. 12.23. The agreement between the measured results and Eq. (12.34) is quite satisfactory. The coefficient b in the above equation varies only slightly, between 27 and 31 over the experimental range [7], and it is well represented by its average value of 30. The value of a decreases with the increase of parameter E, as shown in Fig. 12.24, and it reflects the fact that the velocity distribution tends to be uneven while the energy for suspending sediment is increasing.

12.3. EFFECT OF SEDIMENT MOTION ON ENERGY DISSIPATION

12.3.1. Effect of boundary change and sediment motion

At the beginning of this chapter, the two feedback systems between sediment-laden flow and a movable bed are discussed. Flow in a natural river is just such a synthesis of complex systems. In natural rivers, high sediment concentrations usually occur during floods when the rate of flow is high. The high velocities tend to eliminate such bed configurations as ripples or dunes and form a bed that is smooth and nearly plane. Consequently, at those times, the bed is not rough and the resistance to flow is low. The low resistance is not really the result of the high concentration, it reflects rather the smoothening the bed by the high velocity flow.

High concentration and low resistance are associated with each other in natural rivers, but they are not necessarily cause and effect.

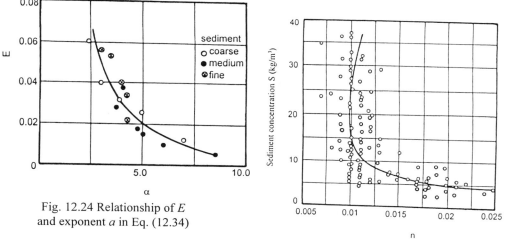

Fig. 12.24 Relationship of E and exponent a in Eq. (12.34)

Fig. 12.25 Relationship of Manning's n and sediment concentration
(Lijin Station on the Yellow River)

In order to clarify this situation, the first author of this book carried out a special analysis of data for the Lower Yellow River [20]. A general tendency is revealed for the roughness to decrease with an increase of concentration, as shown in Fig. 12.25. However, if one sorts these data according to discharge, then within a given range of discharge (i.e., with more or less the same velocity), the trend of n value to vary no longer occurs. Fig. 12.26 is such an example, and it reveals that velocity plays the dominant role in the variation of resistance, rather than concentration.

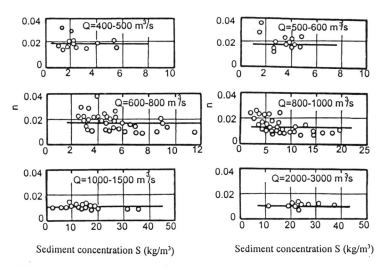

Fig. 12.26 Relationship of Manning's n and sediment concentration for different discharge classes
(Lijin Station on the Yellow River)

Therefore, in the following discussions on the effect of suspended sediment motion on energy dissipation of flow the variation of boundary caused by the change of flow velocity is excluded and in this way experiments with the same boundary conditions are compared.

12.3.2 Effect of bed load

Chapter 9 contains a detailed discussion of how bed load is set in motion, i.e., how a sediment particle lying on a stationary bed is lifted from the bed, enters the flow, is accelerated by flow moving at a higher velocity and then falls and hits other particles on the bed. The trajectory of a saltating sediment particle is shown in Fig. 9.5. The raising of a sediment particle from a stationary position in the bed requires energy from the flow. After jumping up from the bed and encountering flow moving at a higher velocity, the sediment particle is entrained and accelerated by the flow. These processes also take energy from the flow. The process of lifting of a sediment particle into saltation is driven by the surrounding flow, and it takes energy from the flow and converts it into kinetic energy of the horizontal motion of the particle. Then, as the sediment particle falls, it encounters flows with lower velocities. At a certain elevation the horizontal velocity of the particle is just equal to the local flow velocity. Below this point, the surrounding flow decelerates the particle. In other words, the sediment particle drives the surrounding flow and does work on it; thus, the particle returns part of its energy to the flow. Finally, the particle hits the bed with a certain speed. It transfers part of its remaining energy to the stationary particles on the bed, thus causing them to move; nonetheless, most of the energy is converted into heat through the collisions. In summary, the saltation of sediment particles takes energy from the flow and returns a small part of the energy back to the flow; but most of its mechanical energy is transformed into heat. This energy loss is taken from the average motion, instead of from the kinetic energy of the turbulence. For bed load moving in the form of rolling and sliding, the situation is similar in that part of the energy is dissipated through collision and friction. Therefore, bed load motion causes an increase in the energy dissipation of a flow.

Rakoczi carried out experiments similar to those of Vanoni and Nomicos [21]. He conducted preliminary experiments with a movable bed, then used chemicals to fix the bed. Afterwards, he conducted experiments with the same discharge of clear water and measured the resistance. The principlal results, shown in Table 12.3, are different from those of Vanoni and Nomicos. The energy loss of a sediment-laden flow in most of the experiments is larger than that of a flow of clear water over a fixed bed. Rakoczi did not give any explanation for this result. A further analysis of his experimental data, given in Table 12.3, shows that the sediment used by Rakoczi was rather coarse and that all of the corresponding z values ($=\omega/\kappa U_*$) are larger than 5. As shown in Chapter 10, such sediment particles move mainly as bed load. And because of the bed load motion, additional energy dissipation is required. As for sediment sample No.4, which was very coarse, almost no sediment motion occurred;

therefore no difference in energy dissipation between sediment-laden flows with a movable bed and clear water flows with fixed bed took place.

Table 12.3 Comparison of energy loss in sediment-laden flow over movable bed and that in clear water flow (after Ryckoczi, L.)

No. sed.	Mean diameter D	Fall velocity ω	Sediment-laden flow over movable bed			Clear water flow over fixed bed			ω/κU*
	mm	cm/s	slope J	friction vel.U*	friction factor f	slope J	friction vel.U*	friction factor f	
1	0.58	8	0.002080	3.71	0.072	0.001040	2.62	0.0377	5.4
2	1.0	16	0.001234	3.09	0.045	0.000809	2.45	0.0266	12.95
3	3.3	43	0.003065	4.75	0.0925	0.002560	4.35	0.080	22.63
4	3.6	48	0.001790	3.58	0.055	0.001940	3.71	0.058	33.53

12.3.3 Effect of suspended load

Concepts offered in explanations of this occurrence are quite varied; also the existing experimental results show considerable divergence. The data are presented first, and the problem is then discussed.

12.3.3.1 Existing experimental data

1. Data for flume experiments in which the resistance for sediment-laden flows was smaller than that for the flow of clear water.

The aforementioned experiments by Vanoni and Nomicos were carried out under the condition that boundary condition were the same, and the resistance to clear water flow and that to sediment-laden flows (with different concentrations) were compared.

Their experimental results show that the development of dunes plays an important role in resistance and the concentration is a minor factor. The friction factor for flow with or without dunes can differ by factors of 3-6. For the same boundary condition, the existence of suspended load can reduce the friction factor by 4 to 20%.

Zhang collected a large amount of field data in rivers, canals, and pipelines and plotted the value of n against the concentration. He found that n decreases with an increase of concentration. A great number of his data were for values of n smaller than 0.01 (corresponding to the roughness of glass) and for high concentrations of sediment. He concluded that the existence of suspended load reduces the resistance [1]

Table 12.4 Effect of sand wave and suspended load on flow resistance

Set no.	No.	Depth (cm)	Bed condition	Concentration (g/l)	Friction factor		Reduction	
					average f	bed f_b	Δf_b	%
I	1	8.76	dune	3.64	0.074	0.106	0.006	5
	2	8.76		0	0.077	0.112		
II	3	7.52	small dune	4.60	0.0198	0.0211	0.0072	25
	4	7.52		0	0.0246	0.0283		
III	5A	7.86	flat	8.08	0.0165	0.0165	0.0064	28
	6	7.86		0	0.0203	0.0229		
IV	7	7.86	flat	3.61	0.0207	0.0227	0.0035	13.5
	8	7.80		0	0.0230	0.0262		

2. Data for flume experiments in which resistance for sediment-laden flows is the same as for clear water

In the Einstein and Chien experiments (also under the condition of a boundary that is unchanged), the resistances for clear water flow and sediment-laden flow with different concentrations were compared [7]. In their experiments, the concentration was as high as *625kg/m³* (point concentration), and the density gradient was quite large. If suspended load has any effect on resistance, it should be apparent for such extreme experimental conditions.

The results of their comparison of the resistance, given in Table 12.5, include energy slopes, friction factors (wall effect has been eliminated), water depths, velocities, and the maximum concentration measured along the verticals. In runs with nearly the same values of the principal variables, the energy slope and friction factor of clear water flow are almost the same as those for sediment-laden flow. Thus these data indicate no difference in the energy dissipation of sediment-laden flow and clear water flow.

Elata and Ippen also found no significant difference in the resistance coefficients for clear water and sediment-laden flows. The differences did not exceed 5% [4]. They stated that whether the coefficient of resistance was higher for the sediment-laden flow than for clear water flow depended on how the Reynolds number was

Table 12.5 Conditions of experiments conducted by Einstein and Chien

Set-no	Sand	D_{50} mm	Q l/s	h cm	U cm/s	Je	Max.S. g/l	T	R_b cm	κ	f_b
C-1	coarse		78.2×10^3	14.0	185	14.5×10^{-3}	0	23.4	10.0	0.387	0.0334
S-1	coarse		77.8×10^3	13.7	187	13.9×10^{-3}	58	23.6	9.58	0.305	0.0300
			79.2×10^3	13.8	187	14.1×10^{-3}		22.2	9.60	-	0.0298
C-2	coarse		79.2×10^3	13.8	188	14.1×10^{-3}	0	23.3	9.60	0.379	0.0304
C-3	coarse		73.3×10^3	12.3	196	20.6×10^{-3}	0	14.7	9.34	0.403	0.0390
S-2	coarse		73.9×10^3	12.0	204	19.4×10^{-3}	121	17.2	8.69	0.247	0.0329
		1.30	73.6×10^3	12.0	204	19.3×10^{-3}		17.2	8.69	-	0.0316
S-3	coarse		73.3×10^3	11.6	208	20.9×10^{-3}	151	17.2	8.60	0.231	0.0316
			73.6×10^3	11.7	208	20.9×10^{-3}		19.4	8.57	-	0.0324
S-4	coarse		73.0×10^3	11.5	210	23.6×10^{-3}	198	19.2	8.75	0.210	0.0322
			73.3×10^3	11.5	210	23.7×10^{-3}		22.7	8.81	-	0.0367
S-5	coarse		73.3×10^3	11.1	218	25.5×10^{-3}	328	15.5	8.45	0.173	0.0370
			73.0×10^3	10.9	222	25.8×10^{-3}		18.3	8.24	-	0.0354
C-4	coarse		73.3×10^3	12.1	200	18.4×10^{-3}	0	19.4	8.81	0.350	0.0336
C-5	coarse		73.3×10^3	17.8	137	5.5×10^{-3}	0	20.5	10.4	0.350	0.0240
C-6	medium		84.0×10^3	14.9	186	14.7×10^{-3}	0	25.5	10.6	0.398	0.0352
S-6	medium		82.0×10^3	14.3	189	14.7×10^{-3}	28	23.8	9.97	0.295	0.0306
			82.0×10^3	14.3	190	14.3×10^{-3}		26.1	9.97	-	0.0309
S-7	medium		82.3×10^3	14.3	189	14.2×10^{-3}	87	18.3	9.91	0.281	0.0309
			81.4×10^3	14.2	190	14.3×10^{-3}		20.0	9.93	-	0.0301
S-8	medium		79.8×10^3	13.9	190	14.0×10^{-3}	112	23.3	9.67	0.263	0.0322
		0.94	80.3×10^3	13.9	191	15.0×10^{-3}		20.2	9.45	-	0.0300
S-9	medium		80.6×10^3	13.6	195	15.3×10^{-3}	173	22.2	9.45	0.247	0.0283
			79.2×10^3	13.4	195	15.3×10^{-3}		23.3	9.40	-	0.0294
S-10	medium		79.8×10^3	13.2	200	17.3×10^{-3}	263	22.2	9.48	0.248	0.0297
			78.6×10^3	12.9	202	17.3×10^{-3}		26.1	9.24	-	0.0321
C-7	medium		79.2×10^3	13.7	192	15.4×10^{-3}	0	22.4	8.67	0.395	0.0306
C-8	medium		74.7×10^3	12.2	202	20.2×10^{-3}	0	14.7	9.03	0.391	0.0348
C-9	medium		73.3×10^3	12.1	200	19.3×10^{-3}	0	15.3	8.88	0.410	0.0336
C-10	fine		77.5×10^3	13.3	192	12.7×10^{-3}	0	18.3	8.66	0.400	0.0233
C-11	fine		77.0×10^3	13.4	190	12.7×10^{-3}	0	19.2	8.75	0.406	0.0241
S-11	fine		77.5×10^3	13.3	193	13.1×10^{-3}	31	18.3	8.72	0.398	0.0240
			77.5×10^3	13.3	192	13.1×10^{-3}		21.4	8.81	-	0.0244
S-12	fine		77.2×10^3	13.2	193	12.2×10^{-3}	205	20.5	8.45	0.274	0.0217
			77.2×10^3	13.2	193	12.4×10^{-3}		21.7	8.45	-	0.0221
C-12	fine		76.7×10^3	13.2	192	13.2×10^{-3}	0	17.2	8.72	0.391	0.0245
S-13	fine	0.274	77.0×10^3	13.4	189	12.6×10^{-3}	252	19.2	8.81	0.237	0.0242
			76.3×10^3	13.3	189	12.8×10^{-3}		21.7	8.90	-	0.0249
S-14	fine		76.3×10^3	12.3	205	17.4×10^{-3}	386	18.3	8.69	0.48	0.0282
			76.3×10^3	12.4	203	17.0×10^{-3}		20.5	8.72	-	0.0282
S-15	fine		76.3×10^3	12.4	202	16.9×10^{-3}	625	18.1	8.75	0.168	0.0283
			76.3×10^3	12.4	203	16.7×10^{-3}		18.9	8.69	-	0.0276
S-16	fine		74.2×10^3	11.9	206	18.6×10^{-3}	621	16.1	8.51	0.182	0.0291
			74.2×10^3	11.9	206	18.7×10^{-3}		17.7	8.57	-	0.0291
C-13	fine		74.2×10^3	11.7	210	18.9×10^{-3}	0	15.8	8.24	0.427	0.0276

formulated. If the viscosity of clear water was used, the resistance coefficient of sediment laden flow was higher than that for the clear water. If the viscosity of sediment laden water was used in the formation of Reynolds number, then the resistance coefficients of both the sediment-laden and the clear water flows were approximately the same.

3. Data from flume experiments in which resistance for sediment-laden flow is larger

Later, Montes and Ippen continued the MIT flume experiments. Instead of plastic balls, they used quartz sand (D_{50} = 0.365 mm), natural sand (D_{50} = 0.132mm), and mixtures of the two. The walls were smooth as before, but the bed was not horizontal; rather, it had a range of steep slopes, with a maximum value of 1%. Experimental results showed that the friction factor of sediment-laden flow was clearly larger than that of clear-water flow, as shown in Fig. 12.27. In the figure, the concentration close to the bed, S_{vm} is taken as a third parameter, and the data points arrange themselves in belts. The higher the concentration, the larger the friction factor is.

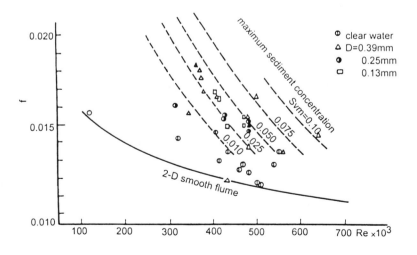

Fig. 12.27 The relationship between friction factor and sediment concentration

556

Wan et al. conducted experiments with hyperconcentrated flows in a tilting flume 0.4m wide, 0.4m high and 24m long. The walls and bottom were made of glass. The sediment used in the experiments was rather fine (D_{50} was 0.042mm), and the maximum concentration was over 900 kg/m^3 [22]. The experimental relationship between the friction factor and the Reynolds number is shown in Fig. 12.28. The effective viscosity of the sediment-laden flow, μ_e, was used to form the Reynolds number:

$$\mu_e = \eta\left(1 + \frac{\tau_B h}{2\eta U}\right) \tag{12.35}$$

in it, h and U are water depth and velocity, τ_B the Bingham yield stress, and η the rigidity of the sediment-laden flow. The figure indicates that, for a given Reynolds number, the friction factor of the hyper-concentrated flow is close to that of clear water flow or slightly smaller. But if a given hydraulic radius (or water depth h in a two-dimensional case) and a given average velocity are taken as the basis of comparison the friction factor of hyper-concentrated flow is definitely larger than that of clear water flow; the cause is the much larger effective viscosity μ_e of sediment-laden flows.

4. Data for experiments in pipelines

Resistance to flow in a pipeline is usually written in the form

$$\Delta P = f_m \frac{1}{d} \frac{\rho_m U^2}{2} \tag{12.36}$$

in which ΔP is the pressure drop per unit length, ρ_m density of the water-sediment mixture, f_m friction factor of a sediment-laden flow, U average velocity and d pipe diameter. If head loss is the drop in a unit length of column of the mixture Δh_m, it reads:

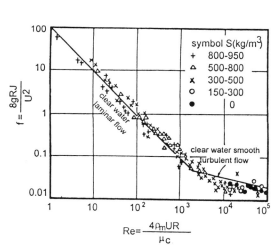

Fig. 12.28 Comparison of resistance coefficients in hyperconcentrated flow and clear water flow

$$\Delta h_m = \frac{\Delta P}{\rho_m g} = f_m \frac{1}{d} \frac{U^2}{2g} \tag{12.37}$$

The corresponding head loss for clear-water flow in the pipeline is,

$$\Delta h = \frac{\Delta P}{\rho g} = f \frac{1}{d} \frac{U^2}{2g} \tag{12.38}$$

Abundant experimental data are available for hydrotransport in pipelines. If sediment moves as bed load or if sediment particles settle out and form ripples on the bottom, the resistance is clearly larger than that for clear-water flow. This result is generally accepted. If all sediment particles are in suspension so that no bed load and no ripples appear on the bottom, the experimental results of the resistance for sediment-laden flow and that for clear-water flow diverge. The conditions for the different experiments are presented in detail in the chapter on hydrotransport (Chapter 17). The experimental results can be classified into three categories:

(1) The head loss of sediment-laden flow per unit length (Δh_m, expressed as the height of a column of the mixture) is equal to that of clear-water flow under the same conditions (Δh is the height of clear-water column). From Eq. (12.37) and Eq. (12.38) one obtains

$$f_m = f$$

i.e., the friction factor for sediment-laden flow is equal to that for clear-water flow.

(2) The head loss for sediment-laden flow and that for clear-water flow, both expressed in terms of the height of a clear water column are equal to each other. If the height of the clear-water column is transformed into the height of mixture column it is smaller, i.e., the head loss of sediment-laden flow is smaller.

(3) The head loss of sediment-laden flow in terms of height of column of mixture is smaller than the head loss of clear-water flow in terms of height of a clear-water column. In this instance, the hydraulic loss of sediment-laden flow is smaller.

Most experimental results belong to the second or third categories, nonetheless, quite a few researchers have reached the conclusion that the resistance to sediment-laden flow is equal to that to clear-water flow as in the first category.

Bruhl and Kazanskij observed the following phenomenon [3]: if coarse particles with a median diameter 0.3 mm are transported in suspension (volumetric concentration S_v=24.5%), the head loss (in terms of height of a clear water column) was clearly larger than that for clear water flow. If fine particles with a median diameter of less than 0.01 mm were added, the head loss was less than that in the former case, and the head loss went down as the concentration of fine particles went up. The head loss decreased as fine particles were added until the concentration of fine particles S_{v2} reached 6.6%. For still higher velocities, in the range of 4 to 5.5 m/s, the addition of fine particles produced no obvious effect. All their experimental results are shown in Fig. 12.29.

12.3.3.2 Discussion on effect of suspended sediment motion on energy dissipation

Experimental data presented in sections 12.3.3.1 reveal such a wide divergence that a preliminary discussion of existing knowledge is necessary. Three points merit prior discussion.

First, in studies of the effect of suspended sediment motion on energy dissipation of flow, one must eliminate the variation of boundary condition, as emphasized in section 12.3.1. As shown in the Vanoni and Nomicos experiments,

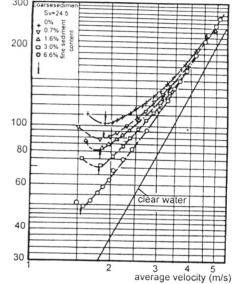

Fig. 12.29 Effect on friction loss by adding fine particles in pipe experiments(after Kazanskij, I. And H. Bruhl)

the friction factor in flow with dunes is much greater than that without dunes, increasing by factors of 3 to 6; but with the same boundary condition, the friction factor in flow with suspended load is slightly smaller than that in clear water flow, decreasing only 4~20%. Therefore, the influence of suspended load is much smaller than that due to boundary conditions. Because of this, all researchers planned their experiments so as to keep the boundary condition unchanged in order to give prominence to the effect of the suspended sediment. However, the task of keeping boundary condition strictly unchanged is not an easy one. Ippen pointed out that in some of the Vanoni and Nomicos experiments the fixed bed was made of sediment particles coarser than the suspended sediment, and some fine suspended particles were likely to have settled into the voids among coarser particles and in this way made the bed smoother and friction loss smaller [18]. Ippen experiments were conducted with a smooth boundary. Nevertheless, the authors stated that sometimes stripped sand belts appeared on the bed. Thus, the increase of friction factor of sediment laden flow shown in Fig. 12.27 possibly resulted from a coarsening of the bed due to the deposition of sediment particles. In the Einstein and Chien experiments, the boundary conditions were kept more nearly constant. On the one hand, the sediment particles glued on the bed were the same size as the suspended ones; on the other, high rates of flow were used throughout to assure that sediment particles did not settle out. For field data in natural rivers, a change of discharge might cause a change in the boundary condition, as has been mentioned. Even at the same discharge, a higher concentration would increase the fluid viscosity, and it could cause a change of bed configuration, as discussed in detail in Chapter 6. In such a way, any change in the friction factor is caused by the change of bed configuration resulting from the presence of the sediment.

Second, in many of the experiments, not all of the sediment particles were really in suspension; some particles moved as bed load. As bed load and suspended load have different effects on energy dissipation, their co-existence complicates the interpretation. For instance, in the Bruhl and Kazanskij pipeline experiments (Fig. 12.29) a large part of the coarse particles (0.3 mm) may have moved as bed load and at rather low velocities. The addition of 0.01 mm fine particles would increase the fluid viscosity, and their fall velocity would be low; consequently, more coarse particles were converted into suspended load with a consequent reduction of resistance. With the high velocities, the coarse particles were rather uniformly suspended, and any reduction of drag resistance caused by the addition of fine particles would not be large.

Third, the damping of turbulence is not equivalent to a reduction of energy dissipation, as discussed in Chapter 4. The effective potential energy of flow is transferred to the zone near the boundary through shear where it produces turbulent eddies. These eddies diffuse into the main flow region and decompose into smaller eddies in a cascade. Finally turbulence energy dissipates through viscosity to heat. The turbulence kinetic energy is taken from the effective potential energy of flow and is finally converted into heat. Suspended particles are supported by turbulence, i.e., their suspension takes part of the energy from the kinetic energy of turbulence. Consequently, the existence of suspended load damps turbulence. Most laboratory experiments and field observations reveal such effects. But the kinetic energy of turbulence is intermediary in the processes of energy transformation. The use of a part of the turbulence energy for supporting sediment suspension does not mean a comparable reduction in the amount of effective potential energy converted into turbulent kinetic energy (i.e., a reduction of energy loss). Therefore, the fact that the existence of suspended load damps turbulence does not mean a comparable reduction of energy dissipation.

These three points provide a basis for the following discussion of various points of view on the effects of suspended load on energy dissipation.

1. An analysis of the decrease of energy dissipation of flow caused by suspended sediment is approached in two ways.

(1) Change of flow field caused by the presence in part of the flow field of solid particles

Hino pointed out that a part of the space of a sediment-laden flow is occupied by solid particles, resulting in less space (or fluid volume) in which turbulence kinetic energy can be converted into heat [10]. From this point of view, the existence of solid particles reduces energy dissipation. However, the presence of solid particles also increases energy dissipation because of their relative motion and rotation.

As to the first point, the average energy dissipation rate W in the space replaced by a solid particle would have been

$$\overline{W} = \left(\frac{\pi}{6}D^3\right)gUJ = \left(\frac{\pi}{6}D^3\right)\frac{U_*^3}{h}\sqrt{\frac{2}{C_f}}\left(1 - \frac{1}{\kappa}\sqrt{\frac{C_f}{2}}\right) \qquad (12.39)$$

in which D is the diameter of a sediment particle, $C_f=2(U_*/u_{max})^2$ and u_{max} is the velocity at the water surface.

For the second point, no valid expression for the additional energy dissipation caused by a solid particle has been presented. If the effect of rotation is neglected, the additional energy dissipation rate because of their relative motion is

$$W_p \approx \left(\frac{\pi}{6}D^3\right)\frac{3\rho\left(2\frac{\rho_s}{\rho}+1\right)}{2}\frac{\overline{u'}^2}{U_*^2}\frac{U_*^3}{h} \qquad (12.40)$$

A comparison of Eq. (12.39) and Eq. (12.40) for a given flow condition shows that the reduction of energy dissipation caused by the effect of a sediment particle on the viscosity is 10 to 12 times larger than the additional energy dissipation. Therefore, the latter can be neglected. Hino's study suggests that the presence of suspended particles reduces energy dissipation. In his analysis, the density of the particles and the density gradient caused by an uneven concentration distribution did not play important roles.

(2) Difference in the viscosities of sediment-laden flow and clear-water flow

The existence of sediment particles increases the viscosity of a sediment-laden flow and thus thickens the laminar sublayer. If the flow boundary for clear-water flow is hydraulically rough or in the transition zone, a thicker laminar sublayer would make the boundary relatively smoother. The effect of an increase of fluid viscosity is more pronounced for fine sediment particles and, hence, so is its effect on the resistance. But if the boundary for clear water flow is hydraulically smooth, the change of viscosity and of the laminar sublayer in sediment-laden flow has no effect on resistance.

If the content of fine particles exceeds a certain critical value, a sediment-laden fluid becomes non-Newtonian, and the local energy dissipation is then significantly less. When a non-Newtonian fluid encounters protruding obstacles in its flow course, it cannot as easily form the rotating vortexes behind obstacles of the kind that usually appear in turbulent flows. Shishenko observed that the flow of a non-Newtonian fluid in an abrupt expansion zone was essentially steady, that no vortexes formed [24]. Therefore for a non-Newtonian fluid, the effect of surface roughness on turbulence properties is not as marked as is that for a Newtonian fluid; consequently, the local resistance in a non-Newtonian fluid is less.

Based on a simple model, Gyr stated that due to the higher viscosity of sediment-laden flow the size of the small eddies formed near the boundary must be larger [25]. Thus, from the term for energy dissipation, the larger the eddy is, the smaller the velocity gradient and the less the energy dissipation. Hence, in comparison with clear-water flow, the smallest eddies in a sediment-laden flow are larger and the energy dissipation is less. Muller carried out experiments in a pipeline to verify this concept [26]. His experiment showed that the size of smallest eddies increased with increasing concentration; this trend was more evident in the central part of the pipe and decreased towards the boundary.

2. Energy dissipation increases due to suspended motion of sediment particles.

As already mentioned, comparisons of energy dissipation in sediment-laden and clear-water flows should be based on a consistent standard. Comparisons based on the same Reynolds number or on the same velocity (and same depth) may lead to different conclusions. The well-known relationship between friction factor and Reynolds number is different in the three regions:

(1) Laminar flow – in which the friction factor is inversely proportional to the Reynolds number;

(2) Smooth boundary turbulent flow – in which the friction factor is inversely proportional to 0.25 power of Reynolds number;

(3) Rough boundary turbulent flow – in which the friction factor does not vary with Reynolds number.

For the same Reynolds number, the friction factor for sediment-laden flow may be equal to or less than that in clear water flow. However, in engineering practice, the problem normally encountered is the following: for flow of a given discharge passing a certain section, does a sediment-laden or a clear-water flow require the larger slope? Such a comparison is carried out under the condition that the flows have the same velocity and the same depth. Since the viscosity of a sediment-laden flow is much higher than that of clear-water flow, the Reynolds number of the sediment-laden flow would be much smaller than that of the clear-water flow. In the laminar flow region, particularly, and in the turbulent flow region for a smooth boundary, the friction factor of the sediment-laden flow would be larger than that for the clear water flow. Such a situation is likely to occur for flow with a high content of fine particles. In such a case, a given flow could be turbulent flow for a clear water flow and laminar for a sediment-laden one. The experiments with hyper-concentrated flows by Wan et al. showed just such an occurrence.

3. *Suspended sediment motion may cause different changes of energy dissipation.*

Based on data of Roberts, Kennedy, and Ippen experiments, Mih stated that, depending on the net effect of two contrary tendencies, the energy dissipation in sediment-laden flow can be larger than, equal to or smaller than that in clear water flow [27,28].

Mih started from the equation for the logarithmic velocity profile

$$\frac{u}{U_*} = \frac{1}{\kappa} \ln \frac{y}{y_o} \tag{7.34}$$

For turbulent flow passing a smooth boundary,

$$y_o = \frac{C\nu}{U_*} \tag{12.41}$$

the friction factor in a smooth pipe can be expressed as Karman-Prandtl formula in the form

$$\frac{1}{\sqrt{f}} = \frac{0.353}{\kappa} \left(\ln \frac{R_e \sqrt{f}}{C} - 3.23 \right) \tag{12.42}$$

in which

$$\text{Re} = \frac{Ud}{\nu} \tag{12.43}$$

d is pipe diameter. For clear-water flows, Mih used the values $\kappa = 0.406$ and $C = 0.10$, and then transformed Eq. (12.42) into

$$\frac{1}{\sqrt{f}} = 2\log \mathrm{Re}\sqrt{f} - 0.8 \qquad (12.44)$$

With neutral-density particles suspended in the flow, as for a series of experiments in a smooth pipe carried out at MIT, κ and C vary with volumetric concentration of particles, as shown in Fig. 12.30 [28, 29, 30]. In the range of experiments,

$$\kappa = 0.4(1 - 0.95S_v) \qquad (12.45)$$

Thus, the coefficient C, which reflects the thickness of laminar sublayer, increases with an increase of S_v. Along with the increase of concentration, κ and C in Eq. (12.42) have opposite trends of development. Hence, different trend for the energy dissipation in sediment-laden flow and in clear-water flow can occur. Fig. 12.31 shows the experimental results obtained by Roberts. The experiment was carried out in a smooth pipe 50 mm in diameter. The fluid was a sugar solution with a concentration of 4%, and the suspended particles were plastic cubes with a side length of 2.74 mm and a neutral density. The figure shows that friction factor for a volumetric concentration of plastic particles of 30% is consistently above that for a pure liquid. The values for 10% and 20% volumetric concentration are close to the curve for small Reynolds number but below it for Reynolds number above 7×10^4.

Mih's analysis is for a uniformly suspended sediment in a smooth pipe. Further

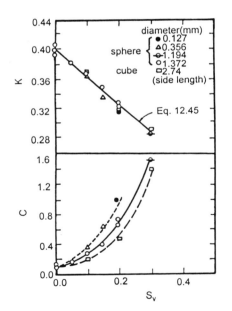

Fig. 12.30 Variation of C and κ with sediment concentration with neutral-density particles suspended in the flow (MIT experimental results)(after Daily, J. W. and, R.W. Hardison)

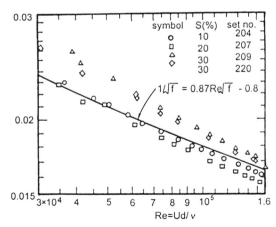

Fig. 12.31 Comparison of the resistance coefficient of pure fluid flow and that of flow with suspended neutral-density cubes (after Roberts, C.P.R.; J.F. Kennedy; and A. T. Ippen)

study is needed to establish trends for other conditions.

In summary, the process of flow energy transformation from potential energy to heat is different for flows with suspended sediment. This complex change has not yet been clearly explained in detail, so that no general theory can be presented to explain various related phenomena concerning the resistance. In alluvial rivers the effect of concentration on the resistance is much less than the effect of bed configuration on the resistance. For high concentrations, particularly of fine particles, the rheological properties of the flow may change, so that the situation is different. In such cases, one must determine which standard of comparison to adopt in comparing sediment-laden and clear-water flows. Hyperconcentrated flow in the Middle and Lower Yellow River may require larger slopes than do clear water flows for the same depth and same velocity.

12.4 THE FEEDBACK EFFECT

The existence of sediment affects flow conditions, and the resulting changes naturally affect the sediment motion in turn. Studies of this aspect are quite incomplete. The fragmentary results that are available are introduced briefly in this section.

12.4.1 Variation of velocity profile induced by the existence of concentration gradient near bed and its effect on the concentration profile

If the concentration gradient near the bed is large, the vertical distribution of suspended sediment is different from that given by Eq. (10.16). As shown in Fig. 12.15, the data points near the bed no longer follow a straight line on semi-logarithmic paper. Chien and Wan analyzed this divergence [8] and verified it by using Einstein and Chien's data [7].

First, they determined the effect of concentration on fall velocity of particles from the following formula, introduced in Chapter 3:

$$\omega/\omega_o = \left(1 - S_v\right)^m \tag{3.38}$$

The corresponding m values for sediments used in the experiments are as follows:

Coarse sand	*D=1.3mm*	*m ≈ 2.5*
Medium sand	*D=0.94mm*	*m ≈ 3.0*
Fine sand	*D=0.27mm*	*m ≈ 4.5*

Second, they adopted Eq. (12.34), in which the effect of concentration is included, for the velocity distribution, and expressed the velocity gradient as follows:

$$\frac{du_r}{d\eta} = \frac{b}{a}\eta^{\frac{1}{a}-1} = b'\eta^{-\beta'} \tag{12.46}$$

in which: $b'=b/a$; $\beta'=1-(1/a)(0<\beta'<1)$. In accordance with its definition, the momentum exchange coefficient is:

$$\varepsilon_m = \frac{\tau}{\rho\dfrac{du}{dy}} = \frac{\rho U_*^2(1-\eta)}{\rho U_* \dfrac{du_r}{h}\dfrac{}{d\eta}} = \frac{U_* h}{b'}\eta^{\beta'}(1-\eta) \tag{12.47}$$

Next, from the basic equation of diffusion theory,

$$\varepsilon_s \frac{dS_v}{dy} + S_v\omega = 0 \tag{10.7}$$

in Eq. (10.7), ω is replaced by $\omega_o(1-S_v)^m$. The simplification $\varepsilon_s=\varepsilon_m$ is used, as before. Substituting Eq. (12.47) into Eq. (10.7), rearranging and integrating both sides, one obtains:

$$\int\frac{dS_v}{S_v(1-S_v)^m} = -\frac{\omega_o b'}{U_*}\int\frac{d\eta}{\eta^{\beta'}(1-\eta)} + const. \tag{12.48}$$

If F and G are as follows:

$$\ln G(S_v,m) = \int\frac{dS_v}{S_v(1-S_v)^m}, \quad F(\beta',\eta) = \int\frac{d\eta}{\eta^{\beta'}(1-\eta)} \tag{12.49}$$

Eq. (12.48) can be rewritten as:

$$\log G(S_v,m) = -\frac{1}{\zeta}F(\beta',\eta) + const. \tag{12.50}$$

in which

$$\zeta = \frac{2.3U_*}{\omega_o b'} \tag{12.51}$$

Eq. (12.50) is the equation for the profile of suspended sediment concentration, in which the effect of concentration (and of its gradient) has been taken into consideration. On semi-logarithmic paper with $G(S_v, m)$ as ordinate and $F(\beta', \eta)$ as abscissa, Eq.(12.50) is a straight line. Both of these functions can be integrated and tabulated for convenience. Like the equation of the concentration profile from diffusion theory, Eq. (12.50) gives only the relative distribution of concentration, not the absolute values.

From measured data of concentration [7] and the a value presented in the section on velocity profile, experimental values of $G(S_v, m)$ and $F(\beta', \eta)$ can be calculated and compared with Eq. (12.50). The results for fine sand shown in Fig. 12.32 are quite well distributed along straight lines. The values of ζ can be calculated from the slopes of the lines. Experimental and theoretical values of ζ are compared in Fig. (12.33). The agreement between them is only fair. This result is similar to that for the values of z for which the theoretical values do not coincide well with the experimental ones for low concentrations.

Recently, Lavelle and Thacker noticed the effect of concentration on fall velocity summarized by Maude [31]. Referring Maude's summary, they used an equation similar to Eq. (3.38) to show the effect of concentration on fall velocity. In their analysis, they used a constant of 5 for the exponent m in Eq. (3.38), and they used the following momentum exchange coefficient:

$$\varepsilon_s = (\alpha + \beta y)(h - y)/h \qquad (12.52)$$

in which α and β are parameters.

If one substitutes Eq. (13.38) and Eq. (12.52) into Eq. (10.7), the basic formula of diffusion theory, and integrates it, one obtains an equation for the concentration profile in which the parameter S_o, the concentration at a reference elevation, appears, in addition to the parameters α and β.

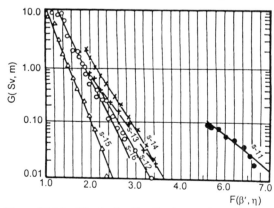

Fig. 12.32 Vertical distribution of sediment concentration plotted according to Eq. (12.50) (fine sediment set)

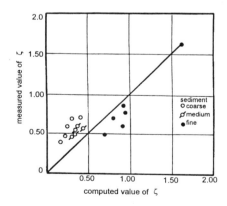

Fig. 12.33 Comparison of computed and measured values of ζ

Lavelle and Thacker determined the parameters α and β in their equation for the concentration profile by using six sets of data from experiments with fine sand conducted by Einstein and Chien (sets *S-11* to *S-16*). They took the bed surface as the reference elevation in their determination of S_o; the values thus obtained for the various sets were close to 0.48, a value that is equal to the concentration of densely packed natural quartz sand. Thus they suggested that the bed surface could be taken as the reference elevation and the concentration of densely packed sand could be taken as the reference concentration; these values are clearly independent of the flow conditions. However, the authors explained in Chapter 10 why the diffusion theory should not be applied to the region near the bed. Therefore, the suggestion of Lavelle and Thacker may not be theoretically acceptable. Actually, the results from their method for data set *S-11*, for which the concentration was low, did not fit well.

12.4.2. Effect of fine particles on the motion of coarse particles

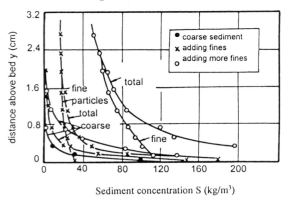

Fig. 12.34 Effect on vertical distribution of sediment concentration by adding fine particles (after Kikkawa, H., and S. Fukuoka)

The results obtained by Bruhl and Kazanskij on hydrotransport have been discussed. The presence of fine particles (D=0.01 mm) in a pipe flow carrying coarse particles (0.3 mm) caused the head loss to be significantly lower [3]. The addion of the fine particles also reduced the turbulence intensity markedly.

Kikkawa and Fukuoka also conducted experiments specially planned to study the effect of fine particles on the transport of coarse particles [32]. They used a recirculating glass flume 8 m long and 0.4 m wide. First, they measured the velocity profiles for clear water. Then they added 0.18 mm coarse particles. Once equilibrium was reached, they measured the velocity profiles, concentration profiles and average concentrations. Afterwards, they repeated the measurements after adding fine particles with diameters of 0.05 mm (or 0.015 mm). Finally, they added still more fine particles and repeated the measurements. During the entire experiment, all fine particles were in suspension.

The results revealed that the presence of fine particles made both the velocity and concentration gradients larger (i.e., the Karman constant κ was less). Also, the rate of transport of coarse particles was markedly higher, as shown in Fig. 12.34.

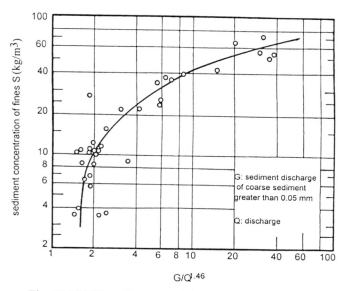

Fig. 12.35 Effect of content of fine particles (finer than 0.01 mm)
on the transport capacity of coarse particles

The fine particles affect the transport rate of coarse particles primarily through their effect on the viscosity of the suspension. As stated in Chapter 11, the sediment transport capacity of flow generally increases with a decrease of water temperature and the corresponding increase of viscosity. In rivers heavily laden with sediment, the effect of sediment concentration on viscosity is much more significant than is the effect of water temperature. Applying the revised Einstein bed load function, Chien and Zhang calculated the bed material load for the Huayuankou Reach of the Lower Yellow River. The calculations indicated that the sediment transport capacity of bed material load for bankful discharge increased by 50% as water temperature dropped from 30 to 0°C; in contrast, the sediment transport capacity of bed material load can increase by a factor of 10 as the concentration in the oncoming flow increases from 100 kg/m^3 to 500 kg/m^3 [33]. From field data from the Yellow River, Fang and Qi also plotted a relationship between sediment-carrying capacity for coarse particles (larger than 0.05 mm) and the concentration of fine particles (smaller than 0.01 mm), Fig. 12.35. [1] The figure shows that the higher the concentration of fine particles, the larger the capacity of the flow is to transport coarse particles. This phenomenon is discussed in detail in the chapter on hyperconcentrated flows (Chapter 13).

[1] Fang, Zongdai, Pu Qi. *Effect of very fine sediment on sediment-carrying capacity and fluvial processes.* Research Institute of Hydraulic Technology, Yellow River Water Conservancy Commission, Oct. 1978.

REFERENCES

[1] Zhang, Ruijin. "On Gravitation Theory as Well as Motion Processes of Suspended Load." *Journal of Hydraulic Engineering*, 1963, No. 3, p. 11-23.

[2] Bagnold, R.A. "Some Flume Experiment on Large Grains but Little Denser than Transporting Fluid, and Their Implication." *Proc., Inst. Civil Engrs.*, 1955, pp. 174-205.

[3] Bruhl, H. and I. Kazanskij. "New Results Concerning the Influence of Fine Particles on Sand-Water Flows in Pipes." *Proc. of Hydrotransport, No. 4*, 1976, pp. B2 19-25.

[4] Elata, C., and I.T. Ippen. "The Dynamics of Open Channel Flow with Suspensions of Neutrally Buoyant Particles." *Tech. Rep. No. 45*, Hydrodynamics Laboratory, Massachusetts Inst. Tech., January 1961.

[5] Muller, A. "Turbulence Measurements over A Movable Bed with Sediment Transport by Laser-Anemometry." *Proc. 15th Congress of Intern. Assoc. Hyd. Res.*, 1973.

[6] Bohlen, W.F. "Hotwire Anemometer Study of Turbulence in Open Channel Flow Transporting Neutrally Buoyant Particles." *Report No. 69-1*, Experimental Sedimentology Laboratory, Dept. of Earth and Planetary Sciences, Massachusetts Inst. Tech., 1969.

[7] Einstein, H.A., and Ning Chien. "Effects of Sediment Concentration Near the Bed on the Velocity and Sediment Distribution." *M.R.D. Sediment Series No. 8*. Missouri River Div., Corps Engrs.,1955,p.76.

[8] Chien, Ning, and Zhaohui Wan. "Preliminary Study on The Effect of Hyperconcentrated Layer Near Bed on Flow and Sediment Motion." *Journal of Hydraulic Engineering*, 1965, No. 4, p. 1-20. (in Chinese)

[9] Yalin, M.S. *Mechanics of Sediment Transport*. Pergamon Press, 1972.

[10] Hino, M. "Turbulent Flow with Suspended Particles." *J. Hyd. Div., Proc. Amer. Soc. Civil Engrs.*, Vol. 89, No. HY4, pp. 161-185.

[11] Monin, A.S. "Turbulence in Shear Flow with Stability." *J. Geophys. Res.*, Vol. 64, No. 12, 1959. pp. 2224-2225.

[12] Shaver, R.G., and E.M. Merrill. "Turbulent Flow of Pseudoplastic Polymer Solutions in Straight Cylindrical Tubes." *J. Amer. Inst. Chem. Engrs.*, Vol. 5, No. 2, 1959, pp. 181-188.

[13] Kazanskij, I., and H. Bruhl. "Influence of High Concentrated Rigid Particles on Macroturbulence Characteristics of Pipe Flow." *Proc., Hydrotransport, 2*, 1972, pp. A2 15-30.

[14] Vanoni, V.A., and G.N. Nomicos. "Resistance Properties of Sediment-Laden Streams." *Trans., Amer. Soc. Civil Engrs.*, Vol. 125, 1960, pp. 1140-1175.

[15] Kalinske, A.A., and C.H. Hsia. "Study of Transportation of Fine Sediments by Flowing Water." Univ. Iowa Studies in Engin., *Bull. No. 29*, 1945, p. 30.

[16] Tsulaki. "Effect of Suspended Load on Flow Properties." *J. of Sediment Research*, Vol. 1, No. 1, 1956, p. 19-28.

[17] Barton, J.R., and P.N. Lin. "Roughness of Alluvial Channels." *Report 55-JBR-2*, Civil Engin. Dept., Colorado A. and M. College, 1955, p. 43.

[18] Ippen, A.T. "The Interaction of Velocity Distribution and Suspended Load in Turbulent Streams." *Proc., Intern. Symp. on River Mech.*, Vol. 4, Asian Inst. Tech., Bangkok, Thailand.

[19] Oda, H., J. Gruat, and F. Valentian. "Variation of Characteristic Length of the Turbulence in Pipes Flows with Solid Particles in Suspension." *Proc. Hydrotransport 2*, 1972, pp. A3 31-44.

[20] Chien, Ning. "Roughness of the Lower Yellow River." *Journal of Sediment Research*, Vol. 4, No. 1, 1959, p. 1-15. (in Chinese)

[21] Ryckoczi, L. "Experiment Study of Flume Bed Roughness." *Proc., 12th Cong., Intern. Assoc. Hyd. Res.*, Vol. 1, 1967, pp. 181-186.

[22] Montes, J.S., and A.T. Ippen. "Interaction of Two-Dimensional Turbulent Flow with Suspended Particles." *Rep. No. 164*, Lab. Water Resources and Hydrodynamics, Massachusetts Inst. Tech, Jan. 1973.

[23] Wan, Zhaohui, Yiying Qian, Wenhai Yang, Wenlin Zhao. "An Experimental Study on Hyperconcentrated Flow." *Yellow River*, 1979, No. 1, p. 53-65. (in Chinese)

[24] Shishenko, R.E. *Hydraulics of Slurry*. Petroleum Industry Press, 1951. (in Russian)

[25] Gyr. A. "The Behavior of the Turbulent Flow in A Two-Dimensional Open Channel in Presence of Suspended Particles." *Proc., 12th Cong., Intern. Assoc. Hyd. Res.*, Vol. 2, 1967.

[26] Muller, A. "Measurement of the Influence of Suspended Particles on the Size of Vortices." *Proc., 12th Cong., Intern. Assoc. Hyd. Res.*, Vol. 4, 1967.

[27] Mih, W.C. "Solid-Liquid Suspension Flow in Pipes." *Sedimentation.* Edited by H.W. Shen, Colorado, 1972, pp. 24, 1-23.

[28] Roberts, C.P.R., J.F. Kennedy, and A.T. Ippen. "Particle and Fluid Velocities of Turbulent Flows of Suspensions of Neutrally Buoyant Particles." *Tech. Rep. 103*, M.I.T. Hydrodynamics Lab., 1967, p.45.

[29] Daily, J.W., and T.K. Chu. "Rigid Particle Suspensions in Turbulent Shear Flow: Some Concentration Effects." *Tech. Rep. 48*, M.I.T. Hydrodynamics Lab., 1961.

[30] Daily, J.W., and R.W. Hardison. "Rigid Particle Suspensions in Turbulent Shear Flow: Measurements of Total Head, Velocity and Turbulence with Impact Tubes." *Tech. Rep. 67*, M.I.T. Hydrodynamics Lab., 1964.

[31] Lavelle, J., and W.C. Thacker. "Effects of Hindered Settling on Sediment Concentration Profiles." *J. Hyd. Res., Intern. Assoc. Hyd. Res.*, Vol. 16, No. 4, 1978, pp. 347-356.

[32] Kikkawa, H., and S. Fukuoka. "The Characteristics of Flow with Wash Load." *Proc., 13th Cong., Intern. Assoc. Hyd. Res.*, Vol. 2, 1969, pp. 223-240.

[33] Chien, Ning, Ren Zhang, Jiufa Li, and Weide Hu. "Preliminary Study on Adjustment Mechanism of Sediment-Carrying Capacity in Heavily Sediment-Laden Rivers." *Report No.80-2* of Sedimentary Laboratory, Department of Hydraulic Engineering, Tsinghua University, 1980.

CHAPTER 13

HYPERCONCENTRATED AND DEBRIS FLOWS

Sediment concentrations in many of the rivers in China are the highest in the world. Whenever sediment concentrations exceed a certain critical value, the properties of the flow change significantly; they became different from ordinary sediment-laden flows both in their motion and in the manner sediment is transported. Dynamic (or hydraulic) debris flow is also a kind of hyperconcentrated flow. Although debris flow contains coarser particles with a wider range of sizes and higher specific weights than those for hyperconcentrated flow, the two flows have similar mechanisms. The volume concentration of hyperconcentrated flow does not exceed 0.60, a value that corresponds to a specific weight of 2.0 t/m³, but the volume concentration of debris flow in Jiangjia Gully reaches 0.76, a specific weight of 2.25 t/m³. These two subjects are discussed separately because of their differences, but they still fit well within the same chapter.

13.1 HYPERCONCENTRATED FLOW [1]

13.1.1 Introduction

Sediment concentrations in most tributaries in the middle reaches of the Yellow River are extremely high during the flood season. The average annual concentrations in some tributaries approach 500 kg/m³ , and concentrations of 1500-1600 kg/m³ have been recorded. In that region, the tributaries that originate from desert areas possess lower concentrations, so that only the storm-induced floods originating in the loess region are likely to be hyperconcentrated. Such floods, which carry some clay and fine silt, also have more coarse particles than do those originating in the desert regions.

Loess composition in Northwestern China, that is, in the middle reaches of the Yellow River, becomes finer and finer moving from west to east and from north to south. The coarser the particles in the river basin are, the higher the recorded maximum concentration in the river. The relationship between the content of particles coarser than 0.05 mm and the recorded maximum concentration for rivers in the upper and middle reaches of the Yellow River is shown in Fig. 13.1 [1] . It shows that the coarser the particles in the basin are, the higher the recorded concentration of the sediment. The variations in one river or within one region are similar; i.e., the higher the concentration, the coarser the sediment that is carried. Relationships between concentrations and median diameters for the Wudin River at Dingjiaguo and Weihe

---

[1]   Research Group on Soil-Water Conservation, Planning Office of Yellow River Water Conservancy Committee. *Analysis of Sediment Entering the Yellow River Basin.* Dec. 1977, p. 23.

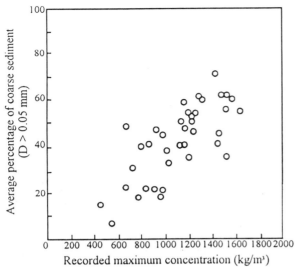

Fig. 13.1 Relationship between coarse sediment percentage and measured maximum concentrations in the Yellow River

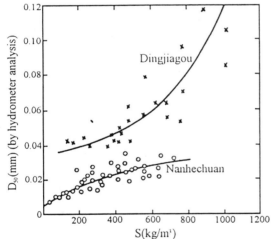

Fig. 13.2 Relationship between concentration and median diameter for sediment carried by hyperconcentrated flow

River at Nanhechuan are shown in Fig. 13.2. The percentage of coarse particles increases, and median diameter increases correspondingly, with the increase of concentration. In this respect, the characteristics of hyperconcentrated flow are completely different from those of a common sediment-laden flow, and they have provided important clues for understanding the mechanism of hyperconcentrated flow.

Specific weights and viscosities are much higher for hyperconcentrated flows than are those for ordinary flows. A certain amount of clay and fine silt in hyperconcentrated flows causes their rheological properties to change qualitatively. Because of these changes of rheological properties, the fall velocity of sediment particles in a hyperconcentrated flow also shows specific features that are quite different from those in clear water. All of these effects are described in Chapters 2 and 3 in some detail and are not repeated here.

13.1.2 Characteristics of motion

13.1.2.1 Phenomenon of motion---two flow patterns

The intensity of turbulence in hyperconcentrated flow is normally comparatively weak, and turbulence at small scales is almost non-existent. Although the flow velocity of a hyperconcentrated flow can be rather high, the water surface appears to be calm and smooth to the extent that surrounding scenes are clearly reflected in it. However, cloud-like patterns form within the flow, and these patterns twist and deform as they move along the channel. Wakes that form downstream of projecting objects along irregular banks remain visible.

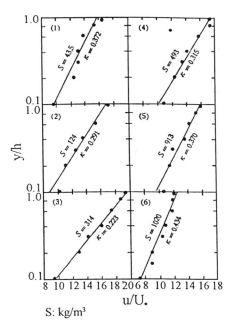

S: kg/m³

Fig. 13.3 Vertical velocity profiles for various concentrations at Dingjiagou Station, Wuding River

Hyperconcentrated flows can become laminar if the concentrations are extremely high, particularly in small canals. In such cases, a thin layer of clear water several millimetres thick exists at the surface and thread-like stream filaments are visible. They are called "flow-induced threads" by local people. Different from ordinary laminar flows, those of hyperconcentrated fluid sometimes possess a degree of plasticity. If a straight line of white powder were spread on the surface perpendicular to the flow, the line would keep its form as it moves with the flow, and the powder would not mix. Observations reveal that a "plug" forms, so that no relative motion or shear exists in the central part of the flume. From the vertical velocity profiles, the central "plug" also extends downward through most of the depth [2]. Sometimes two types of flow pattern exist simultaneously in a river reach, with turbulent flow in the main channel and laminar flow in the by-channels.

13.1.2.2 Velocity profile

1. Velocity profile in turbulent flow

In turbulent hyperconcentrated flow, the velocity distribution still follows the logarithmic law (Fig.13.3)[3]. For high sediment concentrations, the velocity distribution is more uneven; then if the sediment concentration exceeds a critical value, the velocity distribution tends to be uniform. The variation of the velocity distribution is clearly revealed by the relationship between the Karman constant and the sediment concentration (Fig.13.4). For the sediment carried in the Wuding River, the Karman constant had a minimum value of about 0.27 for a sediment concentration of about *300* kg/m³.

In Chapter 12 the relationship between κ and E is described, with E being the ratio of suspension energy w_d to potential w_s provided by the flow in unit time. For hyperconcentrated flow

$$w_s = [\gamma + (\gamma_s - \gamma)S_v]UJ \qquad (13.1)$$

$$w_d = (\gamma_s - \gamma)S_v(1 - S_v)\omega \qquad (13.2)$$

575

The relationship between $E = w_d/w_s$ and Karman constant κ is plotted in Fig. 13.5; the solid line is for common sediment-laden flow, and it is based on both field data and laboratory data, the dotted lines are the envelopes of the data points. Field data for hyperconcentrated flow at Dingjiaguo Gauging Station on Wuding River are also plotted in the figure. Fall velocities at different concentrations for the latter flows were not available, and hence were roughly estimated in two different ways. In Fig. 13.5a, fall velocities were calculated according to Eq. (3.38):

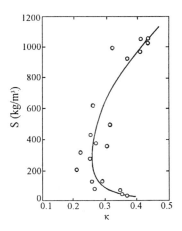

$$\frac{\omega}{\omega_0} = (1 - S_v)^m \qquad (3.38)$$

Fig. 13.4 Relationship between concentration and Karman constant κ, at Dingjiagou Station, Wuding River

This formula is suitable for uniform sediment; the effect of fine particles on fall velocities of coarser particles is not taken into consideration, thus giving larger values of ω/ω_0. In Fig. 3.27, the experimental curves obtained in the Institute of Hydraulic Research, Yellow River Conservancy Commission (IHR, YRCC) are plotted. From experience with high concentrations at the Luohui Irrigation District, ω/ω_0 values obtained from these curves are probably smaller than the correct values. Correspondingly, values interpolated from Eq. (3.38) and the experimental curve for sample 1 in Fig. 3.27 were used for estimating the fall velocity at different concentrations in Fig. 13.5.

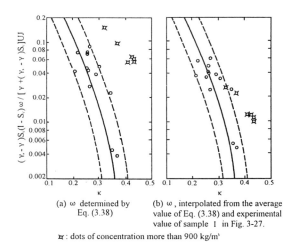

(a) ω determined by Eq. (3.38)

(b) ω, interpolated from the average value of Eq. (3.38) and experimental value of sample 1 in Fig. 3-27.

⋈ : dots of concentration more than 900 kg/m³

Fig. 13.5 Relationship between the Karman constant κ and E, for hyperconcentrated flow at Dingjiagou Station, Wuding River

576

From the distribution of data in Fig. 13.5, turbulent hyperconcentrated flow follows the same law as do those for ordinary sediment-laden flows if the effect of concentration on fall velocity is included. Data points for concentrations close to 1,000 kg/m^3 or higher have higher κ values; correspondingly, velocity profiles of such flows are more uniform than are those for turbulent flow of clear water. In such cases, the turbulent hyperconcentrated flow has probably transformed into laminar because of the hyperconcentration.

2. Velocity profile in laminar flow

As discussed in Chapter 2, the hyperconcentrated flows containing silt and clay usually behave as a non-Newtonian fluid. For a Bingham fluid, and starting from the rheological equation

$$\tau = \tau_B + \eta \frac{du}{dy} \tag{13.3}$$

one can easily deduce the velocity profile of a Bingham fluid in the region outside of the plug

$$u = \frac{y}{2\eta}(2\gamma_m hJ - \gamma_m yJ - 2\tau_B) \tag{13.4}$$

In it, η and τ_B are the rigidity and the yield stress of a Bingham fluid, respectively, and γ_m is the specific weight of the hyperconcentrated flow. Written in the form of velocity difference, Eq. (13.4) becomes

$$\frac{u_{max} - u}{u_{max}} = (1 - \frac{\gamma_m yJ}{\gamma_m hJ - \tau_B})^2 \tag{13.5}$$

in which U_{max} is the velocity at the surface. Such a velocity profile has a unusual feature; in the region

$$y \geq h - \frac{\tau_B}{\gamma_m J} \tag{13.6}$$

the shear stress is less than the Bingham yield stress, and therefore, no relative motion exists. In this region the flow moves forward as an entity, i.e., in the form of a "plug," as already described, and the expression for the velocity is

$$u_p = \frac{1}{2\eta\gamma_m J}(\gamma_m hJ - \tau_B)^2 \tag{13.7}$$

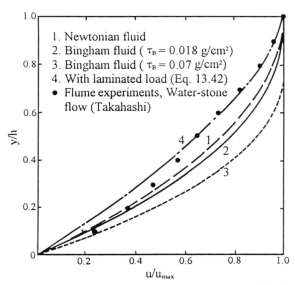

1. Newtonian fluid
2. Bingham fluid ($\tau_B = 0.018$ g/cm²)
3. Bingham fluid ($\tau_B = 0.07$ g/cm²)
4. With laminated load (Eq. 13.42)
• Flume experiments, Water-stone flow (Takahashi)

Fig. 13.6 A comparison of various velocity profiles in laminar flow and in flow with laminated load

Fig. 13.6 shows the velocity profile at the Dingjiagou Station of Wuding River for a concentration of about 1,000 kg/m³ (h=1.3 m and J=13×10⁻⁴); the flow was laminar. In the calculation, two different values were assumed for the Bingham yield stresses. From the figure, the larger the value of τ_B, the more pronounced is the plug, and the more the velocity profile deviates from that of a Newtonian laminar flow.

A significant point is that the plug can exist in other non-Newtonian flows as well as in a Bingham fluid. In a pseudo-plastic fluid with a large value of m (plastic index in Table 2.8), the velocity near the surface also varies only slightly. Since velocity profiles of non-Newtonian fluids are usually measured in pipelines, this comparison is discussed further in Chapter 17. In another case, laminar hyperconcentrated flow might occur in the form of Newtonian fluid carrying a laminated sediment load. In this latter case, the velocity profile differs from that for Eq. (13.4). These results are discussed in some detail in the section on non-viscous debris flow.

13.1.2.3 Resistance to flow

1. Smooth boundary

From experience with hyperconcentrated hydrotransport for Bingham fluid [4], one can base the relationship between the friction factor and Reynolds number on either the effective viscosity μ_e or the rigidity η, as follows:

First, effective viscosity μ_e is used to form the Reynolds number [5].

The integration of Eq. (13.4) provides a relationship between discharge per unit width q, hydraulic factors and the physical properties of the fluid, as follows:

$$q = Uh = \frac{\gamma_m J h^3}{3\eta}[1 - \frac{3}{2}(\frac{\tau_B}{\gamma_m hJ}) + \frac{1}{2}(\frac{\tau_B}{\gamma_m hJ})^3] \qquad (13.8)$$

The term $\left(\dfrac{\tau_B}{\gamma_m hJ}\right)^3$ in the above equation is usually negligible, and Eq. (13.8) then reduce to

$$q = Uh = \frac{\gamma_m Jh^3}{3\eta}[1 - \frac{3}{2}(\frac{\tau_B}{\gamma_m hJ})]$$

(13.9)

hence,

$$J = \frac{h_f}{L} = \frac{3U\eta}{\gamma_m h^2} + \frac{3}{2}\frac{\tau_B}{\gamma_m h}$$

in which h_f is the head loss in distance L. With additional transformations

$$h_f = \frac{96}{4\rho_m Uh} \frac{L}{4h} \frac{U^2}{2g}$$

(13.10)

$$\eta(1 + \frac{1}{2}\frac{\tau_B h}{\eta U})$$

If μ_e denotes the effective viscosity

$$\mu_e = \eta(1 + \frac{1}{2}\frac{\tau_B h}{\eta U})$$

(13.11)

and the Reynolds number has the form

$$\text{Re} = \frac{4\rho_m Uh}{\mu_e}$$

(13.12)

then

$$h_f = \frac{96}{\text{Re}} \frac{L}{4h} \frac{U^2}{2g}$$

(13.13)

In terms of the friction factor

$$f = \frac{8ghJ}{U^2} = \frac{8gh}{U^2} \frac{h_f}{L}$$

(13.14)

A comparison of Eq. (13.13) with Eq. (13.14), yields the relationship

579

$$f = \frac{96}{\text{Re}} \qquad (13.15)$$

This is the relationship between the friction factor and the Reynolds number for ordinary laminar flow. That is, if Reynolds number is written in the form of Eq. (13.12), the relationship between friction factor and Reynolds number for a Bingham laminar flow is the same as it is for a Newtonian laminar flow.

The usual relationship between f and Re for experimental data in a flume is plotted in Fig. 12.28. In the figure, the data for laminar flow follow the line for Eq. (13.15), and those points for turbulent flow are slightly lower than the line for flow in a smooth channel.

The flow discussed in the preceding paragraphs is nearly two-dimensional. Experimental data for three-dimensional flow obtained at Northwest Hydrotechnical Scientific Research Institute (NWHRI) and Northwest Agricultural College (NWAC) indicated that the coefficient in Eq. (13.15) varies with the shape of cross section. For U-shaped and rectangular flume, the constant is 84, and for a trapezoidal flume with a side slope of 1:1, it is 75 [6,7].

Second, if rigidity is used in the formation of Reynolds number, Eq. (13.8) can be written in dimensionless form as follows:

$$\frac{1}{\text{Re}} = \frac{f}{96} - \frac{1}{8}\frac{He}{\text{Re}^2} + \frac{8}{3f^2} \cdot \left(\frac{He}{\text{Re}^2}\right)^3 \qquad (13.16)$$

in which

$$\text{Re} = \frac{4\rho_m Uh}{\eta} \qquad (13.17)$$

$$He = \frac{16\rho_m \tau_B h^2}{\eta^2} \qquad (13.18)$$

Fig. 13.7 shows relationships between friction factor and Reynolds number for various values of the Hedstrom number He. In the laminar flow region, the measured data scatter but are near the theoretical curves for the corresponding Hedstrom numbers; the data for turbulent flow with a smooth boundary fall a little below the curve for clear water.

From a synthesis of the two different expressions, one can draw the following conclusions:

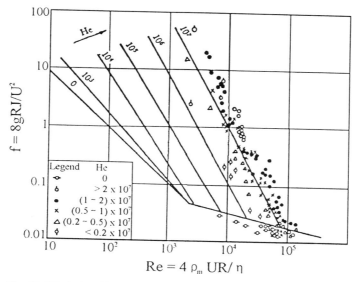

Fig. 13.7 Relationship between friction factor for hyperconcentrated flow
and Reynolds number with Hedstrom number as parameter

(1) For turbulent flow with a smooth boundary, the data fall below the curve for clear water. Thus, for the same Reynolds number, the friction factor of a hyperconcentrated flow is near or a little less than that for clear water flow. But for the same velocity and depth, the absolute value of head loss in a hyperconcentrated flow is larger than that of a clear water flow. This result is explained in more detail in Chapter 12.

(2) For the laminar flow region, friction factor increases greatly with the decrease of the Reynolds number. For the same velocity and depth, the head loss of a hyperconcentrated flow is much higher than that of a clear water flow.

(3) If the Reynolds number has the form of Eq. (13.12), the critical Reynolds number at which a laminar hyperconcentrated flow transforms into a turbulent one is the same as that for clear water flow. That is, the transition occurs at $Re = 2,000$. If the Reynolds number has the form of Eq. (13.17), and depends on Hedstrom number, the critical Reynolds number varies within the range from 2,000 for $He = 0$ to 10^5 for $He = 10^7$. In contrast to clear water flow, hyperconcentrated flow transforms between turbulent and laminar at a higher velocity and a larger depth.

2. Rough boundary

As for clear water flow, the friction factor for hyperconcentrated flow does not vary with the Reynolds number; it is a function only of relative roughness Δ/h. From the experimental results of Zhang and Reng for fully developed turbulent flow[8],

$$\frac{1}{\sqrt{f}} = 2\log\frac{h}{\Delta} + 2.12 \quad (13.19)$$

Since the viscosity of a hyperconcentrated flow and the corresponding thickness of the laminar sublayer are much greater than those of clear water flow, the absolute roughness for hyperconcentrated flow to become turbulent is also much larger.

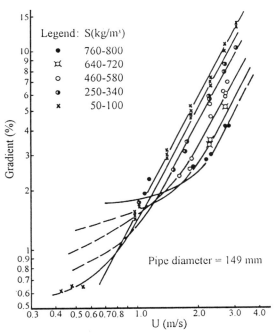

The resistance loss for a hyperconcentrated flow with a smooth boundary is larger than that for a clear water flow, as mentioned above. The situation for a hyperconcentrated flow with a rough boundary is the opposite. Because of the siltation of the slurry on the boundary and thus make it smoother, the head loss of a hyperconcentrated flow is less than that of a clear water flow. Fig. 13.8 presents the

Fig. 13.8 Comparison between pressure gradient for clear water in a rough pipeline with that for hyperconcentrated flow with sediment particles fully suspended

experimental results for flow in a rough pipeline obtained by NWHRI [1]. For a given velocity and pipe diameter, the higher the concentration, the lower the pressure gradient that is required. The gauging section at the Baiyangshu in Baojixia Irrigation District is located in a concrete-lined trunk canal with some deposits of coarse bed load material. The Manning n for the canal was 0.03 for the flow of clear water. As the concentration in the flow increased to 340 kg/m^3, a layer of smooth silt deposited on the bottom, and the Manning n dropped to 0.014 [2].

In natural rivers, the sediment size increases with an increase of concentration. If the longitudinal slope of a canal is not steep enough for all of the coarse particles to be conveyed, some coarse particles deposit and dunes form on the canal bed, and they cause a large increase in the roughness of the canal. In Luohui Irrigation District, the

---

[1] Jiang, Suqi, Dongzi Sun. "Study and Basic Characteristics of Hyperconcentrated Hydrotransport in Pipeline and Its Basic Characteristics." *Report of Northwest Hydrotechnical Scientific Research Institute*, 1979, p. 16.

[2] Baojixia Weihui Irrigation Bureau, Shaangxi Province. *Progress Report, Study of Sediment Transport in Hyperconcentrated Canal Flow in Baojixia Weihui Irrigation District*. 1978, p. 21.

following situation was observed: with a hyperconcentrated flow in an earth canal, the roughness increased from 0.017 to 0.025 because of the deposition of coarse particles [9].

13.1.2.4 Flow instability

In flume experiments, as the concentration rises beyond a critical value, the rigidity or stickiness coefficient increases sharply, Fig. 2.29. If the flow transforms into a laminar flow of a homogeneous slurry, the stages at various cross sections often fluctuate periodically. The period can be short, of the order of 3 to 5 minutes, or as long as half of an hour to an hour. Such pulsating flows are closely related to the formation, development and destruction of a stagnation layer near the bottom of flume. Fig. 13.9 is an example of a pulsating flow that are observed at the IHR, YRCC [2,5]. If the flow was not high, a layer of slurry often stagnated on the bottom. Differing from the deposition in a common sediment-laden flow, this layer did not form a rigid bed but maintained a certain fluidity. Because of the development and thickening of the stagnant layer, the water level rose accordingly. The thickening of the stagnant layer and the raising of the water level happened mainly in the upper and middle reaches of the flume. The result was a steepening of the surface slope and a slight increase in velocity. Once it reached a certain thickness, the stagnant layer collapsed rather suddenly, and the water level dropped accordingly. The surface slope was then flatter and the velocity was slightly less. Subsequently, a new stagnant layer formed and the cycle repeated itself, thus causing the periodical fluctuation of the water level.

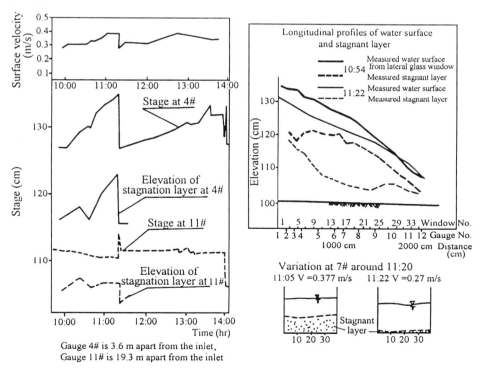

Fig. 13.9 Pulsating phenomenon of hyperconcentrated flow as observed in a flume

583

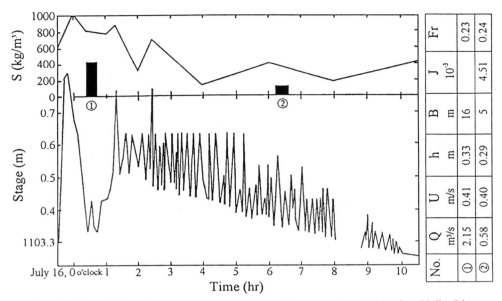

Fig. 13.10 Instability of a hyperconcentrated flow at Lanxipu Gauging Station, Heihe River (tributary of Puhe River)

In some small tributaries of the Yellow River, such unstable phenomena in hyperconcentrated flows have also been observed. Fig. 13.10 shows a periodical fluctuation of the water level that was recorded at Lansipo, Heihe River in July of 1967 [3]. The period of its cycle was 8 to 10 minutes and the amplitude was 15 to 26 cm. The corresponding Froude number was only about 0.23. These regular fluctuations reveal an inherent instability of the flow.

Study of the formation of unstable flow is based on the hypothesis that a small disturbance appears on the water surface at first; then one can analyse whether the disturbance develops or is attenuated during propagation. In the former case, a series of periodical rolling waves finally form. From this concept, Dracos and Glenne analysed the stability of various kinds of open channel flows, such as steady and unsteady, uniform and nonuniform flows [10]. A steady, uniform flow in an open channel that is wide and shallow loses its stability if the Froude number (Fr) has the value

$$Fr = \frac{p}{1+q} \qquad (13.20)$$

in which p and q are exponents in the appropriate equation of motion

$$Fr = U / \sqrt{gh}$$

$$J \propto U^p / h^{1+q}$$

584

From the law of resistance for hyperconcentrated flow with laminar pattern, depending on Hedstrom number, p and q vary in the following ranges:

$$p=0.05 \sim 1 \qquad q=1 \sim 1.95$$

Hence, a hyperconcentrated flow can lose its stability for Froude numbers in the range of 0.02~0.5. The larger the Hedstrom number, the more readily the pulsating phenomenon occurs. For turbulent flow in a smooth conduit,

$$p=1.76 \qquad q=0.24$$

The corresponding Froude number is 1.42. Thus, turbulent hyperconcentrated flow is more stable than laminar hyperconcentrated flow.

13.1.3 Characteristics of sediment motion

13.1.3.1 Two extreme modes of hyperconcentrted flow

Hyperconcentrated flow can be classified according to the size distribution of sediment particles transported. Two extreme modes are the following:

1. Mode I — Hyperconcentrated flow consisting mainly of fine particles such as silt and clay.

Hyperconcentrated flows that consist primarily of fine particles behave like a non-Newtonian fluid, usually of the type called a Bingham fluid. As concentration increases, flocculation structure among sediment particles forms rapidly and the viscosity increases sharply. During settling, no sorting occurs and the particles move as an integral mass; as they settle a clear interface forms between the slurry and the clear water. The settling velocity of the interface may be only one hundredth or even as low as one thousandth of the fall velocity of a single particle. In fact, the whole fluid becomes a homogeneous slurry. According to the experimental data of Wan, Chien, et al., for a sample with $D_{50} = 0.0015$ mm, in which 85% of particles were finer than 0.01 mm, the fluid became a homogeneous slurry at a concentration only *90* kg/m^3 [2].

An important question is, by which force are the weight of sediment particles supported? If a particle settles in a fluid with flocculation structure, the flow pattern around the particle is quite different from that in clear water, as shown in Fig. 13.11 for laminar flow [11]. The region with curved streamlines is limited to a small zone near the particle; outside the region marked by the dashed line, the fluid is undisturbed. As the ambient flow velocity increases, even if eddies form behind the body, the rotating speed of the eddies is much less than it would be in clear water. Shishenko suggested that the dashed line marks the surface (of a rotator) on which the shear stresses in the fluid are equal to the Bingham yield stress [12]. He assumed that, for a sphere in a state of balance in a fluid with structure, the maximum shear stress acting on the surface of the sphere in such a fluid was proportional to the submerged weight of the sphere and inversely proportional to the surface area of the sphere; that is,

$$\tau = k \frac{W'}{\pi D^2} = \frac{kD(\gamma_s - \gamma_f)}{6}$$

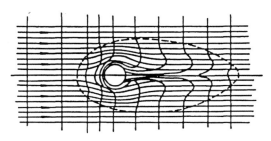

Fig. 13.11 Flow pattern of a structured liquid passing around a sphere(after Frisch, H. L., and R. Simha)

in which γ_s and γ_f are the unit weights of sphere and fluid, respectively, and k is a coefficient. At equilibrium, $\tau = \tau_B$, so that

$$\tau_B = \frac{kD_0(\gamma_s - \gamma_f)}{6}$$

In other words, the maximum diameter of a sphere that does not settle in a fluid with structure is given by

$$D_0 = \frac{6\tau_B}{k(\gamma_s - \gamma_f)} \qquad (13.21)$$

From this experimental data, k is a function of sphere diameter and varies in the range of 0.3~0.6. According to studies at IHR, YRCC and at NWHRI, k is about 1.05 [2]. If the coarsest sediment particle in a fluid is smaller than the size given by Eq. (13.21), the sediment-laden water will continue to move as long as the longitudinal slope or pressure gradient is large enough to overcome the resistance. The capacity of sediment transport that is commonly referred to in ordinary sediment-laden flows is meaningless in such hyperconcentrated flows.

2. Mode II — Hyperconcentrated flow consisting mainly of particles coarser than fine sand.

Hyperconcentrated flow without fine particles that form a structure maintains the properties of Newtonian fluid except for those with extremely high concentrations[1] . In

---

[1] Studies of this phenomenon give rise to some confusion. Many experimental data seem to show that a mixture of water and hyperconcentrated coarse particles also behaves as a Bingham fluid, one with a rather high yield stress. These data contradict the results observed by Chien and Ma, and those by Migniot cited in Chapter 2. According to the latest research of Fei, the conduct of rheological measurement of a mixture with a number of coarse particles by means of a capillary viscometer requires a strong stirring in order to maintain a uniform concentration of coarse particles. Such stirring causes a rather large local dissipation of energy. If the energy loss is not properly accounted for, the rheogram would indicate too large a yield stress. However, a corrected rheogram would indicate the properties of a

this extreme case, although there is a Bingham yield stress, its absolute value is usually comparatively small. The fall velocity of sediment particles decreases as the concentration increases, but compared with the above-mentioned case the decrease is much smaller. Sorting occurs during the settling of coarse particles. In this sense, the flow is a two-phase one, and sediment particles move both as bed load and suspended load. As the concentration increases, the intensity of turbulence decreases, and the shear-induced dispersion force among particles becomes larger and larger. Finally, the turbulent flow is transformed into laminar flow, and the weight of the sediment particles is supported entirely by the dispersive force. The vertical distribution of concentration then becomes nearly uniform. The motion is just the laminated load described in Chapter 5; Bagnold conducted special experiments on this subject[13].

Bagnold conducted experiments in a small flume using spheres with a diameter of 1.36 mm that were made of a mixture of paraffin wax and lead stearate. The specific weight of the spheres was only 1.004 t/m^3. In the experiments, the water depth was kept at a constant value of 60 mm, and the shear stress was increased gradually by increasing the slope. Since sediment particles were rather coarse, the fluid remained Newtonian and no Bingham yield stress occurred until volume concentration of sediment reached 60%. His experimental results are shown in Fig. 13.12. In the figure, turbulence in region cd was not completely damped, therefore the transport rate was proportional to 3/2 power of shear stress. As ϕ_T increased gradually, the region of concentrated sediment motion had to expand into the main flow region because the water depth was limited and so was the concentration near the bed. Generation of turbulence was more and more restricted and the kinetic energy of turbulence already produced was exhausted in overcoming the concentration gradient. Thus, in the region de, the experimental data deviate from the trend represented by the curve $abcd$. A certain distance among particles was maintained in the existence of turbulence. When turbulence gradually diminished and particles kept on their motion supported by dispersive force, the distance among particles might be further reduced and the concentration further increased consequently. The tendency is shown in Fig. 13.12, that is, the rate of increase of the intensity of sediment motion increased even more sharply than that in the earlier stage when the turbulence had not been fully damped. When the mean volume concentration reached 25%, the general random turbulence began to be replaced by a more regular secondary flow in which water ascended near the side walls and descended in the middle of the flume. At a mean volume concentration of about 30%, the turbulence appeared to be much weaker. Simultaneously the vertical concentration profile became uniform. At a mean volume concentration of about 35%, both the turbulence and secondary flow disappeared entirely, and uniformly distributed sediment particles moved forward in a laminar flow. Due to the obviously much higher

---

Newtonian fluid. The details are contained in the report: "Rheological Measurement of the Slurry of Mining Tail from Qingjiang Mine." *Progress Report 80-1 of Sedimentation Lab*, Department of Hydraulic Engineering, Tsinghua University, 1980.

fluid viscosity, the velocity was lower. Therefore, although the concentration would further increase with the increase of shear stress, the parameter ϕ_T in terms of transport rate would decrease as $\tau/[(\gamma_S-\gamma)D]$ increased. In Fig. 13.12, the curve in the region *fgh* reverses. At still higher concentrations, the spacing between the particles was so small that they could no longer move; finally the phenomenon of "freezing" occurred, and sediment particles blocked the whole flume. The flow was quite unstable in the latter stages, and the limiting concentration varied over a wide range, reaching about 60% at its maximum.

13.1.3.2 Modes of hyperconcentrated flow in Northwest China

The properties of the hyperconcentrated flows in Northwest China place them somewhere between the afore-mentioned modes. The flows contain fine particles that form a structure and also a significant proportion of coarse particles. The further north the rivers are located, like the Huanfuchuan River, Kuye River, Wuding River, the coarser the particles are.

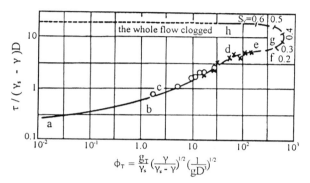

Fig. 13.12 Relationship between flow condition and sediment transport rate, including laminated load (after Bagnold, R. A.)

Whenever the concentration increases to some limiting value, the fine particles and water form a homogeneous slurry, and the coarse particles in the slurry settle out freely. The whole flow still maintains features of two-phase sediment-laden flow. The only difference is that the liquid phase is a slurry, consisting of a mixture of fine sediment particles and water, rather than clear water alone. For still higher concentrations, more and more coarse particles join the flocculent structure and become a part of the slurry. For concentrations above a certain critical value, the whole flow is a homogeneous slurry. As pointed out in Chapter 3, the coarser the particles are, the higher the critical concentration will be.

The occurrence of laminated load leads to large amounts of bed load, on the one hand, and to an expanded region for the bed load, on the other. For pipe flows, the resulting increase of shear stress is overcome by a larger pressure gradient because the depth of the flow can not increase, as in open channel flow. Ultimately the bed load can occupy the whole space, and the flow is then laminar flow carrying laminated load. In natural rivers, the slope usually varies slightly, and the increase of shear stress is caused by an increase in the depth. Therefore, the situation that the bed load occupy the whole space and the whole flow is laminar can hardly occur. However, in the area that supplies coarse sediment to the Yellow River, the liquid phase is a slurry with a high viscosity that favours the transition from turbulent to laminar flow. Then the coarse particles move forward as laminated load.

13.1.3.3 Suspended load movement

1. Vertical distribution of concentration of suspended load

The suspended load generally dominates transport of sediment in two-phase turbulent flows with hyperconcentration. Vertical profile of concentration in such flows can be described by diffusion theory [Eq. (10.16)]. Five sets of data obtained at Dingjiaguo Gauging Station, Wuding River are plotted in Fig. 13.13. The ordinate is *(h-y)/y* and the abscissa is total sediment concentration, both in logarithmic scales. The data are well distributed along straight lines. Fig. 13.14 shows the profiles of the concentration of various grain sizes for two sets of data. These partial concentrations can also be described in terms of diffusion theory. Because high concentrations significantly reduce the fall velocity, the vertical concentration profile at hyperconcentration is extremely uniform.

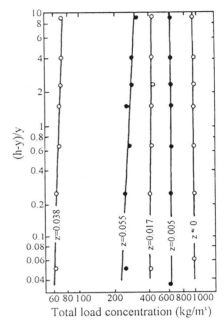

Fig. 13.13 Total load concentration profiles at Dingjiagou Station, Wuding River

Fig. 13.15 is a comparison of the measured and the theoretical values of exponent *z*. In the calculation, ω is taken as the gross fall velocity corresponding to median diameter of nonuniform sediment, which is the reduced fall velocity due to high concentrations. The data, except those corresponding to concentration higher than 900 kg/m³, scatter on both sides of the 45° line. The vertical concentration distribution is rather uniform for concentration exceeding 900 kg/m³. Very likely, at that time the flow was just transforming from turbulent to laminar.

2. Variation of suspension energy with sediment concentration

Turbulent flow has to spend a part of its energy for supporting the sediment suspension. The energy spent per unit time to support the suspension of sediment particles is called suspension energy, and it is expressed in Eq. (13.2).

Since the gross fall velocity decreases with the increase of sediment concentration, suspension energy does not always increase with an increase of concentration. For uniform sediment, if Eq. (3.38) is used to express the effect of concentration on the fall velocity, Eq. (13.2) can be written as

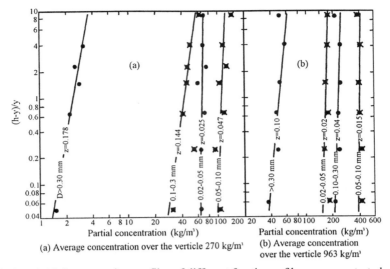

Fig. 13.14 Concentration profiles of different fractions of hyperconcentrated
flow at Dingjiagou Station, Wuding River

$$\omega_d = (\gamma_s - \gamma)S_v(1 - S_v)^{m+1}\omega_0 \qquad (13.22)$$

If the expression for ω_d is differentiated with respect to S_v, the derivation is set equal to zero, one obtains the concentration at which the suspension energy is a maximum:

$$S_{vo} = \frac{1}{m+2} \qquad (13.23)$$

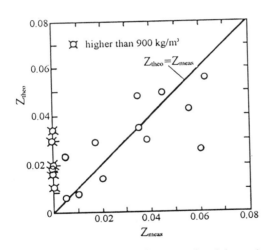

Fig. 13.15 Comparison of measured and theoretical
suspension indexes
(at Dingjiagou Station, Wuding River)

The value of critical concentrations are listed in Table 13.1 for various sizes of uniform sediment.

From the table, the coarser the particles are, the higher is the concentration at which the suspension energy reaches maximum. In natural rivers most suspended sediment particles are finer than 0.10 mm; for them, the critical concentration varies within a narrow range, for which it is about 400 kg/m³. At concentrations higher than 400 kg/m³, the addition of more sediment particles would not increase the energy consumption of the flow.

Table 13.1 Critical concentration for uniform sediment of different sizes, at which the suspension energy is a maximum

D (mm)	m	Concentration, at which the suspension energy is a maximum	
		in volume	kg/m³
≤0.05	4.91	0.145	385
0.10	4.80	0.147	390
0.25	4.38	0.157	416
0.50	3.57	0.179	475
1.00	2.92	0.203	538
≥2.86	2.25	0.235	612

In nature, the sediment size in a hyperconcentrated flow becomes coarser as the concentration increases, Fig. 13.2. Therefore, the critical concentration at which the suspension energy reaches its maximum should be higher than that for uniform sediment. The critical concentration should be determined in each specific application.

3. Sediment-carrying capacity of suspended load

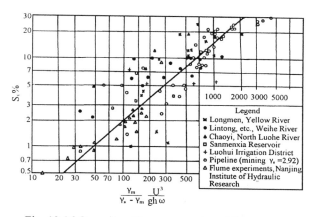

Fig. 13.16 Capacity of hyperconcentrated flow to carry
coarse sediment in suspension

As pointed out in Chapter 10, a commonly used relationship for suspended-sediment transport is an expression of the concentration of suspended load as a function of the flow intensity index $U^3/(gH\omega)$. Scientists of NWHRI plotted the relationship between the volume concentration of the bed material load in the suspended load of a turbulent hyperconcentrated flow and the parameter

$$\frac{\gamma_m}{\gamma_s - \gamma_m} \frac{U^3}{gh\omega}$$

in which γ_m is the specific weight of the sediment-laden flow. Only if the effects of sediment concentration on both fall velocity and specific weight have been taken into account, does the sediment carrying capacity of bed material load follows the same law, whether the concentration is high or low (Fig. 13.16).

13.1.3.4 Bed load movement

Bed load movement in hyperconcentrated turbulent flow is essentially bed load movement of coarse particles carried by slurry. A slurry differs from clear water in the following ways:

1. Specific weight of slurry is larger than that of clear water. Hence, the drag force along flow direction is larger, and so is the force due to buoyancy;

2. Viscosity of a slurry is higher than that of clear water. Once sediment particles are picked up from the bed by a slurry, their fall velocities are smaller than those in clear water;

3. Turbulence is less intense in a slurry than in clear water. Consequently, on the one hand, the effective force acting on sediment particles lying on bed is reduced. On the other hand, the resistance for a flow with a rough boundary is less because the laminar sublayer is thicker, and therefore the relative roughness is less.

The overall effect of these three factors is to make the bed load carrying capacity of slurry flow larger than that of clear water flow.

Fleshman put concrete blocks and spheres of different sizes on fixed beds and conducted experiments with both water flow and slurry flow; he then compared the carrying capacity of the two flows for a given velocity [15]. If D_c and D_s denote the maximum size of concrete blocks that can be moved by water flow and by slurry flow, respectively, then the relationship between the ratio D_c/D_s and the flow velocity is as shown in Fig. 13.17. The maximum weight of concrete blocks that can be moved by the mixture of fine sediment particles and water is dozens of times larger than that by the clear water. The larger the effective viscosity of the slurry and the lower the flow velocity, the larger this ratio is. In the field data for Baojixia Irrigation Districts, the bed load carrying capacity of hyperconcentrated flow was 45-87 times larger than that of low concentration flows.

Fleshman experiments demonstrated the ability of flow to move the large stones that protrude from the bed. Rabkova conducted bed load experiments with slurries at concentrations up to 450 kg/m^3 flowing over a sand bed [16]. His results revealed that the size composition of bed material varied because some sediment particles finer than 0.02 mm filled the interstices among the sand particles in the bed and even covered parts of the bed. The consequences were the cessation of dune movement and the formation of a plane bed. This process led to a reduction of bed roughness, a more uniform velocity profile, an increase in velocity near the bed, and a corresponding increase in the capacity to carry the bed load. In some cases, part of fine silt and clay carried by the hyperconcentrated flow deposited among the sand particles on bed, and possibly, the cohesion of the surface bed material may have increased and the intensity of the bed erosion reduced in consequence. Wright observed and compared bed erosion by clear water and that by a slurry consisting of kaolinite and water [17]. He used two kinds of bed in his experiments. The experimental results showed that, for the same rate of erosion, the velocity of a slurry should be 10~25% larger than that of clear water.

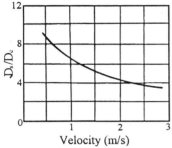

Fig. 13.17 Comparison of the maximum sizes of bed load particles carried by mud flow and by clear water flow

With a gradual increase of the concentration and the consequent gradual attenuation of turbulence, sediment motion is transformed from normal suspension toward lamination. This process is discussed in the following section on water-debris flow.

13.1.3.5 Clogging phenomenon

If the sediment concentration exceeds a certain limit, a sudden drop in the flood discharge and flow velocity may sometimes cause the flow to stop moving and stagnate locally. People in Northwestern China refer to this type of blockage in hyperconcentrated flow as clogging.

On June 17, 1963, this type of clogging was observed at Lijiahe Guaging Station, Xiaoli River. The flood peak reached the gauging station at 19:45. As the flood receded, the maximum concentration reached 1,220 kg/m³, the highest concentration recorded at that gauging station. The velocity of the flow became smaller and smaller. Finally it stopped moving altogether, and no noise of flowing water could be heard. The clogging, started at 21:48 on the 17th and ended at 16:18 on the 18th; thus it lasted for eighteen and one-half hours. The flow did not stop for the entire period; rather, it stopped and started several times, or it became somewhat larger or smaller intermittently. Suddenly the flow increased dramatically for a period, and a bore moved through the region in the downstream direction. The bore broke up the stagnated mud at the upstream bend with a loud noise. After a while, the bore reached the location of the gauging station and the water level there rose. A little later, it reached a bend downstream and broke the stagnated mud there, with more loud noise. After a while, the noise level diminished gradually, as did the velocity of the flow. Once again, clogging occurred. Intermittent changes in the flow occurred six times between 21:48 and 24:10. The water level rose some 0.1 to 0.5 m and then dropped down again during the bore cycles; times for the rising and dropping of water level were two to four minutes.

The same kind of intermittent flow has also been demonstrated in flume experiments. During clogging, the longitudinal profile had a convex shape. After the slurry stopped moving, the continuous supply of slurry caused the water depth at the entrance of the reach to increase gradually and the slope of the water surface to increase consequently. After a while, the stagnated slurry started to flow again. The flow restarted at the entrance of the flume and the moving slurry there forced the stagnated slurry downstream to move as well, with the initiation of flow propagating from upstream to downstream in a kind of wave. After a while, the water level in the flume was lower and the slope flatter, and then the flow ceased again. The cycle of clogging and flowing repeated itself several times.

In the clogging stage, the concentration was so high that the sediment motion could be classified as either mode I (the motion of a homogeneous slurry), or mode II (laminated flow). In the former case, the flow could not be maintained if the shear stress acting on bed was less than the Bingham yield stress. From Eq. (13.8), if $\gamma_m\,hJ = \tau_B$, then q is zero. In the latter case the porosity among particles was so low that the water and sediment could not sustain an ordinary motion. It was not a problem of sediment carrying capacity. Using glass fragments of size 0.21~0.70 mm as sediment, Batin and Streat conducted experiments in a pipeline. Their results revealed that the maximum volume concentration of solid material that could be carried by the flow was 59% [18] , a value that was just between the concentration for loose contact after deposition (= 57%) and that for compact packing (= 62.4%).

No matter which mode of sediment motion the clogging belongs to, the coarser the sediment composition, the higher the critical concentration at which clogging occurs, or the maximum concentration that can be carried by the flow, Fig. 13.1. From the viewpoint of the Bingham yield stress for mode I, the more fine particles in the sediment, the larger the Bingham yield stress is; correspondingly, for given flow conditions, the lower the concentration at which clogging occurs. For mode II, for a given concentration, the finer the sediment, the smaller are the spaces among sediment particles; consequently, the whole flow clogs at lower concentrations.

13.1.3.6 Tearing bottom phenomenon

At Longmen Station on the main stem of the Yellow River and at Lintong Station on the tributary Weihe River, drastic bed scouring sometimes took place during the passage of hyperconcentrated floods. Large pieces of the river bed as much as a meter thick were lifted so that they towered over the water surface and then collapsed back into it, or a piece of deposited material was rolled up like a carpet and then carried away by the flow. The river bed could be eroded in this manner to depth of several meters or even up toward ten meters during a large flood. Such intensive scouring with hyperconcentration is called the "tearing bottom" phenomenon by the local people.

The process of lifting large chunks of river deposits is similar to the mechanism of intensified bed load motion by hyperconcentration that is already described in Section 13.1.3.4. The drag force acting on river bed equals $\gamma_m hJ$, and the specific weight of hyperconcentration flow is much larger than that of clear water. Therefore the drag force of a highly concentrated flow is proportionally much larger than that of a clear water flow. Generally the dry specific weight of river deposits is about 1.4 t/m³, corresponding to a saturated specific weight of 1.87 t/m³. Hence, for flow with a concentration of about 1,400 kg/m³, the deposits could float on the slurry surface. For the hydraulic parameters at Longmen Station, the equivalent diameter of river deposits that can be lifted up can be calculated from the Shields curve for the threshold drag force, Fig. 8.6. From it, the maximum equivalent diameter of river deposits that can be eroded during the peak discharge is about 1.3 m, and the tearing bottom phenomenon occurs just at the moment, the equivalent diameter reaches its maximum. Because the flow is hyperconcentrated, the eroded sediment can be easily carried away by the flow without increasing unduly the burden of the flow. The analysis also shows that the tearing bottom occurs more readily in a homogeneous hyperconcentrated flow (mode I) [41].

With the tearing of the bottom, sediment is eroded in large chunks rather than in individual grains. Perhaps the amount of fine particles in the bed deposits is sufficient to make the deposits cohesive. During the processes of deposition, local weak spots could form. At present, however, we lack the data necessary to prove such an hypothesis.

13.1.3.7 Difference in hyperconcentrated flows from areas of coarse and fine sediment

In the Yellow River basin, the maximum concentration of hyperconcentrated flow that originate from a source area of coarse sediment may reach 1,600 kg/m³, whereas that from source areas of fine sediment usually is less than 1,000 kg/m³. The difference in the mechanism of hyperconcentrated flow motion originating from two source areas arises from different aspects of sediment motion in hyperconcentrated flow.

As already mentioned, if hyperconcentrated flow approaches or actually becomes laminar, two different modes of motion can occur. The key point for the first mode is the concentration at which the whole flow turns into a single-phase homogeneous slurry. From Eq. (13.21), this critical concentration S_c depends on the size composition of the sediment. From experiments on settling in quiescent water, Zhang et al. suggested the following empirical formula for mine tailings and loess in Northwest China [6]:

$$S_v = 390(D_{50} \cdot \Delta p)^{0.61} \qquad (13.24)$$

where: S_c is critical concentration, at which a homogeneous slurry is formed, in kg/m³; D_{50} median diameter of the sediment carried by the flow, in mm; Δp percentage finer than 0.007 mm for sediment carried by the flow.

For hyperconcentrated flow with shear, the structural strength caused by the existence of fine particles is less than that it is if hyperconcentrated slurry is stationary [19]; therefore, the critical concentration for which a homogeneous slurry forms should be larger than that given by Eq. (13.24). In contrast, the maximum concentration for sediment motion mode II depends on the fluidity of the water-sediment entity. Researchers studying debris flow in China suggested a concept of free porosity [20] :

$$e_r = 1 - \frac{N_s(1 - N_e)}{N_e(1 - N_s)} \qquad (13.25)$$

in which N_e is fluid porosity, i.e., volumetric percentage of the water in debris flow (the volume occupied by air is considered to be neglected); N_s porosity under the condition of loose contact among particles after deposition, by volumetric percentage.

Table 13.2 Free porosities of solid particles for the limiting
sediment concentration a flow can carry

Author	Sediment used in experiments	Measured S_{vm}	N_e	N_s	e_r
Bagnold	1.33 mm balls (paraffin wax-lead stearate)	0.60	0.40	0.42[(1)]	-.086
Bartin & Streat	0.21~0.70 mm glass slack	0.59	0.41	0.424[(2)]	-.060
Cloete et al.	0.36 mm sand	0.539	0.461	0.489[(2)]	-0.12
Cloete et al.	0.36 mm glass balls	0.620	0.38	0.439[(2)]	-0.28

Remark:1. The figure is deduced from Table 2.7 (loose porosity for wet lead ball) 2. The porosity of deposit of particles freely setting in quiescent water.

Free porosity reflects the compactness of sediment particles. The larger the e_r is, the more the degree of freedom of solid particles. If sediment particles that are already in loose contact with each other while moving, the free porosity is zero. The free porosities at the moment of blockage or for flows approaching blockage are listed in Table 13.2 from flume experiments conducted by Bagnold, pipeline experiments conducted by Bantin and Streat and pipeline experiments conducted by Cloete [21]. They show that the free porosities are negative if the sediment concentration is near its maximum value.

Eqs. (13.24) and (3.25) can apply to identify quantitatively the mode of sediment motion at hyperconcentration, at least approximately. These equations have been applied to Huangfuchuan River, representative of a hyperconcentrated flow originating from an area of coarse sediment, and to the Weihe River, representative of a hyperconcentrated flow originating from area of fine sediment. The results are shown in Table 13.3. For the Weihe River, the entire flow approaches a homogeneous slurry if the concentration is about 800 kg/m³, but for this value the solid particles in the slurry are not in direct contact with each other. The inference then is that hyperconcentrated flow originating from a source of fine sediment behaves as mode I type of sediment motion. For the Huangfuchuan River, in contrast, the whole flow was not transformed into a homogeneous slurry for the much higher concentration of 1,500 kg/m³. [1) At this concentration, solid particles do directly contact each other. The inference in this case is that hyperconcentrated flow originating from a source of coarse sediment behaves as a mode II type of sediment motion.

Table 13.3 Free porosities of hyperconcentrated flow in the Weihe and Huangfuchuan Rivers, and the limiting concentration to form a homogeneous slurry

River	Gauging station	Measured S		Sediment		$N_e^{(1)}$	$S_c^{(2)}$	e_r
		kg/m³	S_v	D_{50}	ΔP		kg/m³	
				mm	%			
Weihe	Nanhechuan	735	0.277	0.032	84.4	0.330	710	0.222
Huangfuchuan	Huangfu	1500	0.566	0.384	96.8	0.556	3539	-0.04

Remark: 1. determined by the method in reference [22]; 2. calculated according to Eq. (13.24).

13.1.3.8 Canal design for the conveyance of hyperconcentrated flow

In large irrigation districts like the Jinghui, Luohui, and Weihui Districts of Northwest China, for a long time, water diversion was not permitted, or interrupted, if the concentration in the river exceeded 15% (by weight), i.e., 167 kg/m³, in order to avoid siltation in canals. In recent years, the limiting concentration was raised gradually in these irrigation districts because of the urgent need for water. The practice showed that high flow velocity is not necessary to convey hyperconcentrated flow. In an ordinary canal with a depth of about 1 m, an average velocity of 0.8 to 0.9 m/s is high enough to convey hyperconcentrated flow at a concentration of 50% by weight, without siltation [9,23]. Similar phenomena have been observed in other countries for flumes conveying tailings [24].

---

1) In this instance, sediment particles are nearly uniformly distributed in the flow and their fall velocities are extremely low. But the mechanism is different from that expressed by Eq. (13.24).

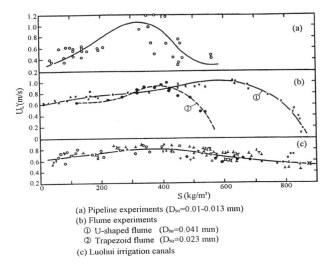

(a) Pipeline experiments (D₅₀=0.01-0.013 mm)
(b) Flume experiments
    ① U-shaped flume (D₅₀=0.041 mm)
    ② Trapezoid flume (D₅₀=0.023 mm)
(c) Luohui irrigation canals

Fig. 13.18 Relationship between non-siltation velocity and concentration

The relationship between suspension energy and concentration in Section 3.1.2.2 can be used to make predictions: for a certain slope, the velocity required for conveying hyperconcentrated flow decreases with an increase of concentration if the concentration is more than a certain limit. That is, a maximum value exists for the velocity in the relationship between non-siltation velocity and concentration. Measured data in flumes, pipelines, and field canals are shown in Fig. 13.18 [7,25]. In Fig. 13.18(c), data obtained from different canals are marked by different symbols, and they show that such a limiting value for the velocity does exist. The finer the sediment composition is, the lower the concentration for which the velocity reaches its maximum. The field and the laboratory results are significantly different. In the laboratory, the sediment composition does not change during experiments. But in the field, the sediment diverted from the rivers becomes coarser and coarser as sediment concentration increases(Fig.13.2). Hence, in canals, the variation of non-siltation velocity with concentration is not as large as it is in laboratory flumes. For this reason, the data range is so narrow that only left limb of the curves occurs. That is, the non-siltation velocity U_L' is an exponential function of concentration with the following form:

$$U_L' = \alpha S_w^\beta \qquad (13.26)$$

where U_L' is in m/s and S_w in percentage by weight. The coefficient α and the exponent β are 1.30 and 0.67, respectively, for Luohui Irrigation District; the corresponding values are 1.19 and 0.45 for Weihui Irrigation District.

Hydraulic conditions for preventing siltation in canals can be expressed in terms of parameters other than the velocity, as in the following relationship obtained for the Luohui Irrigation District: [1]

$$Q = A(\frac{n^2}{J})^{1.7} \qquad (13.27)$$

in which the terms used are: Q — diverted discharge, in m³/s; n — Manning friction factor; J — slope of the canal; A — coefficient, $A = 7.56 \times 10^6$ with $S_W = 20$-30%; $A = 25.2 \times 10^6$ with $S_W > 50\%$.

Fig. 13.18 reflects the fact that accumulated quantitative changes cause a qualitative change. In fact, the design of canals for conveyance of hyperconcentrated flow shows that two qualitative changes were caused by accumulated quantitative changes [2], and these are indicated by the particle fall velocity and by the flow patterns.

A flow spends a certain amount of energy in carrying sediment. For a suspended load, the work done to suspend sediment depends on the concentration and the fall velocities of the sediment particles. The trend with increasing concentration is for the concentration to be the dominant factor initially, so that the work done increases with the increase of concentration. Once the concentration exceeds a critical value (Table 13.1), the reduction in the fall velocity of particles induced by the further increase of sediment concentration becomes the dominant factor. An increase of concentration then reduces the work done for suspension instead. This is the first qualitative change caused by the accumulated quantitative changes with increase of concentration, and it shows as a peak in the relationship between non-siltation velocity and concentration, as shown in Fig. 13.18.

As concentration increases still further, the turbulence in the flow is progressively damped and the turbulent flow finally becomes laminar. Chapter 12 and this chapter have stated repeatedly that the resistance to flow for a smooth boundary is much larger once the turbulent flow becomes laminar. In order to prevent the occurrence of laminar flow, the flow velocity must be higher for concentrations in this range; that is, for this condition, the accumulated quantitative changes cause the second qualitative change in

---

[1] Louhui Irrigation Bureau. "Summary of Hyperconcentrated Irrigation in 1979," February 1980, p. 18.

[2] Chien, Ning, and Zhaohui Wan. "Discussion on The Suggestion of Regulating Sediment by Xiaolangdi Reservoir and Releasing Sediment in The Form of Hyperconcentrated Flow for Solving Aggradation Problem of Lower Yellow River." *Report of Institute of Water Conservancy and Hydro-Power Research*, 1979, p. 21.

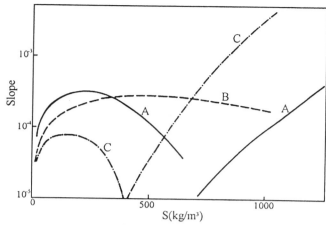

Fig. 13.19 Canal slope required for transport of
hyperconcentrated flow at various concentrations

that a minimum value of velocity occurs in the relationship between velocity and concentration.

Fig. 13.19 was used in the design of a concrete canal for the conveyance of hyperconcentrated flow. The flow velocity at a certain discharge was kept constant and the slope required for the canal to convey hyperconcentrated flow at different concentration was investigated. In the figure, curve A corresponds to hyperconcentrated flow with 12% of sediment particles finer than 0.01 mm. At first, the required slope gradually increases with increasing concentration, and it reaches its first extreme value (a maximum) for a concentration of about 300 kg/m³. As the concentration increases further, the required slope decreases. For concentrations greater than 700 kg/m³, the flow is laminar. In order to avoid this occurrence, a larger slope must be used; consequently, the curve for the required slope displays a second extreme value (a minimum). The shape of curve A depends greatly on the percentage of fine particles in the sediment. The figure also shows curves for 4% of fine particles (curve B) and for 36% of fine particles (curve C). With little fine material in the diverted sediment, the viscosity of the flow is low so that laminar flow is less likely to occur. Consequently, in the concentration range encountered in the design, the second critical condition does not occur. In contrast, if the content of fine particles in the diverted sediment is large, the flow easily becomes laminar, and the required slope is then comparatively large.

13.2 DEBRIS FLOW

13.2.1 General characteristics

Debris flow is a two-phase flow that is saturated by containing solid particles ranging from fine clay to large cobbles, and it originates in gullies or on steeply sloping land. A debris flow might form under the following conditions: (1) a large amount of loose solid material ready to be moved; (2) a steep slope; (3) intensive precipitation.

The solid material carried by debris flow may be the colluvial deposits of former landslides, landslides induced by earthquakes, deposits carried and deposited by glaciers, or products of intensive rock weathering. In some mountainous basins, an armouring layer formed by long-term flow action protects the surface, and the material underneath it can be much finer. If a large flood erodes this surface layer, the material underneath, which has a low resistance, can be deeply eroded and thus form a debris flow. The specific weight of debris flow formed in this way can reach 1.9~2.1 t/m³. [1)]

China is a vast country that includes large mountainous areas and has complex geological, geomorphological, and climatological conditions that favour the formation of debris flows. Thus it is one of the countries prone to the formation of debris flows. These can occur in the foot hills of the many mountain areas: Tianshan, Qilian, Kunlun mountains, Qinglin, Taihan mountains, West mountain area of Beijing, mountainous area in west Liaoning, and Changbei Mountain in Jilin. In addition, Southeastern Tibet, Transverse Crossing Mountains, West and Northeastern Yunan, and the Western mountainous area of Sichuan are typical areas prone to frequent and serious disasters caused by debris flow [26]. For example, in Xiaojiang basin in Northeastern Yunan with a catchment area of 3,120 km², debris flow takes place 500 to 1,000 times annually. In some years debris flows can occur as many as two to three thousand times [27]. Carrying huge stones and cobbles, debris flows sometimes move with high velocities. The maximum velocity recorded in China is 13.4 m/s; but reports indicate that the maximum velocity may have reached 20 m/s. Debris flows with such high velocities are extremely destructive. The measured impact force has reached 92 t/m² [28]. As it runs out of a gorge, a debris flow deposits a large amount of material that can form a big fan-like stone "sea"; bury villages and farmland near the creek; destroy roads, bridges, and hydraulic structures; block rivers; form shoals; and cause dam-breach floods that seriously threaten regions downstream.

13.2.2 Classification of debris flow

Debris flow can be classified in various ways: in terms of the geological and morphological characteristics of debris flow regions; according to its activity and intensity of destruction; according to its solid material it carries; by dynamic conditions of its formation; and according to its physical and mechanical properties. The last three classifications are the most significant ones for this study.

[1)] Feng, Qinghua and Daoming Xu. *Case Study of Debris Flow Action in a Small Mountainous Basin---Debris Flow in Hanlou Gully*. Langzhou Institute of Glacial, Frozen Soil and Desert Research, Academia Sinica, 1978, p. 27.

13.2.2.1 Classification according to the composition of debris flow

Debris flow can be classified in three categories according to the material they carry:

1. Mud flow — carries mainly sediment finer than coarse sand, but sometimes also a small amount of rock fragment. The hyperconcentrated flow at rather high concentration mentioned in Section 13.1 is mud flow.

2. Viscous debris flow — consists of large amount of fine particles (silt, clay, etc.) and also carries large stones, boulders, and cobbles.

3. Water-borne debris flow (non-viscous debris flow) — consists mainly of large boulders carried by water or dilute slurry. Such debris flow might form under conditions in which no fine particles or very few are present. [1]

Debris flows with different size compositions behave in special ways; one must study the composition of debris flows in order to understand the mechanisms of debris flow motion. In studies made abroad, the specific weight of water-debris flow did not exceed $1.1 \sim 1.3$ t/m^3. By contrast, in Hunshui Gully, a tributary of Dayinjiang River in Yunan Province, the measured specific weight of water-borne debris flow was 2 t/m^3, corresponding to a volume concentration of 0.61 [2] ; that is, the content of sand and boulders was 1,606 kg/m^3. The specific weight of a water-borne debris flow, consisting only of clear water, large stones and boulders, which once happened in Dahongluo mountainous area, Jingxi County, Liaoning Province also reached 2 t/m^3.

13.2.2.2 Classification according to dynamic conditions for formation of debris flow

Two categories of dynamic force can cause material stored on the ground to move as debris flow:

First, hydraulic debris flow — formed because of the abundance of solid material entering gullies following intensive erosion by surface runoff or dam breaching.

Second, gravitational debris flow — formed because over-saturated soil loses its stability and slides along the ground surface.

---

[1] Some researchers suggest that the minimum volume concentration of fine particles in hydraulic debris flow is 1% [30].

[2] South-Western Division of Railway Research Institute, Kunming Railway Bureau of Railway Ministry and Chengdu Geography Institute of Academia Sinica. *Synthesised Comment on Investigation, Survey and Experiments of Debris Flow in Santan along the Chengdu-Kunming Railway*. 1980, p.20.

The study of the first is within the field of the mechanics of sediment motion. The study of the second is in the field of soil mechanics. This section is limited to the discussion of hydraulic debris flow.

13.2.2.3 Classification according to physico-mechanical properties

Just as hyperconcentrated flow can be classified into a turbulent or a laminar, according to its physico-mechanical properties, so can debris flow be classified into these two basic flow patterns. The only difference is that the former is called non-viscous debris flow, and the latter viscous debris flow, or structural debris flow. Table 13.4 presents the classification for different regions according to this standard. The table shows that for various patterns of debris flow, the specific weight varies within certain ranges. The variation is caused by the difference in the criteria adopted for judging the flow patterns and by the difference in the composition of the local solid material. In order to eliminate this subjectivity, Chinese researchers of debris flow suggested the use of free porosity, Eq. (13.25), as the criteria for judging flow patterns [31], as in Table 13.4.

Table 13.4 Classification of patterns of debris flow

Category	Flow pattern	Specific weight (t/m³) in the following region				e_r
		Tibet Plateau	Xiaojiang Basin	Hunshui Gully	Beijing Mountain	
Sediment-laden flow	intense turbulent	—	—	1~1.35	—	0.7~1
Non-viscous debris flow	intense turbulent	1.10~1.80	1.50~1.90	1.35~1.69	1.60~1.90	0.37~0.7
Transitional	weak turbulent	1.80~2.00	—	1.69~1.94	—	0.1~0.37
Viscous debris flow	laminar	>2.0	2.0~2.4	1.89~2.24	1.90~2.30	<0.1

13.2.3 Characteristics of debris flow

13.2.3.1 Flow patterns and characteristics of deposition [26, 32]

First, non-viscous debris flow

Non-viscous debris flow is a two-phase turbulent flow. The liquid phase is a slurry, consisting of water and fine particles. The solid phase is the coarse particles: coarse sand and pebbles in the size range of 0.2 to 2 mm that are carried in suspension. Coarser particles roll and jump with velocities much lower than that of the slurry, the

Fig. 13.20 Overall view of the frontal
part of a glacial debris flow

Fig. 13.21 Close picture of the frontal part of a
glacial debris flow in Dongchuan, Yunan
(Chengdu Geography Institute, Academia Sinica)

noise of the colliding particles can easily be heard. With surface waves and splashing, the whole flow is obviously turbulent. Sometimes the motion is intermittent. But the wave amplitude of the intermittent flow is low (20 to 50 cm), and usually the flow is not completely interrupted between the waves.

Non-viscous debris flow normally runs smoothly. Only in reaches with sudden changes in the gully bed are concentrations of stones deposited, usually in crescent-shaped piles along the concave bank. Sometimes stones stop moving because of some local resistance. More and more stones then pile up, and the flow is forced to disperse and run in the depressions between piles. Deposits of stones quickly rise above the surface, like a dam suddenly arising from the bed. As the flow runs out from a gorge, the velocity of the non-viscous debris flow quickly drops and the solid material then deposits. Water and slurry continue to flow, but more slowly, and the coarse particles spread out like a fan with a flat surface.

Second, viscous debris flow

Generally, viscous debris flow are characterised by an intermittent pulsing. A single event can include from 10 to several dozens such pulses. The hydrograph of a debris flow in Jiangjia Gully, Dongchuan County, Yunan Province on July 23, 1966 is shown in Fig. 13.22 [32]. The front of the intermittent flow was high and steep. It consisted mainly of big stones, and it is called the "dragon head" by the local people; it has a convex tongue-shaped front with a steep slope. Fig. 13.20 is a view of the whole front and Fig. 13.21 is a close-up of the frontal part. This part, roaring as it flows, has a height ranging from a few meters to more than 10 meters as it rushes out of a gorge. As it approaches, the area is draped in a veil of dust. The slurry splashes, and the movement makes a loud noise; it can even cause the ground to tremble. Sometimes, at the beginning of an intermittent debris flow, wherever the debris flow passes, a layer of fresh slurry adheres to the deposits of prior debris flows. The original surface is dry and rough, but the debris flow makes the bed smooth. As the flow loses mass along the way, it becomes thinner and slower; finally it ends as a deposit on the bed of the gully with a leaf-shaped outline. Soon, the second intermittent flow comes, then the third one .... Each pushes further along than the former one. This process is called "bed paving". The period of these intermittent flows is in the range of 10 to 40 seconds. Their lengths vary from 50 m to 300 m. After the initial bed-paving process, succeeding debris flows run over the smoothened bed with higher velocities.

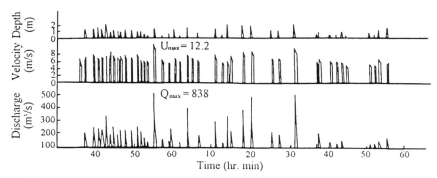

Fig. 13.22 Hydrograph of a debris flow in Dongchuan, Yunan, July 23, 1966

At a first glance, solid grains of various sizes in viscous debris flow appear to adhere to each other, forming a viscous entity, and moving forward with the same velocity. But with closer observation, one discovers that only particles finer than 10 cm move in this way. Coarser stones move intermittently, as they roll or jump along the bed. Big stones, some as large as 5 m in size, tend to concentrate in the front part and move forward in a kind of dynamic floatation.

Because of its high velocity and large inertia, viscous debris flow superelevates strongly in bends, and this action can cause cut-offs to form. A viscous debris flow will climb an adverse slope and still move strongly forward. In some cases, a debris flow will climb a terrace, cliff or guiding dike 5 to 10 m high and continue to run forward. If a viscous debris flow suddenly encounters an obstruction, its kinetic energy transforms into potential energy; the slurry and stones then splash to heights indicated in the following formula:

$$\Delta h = 1.6 \frac{U_c^2}{2g} \qquad (13.28)$$

in which U_c is the velocity of the frontal part of a debris flow [32]. According to observations in Jiangjia Gully, the values of Δh is 3 to 5 times the height of the frontal part.

In regions in which the content of fine particles is low, continuous viscous debris flow can also occur. In such cases, neither a definite frontal part forms nor does a pulsing motion occur. Such debris flows have been observed in Hunshui Gully, Dayinjiang River, Yunan Province[31].

Deposition of viscous debris flow can still maintain the dynamic structure formed during its motion and create rows of pebbles that are parallel to the flow direction. The piles of pebbles, which can be tongue-shaped, dune-shaped or island-shaped, usually have a flat or slightly convex surface, steep lateral slopes, and a steep front. Widely spread, isolated islands, in a vast "stone sea," are usually the deposited frontal parts of prior flows. These special movements and deposition patterns cause the ground surface in the alluvial fans to undulate; their patterns are completely different from the alluvial fans formed by floods or glacial deposits[26].

13.2.3.2 Sorting of debris flows carrying solid particles

Table 13.5 gives the averaged size compositions of various debris flows and original soils in Hunshui Gully. Fig. 13.23 shows the compositions of debris flow with different specific weights in Jiangjia Gully. They are similar to those of hyperconcentrated flow; that is, the more solid particles they contain, the coarser they are. The less the specific weight, the more pronounced is the sorting of the flow. For a laminar debris flow, all of the solid material on the ground surface moves in the form of laminated load; therefore, the size composition of debris flow does not differ from that of the original soil. One of the distinct characteristics of viscous debris flow is the large range of their size compositions, consisting of both coarse and fine particles. One consequence is that their specific weight can attain such large values.

Table 13.5 Average size compositions of various debris flows and of the original soils in Hunshui Gully

Category	Percentage of different sizes (mm)				D_{50} (mm)	$\sigma = (D_{84}+D_{16})/2$ (mm)
	>2	0.05--2	0.005--0.05	<0.005		
Sediment-laden flow	1.7	54.2	28.6	16.1	0.07	0.11
Non-viscous debris flow	6.0	61.7	19.4	12.9	0.18	0.49
Transitional debris flow	35.1	51.2	7.7	6.0	1.22	2.25
Viscous debris flow	52.1	37.5	5.9	4.5	2.50	4.97
Original soil	51.5	39.7	6.1	2.7	2.10	5.09

13.2.3.3 Longitudinal variation of the discharge of debris flow

Once triggered by precipitation, a debris flow possesses powerful erosive ability at the very beginning, and it can cut the river bed intensively. With the addition of the sediment thus eroded and the water in voids among sediment particles, the discharge of a debris flow increases along its course. The maximum discharge of a debris flow can be much larger than the initial discharge of clear water caused by precipitation. In a debris flow observed in former USSR, the clear water discharge that came directly from precipitation was only 49 m^3/s. After passing through a reach of 10 km and eroding the river bed, the flow attained a maximum discharge of 490 m^3/s, 10 times of the initial flow. The larger discharge greatly increases the harmful effects of the debris flow. Of the 441 m^3/s added from the river bed, solid particles and water constitute 54% and 46% respectively.

13.2.4 Mechanism of water-borne debris flow

Non-viscous debris flow is a homogeneous slurry carrying coarse particles. The homogeneous slurry consists of water, clay and silt if the specific weight of debris flow is low. The homogeneous slurry consists of water, clay, silt, and sand if the specific weight of debris flow is high.

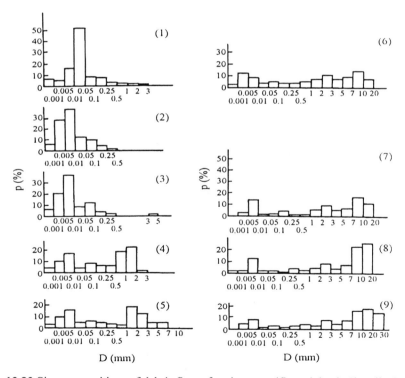

Fig. 13.23 Size compositions of debris flow of various specific weights in Jiangjia Gully

Explanations offered to describe the mechanism of motion of a hydraulically viscous debris flow diverge greatly. Briefly, the divergence results from the two modes of hyperconcentrated flow mentioned in the preceding paragraph. One concept indicates that the coarse particles are probably supported by the plasticity of the slurry due to the clay particles in the slurry [33]. Another concept is that viscous debris flow is a shear flow of solid particles, and the sediment particles are supported by the dispersive force among them.

As pointed out previously, the specific weight of viscous debris flow can be as high as 2.25 t/m³; for this value, water constitutes only 24% of the unit volume, and the solid particles are in direct contact with each other. Fragments of different sizes are densely packed, and the amount of slurry in the voids among solid particles is quite small. In such a state, the stones can hardly be supported by the slurry in the small voids. A more rational explanation is that, in the shear motion, the weight of stones is transferred directly to the bed through continuous collisions of stones. In fact, from his experiments, Fleshman pointed out that if a rather large body is suspended in structural liquid, the shear stress acting on the surface of the body is so small compared with the static pressure that it can be neglected. Thus, one can doubt that a body in such a suspension would still obey Archimedes law [15]. The more logical explanation for the motion of hydraulic debris flow is that it moves as laminated load. Recently, Takahashi in Japan reached a similar conclusion by analytical means [34].

In order to make the phenomenon more pronounced, let us take water debris flow without fine particles as an example to explain: conditions for forming water debris flow, the maximum content of solid material, the cause of the concentration of large stones in the frontal part, and the mechanism of forming intermittent flow.

13.2.4.1 Conditions for formation of water-borne debris flow [35]

Debris flows form if certain conditions prevail. For example, a flow with depth h_0 is formed after precipitation on ground sloping with an inclination θ, as in Fig. 13.24; the thickness of loose deposits underneath the original ground surface is d, the shear acting on the water and the deposits is τ, and the resistance among solid particles is τ_L. According to the relative magnitude of τ and τ_L, two cases are possible. For the one shown in Fig. 13.24a the shear τ over the depth is larger than τ_L, and therefore, all of the loosely deposited material will move under the action of the runoff. In Fig. 13.24b, only that part of the deposits within the depth range z_0 underneath the primary ground surface will be eroded and carried away. This analysis is based on certain assumptions: (1) the deposit is cohesionless; (2) the loose deposits are saturated when the direct surface runoff is formed; and (3) interstitial water in deposits seeps along the slope so that interstitial pressure does not exist.

The shear at a distance z below the primary bed may be written as:

$$\tau = [S_{v*}(\gamma_s - \gamma)z + \gamma(z + h_0)]\sin\theta \qquad (13.29)$$

where γ_s and γ are specific weights of the solid particles and water, respectively, S_{v*} is the part of the total volume that is occupied by solid particles in deposits in their static state. If loose deposits are acted upon by massive shearing stresses, the stress induced by deformation of the fluid among particles can be neglected. Then

$$\tau_L = \cos\theta[S_{v*}(\gamma_s - \gamma)z]\tan\phi \qquad (13.30)$$

in which ϕ is the angle of internal friction.

The two cases in Fig. 13.24 are the following:

CASE a

No matter how deep the deposit is, the condition for all the deposits to be moved by direct surface runoff is

$$\frac{d\tau}{dz} \geq \frac{d\tau_L}{dz}$$

From Eqs. (13.27) and (13.28),

$$\tan\theta \geq [\frac{S_{v*}(\gamma_s - \gamma)}{S_{v*}(\gamma_s - \gamma) + \gamma}]\tan\phi \qquad (13.31)$$

Bagnold derived the same critical condition early in 1956 [36].

CASE b

The condition that only loose deposit above a limiting depth z_0 can be moved is

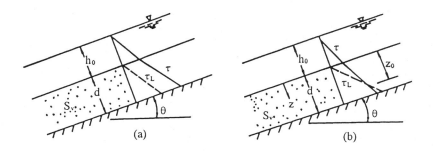

(a)   (b)

Fig. 13.24 Forces acting on loosely deposited material
during surface runoff (after Tamotsu Takahashi)

$$\left.\begin{array}{c} \dfrac{d\tau}{dz} < \dfrac{d\tau_L}{dz} \\[2mm] if \quad z = z_0, \quad \tau = \tau_L \quad \left(z_0 \geq D\right) \end{array}\right\}$$

If both conditions are fulfilled,

$$\frac{S_{v*}(\gamma_s - \gamma)\tan\phi}{S_{v*}(\gamma_s - \gamma) + \gamma(1 + \dfrac{h_0}{D})} \leq \tan\theta < \frac{S_{v*}(\gamma_s - \gamma)}{S_{v*}(\gamma_s - \gamma) + \gamma}\tan\phi$$

The minimum slope at which the sediment underneath the surface is moved as an entity is

$$\tan\theta = \frac{S_{v*}(\gamma_s - \gamma)}{S_{v*}(\gamma_s - \gamma) + \gamma(1 + \dfrac{h_0}{D})}\tan\phi \qquad (13.32)$$

Takahashi stated: for a water-borne debris flow as usually conceived, the moving solid particles will be uniformly distributed in the flow once water debris flow occurs; therefore, a certain thickness of deposit beneath the bed surface must take part in the motion; that is,

$$z_0 \geq kh_0$$

where $k \geq 1$. Hence, Eq. (13.32) should be written as

$$\tan\theta = \frac{S_{v*}(\gamma_s - \gamma)}{S_{v*}(\gamma_s - \gamma) + \gamma(1 + \dfrac{1}{k})}\tan\phi \qquad (13.33)$$

As an example, the values $S_v* = 0.7$, $\gamma_s = 2.6$ t/m³, $\gamma = 1$ t/m³, $k = 0.75$, $tan\,\varphi = 0.8$ are used in Eqs. (13.31) and (13.33), the angle of inclination of the ground surface for triggering debris flow is in the range of

$$14.5° < \theta < 22.9°$$

Viscous debris flow is quite complex because the liquid phase is slurry and it behaves as a Bingham fluid; in this case, γ is the specific weight of slurry, and it consists of water and fine sediment particles. According to field data taken in Hunshui Gully [20], the specific weight of slurry in viscous debris flow can reach 1.75 t/m³ and perhaps even exceed it. If this value is substituted into Eq. (13.33), the inclination angles for the formation of viscous debris flow are smaller,

$$5.8° < \theta < 11.5°$$

This result is consistent with field data obtained in Southwestern China.

13.2.4.2 Maximum content of solid particles in water-borne debris flow

Once a water-borne debris flow forms, it moves toward downstream in the form of a bore with depth h and volume concentration of solid particles S_v. If the loose deposit on a slope (or gully bed) has been saturated before the arrival of the bore, the stress distribution at the time of arrival is as shown in Fig. 13.25. The shear and the resistance at a distance z from the stationary bed are, respectively,

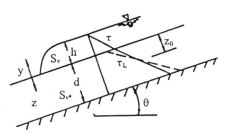

Fig. 13.25 Embryonic form of the bore of a debris flow

$$\tau = \sin\theta[(S_v h + S_{v*}z)(\gamma_s - \gamma) + (h+z)\gamma]$$

$$\tau_L = \cos\theta[(S_v h + S_{v*}z)(\gamma_s - \gamma)]\tan\phi$$

If $\tau = \tau_L$,

$$z_0 = \frac{\gamma\tan\theta - S_v(\gamma_s - \gamma)(\tan\phi - \tan\theta)}{S_{v*}(\gamma_s - \gamma)(\tan\phi - \tan\theta) - \gamma\tan\theta}h$$

(13.34)

the deposit within the depth z_0 will be eroded and move as part of debris flow. The content of solid particles of a water-borne debris flow moving on a slope (or gully) is a maximum as $z_0 = 0$. This maximum denoted by S_{v0}, can be deduced from Eq. (13.34)

$$S_{v0} = \frac{\gamma\tan\theta}{(\gamma_s - \gamma)(\tan\phi - \tan\theta)}$$

(13.35)

in which γ is the specific weight of clear water for water-borne debris flow and that of slurry for viscous debris flow. One can deduce, for a given slope, the maximum content of solid particles of a viscous debris flow formed from deposits with large amount of fine particles, is much larger than that of a water-borne debris flow formed from deposits without fine particles.

This phenomenon can also be approached in another way — by starting from the Bagnold basic formula for bed load. According to Eq. (9.16),

$$w_b'\bar{u}_b\tan\alpha = \tau_0 U e_b$$

(9.16)

where w_b' — submerged weight of bed load on unit bed area; $\bar{u}_b$ — average velocity of bed load; α — dynamic friction angle caused by collision among particles in bed load motion; τ_0 — shear acting on the bed; U — average flow velocity; e_b — efficiency of conveying bed load by the flow.

For viscous debris flow, turbulence and suspended load no longer exist, all the moving sediment particles belong to laminated load — a kind of bed load motion (details are in Chapter 5). In such flows, the shear induced by the deformation of fluid in the interstices can be neglected. Because of the extremely large content of solid particles, no relative motion occurs between the liquid and solid phases. Hence the efficiency of sediment transport approaches one.

$$w_b' = S_v(\gamma_s - \gamma)h$$
$$\tau_0 = [S_v\gamma_s + (1 - S_v)\gamma]h\tan\theta$$
$$\bar{u}_b = U$$
$$e_b = 1$$

Combining these formulas with Eq. (9.16), one obtains an expression for the maximum content of solid particles in water-borne debris flow [37]

$$S_{v0}' = \frac{\gamma\tan\theta}{(\gamma_s - \gamma)(\tan\alpha - \tan\theta)} \tag{13.36}$$

Eq. (13.36) can also be deduced by solving Eq.(13.39) and Eq.(13.40).

From a comparison of Eqs. (13.35) and (13.36), one can deduce that $S_{v0}' > S_{v0}$, since $tan\ \alpha$ is smaller than $\tan\varphi$ ($\tan\ \alpha$ is taken as 0.63 by Bagnold). From the physical point of view, one can conclude:

First, along with erosion of the slope (or gully bed), the maximum content of solid particles of a water-borne debris flow can be further enlarged if other mass supply (for instance, from bank slides) is available. For a water-borne debris flow with concentration S_{v0}, moving along a slope (or gully) with inclination angle $\tan\theta$, the maximum content of solid particles of a water-borne debris flow can be further enlarged for

$$\frac{\tan\phi - \tan\alpha}{\tan\alpha - \tan\theta}$$

times.

Second, if a water-borne debris flow with concentration S_{v0}, moving along a slope (or gully), enters a region with a gentler slope, deposition will not take place if the latter slope is larger than

$$\frac{\tan\alpha}{\tan\phi}\tan\theta$$

13.2.4.3 Concentrating of coarse particles in the frontal part of a debris flow

In the debris flow, the shear T and the dispersive force P, induced by collision and friction among particles, play an important role. In Fig. 5.9, the experimental results of Bagnold are presented; he established a relationship between T, P and the following parameter

$$N = \frac{\lambda^{\frac{1}{2}} \rho_s D^2 \, du / dy}{\mu}$$

in which ρ_s is the density of sediment, du/dy is velocity gradient in the direction perpendicular to the shear surface, μ is the viscosity of water (for water-borne debris flow) or slurry (for viscous debris flow),

$$\lambda = \frac{\text{diameter of sediment particles}}{\text{distance among sediment particles}} = \left[\left(\frac{S_{v*}}{S_v} \right)^{1/3} - 1 \right]^{-1}$$

Although no turbulence exists in viscous debris flow in the usual sense, granular shear flow still can be classified into two regions in which either viscous or inertial force dominates according to the magnitude of N [34],

$$T = c\rho_s (\lambda D)^2 (\frac{du}{dy})^2 \tag{13.37}$$

$$P = \frac{T}{\tan \alpha} \tag{13.38}$$

The quantity c is a constant, taken to be 0.013 by Bagnold, $\tan\alpha$ is the dynamic friction factor as in the foregoing.

From Eqs. (13.37) and (13.38), the dispersive force is proportional to the square of particle diameter. Therefore, in debris flow, the coarser the particles, the larger are the dispersive forces that act on them. These forces act in the vertical direction in such a way that coarse particles gradually move towards the water surface. As the velocity at the water surface is the largest, large stones continuously overtake other parts of the flow and concentrate in the frontal part [37]. Takahashi derived a formula to express the preceding concept and established experimentally the following phenomenon — if non-uniform sediment is moving as laminated load, particles that are larger than average will move towards the water surface, and those smaller than average will move towards the bottom [34].

13.2.4.4 Mechanism of the formation of intermittent flow

The intermittent flow of viscous debris flow is different from that of hyperconcentrated flow. The intermittent property of the viscous debris flow is obviously the result of the frequent stagnation of debris flow during its course of motion. Such stagnations result partly from blockage at constricted reaches, abrupt bends and meanders along the gully, and partly caused primarily by the internal mechanics of the debris flow. Pulsations are observed more often at gauging stations that are further downstream because debris flow originating from different gullies may not converge at same time.

As pointed out already, the clogging of hyperconcentrated flow can be explained in two ways according to the mechanism of hyperconcentrated flows. The situation for debris flow is similar. For instance, some researchers attribute the stagnation of debris flow in a straight gully to the fact that the shear on the bed surface is insufficient to overcome the Bingham yield stress; hence the debris flow disperses and become thinner and thinner due to some sort of resistance encountered in the process of motion [38]. As the stagnation layer thickens continuously and reaches a certain thickness so that it enlarges the shear on bed, then the stagnation layer starts moving. Although such a situation may arise, a different mechanism is also possible.

The bed paving process caused by intermittent debris flow in Jiangjia Gully, Yunan Province, has been described. Such a process gradually propagates downstream. On August 8, 1975, a rainstorm occurred in the upstream basin and intermittent light rain fell further downstream in the area near the gauging station. A viscous debris flow began on a dry bed. During propagation, some of the water in the debris flow infiltrated into the bed so that the percentage of solid particles in the debris flow became larger and larger and the porosity of the flow smaller and smaller. Finally, the flow lost its fluidity and stagnated. At the places that had been reached by preceding debris flows, the bed was saturated. The process of losing water to the bed thus propagated downstream. Correspondingly, the place of stagnation also propagated downstream.

13.2.5 Velocity and resistance in debris flow

13.2.5.1 Velocity profile of water debris flow [35]

If the specific weight of water debris flow is so large that all the sediment particles move in the form of laminated load, the dispersive force among particles at a distance y above the bed is equal to the submerged weight of all the sediment particles in the water column between elevation y and the surface

$$c\rho_s (\lambda D)^2 (\frac{du}{dy})^2 \cot \alpha = S_v (\gamma_s - \gamma)(h - y)\cos\theta \qquad (13.39)$$

If the shear induced by deformation of liquid among particles can be neglected, then

$$c\rho_s(\lambda D)^2 (\frac{du}{dy})^2 = [S_v(\gamma_s - \gamma) + \gamma](h - y)\sin\theta \qquad (13.40)$$

Integrating Eq. (13.40) with the boundary condition $y = 0$, $u = 0$, one obtains

$$u = \frac{2}{3D}\left\{\frac{g\sin\theta}{c}\left[S_v + \frac{\gamma}{\gamma_s}(1 - S_v)\right]\right\}^{\frac{1}{2}}\left[(\frac{S_{v*}}{S_v})^{\frac{1}{3}} - 1\right][h^{\frac{3}{2}} - (h - y)^{\frac{3}{2}}] \qquad (13.41)$$

$u = u_{max}$ at $y = h$,

$$\frac{u_{max} - u}{u_{max}} = (1 - \frac{y}{h})^{\frac{3}{2}} \qquad (13.42)$$

In Fig. 13.6, Eq. (13.42) is compared with the results of experiments conducted by Takahashi, in which non-uniform sediment was used with $D_{av} = 9.87$ mm and $\sqrt{D_{84}/D_{16}} = 3.46$; and the angle of inclination of the flume was $18°$; the depth of the water debris flow was kept constant at 10 cm. The figure shows that the theoretical result essentially coincides with the experimental data. If all the sediment particles move in the form of laminated load, the velocity profile is more non-uniform than it is for laminar flow of a Newtonian fluid, and the difference between the velocity profile in laminated flow and that in laminar flow of Bingham fluid is even larger.

Since

$$U = \frac{1}{h}\int_0^h u\,dy$$

If θ is not too large

$$\sin\theta = \tan\theta$$

If Eq. (13.41) is substituted into the preceding equation and integrated, the result is

$$U = \frac{2}{5D}\left\{\frac{g}{c}[S_v + \frac{\gamma}{\gamma_s}(1 - S_v)]\right\}^{\frac{1}{2}}[(\frac{S_{v*}}{S_v})^{\frac{1}{3}} - 1]J^{\frac{1}{2}}h^{\frac{3}{2}} \qquad (13.43)$$

The above formula shows that the velocity decreases with increases in the content of solid particles and in the size of the sediment particles, and it increases with an increase of the specific weight of the liquid phase. Besides, the velocity is proportional to a high power of the water depth. Such a relationship was also found in the experiments on hyperconcentrated flow conducted in IHR, YRCC at Zhengzhou.

For viscous debris flow, the liquid phase is a slurry consisting of fine particles and water. No one has studied the velocity profile for the condition that all the coarse particles move in the form of laminated flow.

13.2.5.2 Resistance to debris flow

Table 13.6 Roughness of debris flow

Category	Characteristics of channel	Slope	Values of n for various depths (m)			
			0.5	1.0	2.0	4.0
Non-viscous debris flow	Narrow and steep channel with steps and contractions; bed material is 0.5-2m stones.	0.15~0.22	0.20	0.25	0.33	0.50
	Channel with many bends and steps; bed material is 0.3-0.5 m stones.	0.08~0.15	0.10	0.125	0.167	0.25
	Wide and straight channel; bed material is 0.3 m stones, sand and gravel.	0.02~0.08	0.056	0.071	0.10	0.125
Viscous debris flow	Narrow, steep and meandering channel; bed material is big stones, sand and gravel, forming blockages and steps.	0.12~0.16	0.056	0.067	0.083	0.10
	Comparative straight channel, bed material is stones, sand and gravel.	0.08~0.12	0.036	0.042	0.05	0.06
	Wide and straight channel, bed material is stones finer than 0.3 m, sand and gravel.	0.04~0.08	0.029	0.036	0.042	0.05

No systematic study of this topic has been conducted. Only a few data on roughness are available. The roughness for debris flow is large. The Manning friction factor for glacial debris flow can exceed 0.45 [38]. Another characteristics of the roughness is that Manning friction factor is proportional to the water depth. The following relationship was based on field data of Jiangjia Gully, Dongchuan County, Yunan Province:

$$n = 0.035h^{0.34} \tag{13.44}$$

In the process of propagation, the discharge of debris flow increases along its course. Hence, the greater the depth of a debris flow, the larger is the amount of the solid particles that are carried; consequently, the more is the additional potential energy dissipation caused by bed load motion. As a result, the Manning roughness increases with an increase of water depth. However, the Manning formula should be used only for flow in the turbulent, rough-boundary region. And since most debris flows in Jiangjia Gully are laminar or close to laminar, the use of a Manning coefficient of roughness for describing the resistance of the flow is not appropriate.

Roughness for debris flow is closely related to the boundary conditions of the flow. In Table 13.6, Manning friction factors are given for different types of debris flows running through various types of channels. This table is based on the experience accumulated by Chinese researchers over many years [40]. It shows that the more irregular the outline of the channel and the coarser the bed material, the larger is the roughness for otherwise similar debris flows. Viscous debris flow is a laminar flow, and therefore the resistance should not vary with the boundary condition. Table 13.6 shows that the dependence of the Manning friction factor on channel conditions is not as strong as that for non-viscous debris flow; nonetheless, some interdependence still exists. After the bed-paving process of intermittent debris flow has taken place, the Manning friction factor is about half of what it was before [32]. Such a discrepancy might have been caused by the following factors: (1) Manning formula is not suitable for determining the velocity of viscous debris flow; (2) channel characteristics listed in Table 13.6 refer not only to the boundary roughness, but also to its shape in plan; hence they involve losses due to form resistance; (3) some of selected data are in the transition region between turbulent and laminar flows.

REFERENCES

[1] Chien, Ning, Zhaohui Wan, and Yiying Qian. "Hyperconcentrated Flow on The Yellow River." *Science Bulletin*, Vol. 24, No. 8, 1979, pp. 368-371; Journal of Tsinghua University, Vol. 19, No. 2, 1979, pp. 1-17. (in Chinese)

[2] Wan, Zhaohui, Yiying Qian, Wenhai Yang, and Wenlin Zhao. "Laboratorial Research on Hyperconcentration Flow." *Yellow River*, No. 1, pp. 53-65, 1979. (in Chinese)

[3] Chien, Ning. "Preliminary Study on The Mechanism of Hyperconcentrated Flow in North-Western Region of China." *Selected Papers of The Symposium on Sediment Problems on The Yellow River*, Vol. 4, pp. 244-267. (in Chinese)

[4] Hedstroelm, B.O.A. "Flow of Plastic Materials in Pipes." *Indus. Engin. Chem.*, Vol. 44, 1952, pp. 651-656.

[5] Qian, Yiying, Wenhai Yang, Wenlin Zhao, Xiuwen Cheng, Longrong Zhang, and Wengui Xu. "Basic Characteristics of Hyperconcentration Flow of Sediment." *Proc. of The International Symposium on River Sedimentation*, Vol. 1, Guanghua Press, 1980, p. 175-184. (in Chinese)

[6] Zhang, Hao, Zenghai Ren, Suqi Jiang, Dongzhi Sun, and Naishi Lu. "Settling of Sediment and the Resistance to Flow at Hyperconcentrations." *Proc. of the International Symposium on River Sedimentation*, Vol. 1, Guanhua Press, 1980, p. 185-194. (in Chinese)

[7] Chi, Yaoyu. "Transport of Hyperconcentrated Flow in Canal." *Journal of North-Western Agriculture College*, 1979, No. 3. (in Chinese)

[8] Zhang, Hao, and Zenghai Ren. "Experiments on Resistance to Hyperconcentrated Bingham Fluid in Channel with Rectangular Cross Section." *Science Bulletin No. 13*, 1981, pp. 811-814. (in Chinese)

[9] Yang, Tingrui, Zhaohui Wan, Yaoyu Chi, Yian Xu, Zaiyang Wang, and Naixiong Zhao. "Problems of Utilization of Muddy Water with Hyperconcentration of Sediment." *Proc. of the International Symposium on River Sedimentation*, Vol. 1, Guanhua Press, 1980, pp. 93-102. (in Chinese)

[10] Dracos, T.A., and G. Glenne. "Stability Criteria for Open Channel Flow." *J. Hyd. Div., Proc., Amer. Soc. Civil Engrs.*, Vol. 93, HY6, 1967, pp. 79-101.

[11] Frisch, H.L., and R. Simha. "The Viscosity of Colloidal Suspensions and Macromolecular Solutions." In *Rheology, Theory and Applications*. Edited by F. R. Eirich, Academic Press Ins., 1956, pp. 525-613.

[12] Shishchenko, R.E. *Hydraulics of Slurry*. Press of Petroleum Industry. 1957, p. (in Russian)

[13] Bagnold, R.A. "Some Flume Experiments on Large Grains But Little Denser Than the Transporting Fluid and Their Implications." *Proc., Inst. Civil Engrs.*, Part 3, April 1955, pp. 174-205.

[14] Northwest Hydrotechnical Scientific Research Institute. "Study on Sediment-Carrying Capacity of Hyperconcentrated Flow." *Selected Papers of The Symposium on Sediment Problems on The Yellow River*, 1975, Vol. 2, pp. 249-258. (in Chinese)

[15] Fleishman, S.M. *Debris Flow and Design of Road in Regions Prone to Debris Flow*. Press of Transportation Industry, Moscow, 1955.

[16] Rabkowa, E.K. "Debris Flow Motion." *Hydrotechnics and Soil Improvement*, 12, 1955, p. 14-23. (in Russian)

[17] Wright, C.A. "Experimental Study of the Scour of A Sandy River Bed by Clear and Muddy Water." *J. Res., Bureau of Standards*, Vol. 17, No. 2, 1936, pp. 193-206.

[18] Bantin, R.A., and M. Streat. "Dense Phase Flow of Solid-Water Mixtures in Pipelines." *Proc. Hydrotransport 1*, 1970, pp. G-1 to G-24.

[19] Hampton, M.A. "Competence of Fine Grained Debris Flows." *J. Sedim. Petro.*, Vol. 45, No. 4, 1975, pp. 834-844.

[20] Zhang, Xinbao, and Shuifang He. "Preliminary Study on the Composition of Debris Flow in Hunshui Gully." *Proc. of Debris Flow*, Chengdu Institute of Geography, Academia Sinica, 1980, p. 155-164. (in Chinese)

[21] Cloete, F.L.D. et. Al. "Dense Phase Flow of Solids-Water Mixtures Through Vertical Pipes." *Trans., Inst. Chem. Engrs.*, Vol. 45, No. 10, 1967, pp. 392-400.

[22] Chien, Ning. "Some Problems of Hyperconcentrated Flow." *Yellow River*, 1981, No. 4, p. 1-9. (in Chinese)

[23] Research Group of Hyperconcentrated Irrigation in Luohui Irrigation District. "Hyperconcentrated Irrigation in Luohui Irrigation District." *Selected Papers of The Symposium on Sediment Problems on The Yellow River*, 1975, Vol. 1, Part. 1, pp. 139-157. (in Chinese)

[24] Kleiman, P. "Hydraulic Transport of Copper Tailings." *J. Hyd. Div., Proc., Amer. Soc. Civil Engrs.*, Vol. 102, No. HY10, 1976, pp. 1589-1595.

[25] Dai, Jilan, Zhaohui Wan, Wenzhi Wang, Wukui Chen, and Xijun Li. "Experimental Study of Slurry Transport in Pipes." *Proc. of The International Symposium on River Sedimentation*, Vol. 1, Guanghua Press, 1980, pp. 195-204. (in Chinese)

[26] Ganshu Institute of Glacial, Frozen Soil and Desert Research, Academia Sinica. *Debris Flow*. Science Press, 1973, pp. 82. (in Chinese)

[27] Li, Xie. "Debris Flow in Xiaojiang Basin, North-Eastern Yunnan." *Proc. of Debris Flow*, Chengdu Institute Geography of, Academia Sinica, 1980, p. 34-42. (in Chinese)

[28] Zhang, Shucheng, and Jianmo Yuan. "Impact of Debris Flow and Its Measurement." *Proc. of Debris Flow*, Chengdu Institute of Geography Institute, Academia Sinica, 1980, pp. 187-192. (in Chinese)

[29] Yang, Kaishu. "Comment on Classification of Debris Flow and Suggestion of Criterion for Classification." *Report of South-Western Division*, China Railway Academy, 1980, p. 17. (in Chinese)

[30] Tian, Lianquan, Singbao Zhang, and Jishan Wu. "The Formation of Debris Flow." *Proc. of Debris Flow*, Chengdu Institute of Geography, Academia Sinica, 1980, pp. 69-73. (in Chinese)

[31] First Department, Chengdu Institute of Geography, Academia Sinica. "Characteristics of Debris Flow in Hunshui Gully of Daiyinjiang River, Yunnan Province." 1979, p. 120. (in Chinese)

[32] Kang, Zhicheng. "Flow Pattern Characteristics of Debris Flow in Jiangjia Gully, Dongchuan County, Yunnan Province." *Proc. of Debris Flow*, Chengdu Institute of Geography, Academia Sinica, 1980, p. 165-178. (in Chinese)

[33] Middleton, G.V., and M.A. Hampton. "Subaqueous Sediment Transport and Deposition by Sediment Gravity Flows." In *Marine Sediment Transport and Environmental Management*, John Wiley and Sons, Inc., N. Y., 1978.

[34] Takahashi, T. "Debris Flow on Prismatic Open Channel." *J. Hyd., Proc., Amer. Soc. Civil Engrs.*, Vol. 106, No. HY 3, 1980, pp. 381-396.

[35] Takahashi, T. "Mechanical Characteristics of Debris Flow." *J. Hyd. Div., Proc. Amer. Soc. Civil Engrs.*, Vol. 104, No. HY8, 1978, pp. 1152-1169.

[36] Bagnold, R.A. "The Flow of Cohesionless Grains in Fluids." *Phil. Trans.*, Royal Soc. London, Ser. A, Vol. 249, 1956, pp. 235-297.

[37] Bagnold, R.A. "Deposition in the Process of Hydraulic Transport." *Sedimentology*, Vol. 10, No. 1, 1968, pp. 45-56.

[38] Du, Ronghuan, and Shucheng Zhang. "Characteristics of Glacial Debris Flows in South-Eastern Qinghai-Tibet Plateau." *Proc. of Debris Flow*, Chengdu Institute of Geography, Academia Sinica, 1980, pp. 27-33. (in Chinese)

[39] Kang, Zhicheng. "Preliminary Analysis on Velocity of Pulsing Viscous Debris Flow in Jiangjia Gully, Dongchuan County, Yunnan Province." *Proc. of Debris Flow*, Chengdu Institute of Geography, Academia Sinica, 1980, pp. 132-139. (in Chinese)

[40] Xu, Daoming, and Qinghua Feng. "Table of Roughness of Debris Flow Channel." *Abstract of papers of First National Symposium on Debris Flow*, 1979, p. 51-52. (in Chinese)

[41] Zhang Ren, Ning Chien, and Tilu Cai. "An Analysis on the Conditions of Transporting Flow with Hyperconcentration over a Long Distance." *J. of Sediment Research*, No. 3, 1982, pp.1-12. (in Chinese)

CHAPTER 14

DENSITY CURRENTS

A density current is a relative motion that takes place between two fluid layers that have slightly different densities; it is caused by the density difference. Under certain conditions, a sediment-laden flow dives under the clear water and continues forward flowing as a density current, because of its larger density. Thus, a density current is a special type of sediment-laden flow.

At the end of the nineteenth century, Swiss scientists noticed that after the River Roan and River Ryan flowed into Lake Constance and Lake Geneva, the muddy and cold river water did not mix with the clear and warm lake water; instead, it dove to the bottom of the lake and continued to move as an entity. The Hoover Dam on the Colorado River in the United States was built in 1935 and the backwater region was 110 km long. Observations showed that during the flood season, muddy water flowed the entire length of the reservoir in the form of a density flow and then appeared in the flow being released at the dam. These occurrences indicate that such density currents can carry large amounts of sediment over long distances without mixing appreciably with the adjacent clear water; this process can reduce deposition in the reservoir and extend the life of the reservoir. Since then, many scientists have studied density currents, and they have found that they occur often in nature. The process involves not only reservoir deposition, but also many other types of flow, it can affect a variety of engineering processes:

1. Density currents formed by sediment-laden water play an important role in the release of sediment from reservoirs, settling basin design and deposition both in navigation channels and downstream of tidal guard sluices. Geologists and geomorphologists have reported that after a land slide in the sea bed following an earthquake, density currents with high velocity, large thickness, and high concentration can form along the sea bottom. Oceanographers call them turbidity flows. They can also play an important role in moulding coast lines.

2. Density currents that form because of the salt water in estuaries can greatly affect the mixing of the salt water with the clear water there. In estuaries in regions with low tides, salt-water wedges can extend rather long distances upstream from the river mouth, and they can have significant effects both on irrigation and on sediment deposition in the region of the river mouth.

3. Density currents due to temperature differences not only can cause problems at intakes for industrial cooling water but can also create large scale flows in the ocean. A cold front in atmosphere, which is the concern of meteorologists, is also a density current that is due to a temperature difference.

4. The two-phase flow formed by petroleum and water has the same characteristics as a density flow, and it is an important process in the oil extraction industry.

Because of the wide variety of density currents, they have been studied as part of the sciences of water conservancy, geography and geology, meteorology, oceanography, petroleum extraction, and others. Hence, a summary of all of the published works is impossible within this book, one can find more complete information in references [1, 2], and reviews of theoretical studies of density currents have been made by Yih[3] and Turner[4]. The basic properties of density currents caused by sediment-laden flows are presented in this chapter. Some studies of salt water are included because many of the experimental studies of density currents have been conducted using salt solutions as the heavier fluid.

14.1 FORMATION AND MOVEMENT OF DENSITY CURRENTS

14.1.1 Similarities and differences between density currents and ordinary open-channel flows

The movement of a density flow is quite similar to that of ordinary open-channel flow. In fact, flow in an open channel can be thought of as a particular case of a density current; in it, the atmosphere serves as the upper layer of liquid. Almost all of the phenomena that occur at a free surface can also occur at the interface between flows of different densities.

The density flows encountered in hydraulic projects generally involve only slight differences in the densities of the upper and lower layers, and their inertial forces and densities are often considered to be the same. Furthermore, since the density difference is small, the upper liquid creates a large buoyancy effect within the lower liquid, so that the effective gravity of the lower liquid is greatly reduced, if g' is defined as follows:

$$g' = g \frac{\Delta \rho}{\rho} \qquad (14.1)$$

in which g is the usual gravitational acceleration, $\Delta \rho$ is the density difference between the upper and lower liquids, and ρ is the density of the upper liquid. Many formulas describing open-channel flow apply also to density currents once g is replaced by g'. For example, the flow pattern in an open channel depends greatly on the Froude number of the flow; in a density current, the Froude number remains the key parameter but its form is modified:

$$Fr' = \frac{U_c}{\sqrt{g'h'}} \qquad (14.2)$$

in which U_c is the relative velocity between the two liquid layers, h' is the thickness of the density current. As pointed out in Chapter 12, an important parameter for a flow field having a density gradient is the Richardson number:

$$Ri = \frac{-\dfrac{g}{\rho}\dfrac{d\rho}{dy}}{(\dfrac{du}{dy})^2} \qquad (14.3)$$

If the upper and lower liquids move with a relative velocity of U_c because of the density variation $\Delta\rho$,

$$\frac{d\rho}{dy} \quad \sim \quad -\frac{\Delta\rho}{h'}$$

$$\frac{du}{dy} \quad \sim \quad \frac{U_c}{h'}$$

Thus

$$Ri = \frac{g\dfrac{\Delta\rho}{\rho}h'}{U_c^2} = \frac{1}{Fr'^2} \qquad (14.4)$$

Although the physical meanings of the two parameters are different, they are equivalent.

Of course, density currents and general open-channel flows also display some differences, the three primary ones are the following:

1. As the effective gravitational acceleration g' is much less than g, Froude numbers of density currents are often much larger than those for open-channel flows. Even if the bed slope is rather gentle, the density current can still be supercritical.

2. If a density gradient exists in a flow field, the density generally increases in the downward direction. That is, $d\rho/dy < 0$. Consequently, vertical exchanges within the flow are greatly restricted.

If a water element with a volume V and a density of $\rho(y_0)$ moves from level y_0 to y_1, where the local density is $\rho(y_1)$, the work done in moving this water element is

$$W = V\int_{y_0}^{y_1}[\rho(y_0) - \rho(y_1)]g\,dy$$

if $y_0 < y_1$, $\rho(y_0) > \rho(y_1)$

                      $y_0 > y_1$, $\rho(y_0) < \rho(y_1)$

In both circumstances, W is positive; that is, if the two water elements, each with volume V, exchange places, the work done by the flow is $2W$. In contrast, if the density in the flow field is the same everywhere, so that $\rho(y_0) = \rho(y_1)$, the exchange of two water elements takes place with a net work of zero. The water element moving from the lower to the upper layer takes a certain amount of energy from the flow whereas the body moving from the upper to the lower place releases an equal amount of energy.

In the physical sense, the vertical exchanges that occur in a flow field having a density gradient are much less extensive than are those in a fluid of uniform density. This point has far-reaching effects. For example, in drawing water from different levels of a reservoir without a density gradient, water at any level could be drawn out through the outlet as readily as any other, as shown in Fig. 14.1a. In contrast, if a density gradient occurs in the reservoir, vertical movement is constrained so that the water discharging from the outlet tends to come from a region of smaller vertical extent; thus its velocity is distributed more as shown in Fig. 14.1b[5]. As a result, if a density gradient exists, water can be drawn selectively from different levels. This point is discussed in more detail in Paragraph14.2 .

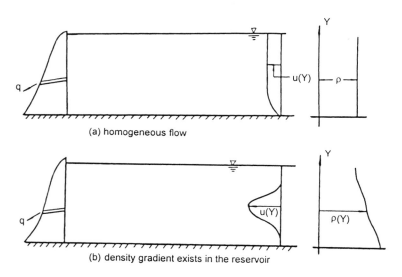

(a) homogeneous flow

(b) density gradient exists in the reservoir

Fig. 14.1 Water drawing through an outlet in a
reservoir with and without density gradient

3. The inertia of the upper layer in open-channel flow is usually negligible because the fluid in it is air. But the inertia of the upper layer of a density flow is not

negligible, and the densities of both layers must be included in an analysis. Also, the resistance the air exerts on the flowing water is relatively small and is rarely included in calculations. With density currents, the resistance at the interface is far greater. Though it is usually smaller than that at the river bed, it can sometimes even be larger. Finally, a free surface is generally clearly discernible, whereas the mixing of the upper and lower liquids in a density flow often forms a transition layer whose density and velocity change continuously; hence they do not form a clearly defined interface.

The foregoing differences all make the analysis of density flows more complex than that of open-channel flows.

14.1.2 Conditions for formation of density flows

Fig. 14.2 is a schematic diagram of the transition from a sediment-laden open-channel flow to a well defined density flow in a reservoir. After the sediment-laden flow enters the backwater region of the reservoir at point A, the sediment carried by the flow settles down steadily toward the river bed. The velocity and concentration at the water surface gradually tend toward zero because of the increasing depth and the decreasing water velocity. The coarser particles tend to deposit first and form a delta near the entrance to the reservoir, whereas the finer particles can remain

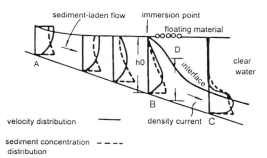

Fig. 14.2 The longitudinal change of velocity and sediment concentration during the transition from a sediment-laden open-channel flow to a density current

suspended because of their smaller settling velocities. Beyond point B, the surface water becomes clearer and a distinct interface forms between clear water and sediment-laden water. From then on, two kinds of flow with different densities are taking place. Because of the action of gravity on the remaining suspended material, the submerged flow continues to move. Thus, it forms a density flow between point B and C in Fig. 14.2. During the movement, the density flow would drag some of the clear water above the interface along with it; other parts of the clear water in the upper region would then flow backwards, to maintain a water volume balance, and this flow pushes any floating material toward point B. The presence of such material is the most reliable signal that a density flow has formed. The patterns of flow and sediment at the immersion point can be considered as the condition for the occurrence of a density flow.

Between points B and C, the interface between clear water and sediment-laden flow changes progressively. At some point D, downstream of the immersion point, the slope of the surface dh'/dx tends toward $-\infty$. For a non-uniform density flow, the rate of change in water depth along the course of the flow is inversely proportional to

$$1 - \frac{U'^2}{\frac{\Delta\rho}{\rho'}gh'}$$

in which U', h', ρ' are the velocity, thickness and density of the denser liquid, respectively; $\Delta\rho$ is the difference in density between the clear water and sediment-laden flows. Then, the following hydraulic condition exists at the inflection point D of the interface:

$$\frac{U'^2}{\frac{\Delta\rho}{\rho'}gh'} = 1$$

Because the water depth h_0 at the immersion point is larger than that of the denser liquid h' at point D, the following relationship should exist at the immersion point

$$\frac{U_0^2}{\frac{\Delta\rho}{\rho'}gh_0} = k < 1$$

Data from both flume experiments and field observations show that the value of k is about 0.6 [6]. Thus the condition for the formation of a density current is:

$$\frac{q^2}{\frac{\Delta\rho}{\rho'}gh_0^3} = 0.6 \qquad (14.5)$$

in which q is the discharge per unit width at the immersion point. From the above equation, if the water level upstream of the dam remains constant, an increase of the inflow discharge would cause the immersion point to move downstream; an increase in the sediment concentration of the inflow would cause the point to move upstream.

The condition for the formation of a density current should be distinguished from the condition required to maintain it; the latter is the condition required to keep the density flow moving toward the dam after it forms. The condition for maintaining a density flow has two parts. One is a continuous supply of sediment-laden liquid. If the inflow of sediment-laden liquid ceases to supply water and sediment so that it no longer forms a density current at the immersion point, the already formed density current downstream would soon stop moving, and it would then disappear. The other is the condition to keep the flow moving; that is, if a density current is to move stable, it must be able to overcome any resistance it encounters. Since the energy to maintain the flow comes from the sediment suspended by the flow, some minimum concentration is clearly required. This concentration is much higher than that required to form a density current. According to field data from Shaver Lake, U.S.[7], a density current will form if the density difference between river water and lake water is 0.0008; that is, if the concentration is about 1.28 kg/m^3. The yearly average concentration of sediment in the Yangtze River in China is only 1 kg/m^3, density currents can form and cause depositions in the Qingshan Channel and in the navigation channel of the Gezhouba Dam. However, the data of Guanting Reservoir in Yongding River indicates that only if the sediment concentration in a density current is more than 20 kg/m^3, can the current continue to flow and finally reach the dam. Considering the maintenance of a stable density current, Levy derived the following relationship[8]:

$$(\frac{\Delta\rho}{\rho'})^{1/2} - \frac{0.8C\alpha(\frac{J_0}{\lambda_c}\frac{\Delta\rho}{\rho'})^{1/6}}{1-\alpha(\frac{\lambda_c}{J_0}\frac{\rho'}{\Delta\rho})^{1/3}} = 0.03 \tag{14.6}$$

in which

$$\alpha = (\frac{q^2}{gh^3})^{1/3}$$

$$\lambda_c = \lambda_0 + \lambda_i \frac{H}{H-h'}$$

H water depth in the reservoir;

λ_0, λ_i — resistance factors at bottom of reservoir and at interface of density current;

J_0 bottom slope of reservoir;

C ratio of velocity at interface to the average velocity of the density current.

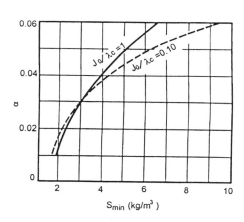

Fig. 14.3 The minimum sediment concentration for maintaining a density current

The minimum sediment concentration to maintain a density current is shown in Fig. 14.3 for $C = 1$ and for various values of α and J_0/λ_c. As α increases, the minimum concentration increases. It also varies with J_0/λ_c but not as much.

14.1.3 Movement of a density current

14.1.3.1 Velocity of the front

As a density current moves through a reservoir, some clear water must also be pushed aside. Accordingly, the force that pushes the head of a density current forward must be, and is, larger than that required for the following underflows alone. As a result, the head of a density flow is thicker than the stable submerged flow behind it, about twice as thick. The head of a density current is relatively short and the transition to the following underflow is rather abrupt. The head of a density current is a little like the front of a debris flow. While progressing, the shape and velocity of the head remain essentially stable.

The motion of the head of a density current, moving through quiescent fluid at a velocity U_f', can be transformed into a steady flow by the usual device of having the observer move with the head of the current. Then the flow field is as shown in Fig. 14.4a, and the density current no longer appears to the observer to move; instead the water in the reservoir, which was still before, now appears to move from right to left with the velocity U_f'. If the thickness of the steady underflow is h', the water depth in the reservoir is H and the density current and the reservoir have density distributions, $\rho'(y)$ and $\rho(y)$, with reference to Fig. 14.4a, the Bernoulli Equation can now be written between points (1) and (2), as follows:

$$\int_0^H \rho(y)g dy + \rho_0 \frac{U_f'^2}{2} = \int_0^{h'} \rho'(y)g dy + \int_{h'}^H \rho(y)g dy$$

It can be simplified to

$$\rho \frac{U_f'^2}{2} = \int_0^{h'} (\rho' - \rho)g dy \qquad (14.7)$$

Eq. (14.7) can be solved for various special conditions[9]:

1. $\rho(y) = \text{constant} = \rho_0$, $\rho'(y) = \text{constant} = \rho_0'$

$$\Delta\rho = \rho_0' - \rho_0$$

then

$$U_f' = \sqrt{\frac{\Delta\rho}{\rho} 2gh'} \qquad (14.8)$$

obtained in 1940 by Von Karman [10].

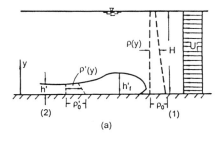

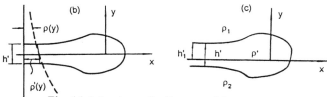

Fig. 14.4 A schematic diagram of front velocity of bottom
and intermediate density current

2. Both densities vary linearly in the vertical direction:

$$\rho(y) = \rho_0 - \beta\rho_0 y$$

$$\rho'(y) = \rho_0' - \beta'\rho_0 y$$

Substituting these values into Eq. (14.7), one obtains

$$U_f' = [2\frac{\Delta\rho}{\rho}gh' + (\beta - \beta')gh'^2]^{1/2} \qquad (14.9)$$

3. The density of the lower layer varies linearly, whereas that in the reservoir is constant, so that

$$\beta = 0, \ \beta' > 0$$

Then, formula Eq. (14.9) can be simplified to

$$U_f' = [2\frac{\Delta\rho}{\rho_0}gh' - \beta'gh'^2]^{1/2} \qquad (14.10)$$

In this case, the density current, after submerging in the reservoir, moves a little more slowly than does a homogeneous density current.

4. Density current has a constant density, and that in the reservoir varies linearly so that

$$\beta' = 0, \quad \beta > 0$$

Then Eq. (14.9) becomes

$$U'_f = [2\frac{\Delta\rho}{\rho_0}gh' + \beta g h'^2]^{1/2} \tag{14.11}$$

If the density in the reservoir varies, the speed of the density current is greater than it is in a homogeneous reservoir, after submerging in the reservoir.

If the density of the density current falls between those of a lower layer of water and an upper one in the reservoir, the current flows between the layers (an intermediate flow). If both the densities of the current and the density of the water in the reservoir are distributed linearly along the vertical, and the coordinate system is as shown in Fig. 14.4b, then

$$\rho'(y) = \rho(y)$$

$$U'_f = \left[(\beta - \beta')\frac{gh'^2}{4}\right]^{1/2} \tag{14.12}$$

The speed of such a density current is less than that of a density current on the bottom under the same circumstances, Eq. (14.9).

Another special case is a homogeneous density current (density ρ') that flows between two fluid layers with different densities; the density and thickness of the upper layer are ρ_1 and h_1, those of the lower layer are ρ_2 and h_2, $\rho_1 < \rho' < \rho_2$, as shown in Fig. 14.4c. From a vertical force balance, one obtains

$$\rho'gh' = \rho_1 g h'_1 + \rho_2 g(h' - h'_1)$$

This formula together with Eq. (14.7) lead to the result

$$U'_f = (2\frac{\Delta\rho}{\rho_2}\beta gh')^{1/2} \tag{14.13}$$

in which $\beta = \dfrac{\rho_2 - \rho'}{\rho_2 - \rho_1}$

$$\Delta\rho = \rho' - \rho_1$$

Density currents can also flow at the surface, and they follow Eq. (14.13) with ρ_2 and ρ' being nearly the same and much larger than ρ_1, then

$$U'_f = (2\frac{\rho_2 - \rho'}{\rho_2}gh')^{1/2}$$

The equation is the same as Eq. (14.8), so that the Von Karman solution is suitable for density currents both on the bottom and on the surface.

The foregoing analysis is approximate because it does not include the resistance. If a term for the resistance $\lambda \rho_0 U_f'^2 / 2$ is introduced into the Bernoulli Equation, in which λ is a resistance coefficient, then with no density gradient[11],

$$U_f' = \sqrt{\frac{2}{1+\lambda} \frac{\Delta \rho}{\rho_0} gh'} \qquad (14.14)$$

For a surface flow, the resistance is negligible, that is $\lambda \approx 0$. For a density current on the bottom, Abraham and Vreugdenhil used $\lambda = 0.6$[12] and Barr used $\lambda = 1.0$ [13]. If $\lambda = 0.6$,

$$U_f' \approx 1.10 \sqrt{\frac{\Delta \rho}{\rho_0} gh'} \qquad (14.15)$$

which corresponds to the experimental data of Wilkinson and Wood for large values of the Reynolds number[14].

In fact, Eq. (14.14) is only suitable for density currents with a thickness h' that is much smaller than the water depth H. If h' and H are of the same order, then[11]

$$U_f' = [\frac{\Delta \rho}{\rho_0} g \frac{h'}{H} \frac{(H-h')(2H-h')}{H+h'+\lambda(H-h')}]^{1/2} \qquad (14.16)$$

If $h' \ll H$, Eq. (14.16) approaches Eq. (14.14).

To summarize, the speed of the head of density current has no relationship with the slope of the reservoir bottom. Middleton found through experiments that the foregoing conclusion is correct if the gradient is less than 4%. If it is more than this value, the coefficient in Eq. (14.15) is no longer a constant with a value of 1.10 but a variable that increases with the gradient[15]. Some results indicate that the front velocity is directly proportional to the square root of the bottom slope[16]. This front velocity denotes the velocity of the steady uniform underflow behind the head. In Fig. 14.4a, the velocity of the density current head U_f' is taken to be equal to the velocity of the density current U'; however, because of the disturbance caused by the movement of the density current and the mixing of clear and sediment-laden flow, the two are not necessarily equal. Field data show that, although the ratio between the two velocities is close to unity, if the Froude number is less than 1, the ratio decreases as the Froude number increases, as shown in Fig. 14.5[15].

14.1.3.2. The interfacial boundary layer

The relative motion implicit in the flow of a density current, from the point of submergence on downstream, causes a boundary layer to form along the interface; its thickness grows with distance traveled. This phenomenon is like the flow of water over a plate, the only difference is that the boundary layer on the interface develops in two directions — both upward and downward. That is, the layer grows into both the sediment-laden and the clear water regions. Only after the boundary layer extends over the entire depth, can the flow be considered to be fully developed.

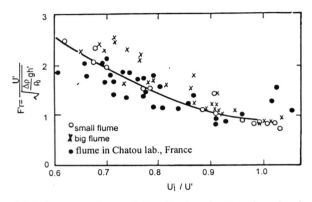

Fig. 14.5 A comparison of the front velocity of a density current U_f' with its average velocity U' (after Middleton, G.V.)

An idealized case is a two-dimensional flow with density ρ, viscosity μ, and velocity U that flows over a parallel two-dimensional flow with values ρ', μ', U', and the two flows extend to infinity upward and downward from the interface. The interface of the two layers is taken as the point of reference, the direction along the interface as x and the perpendicular direction as y (Fig. 14.6). If U and U' are sufficiently small, the flow in the vicinity of the interface is laminar. Then at the interface, the only shearing stresses are in the x-direction; with no force in the y-direction, the interface remains horizontal. The velocity at the interface is a constant u_i.

The thickness of the boundary layer is zero at the initial point and increases in the x-direction.

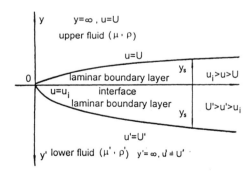

Fig. 14.6 Laminar boundary layer on the interface of two flows

According to the Prandtl boundary-layer theory for the two dimensional flow of an incompressible viscous fluid, the equation of motion of the and the conditions for static pressure and continuity can be written as follows:

632

$$\left.\begin{array}{l} u\dfrac{\partial u}{\partial x}+v\dfrac{\partial u}{\partial y}=v\dfrac{\partial^2 u}{\partial y^2}-\dfrac{1}{\rho}\dfrac{\partial p}{\partial x}\\[2mm] \dfrac{1}{\rho}\dfrac{\partial p}{\partial y}+g=0\\[2mm] \dfrac{\partial u}{\partial x}+\dfrac{\partial v}{\partial y}=0 \end{array}\right\}$$

(14.17)

in which u and v are the x and y components of the velocity and p is the static pressure.

Keulegan obtained a solution for the case in which the upper layer is at rest ($U = 0$) using Pohlhausen's method for solving the boundary layer equation[17]. The thicknesses of the two parts of the boundary layer are as follows:

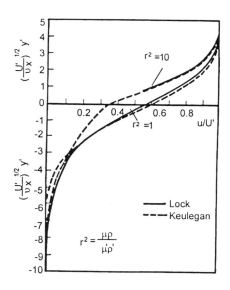

Fig. 14.7 Velocity distribution in a laminar boundary layer

$$\left.\begin{array}{l} y_s = m\sqrt{\dfrac{vx}{U'}}\\[3mm] y'_s = m'\sqrt{\dfrac{v'x}{U'}} \end{array}\right\}$$

(14.18)

The value of m and m' are functions of the parameter

$$r = (\frac{\mu\rho}{\mu'\rho'})^{1/2}$$

as shown in Table 14.1. For a velocity of the density current of about 30 cm/s and a value of r^2 of 0.01, with every kilometer the density current moves forward, the boundary layer extends an additional 25 cm into the sediment laden flow region, and 37.5 cm into the clear water region. The following formula expresses the shear stress at the interface,

$$\tau_i = \lambda_i\sqrt{\frac{v'}{U'x}}\frac{\rho'}{2}U'^2$$

(14.19)

The friction factor λ_i is also a function of r^2 (Table 14.1). Fig.14.7 shows the velocity distribution in the upper and lower boundary layers for $r^2 = 1$ and 10.

Lock also obtained solutions to Eq. (14.7) for the conditions that $U = 0$ and $U \neq 0$[18]; and they are nearly the same as those of Keulegan, as shown in Table 14.1 and Fig. 14.7.

Keulegan and Lock treated flows in which the thickness of both the clear water and the muddy layers are large, so that the boundary layer is not affected by either the water surface or the river bottom. Sometimes, however, the depth of the muddy flow is much smaller than that of the clear water. The lower boundary layer soon extends to the river bed even though the upper one has not reached the water surface. Bata studied this case, and he found that the resistance on the interface is not only related to the Reynolds number and the parameter r, but it is also affected by the ratio of density current thickness h' to distance x[19]:

$$\frac{(384 - \lambda_i \, \text{Re}')^{3/2}}{4\lambda_i \, \text{Re}' - 384} = \frac{31.2}{\sqrt{\text{Re}' \, M}} \qquad (14.20)$$

in which

$$\text{Re}' = \frac{U'h'}{v'} \qquad (14.21)$$

$$M = \frac{h'}{x} \frac{\mu\rho}{\mu'\rho'} \qquad (14.22)$$

The forgoing study indicates that the resistance on the interface is larger before the density current and the upper clear water layer have fully developed than it is afterwards. Usually, the course over which a density current flows is relatively long, and flows in both the upper and lower layers do develop fully; thus the phenomenon is no longer in the process of development and the flows can not be considered as boundary layers. The following sections are for such fully developed flows. The analyses by Keulegan and others are all for laminar boundary layers; although they may be turbulent in practice, no one has obtained a result for this case as yet.

14.1.3.3. Velocity distribution

1. The velocity distribution for laminar flow

Ippen and Harleman used the velocity distribution of flow between two parallel plates for that of a density current with laminar flow. If one of the plates is fixed, so as

Table 14.1 Relationships between m, m' and λ_i, r^2

γ^2		0	0.01	0.10	0.316	1.0	3.16	10	100	∞
m'		-	4.34	4.47	4.58	4.75	4.95	5.16	5.53	6.05
m		6.29	6.47	7.00	7.47	8.19	9.26	10.72	15.09	-
λ_i	Keulegan	-	0.078	0.20	0.29	0.39	0.43	0.55	0.63	0.66
	Lock	-	-	-	-	0.40	-	0.56	0.64	-

to represent the river bottom, and the other plane moves at a speed u_i[20] then the velocity distribution can be written as follows:

$$u = u_i \frac{y}{h'} + \frac{y^2 - yh'}{2\mu'} \frac{\partial}{\partial x}(p + \rho'gz_i) \tag{14.23}$$

The average velocity is

$$U' = \frac{u_i}{2} - \frac{h'^2}{12\mu'} \frac{\partial}{\partial x}(p + \rho'gz_i) \tag{14.24}$$

if

$$\xi = \frac{Fr'^2}{Re'\, J} \tag{14.25}$$

in which J is the slope of the density current, the Reynolds number Re' is as expressed in Eq. (14.21), and the Froude number Fr' has the following form:

$$Fr' = \frac{U'}{\sqrt{\dfrac{\Delta\rho}{\rho'} gh'}} \tag{14.26}$$

One notices that

$$\frac{\partial z_i}{\partial x} = -J$$

$$\frac{\partial p}{\partial x} = \rho g J$$

$$y = h' \qquad u = u_i$$

$$y = y_m \qquad u = u_{max}, \qquad \frac{du}{dy} = 0$$

then

$$\frac{y_m}{h'} = 2\xi + \frac{1}{3} \tag{14.27}$$

$$\frac{u}{u_{max}} = 2\frac{y}{y_m}\left(1 - \frac{y}{2y_m}\right) \tag{14.28}$$

635

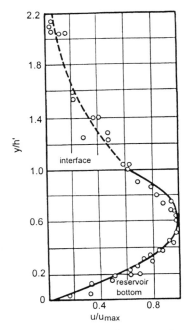

y/h'

2.2

1.8

1.4

interface

1.0

0.6

0.2

0

0.4   0.8
u/u$_{max}$

reservoir
bottom

Fig. 14.8 Comparison of measured and
theoretical velocity distribution for a laminar
density current   (after Rayraud, J.P.)

Raynaud assumed that the velocity distribution was parabolic and derived a result like that of Eq. (14.28) [21]. Fig. 14.8 shows the data from flume experiments for the case that $\Delta\rho/\rho'$ changes very little over the depth; the measured data follow Eqs. (14.27) and (14.28) quite well for the value $\xi = 0.15$ [1]. Substituting this value into Eq. (14.25), one obtains

$$U' = 0.375(Re')^{1/2}\sqrt{\frac{\Delta\rho}{\rho}gh'J}$$

(14.29)

and it gives the average velocity in a laminar density current. It has the form of the Chezy formula and can be rewritten as

$$h' = [7.15\frac{\mu'}{\Delta\rho g}\frac{q}{J}]^{1/3}$$

(14.30)

Hence, in a given reservoir, the thickness of a laminar density current is directly proportional to the cube root of unit discharge; and, if the unit discharge is fixed, the thickness is inversely proportional to the cube root of the gradient.

2. Velocity distribution for turbulent flow

The vertical distributions of velocity, sediment concentration, and shear stress in a turbulent density current are indicated in Fig. 14.9. An inflection point A occurs in the upper part of the velocity curve, and the plane passing through this point effectively marks the interface between clear and muddy water. The point at which the maximum velocity u_{max} occurs can be used to divide h' into two regions with thicknesses h_2' and h_1'. The velocity distributions within them obey different laws[16,22].

For the region below the level of maximum velocity, the flow is not affected by the interfacial resistance; hence the velocity distribution is like that in ordinary open-channel flow. From the experimental data obtained by Geza and Bogich [23] at the Chatou laboratory in France, the velocity distribution essentially obeys the logarithmic law near the bottom for a smooth boundary,

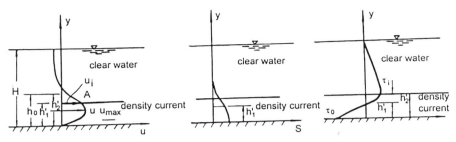

(a) Velocity profile   (b) Sediment concentration profile   (c) shear stress profile

Fig. 14.9 A schematic diagram of the vertical velocity, sediment concentration and shear stress profiles for turbulent density current

$$\frac{u - u_{max}}{U_*} = \frac{1}{\kappa} \log \frac{y}{h_1'} \quad (14.31)$$

in which $U_* = \sqrt{\dfrac{\tau_0}{\rho'}}$

τ_0 is the shear stress on the bottom; the Karman constant κ for sediment-laden flow is less than the value for clear water of 0.4. The value of κ approaches 0.4 as the average velocity of the sediment-laden liquid increases, but the trend is not well established by the data.

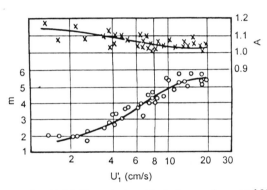

Fig. 14.10 The relationship between A, m and U_1' (after Geza, B., and J. Bogich)

The velocity distribution for a density current in a flume with a smooth bottom can also be expressed by an exponential formula:

$$u = A u_{max} \left(\frac{y}{h_1'}\right)^{1/m} \quad (14.32)$$

in which the coefficient A and m in the exponent are functions of U_1', as shown in Fig. 14.10; then

$$\frac{U_1'}{u_{max}} = A \frac{m}{m+1} \quad (14.33)$$

In the interfacial region between the density current and clear water ($h_0>y>h_1'$), the velocity distribution has an inflection point at A as the velocity drops from u_{max} to zero. The form of the curve is similar to that for the velocity distribution of a turbulent jet (Fig.14.11). It follows the Gaussian normal error distribution [24],

$$\frac{u}{u_{max}} = e^{-\frac{1}{2}\left[\frac{y-h_1'}{\sigma}\right]^2} \qquad (14.34)$$

in which σ is the distance from the point of maximum velocity to the inflection point; the local velocity u_i there is equal to $0.606u_{max}$. Fig. 14.12 is a comparison of field data with the formula, and the two correspond well. The average velocity U_2' in this region is $0.86\ u_{max}$ from Eq. (14.34).

Recent studies show that the interfacial

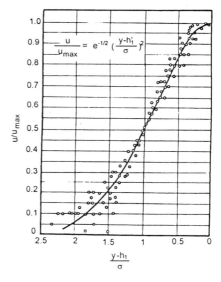

Fig. 14.12 A comparison of velocity distribution between field data and theoretical formula in transitional region of turbulent density current (after Albertson, M.L. et al)

mixing in a supercritical density current differs greatly from that in a subcritical one, and it may affect the velocity distribution. Fig. 14.13 is a comparison of the velocity distributions for a supercritical density current with that for a subcritical one. Several of the velocity distributions for subcritical density currents correspond with the

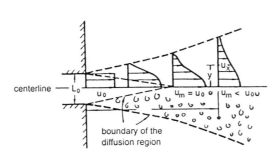

Fig. 14.11 A schematic diagram of velocity distribution of a turbulent jet in diffusion

velocity distribution shown in Fig. 14.9 (a) [25]. For distributions in supercritical density currents, the elevation of the point of maximum velocity is significantly lower. From the data of Chatou, Middleton found that the location of the point with maximum velocity (represented by the ratio h_2''/h_1') in Fig. 14.9a is probably a function of the Froude number of the density current.

14.1.3.4 Resistance

1. Resistance in a laminar density current

For the free body ABCD in a two-dimensional density current, as shown in

Fig. 14.14, the forces acting on the body are:

$$p_1 = (\rho g h + \frac{1}{2}\rho' g h')h'$$

$$P_2 = [\rho g (h + \Delta x \cdot J) + \frac{1}{2}\rho' g h']h'$$

$$W = \rho' g h' \Delta x$$

$$\tau_i = \lambda_i \frac{\rho' U'^2}{2}$$

$$\tau_0 = \lambda_0 \frac{\rho' U'^2}{2}$$

From a force balance,

$$\lambda_i \frac{\rho' U'^2}{2} \Delta x + \lambda_0 \frac{\rho' U'^2}{2} \Delta x = \rho' g h' \Delta x J + (\rho g h + \frac{1}{2}\rho' g h')h'$$

$$- [\rho g (h + \Delta x J) + \frac{1}{2}\rho' g h']h'$$

The result can be simplified to

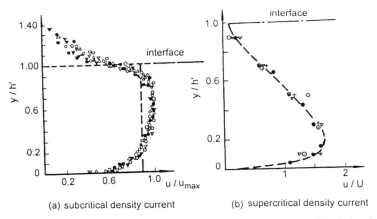

(a) subcritical density current   (b) supercritical density current

Fig. 14.13 A comparison of velocity distribution for a supercritical density current with that for a subcritical one (after Georgiev, B.V., and Wilkinson, G.V.)

$$U'^2 = \frac{2}{\lambda_0}\frac{\Delta\rho}{\rho'}g\frac{h'}{1 + \lambda_i / \lambda_0}J = \frac{2}{\lambda_0}\frac{\Delta\rho}{\rho'}g\frac{h'}{1 + \alpha}J \qquad (14.35)$$

in which

$$\alpha = \frac{\lambda_i}{\lambda_0} = \frac{\tau_i}{\tau_0} \qquad (14.36)$$

In two-dimensional uniform flow, the shear stress is linearly distributed, from τ_0 at the river bottom, to zero at $y = y_m$ $(u = u_{max})$, and on the interface its value is τ_i, hence

$$\alpha = \frac{h' - y_m}{y_m} = \frac{\dfrac{1}{3} - \xi}{\dfrac{1}{6} + \xi} \tag{14.37}$$

in which ξ is defined in Eq. (14.25). From experimental data, $\xi = 0.14$, and substituting this value in Eq. (14.37) , one obtains $\alpha = 0.63$; thus the shear stress on the

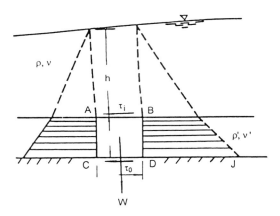

Fig. 14.14 External forces acting on a free body ABCD in density current

interface is about 63% of that at the bottom.

A comparison of Eqs. (14.25) and (14.35) yields

$$\lambda_0 = \frac{2}{1 + \alpha\,\xi} \frac{1}{Re'} \tag{14.38}$$

With the substitution of the values for α and ξ, the relationships for the resistance in a laminar density current are

$$\lambda_0 = \frac{8.75}{Re'}$$
$$\lambda_i = \frac{5.50}{Re'} \tag{14.39}$$

Fig. 14.15 is a comparison of observed data with the theoretical result, and the data follow closely the theoretical line.

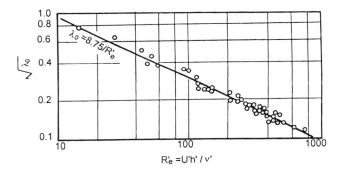

Fig. 14.15 Resistance to laminar density current
(experiment results) (after Ippen and Harleman)

Raynaud treated the same problem but with the difference that he neglected α in his derivation; that is, he neglected the resistance on the interface. As a result, the friction factor is 1.63 times that in Eq. (14.38) or Eq. (14.39). The experimental data of Raynaud are shown in Fig. 14.16, and in the laminar region [21]

$$\lambda_i = \frac{14}{Re'} \tag{14.40}$$

If the Reynolds number exceeds 1600, the experimented data deviate from the straight line and indicate a transition region leading to a constant friction factor for turbulent flow.

The foregoing experiments were all carried out using clear water and a sediment-laden flow. Geza and Bogich also performed an experiment in which they used clear water and various kinds of oil. They found that the factor in Eq. (14.40) depended on the viscosity of the oil in the upper layer. In their experiment, its value varied within the range of 12 to15 [23].

2. Resistance in a turbulent density current

As already mentioned, a turbulent density current can be divided into two regions (Fig. 14.9), one is the near-bottom region $(0<y<h_1')$, the other is the mixing region $(h_1'<y<h')$; the average velocities in the two regions are U_1' and U_2', respectively. If the resistance factors for the bottom and for the interface are expressed in the following forms,

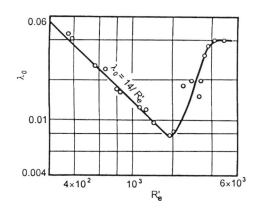

Fig. 14.16 Resistance to laminar and transitional density current (after Raynaud, J.P.)

641

$$\left.\begin{array}{l}\lambda_0 = \dfrac{2\tau_0}{\rho U_1'^2} \\[4mm] \lambda_i = \dfrac{2\tau_i}{\rho U_2'^2}\end{array}\right\}\qquad(14.41)$$

in which τ_0 and τ_i are the shear stress at the bottom and that at the interface; then from the results of studies at the Chatou laboratory [16,22], λ_0 is a function of the relative roughness and the Reynolds number of the flow in the near-bottom region; the latter parameter is defined as

$$\mathrm{Re}_1' = \frac{h_1' U_1'}{v'}\qquad(14.42)$$

If the river bed is smooth, roughness is not a factor, and λ_0 is a function only of Re_1'. Figs. 14.17 and 14.18 present the results of observations in a flume and in the field, respectively. They show that λ_0 is essentially constant in the turbulent range with a value of about 0.08. If the flume is rough, λ_0 depends on the relative roughness, and experimental data for this case are shown in Fig. 14.19. As the absolute roughness did not change during the experiment, the quantity $1/h_1'$ is proportional to the relative roughness. The friction factor rises sharply with increasing relative roughness. In the mixing region, experimental results indicate that the friction factor λ_i varies with the Froude number, defined as follows:

$$Fr_2' = \frac{u_{max}^2}{g\,\dfrac{\Delta\rho_m}{\rho}\,h_2'}\qquad(14.43)$$

in which $\Delta\rho_m$ is the difference between the density at the point of maximum velocity and that at the water surface. Fig. 14.20 shows the relationship between λ_i/J and Fr_2'; the data include tests in flumes, canals and reservoirs, and all the points fall along a straight line, for which the formula is

$$\frac{\lambda_i Fr_2'}{J} = 6\qquad(14.44)$$

Macagno and Rouse studied interfacial resistance[28]. So that the formulation of interfacial turbulence would be caused primarily by the instability of the interface itself, they arranged for two flows with the same density, thickness, and average velocity to move in opposite directions in a rather long air-tight container; a section is shown in Fig. 14.21. Since the properties of the interface depend mainly on characteristics of flow near the interface, they chose U_0 and h_0, shown in the figure, as

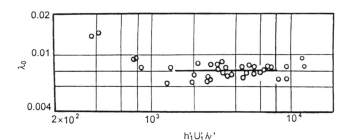

Fig. 14.17 Resistance to turbulent density current (experiment result of Chatou laboratory,France, with smooth flume bottom)(after Michon, X. et al)

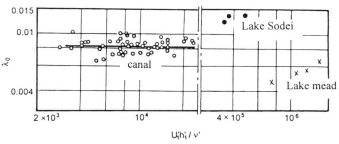

Fig. 14.18 Resistance to turbulent density current (field data in concrete channels and reservoirs) (after Bonnefille, R., and J. Goddet)

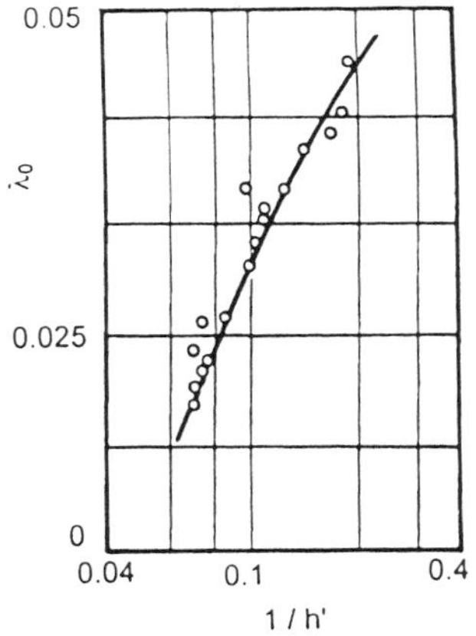

Fig. 14.19 λ_0 vs. $1/h'$

the characteristic velocity and length. Appropriate forms for the Reynolds and Froude numbers are then as follows

$$Re'_0 = \frac{U_0 h_0}{v} \qquad (14.45)$$

$$Fr'_0 = \frac{U_0}{\sqrt{\dfrac{\Delta\rho}{\rho} g h_0}} \qquad (14.46)$$

The experimental results show that the ratio of the shear force τ_i on the interface and the corresponding shear force $\mu(\dfrac{du}{dy})_0$ for laminar flow is a function of Re_0' and Fr_0', as shown in Fig. 14.22a, in which the curve $\tau_i / [\mu (du/dy)_0] = 1$ denotes the laminar shear stress. The quantity $(du/dy)_0$ is the velocity gradient at the interface. Only

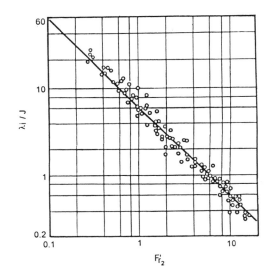

Fig. 14.20 λ_i / J vs. Fr_2' (after Macagno, E.S. and H. Rouse)

643

if the value of τ_i is rather large is the ratio τ_i / $[\mu \ (du/dy)_0]$ a function of the Froude number alone; that is, does not depend on the Reynolds number.

Recently, more data on the resistance at a density current interface have been obtained from both laboratory and field measurements. Bo Pedersen made a comprehensive analysis of these data and presented a universal equation for the interfacial resistance, for both laminar and turbulent flows [29].

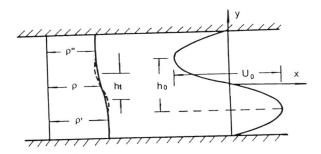

Fig. 14.21 A schematic diagram of density and velocity distribution when the upper flow and lower flow move in opposite direction (experiment of Macagno, E.S. ,and H. Rouse)

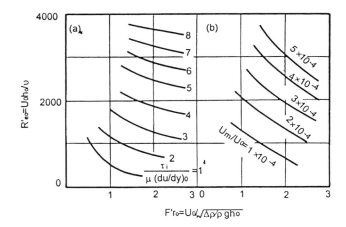

Fig. 14.22 Relationship between shear stress, the mixing factor at the interface and Fr_0', Re_0' (experiment of Macagno, E.S., and H. Rouse)

$$\sqrt{\frac{2}{\lambda_i}} = 2.45[\ln(Re_2'\sqrt{\frac{\lambda_i}{2}}) - 1.3]$$ (14.47)

in which

$$\lambda_i = \frac{2\tau_i}{\rho'(u_{max} - u_i)^2}$$ (14.48)

$$\text{Re}_2' = \frac{(u_{\max} - u_i)h_2'}{v'} \tag{14.49}$$

The quantities used are as defined in Fig.14.9. This equation is applicable within the range

$$500 < Re_2' < 10^7$$

In a laminar density current, the ratio of the interfacial friction coefficient λ_i to that for the flume bottom λ_0 is a constant, with a value of 0.63. For a turbulent density current, this ratio is affected by the Froude number and the relative roughness of flume bottom. From limited flume data, Middleton plotted λ_i/λ_0 vs Fr', as shown in Fig. 14.23[27]; the data points are rather scattered. The flume bottom was smooth, so that the effect of relative roughness was not investigated.

Some researchers have treated the interfacial shear stress and bottom shear stress together, instead of separately; they used a generalized resistance to represent them. For example, Levy suggested the use of the following unified resistance coefficient

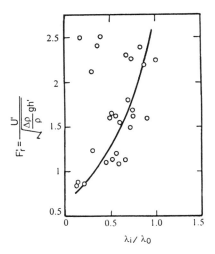

Fig. 14.23 λ_i / λ_0 vs. Fr' (after Middlton, G.V.)

$$\lambda_c = \frac{\lambda_0}{2} + \frac{\lambda_i}{2}\frac{H}{H - h'} \tag{14.50}$$

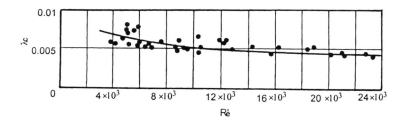

Fig. 14.24 λ_c vs. Re' (experiment of Levy)

The relationship of λ_0 to Re' is shown in Fig. 14.24 [8]. Fan suggested the use of the following expression [6]:

$$\lambda_m = \frac{BH - 2(H - h')h'}{(H - h')(2h' + B)} \lambda_i + \lambda_0 \tag{14.51}$$

The data from experiments in a smooth flume and from the Guanting Reservoir indicate that λ_m is in the range 0.0050 to 0.0075 for turbulent flow. The values presented for the friction factors are rather different, probably because the investigators used different forms for them.

14.1.3.5 Local losses

During the movement of a density current, local energy losses occur wherever the local boundary conditions change. At IWHR, an analysis of local losses was carried out for three types of boundary change: sudden enlargement, sudden contraction and one type of bend[30]. The relevant variables are the velocity U_1', thickness h_1', density ρ_1', and width B_1 before the flow enters the region of change. The thickness h_2' downstream of the abrupt change can be calculated from the following equations:

sudden enlargement

$$\frac{U_1'^2}{\frac{\Delta\rho_1}{\rho_1'} g h_1'} = \frac{1 - (\frac{h_2'}{h_1'})^2}{2(\frac{h_1'}{h_2'}\frac{B_1}{B_2} - 1)\frac{B_1}{B_2}} \tag{14.52}$$

sudden contraction

$$\frac{U_1'^2}{\frac{\Delta\rho_1}{\rho_1'} g h_1'} = \frac{1 - (\frac{h_2'}{h_1'})^2}{2(\frac{h_1'}{h_2'} - \frac{B_1}{B_2})} \tag{14.53}$$

bend

$$\frac{U_1'^2}{\frac{\Delta\rho_1}{\rho_1'} g h_1'} = \frac{1 - (\frac{h_2'}{h_1'})^2}{2(\frac{h_1'}{h_2'} - 1)} \tag{14.54}$$

If the thickness of the density current downstream of local change is known, one can calculate the density of the heavier layer in accordance with the equation of non-uniform flow as introduced in the following paragraphs. Also, from that density difference, one can estimate the quantity of sediment deposited in the region of local change.

The local losses can also be expressed in terms of a local loss coefficient λ_L, which is defined as

$$\lambda_L = \frac{h_f}{2\frac{\Delta\rho'}{\rho'}g} \tag{14.55}$$

in which h_f is the head loss caused by the local change in boundary. Fig. 14.25 shows the relationship between these loss coefficients and the unit discharge for a bend and a sudden enlargement.

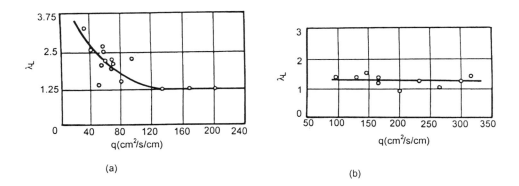

(a)   (b)

Fig. 14.25 Local loss coefficient for density current

14.1.3.6 Hydraulic jump

If a hydraulic jump forms in normal open-channel flow, some flow energy is lost, i.e., changed into heat, in the disturbed region. The same phenomenon occurs in density currents; the pressure change at a front in the atmosphere is another similar phenomenon.

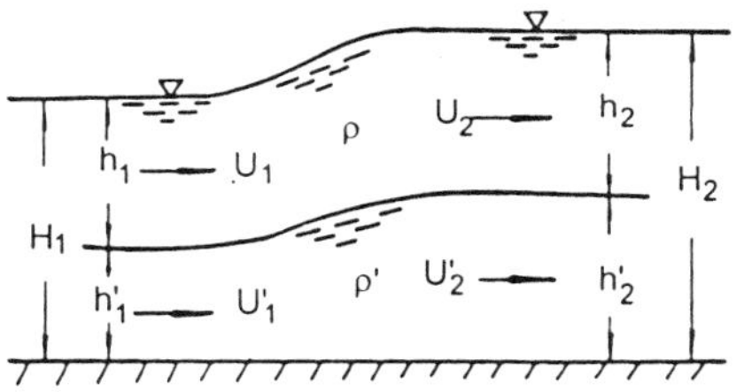

Fig. 14.26 Hydraulic jump
in density current

With reference to the stratified density flow shown in Fig. 14.26, the upper layer has a density ρ, a unit discharge q, the water depths upstream and downstream are h_1 and h_2, and the corresponding velocities are u_1 and u_2. The comparable variables for the lower layer are indicated by a prima. The water depths before and after the jump are called conjugate depths. If the two depths have the same value, that depth is called the critical depth h_c, and the jump that forms in this case is extremely weak.

If a jump forms in the two-layer flow, some mixing usually occurs and the local density changes. In the following calculation, however, the mass exchange is neglected, so that the density after the jump is considered to be unchanged. Komar analyzed the density change after a jump in a turbidity current on a sea bed[31].

Tepper obtained the following solution for the condition that the jump does not affect the free surface, that is, for $H_1 = H_2$,

$$\frac{x+1}{2} - \frac{Fr^2}{x} = 0 \tag{14.56}$$

For the various flow conditions, the values of x and Fr in this equation are as defined in Table 14.2.

Table 14.2   Values of x and Fr in Eq. (14.56)

Flow conditions	x	Fr
Open-channel flow($\rho' \gg \rho$)	h_2' / h_1'	$U_1' / \sqrt{gh_1'}$
Upper layer is at rest	h_2' / h_1'	$U_1' / \sqrt{\dfrac{\Delta\rho}{\rho'} gh_1'}$
Both upper and lower layers are in motion	$\dfrac{h_2' / h_1'}{1 + 2\dfrac{\rho h_1}{\rho' h_1'}\dfrac{U_1}{\sqrt{\dfrac{\Delta\rho}{\rho'} gh_1}}}$	$\dfrac{U_1' / \sqrt{\dfrac{\Delta\rho}{\rho'} gh_1'}}{1 + 2\dfrac{\rho h_1}{\rho' h_1'}\dfrac{U_1}{\sqrt{\dfrac{\Delta\rho}{\rho'} gh_1}}}$

The hydraulic jump in a density current can be either normal or inverse, as shown in Fig. 14.27. Tepper proved that the inverse jump can form only if the upper and lower layers flow in the same direction and the ratio of the velocities satisfies the relationship

$$\frac{U_1}{U_1'} > \frac{h_2(h_1' + h_2')}{h_2'(h_1 + h_2)}\left[\frac{(h_1' - h_2')^2 \dfrac{h_2}{h_1}}{2\dfrac{\rho}{\Delta\rho}\dfrac{U_1^2}{g}(h_1' + h_2')} + 1\right] > 1 \tag{14.57}$$

648

Yih and Guna analyzed further the characteristics of conjugate water depths for jumps in density currents[33]. If the interfacial normal stress is negligible, the pressure exerted on the jump by the upper layer is effectively the static pressure, so that the average pressure head is $\frac{1}{2}(h_1 + h_2)$; then, from the momentum relationship for the lower layer, in the horizontal direction,

$$\rho'q'^2(\frac{1}{h_2'} - \frac{1}{h_1'}) = h_1'h_1\rho g + \frac{1}{2}h_1'^2\rho'g + \frac{1}{2}(h_1 + h_2)(h_2' - h_1')\rho g$$

$$- h_2'h_2\rho g - \frac{1}{2}h_2'^2\rho'g$$

in which q' is the discharge per unit width for the lower layer. If

$$a' = q'^2 / g$$
$$\theta = \rho / \rho'$$

then the preceding equation can be rewritten in the form

$$2a'(h_1' - h_2') = h_1'h_2'(h_1' + h_2')[\theta(h_1 - h_2) + (h_1' - h_2')] \qquad (14.58)$$

Also, from the momentum relationship applied to the upper layer,

$$2a(h_1 - h_2) = h_1h_2(h_1 + h_2)[(h_1 - h_2) + (h_1' - h_2')] \qquad (14.59)$$

in which

$$a = q^2 / g$$

and q is the discharge per unit width for the upper layer. If h_1 and h_1' are known, one can solve Eqs. (14.58) and (14.59) so as to obtain the conjugate depths h_2 and h_2'.

From the analysis by Yih and Guna, three conjugate states can occur at most. If

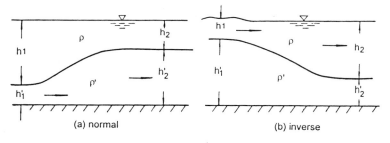

Fig. 14.27 Normal hydraulic jump and inverse hydraulic jump

649

the Froude number of one of the layers is very large, then only one conjugate state is possible, and the conditions downstream of the jump can be completely determined. With more than one conjugate state, the flow situation after the jump can not be fully determined from the momentum relationships because it also depends on the type of downstream control.

If the upper layer is at rest, $a = 0$, and Eq. (14.59) can be rewritten,

$$\frac{h_2'}{h_1'} = \frac{1}{2}(\sqrt{1+8Fr'^2} - 1) \tag{14.60}$$

Only if $F_1'>1$ and $h_2'>h_1'$, can a normal jump take place. Also, from Eq. (14.59), for this situation

$$h_1+h_1' = h_2+h_2'$$

which means that the free surface is horizontal. If the lower layer is still and the upper layer moves, so that, $a'= 0$, one sees from Eq. (14.58) that

$$\theta h_1 + h_1' = \theta h_2 + h_2'$$

and the free surface is no longer horizontal. But if the density difference between the lower and upper layers is small, θ is close to 1, and the change in the level of the free surface is not significant.

If the water depth is equal to the critical depth, that is, if the flow is everywhere critical, (h_2-h_1) and $(h_2'-h_1')$ are infinitesimal, and they can be replaced by dh_1 and dh_1' respectively. Then Eq. (14.58) and Eq. (14.59) have the following differential forms:

$$\frac{dh_1'}{dh_1} = \frac{\theta h_1'^3}{a' - h_1'^3} \qquad\qquad \frac{dh_1'}{dh_1} = \frac{a - h_1^3}{h_1^3}$$

and the critical condition is the following:

$$h_{1c}^3 = \frac{a(a' - h_{1c}'^3)}{a' - (1-\theta)h_{1c}'^3} \tag{14.61}$$

Yih and Guna verified Eq. (14.60) by conducting experiments in a flume. The result showed that the equation essentially reflects the conditions for a jump in a density current, but because the interfacial resistance was neglected in the derivation, the measured values of h_2' are a little smaller than the theoretical ones. As the Froude number and density ratio θ increase, the interfacial resistance becomes more and more important, and the deviation becomes larger and larger.

Eq. (14.60) can also be applied to a hydraulic jump in an open-channel. Experiments show that the profile of a jump for a density current is similar to one in ordinary open-channel flow; however, for the same Froude number in the approach flow, the jump in a density current goes to a higher lever and is longer than one in ordinary open-channel flow, as shown in Fig. 14.28 [34]. The difference is related to the change in density across the jump caused by violent mixing.

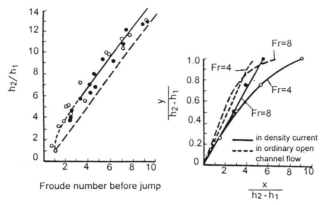

Fig. 14.28 A comparison of geometrical contour of hydraulic jump in density current and that in ordinary open-channel flow (after Wood, I.R.)

14.1.3.7 Non-uniform flow

If a reservoir is not long, the motion of a density current in the reservoir may not become uniform. In flume experiments, particularly, the flow has distinctly non-uniform characteristics.

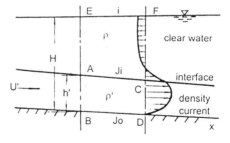

Fig. 14.29 Non-uniform density current with a reverse slope on free surface

Deriving the equation of non-uniform motion for a density current, Levi considered the shear at the interface between a density current and the clear water in the upper layer. The shear induces a circulation in the upper layer and consequently a reverse slope occurs on the free surface[8]. If the slopes of the free surface, interface and reservoir bottom are $-i$, J_i and J_0, then the momentum equation for the free body of unit length ACDB in Fig. 14.29 is

$$(\rho' - \rho)gh'(J_0 - \frac{dh'}{dx}) = \lambda_0 \frac{\rho'U'^2}{2} + \lambda_i \frac{\rho'U'^2}{2} + \rho gh'i + \rho'g\alpha h' \frac{d}{dx}(\frac{U'^2}{2g}) \qquad (14.62)$$

in which α is a correction factor for the velocity distribution.

Next, the free body EFBD can be used in an analysis of the density current. Because the velocity of the clear water is small, the effect of kinetic energy in the clear water can be reflected by the change in the velocity distribution factor (from α to α'). The corresponding momentum equation is

$$(\rho' - \rho)gh'(J_0 - \frac{dh'}{dx}) = \lambda_0 \frac{\rho'U'^2}{2} + \rho gHi + \rho'g\alpha'h' \frac{d}{dx}(\frac{U'^2}{2g}) \tag{14.63}$$

To simplify, one can assume

$$\alpha = \alpha'$$

Solving Eq. (14.62) and Eq. (14.63) simultaneously, one obtains

$$i = \frac{\rho'}{\rho} \frac{\lambda_i}{2g} \frac{U'^2}{H - h'} \tag{14.64}$$

Combining this with Eq. (14.62) and simplifying, one gets

$$\frac{\Delta \rho}{\rho'} h'(J_0 - \frac{dh'}{dx}) = \frac{U'^2}{2g}(\lambda_0 + \lambda_i \frac{H}{H - h'}) + \alpha h' \frac{d}{dx}(\frac{U'^2}{2g}) \tag{14.65}$$

For two-dimensional flow, the equation of continuity is

$$q = h'U' = constant$$

If the critical depth h_c' is defined as

$$h_c' = \left(\frac{\alpha q^2}{\frac{\Delta \rho}{\rho'} g} \right)^{1/3} \tag{14.66}$$

then Eq. (14.65) becomes

$$\frac{dh'}{dx} = \frac{J_0 - \frac{1}{2\alpha}(\frac{h_c'}{h'})^3(\lambda_0 + \lambda_i \frac{H}{H - h'})}{1 - (\frac{h_c'}{h'})^3} \tag{14.67}$$

If $dh'/dx = 0$, the flow is uniform, and the normal water depth h_n' is defined by the following equation

$$h'_n = \sqrt[3]{\frac{q^2(\lambda_0 + \lambda_i \dfrac{H}{H - h'})}{2\dfrac{\Delta\rho}{\rho'}gJ_0}} \qquad (14.68)$$

Substituting Eq. (14.68) into Eq. (14.67), one obtains

$$\frac{dh'}{dx} = \frac{J_0[1 - (\dfrac{h'_n}{h'})^3]}{1 - (\dfrac{h'_c}{h'})^3} \qquad (14.69)$$

This form is known as the Bresse Equation, and it is often used in calculating backwater curves in ordinary open-channel flows. Consequently, all of the back water curves derived for open-channel flows can also occur along the interface of a density current.

14.1.3.8 Unsteady flow

If the inflow to a reservoir forms a density current during the passage of a flood, the flood flow itself is usually unsteady, so that any density current involved would also be unsteady; the velocity, thickness and density then vary with time. For such a situation, the following equation of unsteady motion should be used:

$$\Delta\rho(J_0 - \frac{\partial h'}{\partial x}) = \frac{\rho'}{2}\frac{U'^2}{gh'}(\lambda_0 + \lambda_i \frac{H}{H - h'})$$

$$+ \frac{\alpha U'}{g}\frac{\partial}{\partial x}(\rho'U') + \frac{1}{g}\frac{\partial}{\partial t}(\rho'U') \qquad (14.70)$$

This equation can be solved using methods developed for flows in open-channels; they include the solving of differential equations, the method of characteristics, etc. Fan used the method of characteristics to solve Eq. (14.70) for the special condition that the sediment concentration does not change with time, and he verified the result he obtained experimentally[6].

14.1.3.9 Transverse diffusion

After a density current flows into a wide body of water, it grows laterally. Fietz and Wood assumed that the density current came from a point source, and studied the transverse diffusion of a three-dimensional outflow from an orifice in quiet water in the laboratory [35]. Fig. 14.30 shows the relationship of diffusing angle to Richardson number Ri,

653

$$Ri = (\frac{r^5 g\Delta\rho}{Q^2 \rho_0})^{1/3} \qquad (14.71)$$

in which: r — the radius of the orifice;

Q — the discharge of a density current from the outlet;

$\Delta\rho$ — density difference between the density current and the water in reservoir at the outlet;

ρ_0 — density of water in reservoir.

In this form, the Richardson number is simply the Froude number raised to the power -2/3.

$$Ri = 0.93 Fr'^{(-2/3)} \qquad (14.72)$$

in which

$$Fr' = \frac{U_c}{\sqrt{\frac{\Delta\rho}{\rho} gd}}$$

U_c is the velocity at the outlet, d is the diameter of the orifice.

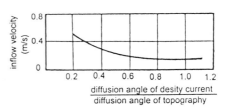

Fig. 14.31 Relationship between transverse diffusion of density current, velocity and topography in Guanting reservoir

In a reservoir, the diffusing angle of the density current is affected by the topography of the reservoir, and the latter can also be described schematically by an angle of expansion. From observations in the Guanting Reservoir, shown in Fig. 14.31, the ratio of these two angles is a function of the mean velocity at the inflow [30]. The curve is flat, especially for large ratios. It shows that the use of inflow velocity as the primary variable does not reflect the entire process well.

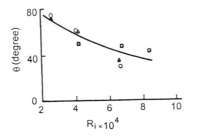

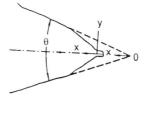

Fig. 14.30 Transverse diffusion of a three-dimensional density current outflow from an quiet water (after Fietz, T.R., and I.R. Wood)

654

14.1.3.10 Formation of a wave at an interface

If two liquid layers move at different velocities, waves will form on the interface once the relative velocity exceeds a certain value. Wind-generated water waves are the most common example of this phenomenon. Two kinds of waves are encountered: oscillatory waves and solitary waves.

1. Oscillatory waves

Fig.14.32a shows the situation in which oscillatory waves commonly occur, both at an interface and on a free surface. The wave celerity at the interface is [2]

$$c = \frac{RU' + TU}{R+T} \pm \left[\frac{g(\rho' - \rho)}{k(R+T)} - \frac{RT}{(R+T)^2}(U'-U)^2 \right]^{1/2} \tag{14.73}$$

in which

$$\left.\begin{array}{l} k = \dfrac{2\pi}{\lambda} \\ R = \rho' \coth kh' \\ T = \rho[\coth kh - \dfrac{b/a}{\sinh(kh)}] \end{array}\right\} \tag{14.74}$$

λ is the wave length and T is the wave period. The ratio of the amplitude of the wave at the interface to that on the free surface is

$$\frac{a}{b} = \cosh kh = \frac{g \sinh kh}{k(U-c)^2} \tag{14.75}$$

Generally, for a density current in a reservoir, the clear water in the upper layer is hardly moving ($U \approx 0$); because the clear water in the reservoir is quite deep, the interfacial wave is not large enough to induce significant waves at the free surface. For that case,

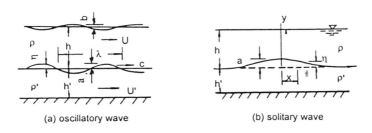

(a) oscillatory wave   (b) solitary wave

Fig. 14.32 Wave phenomena on the interface

$$c = \frac{\rho' U' \coth kh'}{\rho' \coth kh + \rho \coth kh} \pm \left[\frac{g(\rho' - \rho)}{k(\rho' \coth kh' + \rho \coth kh)} \right.$$

$$\left. - \frac{\rho' \rho \coth kh' \coth kh}{(\rho' \coth kh' + \rho \coth kh)^2} U'^2 \right]^{1/2} \tag{14.76}$$

If river water flows into the sea at the mouth of a river and forms a density current on the sea surface, the potential energy of the wave at the interface is relatively small; hence, under certain conditions, large waves can be easily induced there. For a density current on the water surface of an estuary, velocities of both the upper and lower layers can be neglected ($U' = U = 0$), the depth of the upper layer is comparatively small, the depth of the lower layer can be assumed to approach ∞ (so that $coth\ kh' \rightarrow 1$); then, Eq. (14.73) and Eq. (14.75) can be written as follows:

$$c^2 = \frac{g(\rho' - \rho)}{k(\rho' + \rho \coth kh)} \tag{14.77}$$

$$\frac{a}{b} = -\frac{\rho}{\rho' - \rho} e^{kh} \tag{14.78}$$

If $\rho' \rightarrow \rho$ or if kh is large, that is, if the wave is relatively short, $a >> b$; for these conditions, large waves can form on the interface between clear and salt water and still remain hidden under a nearly calm sea surface. Such waves can cause a tremendous resistance to the motion of ships. In nautical terminology, they are called "dead water" [36].

In the open sea, thermal layers are common, and extremely long oscillatory waves can form at the interface between them. In this case, kh ($= \frac{2\pi}{\lambda}h$) is small, coth kh can be replaced by $\frac{1}{kh}$, $U' = U = 0$, and $\rho' \rightarrow \rho$. Hence, Eq. (14.73) becomes

$$c^2 = \frac{gh'h(\rho' - \rho)}{(h' + h)\rho'} \tag{14.79}$$

Hourwitz introduced the effect of the rotation of the earth into the above equation, and he found that the period of such interfacial waves is sometimes close to the period of a tidal wave [37].

2.Solitary wave

A solitary wave at an interface is shown in Fig. 4.32b. Such a wave differs from an oscillatory wave in that a moving wave causes water molecules to be displaced a certain distance in the direction of wave motion after the wave has passed. Keulegan derived the following expression for the celerity of a solitary wave in terms of its amplitude a:

$$c^2 = \frac{gh'h(\rho' - \rho)}{(h' + h)\rho'}\left(1 + \frac{h - h'}{hh'}a\right)$$

(14.80)

The shape of the wave is given by

$$\eta = a \sec h^2\left[\left(\frac{1}{2}\frac{h - h'}{h}\frac{a}{h'}\right)^{1/2}\frac{x}{h'}\right]$$

(14.81)

Eqs. (14.80) and (14.81) show that the relative thickness of the upper and lower layers affect the formation of the solitary wave significantly. If $h' < h$, the solitary wave is a positive wave and the interface is displaced upward; if $h' > h$, the solitary wave is a negative wave and the interface is displaced downward. If the thicknesses of the two layers are the same, and the difference in the densities is small, no solitary wave can form at the interface. Experiments confirm predictions based on Eqs.(14.80) and (14.81)[39].

14.2 SELECTIVE WATER DIVERSION

If the fluid in a reservoir is stratified because of differences in density, one may sometimes seek to divert water from one layer and not from another. For example, if a density current has formed in a reservoir, siltation in the reservoir can be alleviated by the placement of bottom sluices at the dam site so that the density current can be discharged from the reservoir. In this way, water from the upper layer (clear water) is not diverted and remains impounded in the reservoir. Another example is a density current that has formed in a cooling pond for a hydroelectric plant; water from the lower cooling layer is diverted for the hydroelectric plant and the heated water above is to be excluded. As pointed out at the beginning of this chapter, if a stable density gradient exists, vertical exchange is sufficiently reduced that such a requirement can be satisfied by appropriate engineering measures. Actually, questions of selective water diversion arise in many circumstances. They require answers to specified questions: (1) How should orifices be arranged, what outflow is critical? The latter question means that if the discharge of the outflow is larger than this value, some undesirable water will be released. (2) Which region does the outflow come from for a certain elevation of the outlet?

14.2.1 Selective diversion for various specific cases

Much research has been devoted to selective diversion [40], and Table 14.3 contains some of results. The outlet arrangements include diversions from both side wall and bottom openings and by siphons. The density stratification of the water include the

Table 14.3 Some results of selective outflow

Series	Position of outlet	Flow condition	Author	Consideration of theoretical analysis	Critical condition or outflowing range	Reference
1		Lower layer keeps still. Upper layer flow out through outlet. Lower pressure area forms near outlet. Lower layer is abstracted to upraise. In critical condition, lower layer begins to flow out.	Craya and Gariel	Flow viscosity is not considered	$\dfrac{U_c}{\sqrt{\frac{\Delta\rho}{\rho'}gh_L}} = 3.25\left(\dfrac{h_L}{d}\right)^2$ circular outlet $\dfrac{U_c}{\sqrt{\frac{\Delta\rho}{\rho'}gh_L}} = 1.54\dfrac{h_L}{d}$ two dimensional oulet	[41,42]
			Bohan and Grace		$\dfrac{U_{c\,i}}{\sqrt{\frac{\Delta\rho}{\rho'}gh_L}} = \dfrac{h_L^2}{A_0}$ A_0--area of outlet	[43]
2		Density distribution is linear.	Stretch from series 7	Two dimensional outlet. Flow viscosity is not considered	$\delta = 2.7\left(\dfrac{q^2}{g\beta}\right)^{1/4}$	[40]
		Outflow discharge is q (to two dimensional outlet) or	Gelhar and Mascolo	Two dimensional outlet. Flow viscosity is considered	$\delta = 4.74\left(\dfrac{qx\nu}{g\beta}\right)^{1/5}$ $u_{max} = 0.416\left(\dfrac{q^4\,g\beta}{x\nu}\right)^{1/5}$	[44]
		Q(to three dimensional outlet). Depth of outflowing layer is δ which is dependent on x.	Koh	Two dimensional outlet. Flow viscosity and molecular diffusion in interface are	$\delta = 7.14\dfrac{x^{1/3}}{\alpha_0}$	[45]

3						
		Mean velocity during flowing is U.	Koh	considered Three dimensional outlet. Other conditions as proceeding	$u_{max} = 0.284 \dfrac{\alpha_0 q}{x^{1/6}}$ $\alpha_0 = \left(\dfrac{\beta g}{\varepsilon \nu}\right)^{1/6}$ ε--coefficient of molecular diffusion	[40]
		Maximum velocity in central line is Umax.	Koh	Two dimensional outlet. Turbulent diffusion in interface is considered	$\delta = 5.8 \dfrac{r^{1/3}}{\alpha_0}$ $u_{max} = 0.35 \dfrac{Q}{2\pi r} \dfrac{\alpha_0}{r^{4/3}}$ r— radial distance $\dfrac{\delta}{\alpha} = 8.4 \left(\dfrac{\varepsilon_m x}{q a}\right)^{0.245}$ $2.7 < \delta/\alpha < 13.7$ $\dfrac{\delta}{\alpha} = 7.14 \left(\dfrac{\varepsilon_m x}{q a}\right)^{1/3} \cdot \ \delta/\alpha > 13.7$ ε_m--coefficient of turbulent mass exchange, $a = \left(\dfrac{q^2}{g\beta}\right)^{1/4}$	[40]
		Lower layer outflow through bottom outlet. Upper layer is about to outflow.	Harleman, Morgan and Purple	Flow viscosity is not considered	$\dfrac{U_c}{\sqrt{\dfrac{\Delta\rho}{\rho'}gh_L}} = 2.05\left(\dfrac{h_L}{d}\right)^2$	[46]

#	Diagram	Description	Author	Consideration	Equation	Ref.
4		Two layers both stretch to infinite distance. In cirtical condition, lower layer is about to be diverted.	Davidian and Glover	Flow viscosity is not considered	$$\frac{U_c}{\sqrt{\frac{\Delta\rho}{\rho'}gh_L}} = 5.70\left(\frac{h_L}{d}\right)^{3/2}$$	[47]
5		Lower layer outflow right through submerged bottom outlet. In cirtical condition, upper layer is about to outflow together	Harleman, Gooch and Ippen	Flow viscosity is not considered	$$\frac{U_c}{\sqrt{\frac{\Delta\rho}{\rho'}gh''}} = f\left(\frac{h'}{h''}, \alpha\right)$$ Where functional relationship is shown as Fig.14-33. α-- correction factor of nonuniform velocity distribution near outler	[48]
6		Two layers are still at first. Then lower layer outflow through the two dimensional outlet in the corner. In cirtical condition, upper layer is about to outflow.	Huber	Flow viscosity is not considered	$$\frac{U_c}{\sqrt{\frac{\Delta\rho}{\rho'}gh'}} = 1.66$$ (theoretical results) $=0.87$ (experimental results)	[49,50]
7		Vertical density distribution is linear. Outflowing through the bottom outlet is limited to an area of δ. In cirtical condition, upper layer is about to outflow.	Walesch and Monkmeyer	Two dimensional outlet. Flow viscosity is considered	$$\delta = 2.47\left(\frac{qx\nu}{g\beta}\right)^{1/5}$$ $$u_{max} = 0.556\left(\frac{q^4 g\beta}{x\nu}\right)^{1/5}$$	[51]
			Yih and Debler	Consider outflowing condition of upper layer	$$\frac{q}{H^2\sqrt{g\beta}} = 032 \text{ (theoretical results)}$$ $=0.28$ (experimental results)	[52,53]

cases of water in two layers and a variable density with a linear distribution in the vertical. In theoretical analyses, the following assumptions are generally adopted:

1. If the velocity field near the outlet is the primary concern, the water there accelerates, and the inertia and gravity forces are the primary factors; hence, the viscosity force is usually neglected. However, for the velocity field far from the outlet, viscosity forces must be included.

2. Because the variation of density is small, variations of the viscosity can be neglected; diffusion and mixing at the interface is generally not included. However, if the Froude number is large, some researchers took the effect of diffusion into consideration.

3. The vertical dimension of the flowing layer is much smaller than the whole water depth, so the inertia term can be neglected.

4. Because of mathematical considerations, studies are limited to two-dimensional analyses; still, some three dimensional results can be obtained from the two-dimensional analyses.

In addition to theoretical analyses, many indoor experiments of small scale have been conducted. Because of practical limitations on the scale of such studies, the Reynolds number is small and the flow is often laminar. However, turbulence of some intensity can exist in the field. Consequently, effects of scale on a flow pattern should be fully explored if the results of model studies are to be applied to practical projects.

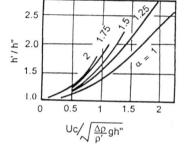

Fig. 14.33 $U_c / \sqrt{\dfrac{\Delta\rho}{\rho} gh''}$ vs. h'/h'' and α

(after Harleman, D.R.F. et al)

As shown in Table 14.3, the critical condition for two layers of different density so that both layers are diverted from the outlet is

$$Fr' = \frac{U_c}{\sqrt{\dfrac{\Delta\rho}{\rho} gh_L}} = m(\frac{h_L}{d})^n \tag{14.82}$$

in which

U_c — sluice flow velocity;

h_L — distance between undisturbed layer surface and outlet;

d — size of the outlet.

The exponent n varies between 1 and 2. The coefficient m is related to the shape of the outlet. With a density gradient, no matter whether liquid is diverted from a bottom outlet or from some other elevation, the outflowing water comes only from a region, the outer boundaries of which are $\delta/2$ from the outlet axis. The farther the liquid is from the outlet, the greater the value of δ. The experimental results of different researchers show that δ is proportional to the distance x raised to a power of 1/5 to1/3.

14.2.2 Quantity of diversion from a density current in a reservoir

A density current forms a wave head and proceeds along the bottom of a reservoir, and it rises up when it meets the dam. If the velocity of the heavier layer is high, some of the turbid water forms a reflecting surge. If the velocity is low, a long wave forms. If the velocity is very low, the long wave disappears; the result is simply that the depth of turbid water gradually increases, and the liquid flows out through the outlet once the level reaches a certain elevation.

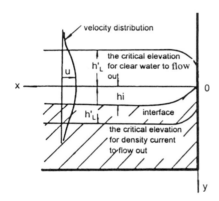

Fig. 14.34 A schematic diagram of a density current passing through an outlet

The rise of a density current when it meets the dam is the result of a conversion of its kinetic energy to potential energy. Because of the buoyant effect of the upper clear water on the turbid water, the potential energy can cause a density current to rise large vertical distances, even though the corresponding rise for the same potential energy is quite small in normal open channel flow. On the other hand, because the velocity of density current in reservoir is not large, and its kinetic energy is therefore small, if the outlet is some distance above it, the density current may not flow out. The situation is not like that in the atmosphere. In the latter case while proceeding along the ground surface, a density current formed by cold air mass can easily climb up hills and continue its way because of its high velocity. Besides, like a sunk, the flow passing through an outlet creates a region of lower pressure nearby, and the turbid water can be raised by the suction caused by the low pressure.

When turbid water reaches a critical height, shown in No. 1 of Table 14.3, it begins to flow out. The sediment concentration of the outflow varies with the relative position of the interface and the elevation of the outlet. If the interface is on the same level as the outlet, about half of the flow region upstream of the outlet is occupied by clear water, and the other half by sediment-laden flow water. If the elevation of the interface is higher than the outlet, the ratio of turbid water in the outflow is greater. Finally, if the interface is high enough, another critical condition is reached, one at which the sediment-laden flow dominates the outflow. The concentration of the outflow is equal to the concentration of the density current. This elevation is called the critical elevation for any clear water to flow out (shown as Fig. 14.34). If the interface is higher than this critical elevation, the concentration of the outflow can increase no further. The discussion on the concentration of outflow in different cases is as following [6].

If the outlet is higher, the velocity distribution can be considered as symmetric to the outlet center. To the bottom outlet, velocity distribution in front of the outlet is approximately the upper half of symmetric velocity distribution. The velocity variation in front of the two-dimensional outlet, except the area near the outlet, can be represented as $u = q/(\pi r)$. The velocity varies slightly in cross section when it is far from the outlet.

In the terminology used, the concentration of the density current reaching the dam is S', in weight per unit volume, the concentration of released water S_0, the concentration at a distance y from the orifice center S, the distance between the interface and the orifice center h_i. The flow passing through the orifice comes from the region $h_L' < y < h_L$ and it has an average velocity U; then

$$\frac{S_0}{S'} = \frac{1}{U(h_L + h_L')S'} \int_{h_i}^{h_L} uSdy \tag{14.83}$$

The boundary conditions are

$$y = h_L \qquad\qquad S_0 = 0$$

$$y = -h_L \qquad\qquad S_0 = S'$$

$$y \to 0 \qquad\qquad S = \frac{1}{2}S'$$

For a bottom sluice, $h_L = 0$, and therefore

$$\frac{S_0}{S'} = \frac{1}{Uh_L'S'} \int_{h_i}^{0} uSdy \tag{14.84}$$

For a homogeneous density current, such as one containing salt, the concentration is nearly uniformly distributed, $S = S'$, and far away from the orifice, the velocity profile is also nearly uniform. Therefore Eq. (14.83) can be simplified,

$$\frac{S_0}{S'} = \frac{h_L - h_i}{h_L + h_L'} \tag{14.85}$$

If the approximation $h_L = h_L'$ is introduced, substitution of the critical condition from No.1 in Table 14.3 into the preceding formula for a two-dimensional orifice yields

$$\frac{S_0}{S'} = \frac{1}{2}\left[1 - K_1\left(\frac{\Delta\rho}{\rho}\frac{gh_i^3}{q^2}\right)^{\frac{1}{3}}\right] \tag{14.86}$$

for a three-dimensional orifice:

$$\frac{S_0}{S'} = \frac{1}{2}\left[1 - K_2\left(\frac{\Delta\rho}{\rho}\frac{gh_i^5}{Q^2}\right)^{\frac{1}{5}}\right] \tag{14.87}$$

For a bottom sluice:

$$\frac{S_0}{S'} = \frac{h_i}{h_L'} \tag{14.88}$$

If h_L' can still be expressed as in No. 1 of Table 14.3, then

for a two-dimensional orifice: $\qquad \dfrac{S_0}{S'} = K_3\left(\dfrac{\Delta\rho}{\rho}\dfrac{gh_i^3}{q^2}\right)^{1/3}$

$$\tag{14.89}$$

for a three-dimensional orifice: $\qquad \dfrac{S_0}{S'} = K_4\left(\dfrac{\Delta\rho}{\rho}\dfrac{gh_i^5}{Q^2}\right)^{1/5}$

The coefficients in Eqs. (14.86), (14.87), and (14.89) must be determined from experiments.

For turbid density currents, the concentration is non-uniformly distributed due to the settlement of sediment particles in a backwater region and to the effect of the velocity distribution near the orifice; therefore the average concentration S near the point of release differs from the average concentration of the density current S', where it reaches the dam. For a two-dimensional orifice placed at a rather high level

$$\frac{S_0}{S'} = \frac{1}{2}\left[K_5 - \left(\frac{\Delta\rho}{\rho}\frac{gh_i^3}{q^2}\right)^{\frac{1}{3}}\right] \tag{14.90}$$

Table 14.4 Concentration of Sluice Flow of Density Current

Density current	Size of outlet (cm)	Number of outlets in 50cm in width	Distance between flume bottom and lower brim of outlet	Variation range of S_0/S'	Corresponding coefficients
Saline wedge	50 × 0.2	1	6, 15, 20, 28	0.5~1.0 0~0.5	K_1=1.55 K_1=1.33
	2 × 2 1 × 1	4 13	5, 11, 24 9, 5, 24, 5	0.5~1.0 0~0.5	K_2=1.67 K_2=1.33
Turbid water	50 × 0.6	1	13 26 39	0~0.5 0~0.5 0.5~1.0	K_1=1.84 K_1=1.26 K_5=1.30
	2 × 2	7	7 24 39	0~0.5 0~0.5 0.5~1.0	K_2=1.43 K_2=0.80 K_6=1.15
	2 × 2 3 × 2	7 4	0 0	– –	K_4=0.47 K_4=0.90

For a three-dimensional orifice:

$$\frac{S_0}{S'} = \frac{1}{2}\left[K_6 - \left(\frac{\Delta\rho}{\rho}\frac{gh_i^5}{Q^2} \right)^{\frac{1}{5}} \right]$$
(14.91)

For an orifice located at a low level with a density interface that is located above the orifice, S and S_i do not differ much, and the concentration of the outcoming flow still obeys Eqs. (14.86), (14.87), and (14.89); only the coefficients are different. Fan conducted flume experiments on the concentration in outflows under various conditions, and his results are shown in Table 14.4[6].

14.3 DISPERSION, TRANSPORT, AND DEPOSITION OF DENSITY CURRENT

Conditions for maintaining a density current are such that the upper and lower layers of liquid do not mix completely, and some effect of gravity remains. Under certain conditions, the stratified flow loses its stability and diffusion through the interface occurs. If water in the upper layer has a certain velocity and turbulence, the added material, such as salt or sediment, disperses from the lower layer to the upper layer, and it can then be transported in the flow direction. For sediment-induced density currents, part of the sediment carried by the density current may deposit during the processes of diffusion and transport.

The mixing of two liquids with different densities consists of local mixing in some regions and a more general mixing at the interface. Where a river empties into a reservoir, part of its energy is lost because of the rapid reduction in momentum. At the plunge point, the incoming flow collides with the reverse flow on the surface of the reservoir. Consequently, the density current loses part of its energy there. Both processes lead to strong local mixing; even though this mixing may not cause the density current to disappear, it dilutes the density current and increases its volume. Field data from Shaver Lake show that such local mixing at the entrance to a reservoir can make the volume of a density flow 2 to 5 times as large as its original size [7]. In addition, where a density current meets a large obstacle, local mixing may also take place. Local losses for a density current passing through sudden contractions, sudden enlargements and sharp bends are discussed in Section 14.1.3.

A number of studies have been made of mixing at an interface. Critical conditions for mixing, the processes of diffusion and transport after loss of interfacial stability and the deposition of a muddy density current are introduced in the following sections.

14.3.1 The interfacial stability of a density current

14.3.1.1 Phenomenon of interfacial mixing

Conditions at an interface between two liquids with different densities vary with the relative velocity of the two layers. If the velocity of a density current in a reservoir is low, only molecular diffusion occurs, and the interface remains clear and plane. If the velocity reaches a certain level, however, waves appear at the interface; a schematic representation of the development is shown in Fig. 14.35a[54] . The tops of interfacial waves are often cut off by the clear water in the upper layer, just as wind blowing over the ocean cuts off the tops of sea waves.

If the velocity of a density current is quite high and the flow is supercritical, a series of eddies appears at the surface, instead of well-defined abrupt waves. The occurrence is similar to the eddies produced on both sides of a submerged jet, as shown in Fig. 14.35b[29] . Either breaking waves or eddies can cause the density current to mix with the clear water at the interface.

Clear water entrained by a density current through mixing comes from a low velocity region; therefore, energy must be drawn from the density current to accelerate it. This action retards the density current near the interface. Still, the turbid water that enters the clear water region cannot penetrate deeply into the clear water region. Hence, an intermediate layer forms that has a density a little higher than that of the clear water in the upper layer. Flow in this new layer gets its momentum from the turbid water that comes from the lower layer with its higher velocity. As a result, the

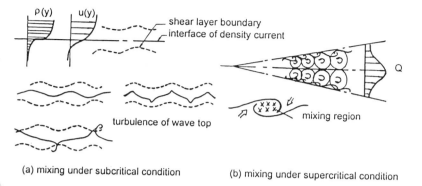

Fig. 14.35 Two types of mixing when the interface of density current loses its stabilit‐

fluid in the intermediate layer moves downstream. That is, the flow forms three layers: the original density current, which is not affected by mixing and moves on the bottom with the original velocity; a second layer, which is the another density current caused by interfacial mixing; an upper layer which is the original clear water in the reservoir and is not affected by the mixing.

The forgoing phenomenon is the first stage of interfacial mixing, and it occurs soon after the turbid water submerges in a reservoir. After the density current has moved some distance, the interface is no longer sharp. Instead, a transitional region, with both a density gradient and a velocity gradient, exists between the upper and lower layers.

14.3.1.2 Critical conditions for the interface losing its stability if the viscosity is neglected

One way of studying interfacial stability is based on the equation for the celerity of an interfacial wave. The wave is stable if the celerity is a real number, and the wave will break if the celerity is an imaginary number. From Eq. (14.76), the celerity is imaginary if

$$\frac{g(\rho'-\rho)}{k(\rho'\coth kh' + \rho\coth kh)} = \frac{\rho'\rho\coth kh' \cdot \coth kh}{(\rho'\coth kh' + \rho\coth kh)^2}U'^2 < 0$$

If ρ' and ρ differs only lightly, the critical condition can be written as

$$\frac{U'_c}{\sqrt{g\frac{\Delta\rho}{\rho}\frac{\lambda}{\pi}}} = \left[\frac{1}{2}\left(\tanh\frac{2\pi h}{\lambda} + \tanh\frac{2\pi h'}{\lambda}\right)\right]^{1/2} \tag{14.92}$$

667

If the wave length is also small, Eq. (14.92) can be further simplified,

$$\frac{U_c'}{\sqrt{g\frac{\Delta\rho}{\rho}\frac{\lambda}{\pi}}} = 1 \qquad (14.93)$$

Eqs. (14.92) and (14.93) are shown in Fig. 14.36a[20]; the abscissa and the ordinate are

$$Fr' = \frac{U_c'}{\sqrt{g\frac{\Delta\rho}{\rho}h'}} \quad \text{and} \quad \frac{2\pi h'}{\lambda}$$

To the right of the curve is an unstable region, and to the left is a stable one. The figure shows that the ratio h/h' plays a role only if the Froude number is larger than 1, Eq. (14.92) and Eq. (14.93) give the same values if the Froude number is less than 1. Experimental results tend to concentrate in the region for $Fr' \cong 1$. Because viscosity was neglected in the derivation, most of the points fall to the right of the curves. In ordinary open channel flow, the specific energy of the flow is a minimum if the Froude number is 1. In this case, the flow is unstable and a series of waves appears on the water surface. Apparently a similar phenomenon occurs at the interface of a density current. For $Fr' = 1$ and from Eq. (14.93), one obtains

$$\lambda = \pi h' \qquad (14.94)$$

Fig. 14.36b is a comparison of the measured wave length and $\pi h'$ at the moment a wave breaks and the upper and lower layers starts to become mixed. The points are well-distributed along the straight line representing Eq. (14.94).

Like the formation of sand waves on a bed discussed in Chapter 6, the appearance of interfacial waves in a density current can be treated as a stability problem in fluid mechanics[55]. In the analysis, the interface is considered not to be a distinct, clear-cut boundary between two layers, but rather a transitional region between them; also, the distributions of both velocity and density are taken to be linear. The quantities R_{it} and α are defined then as follows:

$$Ri_t = \frac{g(\frac{\Delta\rho}{\rho})h_t}{(\Delta u)^2} \qquad (14.95)$$

$$\alpha = \frac{1}{2}kh_t \qquad (14.96)$$

in which h_t is the thickness of the transitional region; Δu, $\Delta \rho$ are the differences of velocity and density, respectively, between the layers above and below the transitional region; $k = 2\pi/\lambda$, in which λ is the wave length of the interfacial waves. The stable and unstable regions for R_{it} are shown in Fig. 14.37a[56], in which R_{it} is shown as a function of α. If R_{it} is greater than 1/4 which also means the Froude number expressed by Eq. (14.95) is less than 2, any perturbation attenuates and the interface remains stable; if R_{it} is less than 1/4, the wave that is most likely to develop is the one with a wavelength 7.5 times the thickness of the transitional region.

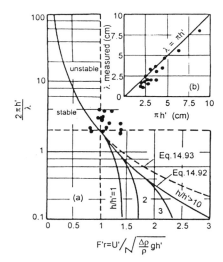

Fig. 14.36 Stability criterion for waves at the interface of density current (treated as an ideal flow)

The stability condition for the interface also depends on the velocity and density distributions within the transitional region. If the two profiles follow either a hyperbolic tangent function or an error function, the two demarcations of the stable and unstable regions are different, as shown in Fig. 14.37b[57]. The maximum value of R_{it}, for which the interface can be unstable remains 1/4; however, it is the wave with

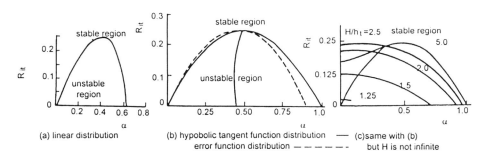

(a) linear distribution

(b) hypobolic tangent function distribution —— (c)same with (b)
error function distribution – – – – – but H is not infinite

Fig. 14.37 Stability condition for the interface when velocity and density current in interface region (transitional region) distribute differently (after Turner, J.S.)

wavelength $2\pi h_t$ that is most likely to develop.

In the two foregoing cases, the thickness of both the density current and the upper layer are assumed to be infinite. If the water surface and the river bed are physically close to each other, and if $H/h_t < 5$, the curves in Fig. 14.37b take other forms like those shown in Fig. 14.37c[4]. Longer waves tend to be less stable and shorter waves more stable. If H/h_t is less than 1.25, any wave is stable.

14.3.1.3 Critical conditions for an interface to lose its stability if effect of viscosity is included

In the preceding analysis, the fluid is treated like an ideal one; that is, the fluid has no viscosity. Jeffreys included the effect of viscosity in his treatment of wind-induced waves. Following that work, Keulegan proposed the following criterion for the stability of the interface in a density current[58]:

$$K_u = \frac{(v'g\frac{\Delta\rho}{\rho})^{1/3}}{U'} \tag{14.97}$$

K_u is called the Keulegan number, and it is actually a function of both the Froude number and the Reynolds number:

$$K_u^3 = \frac{1}{Fr'^2 \, Re'} \tag{14.98}$$

The expressions for Fr' and Re' are those in Eq. (14.26) and Eq. (14.21), respectively.

According to the experiments of Keulegan and those at MIT[20], the relationship between K_u and Re' is as shown in Fig. 14.38. For laminar flow,

$$K_{uc} = 1/Re'^{1/3} \tag{14.99}$$

From Eq.(14.98) and Fig.14.36a, one can see that the interface loses its stability at $Fr' = 1$ if the effects of viscosity are negligible. If the Reynolds number is larger than 1,000, the flow is turbulent, and then

$$K_{uc} = 0.178$$

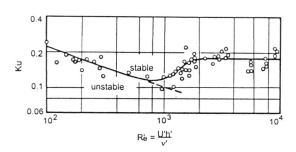

Fig. 14.38 Stability index for the interface of a density current when the viscosity is considered

If the density varies gradually in a transitional region between the upper and lower layers, the stability index K_u may still be a function of the Froude number and the Reynolds number,

$$K_u = f(F_{r_t}, R_{et})$$

in which

$$Fr_t = \frac{U'}{\sqrt{\dfrac{\Delta\rho}{\rho} gh_t}}$$

$$\mathrm{Re}_t = \frac{U'h_t}{v'}$$

h_t is the thickness of the transitional region. According to Schlichting, an approximate form for the function is[54]:

$$\mathrm{Re}_t\, Fr_t^{3/2} = 77,000 \qquad\qquad (14.100)$$

14.3.1.4 Effects of turbulence formed near the bed on the stability of the interface

The preceding discussion is limited to the instability of the interface caused by perturbations originating at the interface itself, but such perturbations can also come from the bed. If the velocity of a density current reaches some level, turbulence may originate there, and vortices thus formed may be lifted and enter the main flow region. If vortices should pass through the interface and enter the clear water region, they would bring turbid water with them into the clear water region, and in this way they can cause mixing of clear water and turbid water.

Nonetheless, turbulence originating from the bed does not usually affect the stability of the interface for three reasons: (1) If the density gradient in the flow field is stable, it damps any vertical exchange so that only the flow field near the interface can affect the stability of the interface; (2) If the density current carries a significant amount of sediment finer than 0.01 mm, the whole fluid could be a Bingham fluid with a yield stress τ_B; a plug then forms in which the shear stress is less than τ_B and no relative motion occurs, as discussed in Chapter 13. Turbulence originating from the bed could not penetrate such a plug[60] ; (3) If the interface is some distance from the bed, vortices lifted from the bed decompose into smaller vortices and lose their kinetic energy due to viscous deformation; mixing caused by such vortices is usually much weaker at the interface. Consequently, few studies have been made of the effect of turbulence originating from the bed on the interfacial instability.

French did some experiments in a flume with two layers of liquids with different densities, salt water and clear water that were set in motion. From the free surface to the bottom of the flume, the velocity in the two layers varied continuously over the entire depth. The average velocity was U. A density difference occurred at the interface, even though the velocity varied continuously there. Perturbations could originate only from the bed. A related parameter is

$$Ri = \frac{g \dfrac{\Delta\rho}{\rho} H}{U_*^2}$$

in which U_* is the shear velocity; apparently the interfacial stability is related to this Richardson number and to the Keulegan number given in Eq. (14.97), as shown in Fig. 14.39[61]. The points are rather scattered, but the stability appears to be related also to the boundary resistance, that is to U/U_*, to some degree.

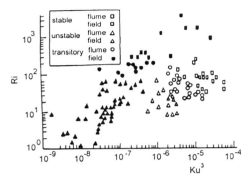

Fig. 14.39 Influence of turbulence originating from flume bottom on the stability of the interface of density current (after French, R.H.)

14.3.2 Diffusion and transport after a density current has lost its stability

After the interface has lost its stability, clear water and turbid water penetrate the interface and mix with each other continuously. However, if there were no turbulence in the field, the turbid water penetrating into the clear water region would settle and return to the interface due to the effect of gravity; also, the clear water penetrating into the turbid water region would return to the interface due to its buoyancy. In other words, only diffusion of mass, not transport of mass, would occur. If turbulence occurs in one of the layers, material coming from another layer is also transported.

Rouse and Dodu put two layers of liquids with different densities into a closed cylinder in which they had installed a set of grids with square holes 10 cm above the interface of the two liquids. They then set the grids in simple harmonic oscillation with various amplitudes and frequencies[62]. Their experiments showed that turbulence induced in the upper layer by the oscillation of grids did not penetrate through the interface into the lower layer so as to form a transitional region. Instead, the disturbance formed irregular waves at the interface; elements of the heavy liquid at the tip of the waves were lifted up and mixed with the lighter liquid. They deduced that only if the turbulence was induced by shear at the interface a transitional region with gradually varying density was formed between the two layers. In their experiments, they used the dimensionless parameter D_f to reflect the diffusion rate

$$D_f = \frac{U_m h}{\varepsilon} \tag{14.101}$$

672

in which U_m is the rate of diffusion discharge per unit area of interface, h the thickness of the upper layer, ε a turbulence mixing coefficient. A relationship between D_f and the Froude number as well as the Reynolds number exists as follows:

$$D_f = 5 \times 10^4 Fr^{5/2} Re \qquad (14.102)$$

In the discussion of interfacial resistance, Macagno and Rouse's experiments are introduced[28]. From those experiments, a relationship between U_m/U_0, which denotes a mixing coefficient at the interface, and R_{e0}' and F_{r0}' exists, and it is shown in Fig. 14.22b. The curves in Fig. 14.22a and b follow similar trends. The reason is evident. The shear at the interface actually represents the diffusion of momentum through the interface, and U_m is a variable reflecting the diffusion of mass.

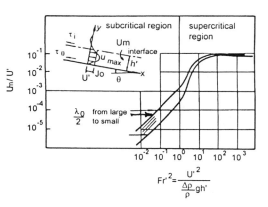

Fig. 14.40 Transport once the interface for a density current has lost its stability (after Peterson, B.)

The preceding result cannot yet be applied directly in solving practical problems relating to density currents. Pederson suggested that when clear water enters a density current region and moves forward with some velocity, its potential energy and kinetic energy vary. These two parts of the energy must be drawn from the average motion — that is, from the turbulence energy of the density current. Since mixing and transport cause the potential energy and the kinetic energy to vary, the ratio of the variation of these energies to the turbulence energy is simply the Richardson number, as mentioned at several places in this chapter. For a subcritical density current, only the variation of the potential energy needs to be considered; for a supercritical flow, the variation of both the potential energy and the kinetic energy need to be considered. Based on such concepts, Pederson established the relationship shown in Fig. 14.40[20]. The figure shows that the Froude number is a basic parameter; in addition, if the flow is subcritical, the friction factor of the boundary λ_0 also plays role. In supercritical flow, the interfacial resistance becomes more and more important as the velocity increases, and the effect of λ_0 decreases. U_m/U' approaches a constant 0.10 if the Froude number exceeds 10.

Pederson also proposed an approximate formula for use in engineering practice

$$\frac{U_m}{U'} = 0.072 J_0 \qquad (14.103)$$

in which J_0 is the slope of the bed.

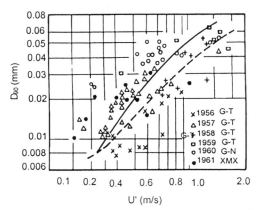

Fig. 14.41 Relationship between average velocity of density currents U' and grain size D_{90} of coarsest particles carried by them

14.3.3 Deposition of sediment carried by a density current

If a turbid density current flows into clear water, the two layers do not form a homogeneous mixture so that the sediment does not settle out near the point of entry.

On the contrary, once it mixes with clear water, a density current drops part of the sediment and then remains stratified as it moves. Thus a second density current forms at the interface and plays the role of a "buffer" that impedes the mixing of the density current with the clear water. A density current can maintain its motion over a rather long distance in a reservoir. The mixing with clear water and the dropping of part of the sediment cause the density of the density current to reduce gradually; as a consequence its velocity reduces also.

How much of the sediment carried by a density current will deposit? The answer depends on the sediment-carrying capacity of the density current at a given velocity. In most cases, the concentration in a density current is not high, and the sediment particles carried by it are small; hence, the sediment-carrying capacity of a density current does not usually need to be considered. The existing data provide limited empirical relationships between the average velocities of density currents U' and the grain sizes of the coarsest particles D_{90} carried by them, as shown in Fig. 14.41[68]. Such a relationship can be used for rough predictions of the grain size of sediment that will deposit in a density current with a given velocity.

The process of deposition of sediment carried by density currents varies from case to case. One example is a hydraulic project in a navigable river in which an artificial access canal is excavated; one end of the artificial canal would be connected to the river and the other end, at the lock, would be closed. In such a case, the turbid water in the river flows underneath the clearer water in the dead water reach as a density current. The sediment it carries deposits along the way and the clear water that forms flows continuously back to the river. In time, the artificial canal can aggradate significantly. Fan proposed a model for calculating the siltation in such a reach[64].

In the preceding paragraphs, the following factors related to density currents are treated: the plunge point of a density current, the transverse diffusion along its course, both boundary shear and local resistance, velocity of a density current, mixing of turbid water with clear water at the interface and consequent retardation of the flow and deposition of the sediment, the rise of a density current as it reaches a dam and the release discharge through an outlet there. Hence, one is able to predict the amount of sediment released from a reservoir in the form of a density current from the hydrograph of discharge and the concentration at the entrance of the reservoir. Such a method has been calibrated using data from Guanting Reservoir in China and Lake Mead in US[65].

14.4 HYPERCONCENTRATED DENSITY CURRENT

Because of the density difference, a density current can maintain its motion underneath the clear water in a reservoir. The discussion so far is limited to the cases in which the concentration of the density current is not high and the density difference is correspondingly small. However, hyperconcentrated flows can also occur in nature. If hyperconcentrated flows meet water with less density, they form density currents that are also hyperconcentrated. Their characteristics are quite different from those of ordinary density currents.

14.4.1 Hyperconcentrated density current in rivers

Density currents in rivers frequently occur in estuarial regions. Sea water flows underneath the fresh water in a river in the form of a salt-water wedge. Turbid density currents also occur in "caecum" reaches connected to rivers, as mentioned in 14.3. In addition, as observed in the Yellow River basin, if a flow with a high concentration flows under the clearer water in a river with a much lower sediment concentration, a density current forms and it can then move downstream a rather long distance.

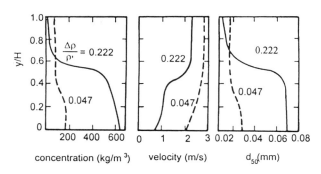

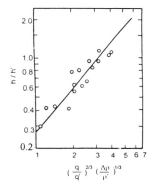

Fig. 14.42 Hyperconcentrated density current formed at confluence of Weihe and Yellow rivers

Fig. 14.43 Location of the interface of density currents in rivers

The Weihe River joins the Yellow River at Tongguan. Vertical profiles of velocity, concentration, and median diameter of suspended load in a density current, which formed when the concentration of the incoming flow from the Weihe River was high, are shown in Fig. 14.42[66]. The figure clearly shows that for $\Delta\rho/\rho' = 0.222$, a distinct interface existed and abrupt changes in concentration, velocity, and median diameter occurred in the vertical direction. At Tongguan, the maximum velocity of such a hyperconcentrated density current was measured to be in the range 0.7 to 1.41m/s, and sediment particles with a diameters of 0.10 mm and even larger were transported.

At the confluence of the Weihe and the Yellow River, the condition for the formation of a hyperconcentrated density current is essentially in accordance with Eq. (14.5); that is,

$$\frac{q^2}{\frac{\Delta\rho}{\rho}gh^3} = 0.6$$

In the calculations, the unit discharge q, density ρ, and water depth h are the measured values at the cross section in Weihe River just beyond the plunge point, $\Delta\rho$ is the density difference between the Weihe River and the Yellow River. The location of the interface can be predicted according to Fig. 14.43, in which parameters without primes are for the upper layer (the flow from the Yellow River), and parameters with primes are for the lower layer (the flow from the Weihe River).

14.4.2 Hyperconcentrated density currents in reservoirs

As a hyperconcentrated flow enters a reservoir, it forms a density current. Due to the high concentration and greatly reduced fall velocity, particularly for the fine particles, a hyperconcentrated density current possesses a strong sediment carrying capacity, can maintain a high velocity and can convey particles as coarse as 0.10 mm. Such a hyperconcentrated density current behaves quite differently from an ordinary density current.

As an ordinary density current reaches a dam, its kinetic energy can be transformed into potential energy. Because of the buoyancy effect due to the small density difference, this transformation of energy enables the interface to rise a considerable distance. If the density current reaching the dam is not released and the turbid flow continues to flow in, turbid water accumulates in the region upstream of the dam and the interface between the two layers rises steadily. Meanwhile, some sediment settles out under the action of gravity. Also, the turbid water mixes with the clear water because of turbulence generated by the reflection of the flowing turbid stream at the dam. Thus, the concentration of the accumulating turbid water decreases substantially. If a sluicing facilities has been installed and has been opened before the density current reaches the dam, part of the turbid water can be released from the reservoir. The hydrograph of the outflow concentration is rather like that of the inflow concentration, but its peak concentration is lower, and there is a time lag between their occurrences, as shown in Fig. 14.44a.

For a hyperconcentrated density current, the situation is quite different. After a hyperconcentrated density current reaches a dam and forms a region of turbid water at the bottom of the reservoir, sediment particles settle extremely slowly due to the high concentration. Besides, the mixing at the interface is greatly damped due to the large density difference between clear water and turbid water. As a result, the high

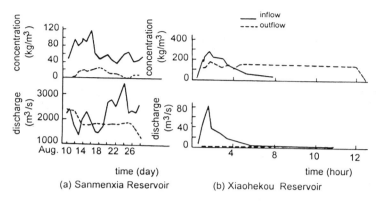

Fig. 14.44 Two cases of releasing sediment as density current

concentration of the turbid water in the denser region can be maintained for a long time[67]. If the density current is being released through sluice gates, and even if the incoming flow has stopped, turbid water can still be supplied from the region of turbid water. Consequently, a high concentration can be maintained in the outflow for a much longer time than it is in the inflow, Fig. 14.44b. Therefore, sluicing sediment in the form of a density current in a reservoir on a river heavily laden with sediment is much more effective than it is in a river with a more normal concentration. For example, in the early stage of the operation of Lake Mead in the U.S., sediment was sluiced in the form of a density current four times, and 27% of the sediment inflow was released. In Liujiaxia, Hesonglin, and other reservoirs in China, 55 to 61% of the incoming sediment was released[63].

14.4.3 Hyperconcentrated turbidity current on sea bottom

Topographic surveys and samples of deposits in deep sea areas have led to significant advances in the recent decades. The knowledge of sea bed geomorphology and geology has been greatly enriched. In the past, the formation of gorges and the discovery of coarse deposits in the deep sea region were not fully understood.

On the submerged continental shelves, both giant gorges and deep channels are found. The geomorphologies of the two forms are different. The gorges, shown in Fig.14.45a, are similar to formations on the surface of the earth. Numerous channels divide and disperse along the surfaces of alluvial cones, and they can have lengths of several thousand kilometers. The gorges are actually valleys that originally formed on the earth's surface and were later submerged. They can be submerged as deeply as 800 m. The situation shown in Fig.14.45b has a completely different morphology. A typical valley is V-shaped with side slopes of about 1:4.5. The longitudinal slope of the channels reduces in the downstream direction, and it is directed along the principal slope of the ocean bottom. The valleys extend over the entire length of the slope, with none of them ending at an intermediate point[68].

In large-scale surveys of deeper regions of the sea, coarse particles are frequently found on the bottom in the deeper regions[69]. Most of these particles are in the range between fine and median sand; few of them are pebbles. Usually the sand is deposited in layers separated by other layers of clay and silt. Sorting occurs along the flow direction. The closer a region is to the continent, the coarser the particles are.

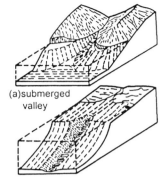

(a)submerged valley

(b) gorge formed by density currents

Fig. 14.45 Two types of gorges on sea bottom

The discovery of such a phenomenon indicates that there must have been flows involved that were powerful enough to cut gorges more than 500 km long and to convey coarse particles to regions of the sea bottom where the water is several kilometers deep. Such flows were not continuous, but they could have been repeated from time to time.

A large earthquake occurred in Grand Bank, in the southern part of New Finland in 1929. After the earthquake, numerous cables were broken sequentially[70]. The level of the earthquake was 7.2, and its epicenter was located on the continental slope where the depth is some 1,800 to 3,600 m and a large number of cables had been laid. During the earthquake, six marine cables located at depths of 275 to 3,300 m were broken simultaneously. Then, in the direction of increasing water depth, which is southwards, several other marine cables were broken in sequence. The last one was located 480 km south of the epicenter and it was broken 13 hours and 17 minutes after the earthquake occurred. From the distances involved and the timing of breaks, the propagation speed was greater on the continental slope in the region closer to the epicenter; further away, the slope was gentler and the rate of propagation was lower. The region in which the cables were broken was limited to the deepest part of the ocean and had a fan shape. It was narrow to the north and wide to the south, and the epicenter was also the center of the fan. This pattern indicates that the cable breaking was induced by the earthquake. The action originated at the epicenter and spread along the continental shelf. As it propagated, the region it affected gradually widened. The speed of propagation became less as the slope became gentler. Apparently, the details of the preceding phenomenon can best be explained by attributing it to a density current moving in the deep sea. From studies during recent decades, the phenomenon of marine density currents has come to be generally accepted. Geologists and geophysicists call them "turbidity currents."

678

14.4.3.1 Mechanism of formation of hyperconcentrated turbidity currents

From hydrodynamics, Plapp and Mitchell demonstrated the possibility of the occurrence of a hyperconcentrated turbidity current at Grand Banks[71]. Bagnold gave an excellent explanation of this phenomenon [72]. He stated that the initial energy of the landslides caused by the earthquake was large, and the volume of earth set in motion was also extremely large. A large amount of sediment was concentrated in a layer with a thickness of several hundred meters in the region, directly above the sea bottom. In addition, the gradient of the continental slope was large. As a result, when the sediment-laden flow moved forward, the kinetic energy that was transformed from the potential energy was large enough to overcome the bottom and interfacial resistance and maintain the motion along the continental slope for a long distance.

14.4.3.2 Velocity of hyperconcentrated turbidity current

According to the data on the progression of cable breaking after the Grand Banks earthquake on the continental slope, where the gradient was 0.006, the average velocity of the density current reached 20 m/s; on the flat bottom in the deeper part of the sea, where the gradient was 0.001, the average velocity was 9.8 m/s. Because of the extremely large thickness of the hyperconcentrated turbidity current, such high velocities are feasible.

Through experiments, Middleton proved that the velocity of a hyperconcentrated density current follows the same law as does that of an ordinary density current. If the thickness of the head of a density current h_f' (Fig. 14.4a) is used as the length in the Froude number, the coefficient in Eq. (14.15) would be 0.75[15]. Fig. 14.46 shows the frontal velocity of a turbidity current on the sea bottom as a function of h_f' and $\Delta\rho/\rho_0$. If the thickness of the head of such a current was grater than 100 m, its velocity could surely have reached 20 m/s.

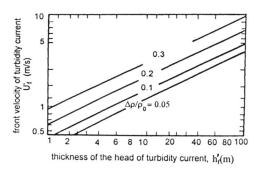

Fig. 14.46 Possible front velocity of turgidity current on sea bottom under different conditions

14.4.3.3 Coarsest particles carried by hyperconcentrated turbidity currents

From Eq. (14.35), the shear exerted on the bed by a hyperconcentrated turbidity current is

$$\tau_0 = \frac{(\rho' - \rho)gh'J}{1 + \alpha} \tag{14.104}$$

in which the ratio of the interfacial resistance to the bed resistance α is a function of the Froude number. As the thickness of a hyperconcentrated turbidity current is extremely large and the density difference of a turbidity current is much larger than that of a common density current, the shear acting on the bed can also be quite large. According to Komar's estimate, based on Shields' threshold curve (Fig. 8.6), a turbidity current like that at the Grand Banks might even be able to pick up small to medium-sized pebbles [73].

14.4.3.4 *Bed form while a turbidity current is passing*

As pointed out in Chapter 6, if the flow intensity is rather high, sand waves or even "chutes and pools" may form on the bed (Fig. 6.1). These unusual dynamic bed forms occur with the high intensities of flow that can take place in flume experiments, but they are seldom seen in nature because the flow in alluvial rivers is not usually supercritical. These sand waves on the bed, which are composed of loose granular material, may, however, occur under hyperconcentrated turbidity currents; the high velocities of hyperconcentrated turbidity currents are often supercritical. Using coal powder as bed material, Hand easily produced these special bed forms in a channel with a gradient of 0.08-0.10 [74]. Kennedy proposed the following formula for the wavelength of sand waves in open channel flow

$$\lambda = \frac{2\pi U^2}{g} \tag{6.24}$$

in which λ is the wavelength. For a density current, the preceding formula should be notified as follows[74]

$$\lambda = \frac{2\pi U'^2}{g} \frac{\rho + \rho'}{\Delta\rho} \tag{14.105}$$

The density ratio term causes the sand waves formed by density currents to be much longer than those in ordinary open channel flows. They can be dozens of meters, even 100 m long.

REFERENCES

[1] Chien, Ning, and Jiahua Fan, et. al. *Density Current*. Water Conservancy Press, 1957, p. 215. (in Chinese)

[2] Harleman, D.R.F. "Stratified Flow", In *Handbook of Fluid Dynamics*, Edinted by V.L. Streeter, McGraw-Hill Book Co., 1961, p. 21.

[3] Yih, C.S. *Dynamics of Non-homogeneous Fluids*. Macmillan Co., N. Y. 1965.

[4] Turner, J.S. *Buoyancy Effects in Fluids*. Cambridge Univ. Press, 1973, p. 367.

[5] Koh, C.Y. "Viscous Stratified Flow Towards A Line Sink." *Report KH-R-6*, Keck Lab. of Hyd. and Water Resources, Calif. Inst. Tech., 1964, p. 172.

[6] Fan, Jiahua et al. "Experimental Study on Density Current." *Journal of Hydraulic Engineering*, 1959, Vol. 5, pp. 30-48

[7] Bell, H.S. "The Effect of Entrance Mixing on the Size of Density Currents in Shaver Lake." *Trans., Amer. Geophys, Union*, Vol. 28, No. 5, 1947, pp. 780-791.

[8] Levy, E.E. "Theory of Underflows in Storage Reservoirs." *Proc., Intern. Assoc. Hyd. Res.*, Vol. 2, 1959, p. 26

[9] Kao, Tivothy W. "Density Currents and Their Applications." *J. Hyd. Div., Proc., Amer. Soc. Civil Engrs.*, Vol. 103, No. HY5, 1977, pp. 543-555.

[10] Von Karman, T. "The Engineer Grapples with Non-Linaar Problems." *Bull.*, Amer. Math. Soc., Vol. 46, 1960., pp. 615-683.

[11] Kranenburg, K. "Internal Fronts in Two-Layer Flow." *J. Hyd. Div., Proc., Amer. Soc. Civil Engrs.*, Vol. 104, No. HY10, 1978, pp. 1449-1453.

[12] Abraham, G.B., and C.B. Vreugdenhil. "Discontinuities in Stratified Flows." *J. Hyd. Res.*, Intern. Assoc. Hyd. Res., Vol. 9, No. 3, 1971, pp. 1449-1453.

[13] Barr, D.I.H. "Densimetric Exchange Flow in Rectangular Channels." *La Houille Blanche*, Vol. 18, No. 7, 1963, pp. 739-766.

[14] Wilkinson, D.L., and I.R. Wood. "Some Observations on the Motion of the Head of A Density Current." *J. Hyd. Res.*, Intern, Assoc. Hyd. Res., Vol. 10, No. 3, 1972, pp. 305-324

[15] Middleton, G.V. "Experiment on Density and Turbidity Currents: 1. Motion of the Head." *Canadian J. Earth Sci.*, Vol. 3, No. 4, 1966, pp. 523-546.

[16] Michon, X., J. Goddet, and R. Bonnefille. "Etude Theorique et Experiments des Courants de Densite." *2 vols.*, Lab. Nat. d'Hydralique Chatou, France, 1995.

[17] Keulegan, G.H. "Laminar Flow at the Interface of Two Liquids." *J. Res. Nat. Bureau of Standards*, Vol. 32, 1944, pp. 303-327.

[18] Lock, R.C. "The Velocity Distribution in the Laminar Boundary Layer Between Parallel Streams." *Quart. J. Mech. and Applied Math*, 1951.

[19] Bata, G.L. "Frictional Resestance at the Interface of Density Currents." *Proc., Intern. Assoc. Hyd. Rcs.*, Vol. 2, 1959, p. 16.

[20] Ippen, A.T., and D.R.F. Harleman. "Steady State Characteristics of Subsurface Flow." *Proc., Nat. Bur. Standards Semicentennial Symp. on Gravity Waves*, 1951, pp. 79-93.

[21] Raynaud, J.P. "Etude des Courant d'Eau Boueuse dans les Retenues." *Trans., Cong. on Large Dams, Res.*, Vol. 2, 1959, p. 33.

[22] Bonnefille, R., and J. Goddet. "Etude des Courants de Densite en Canal." *Proc. Intern. Assoc. Hyd. Res.*, Vol. 2, 1959, p. 33.

[23] Geza, B., and K. Bogich. "Some Observations on Density Currents in the Laboratory and in the Field." *Proc., Minnesota Intern. Hyd. Convention*, 1953, pp. 387-400.

[24] Albertson, M.L., Y.B. Dai, R.A. Jensen, and H. Rouse. "Diffusion of Submerged Jets." *Trans., Amer. Soc. Civil Engrs.*, Vol. 115, 1950, pp. 639-697.

[25] Georgiev, B.V. "Some Experimental Investigation on Turbulent Characteristics of Stratified Flows." *Proc., Intern. Symp. on Stratified Flows*, Novosibirsk, 1972, pp. 507-514.

[26] Wilkinson, D.L. "Studies in Density Stratified Flows." *Report No.118*, Water Res. Lab., Univ. New South Wales, Australia, 1970, p. 167.

[27] Middleton, G.V. "Experiments on Density and Turbidity Currents, 2. Uniform Flow of Density Currents." *Canadian J. Earth Sci.*, Vol. 3, No. 5, 1966, pp. 627-637.

[28] Macagno, E.S., and H. Rouse. "Interfacial Mixing in Stratified Flow." *J. Engin. Mech. Div., Proc., Amer. Soc. Civil Engrs.*, Vol. 87, No. EM 5, 1961, pp.55-81.

[29] Bo Pedersen, F. "A Monograph on Turbidity Entrainment and Friction in Two-layer Stratified Flow." *Ser. Paper No.25*, Inst. Hydrodynamics and Hyd. Engin., Tech. Univ. Denmark, 1980, p.397.

[30] Sedimentation Engineering Department, IWHR. "Study on Density Current and Its Application." *Report of IWHR, No.15*, Water Conservancy & Electricity Press, 1959, p. 179. (in Chinese)

[31] Komar, P.D. "Hydraulic Jump in Turbidity Current." *Bull.*, Geol. Soc. Amer., Vol. 82, 1971, pp. 1477-1488.

[32] Tepper, M. "The Application of the Hydraulic Analogy to Certain Atmospheric Flow Problems." *U.S. Weather Bur. Res. Paper No. 35*, 1952, p. 50.

[33] Yih, C.S., and C.R. Guna. "Hydraulic Jump in a Fluid System of Two Layers." *Tellus*, Vol. 7, 1955, pp. 358-366.

[34] Wood, I.R. "Horizontal Two-Dimensional Density Current." *J. Hyd. Div., Proc., Amer. Soc. Civil Engrs.*, Vol. 93, No. HY2, 1967, pp. 35-42.

[35] Fietz, T.R., and I.R. Wood. "Three Dimensional Density Current." *J. Hyd. Div., Proc., Amer. Soc. Civil Engrs.*, Vol. 93, No. HY6, 1967, pp. 1-23.

[36] Ekman, V.W. "On Dead Water." *Scientific Results of the Norwegian North Polar Expedition*, Pt. 15, Christiania, 1904.

[37] Hourwitz, B. "Internal Waves of Tidal Character." *Trans.*, Amer. Geophys. Union, Vol. 31, 1950.

[38] Keulegan, G.H. "Characteristics of Internal Solitary Waves." *J. Res., Nat. Bur. Standards*, Vol. 51, No. 3, 1953, pp. 133-140.

[39] Keulegan, G.H. "An Experimental Study of Internal Solitary Waves." *Nat. Bur. Standards, Rep. 4415*, 1955.

[40] Brooks, N.H., and C.Y. Koh. "Selective Withdrawal from Density-Stratified Reservoirs." *J. Hyd. Div., Proc., Amer. Soc. Civil Engrs.*, Vol. 95, No. HY4, 1969, pp. 1369-1400.

[41] Craya, A. "Recherches Theoriques sur l'Ecoulement de Couches Superposes de Fluides de Densites Diffenentes." *La Houille Blanche*, Jan.-Feb. 1949, pp. 56-64.

[42] Gariel, P. "Recherches Experimentales fur l'Ecoulement de Coches Superposes de Fluides de Densites Differentes." *La Houille Blanche*, Jan.Feb. 1949, pp. 56-64.

[43] Bohan, J.P., and J.L. Grace. "Mechanics of Stratified Flow Through Orifices." *J. Hyd. Div., Proc., Amer. Soc. Civil Engrs.*, Vol. 96, No. HY12, 1970, pp. 2401-2416.

[44] Gelhar, L.W., and D.M. Mascolo. "Non-Diffusive Characteristics of Slow Viscous Flow Towards A Line Sink." *Report No. 88*, Hydrodynamics Lab., Mass. Inst. Tech., 1966.

[45] Koh, C.Y. "Viscous Stratified Flow Towards a Sink." *J. Fluid Mech.*, Vol. 24, Pt. 3, 1966, pp. 553-575.

[46] Harleman, D.R.F., R.L. Morgan, and R.A. Purple. "Selective Withdrawal from A Vertically Stratified Fluid." *Proc., 8th Cong.*, Intern. Assoc. Hyd. Res., Vol. 2, 1959, p. 16.

[47] Rouse, H. "Seven Exploratory Studies in Hydraulics." *J. Hyd. Div., Proc., Amer. Soc. Civil Engrs.*, Vol. 82, No. HY4, 1956.

[48] Harleman, D.R.F., R.S. Gooch, and A.T. Ippen. "Submerged Sluice Control of Stratified Flow." *J. Hyd. Div., Proc., Amer. Soc. Civil Engrs.*, Vol. 84, No. HY2, 1958, p. 15.

[49] Huber, D.G. "Irrotational Motion of Two Fluid Strata Towards A Line Sink." *J. Engin. Mech. Div., Proc., Amer. Soc. Civil Engrs.*, Vol. 86, No. EM4, 1960, pp. 71-86.

[50] Huber, D.G., and T.L. Reid. Experimental Study of Two-Layered Flow Through A Sink." *J. Hyd. Div., Proc., Amer. Soc. Civil Engrs.*, Vol. 92, No. HY1, 1966, pp. 31-42.

[51] Walesh, S.G., and P.C. Monkmeyer. "Bottom Withdrawal Viscous Stratified Fluid." *J. Hyd. Div., Proc., Amer. Soc. Civil Engrs.*, Vol. 99, No. HY9, 1973, pp. 1401-1420.

[52] Yih, C.S. "On the Flow of a Stratified Flow." *Proc., 3rd U. S. Nat. Cong. Applied Math.*, 1958.

[53] Debler, W.R. "Stratified Flow into a Line Sink." *J. Engin. Mech. Div., Proc., Amer. Soc. Civil Engrs.*, Vol. 85, No. EM3, 1959, pp. 51-66.

[54] Browand, F.K., and Y.H. Wang. "An Experiment on the Growth of Small Disturbances at the Interface Between Two Streams of Different Densities and Velocities." *Proc., Intern. Symp. on Stratified Flows*, Novosibirsk, 1972, pp. 491-498.

[55] Drazin, P.G., and L.N. Howard. "Hydrodynamic Stability of Parallel Flow of Inviscid Fluid." *Advances in Appl. Mech.*, Vol. 9, 1966, pp. 1-89.

[56] Miles, J.W., and L.N. Howard. "Note on a Heterogeneous Shear Flow." *J. Flu. Mech.*, Vol. 20, 1964, pp. 331-336.

[57] Hazel, P. "Numerical Studies of the Stability of Inviscid Stratified Shear Flow." *J. Flu. Mech.*, Vol. 51, 1972, pp. 39-61.

[58] Keulegan, G.H. "Internal Instability and Mixing in Stratified Flows." *J. Res., Nat. Bur. Standards*, Vol. 43, 1949, pp. 487-500.

[59] Keulegan, G.H. "Wave Motion." In *Engineering Hydraulics*. Edited by H. Rouse, John Wiley and Sons, 1950.

[60] Einstein, H.A. "The Viscosity of Highly Concentrated Underflows and Its Influence on Mixing." *Trans.*, Amer. Geophys. Union, Vol. 22, Pt. 3, 1941, pp. 597-603.

[61] French, R.H. "Interfacial Stability in Channel Flow." *J. Hyd. Div., Proc., Amer. Soc. Civil Engrs.*, Vol. 105, No. HY8, 1979, pp. 959-967.

[62] Rouse, H., and J. Dodu. "Diffusion Turbulente a Travers Une Discontinuite de densite." *La Houille Blanche*, No. 4, 1955, pp. 522-529.

[63] River & Canal Department, Northwest Hydro-technical Scientific Research Institute and Sediment Laboratory, Tsinghua University. *Reservoir Sedimentation.* Water Conservancy & Electricity Press, 1979, pp. 95-118. (in Chinese)

[64] Fan, Jiahua. "Analysis on Deposition Caused by Density Current." *Scientia Sinica*, 1980, Vol. 1, pp. 82-89. (in Chinese)

[65] Fan, Jiahua, Shoubai Sheng, and Deyi Wu. "Approximate Calculation of Density Current in Reservoir." *Proc. IWHR*, Vol. 2, 1963, pp. 135-145. (in Chinese)

[66] Tongguan Hydrological Station. "Hyperconcentrated Density Current in River at Tongguan Section." Vol. 2, 1975, pp. 135-145

[67] Irrigation District Downstream Xiaohekou Reservoir, Yecheng, Shanxi, Institute of Hydraulic Research, Shanxi and Hydraulic Engineering Department, Tsinghua University. Vol. 3, 1976, pp. 138-148

[68] Daly, R. A. "Origin of Submarine Canyons." *Amer. J. Sci.*, Vol. 231, 1936, pp. 401-420.

[69] Shepard, E.P. "Transportation of Sand into Deep Water." *Soc. Economic Paleontologists and Mineralogists*, Sp. Pub. No. 2, 1951, pp. 53-65.

[70] Heezen, B.C., and M. Ewing. "Turbidity Currents and Submarine Slumps, and the 1929 Grand Banks Earthquakes." *Amer. J. Sci.*, Vol. 250, 1952, pp. 849-873.

[71] Plapp, J.E., and J.P. Mitchell. "A Hydrodynamic Theory of Turbidity Currents." *J. Geophys. Res.*, Vol. 65, No. 3, 1960, pp. 983-992.

[72] Bagnold, R.A. "Auto-suspension of Transported Sediment; Turbidity Currents." *Proc.*, Royal., Soc. London, Ser. A, Vol. 265, No. 1322, 1962, pp. 314-319.

[73] Komar, P.D. "The Competence of Turbidity Current Flow." *Bull.*, Geol. Soc. Amer., Vol. 81, No. 5, 1970, pp. 1555-1560.

[74] Hand, B.M. "Supercritical Flow in Density Currents." *J. Sedim. Petro.*, Vol. 44, No. 3, 1974, pp. 637-648.

CHAPTER 15

MOVEMENT OF WIND-BLOWN SAND

Sediment movement in open channel and in reservoir has been treated in the preceding chapters. Many of the principles and laws presented are also applicable to sediment movement under other dynamic conditions. In the following three chapters the movement of wind-blown sand and wave-induced sediment, and hydrotransport of granular material in pipeline are briefly discussed.

Deserts in China cover an area of 1.308×10^6 km^2, constituting 13.6% of the total territory. To prevent further expansion of the desert threats to bury roads, farmland, and even urban areas, large amounts of money and labor have been and are being spent. A sound understanding of the mechanism of wind-blown sand will benefit such works by guiding the activities along a scientifically sound course.

15.1 BASIC FORMS OF WIND-BLOWN SAND MOVEMENT AND PRINCIPLE DIFFERENCES BETWEEN IT AND ALLUVIAL SEDIMENT MOVEMENT

Saltation, suspension, and surface creep are the three primary patterns of wind-blown sand movement

15.1.1 Basic forms of wind-blown sand

15.1.1.1 Saltation

Sand particles are accelerated by wind once lifting forces have dislodged them from the bed surface, and in this way they take momentum from the airflow. Because the specific gravity of air is much smaller than that of sand particles, the wind exerts only a small force on the moving particles. Still, when particles drop back down to the bed, they may possess a considerable amount of momentum. If the bed is sufficiently rigid, the particles can bounce up again after impingement, just like a ball, and thus continue to be as saltating grain. If the bed is composed of loose granular particles, not only can falling particles rebound, they can also initiate saltation of other particles near the point of impingement. Thus, once a saltation begins, it induces a series of chain reactions that cause sand movement of considerable intensity. Sand particles moving in this way are called saltation load and it is the primary pattern for the movement of wind-blown sand.

Sand partcles with a diameter of 0.1-0.15 mm are the ones that move most easily by saltation. In desert areas, large quantities of sand particles passed over the ground surface within a limited distance from ground level. They look like a sand cloud moving along the ground-- almost like a moving carpet. Sometimes the thickness of the sand cloud is rather large, but the majority of sand particles move quite near the ground surface. In those regions with intensive wind-blown sand, columns and

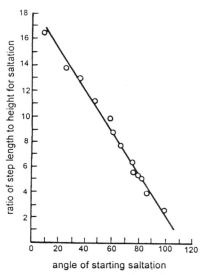

Fig.15.1 Ratio of step length to height for saltation versus angle of starting saltation

boulders protruding above the surface are effectively "sand blasted" and become very smooth due to abrasion, but such traces of wind abrasion extend only to a height of 50 cm above the surface of the ground. Observations carried out in desert areas provide ample evidence that more than 90% of the saltation load moves within 30 cm of the ground, and about half of it moves within 5 cm of the ground [1].

The height of saltation of sand particles can vary considerably, but the angle of impingement against the ground surface varies within only a small range, mostly between 10° and 16° [2]. The angle at which the saltation begins also varies widely. Ling and Wu used high-speed photography in wind tunnel tests, and they found that about 40% of sand particles start saltation at angles of 30°-50°, and another 28% at angles of 60°-80° [3]. Using coarse sand diameters of 1.44 to 2.25 mm and seeds with specific gravity of 1.15, Tsuchiya found that the initial angle for saltation was about 40°, and that it was independent of the velocity of the air [4]. The ratio of step length to height for saltation is related to the angle at which saltation starts, as shown in Fig. 15.1. Both the step length and the height of saltation increase with the angle. But since the latter increases more than the former, the ratio of step length to height decreases with the increase of angle of starting saltation [3].

During saltation, sand particles swirl continuously with rates in the range of 100 to 1,000 rpm.

15.1.1.2 Suspension

For sand particles of $D>0.1$ mm, the fall velocity is usually smaller than the *upward movements* in the turbulent airflow, and they can therefore move in suspension. The behaviour of suspended wind-blown sand particles depends upon the airflow structure at higher elevations. Sometimes they reach heights of several hundred meters and can even change the colour of the sky; the so-called dust storms, seen in north and northwest China, are nothing other than the suspension of "blowing yellowish sand." Both saltation load and surface creep load move near the ground surface. As wind direction on the ground changes often, saltation load and surface creep load usually move to-and-fro within any local region; thus, the speed of migration is quite slow. But for suspended load in high altitude, the situation is totally different. A dust storm can move large quantities of fine sand over large distances. For example, in 1934 the central and southern parts of the United States experienced a severe and long-lasting drought. Huge dust storms occurred, and they blew large amounts of sand across the continent to the Pacific Ocean. As the dust storm moved, the sky was darkened for 47 days during a period of two months. Over a large area, 43% of the farmland suffered from serious wind erosion. Over a vast area (65,000 km²) near the centre of the dust storm, more than 80% of the farmland lost its value for cultivation because so much top soil had been removed.

Karman evaluated the time duration t and the travel distance L for sand particles blown by wind

$$i = \frac{40\varepsilon\mu^2}{\rho_s^2 g^2 D^4}$$ (15.1)

$$L = \frac{40\varepsilon\mu^2 U}{\rho_s^2 g^2 D^4}$$ (15.2)

in which μ — viscosity of air; U — mean velocity of wind; ρ_s — density of sand particles; ε — coefficient of turbulence exchange. For a strong wind ε can be taken as $\varepsilon = 10^4\text{-}10^5$ cm²/s. According to Eq. (15.1) and Eq. (15.2), the travel distances and heights for sand particles blown by wind at a mean velocity of 15 m/s are listed in Table 15.1 [5]. Particles finer than silt can travel long distances, even across oceans. For this reason, dunes located in a desert lack fine material, particularly material with $D < 0.06$ mm, and eolian deposits of this fine material can be found on sea beds large distance away.

Table 15.1 Height and distance for saltation particles
(mean wind velocity $U = 15$ m/s) (after Malina, F. J.)

Grain size (mm)	Fall velocity (cm/s)	Duration of suspension in air flow	Distance	Height
0.001	0.0083	0.95-9.5 years	$4.5\times10^5\text{-}4.5\times10^6$ km	7.75-77.5km
0.01	0.824	0.83-8.3 hours	45-450km	78-775m
0.10	82.4	0.3-3 sec	4.5-45m	0.78-7.75m

15.1.1.3 Surface creep

Wind-blown sand particles that return to the ground surface still possess a considerable amount of momentum; they cannot only cause the dispersion of other particles and put them in saltation, but they can also push other particles forward as a result of the impact already. If the wind velocity is large enough, the whole surface appears to creep forward. Of course, surface creep is quite different from saltation. The latter acquires its momentum from wind pressure, while the former is not directly affected by the wind, but gains its momentum from the impingement of grains in saltation.

Sand particles that are too coarse to be picked up by wind pressure nevertheless can move because of the impingement that comprise the saltation load of the much finer particles. Observations show that sand at comparatively high speeds in saltation can push forward sand particles in the surface layer with diameters six times larger, that is, with weights that are more than two hundred times as great [2]. All sand particles in the range of $D = 0.5$-1.0 mm belong to the category of surface creep load.

Since the momentum of surface creep load comes from saltation load, the transport rate of the two kinds of loads are logically related to each other. This can be expressed in terms of the ratio of the two kinds of load or the weight of particles in the surface creep load as a fraction of the total wind load (not including suspension load), the ratio varies within a narrow range. Results of measured data by various authors are given in Table 15.2; they show that the weight factor varies between 0.065 and 0.25 with an average value of 0.20 [7]. The coarser the sand particles are, the larger the weight factor for the surface creep load.

Table 15.2 Weight percentage of surface creeping load in the total
wind driven sand load (after Cooke, R.U. and A.Warren)

Researcher	Weight percentage of surface creeping load
R.A. Bagnold	0.26
W.S. Chipil	0.157-0.25
	0.25-0.83
Ishihara and Iwagaki (Field data)	0.065-0.166
Kigoshi Horikawa and H.W. Shen	0.20

Among the three basic forms of wind-blown sand, saltation is the most important. Not only does the saltation load constitute the major part of the movement, but also, surface creep and suspension are both related to it. Surface creep load acquires momentum directly from saltation load, as already mentioned. Also suspension load, composed of fine sand particles, can hardly be lifted directly by wind forces, because they exist in eolian deposits on the ground surface where they are sheltered in the laminar sublayer; also, there can be a cohesion among fine particles. But once these particles are dislodged due to impingement of saltation load, eddies in airflow can easily raise them to high altitude and transport them to far-away places. Therefore, the crux of the prevention of wind erosion is the effective control of saltation. Because chain reactions are induced by saltation, priority should be given to protective measures in those small-sized districts where wind erosion is apt to begin. Such measures often result in the alleviation, even the elimination of wind erosion over a much larger areas. In view of the limited height at which the saltation load moves, the planting of wind-barriers or shelter belts can be effective.

Next, because fine sand is mostly carried some distances from the local region, the sand particles that form dunes move as saltation and surface creep, and those in surface creep cannot be too coarse to move. Hence the compositions of wind formed dunes that are spread widely over vast areas in the world have a degree of uniformity. For example, the size of the most common wind-blown sand particles in Kansas, U.S., is 0.16 to 0.35 mm, with a mean diameter of 0.25 mm [8]. In west Canada and in Central Asia of the former USSR, the median diameter of wind blown dunes is also 0.25 mm. In China, dunes in the Taklamakan Desert are composed of sand with D in the range 0.06 to 0.19 mm. The median diameter of the eolian sand in Ulan Buh and BadainJaran Deserts are in the range of 0.10 to 0.24 mm.

15.1.2 Principal difference between the movement of alluvial sediment and wind-blown sand

Prior to the introduction of the saltation for wind-blown sand, a discussion of the similarities and difference of sediment movement in airflow and in water flow is useful.

The differences of sediment movement in water and air arise from the following aspects:

1. Difference in the forms of movement

Chapter 5 points out that saltation plays a predominant role in wind-blown sand, whereas most sediment particles in water flow move in suspension. This difference is caused by the fact that the density of water and that of air are not of the same order. With the buoyancy effect, the ratio of density of air to that of sand is 1:2000, whereas the ratio of the density of water to that of submerged sediment is 1:1.65. Because of this large disparity, a given uplift force can cause a sand particle to move up and down much higher levels in air than in water, and consequently, saltation is much more important in air than in water. This difference in form of movement leads to a series of other disparities.

2. Difference in forces acting on sediment particles

When sand particles fall from airflow at high altitude onto the ground surface with a considerable momentum, the impinging force against particles on the surface is also large. From data observed by Lin and Wu using high-speed photography, the impinging force can be several times the particle weight, even several thousand times. The drag force may equal or exceed the weight, and the uplift force amounts to only a small fraction—one-tenth to a few hundredths of the weight [3]. The distribution of forces acting on sand particles resting on the ground surface can differ from the amounts, but the impinging force is surely the dominant one. For the movement of sediment in water flow, in contrast, the primary. forces acting on the particles resting on the bed surface are the forces of drag and uplift, as treated in Chapter 5.

3. Difference in resistance to fluid media

The large difference in the ratios of density between sand to water and sand to air also results in a difference in the interaction between the sediment particle and the surrounding fluid media. Saltation load acquires momentum from that of the surrounding fluid. For sand particle at rest to acquire a given velocity from air flow, the air must lose momentum from a volume 2000 times the size of the sand particle. The same process in water requires the momentum from a volume only 1.65 times greater. On the one hand, when sediment particles move in water flow, starting from the entrainment until they reach the highest point of their trajectory during saltation, the velocity of sediment particles rapidly approaches or even reaches the local flow velocity. However, for wind-blown sand, the velocity of the sand is much smaller than that of the air. Velocities of sand particles with $D = 0.15$ mm at various stages of saltation are listed in Table 15.3 for a wind velocity of 18 m/s [3]. The table shows that not only is the velocity of surface creep load much smaller than the wind velocity, but also the velocity of saltation load as it moves horizontally is only about one-sixth that of wind velocity.

Table 15.3 Velocity of sand particles moving in different ways
at wind velocity of *18* m/s, in m/s

	Height(cm)	0-1	1-2	2-3	3-4
Surface creep	Rolling		0.93		
	Sliding		1.29		
Saltation	Starting period	Angle of saltation increases from 10° to 120° , velocity decreases from 1.13 m/s to 0.64 m/s			
	Ascending period	0.90	2.08	2.09	2.63
	Horizontal travel	1.37	2.24	2.83	3.20
	Descending period	Angle of impingement against surface 10°-30° , with mean velocity of 2.10 m/s			

690

On the other hand, according to the law of momentum, the loss in momentum for the fluid is the resistance caused by the transport of the saltation load, a resistance that can be called a saltation resistance. For wind-blown sand, the saltation resistance is so large that the resistance produced at the ground surface (such as grain resistance, form resistance, etc.) is negligible in comparison. The situation is the reverse of that for sediment movement in open channel flow; the effect caused by saltation is negligible and the resistance to flow depends mainly upon the shear at the bed. Obviously, the difference in the mechanism of resistance affect also the patterns of fluid flow.

4. Difference in effect of turbulence on sediment movement

The importance of eddy size relative to the size of sediment particles for sediment movement in suspension has been discussed at several points. Eddies that are too small are unable to carry sediment particles in suspension because of the ever-present relative motion of sediment as it settled through the surrounding fluid. Bagnold explained this condition from another viewpoint [10]. The discussions of the trajectory of saltation load and the forces acting on particles in Chapter 5 pointed out that when sand particles reach the highest level and start to drop down, a relative velocity U_r occurs between the particle and the surrounding fluid along the direction of motion. Because of the resistance, U_r decreases continuously while the particle falls. Bagnold called the distance travelled by the sand particle when U_r has decreased to $U_r/2$ the "penetration distance." He stated that when a particle of saltation load encounters an eddy, if the eddy size is smaller than the penetration distance, the sand particle will pass through the eddy and escape its influence before the eddy can affect its motion. Hence, only those eddies larger than the penetration distance can cause the sand particle to deviate from its original trajectory and bring it into suspension. After some straightforward manipulation, Bagnold found that the penetration distance is proportional to density ratio (ρ/ρ_s). Then, since ρ/ρ_s of water is quite small, sediment particles cannot attain high altitude by saltation, but can be readily entrained by eddy currents and become suspended after moving only a limited distance, the order of several times the sediment diameter. In contrast, the penetration distance in the atmosphere is quite large, therefore, the small eddies near the ground surface are ineffective for sand transport. Only after travelling a considerable distance, the order of several hundred to a few thousand times the sediment diameter, are these particles caught by relatively large eddies and changed from saltation to suspension. In other words, turbulence in water flow exerts a much stronger effect than it does in airflow. Besides, the large amount of saltation load has partially transmitted the shear force of the fluid and a relatively large density gradient exists near the bed. Hence, the turbulence intensity in airflow is smaller than that in water flow for otherwise similar conditions.

15.2 WIND VELOCITY DISTRIBUTION OVER DESERTS

15.2.1 Wind velocity distribution with no sand movement

Field observations and wind tunnel experiments show that the wind velocity profile near the ground follows the logarithmic law for a stable surface, i.e., with no movement of sand:

$$\frac{u}{U_*} = \frac{1}{k}\ln(\frac{y}{y_0}) \tag{7.34}$$

The parameter y_0 is related to the air viscosity for a smooth bed, and it is some characteristic length for a rough bed. For rough beds, Bagnold obtained the same results for air that Keulegan obtained from experiments with water [2]

$$y_0 = \frac{D}{30} \tag{15.3}$$

in which D is the sediment diameter of the particles on the bed surface. Zingg found from wind tunnel experiments that

$$y_0 = 0.081\log\frac{D}{0.18} \tag{15.4}$$

in which y_0 and D are both in mm.

Recent research results reveal that the Karman constant κ in Eq. (7.34) maintains its constant value of 0.4 only if the temperature near ground surface is constant. If the temperature at the higher elevations is higher than that near ground surface, $\kappa < 0.4$; if it is lower, $\kappa > 0.4$, Fig. 15.2 [12]. $T_{0.1}$ and $T_{1.1}$ denote the temperature at 0.1 m and 1.1 m above the ground surface, respectively.

In fact, if a temperature gradient exists, the logarithmic law is no longer valid for the wind velocity distribution; instead, it forms a curve, either upward or downward, on a semi-logarithmic plot, Fig. 15.3 [13]. As pointed out in Chapter 7, with a density gradient in the velocity field, many of its features are related to the relevant parameter — the Richarson number. If the density gradient is induced by changes in temperature, Deacon suggested the use of a Richardson number in the form of

$$R_i = g\frac{(\dfrac{d\ln T}{dy})}{(\dfrac{du}{dy})_{y=y_1}} \tag{15.5}$$

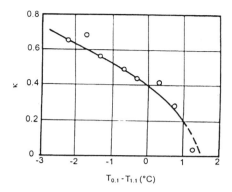

Fig. 15.2 Relationship between Karman constant in wind velocity distribution and temperature gradient (after Sheppard, P. A.)

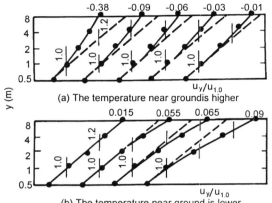

(a) The temperature near ground is higher

(b) The temperature near ground is lower

(Dotted lines represent wind velocity distribution without temperature gradient)

Fig. 15.3 Wind velocity distribution on ground surface if a temperature gradient exists (after Deacon, E. L.)

in which T is in $°F$, and y_l denotes a reference plane. If the effects of Richardson number are included, the wind velocity profile can be expressed in the following form

$$\frac{u}{U_*} = \frac{1}{k(1-\beta)}[(\frac{y}{y_0})^{1-\beta} - 1]$$ (15.6)

in which the Karman constant κ has the value 0.4, and β is a function of the Richardson number R_i as shown in Fig.15.4. If the temperature at high elevations is lower than that at ground surface, $R_i < 0$, $\beta > 1$; if it is higher, $R_i > 0$, $\beta < 1$; with no temperature gradient, $R_i = 0$, $\beta = 1$. Eq. (15.6) can be expended into a series

$$\frac{u}{U_*} = \frac{1}{k}[\ln\frac{y}{y_0} + \frac{(1-\beta)}{2!}\ln^2(\frac{y}{y_0}) + \cdots]$$ (15.7)

if $\beta = 1$, Eq. (15.7) is the logarithmic law of wind velocity distribution. Field data show that during the daytime, the wind velocity profile follows the logarithmic law more than 90% of the time; only during the night does a temperature gradient occurs, and then Eq. (15.6) is more reliable than the logarithmic expression.

As pointed out in Chapter 12, if the gradient of sediment concentration near the bottom is large, the velocity distribution can

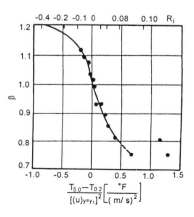

Fig. 15.4 Relationship between parameter β in wind velocity distribution and temperaure gradient (after Deacon, E.L.)

also be expressed in exponential form [Eq. 12.34]; both the coefficient and the exponent are functions of E (Eq. 12.19), which is another form of expression for Richarson number. This example shows that problems in different disciplines in nature tend to follow universal laws. To link the problems of different disciplines may help to provide a deeper understanding of natural phenomena.

15.2.2 Wind velocity distribution with wind-blown sand

If the wind is causing sand to move over the surface of the ground, the velocity distribution differs appreciably from that over a rigid bed. Bagnold found that the wind velocity distribution follows the logarithmic law, but because of the large saltation resistance (due to the intensification), the wind velocity near the ground does not increase with an increase of the wind velocity. A plot with semi-logarithmic scale is a series of straight lines that represent different wind-velocity distributions. All the lines converge at a focal point A at a height of Y_t; there the wind velocity has a constant value of u_t, as shown in Fig. 15.5 [14]. For these conditions, the formula of wind velocity distribution can be expressed as follows:

$$u = 5.75 U_* \log \frac{y}{y_t} + u_t \qquad (15.8)$$

From his experimental results, Bagnold found for uniform fine sand that y_t is about 0.3 cm; for the sand particles normally found in sand dunes, y_t is about 1 cm. The wind velocity at such a height corresponds to the threshold velocity on the bed surface due to impingement. Zingg found from his wind tunnel experiments that

$$y_t = 10D \qquad (15.9)$$

$$u_t = 8.49D \qquad (15.10)$$

in which y_t and D are in mm, u_t in m/s.

As to the wind velocity distribution following the form of Eq. (15.9), the following three questions need to be clarified.

1. The intrinsic mechanism of convergence at one focal point for wind velocity profile

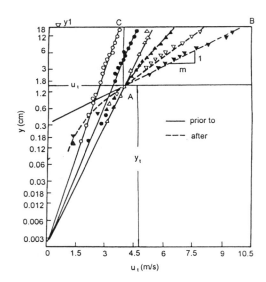

Fig.15.5 Velocity distribution prior to and after the threshold of sand movement (after Kiyoalu Horikawa, and H. W. Shen)

694

of different intensities.

Chapters 5 and 9 introduce Bagnold's basic concept for bed load movement: that the shear force at elevations near the bed is composed of two parts: the grain shear T and the liquid shear τ'. If sediment particles on a river bed are not to be carried away layer by layer, the shear force acting on the motionless bed surface should just balance the threshold force τ_c for particles on the bed surface, Eq. (9.65). Bagnold suggested that a similar mechanism should occur for wind-blown sand [15]. In order for airflow to maintain a steady movement of sand, the saltation resistance to airflow should make the wind velocity near ground surface equal to the threshold wind velocity due to impingement of saltation load. Since the wind velocity near the ground depends only upon sand properties and is unrelated to the intensity of wind force, the velocity distributions for various wind intensities should converge at some point near the ground.

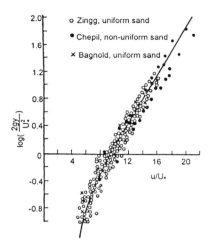

Fig. 15.6 Wind velocity distribution according to Owen formula (after Owen, P.R.)

Different interpretations have been made of the physical picture suggested by Bagnold. Owen stated that, if the saltation resistance is large enough, the characteristic length y_0 in Eq. (7.34) denoting the bed roughness and the saltation height must be about the same. According to Bagnold's data, the upward component velocity of a sand particle at the moment of starting saltation is equivalent to the friction velocity U_*, thus

$$h \sim U_*^2/2g$$

If this value is substituted into Eq. (7.34), one obtains [16]

$$\frac{u}{U_*} = \frac{1}{k}\ln\frac{2gy}{U_*^2} + \text{constant}$$

Fig. 15.6 is a comparison of the above equation with the experimental data. All data points fall along a straight line that can be expressed by

$$\frac{u}{U_*} = 2.5\log\frac{2gy}{U_*^2} + 9.7 \tag{15.11}$$

This equation has the term U_* on the left-hand side and a function of the wind velocity on the right-hand side; their signs are opposite and they interact so that within a certain range of y, the wind velocity does not vary with the wind force (U_*), if

$$\partial u / \partial U_* = 0$$

it follows that

$$2.5\log(2gy_t) = 5\log U_* - 4.7$$

For Bagnold's experiments with uniform sand, the averaged value of U_* is about 60 cm/s, and the corresponding value for y_t, from the above equation, is about 2.9 mm. In other words, wind velocity profiles of various wind intensities converge at a point about 3 mm above the ground surface.

2. Variation of the Karman constant κ

In formulas proposed by Bagnold and Owen, the Karman constant is taken to be 0.4, even with the presence of sand particles. In fact, however, velocity profiles for the case of sediment transport in water flow at high intensity show that the Karman constant decreases with an increase of sediment concentration (Fig. 12.12). The contradiction comes from the fact that airflow has no free surface, as does water flow, so the friction velocity can be deduced only indirectly from the wind velocity distribution. Above the saltation zone, the wind velocity distribution follows the logarithmic law (Fig.15.5). If wind velocities are measured at a fixed height, y_1, the slope of straight line is

$$m = \frac{BC}{AC} = \frac{u - u_t}{\log \dfrac{y_1}{y_t}}$$

For a given sand, u_t and y_t are nearly constant. The expression for the logarithmic distribution of wind velocity would then take the basic form of Eq. (7.34),

$$\frac{U_*}{k} = \frac{1}{2.3}m$$

if κ is taken to be 0.4

$$U_* = m/5.75$$

In other words, if U_* is determined in this way, for the region above the saltation zone, the Karman constant should vary with sediment concentration in the logarithmic distribution of wind velocity. But if the Karman constant is fixed, the effect introduced is included in the computation of friction velocity.

3. Wind velocity distribution in saltation zone

Within the saltation zone the density has a large gradient, and the wind velocity distribution should be concave downward on a semi-logarithmic plot, as in Fig. 12.12. Such a predictable tendency actually exists, as shown in Figs. 15.5 and 15.6 in the region where $y < (U_*^2/2g)$.

15.3 LAWS OF WIND-BLOWN SAND

15.3.1 Incipient motion of sand particles

15.3.1.1 Threshold conditions for coarse sand particles

Whether the wind causes sand movement depends upon whether a saltation load enters the region from upward,. This is the essential difference between wind-driven sand movement and incipient motion of sediment particles in flowing water. The situation can be best explained on the basis of a wind tunnel experiment.

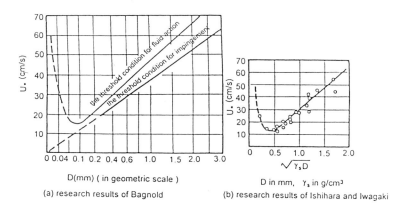

(a) research results of Bagnold

(b) research results of Ishihara and Iwagaki

Fig.15.7 Threshold condition for sand movement by wind
(after Bagnold, R. A.)

First, a thick layer of sand is placed on the floor of the wind tunnel. Then the fan is turned on to produce airflow over the tunnel bed at a given velocity. If the velocity is low, sand particles on the bed remain at rest. Only if the velocity exceeds a certain limit, do sand particles begin to move. Since the sand movement is induced directly by the action of the air flow on the particles, the critical condition is properly called the threshold for fluid action.

In the next stage of the experiment, a relatively low wind velocity is maintained and sand is supplied continuously at the entrance to the test section. The sand particles then gain a certain amount of momentum from the airflow as they fall to the bed, and their impingement produces surface creep and saltation of the particles on the bed. However, if the wind velocity is not high enough, the energy supplied by the flowing are compensated for the energy loss due to saltation resistance. Hence, not far from the entrance, the disturbance due to the sand flux at the entrance disappears. Further downstream, the sand does not move. If the wind velocity is then increased until it reaches a critical value, the energy gain and energy loss reach a balance and the sand movement continues steadily throughout the entire length of the test section. Since the sand movement is caused mainly by the impingement of saltation load, this critical condition is called the threshold condition for impingement.

Bagnold's results are plotted in Fig. 15.7a. Here, the friction velocity is used because the mean value of wind velocity does not have a clear definition for wind-blown sand, and the shear force on the bed is not easy to measure. Still, the friction velocity can be obtained from the wind velocity distribution, as described in the preceding paragraph.

In analogy to the critical threshold condition in water flow, the threshold condition for initial movement by fluid action can be expressed in the basic form

$$\frac{\tau_c}{(\gamma_s - \gamma)D} = f(\frac{U_* D}{v}) \qquad (8.12)$$

Substituting $\tau_c = U_{*c}^2 \rho$ into Eq. (8.12) yields

$$\frac{U_{*c}}{\sqrt{\frac{\gamma_s - \gamma}{\gamma} gD}} = f_1(\frac{U_* D}{v}) \qquad (15.12)$$

For the region in which $U_* D/v > 3.5$ (equivalent to incipient motion of sand with $D > 0.25$ mm), the right-hand side of the expression is nearly constant. From experimental results with uniform sand, Bagnold obtained a constant value of 0.10; Chepil found that the value should vary within the range of 0.09 to 0.11 [17]. In Chapter 8, the threshold of sediment movement in water flow is shown to follow the Shields curve (Fig. 8.6). If the grain Reynolds number is large, the curve reaches a constant value between 0.04 to 0.06. If the Shields criteria are converted to the form of Eq. (15.12), the constant should vary between 0.20 to 0.25, a value larger than those Bagnold and Chepil obtained for wind-blown sand movement. According to the latest experimental results of wind-blown sand movement obtained by Lyles and Woodruff [18],

$$\frac{U_{*c}}{\sqrt{\dfrac{\gamma_s - \gamma}{\gamma} gD}} = 0.17 - 0.20$$

which is closer to the results obtained for water flow. Such a big disparity in the experimental results obtained by various authors may be related to the following aspects. On the one hand, no strict standard has been formulated for determining the threshold condition for sand movement; on the other hand, the methods used by various authors to compute the friction velocity are also different.

According to field observation in desert areas in China, the relation between sand diameter and the threshold wind velocity is as shown in Table 15.4 [9]. The threshold wind velocity is nearly proportional to the square root of grain diameter (for wind velocity measured at a height of 2 m above ground surface).

Table 15.4 Observed threshold wind velocity for various sand diameters, as measured in desert areas in China

Grain size (mm)	Wind velocity for incipient motion (m/s) (measured *at* 2 m above ground surface)
0.1-0.25	4.0
0.25-0.5	5.6
0.5-0.10	6.7
>1.0	7.1

In Fig. 15.7a also, the threshold condition for impingement is lower than that of fluid action. For coarse sand particles $D > 0.25$ mm with the impingement of saltation load, the constant in the right-hand side of Eq. 15.12 should be 0.08 rather than 0.1 [2]. Particles larger than coarse sand are not easily set in motion; however, under certain specific conditions, for example on the surface of glacier, because of small friction, coarse particles like gravel can also be moved by wind, but if the wind velocity is large enough. Schumm found that a stone with dimensions of 4.9×3.9×2.7 cm and a weight of 56g would be moved by wind with a velocity of 66 Km/h; once the motion starts, it continues, even for a lower wind velocity [19].

15.3.1.2 Threshold condition of fine sand particles

For smaller sand particles, the laminar sublayer begins to play a role because of the shading effect. Fine sand particles often absorb moisture from the atmosphere, which leads to the occurrence of cohesion among sand particles. Thus the threshold conditions for fine sand particles are qualitatively similar to those of alluvial sediment in water flow; they show a saddle-shaped curve, as in Fig. 15.7a. Fig. 15.7b shows the threshold condition for fluid action, plotted from the work of Ishihara and Iwagaki [14]. The curve displays the same tendency as that given by Bagnold. Fletcher derived a generalized formula for the friction velocity at threshold condition--one that is suitable for both coarse and fine sand--on the basis of dimensional analysis and a series of experimental result [20].

$$U_{*c} = (\frac{\gamma_s - \gamma}{\gamma})^{1/2} \left[0.13(gD)^{1/2} + 0.057(\frac{c}{\rho_s})^{1/4} (\frac{v}{D})^{1/2} \right]$$
(!5.13)

in which ρ_s — density of sand particle; c — cohesion between particles. If c is small, Eq. (15.13) reduces to the formula proposed by Bagnold and Chepil. The only difference is that the constant Fletcher adopted is slightly larger.

The threshold condition for the impingement for fine sand particles is not yet clear. Once such particles enter the airflow, they move in the form of suspension, not saltation. If some coarser saltation load falls onto a bed composed of fine sand, the coarse grains may sink among the fine particles, and gradually change the nature of the bed. According to available information, the threshold condition for the impingement of relatively fine particles seems to approach that of fluid action.

The fact that fine sand particles are not apt to be set in motion is significant in regard to the morphological feature of the loess areas in China. Since the grain size on the ground surface of loess areas is quite fine, they cannot be blown even by strong winds; but once they are disturbed and dislodged, they are easily blown away. Paths near villages are frequently disturbed by blocks of sheep flock and carts. The effect of wind blowing is then different from that on the neighbouring land surface. Consequently, the paths are eroded more and more and eventually make the landscape steep and gorge-like.

Finally, wind erosion of soils is much more complicated than that for the threshold of sand particles motion; it involves a number of factors like vegetation cover, cultivation and physico-chemical properties of soil. Details are contained in references [21, 22, 23]

15.3.2 Bed load movement

Just as for sediment movement in flowing water, saltation load and surface creep load together are called bed load under wind action.

15.3.2.1 Distribution of saltation load along a vertical

In sediment-laden flow, bed load movement takes place within a zone that is several sediment diameters thick in water flow. The investigation of how bed load is distributed in such a small layer is meaningless in practice. For wind-driven sand, however, the situation is quite different; saltation can reach heights of several hundred or even a thousand times the sand diameter. From the distribution of saltation load along vertical and the wind velocity distribution, one can calculate the sand transport rate due to the saltation load.

Field data indicate that the distribution of saltation-load concentration along a vertical appears to be logarithmic. Fig. 15.8 contains the concentration profiles in airflow at three wind velocities, as measured at Suoche of Xingjiang Uygur Autonomous Region [9]. The data points follow straight

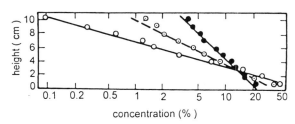

Fig. 15.8 Sand concentration distribution in airflow at various wind velocities (measured at Suoche, Xingjiang Uygur Autonomous Region, China)

lines on semi-logarithmic plots. Zingg obtained an empirical formula using experimental data from wind tunnel tests [11].

$$g_y = (\frac{\alpha}{y+b})^{\frac{1}{n}} \qquad (15.14)$$

in which g_y — transport rate of saltation load at height y, in g/cm³/s; b — function of sand diameter; α, n — vary with sand diameter and wind velocity.

Kawamura studied this problem theoretically and obtained the following expression for saltation-load concentration along a vertical [24]

$$S(y) = \frac{2W_0}{\sqrt{gh}} K_0(\sqrt{\frac{2y}{h}}) \qquad (15.15)$$

in which

$S(y)$ — concentration of saltation load at height y in g/cm³ ; W_0 — amount of sand particles falling on unit area per unit time in g/cm²/s; h — mean value of the max-height of saltation; K_0 — Bessel function of zero order.

Kawamura approximated the wind velocity distribution by an exponential distribution

$$u(y) = ay^{1/2} \qquad (15.16)$$

Then, the distribution of saltation load per unit area along a vertical can be expressed as

701

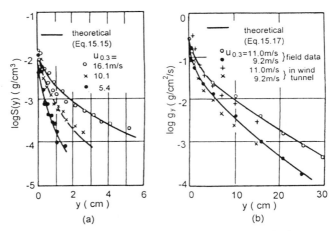

Fig. 15.9 Concentration profile of saltation load along vertical — comparison of theoretical formula against experimental and field data (after Kawamuta, R.)

$$g_y = W_0\{2\sqrt{2}\eta[K_0(\xi) - \beta\sqrt{\frac{h}{g}}K_1(\xi)]$$

$$+ \frac{1}{\sqrt{2}}\frac{a\beta\sqrt{0.75h}}{g}\cdot\xi^2[K_0(\xi)+K_2(\xi)]\} \tag{15.17}$$

in which

$$\xi = \sqrt{\frac{2y}{h}}$$

$$\eta = \frac{\bar{u}_1}{\sqrt{2gh}}$$

$\bar{u}_1$ is the mean horizontal velocity component at which saltation of sand particles begins,

$$\beta = \frac{2\pi\mu D}{m}$$

m — mass of sand particle; K_1, K_2 — Bessel functions of the first and second orders.

Experimental results show that η is close to a constant of 2, and independent of the wind velocity. Fig. 15.9 presents a comparison of Eqs. (15.15) and (15.17) against the measured data; $u_{0.3}$ denotes the wind velocity 0.3 m above the ground surface. These figures reveal that sand particles moving as saltation load concentrate near the ground surface, the thickness of the layer of saltation load gradually increases with the increase of wind velocity.

15.3.2.2 Bed load transport rate

In the following, some representative formula of wind drift bed load are presented along with a comprehensive comparison of the various formulae.

1. Formula of O'Brien and Rindlaub

Early in the 1930s, O'Brien and Rindlaub developed a wind-driven bed load transport rate formula based on field data obtained in the estuarine region of the Columbia River, U.S.[25].

$$g_b = 0.022u_{1.5}^3 \qquad (15.18)$$

in which g_b — unit sediment discharge of bed load in g/s/m; $u_{1.5}$ — wind velocity *at* 1.5 m above ground surface.

Substituting Eqs. (15.9) and (15.10) into Eq. (15.18) yields

$$g_b = 9.96*10^{-5}(U_* + 10.8)^3 \qquad (15.19)$$

in which U_* — friction velocity in cm/s. This is an empirical formula, but the functional relationship that bed load transport rate is proportional to the cube of friction velocity accords well with the formula of Bagnold and Kawamura that were deduced from theoretical backgrounds.

2. Bagnold formula [2].

For a characteristic trajectory of saltation, u_1 denotes the horizontal velocity component of a sand particle during uplift from the ground surface and u_2 the horizontal velocity component when it falls to the ground after travelling a distance of l. The momentum taken from the wind per unit area of bed surface that is spent to support the sand movement is

$$\frac{g_b}{g}\frac{u_2 - u_1}{l}$$

From Newton's law, the rate of momentum loss equals the resistance of sand movement to airflow (the saltation resistance). As mentioned, other forms of resistance on the bed surface can be neglected. The relationship has the form

$$\frac{g_b}{g}\frac{u_2 - u_1}{l} = \tau_0 = \rho U_*^2$$

u_1 is small compared with u_2, and can be neglected; u_2/l is approximately equal to g/v_1, where v_1 is the vertical velocity component during the uplift from ground surface. Thus, the above equation can be rewritten as

$$g_b = \rho U_*^2 v_1$$

Bagnold assumed further that v_1 is proportional to U_*, in which β denotes the constant of proportionality, then

$$g_b = \beta\rho U_*^3 \qquad (15.20)$$

However, the term bed load should include both saltation load and surface creep load. As mentioned before, the ratio of surface creep load to the total bed load is approximately constant, and it is included in the term β in Eq. (15.20).

From experimental results obtained in wind tunnel tests, Bagnold found that g_b appeared to be proportional to the square root of the sand diameter. Hence the bed load formula can be expressed as

$$g_b = c\sqrt{\frac{D}{D_1}}\rho U_*^3 \qquad (15.21)$$

in which D_1 — standard size of sand with $D = 0.25$mm; D — sand diameter under study; c has the following values for difference cases:

nearly uniform sand $\qquad\qquad\qquad\qquad$ *c = 1.5*

sand in natural sand dunes $\qquad\qquad\qquad$ *c = 1.8*

sand with wide range of size distribution $\quad$ *c = 2.8*

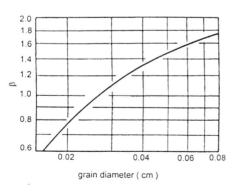

grain diameter (cm)

Fig. 15.10 Relationship between paramenter β in Bagnold bed load formula and grain diameter (after Bagnold, R.A.)

In obtaining Eq. (15.21), Bagnold had not measured directly the shear force acting on the bed surface, but he incorrectly assumed the Karman constant to remain unchanged at the value 0.4, thus he obtained an overestimation of c-values. Zingg measured directly the shear force acting on a bed surface in wind tunnel tests and deduced [11]

$$g_b = 0.72(\frac{D}{D_1})^{0.75}\rho U_*^3 \qquad (15.22)$$

But when he determined the transport rate, he extrapolated the wind velocity distribution to the bed surface and then integrated. In fact, the surface creep load is not included in the formula, and the result was an underestimation of g_b. In 1956, Bagnold conducted a series of experimental studies on sediment transport in water flow and in airflow, and he found the relationship between β and sand diameter shown in Fig. 15.10 [26].

Recently, Hsu developed a formula of the same type as Eq. (15.20) [27].

$$g_b = K(\frac{U_*}{\sqrt{gD}})^3 \qquad (15.23)$$

K is a function of D

$$D = 0.46logK + 0.095 \qquad (15.24)$$

K is in 10^{-4}g/cm/s, D is in mm.

3. Kawamura formula

Starting from Eq. (9.64), Kawamura also separated the shear force acting on the bed into two parts

$$\tau_0 = T + \tau_c \qquad (9.64)$$

in which T is the friction force caused by the impingement. τ_c is the shear force for the threshold condition. Also, T equals the momentum loss in the direction of sand movement due to impingement.

$$T = \frac{W_0}{g}\overline{\left|u_2 - u_1\right|}$$

in which W_0 — weight of sand particles falling on the bed per unit area per unit time.

The bar denotes the average value. Kawamura further assumed

$$\overline{\left|u_2 - u_1\right|} = \xi\overline{\left|v_2 - v_1\right|}$$

and he assumed that the vertical component of sand particle falling on the bed was equal to that of starting saltation, but with opposite signs.

Thus,

$$T = 2\xi \frac{W_0}{g} \overline{v}_1 \qquad (15.25)$$

Substituting Eqs. (9.64) and (15.25) yields

$$2\xi \frac{W_0}{g} \overline{v}_1 = \tau_0 - \tau_c = \rho(U_*^2 - U_{*c}^2)$$

Kawamura obtained from experimental data the relationship

$$W_0 = K_1 \rho g (U_* - U_{*c}) \qquad (15.26)$$

Combining the two equations and solving, he obtained

$$\overline{v}_1 = K_2 (U_* + U_{*c})$$

If the saltation height is proportional to $v_1{}^2/2g$, and saltation distance is proportional to height,

$$\overline{l} = K_3 \frac{(U_* + U_{*c})^2}{g}$$

If g_b denotes the unit sand transport rate passing through a vertical, and all the sand particles fall within a distance $\overline{l}$, then

$$g_b = W_0 \overline{l} = K_4 \rho (U_* - U_{*c})(U_* + U_{*c})^2 \qquad (15.27)$$

The constant K_4 as determined by Kawamura was 2.84×10^{-3}. If $U_* = U_{*c}$, the rate of sand transport is actually zero. In this aspect, Kawamura's formula is more rational than are those of Bagnold and Hsu.

4. Application of the Einstein bed load function to wind-blown sand movement

Kadib applied the Einstein theory in wind-blown sand movement [28]. According to the aforementioned principal differences of sediment movement in water flow and in airflow, the impingement force of saltation load acting on sand particles on bed surface has to be included. If a sand particle falls on another particle with mass m. This causes the latter to jump upward with velocity v_1. The upward impinging force the sand particle receives is

$$F_t = m \frac{v_1}{t}$$

in which t is the time required to saltate with velocity v_1 from rest after impingement. Kadib took the time period, which Einstein considered as the time for

706

the particle on the bed to be picked up by scouring [Eq. 9.35], as t in the above equation, and assumed v_l to be proportional to U_*. In this way he deduced the expression

$$F_i \approx \psi(\frac{\gamma_s^2}{g^3}\frac{\gamma}{\gamma_s - \gamma})^{1/2}(U_*^3 D^{3/2}) \qquad (15.28)$$

The impinging force acts in the same direction as the uplift force and can be combined with the latter by using a correction factor. Then Eq. (9.81) should be rewritten as

$$\psi_* = \frac{\xi}{k}Y(\frac{\beta_.}{\beta_x})^2 \psi \qquad (15.29)$$

in which k is the correction factor for the impinging force.

Following Eq. (15.28)

$$k \propto \frac{1}{U_*^3 D^{3/2}}$$

Based on experimental data of various authors (with D in the range 0.15 to 1.0 mm). Kadib obtained

$$k = \frac{3,280}{U_*^3 D^{3/2}} \qquad (15.30)$$

k is in cm.s. Furthermore, Kadib found also that the shading effect for sediment particles in airflow is slightly different from that in water, and he suggested

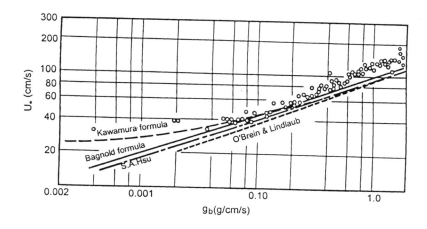

Fig. 15.11 Comparison between formulae for sand transport
rate and experimental data (after Hsu, Shih-Ang)

$$\xi / k = 6.34 \, \psi^{-0.7} \tag{15.31}$$

With adjustment correction, the remainder can be computed by the Einstein theory.

5. Comparison of various formulae

In Eqs. (15.9), (15.20), (15.23), and (15.27), the bed load transport rate is a function of the friction velocity U_* for all, so they can be directly compared with each other. Fig. 15.11 shows such a comparison of the four formulae for the experimental data. Obtained by Shen and Horikawa, if $U_* < 40$ cm/s, Kawamura's formula appears to be more reliable; for $U_* = 40$-70 cm/s, Bagnold's formula gives better results. Horikawa and Shen also pointed out that if the coefficient in Fig. 15.27 is assigned the value 1.02×10^{-3}, then Kawamura's formula would accord better with the experimental results.

15.3.3 Suspended load movement

The diffusion equation for the profile of suspended load concentration along a vertical is contained in Chapter 10, as follows:

$$\varepsilon_y \frac{dS_v}{dy} + S_v \omega = 0 \tag{10.7}$$

It was derived by W. Schmidt in 1925 when he studied dust distribution in the atmosphere [29]. The solution to Eq. (10.7) is

$$\frac{S_v}{S_{va}} = (\frac{h-y}{y} \frac{a}{h-a})^z \tag{10.16}$$

For wind-driven sand the vertical dimension of the airflow h is much larger than y, so that Eq. (10.16) can be rewritten as

$$S_v \sim y^{-z} \tag{15.32}$$

That is, the distribution of suspended sediment concentration with elevation should be a straight line on a logarithmic plot. Fig. 15.12 presents suspended sediment concentration profiles along the vertical. Within 12 m of the ground, the concentration distribution follows the exponential law. From the concentration distribution and wind velocity profile, one can integrate the product of the two along the vertical,

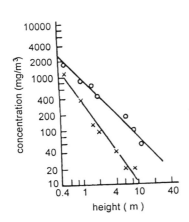

Fig. 15.12 Distribution of suspended load along vertical (after Nickling, W. G.)

and obtain the total sand transport rate of suspended load up to a given height. Statistics from 15 dust-storm events in Slims River valley show that 12.5 to 65.5% of the total load is transported in suspension within a height of 12 m, and of the suspended load alone, 20 to 60% is concentrated within only 0.5 m of the ground surface.

For open channel flow, turbulence originates near the boundary; hence, good correlation exists between suspended load and bed shear force (of friction velocity). In wind-driven sand movement, turbulence structure at high altitudes is greatly affected by both temperature gradient and convection. For such conditions, the friction velocity near the ground will no longer be the only factor involved [23].

15.4 OCCURRENCE AND DEVELOPMENT OF EOLIAN BED FORMS

Under the action of wind, sand particles accumulate as they move and form a variety of morphologic patterns. Because they are easy to observe, the occurrence and development of such bed forms and the intrinsic mechanisms involved are better understood than are those for bed forms in water flow.

15.4.1 Basic types of eolian bed forms

According to field observations by various authors, wind-driven sand waves can have thousands of shapes, and they also behave in many different ways. After detailed analysis, however, all can be classified within no more than three basic types: ripples, dunes, and giant dunes. Wave lengths of the three types have different orders of magnitude: ripples 0.01 to 10 m; dunes 10 to 500 m; giant dunes 500 to 5,000 m.

Fig. 15.13 is a plot of wave length versus sand diameter [30]. The distribution of the data is not continuous, but instead, they fall into three distinct zones. For these wind-blown sand waves, each type has its own equilibrium state; that is, sand waves of one type do not develop continuously from those of another.

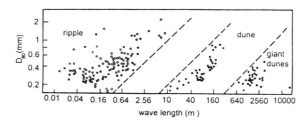

Fig.15.13 Zonal distribution of eolian bed forms according to grain size and wave length (after Wilson, J.G.)

Ripples, dunes, and giant dunes all have longitudinal and transversal faces with different orientations. When longitudinal and transversal sand waves intersect each other, three intersecting patterns can occur, depending on whether there exists a displacement of half wave length in the intersection.

15.4.1.1.Ripples

Fig. 15.14a shows profiles of eolian sand ripples. The windward slope is a slightly convex curve, making an angle of 8 to 10°, with the ground surface; the leeward side is relatively steep near the crest (close to angle of repose of sand particles of about 30°), but its lower part has a gentler angle of about 20°. The cause is the impingement of the saltation load as it falls down directly or impinges against other particles in airflow and then falls at acute angle, Fig. 15.14b.

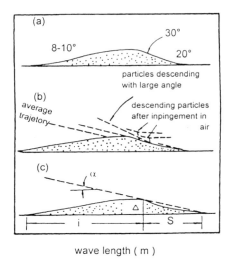

Fig.15.14 Longitudinal profile of eolian ripples(after Bagnold, R. A.)

The wave length of ripples varies between 3 to 20 cm and is generally about 10 cm. Their height falls mostly in the range 0.5 to 1.0 cm. The average ratio of wave length to wave height is about 18, but it can vary considerably. Generally, the ratio for eolian ripples is larger than that for ripples in water flow.

Bagnold indicated that eolian ripples are the direct result of saltation [2]. Sand particles in saltation strike the ground surface at a rather flat angle and impinge against the latter. If the sand particles fell to the ground uniformly, their distribution would be represented by a series of equally spaced parallel lines. But if a small depression forms for some reason, in the region *ABC*, in Fig. 15.15, the sand transport intensity induced by impingement would be proportional to the times of impingement at a point per unit area per second; that is, it would be proportional to the concentration at the point of impingement in Fig. 15.15. On the leeward side of *AB* the impinged points distribute less densely than on the windward side *BC*, hence, the number of impingements is less. Thus, the number of sand particles removed from the windward side is more than those from leeward side, and the original depression tends to enlarge gradually. At the same time, the slope *BC* experiences more impingements than the flat part along the leeward side. At point *C*, the velocity of sand movement from windward side is larger than that toward the leeward side. Therefore, more sand particles accumulate near that point and a second leeward slope *CD* forms where the intensity of sand movement is relatively weak. Again, sand particles moving to the outside from point *D* are more numerous than those moving

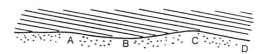

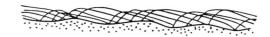

Fig. 15.15 Disparity of impinging intensity on
windward and leeward slopes of eolian ripples
(after Bagnold, R. A.)

Fig. 15.16 Coincidence of wave length of
eolian ripple and characteristic length
of trajectory for saltating particles
(after Bagnold, R. A.)

along slope *CD*, and a second depression is formed. The process repeats itself over and over, and causes the commonly observed ripple formations.

Although the intensity of saltation varies in time and in space, for a given wind condition, the trend of the saltation particles is to have a certain average trajectory, which Bagnold called the characteristic trajectory of saltation movement. Hence, sand particles displaced from one small area will most likely fall on another small area, and the spacing between them is the length of a characteristic trajectory; i.e., the probability of falling there is greater than the probability of falling at other places.

Table 15.5 Comprison of wave length for ripples with average travel distance in
wind-driven saltation of sand particles

Friction velocity (cm/s)	19.2	25.0	40.4	50.5 .	62.5
Averaged travel distance (cm)	2.5	3.0	5.4	8.0	11.6
Observed wave length of ripples (cm)	2.4	3.0	5.3	9.15	11.3

Consequently, sand particles are displaced from a certain point. Then, at another point one characteristic trajectory away, the concentration of impingements will be greater than at other places; and in turn, sand particles dispelled from the second point will reach the ground surface at a third point again about the same distance away. The local increase or decrease of sand transport rate in the lower layer will cause small mounds and depressions to form, and these form the crests and troughs with a wave length equal to the length of a characteristic trajectory (Fig. 15.16). In Table 15.5, the travel distance of saltation load and the measured wave length of sand ripples are compared for various friction velocities. The wave length of the ripples are essentially equal to the average travel distance of the saltating particles.

Although Bagnold's explanation of the formation of wind-blown sand ripples is widely accepted, Sharp has raised some questions about it. He suggested that except the characteristic length of the trajectory, the grain size and the wind velocity should have additional and even more direct effects on the formation and development of ripple [31]. He stated that the constitution of ripples is mainly due to surface creep load. In Fig. 15.14c, either the length of the shaded zone S or the length of impingement zone i is related to the wave height of the ripples $\varDelta$ and the angle of impingement of saltation load α; and with the increase of wind velocity $\varDelta$ and α tends to decrease; and wave height of ripple $\varDelta$ increases as the grain size becomes coarser. According to the available data, S and i both increase with an increase of grain size; if the wind velocity increases, i tends to decrease, and S tends to increase. These results indicate that wave length of ripples is directly affected by both sand size and wind velocity.

15.4.1.2 Sand ridge

A sand ridge is a kind of ripple. It occurs in desert areas where erosion has occurred and the supply of coarse particles is available. During wind erosion, coarse sand gradually accumulates in a layer that can migrate as surface creep load along the windward slope because of the impingement of saltating particles. After passing over the crest, the coarse particles settle in the wind-shaded zone on the leeward slope. If these particles are coarse enough

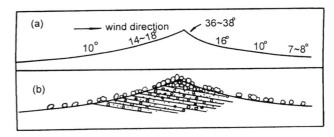

Fig. 15.17 Profile of sand ridge (after Sharp, R.P.)

that they cannot be moved directly by the wind, they will accumulate near the crest, and produce an extension of the sand ripple; eventually they form a sand ridge.

Sand ridges belong to a bed form that is larger in scale than ripples. The wave length of sand ridges is usually about 60 to 100 cm, and their wave height is 5 to 8 cm; the average ratio of wave length to wave height is about 15. In the deserts of Libya, huge sand ridges form, that they can be as much as 20 m long and 60 cm high.

Sand ridges are more stable and can be preserved for much longer than can sand ripples. If the local wind changes direction frequently, the base of a sand ridge will be relatively symmetric; only the upper part near the crest is asymmetric, reflecting the wind direction at the time (Fig. 15.17)[31]. In regions where one wind direction prevails, sand

Fig. 15.18 Plan view of sand strips and local circulation of airflow

ridges show an asymmetric form. In Fig. 15.17b, the distribution of different sizes of sand particles at a sand ridge is also shown. On the bed surface in the trough of a sand ridge, coarse particles do not exceed 10 to 20%, but near the crest, they constitute 50 to 80%. The coarsest particles are as big as 2 to 5 mm, and they can even exceed 1 cm. The crest of sand ripples and sand ridges are perpendicular to the wind direction and present a continuous wave front, so that they have a rather regular appearance.

15.4.1.3 Sand strip

A sand strip is a kind of longitudinal ripple. It can form in the open desert when ever a strong wind carries large amounts of sand over a ground surface covered with gravel. If, occasionally, a small amount of sediment is concentrated at a certain place, the transient pile of sand covers the bed surface and intensifies the local sand movement and the corresponding saltation resistance. Then the wind velocity near the sandy bed is less than that over the gravel bed, The difference in wind velocity along the line making the different kinds of ground cover leads to the formation of longitudinal vortices which push sand particles from the sides to the centre and form parallel sand strips, as in Fig. 15.18. Bagnold reported that once he saw 200 and more of these sand strips in the desert of the southern Egypt. The strips 1 to 3 m wide, 1 to 2 cm thick and about 500 m long, on average; and their spacing was 40 to 60 m [2]. If the wind velocity gradually decreases, the intensity of sand movement and saltation

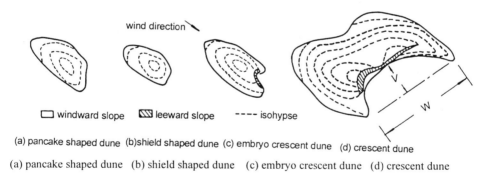

(a) pancake shaped dune (b)shield shaped dune (c) embryo crescent dune (d) crescent dune

(a) pancake shaped dune   (b) shield shaped dune   (c) embryo crescent dune   (d) crescent dune

Fig. 15.19 Configuration of sand dunes in different stages of development

resistance also decrease and no longer play an important role. Consequently, the velocity of wind blowing over gravel-covered bed with larger roughness is lower than that blowing over sand-covered bed, then longitudinal vortices appear with an opposite direction as shown in Fig. 15.18, the accumulated sand particles tend to disperse and the sand strips disappear.

15.4.1.4 Sand dunes

In flat denuded desert areas, the most common eolian bed forms are crescent-shaped sand dunes and sand dune chains. Fig. 15.19 shows sketches of sand dunes that were observed on the alluvial fan near the foot of a mountain region, west of oasis in Pishan county, China. The figure presents different stages of dune development: the pancake-shaped dune, shield-shaped dune, embryo crescent dune and real crescent dune. A crescent dune looks like a round or elliptic pancake with a bite out of it on one side. The bitten side has a discontinuous face, called the sliding face. Wind-blown sand particles passing over the crest stop and accumulate there.

In desert areas in China, crescent sand dunes are the most widely distributed eolian bed form. The dune height is usually 1 to 5 m, rarely exceeding 15 m, the dune width is generally about 10 times the dune height.

15.4.1.5 Longitudinal sand dunes

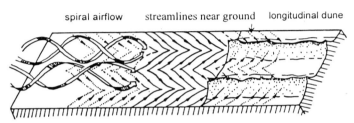

spiral airflow   streamlines near ground   longitudinal dune

Fig. 15.20 Sketch of longitudinal dune induced by spiral airflow (after Mabbutt, J.A.)

The formation of longitudinal dunes is somewhat similar to that for sand strips. It is also the product of a spiral airflow (Fig. 15.20). The spiral is large, possibly induced by a difference in resistances on the ground surface. In some desert areas,

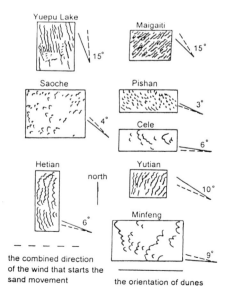

Fig. 15.21 Relationship between direction of dune arrangement and resultant wind direction

sand dunes are widespread, and in other places, the ground surface is simply denuded [1]. Possibly, longitudinal sand dunes are induced by turbulence caused by the radiation of heat to the airflow, or by differential heat radiation from windward and leeward slopes [35].

The height of longitudinal sand dunes is 5 to 30 m, and the width at the base is about 5 to10 times the height. The length is usually about 200 to 500 m, but the dunes can still extend longer .

15.4.1.6 Sand dune belt and sand dune chain

Dunes often accumulate in deserts and develop to dune belts or dune chains;

714

the orientation of the belt or chain is along the direction of the wind that started the sand movement. Fig. 15.21 shows the relationship of sand dunes orientation and wind direction in the Taklamaken Desert [34]. The maximum deviation between the two direction is not over 15°.

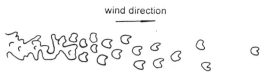

wind direction

Fig. 15.22 Crescent dune belt for monodirectional wind

A sand dune belt (or chain) may have various appearances, including that of the longitudinal sand dunes. If the wind blowing over the desert is effectively mono-directional, crescent dune belts similar to those shown in Fig. 15.22 occur. In the entire belt, those dunes near the origin have an ample supply of sand and can grow to large sizes. Since the rate of migration of dunes is inversely proportional to dune

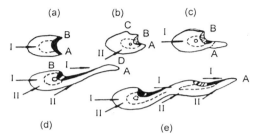

Fig. 15.23 Sketch of transition from crescent dune to chain of longitudinal dunes, wind blows alternately from two directions (after Bagnold, R. A.)

height, large dunes migrate slowly and may come into contact with one another; in this way they combine into one large dune. Those dunes located to the leeward have a smaller supply of sand, and they form dunes that are small and thus move faster and faster. The whole dune group then tends to disperse.

If the wind direction blowing over a desert is a crossed wind, i.e., if it comes alternately from two different directions, a longitudinal dune chain will develop; it is also called a Sief dune. Bagnold suggested that the formation of a chain-shaped dune system starts from a single crescent sand dune [2]. In Fig. 15.23a, wind blowing along direction marked *I* first forms a typical crescent sand dune. When the wind direction changes from *I* to *II*, the right-hand wing develops vigorously to point *A* in Fig.15.23b, because it receives more sand; simultaneously the left-hand wing shrinks and a new wing appears at *C*. In this stage, if the wind direction returns to the more stable direction of *I*, the windward appearance returns to a symmetric form, but the leeward wing *A* will continue to grow and forms a low and small tongue of sand, as shown in Fig. 15.23c. If the wind direction continues for a long time, the overextended part of the wing at *A* disappears, and the sand dune recovers the shape shown in Fig.15.23a. If instead the right-hand wing extends beyond the sheltered zone formed by sand particles that have passed over the crest of crescent dune and reaches point *D* (Fig. 15.23d), and its sand flux comes from the wing end at *B*, then a new sand dune may form at point *D*, because of the ample supply of sand coming from directions *I* and *II* , so that the rear part can grow faster than the combined part. In this way, if the wind continues, a huge longitudinal sand dune chain can form.

In certain areas, strong winds blow toward the dunes from both sides of the principal direction, and they form a pseudo-symmetric Sief dune chain; they do not resemble the crescent dunes (Fig. 15.24). The principal difference between a longitudinal sand chain and a crescent dune belt lies in the fact that the sliding face is parallel to

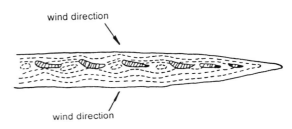

Fig. 15.24 Symmetric chain of longitudinal dunes (after Bagnold, R. A.)

the prevailing wind direction, rather than perpendicular to it. In a symmetric sand dune chain, the sliding face exists as a temporary phenomenon; it is located on the sheltered side, out of the wind.

15.4.1.7 Giant dunes

Giant dunes, called Draa in Arabian, mostly occur in extremely dry desert areas like the Arabian and Sahara Deserts. Fig. 15.25 shows the famous Pur-Pur Giant crescent Dune in Peru. It is 55 m high, 850 m wide and 2 km long. On the surface of the windward slope and along the two extended arms, innumerable secondary crescent dunes have formed. These giant dunes take several thousand years to attain this size.

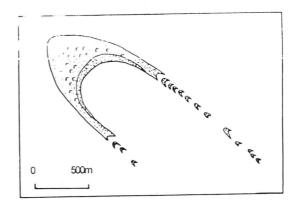

Fig. 15.25 Giant dune in Peru, the Pur-Pur Dune

Giant dunes may have diversified appearances: some look like pyramids with dune heights of up to 150 m, and the perimeter may be 1 to 2 km. They have such a huge scale, comparable with the Pyramids of Egypt, that they are sometimes called pyramid dunes. In Egypt the huge Whale Back with a base 3 km wide and 300 km long is such a giant longitudinal dune.

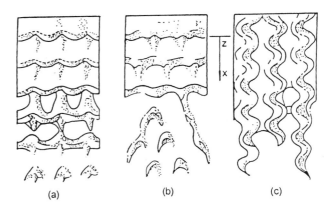

Fig. 15.26 Three basic types of intersection of transverse and longitudinal dunes (after Wilson, J.G.)

716

15.4.1.8 Beam and nest sand dune

The sand dune formed like a beam and nest is a composite configuration formed by the intersection of longitudinal and transverse sand dunes because of winds from different directions. Fig. 15.26a shows a grid-like sandy dune formed by the intersection of longitudinal and transverse sand dunes that has maintained the line of its wave front. If each time the wave front of a longitudinal dune is displaced half a wave length laterally towards both sides, a fish-scale shaped dune forms, as shown in Fig. 15.26b. Each time the wave front of transversal sand dunes is displaced half a wave length along the wind direction, braided sand dunes form, as shown in Fig. 15.26c [30]. If the supply of sand is insufficient, the dunes degenerate like those shown in the lower part of Fig. 15.26.

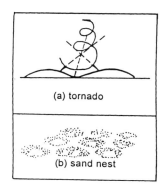

Fig. 15.27 Tornado and sand nest bed form

Also, in hot seasons with no wind, the ground surface of desert areas is strongly heated and may cause convection. The strongly heated air particles lift up in a spiral pattern (Fig. 15.27a), called a tornado. In the northern hemisphere, tornadoes have clockwise swirls, and in the southern hemisphere counter-clockwise ones. Tornadoes can pick up sand particles from the ground surface, and thus form small depressions on the ground. After ascending some distance, the uplifting air current diffuses and the entrained sand particles fall back down. Eolian bed forms causes in this way have no preferred direction, but they create round depressions on the surface with mounds in between. They are called sand nests, in Fig. 15.27b.

These several bed forms are only the most basic of the many eolian geomorphological forms that develop in desert areas. In addition to these, transitional patterns form that correspond to dune development in different stages. If the bed surface is partially covered by vegetation, the situation is still more complex. For example, on land with shrubs, sand dunes can develop branch-shaped sand dune chains, in the same way that a longitudinal sief dune chain can develop from crescent dunes [35]. The wave length is usually about 200 to 500 m, but the dunes can be still longer.

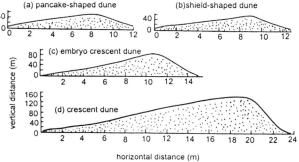

Fig.15.28 Longitudinal profiles of sand dunes with various shapes

15.4.2 Regularity of migration of sand dunes

Fig. 15.19 shows sketches of a sand dune at different stages of development. Fig. 15.28 shows the corresponding profiles of the four types of dunes.

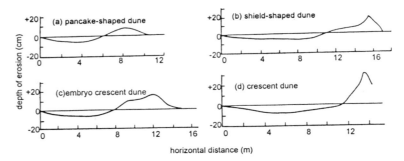

Fig. 15.29 Distribution of scouring and deposition along
the longitudinal profiles of various sand dunes

The profiles of pancake-shaped and shield-shaped dunes are more or less symmetrical; the flow stream lines at the crest of the dune appear not to separate, so the scouring occurs generally on the windward slope and deposition occurs on the leeward slope (Fig. 15.29). Starting from an embryo crescent dune, a steep sliding plane forms; stream lines separate from the sliding plane at the crest and a transverse vortex forms on the leeward slope, where sand particles concentrate and deposit. Although the maximum wind velocity occurs at the crest of dune, the maximum amount of wind erosion occurs midway along the windward slope, as shown in Fig. 15.29.

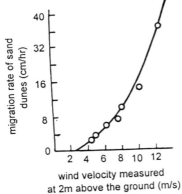

Fig. 15.30 Relationship between wind velocity and migration rate of sand dunes (measured at Taklamaken Desert in China)

Along with the scouring on windward slope and deposition on the leeward slope, the sand dune as a whole migrates slowly in the direction of the wind. Bagnold pointed out that rate of dune migration C can be expressed as [2]

$$C = \frac{g_b}{\gamma_* \Delta} \qquad (15.33)$$

where g_b — sand transport rate per unit width; γ_* — unit weight of eolian deposit; Δ — wave height of dune.

Since the rate of sand transport is proportional to the cube of wind velocity, the rate of dune migration also increases rapidly with increase of wind velocity as shown in Fig. 15.30.

During the migration of dunes, if the direction and velocity of the wind remain unchanged, the configuration of the dunes also remains unchanged. A balance is created among factors of dune configuration: velocity distribution of airflow and the intensity of sand transport. If one can plot the flow net based on a survey map of dune-covered ground so as to compute the rate of sand transport in each grid, then one can obtain the changes in deposition and scouring of various parts of a dune covered area. In computations of sand transport rates, the effect of the local gradient of ground surface on transport rate should be included. In fact, as mentioned above, the maximum amount of wind scour in crescent dunes occurs in the middle part of the windward slope, because the steepest slope is located there. Howard et al. computed the effects of grain size, wind direction, wind velocity, and level of saturation of sand concentration carried by airflow on the configuration of crescent dunes[36]. With an increase in wind velocity and a reduction of particle grain size, Δ/W increase slightly and W/L increases more rapidly (W and L are defined in Fig. 15.19); under such conditions, dunes appear to be steeper and more blunt. If the level of saturation of sand concentration in airflow is low, sand dunes are crescent-shaped. If the extent of saturation of sand concentration in airflow is high, the windward slope looks more like a whale back. and the sliding plane is small. If the wind direction changes frequently, the configuration of dune beds is more blunt than it is for a monodirectional wind.

REFERENCES

[1] Mabbutt , J.A. *Desert Landforms*. Mass. Inst. Tech. Press, 1977, p.340

[2] Bagnold, R.A. *The Physics of Blown Sand and Desert Dunes*. Methuen & Co., London, 1941, p.265.

[3] Ling, Yuquan, and Zhen Wu. "Experimental Study of Wind-Blown Sand Movement by Dynamic Photography." *Acta Geographic*, Vol. 35, No. 2, 1980, pp.174-181.

[4] Yoshita Tsuchiya. "Successive Saltation of a Sand Grain by Wind." *Proc., Coastal Engin. Conf.*, Vol. 2, 1970, pp.1417-1427.

[5] Malina, F.J. "Recent Developments in the Dynamics of Wind Erosion." *Trans.*, Amer. Geophys. Union, 1941, pp.262-284.

[6] Windom, H.L. "Eolian Contributions to Marine Sediments." *J. Sedim. Petro.*, Vol. 45, 1975, pp.520-529.

[7] Cooke, R.U., and A. Warren. *Geomorphology in Deserts*. B. T. Batsford Ltd., London, 1973, p.374.

[8] Zingg, A.W., and W.S. Chepil. "Aerodynamics of Wind Erosion." *Agri. Engin.*, Vol.31, No.6, 1950, pp.279-282.

[9] Zhu, Zhengda, Zhen Wu, and Shu Liu, et al. *Introduction to Deserts in China. Scientific Press*, 1980, p.107.

[10] Bagnold, R.A. "The Movement of A Cohesionless Granular Bed by Fluid Flow over It." *British J. Applied Phys.*, Vol.2, No.2, 1951.

[11] Zingg, A.W. "Wind Tunnel Studies of the Movement of Sedimentary Material." *Proc., 5th Hyd. Conf.*; Univ. Iowa, 1953, pp.111-135.

[12] Sheppard, P.A. "The Aerodynamic Drag of the Earth's Surface and the Value of von Karman's Constant in the Lower Atmosphere." *Proc.*, Royal Soc. London, Ser. A, Vol. 188, 1947, pp.208-222.

[13] Deacon, E.L. "Vertical Diffusion in the Lowest Layers of the Atmosphere." *Quat. J., Royal Meteorological Soc.*, Vol. 75, No. 323, 1949.

[14] Kiyoshi Horikawa, and H.W. Shen. "Sand Movement by Wind Action (On the Characterisitics of Sand Traps)." *U.S. Beach Erosion Board*, Tech. Mem., No. 119, 1960, p. 51.

[15] Bagnold, R.A. "The Nature of Saltation and of Bed Load Transport in Water." *Proc.*, Royal Soc. London, Ser.A, Vol. 332, 1973, pp.473-504.

[16] Owen, P.R. "Saltation of Uniform Grains in Air." *J. Fluid Mech.*, Vol. 20, Pt. 2, 1964, pp. 225-242.

[17] Chepil, W.S. "Dynamics of Wind Erosion-3.The Transport Capacity of the Wind." *Soil Sci.*, Vol. 60, 1945, pp.475-480.

[18] Lyles, L., and N.P. Woodruff. "Boundary Layer Flow Structure, Effects on Detachment of Noncohesive Paticles." In *Sedimentation*. Edited by H.W. Shen, 1972, p.16.

[19] Schumm, S. "The Movement of Rocks by Wind." *J. Sedim. Petro.*, Vol. 26, No. 3, 1956, pp. 284-286.

[20] Fletcher B. "The Incipient Motion of Granular Materials." *J. Phys. D: Applied Phys.*, Vol. 9, No.17, 1976, pp. 2471-2478.

[21] Chepils, W.S. *Properties of Soil which Influence Wind Erosion*:
    (1) "The Governing Principle of Surface Roughness." *Soil Sci.*, Vol. 69, No. 2, 1950, pp.149-162.
    (2) "Dry Aggregate Structure As An Index of Erodibility." *Soil Sci.*, Vol. 69, No. 5, 1950, pp.403~414.
    (3) "Effect of Apparent Density on Erodibility." *Soil Sci.*, Vol. 71, No. 2, 1951, pp.141-153.
    (4) "State of Dry Aggregate Structure." *Soil Sci.*, Vol. 72, No. 5, 1951, pp.387-401.
    (5) "Mechanical Stability of Structure." *Soil Sci.*, Vol. 72, No. 6, 1951, pp.465-478.

[22] Chepil, W.S., and N.P. Woodruff. "The Physics of Wind Erosion and Its Control." *Advances in Agronomy*, Amer. Soc. Agronomy, Vol.15, 1963, pp.211-302.

[23] Nickling, W.G. "Eolian Sediment Transport During Dust Storms: Slims River Valley, Yukon Territory." *Canadian J. Earth Sci.*, Vol. 15, No. 7 1978, pp.1069-1084.

[24] Kawamura, R. "Study on Sand Movement by Wind." *Rep.*, Inst. Sci. & Tech., Univ. Tokyo, Vol. 5, No. 3/4, 1951.

[25] O'Brien, M.P., and B.D. Rindlaub. "The Transportation of Sand by Wind." *Civil Engin.*, May 1936.

[26] Bagnold, R.A. "The Flow of Cohesionless Grains in Fluids." *Phil. Trans.*, Royal Soc. London, Ser.A, Vol. 249, No. 964, 1956, pp.235-297.

[27] Hsu, Shih-Ang. "Wind Stress Criteria in Eolian Sand Transport." *J. Geophys. Res.*, Vol.76, No.36, 1971, pp.8684-8686.

[28] Kadib, Abdel-Latif A. "Mechanism of Sand Movement on Coastal Dunes." *J. Waterways and Harbor Div., Proc., Amer. Soc. Civil Engrs.*, Vol. 92, No. WW2, 1966, pp.27-44.

[29] Schmidt, W. "Der Massenaustauch in Freier und Verwandte Erscheinugen." *Problems der Kosmischen Physik*, Pt. 7, 1925.

[30] Wilson, J.G. "Aeolian Bedforms-Their Development and Origins." *Sedimentology*, Vol.19 , No.3/4, 1972, pp.171-210

[31] Sharp, R.P. "Wind Ripples." *J. Geol.*, Vol. 71, No. 5, 1963, pp.617-636.

[32] Zhu, Zhengda. " Preliminary Study of Some Aspects of the Dynamic Process of Sand Dune Movement under Wind Action." *Special issue of Georaphy*, No.5, 1963, pp.58-78.

[33] Bagnold, R.A. "The Surface Movement of Blown Sand in Relation to Meteorology." *Proc. Intern. Symp. on Desert Res.*, Israel, 1953, pp. 89-93.

[34] Zhu, Zhengda , Hengwen Gao, and Gongcheng Wu. " Investigation of Migration of Sand Dunes near the Oasis in Southwestern Region of Taklamaken Desert." *Acta Geographica*, Vol. 30, No. 1, 1964, pp. 35-50. (in Chinese)

[35] Chen, Zhiping. "Basic Characteristics of the Gurbantunggut Desert in Junggar Basin." *Special issue of Geograph*, No. 5, 1963, pp. 79-90

[36] Howard, A.D. et al. "Sand Transport Model of Barchan Dune Equilibrium." *Sedimentology*, Vol. 25, No. 3, 1978, pp. 307-338.

CHAPTER 16

SEDIMENT MOVEMENT DUE TO WAVE ACTION

This chapter is mainly an elucidation of the physical processes of sediment movement due to wave action. As the intensity of waves increases, patterns of flow and of sediment movement near the bed undergo a series of changes[1].

1. An oscillatory laminar sublayer develops at the bed surface.
2. The laminar sublayer loses its stability and becomes a turbulent sublayer.
3. Motion of sediment particles is initiated (fine particles may begin to move even when the sublayer is still laminar).
4. General movement of sediment particles occurs.
5. Ripples appear on the originally plane bed.
6. Sediment is transported in the direction of the wave advance.
7. Turbulent eddies gradually diffuse upward from the bed surface.
8. Sediment is suspended near the bed and is then transported in the direction opposite the wave advance.
9. Ripples become longer and lower and finally disappear.

These various phenomena may not always follow just this sequence. In special cases, some of the stages may not occur.

As mentioned, sediment can move in the direction of waves or opposite it. If waves approach the coastline at some angle, littoral currents form and induce drifts along the coastline. A large part of the seashore of North China consists of mud (fine silt and clay), and it displays movement characteristics due to wave action that differ from those for coarse particles. These two topics are discussed in separate sections of the chapter.

16.1 GENERAL DESCRIPTION

16.1.1 Nature of the issue

Waves have two unique characteristics with regard to sediment movement that differ from those of unidirectional flows.

First, waves cause periodic oscillations of velocity and pressure and therefore the force and velocity acting on sediment particles on the bed are near their maximum values for only a short part of each period; their values are much smaller at other times so large acceleration occur. Consequently, the inertia of sediment particles plays a more important role with waves[2].

Second, the shear stress on the bed affects the entire velocity field in open channel flows, whereas wave motion is primarily an exchange between potential energy and kinetic energy. As an approximation, wave motion can be viewed as non-resistant. The shear stresses on the bed arise solely from the fact that the velocity at the bed must be zero. Some energy is taken from the waves to create the intermittent boundary layers that form at the bed, but the energy involved is only a small fraction of the energy in waves. The advance of the fluid mass as a whole is generally small if the water is comparatively deep. Only when the waves are in shallow water does a significantly large movement of fluid mass near the bottom take place so the boundary affects the wave patterns. The formation and development of the boundary layer not only determines the magnitude of the shear stress and velocity distribution near the bed, these actions of the layer are usually also the only source of turbulence. Only in the breaker zone does the main flow become turbulent. Therefore, the boundary layer has a profound effect on sediment movement. The thickness of the boundary layer for short-period waves is generally about one percent of the water depth; i.e., the sediment movement due to wave action is confined in an extremely thin layer close to the bottom. Nevertheless, many relationships between sediment movement and shear stress in unidirectional flows can be applied to studies of sediment movement due to wave action on the basis of comparable patterns of shear stresses.

16.1.2 Laboratory test technique

Observation and measurement of waves and sediment movement near the vast area in the shorelines are more difficult than is hydrometry in rivers. Therefore, observations and measurements of waves and sediment movement are primarily those conducted in laboratories. With the development of special techniques, the three different methods used in these experiments are as follows [3].

The first involves the use of wave tanks. Sediment is placed on the bottom of a tank at a certain slope. Waves generated by a wave maker at one end of the tank propagate along it toward the sediment-covered shore. Then the flow structure and sediment movement near the bottom at the sloping surface are observed and measured. The flow field generated in this way is similar to that in nature except for disturbances produced by the reflections of waves at both the sloping shore and the wave maker. However, the strength of the waves in wave tanks is much smaller than that in nature. The periods of natural waves are usually 5 to 15 seconds, and the boundary layer that forms is turbulent; but in wave tanks the period is usually only 2 to 3 seconds and the boundary layer is often laminar.

The second method is to produce the relative motion by locating a plate covered with sediment in a flume and moving it horizontally in simple harmonic motion with controlled period and amplitude. The horizontal movement of water particles near the bed is essentially simple harmonic motion if the wave height is much smaller than the water depth. This method is not restricted as to the wave intensity, but it does not reproduce the periodic acceleration of the water and the variation of static pressure near the bed; therefore, the flow structure differs somewhat from than of actual waves. Also, the process cannot simulate the mass movement of water along the slope.

The third method, only recently developed, is to use alternating flow in a water tunnel instead of a wave tank. The oscillatory movement of water particles is generated by the to-and-fro movement of a piston. The results obtained with this method are probably the most like natural waves, but the facility is comparatively expensive.

The conditions and the relative importance of the role of inertia differ in these three methods. As a result, the results obtained from them must be assessed and interpreted accordingly.

16.2 BASIC CHARACTERISTICS OF WAVES

16.2.1 Generation and propagation of waves

16.2.1.1 Characteristics of wind waves

If wind blows over open water, it can generate waves from almost any disturbance that forms on the surface. The resulting waves are irregular and have the following characteristics.

1. The wind force varies both in time and in space; consequently, wind waves at a certain place may have formed locally or have propagated from other areas. Hence wind waves at any location are a composite of waves with a range of periods and heights. The water surface is often quite irregular as a result of the interference of waves with different periods. Fig. 16.1 shows the superposition of two waves with different wave lengths. If waves combine, the irregularity of the water surface is more pronounced. From a record of a composite wave, each constituent part can be sorted out by statistical methods. If the period and height of each wave are known, they provide the wave spectrum (Fig. 16.2). The width of the wave spectrum curve represents the number of wave trains, and one among them has the largest wave height.

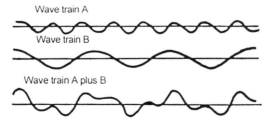

Fig. 16.1 Composite wave formed from two constituent waves
with different periods travelling in the same direction

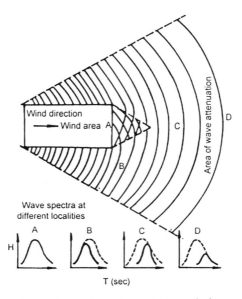

Fig. 16.2 Waves attenuated outside a wind area

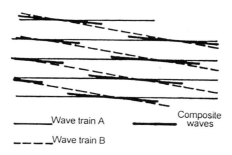

—————Wave train A

– – – –Wave train B

Composite
waves

Fig. 16.3 Interference of two wave trains from different directions,
forming a short-crested wave pattern

2. If waves from different directions meet, they form a short-crested wave pattern as the result of wave interference, as indicated in Fig. 16.3. The larger the angle between the two wave trains, the narrower the composite wave. The water surface is much higher where two wave crests coincide and much lower where two wave troughs coincide. If many waves from various directions meet, the composite crests and troughs are quite complex (Fig.16.4) [4].

3. The profiles of wind waves are asymmetric. The windward slope is gentler than the leeward one. As the wind force increases, the waves become steeper. Finally, further increase causes the waves to break.

Because waves are so complex, one cannot describe fully their mechanical properties in mathematical terms. However, they cause sediment movement only along the coast, and the waves there are generally far from the area where they formed. Consequently, waves near a coast are usually residual waves that have propagated from some distant storms. By the time they reach the coast, they are much more regular than the waves at sea and they are accordingly easier to analyse.

16.2.1.2 Propagation of wind waves

Once generated, wind waves propagate along straight lines as a consequence of gravity and inertia; after some time they enter an area with less wind. The residual waves then spread out in both longitudinal and transverse directions as shown in Fig. 16.2 [4]. If the wind supplies no more energy to them, their heights diminish and their lengths and periods increase because energy is dissipated by air friction. Short period waves tend to disappear first in this process. Besides, during propagation, various waves

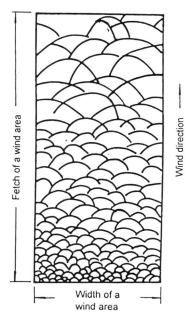

Fig. 16.4 Plan view of wave crests formed by wind within a rectangular area (after Silvester, R.)

of different periods separate from each other; waves with long periods move to the front of a wave group because they travel faster than do waves with short periods. At any point along the path of propagation, long waves are thus the first to arrive, and the shorter waves follow. The width of the wave spectrum is smaller than that in the wind area, the farther they have travelled from the wind area, the longer their wave periods have become, as indicated schematically in Fig. 16.2.

During propagation, the form of the long waves also becomes more regular, and the wave crests tend to form straight lines. Such waves tend to approach the condition of simple oscillatory travelling waves. It is simple because one wave length is predominant; oscillatory because water particles at one point simply move back and forth; and travelling because the wave form propagates forward, in contrast to standing waves. Such waves can be described mathematically. Finally, as they approach the coast, they often cause the sediment at the shore to move.

16.2.2 Characteristics of waves

16.2.2.1 Small amplitude waves, finite amplitude waves, and standing waves

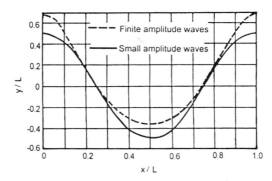

Fig. 16.5 Comparison of wave profiles

Water particle movement, water pressure, and water surface elevation caused by oscillatory waves all vary periodically. They repeat themselves over and over again. Oscillatory waves are often divided into two categories on the basis of the analytical approach used to describe them: small amplitude waves that are sinusoidal in form, and larger amplitude waves that are more nearly trochoidal, as shown in Fig. 16.5. The characteristics of different types of waves are summarised in Table 16.1.

16.2.2.2. Deep water waves and shallow water waves

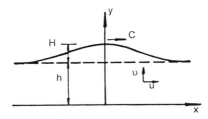

Fig. 16.6 Diagram of a solitary wave
(after Munk, W.H.)

The movement of water particles decreases rapidly with distance below the water surface. At a depth about equal to half of the wave length, the amplitude and velocity of water particles are negligible. For convenience, oscillatory waves are classified into two groups: deep water waves (h/L larger than $1/2$) and shallow water waves (h/L smaller than $1/2$). Such a division is, of course, arbitrary because the characteristics of waves do not change abruptly for this specific condition. Within this chapter, characteristics of deep water waves are denoted by the subscript ∞.

If an oscillatory wave moves from deep water into shallow water, the crest becomes higher, the curvature increases, and the trough becomes flatter. Between two troughs, the characteristics are almost independent of the wave length, i.e., the wave is much like a solitary wave [5], shown in Fig. 16.6. Water moves over the whole depth and the velocity is nearly constant over the entire depth at any location. If the effects of friction are negligible, the form and velocity of a solitary wave do not change. As a solitary wave passes, the local water starts moving and its velocity reaches a maximum when the crest passes. Then, it gradually decelerates and finally becomes quiescent again. The passage of a solitary wave causes particles of water to move a horizontal distance that is proportional to the volume of the wave [6].

726

Table16.1 Characteristics of different waves

Item	Oscillatory wave			Solitary wave	Remark
	Small amplitude wave	Finite amplitude wave			
		Irrotational theory	Theory of trochoidal wave or ellipsoid-trochoidal wave		
1. Relationship between C, L, and T		$L = cT$			–
2. Wave celerity C	$c = \sqrt{\dfrac{gL}{2\pi}\tanh\dfrac{2\pi h}{L}}$ $c_\infty = \sqrt{\dfrac{gL_\infty}{2\pi}}$	$c_\infty^{\,2} = \dfrac{gL_\infty}{2\pi}\Big[1 + \pi^2\Big(\dfrac{H_\infty}{L_\infty}\Big)^2 + \dfrac{1}{2}\pi^4\Big(\dfrac{H_\infty}{L}\Big)^4 + \cdots\Big]$	$c = \sqrt{\dfrac{gL}{2\pi}\tanh\dfrac{2\pi h}{L}}$ $c_\infty = \sqrt{\dfrac{gL_\infty}{2\pi}}$	$c = \sqrt{gh}\Big(1 + \dfrac{H}{h}\Big)^{1/2}$	h-water depth, H-wave height, subscript ∞ represents corresponding values in deep water
3. Amplitude of water particles Horizontal, a Vertical, b	$a = \dfrac{H}{2}\dfrac{\cosh 2\pi(h+y)/L}{\sinh 2\pi h/L}$, $b = \dfrac{H}{2}\dfrac{\sinh 2\pi(h+y)/L}{\sinh 2\pi h/L}$ $a_\infty = b_\infty = \dfrac{H_\infty}{2}e^{2\pi y/L_\infty}$			–	y-distance from still-water surface, positive upwards
4. Velocity of water particles Horizontal, u Vertical, v Tangential, u_r	$u = \dfrac{2\pi a}{T}\sin 2\pi\Big(\dfrac{x}{L}-\dfrac{t}{T}\Big)$, $v = -\dfrac{2\pi b}{T}\cos 2\pi\Big(\dfrac{x}{L}-\dfrac{t}{T}\Big)$ $u_{r\infty} = \dfrac{\pi H_\infty}{T}e^{2\pi y/L_\infty}$			$\dfrac{u}{c} = N\dfrac{1+\cos(M\frac{y}{h})\cosh(M\frac{x}{h})}{\{\cos(M\frac{y}{h})+\cosh(M\frac{x}{h})\}^2}$ $\dfrac{v}{c} = N\dfrac{\sin(M\frac{y}{h})\sinh(M\frac{x}{h})}{\{\cos(M\frac{y}{h})+\cosh(M\frac{x}{h})\}^2}$	t-time, M and N are functions of H/h $\dfrac{H}{h} = \dfrac{N}{M}\tan\dfrac{1}{2}[M(1+\dfrac{H}{h})]$ $N = \dfrac{2}{3}\sin^2[M(1+\dfrac{2}{3}\dfrac{H}{h})]$
5. Wave energy per unit width, E	$E_\infty = \dfrac{\gamma L_\infty H_\infty^{\,2}}{8}$	$E = \dfrac{\gamma L H^2}{8}\Big[1 - A\Big(\dfrac{H}{L}\Big)^2\Big]$ $E_\infty = \dfrac{\gamma L_\infty H_\infty^{\,2}}{8}\Big[1 - 4.93\Big(\dfrac{H_\infty}{L_\infty}\Big)^2\Big]$		$E = \dfrac{8}{3\sqrt{3}}\gamma H^{3/2}h^{3/2}$	$A = \dfrac{\pi^2}{2\tanh^2\dfrac{2\pi h}{L}}$
6. Velocity of mass transport, u_i	$u_i = 0$	$u_i = \dfrac{\pi^2 H^2}{L^2}c\dfrac{e^{4\pi(y+h)/L}+e^{-4\pi(y+h)/L}}{(e^{2\pi h/L}-e^{-2\pi h/L})^2}$ $u_{i\infty} = H^2\sqrt{\dfrac{g\pi^3}{2L_\infty^{\,3}}}e^{4\pi y/L_\infty}$	$u_i = 0$		–

16.2.2.3 Movement of water particles

The propagating velocity of a wave (called wave velocity or wave celerity) is quite different from the velocity of the water particles. Fig. 16.7 shows the orbit of water particles of an oscillatory wave; the dashed lines are path lines. The water particles describe slightly open circular orbits in deep water and slightly open elliptic orbits in shallow water, progressing only a short distance during each orbit. One can compare the situation to that of wind blowing across a wheat field. The wind clearly produces waves that move across the wheat field much like water waves, but each standing wheat stalk can only move back and forth at its original location. Equations for the wave velocity, amplitude, and velocity of water particles are listed in Table 16.1. At the bed, the vertical velocity of the water particles is zero, and the horizontal velocity of the water particles is given by

$$u_0 = \frac{\pi H}{T \sin 2\pi h / L} \sin 2\pi (\frac{x}{L} - \frac{t}{T}) \qquad (16.1)$$

The amplitude of the water particle motion is

$$a_0 = \frac{H/2}{\sinh 2\pi h / L} \qquad (16.2)$$

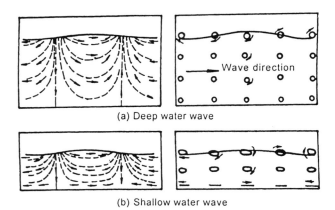

(a) Deep water wave

(b) Shallow water wave

Fig. 16.7 Orbits and path lines of water particles for an oscillatory wave

16.2.2.4 Wave energy and its propagation

The energy in waves is either kinetic or potential. The average kinetic energy over a cross section is a constant while the potential energy is a function of space, it is maximum at the crest or at the trough and a minimum at the still-water line. For a sinusoidal wave in deep water, the kinetic energy and potential energy per unit width in one wave length are equal,

$$E_{p\infty} = E_{k\infty} = \frac{1}{16}\gamma L_{\infty} H_{\infty}^2 \tag{16.3}$$

Therefore, the total energy is

$$E_{\infty} = E_{p\infty} + E_{k\infty} = \frac{1}{8}\gamma L_{\infty} H_{\infty}^2 \tag{16.4}$$

The wave energy for other types of waves is listed in Table 16.1.

A part of the energy in a wave propagates along with the wave. For small amplitude oscillatory waves in deep water, Reynolds showed that the wave energy propagating through a cross section at a distance x from the crest per unit time is as follows.

$$\frac{C_{\infty}}{B}\gamma H_{\infty}^2 \sin\frac{2\pi x}{L_{\infty}}$$

The energy of a wave averaged over the wave length is

$$\frac{C_{\infty}}{16}\gamma H_{\infty}^2$$

that is, half of the total wave energy propagates at the wave celerity. Because the kinetic energy of waves is a constant and the variation of potential energy in the x direction is the same as the above expression, the energy of waves that propagates comes from their potential energy.

The number of waves in a natural train of waves is limited. As a finite train of waves travels through still water, the wave at the front leaves half of its total energy behind as it travels one wave length; the other half of the total energy propagates with the wave. As a result, the wave height of the leading wave diminishes with this transfer of energy. At the rear of the train of waves, half of the total energy creates new waves in the still water. In the middle of the train of waves the total energy remains constant, as does the wave height.

Due to the variation of wave energy during propagation, the group velocity of a wave train, c_g, differs from the wave velocity, c. For a small amplitude oscillatory wave

$$\frac{c_g}{c} = \frac{1}{2}\left(1 + \frac{4\pi h/L}{\sinh 4\pi h/L}\right) \tag{16.5}$$

In deep water, h/L is quite large, so that $c_g/c=1/2$; in shallow water h/L is quite small, so that $c_g/c=1$, thus c_g/c varies between these two extreme values.

16.2.3 Development of boundary layers

As waves travel from deep water into shallow water and the movement of the water particles caused by their motion is affected by the presence of the bottom, the pattern of the motion is affected. Over most of the depth, the flow is still essentially potential motion, but the effect of the viscosity is not negligible within the thin boundary layer near the bottom. The flow characteristics within this layer are quite different from those in the rest of the flow region. Sediment movement near the bottom is directly affected by the shear flow within it.

16.2.3.1 Thickness of boundary layers

According to potential flow theory, the horizontal velocity and amplitude of water particles should follow from Eqs. (16.1) and (16.2). In reality, however, the velocity at the bottom is zero because of boundary shear. As a good approximation, Eqs. (16.1) and (16.2) still hold even quite near the bottom because the boundary layer is thin. If

$$N = 2\pi / T \tag{16.6}$$

then Eq. (16.1) can be written as

$$u_0 = -Na_0 \sin\left(Nt - \frac{2\pi x}{L} \right) \tag{16.7}$$

That is, the water particles move in simple harmonic motion. Because the turbulent boundary layer that usually occurs in nature is difficult to replicate in a wave tank, Bagnold developed a technique of moving a plate in still water to simulate the relative motion near the bottom [7]. Later, Li and Manohar developed this technique further [8,9].

If the bottom surface is the x-z plane and the vertical coordinate y is positive in the upward direction, and if the non-linear terms of inertia, horizontal pressure gradient, and the effect of viscosity are neglected, the equation of motion for the shear flow becomes

$$\frac{\partial u}{\partial t} = v \frac{\partial^2 u}{\partial y^2} \tag{16.8}$$

If the bottom moves in simple harmonic motion

$$u = Na_0 \sin\left(Nt - \frac{2\pi x}{L} \right)$$

and if the water depth is infinite, so that for

$$y \to \infty \qquad u = 0$$

the solution of Eq. (16.8) is:

$$u = u_m e^{-\beta y} \sin\left(Nt - \beta y - \frac{2\pi x}{L} \right) \tag{16.9}$$

in which

$$u_m = Na_0 = \frac{\pi H}{T \sinh 2\pi h / L} \tag{16.10}$$

and

$$\beta = \sqrt{\frac{N}{2\nu}} \tag{16.11}$$

If the wave is quite long, Eq. (16.9) can be simplified:

$$u = u_m e^{-\beta y} \sin(Nt - \beta y) \tag{16.12}$$

Eq. (16.12) represents a horizontal oscillatory wave, propagating from the bottom surface upward with the speed of $N / \beta = \sqrt{2\nu N}$. The amplitude decreases rapidly as the wave propagates upward, as shown in Fig. 16.8. At $\beta y = 4.6$, the amplitude is only about one percent of that at the bottom surface. For practical purposes, the thickness of such a boundary layer can be written as:

$$\delta = 4.6\frac{1}{\beta} = 6.5\sqrt{\frac{\nu}{N}} \tag{16.13}$$

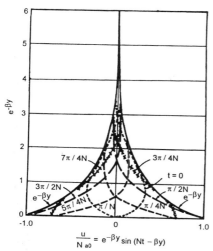

Fig. 16.8 Velocity distribution within the laminar boundary layer on a plate moving laterally in a simple harmonic motion

16.2.3.2 Flow patterns in boundary layers

The ratio of the thickness of a boundary layer to the diameter of bed sediment particle determines the condition of the flow at the boundary. As for unidirectional flows, the boundary is in effect one of three types: smooth, transitional, or rough. From his experiments, Li found that if $\delta/D < 18.5$, the bed is rough; if $\delta/D > 30$, it is smooth; and if it is in between these limits, it is transitional [8].

The laminar boundary layer loses its stability and becomes turbulent once the wave attains a certain strength. The critical value for the transition depends on the relative roughness of bed.

For a smooth bed, Li and Manohar showed that the stability of the boundary layer depends on the Reynolds number for the boundary layer expressed in the form

$$\frac{u_m \delta}{v}$$

After some modification this Reynolds number can be written in the following form:

$$\text{Re}_1 = \frac{N^{\frac{1}{2}} a_0}{v^{1/2}} = \left(\frac{2\pi}{vT}\right)^{1/2} \frac{H/2}{\sinh 2\pi h/L} = 400 \tag{16.14}$$

If Reynolds number of a flow is larger than 400, the boundary layer is turbulent, and if smaller than 400, it is laminar. Because Li and Manohar conducted their experiments by moving a vibrating plate laterally in still water, the flow induced was a little different from the real situation under a wave. Vincent [1] and Collins [10] conducted experiments in water tanks with oscillating flow to study the critical Reynolds number. Collins obtained a critical value for the Reynolds number of 112.5, which is only 30% of the value from Eq. (16.14). As shown in Fig. 16.9, Vincent's value is even smaller. The cause of the difference between Li-Manohar and Collins-Vincent is that the former experiment did

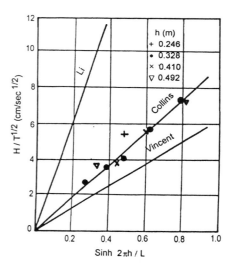

Fig. 16.9 Condition for stability of boundary layer on a smooth bed

not include the effects of periodic acceleration of water particles and pressure variations near the bottom. The result indicates that these two factors increase the instability of the boundary layer.

Li, Manohar and Kalkanis studied the stability of the boundary layer on a rough bed with both two and three-dimensional artificial roughness. Their Reynolds number has the form:

$$\text{Re}_2 = \frac{Na_0 K_s}{v} \qquad (16.15)$$

in which K_s is a measure of bed roughness. The critical Reynolds numbers obtained by Li and by Manohar were later shown to be too large because of some deficiencies in their methods of measurement. After revision by Kalkanis, the critical Reynolds numbers were as follows [11]:

For two-dimensional roughness $R_{e2} = 640$, for $a_0/K_s < 266$,

For three-dimensional roughness $R_{e2} = 104$, for $a_0/K_s < 1630$.

Later, Jonsson [12] and Kamphuis [13] studied the condition of stability of boundary layers more thoroughly. They used a_0/K_s and

$$\text{Re}_3 = \frac{u_m a_0}{v} \qquad (16.16)$$

as the parameters that determine the flow pattern.[1] The results obtained by Kamphuis are shown in Fig. 16.10. He used 10,000 as the critical Reynolds number. Its corresponding value for Re_2 is 100, a value that is close to that given by Collins and Vincent for a smooth boundary.

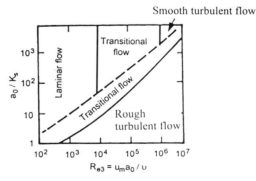

Fig. 16.10 Diagram showing types of flow for boundary
layers formed by wave action (after Kamphuis, J.W.)

---

[1] Kamphuis used the maximum horizontal velocity and amplitude of the wave action at the upper limit of the boundary layer in the two parameters. Since the thickness of the boundary layer is small, these two values differ little from u_m in Eq. (16.10) and a_0 in Eq. (16.2).

16.2.3.3 Velocity distribution

The velocity distribution in a turbulent boundary layer is more complex than that in a laminar boundary layer. Under the guidance of Einstein, researchers at the University of California introduced empirical approaches that deal with this issue [14].

A generalized form of Eq. (16.12) is:

$$u = u_m f_{1*}(y) \sin[Nt - f_{2*}(y)] \tag{16.17}$$

in which

$$
\left.
\begin{aligned}
f_{1*}(y) &= \{1 + [f_1(y)]^2 - 2f_1(y)\cos f_2(y)\}^{1/2} \\
f_{2*}(y) &= \tan^{-1}\left\{ \frac{f_1(y)\sin f_2(y)}{1 - f_1(y)\cos f_2(y)} \right\}
\end{aligned}
\right\} \tag{16.18}
$$

the functions $f_1(y)$ and $f_2(y)$, as determined from the experiments, are presented in Table 16.2.

Fig. 16.11 shows the velocity distributions in boundary layers due to wave action as curves 4 and 5; U is the velocity at the upper limit of the boundary. For comparison, velocity distributions in a boundary layer for unidirectional steady flow are also shown, for them the thickness of the boundary layer corresponds the entire water depth.

Table 16.2 Empirical expressions of $f_1(y)$ and $f_2(y)$ (after Einstein, H.A.)

Boundary conditions		$f_1(y)$	$f_2(y)$
Smooth		$0.3\exp\left(-0.75\dfrac{y}{a_0}\right)$	$1.55\left(\dfrac{y}{\sqrt{2v/N}}\right)^{1/3}$
Rough	Two-D artificial roughness	$\exp\left(-1000\dfrac{y/\sqrt{2v/N}}{a_0 K_s}\right)$	$0.5\left(\dfrac{y}{\sqrt{2v/N}}\right)^{2/3}$
	Three-D artificial roughness	$\exp\left(-133\dfrac{y/\sqrt{2v/N}}{a_0 K_s}\right)$	$0.5\left(\dfrac{y}{\sqrt{2v/N}}\right)^{2/3}$

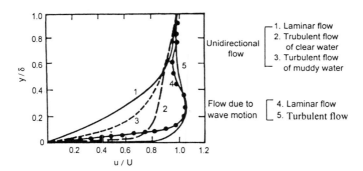

Fig. 16.11 Velocity distribution in boundary layers for unidirectional
flows and for flows due to wave motion (after Teleki, P.G.)

16.2.3.4 Resistance

The maximum shear stress at the bed due to wave action can be expressed as:

$$\tau_{0m} = \frac{1}{2} c_f \rho u_m^2 \qquad (16.19)$$

in which c_f is the resistance coefficient[1].

For laminar flow, the relationship between c_f and the Reynolds number Re_3 is [12],

$$c_f = \frac{2}{\sqrt{Re_3}} \qquad (16.20)$$

For a turbulent boundary layer, Kajiura developed an expression for the resistance coefficient based on the boundary layer theory in unidirectional steady flow [16]. Jonsson and Kamphuis revised Kajiura's equations in accordance with their experimental results to obtain the form given in Table 16.3 [12]. Fig. 16.12 is a plot of the relationship between the resistance coefficient and the Reynolds number presented by Kamphuis; in his work, the representative roughness diameter K_s was taken to be $2D_{90}$.

---

[1] The resistance coefficient, c_f, is the Fanning resistance coefficient, λ, used in Chapter 14. To avoid confusion with the wave length of sand waves, c_f is used in this chapter and the next one.

Table 16.3 Resistance coefficient for turbulent boundary layer due to wave action (after Jonsson, J.G.)

Researcher	Smooth turbulent flow	Rough turbulent flow
Kajiura	$\dfrac{1}{8.1\sqrt{c_f}} + \log\dfrac{1}{\sqrt{c_f}} = -0.135 + \log\sqrt{Re_3}$	$\dfrac{1}{4.05\sqrt{c_f}} + \log\dfrac{1}{4\sqrt{c_f}} = -0.254 + \log\dfrac{a_0}{K_s}$
Jonsson	$\dfrac{1}{4\sqrt{c_f}} + 2\log\dfrac{1}{4\sqrt{c_f}} = -1.55 + \log\sqrt{Re_3}$	$\dfrac{1}{4\sqrt{c_f}} + \log\dfrac{1}{4\sqrt{c_f}} = -0.08 + \log\dfrac{a_0}{K_s}$
Kamphuis	25-30% smaller than Kajiura	$\dfrac{1}{4\sqrt{c_f}} + \log\dfrac{1}{4\sqrt{c_f}} = -0.35 + \dfrac{4}{3}\log\dfrac{a_0}{K_s}$

16.2.4 Variation of waves approaching a coastline

As waves travel from deep water into shallow water during their approach a coastline, they undergo a series of changes. Fig. 16.13 shows a representative profile

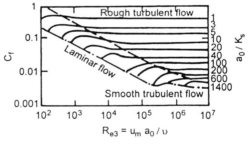

Fig. 16.12 Relationship between resistance coefficient and Reynolds number in a boundary layer due to wave action

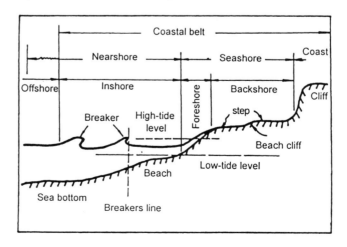

Fig. 16.13 Typical profile for waves near the coast

736

of the sea bottom near the coast and illustrates the terms that are used in the following discussion.

16.2.4.1 Variation of wave elements as the water depth diminishes

If the contours of the sea bottom are straight lines parallel to the coast and if the waves travel perpendicular to them, the wave elements vary as the water depth diminishes in the manner shown in Fig. 16.14. Only the wave period, T, remains unchanged from place to place.

In Fig. 16.14, the three ratios -- C/C_∞, L/L_∞, b_s/a_s -- are all equal to tanh $2\pi h/L$; b_s and a_s represent the vertical and horizontal amplitudes of water particles at the water surface, respectively. As the water depth reduces and becomes less than half of the wave length, the wave velocity and wave length reduce and the orbits of water particles near the water surface change from circles to ellipses, gradually becoming flatter.

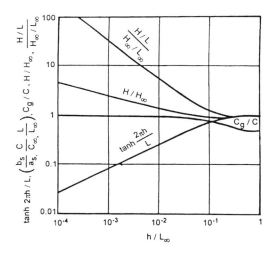

Fig. 16.14 Variation of wave elements as water depth diminishes

The increase of C_g/C implies that more wave energy propagates forward and that the wave height decreases; however, the decrease of C/C_∞ implies an increase in wave height. The former effect is a dominant one, and it leads to a small reduction of wave height as the waves first enter shallow water. The minimum value is 91% of the height in deep water; if h/L_∞ is smaller than 0.15, the effect of C/C_∞ is the more significant one, and the wave height increases. If h/L_∞ is equal to 0.55, the wave height has its original value, then the amplitude increases gradually.

The variation of wave height shown in Fig. 16.14 does not include the effect of wave energy dissipation, which tends to reduce the wave height. Wave energy can be dissipated by the friction of sand ripples [17] or by the seepage flow caused by the variation of pressure along the sea bottom [18]. Generally, the former effect is much more significant than the latter.

16.2.4.2 Distortion of wave shape and wave breaking

As waves approach the coastline, the wave form does not maintain its symmetry; the forward slope becomes steeper and the leeward slope gentler. Finally, the waves become unstable and break. Depending on the wave steepness (H/L) and bottom slope,

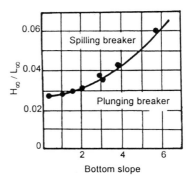

Fig. 16.15 Critical condition for breakers (after Iwagaki, Y., and H. Noda)

two basic types of breakers can occur. If comparatively low waves approach a steep pebbly beach, a plunging breaker occurs in which the crest of the wave tumbles into the trough and encloses a pocket of air. If the waves are fairly steep in deep water and advance over a gently sloping sandy beach, the wave does not lose its form, but the crest becomes sharper until foaming appears there, and the foam spills down the forward slope. This is called a spilling breaker. Most breakers are somewhere in between these two extreme types. From experiments in wave tanks with a rigid bottom, the critical wave steepness and bottom slope vary as shown in Fig. 16.15 [19]. Wave breaking causes strong local turbulence.

As waves enter shallow water, the wave velocity decreases and the velocity of the water particles increases rapidly with the increase of wave height. Once the velocity of water particles surpasses the wave velocity, the wave breaks; the critical water depth at the breaker line (measured from the wave trough) is about 1.28 times the wave height [20]. The position of the breaker line is independent of the wave length. Furthermore, if the wave height is more than one-seventh of the wave length in deep water, the wave also loses its stability and breaks [21], a condition that is independent of water depth. The most common breakers on a beach are situated in between these two extremes; the position of the break point depends on both the wave steepness and water depth. In addition, the beach slope has some effect: the steeper the slope, the farther from the coastline the line of breakers is [22].

16.2.4.3 Flow in breaking zones

After breaking, waves still maintain a certain amount of energy that enables different types of flow to occur in the breaker zone, and they can vary with the magnitude of the beach slopes [23].

On a steep pebbly beach, the breaker forms so close to the shoreline that there is not enough room for another to form; instead a swash sometimes flows up the beach

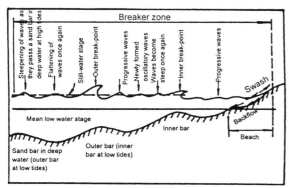

Fig. 16.16 Flow in a zone of breaking waves
(after Russell, R.C.H., and D.H. MacMillan)

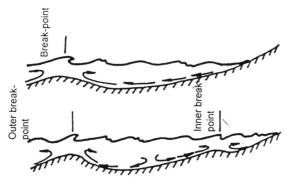

Fig. 16.17 Circulation in a zone of breaking waves
(after Dyhr-Nielsen, M., and T. Sorensen)

and then back to the sea. Much of the wave energy is dissipated during this process. If the backflow is strong enough, it can create reflected waves that propagate offshore and gradually disappear.

On a gently sloping sand beach the foaming area formed as the wave breaks expands and spreads, but the wave does not re-form; if the beach is nearly flat and the break point is far out from the coastline, the situation shown in Fig. 16.16 can occur. After breaking, a new wave may form. It is generally a progressive wave, but sometimes it can be an oscillatory one. The new wave may break again when it enters still shallower water. Such a pattern of breaking-reestablishment-breaking may occur several times before a swash forms.

Circulation can occur in the zone of breakers. Fig. 16.17 shows circulation as formed for either one or two break points [24]. Such circulation often has a profound influence on sediment movement near the shoreline.

16.2.5 Nearshore flow due to wave action[25]

In addition to the tidal flows and turbidity currents that can occur in the near shore zone, wave action can cause various other flows, including mass transport, long shore currents, and rip currents.

16.2.5.1 Mass transport

In normal waves, water particles move forward a short distance during each period, as shown in Fig. 16.18, thus causing a net mass transport in the direction of wave advance. Stokes obtained an equation for the velocity from the potential theory. The average velocity of this mass transport is given in Table 16.1. If the effect of the viscosity of water is negligible, the average velocity of mass transport at y (measured downward from the water surface) is:

$$u_{i\infty} = H_\infty^2 \sqrt{\frac{g\pi^3}{2L_\infty^3}} e^{\frac{4\pi y}{L_\infty}} \qquad (16.21)$$

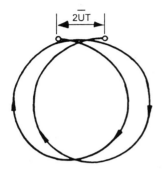

Fig. 16.18 The rotation combined with displacement of a water particle in a oscillating wave

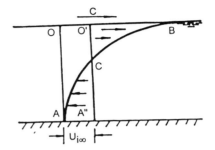

Fig. 16.19 Vertical velocity profile of mass flow in deep water zone if viscosity is neglected

The vertical distribution of u is indicated by the curve OAB in Fig. 16.19. Integrating it along the vertical from $y = 0$ to $y = \infty$, one obtains the discharge mass transport

$$Q_{i\infty} = H_\infty^2 \sqrt{\frac{g\pi}{32 L_\infty}}$$

and the average velocity of mass transport over the vertical is:

$$U_{i\infty} = \frac{Q_{i\infty}}{h} = \frac{H_\infty^2}{h} \sqrt{\frac{g\pi}{32 L_\infty}} \qquad (16.22)$$

The concept of continuity requires that an equivalent flow must exist in the opposite direction along the vertical. Then the actual velocity related to mass transport should be referred to the line $O'A''$, instead of OA in Fig. 16.19, and its expression is:

$$u_{i\infty}' = H_\infty^2 \sqrt{\frac{g\pi}{32 L_\infty}} \left[e^{4\pi y/L_\infty} \sqrt{\frac{16\pi^2}{L_\infty^2}} - \frac{1}{h} \right] \qquad (16.23)$$

Near the water surface the water travels in the direction of the wave advance, whereas near the bottom it travels in the opposite direction.

Due to the existence of a boundary layer, however, the vertical distribution of mass transport is significantly different from that given by Stokes. Such differences from Eq. (16.23) have been found from observations in wave tanks. Bagnold found in some cases that the directions of mass transport were just opposite to those shown in Fig. 16.19 [26]. Other observations showed that the surface water and bottom water moved in the direction of wave advance and the water in the central part moved seaward, resulted in two types of circulation. This contradiction was resolved by the work of Longuet-Higgins in 1953 [27]. His theory was developed for waves propagating in a viscous fluid in water of finite depth. He analyses the case for a boundary layer that is laminar and derived the following expression for the velocity at the bottom,

$$u_{i0} = \frac{5}{4} \frac{\pi^2 H^2}{LT} \frac{1}{\sinh^2 2\pi h / L}$$ (16.24)

and it is always in the direction of wave advance. The types of vertical distribution of velocity due to mass transport depends on the ratio of the wave height to the thickness of the boundary layer. If this ratio is small, the velocity of mass transport at y is

$$u_i = \frac{\pi^2 H^2 / LT}{4 \sinh^2 2\pi h / L} \left\{ 2\cosh\left[\frac{2\pi h}{L}\left(\frac{y}{h}+1\right)\right] + 3 + \frac{2\pi h}{L}\sinh\frac{2\pi h}{L}\left[3\left(\frac{y}{h}\right)^2\right.\right.$$
$$\left. +4\left(\frac{y}{h}\right)+1\right] + 3\left[\frac{\sinh 2\pi h / L}{2\pi h / L}+\frac{3}{2}\right]\left[\left(\frac{y}{h}\right)^2 - 1\right]\right\}$$ (16.25)

in which $\pi^2 H^2/LT$ has the dimension of velocity, and it is the reference quantity for the velocity of mass transport; the remaining terms, which can be represented by $F(y/h)$, determine the variation of the velocity with depth. Fig. 16.20 shows the relationship between $F(y/h)$ and y/h with $2\pi h/L$ as the third parameter. If h/L is small, the mass transport at the surface is in the direction opposite to the wave advance, as observed by Bagnold. If h/L is large, the mass transport near both surface and bottom is in the direction of the wave advance, but in the central part it is seaward. Russell and Osorio verified Eq. (16.24) for values of $2\pi h/L$ between 0.7 and 1.5 by means of experiments in a wave tank. They also found that Eq. (16.25) can be used to describe the vertical distribution of velocities due to mass transport even if the ratio of the wave height to the thickness of the boundary layer is large [28]. Experiments showed that Eqs. (16.24) and (16.25) correspond with the experimental data when the boundary layer becomes unstable. Longuet-Higgins discussed this phenomenon further [29].

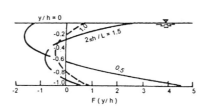

Fig. 16.20 Vertical velocity distribution of mass transport, as presented by Longuet-Higgins (after Longuet-Higgins, M.S.)

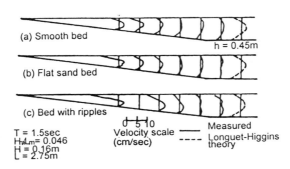

Fig. 16.21 Effect of bottom roughness on vertical distribution of velocity due to mass transport (after Bijker, E.W.; J.P.Th Kalkwijk; and T.Pieters)

Bijker et al. studied mass transport for wave motion with a turbulent boundary layer on a sloping bottom [30]. Their experiment showed that the bottom velocity of mass transport depends primarily on such factors as water depth, wave height, and wave form, and that it is independent of the slope. The vertical velocity distribution of mass transport is, however, strongly affected by the bed roughness, as shown in Fig. 16.21. On a sandy bed, the bottom velocity is less and the surface velocity is more. If ripples appear on the bed, the landward velocity at the bottom is zero, and

can even be directed seaward.

16.2.5.2 Longshore currents

If the line of breakers is at an angle with the coastline, a longshore current is created because the mass transport of water in the direction of wave advance due to wave breaking has a component along the coastline.

The mechanism of the formation of a longshore current has been studied from three different points of view: continuity, energy, and momentum. The first approach combines the longshore current and rip currents (treated in the next section); the water mass moved by the former goes back to the sea through the latter. However, recent studies have shown that the mechanisms of the generation of these two currents are different. The generation of rip currents depends rather on the variation of wave height along the coastline. The second approach is that the component of wave energy along the coastline produces the longshore currents and this energy is dissipated by friction on the bottom during movement. This second approach is not as satisfactory as the third one, because a portion of the wave energy is dissipated as waves break, whereas the momentum remains unchanged. Therefore, the momentum approach is used to derive the velocity of the longshore current.

In the simplest situation, the shoreline is straight, the bottom has a constant slope, J, and α_b is the angle between the wave crests where they break and the shoreline, as shown in Fig. 16.22. For a specific volume of water, $ABCDE$, with a width dx along the beach, if the quantities L_b, C_b, and A are the wave length, celerity and cross-sectional area of the breaking wave crest, respectively, and ρ is the density of the water, then the average momentum per unit surface area is $\rho AC_b / L_b$ [1] , the average momentum flux entering $ABCDE$ is $(\rho AC_b / L_b)C_b\cos\alpha_b dx$, in which the component of momentum flux parallel to the shore is

$$(\rho AC_b / L_b)C_b \cos\alpha_b dx \sin\alpha_b$$

The portion of momentum flux that is transmitted out of $ABCDE$ is

$$U_l(\rho AC_b / L_b)\cos\alpha_b dx$$

in which U_l is the velocity of the longshore current. The difference of the two momentum fluxes,

$$(C_b \sin\alpha_b - U_1)(\rho AC_b / L_b)\cos\alpha_b dx \qquad (16.26)$$

is the dynamic force acting on $ABCDE$ due to wave breaking.

As the longshore current approaches steady state, this dynamic force should be balanced by the friction on the bottom,

$$c_f \rho U_l^2 l dx$$

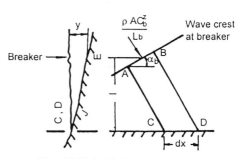

Fig. 16.22 Definition sketch for a longshore current (after Putnam, J.A.; W.H. Munk; and M.A. Taylor)

---

[1] The velocity of the water movement at the line of breakers is approximately the wave velocity.

743

in which c_f is the resistance coefficient, l is the distance between the point of breaking and the shore and the slope $J = h/l$, with h being the depth at the breakers (Fig. 16.22). Equating the two forces gives

$$U_l = \frac{m}{2}\left[\sqrt{1 + \frac{4C_b \sin\alpha_b}{m}} - 1\right]$$ (16.27)

in which

$$m = JA\cos\alpha_b / (c_f hT)$$ (16.28)

If the breaking wave is analysed as a solitary wave, then

$$C_b = \sqrt{g(h + H_b)}$$
$$h = 1.28H_b$$

$$A = 4h^2\sqrt{\frac{H_b}{3h}}$$

Introducing these quantities into Eq. (16.28), one obtains

$$m = \frac{2.61JH_b \cos\alpha_b}{c_f T}$$ (16.29)

From Eqs. (16.27) and (16.28) one finds that the velocity of the longshore current is related to the wave height at the point of breakers, the wave period, the angle between the breaking line and shoreline, and the slope and roughness of the bottom. Both field observations and laboratory experiments confirmed the validity of Eq. (16.27).

Longuet-Higgins also derived an expression for the velocity of the longshore current using the momentum approach and obtained the expression [33]:

$$U_l = \frac{5\pi}{8}\frac{J}{c_f}u_m \sin\alpha_b$$ (16.30)

in which u_m is the maximum horizontal velocity due to mass transport in the breaking zone. Of the two, Eq. (16.30) is the more widely used.

16.2.5.3 Rip currents

A rip current is a concentrated offshore current, as shown in Fig. 16.23. After a rip current passes the line of breakers, it spreads out in a fan shape and gradually loses its identity. It can be more than 800 m long.

Because of wave action, the mean water level near a shore is higher than the mean level in the sea nearby. The superelevation is related to the wave height; the higher the waves, the greater the increase in water level. As the offshore relief can be rather complex, the superelevation of the water surface is not the same all along the shoreline. Thus, currents form from higher-wave breaker zones toward lower-wave breaker zones, and they tend to concentrate and create a seaward flow. Bowen and Inman explained the mechanism of the formation of a rip current in detail, from both theoretical and experimental points of view [34,35].

Nearshore currents induced by waves are complex and often change with the variation of the wave characteristics. Fig. 16.24 shows the currents near a shore that are caused by three different waves [36]. If short-period waves approach the shore perpendicularly, the waves are refracted very little, and the rip currents are more numerous but smaller as shown in Fig. 16.24a. If short-period waves approach the shore obliquely with a rather large angle, a continuous longshore current forms as shown in Fig. 16.24b. If long-period waves approach the shore obliquely, fewer areas with more marked concentration of wave energy occur, and these result in stronger rip currents, as shown in Fig. 16.24c.

The formation of various nearshore currents shown in Fig. 16.24 are closely related to wave refraction, and this process together with wave reflection and diffraction determine the nearshore energy spectrum of the waves [25, 37].

16.3 MECHANISM OF SEDIMENT MOVEMENT DUE TO WAVE ACTION

16.3.1 Incipient motion of sediment particles

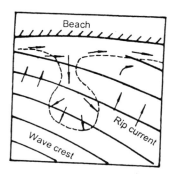

Fig. 16.23 Diagram of a rip current

16.3.1.1 Analysis of incipient motion [38]

If the fluid velocity near a grain of sediment on a bed is u_D, the tractive force acting on the grain is

$$F = C_D \frac{\rho u_D^2}{2} \frac{\pi D^2}{4}$$

The submerged weight of the grain is

$$W' = \frac{\pi D^3}{6}(\rho_s - \rho)g$$

745

If motion is impending

$$F = fW'$$

in which f is the friction coefficient. From these equations,

$$u_{Dc}^2 = \frac{4f}{3} \frac{1}{C_D} \left(\frac{\rho_S - \rho}{\rho} g \right) D$$

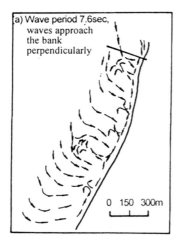

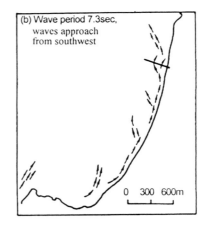

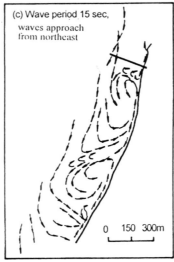

Fig. 16.24 Longshore currents for different waves (after Shepard, F.P., and D.L. Inman)

746

If the flow around the grain is turbulent, the resistance coefficient is constant

$$C_D = k_1$$

Then

$$u_{Dc} = A_1 \left(\frac{\rho_s - \rho}{\rho} g \right)^{1/2} D^{1/2} \tag{16.31}$$

in which

$$A_1 = \sqrt{\frac{4f}{3k_1}}$$

If the flow around the grain is laminar, the resistance coefficient is inversely proportional to the Reynolds number

$$C_D = \frac{k_2}{(u_{Dc} D / v)}$$

Then

$$u_{Dc} = A_2 \left(\frac{\rho_s - \rho}{\rho} g \right) v^{-1} D^2 \tag{16.32}$$

in which

$$A_2 = \frac{4f}{3k_2}$$

If the flow around a grain is in transition between laminar and turbulent, the resistance coefficient is approximately inversely proportional to the Reynolds number used to

$$C_D = \frac{k_3}{(u_{Dc} D / v)^m}$$

in which the power m is also dependent on the Reynolds number. Then,

$$u_{Dc} = A_3 \left(\frac{\rho_s - \rho}{\rho} g \right)^{\frac{1}{2-m}} D^{\frac{1+m}{2-m}} v^{-\frac{m}{2-m}} \tag{16.33}$$

in which

$$A_3 = \left(\frac{4f}{3k_3}\right)^{\frac{1}{2-m}}$$

Now the remaining step is to determine the effective velocity around a grain on the bed due to wave action — a quantity that is related to both the boundary layer and the pattern of flow near the grain.

16.3.1.2 Experiments

16.3.1.2.1 Laminar boundary layer

For a laminar boundary layer, the velocity distribution near the bed is expressed by Eq.(16.21). One can express it in the form

$$u_D = \alpha D \left(\frac{\partial u}{\partial y}\right)_{y=0}$$

in which α is a constant with a value less than 1. If the flow near the bed is oscillating periodically, motion impends at the moment the velocity reaches its maximum value. Therefore, in differentiating u, one can set the sine term in Eq. (16.12) equal to one. The expression for u_D can then be written

$$u_D = 0.7 \alpha D v^{-1/2} N^{3/2} a_0 \qquad (16.34)$$

Using Eq. (16.34) in combination with Eq. (16.31) through Eq. (16.33), one obtains expressions for the angular velocity N_c, for incipient motion of a sediment particle for the three types of boundary layers,

Turbulent

$$N_c = c_1 \left(\frac{\rho_s - \rho}{\rho} g\right)^{1/3} D^{-1/3} v^{1/3} a_0^{-2/3} \qquad (16.35)$$

Laminar

$$N_c = c_2 \left(\frac{\rho_s - \rho}{\rho} g\right)^{2/3} D^{2/3} v^{-1/3} a_0^{-2/3} \qquad (16.36)$$

Transitional

$$N_c = c_3 \left(\frac{\rho_s - \rho}{\rho} g \right)^{\frac{2}{3(2-m)}} D^{\frac{2(2m-1)}{3(2-m)}} v^{\frac{2-3m}{3(2-m)}} a_0^{-2/3} \qquad (16.37)$$

First, laminar flow around a particle

Manohar tested the validity of Eq. (16.35) with glass balls and natural sand for flow around a particle within a laminar boundary layer [9]. From his experiments

$$c_2 = 8.65 \tan^{2/3} \phi$$

in which φ is the angle of repose of sediment particles.

Manohar's study was for a horizontal beach. Eagleson et al. considered the effect of inclining the beach at an angle θ so as to study the equilibrium beach profile and sorting of sediment along a sloping beach [39,40]. They included the inertia effect due to the oscillation of water in formulating their expression of moment balance for a particle. However, in solving it, they neglected the effect of inertia, retaining only the three terms for lift, tractive force, and gravity force. Based on Chapel's experiment in a wind tunnel, they took the ratio of the lift coefficient to resistance coefficient to be approximately constant and used the value 0.85. In their experiments

$$C_D = \frac{19.2}{u_D D / v}$$

from which the derived expression for the angular velocity at incipient motion is the same as Eq. (16.35), with the constant c_2 given by

$$c_2 = 0.373 \left[\frac{\sin(\phi \pm \theta)}{1 + \cos\phi + 0.85 \sin\phi} \right]^{2/3} \qquad (16.38)$$

In the numerator, the positive sign is for landward movement of particles, and the negative sign for seaward movement.

Second, transitional flow around a particle

Goddet studied particles immersed in a boundary layer that has a thickness defined as

$$\sigma = \sqrt{1.3vT} \qquad (16.39)$$

which is less than half of the value from Eq. (16.13). The velocity at the surface of the boundary layer is taken to be the horizontal velocity of water particles near the bottom due to wave action (Eq. 16.7) [41]. He conducted experiments in a wave tank using polystyrene, plexiglass, plastic material, lignite, and natural sand for the sediment, and obtained

$$\frac{u_D}{u_0} = 0.417(D / \sqrt{vT})^{3/4} \tag{16.40}$$

Eq. (16.40) is quite similar to Eq. (16.34), in which u_D is also proportional to $(D / \sqrt{vT})$. He found that the flow around the particles was in the transitional region and that $m = 1/2$. Thus, Eq. (16.33) can be reduced to

$$u_{0a} = A_4 \left(\frac{\rho_s - \rho}{\rho} g\right)^{2/3} D^{1/4} v^{1/24} T^{3/8} \tag{16.41}$$

in which

$$A_4 = 2.4 A_3^{2/3}$$

Fig. 16.25 is a comparison of the experimental data with Eq. (16.41), from which $A_4 = 0.33$.

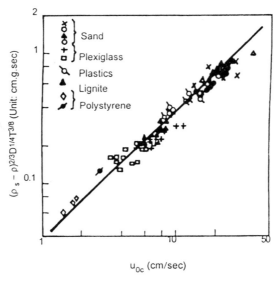

Fig.16.25 Incipient velocity of sediment particles due to wave action
(after Goddet, J.)

If one substitutes

$$u_0 = \frac{2\pi a_0}{T}$$

750

$$T = \frac{2\pi}{N}$$

into Eq. (16.41) and rearranges it, the result is

$$N_c = A_5 \left(\frac{\rho_s - \rho}{\rho} g \right)^{0.485} D^{0.182} v^{0.03} a_0^{-0.727} \qquad (16.42)$$

Using 0.16 and 0.36 mm sand as sediment in a wave tank experiment, Bagnold derived the angular velocity for incipient motion for laminar flow conditions in the form [7]

$$N_c \sim \left(\frac{\rho_s - \rho}{\rho} g \right)^{0.5} D^{0.325} a_0^{-0.75} \qquad (16.43)$$

French researchers modified Eq. (16.43) on the basis of their experimental data using plexiglass and sand from the Seine River

$$N_c \sim \left(\frac{\rho_s - \rho}{\rho} g \right)^{0.525} D^{0.325} v^{-0.05} a_0^{-0.75} \qquad (16.44)$$

If $m = 0.775$, then Eq. (16.37) can be written as

$$N_c \sim \left(\frac{\rho_s - \rho}{\rho} g \right)^{0.544} D^{0.30} v^{-0.088} a_0^{-0.67} \qquad (16.45)$$

which is close to Eq. (16.44). One can infer from these results that the flow around the particles in Bagnold's experiment was in the transitional region, rather than laminar.

16.3.1.2.2 turbulent boundary layer

For coarse particles (D larger than 0.5 mm), the boundary layer is turbulent for the initiation of movement. Manohar studied the initial movement of particles in a turbulent boundary layer by oscillating a plate in still water [9]. In his analysis he considered the lift on a particle, as well as the tractive force. The lift is expressed as

$$F_L = C_L \frac{\rho u_D^2}{2} \frac{\pi D^2}{4}$$

He assumed that the tractive force passed through the center of gravity of the particles and let ϕ be the angle between the vertical and a line connecting the center of gravity and the support point on the neighbouring particle. From a moment equation, he obtained the critical condition for a particle to be lifted

$$C_D \frac{\rho u_D^2}{2} \frac{\pi D^2}{4} \cos\phi + C_L \frac{\rho u_D^2}{2} \frac{\pi D^2}{4} \sin\phi = \frac{\pi D^3}{6}(\rho_s - \rho)g\sin\phi$$

Then, if

$$C_L + C_D \frac{\cos\phi}{\sin\phi} = f\left(\frac{u_D D}{v}\right) \tag{16.46}$$

the critical condition for a particle can be expressed as

$$\frac{\rho u_D^2 D^2 f(\frac{u_D D}{v})}{(\rho_s - \rho)gD^3} = constant \tag{16.47}$$

Manohar assumed that u_D was proportional to the maximum horizontal velocity at the bed as obtained from the potential theory for waves, so that

$$u_D \sim N a_o$$

From experimental data, he obtained

$$f(u_D D / v) \sim \left(\frac{N a_o D}{v}\right)^{0.5}$$

and, together with Eq. (14.47),

$$\frac{\rho N^2 a_0^2 \left(\frac{N a_o D}{v}\right)^{0.5}}{(\rho_s - \rho)gD} = constant$$

his final result is

$$N_c = 14.9\left(\frac{\rho_s - \rho}{\rho}g\right)^{0.40} D^{0.20} v^{0.20} a_0^{-1} \tag{16.48}$$

One cannot readily accept that the viscosity of the water should play such an important role that is indicated by Eq. (16.48). One possibility is that although the boundary layer was turbulent in the range of Manohar's experiment, the flow around the bed particles was still transitional. It is more likely, however, that the result is

752

incorrect since he used only water in his experiments; viscosity was not a proper variable. Whether the relationship between N_C and v in Eq. (16.48) is a true one needs more study.

Rance and Warren conducted experiments on incipient motion of different sediment particles with diameters of 4 to 48 mm as the result of wave action in which the boundary layer and the flow around the particles were fully turbulent [42]. Komar and Miller analysed all of the data-from Manohar, Rance and Warren-and obtained a unified expression for incipient motion in the form [43]

$$\frac{\rho u_{mc}^2}{(\gamma_s - \gamma)D} = 0.55\pi \left(\frac{a_0}{D}\right)^{1/4}$$

(16.49)

in which u_{mc} is the velocity for incipient motion in the form indicated by Eq. (16.10). Eq. (16.49) can be transformed into

$$N_c = 1.31 \left(\frac{\rho_s - \rho}{\rho} g\right)^{1/2} D^{3/8} a_0^{-7/8}$$

(16.50)

Furthermore, for a turbulent boundary layer adopting

$$u_D \sim N a_0$$

as Manohar did, then substituting it into Eq. (16.33) and letting $m = 0$, one obtains the theoretical expression for the critical angular velocity with a turbulent flow around a particle in a turbulent boundary layer in the form

$$N_c \sim \left(\frac{\rho_s - \rho}{\rho} g\right)^{1/2} D^{1/2} a_0^{-1}$$

(16.51)

Clearly, Eq. (16.50) and Eq. (16.51) are quite similar.

All these expressions for the critical velocity are based on data obtained from wave tanks. To extend these results, Sato et al. carried out field measurements along the coast of Japan in which they used radioactive glass sand as tracers to determine the critical condition for motion of the sediment particles. For both laboratory and field usage [44], they proposed the expression

$$\frac{u_m^2}{\left(\frac{\rho_s - \rho}{\rho}\right) gD \sqrt{u_m D / v}} = \frac{0.6}{\sqrt{ND / v}}$$

(16.52)

in which u_m is determined from Eq. (16.10). Eq. (16.52) can be transformed into

$$N_c \sim 1.3\left(\frac{\rho_s - \rho}{\rho}g\right)^{0.5} D^{0.5}a_0^{-0.75} \qquad (16.53)$$

which also is close to Eq. (16.51).

In summary, the basic expression for incipient movement of sediment particles due to wave action is as follows:

$$N_c = constant\left(\frac{\rho_s - \rho}{\rho}g\right)^{a} D^b v^c a_0^d \qquad (16.54)$$

in which the exponents depend on conditions in the boundary layer and the state of flow around the particle, as shown in Table 16.4. With more experimental data, one should be able to determine the relationship between the lift coefficient or the tractive force coefficient and the Reynolds number and unify these various expressions.

Table 16.4 Summary of expressions for angular velocity at incipient motion for sediment particles subjected to wave action

Researchers	Boundary layer	Flow pattern around a particle	$N_c \sim \left(\frac{\rho_s - \rho}{\rho}g\right)^{a} D^b v^c a_0^d$			
			a	b	c	d
Manohar, Eagleson et al	Laminar	Laminar	0.67	0.67	-0.33	-0.67
Bagnold, Martinot-Legarde et al		Transitional	0.525	0.325	-0.05	-0.75
Goddet		Transitional	0.485	0.182	0.03	-0.72
Manohar, Rance et al	Turbulent	Turbulent	0.5	0.375	0	-0.875
Sato et al.		Turbulent	0.50	0.50	0	-0.75

16.3.1.3 Comparison of threshold conditions for wave action and for unidirectional flow

Substituting Eq. (16.19) into Eq. (16.49), one obtains the threshold condition in terms of maximum bed shear stress,

$$\frac{\tau_{mc}}{(\gamma_s - \gamma)D} = 0.275\pi c_f \left(\frac{a_0}{D}\right)^{1/4} \tag{16.55}$$

in which the left side is the parameter used by Shields in his analysis of threshold conditions for unidirectional flow, and the resistance coefficient, c_f, is a function of the Reynolds number $u_m a_0 / \nu$ and the relative roughness a_0 / Ks, as shown in Fig. 16.12.

For sediment movement within a laminar boundary layer, Bagnold's expression [Eq. (16.43)] can be rewritten in the form

$$\frac{\rho u_{mc}^2}{(\gamma_s - \gamma)D} \sim \left(\frac{a_0}{D}\right)^{1/2} D^{0.15}$$

which is not dimensionally consistent because of the term $D^{0.15}$. Komar and Miller showed that this expression is substantially correct even if the D-term is omitted. Then the threshold condition becomes

$$\frac{\rho u_{mc}^2}{(\gamma_s - \gamma)D} = 0.30\left(\frac{a_0}{D}\right)^{1/2} \tag{16.56}$$

This relationship can be rewritten as

$$\frac{\tau_{mc}}{(\gamma_s - \gamma)D} = 0.15c_f\left(\frac{a_0}{D}\right)^{1/2} \tag{16.57}$$

Using the relationship proposed by Jonsson[12]

$$c_f = f\left(\frac{u_m a_0}{\nu}, \frac{a_0}{K_s}\right)$$

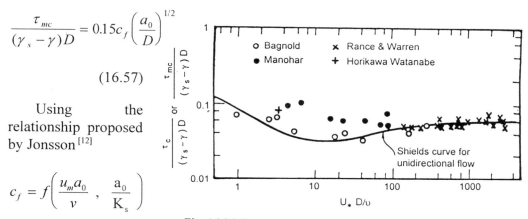

Fig. 16.26 Comparison of threshold conditions due to wave action and the Shields curve for unidirectional flow

755

Komar and Miller analyzed the threshold condition due to wave action on the basis of Eqs. (16.56) and (16.57). In Fig. 16.26, they used the parameter

$\dfrac{\tau_{mc}}{(\gamma_s - \gamma)D}$ as the ordinate and $U*D/v$ as the abscissa so as to compare their results

with the Shields curve in Fig. 8.6. They found that the results for unidirectional flow can be used for wave motion, if the shear stress is used as a reference quantity in this way. The difference between Bagnold's data and Manohar's data is a consequence of the different criteria they used for incipient movement.

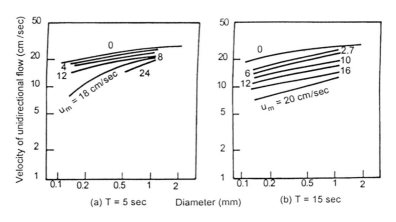

Fig. 16.27 Threshold condition for sediment particles under combined action of waves and unidirectional flow (after Hammond, T.M., and M.B. Collins)

16.3.1.4 Threshold condition for sediment particles under combined action of waves and unidirectional flow

Studies of this topic are scarce. Hammond and Collins carried out four sets of experiments using sediment particles with diameters of 0.14, 0.36, 0.77 and 1.13 mm. The velocity of the unidirectional flow varied from 0 to 27 cm/sec, the wave periods were 5 and 15 sec, and the amplitudes of the velocities of the water particles induced by waves varied from 0 to 35 cm/sec. The result is shown in Fig. 16.27 [45].

16.3.2 Trajectory of moving particles due to wave action

16.3.2.1 Trajectory in deep water

The trajectory of moving particles is closely related to the vortex generated in the troughs of ripples. Fig. 16.28 shows the flow pattern and the kinds of trajectories of particles moving near the bottom at different stages within a half period. Once the oscillatory motion of the water reaches a critical value, vortices are created in the troughs of the ripples. Fig. 16.28a shows the flow pattern as the wave crest passes.

The sediment particles are scoured out from the trough bottom by a vortex of increasing intensity in each trough. The upper vortex, above the ripple in that figure was formed earlier and is carrying sediment particles upward. Fig. 16.28b shows the flow pattern when the velocity due to the waves is zero. Although the water is not moving, the vortex continues to rotate and moves upward from the trough to the crest due to the persistence of the vortex. In this process the ripple crest is at its highest level because sediment deposits near the crest; also some finer sediment particles travel upward with the vortex. Fig. 16.28c shows the flow pattern when motion starts in the opposite direction. The height of the ripple is less because of the horizontal movement of water. New vortexes are being created and the old ones that are ascending carry some suspended sediment.

The trajectories of bed load near the bottom for coarse and fine particles differ, as shown in Fig. 16.29 [46]. The graph to the right of the figure shows the displacement of a water particle during one wave period, and that to the left shows the trajectories of sediment particles at several stages:

Stage a. As the accelerated motion begins, sediment particles on the left side of a ripple are entrained by the water flow and enter the clockwise vortex, which moves to the right side of the ripple crest. The fine particles travel with the vortex while the coarse particles fall onto the bed.

Stage b. As the movement of water decelerates, the angular velocity of the vortex decreases and other sediment particles drop out of the vortex. Once the velocity of the water particles reaches zero, the large vortexes break up into smaller ones. The coarse particles fall back to the same ripple from which they came, but the fine particles may move away from this ripple as they settle down more slowly.

Stage c. When the water particles move in the opposite direction, the fine particles move back and fall out of the original ripple. Meanwhile, the accelerated motion of the water entrains more sediment particles on the right side of the ripple, and they enter the anticlockwise vortex in an upper layer.

Stage d. Deceleration of the water movement makes the vortex disappear and causes the sediment particles to drop to the bed again. The coarse particles fall on the left side of the ripple before the horizontal velocity reaches zero, and the fine particles travel further and are carried back to the original ripple during the next stage (stage a). The process then repeats itself.

16.3.2.2 Trajectory of moving particles in shallow water

In shallow water the landward velocity near the bottom of oscillatory waves is larger than the seaward velocity, making the ripple asymmetric. Under certain conditions, landward bed load transport may take place. Fig. 16.30a shows the form of the ripples and trajectories of particles. After a period of oscillation, more sediment particles are moved to the forward slope of the ripple, causing the ripple to advance in the direction of wave advance. Dyed sediment particles placed on the leeward slope of a ripple show that the ripple moves from position 1 to 2 due to wave action as shown in Fig. 16.30b; the dyed particles also move from the lee slope to the forward slope of the ripple, demonstrating the existence of landward movement of bed load [47].

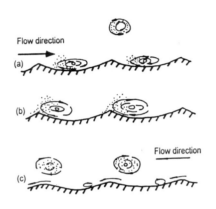

Fig. 16.28 Vortexes and sediment particles moving near the bottom during one wave cycle (after Bagnold, R.A.)

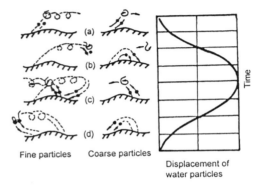

Fig. 16.29 Trajectories of coarse and fine particles near the bottom during one wave cycle (after Bagnold, R.A.)

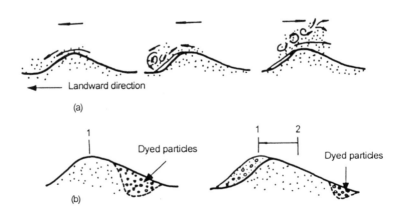

Fig. 16.30 Ripple shape and trajectory of moving particles due to wave action

16.3.3 Bed load motion

When waves pass over a plane bed, sediment particles will travel forward during one half period and travel backward in the other, without consideration of mass transport. The bed load discharge in the following discussion refers to the average bed load discharge during half a period. Many formulas for bed load discharge formulas, originally derived for unidirectional flows, can be transplanted for the transport due to wave action.

16.3.3.1 Einstein bed load function

Starting with the Einstein Bed Load Function in open channel flows, Kalkanis derived a formula for bed load discharge due to wave action [11]. Later, Abou-Seida modified this formula on the basis of additional experimental data [48]. His theory is based on the following conditions:

1. The amplitude of waves is small and the wave length is large ($\frac{2\pi H}{L} \ll 1$). In the flow field outside the boundary layer the linearized flow equations can be used.

2. The boundary layer due to wave action is predominantly turbulent.

3. The fluctuation of the lift force follows the normal error curve.

4. The effect of waves is to induce sediment particles on the bottom to move back and forth. Bed load motion is confined within a layer near the bed with a thickness of $2D$.

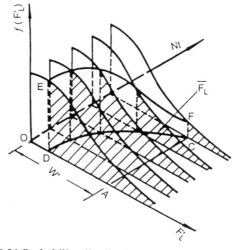

Fig. 16.31 Probability distribution of sediment particle movement
due to wave action during a quarter period

759

The lift force varies periodically because of the wave action, and it also fluctuates because of turbulence. Fig. 16.31 is a three-dimensional diagram that illustrates the probability of a sediment particle being in motion because the lift force acting on it is larger than its immersed weight. The x axis is the value of the fluctuating lift force, F_L', and the y axis is the probability of occurrence of a given F_L', $f(F_L')$, the z axis indicated the phase of the wave. The figure shows the probability distribution for only one-quarter of the period. The total volume under the curves is the total period of time; only part of that time is the fluctuating value of the lift force larger than $W' - \overline{F_L'}$ (i.e., the instantaneous value of the lift force F_L is larger than the immersed weight of the particle, W', $\overline{F_L'}$ is the average value of the lift force). In Fig.16.31 the length OA along the F_L' axis is equal to W', and AC parallel with N_t axis, one can draw a sine curve for F_L along AC. Then, the volume under the right-hand side of the curved surface $DFEC$ corresponds to the period of time during which movement of sediment particles is probable (the dashed area indicating a fraction of the cross section). The ratio of the two periods of time is the probability of particle movement, p.

The expression for p is

$$P = \frac{2}{\pi\sqrt{2\pi}} \int_0^{\pi/2} \int_{B_*\varphi - 1/\eta_0}^{\infty} \exp(-m^2/2)dmd(Nt) \qquad (16.58)$$

In the expression,

$$\psi = \frac{\gamma_s - \gamma}{\gamma} \frac{Dg}{u_a^2}$$

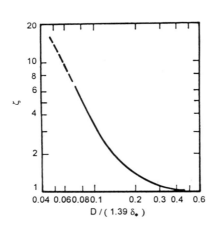

Fig. 16.32 Hiding effect of the boundary layer on sediment particles (after Einstein, H.A.)

in which u_a is the amplitude of the horizontal velocity variation at a distance $0.35D$ above the bed, η_0 is the standard deviation of the fluctuation of the lift force, and m is an integration variable. Moreover, the probability may be derived from the equilibrium condition of which the eroded sediment particles from the bed are equal to the deposited sediment particles from the bed load. Equating it with the value of p in Eq. (16.58) one obtains

$$\frac{A_*\phi}{1 + A_*\phi} = \frac{2}{\pi\sqrt{2\pi}} \int_0^{\pi/2} \int_{B_*\varphi\xi-1/\eta_0}^{\infty} \exp(-m^2/2) \, dm \, d(Nt) \qquad (16.59)$$

This is the Einstein Bed Load Function adapted to wave action, and in it

$$A_* = 13.3 \qquad B_* = 6 \qquad 1/\eta_0 = 2$$

Fig. 16.32 shows the correction factor to account for a hiding effect, ξ, due to the existence of the boundary layer. The thickness of the boundary layer, δ_*, is given by

$$\delta_* = \frac{1100}{4} \frac{\nu}{u_m} \qquad (16.60)$$

in which u_m is the maximum horizontal velocity of water particles near the bed.

The comparison of experimental data with Eq. (16.59) in Fig. 16.33 shows that data points tend to distribute well along the analytical curve.

16.3.3.2 Einstein-Brown formula

In the early 1950s, Brown presented a simplified version of the Einstein bed load discharge formula [49]

$$\Phi = 40\Theta^3 \qquad (16.61)$$

in which Φ is Einstein's dimensionless bed load discharge [Eq. (9.38)], and Θ is the dimensionless flow intensity [Eq. (7.62)]. The Einstein dimensionless bed load discharge due to wave action can be rewritten as

$$\Phi = \frac{\overline{g_{bv}}}{\omega D} \qquad (16.62)$$

in which $\overline{g_{bv}}$ is the average bed load discharge in volume per unit width during a half period and ω is the settling velocity of sediment particles. The dimensionless flow

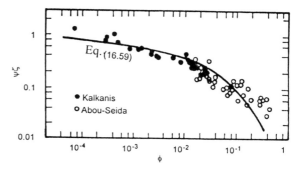

Fig. 16.33 Comparison of the Einstein Bed Load Function for wave action with experimental data

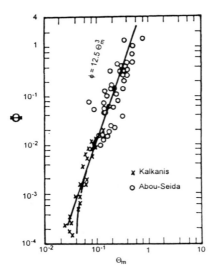

Fig. 16.34 Comparison of Einstein-Brown formula with experimental data
for transport by wave action (after Madsen, O.S., and W.D. Grant)

intensity is written as

$$\Theta_m = \frac{\tau_{0m}}{(\gamma_s - \gamma)D} \tag{16.63}$$

in which τ_{0m} is given by Eq. (16.19). Because the shear stress acting on a particle is larger than the incipient shear stress during only part of the period, Eq. (16.61) should be written in the form of [30]

$$\Phi = 12.5\Theta_m^3 \tag{16.64}$$

Fig. 16.34 is a comparison of experimental data with Eq. (16.64). If the flow intensity is large, they agree.

According to Einstein, if ripples appear on the bed, the flow intensity should correlate with the grain resistance, Θ_m'. [Eq. (7.63)] is the expression for Θ'. Fig. 16.35 is a comparison of Eq. (16.64) and Manohar's experimental data [50]. If Θ_m is replaced by Θ_m' in case, the data for plane bed and rippled beds can be integrated.

16.3.3.3 Meyer-Peter and Muller formula

The Meyer-Peter and Muller formula was written in the form

$$\Phi = (\Theta - \Theta_c)^{3/2} \qquad (9.74)$$

in Chapter 9. Sleath demonstrated that the formula for bed load discharge due to wave action can be expressed as [51]

$$\frac{g_{bv}}{ND^2} = 47(\Theta_m - \Theta_{mc})^{3/2} \qquad (16.65)$$

in which g_{bv} is the bed load discharge, volume per unit width, Θ_{mc} is the threshold value of Θ_m. Eq. (16.65) shows that a form of the Meyer-Peter formula can be used for wave motion.

16.3.4 Suspended load motion

16.3.4.1 Basic features of suspended load motion

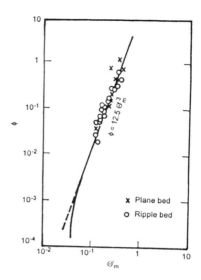

Fig. 16.35 Comparison of Einstein-Brown formula with Manohar's experimental data with and without ripples on the bed

For wave action, the sediment concentration at a point (x,y) in the flow field at time t is composed of three parts:

$$S_w(s, y, t) = \overline{S}_w(x, y) + S_{wp}(x, y, t) + S'_w(x, y, t) \qquad (16.66)$$

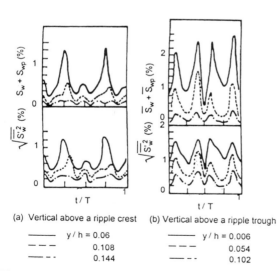

(a) Vertical above a ripple crest   (b) Vertical above a ripple trough

——— y / h = 0.06	——— y / h = 0.006
– – – 0.108	– – – 0.054
— – – 0.144	— – – 0.102

Fig. 16.36 Variation of sediment concentration (% by weight) along different verticals during a wave period for bed with ripples (after Nakato, T. et al)

The first term on the right-hand side is the time-averaged sediment concentration; the second is the periodic change of sediment concentration due to the wave oscillations; and the third term is the fluctuating sediment concentration due to turbulence. Turbulence may be caused by perturbations at the water surface or by the instability of the boundary layer. Except for the intensive turbulence that occurs at the breakers, the effect of turbulence is generally confined in a narrow area near the bed.

The existence of bed ripples

has a large effect on the motion of particles in suspension. Fig. 16.36 shows the variation of sediment concentration $\overline{S_w + S_{wp}}$ and $\sqrt{\overline{S_w'^2}}$ at different heights in verticals above both a ripple crest and a ripple trough during a wave period [32]. At a point near the bed four peaks of sediment concentration occur during a wave period, the nearer the point is to the bed, the more evident the peaks are. The variation of $\overline{S_w + S_{wp}}$ during the first half of the wave period is caused by a vortex that forms on the lee slope of a ripple (Fig. 16.28a). Fine sediment particles moving from the ripple crest fall into the vortex and are kept in suspension by it. As the velocity of water particles diminishes to zero at the middle of the period, the vortex expands and rises, carrying suspended particles with it from the lee slope of the ripple (Fig. 16.28b). The water particles move in the opposite direction during the second half of the wave period, and the vortex is carried by the moving water across the ripple crest (Fig. 16.28c). This action results in four concentration peaks, at $t/T \approx 1/4$ and $t/T \approx 3/4$ as shown in Fig. 16.36a, and at $t/T \approx 3/8$ and $t/T \approx 7/8$ as shown in Fig. 16.36b. Because the displacement of water particles during a half period was 1.6 to 1.8 times the length of a ripple in this experiment, the vortex carrying the suspended sediment passed over the neighbouring ripple crest, and caused concentration peaks at $t/T \approx 1/2$ and $t/T \approx 1$. The two sediment concentration peaks at the ripple trough at $t/T \approx 1/8$ and $t/T \approx 5/8$ are caused by the departure of the vortex from the lee slope of the ripple, Fig. 16.36b. The variation of the mean square deviation of the fluctuating concentration is similar to that of $\overline{S_w + S_{wp}}$, and the two have the same order of magnitude. Hom-ma et al. also reported the occurrence of several concentration peaks during a wave period [35].

16.3.4.2 Vertical distribution of suspended sediment concentration

16.3.4.2.1 Basic differential equation

The basic equation of diffusion of sediment particles, Eq. (10.1), is

$$\frac{\partial S_v}{\partial t} + \frac{\partial}{\partial x}(u_s S_v) + \frac{\partial}{\partial y}(v_s S_v) - \frac{\partial}{\partial y}(\omega S_v) = 0 \qquad (16.67)$$

in which u_S and v_S are the horizontal and vertical velocities of sediment particles, respectively. The effect of the inertia of sediment particles is important during wave action; the velocity of the heavier suspended particles is not the same as the local velocity of water particles and a phase lag between them, t_0, also exists [34]. Their relationship can be expressed as

$$u_s = u - t_0 \frac{\partial u}{\partial t}$$
$$v_s = v - t_0 \frac{\partial v}{\partial t}$$

(16.68)

Substituting Eqs. (16.66) and (16.68) into Eq. (16.67) and averaging the terms in time, one obtains

$$\frac{\partial}{\partial x}\overline{\left[S_{vp}\left(u - t_0 \frac{\partial u}{\partial t}\right)\right]} + \frac{\partial}{\partial x}\overline{\left[S'_v\left(u - t_0 \frac{\partial u}{\partial t}\right)\right]} + \frac{\partial}{\partial y}\overline{\left[S_{vp}\left(v - t_0 \frac{\partial v}{\partial t}\right)\right]}$$
$$+ \frac{\partial}{\partial y}\overline{\left[S'_v\left(v - t_0 \frac{\partial v}{\partial t}\right)\right]} - \frac{\partial}{\partial y}(\omega \overline{S}_v) = 0$$

For a long wave, $\frac{\partial}{\partial x} << \frac{\partial}{\partial y}$, so that the expression can be simplified,

$$\frac{d}{dy}\overline{\left[S_{vp}\left(v - t_0 \frac{\partial v}{\partial t}\right)\right]} + \frac{d}{dy}\overline{\left[S'_v\left(v - t_0 \frac{\partial v}{\partial t}\right)\right]} - \frac{d}{dy}(\omega \overline{S}_v) = 0$$

Integrating this expression, with the condition of zero sediment flux through the water surface, one obtains

$$\overline{S_{vp}\left(v - t_0 \frac{\partial v}{\partial t}\right)} + \overline{S'_v\left(v - t_0 \frac{\partial v}{\partial t}\right)} - \omega \overline{S}_v = 0$$

(16.69)

The vertical velocity of water particles consists of a periodic component, v_p and a fluctuating component, v', i.e.,

$$v = v_p + v'$$

(16.70)

Substituting Eq. (16.70) into Eq. (16.69) for the condition that no relationship exists between S_{vp} and v', or between S_v' and v_p', one obtains

$$\overline{S_{vp}\left(v_p - t_0 \frac{\partial v_p}{\partial t}\right)} + \overline{S'_v(v' - t_0 \frac{\partial v'}{\partial t})} - \omega \overline{S}_v = 0$$

(16.71)

Kennedy and Locher stated, as a rough approximation, that the first item in Eq. (16.71) reflects the influence of the periodic movement of waves and the second term reflects the influence of flow turbulence. For the vertical distribution of suspended sediment in open channel flow, they assumed the second term to be equal to

$-\varepsilon_y \dfrac{d\overline{S_v}}{dy}$, in which ε_y is the diffusion coefficient for sediment particles, so that Eq. (16.71) can then be written as [55]

$$S_{vp}\left(\upsilon_p - t_0 \frac{\partial \upsilon_p}{\partial t}\right) - \omega \overline{S_v} = \varepsilon_y \frac{d\overline{S_v}}{dy} \qquad (16.72)$$

16.3.4.2.2 Vertical distribution of suspended sediment particles near the bed

The effect of the turbulence due to the boundary layer is the principal factor in the zone near the bed; hence, the effect of the periodic oscillation of water particles on diffusion can be neglected. For this condition, Eq. (16.72) can be simplified to the Schmidt formula given in Chapter 10,

$$\varepsilon_y \frac{dS_v}{dy} + \omega S_v = 0 \qquad (10.7)$$

in which S_V is the time-averaged value. The solution of Eq. (10.7) is

$$\frac{S_v(y)}{S_v(a)} = -\omega \int_a^y \frac{dy}{\varepsilon_y} \qquad (16.73)$$

Various researchers have made different assumptions for the vertical distribution of the diffusion coefficient for sediment particles. The most common ones are the following three.

1. Assumption of a constant diffusion coefficient($\varepsilon_y = \varepsilon_0 = constant$)

With this assumption Eq. (16.73) is transformed into

$$S_v(y) = S_v(a)\exp\left[1 - \frac{\omega}{\varepsilon_0}(y - a)\right] \qquad (16.74)$$

which yields a relationship between $S_V(y)$ and y that is a straight line on a semi-log plot. Fig. 16.37 shows the experimental results of Kennedy and Locher, and they fit well with Eq. (16.74). The formula presented by Liu is the same type as Eq. (16.74) [56].

2. Assumption of an exponential variation of ε_y with y

$$\frac{\varepsilon_y}{\varepsilon_a} = \left(\frac{y}{a}\right)^m \tag{16.75}$$

With this assumption, the solution of Eq. (16.73) is

$$\ln\frac{S_v(y)}{S_v(a)} = \frac{\omega a}{(1-m)\varepsilon_a}\left[1-\left(\frac{a}{y}\right)^{m-1}\right] \tag{16.76}$$

Fleming and Hunt used a value of 0.25 for m, based on experimental data [57].

3. Assumption that the diffusion coefficient is equal to the coefficient of momentum exchange of the flowing water

Following the approach used to determine the vertical distribution of suspended particles in open channel flow, Hom-ma et al. assumed that

$$\varepsilon_y = kl^2\left|\frac{du}{dy}\right|$$

in which u and l are determined from potential flow theory. They derived the

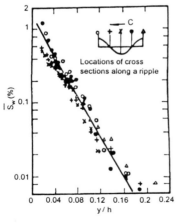

Fig. 16.37 Comparison of Eq. (16.74) with experimental data (after Kennedy, J.G., and G.A. Locher)

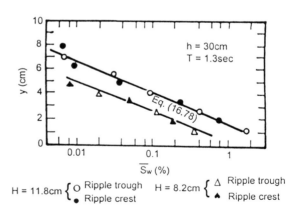

Fig. 16.38 Vertical distribution of suspended sediment near the bed — a comparison of experimental data with Eq. (16.78) (after Das, M.M.)

expressions [53]:

$$\varepsilon_y = \frac{1}{A} \frac{Hc}{\sinh kh} \frac{\sinh^3 k(y+h)}{\cosh^2 k(y+h)} \qquad (16.77)$$

$$\frac{S_v(y)}{S_v(a)} = cxp\left\{-A\frac{\omega}{6H}\frac{\sinh kh}{2k}[F(n_a,kh)-F(n,kh)]\right\} \qquad (16.78)$$

in which

$$F(n,kh) = \frac{\cosh nkh}{\sinh^2 nkh} - \log\left|\tanh\frac{nkh}{2}\right| \qquad (16.79)$$

$$A = 0.161\left(\frac{y+h}{h}\right)^{0.833}\left(\frac{H}{L}\frac{h}{\lambda}\right)^{0.142}\left(\frac{H}{\lambda\sinh kh}\right)^{0.27} \qquad (16.80)$$

$$n = \frac{y+h}{h} \qquad n_a = \frac{h-a}{h}$$

The symbols used are H — wave height, L — wave length, C — wave velocity, k — wave number, λ — wave length of ripples. Eq. (16.78) is the same type of function as Eq.(16.74), and it also follows a straight line on semi-log paper. Fig. 16.38 compares Eq. (16.78) with experimental data [58].

16.3.4.2.3 Vertical distribution of suspended sediment particles within the main flow zone

Within the main flow zone well away from the bed, the effect of turbulence within the boundary layer can be neglected; that is, ε_y in Eq. (16.72) is zero. If t_0 is a constant and using [55]

$$S_{vp} = -K_1\int_0^t v\frac{d\overline{S}_v}{dy}dt \qquad (16.81)$$

one obtains

$$\upsilon = \frac{Nb}{h} y \cos Nt \qquad (16.82)$$

In these equations K_1 is a constant, b is the vertical amplitude of the water particle motion, N is the angular velocity, $N = 2/T$; the solution of Eq. (16.72) is then

$$\ln \frac{S_\upsilon(y)}{S_\upsilon(a)} = \frac{2\omega h^2}{t_0 K_1 b^2 N^2} \left(\frac{1}{y} - \frac{1}{a} \right) \qquad (16.83)$$

Fig. 16.39 is another way of showing the experimental data shown in Fig. 16.37. The experimental data agree with Eq. (16.83) if h/y is smaller than about 15.

16.3.5 Total sediment discharge

How to combine the bed load and suspended load to obtain the total sediment load is an unresolved issue. Bagnold made a preliminary study of this issue following the approach used for unidirectional flow[59].

As pointed out in Chapter 11 for unidirectional steady flow, the total sediment discharge per unit width, in terms of the submerged weight of the sediment, is

$$g_T' = g_b' + g_s' = W \left[\frac{e_b}{\tan \alpha} + \frac{e_s(1 - e_b)}{\omega / u_s} \right] \qquad (9.16) \text{ and } (10.70)$$

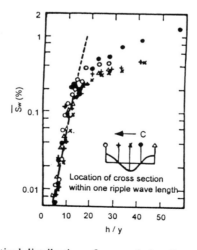

Fig. 16.39 Vertical distribution of suspended sediment in the main flow zone —— a comparison of experimental data with Eq. (16.83)

769

In which e_b and e_s are the efficiencies of bed load and suspended load transport by flow, respectively, $\overline{u_s}$ is the average speed of suspended sediment particles, $\tan\alpha$ is the dynamic friction coefficient of sediment particles, W is the potential energy supplied by the flow per unit time on a unit area of the bed (it is equal to $\tau_0 U$). If an adverse slope ($\tan\theta$) exists, then the expression becomes

$$g'_T = W \left[\frac{e_b}{\tan\alpha - \tan\theta} + \frac{e_s(1-e_b)}{\dfrac{\omega}{\overline{u_s}} - \tan\theta} \right] \qquad (16.84)$$

For waves with a height H, the energy propagating forward per unit area of wave and per unit time is

$$\frac{1}{8}\gamma H^2 c_g$$

in which c_g is the group velocity of waves, Eq. (16.5). The energy loss per a unit distance is

$$\frac{1}{8}\gamma \frac{d(H^2 c_g)}{dx} \approx \frac{1}{8}\gamma H c_g \frac{dH}{dx}$$

If all the energy loss is caused by bed resistance, i.e., if the energy loss due to wave breaking is negligible, the work required for sediment movement from the water per unit area of bed is

$$W = \frac{1}{8}\gamma H c_g \frac{dH}{dx} \qquad (16.85)$$

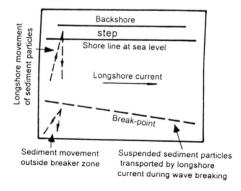

Sediment movement outside breaker zone

Suspended sediment particles transported by longshore current during wave breaking

Fig. 16.40 Three basic patterns of longshore movement of sediment particles

Substituting Eq. (16.85) into Eq. (16.84), one obtains the total sediment discharge per unit width due to wave action. The application of Eq. (16.85) is discussed in the next section.

16.4 LONGSHORE MOVEMENT OF SEDIMENT PARTICLES [60]

16.4.1 Basic patterns

Three basic patterns of longshore movement of sediment particles are shown in Fig. 16.40. Outside the breaker zone, oscillations near the bed due to wave action cause bed particles to move to and fro. Bed load is the main pattern of sediment movement in this zone; suspended load can occur only during periods of strong winds. On the shore side of the breaker zone, a large amount of the sediment is suspended by turbulence and transported by the longshore current. Finally, sediment particles move with the swash and backwash. The latter two patterns are the principal ones. Over 80% of longshore sediment transport takes place inside the breakers. Obviously, the rate of longshore sediment transport at various locations on a beach changes in accordance with the local wave conditions. Fig. 16.41 shows the measured volume of longshore transport of sediment particles on a beach in a coastal area in the former USSR [61]. In the figure, S is the suspended sediment concentration; u_1 is the average surface velocity of the longshore current; the rectangular areas represent the sediment volume passing through the width of the rectangles, in m³/hr; and the height of the rectangle is the sediment volume per unit length, in m²/hr. The maximum concentration occurs in the swash zone, and the largest longshore sediment transport rate occurs at the top of the underwater barrier. During large storms, the position of the largest rate of longshore sediment transport rate shifts seaward.

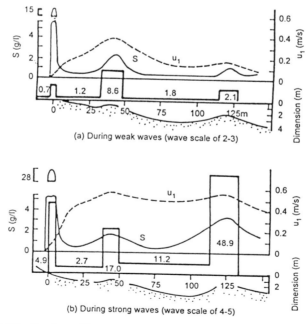

Fig. 16.41 Distribution of longshore movement of sediment particles on a beach

The pattern of longshore movement of sediment particles varies from place to place along a beach. In many coastal areas, the effects of mass transport and rip currents are to transport fine particles seaward and to leave coarse particles on the beach. Komar estimated that 80% of the total longshore transport of sediment moves as bed load [25]. Brenninkmeyer observed that suspended load predominates only in the area where the backwash meets the waves, and that bed load predominates in all other parts of the beach [62].

16.4.2 Sediment transport capacity of longshore currents

Studies of sediment transport capacity of longshore currents over the years have produced many empirical and semi-empirical formulas. They all used one of two approaches. One is to correlate the longshore sediment transport rate with the longshore component of wave energy per unit time, and it is referred to as the wave dynamics approach. The other is based on the concept that sediment movement consumes a certain amount of energy; it then determines the amount of sediment moving due to wave action; finally, the sediment transport is correlated with the longshore current. This method is referred to as the wave energy approach. Although these two approaches appear to be different, they can be unified conceptually.

16.4.2.1 Wave dynamics approach

The wave energy per unit area of water surface, $\frac{1}{8}\gamma H^2$, propagates with a velocity c_g. Hence, referring back to Fig. 16.22, the component of wave energy entering area $ABCD$ per unit time and propagating along the coast is

$$\frac{1}{8}\gamma H_b^2 c_{gb} \sin\alpha_b \cos\alpha_b dx$$

If the value for dx is taken to be unity, the above expression can be rewritten as

$$E_l = \frac{1}{8}\gamma H_b^2 c_{gb} \sin\alpha_b \cos\alpha_b \qquad (16.86)$$

in which E_l is the wave energy propagating along the coast per unit time and per unit length of coast, the subscript b represents the wave elements at the breaker zone. Many researchers have felt that the longshore sediment transport rate should be related to E_l. Fig. 16.42 shows the results from a large amount of laboratory and field data [58]. Komar and Inman proposed a formula that they based mainly on field data because they felt that the longshore movement of sediment particles in laboratory experiments might not be sufficiently well developed [63].

$$G_l' = 0.77 E_l \qquad (16.87)$$

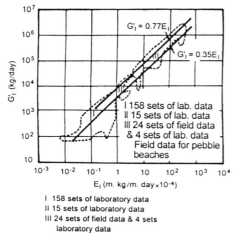

I 158 sets of lab. data
II 15 sets of lab. data
III 24 sets of field data
& 4 sets of lab. data
Field data for pebble
beaches

I 158 sets of laboratory data
II 15 sets of laboratory data
III 24 sets of field data & 4 sets
laboratory data
• Field data for pebble beaches

Fig. 16.42 Comparison of observed longshore
sediment transport rate with Eq. (16.87)
(after Das, M.M.)

in which G_l' is the longshore sediment transport rate by submerged weight of sediment particles in kg/day, and E_l is in kg-m/m-day. If all the laboratory and field data are used, the coefficient in Eq. (16.87) is 0.35. Eq. (16.87) applies only to sand beaches. For pebble beaches the coefficient is probably smaller.

The principal shortcoming of this approach is that to take a component of the wave energy is meaningless because energy is a scalar. Longuet-Higgins studied this issue and gave E_l a new meaning, using momentum flux instead [64].

If the x axis is perpendicular to the coast, y upward from the still-water surface, and z along the coast, then as the waves enter the breaker, the wave energy flux per unit length of coast in the x direction E_x is

$$E_x = \frac{1}{8}\gamma H_b^2 c_{gb} \cos\alpha_b$$

and the momentum flux per unit length of coast along the z axis, S_{xz}

$$S_{xz} = -\int_{-h}^{\xi} \rho u u_e \, dy$$

in which u and u_e are velocities of water particles along the x and z axes, respectively, $y = \xi(x, z, t)$ represents the fluctuating water surface. For small amplitude waves, Longuet-Higgins obtained [33]

$$S_{xz} = \frac{\sin\alpha_b}{c_b} E_x = \frac{1}{8}\gamma H_b^2 \frac{c_{gb}}{c_b} \sin\alpha_b \cos\alpha_b \tag{16.88}$$

From Eq. (16.86) and Eq. (16.88), one obtains

$$E_l = S_{xz} c_b \tag{16.89}$$

i.e., when the momentum of the body of water propagates in some direction, a lateral force parallel to the coastline is created. The longshore sediment transport rate is proportional to this lateral force.

16.4.2.2 Wave energy approach

Bagnold derived a formula for the total sediment load discharge due to wave action by measures of an approach based on wave energy, and it is like Eq. (16.84). If

$$K = \frac{e_b}{\tan \alpha - \tan \theta} + \frac{e_s(1 - e_b)}{\omega / u_s - \tan \theta} \qquad (16.90)$$

then

$$g_T' = KW$$

And if u_t is the velocity of sediment particles in the direction of wave advance due to wave action, then the amount of moving sediment per unit area of the bed by submerged weight, W_T', is as follows:

$$W_T' = g_T' / u_t = KW / u_t$$

Since this amount of sediment has been put in motion by the energy from the waves, sediment movement along any direction can be produced by the longshore current with a velocity U in that same direction, no matter how small the velocity is, that is, even if it is smaller than the incipient velocity of the sediment particles. The sediment transport rate in this direction is

$$g_a' = KW \frac{U_a}{u_t} \qquad (16.91)$$

If u_t is proportional to u_m [Eq. (16.10)], Eq. (16.91) can be rewritten in the form

$$g_a' = K'W \frac{U_a}{u_m} \qquad (16.92)$$

The next step is to apply this concept to the longshore movement of sediment particles inside the breaker zone. As the flow in the breaker zone is complex, only the mean situation is treated here. The wave energy entering the breaker zone per unit time and per unit length along the coast is

$$\frac{1}{8} \gamma H_b^2 c_{gb} \cos \alpha_b$$

Only a fraction of the energy of the wave is used to overcome the resistance — the part that is effective in moving sediment. Therefore, the longshore sediment transport rate (by submerged weight) through a cross section perpendicular to the coast and inside the breakers is

$$G_l' = K'' \frac{1}{8} \gamma H_b^2 c_{gb} \cos \alpha_b \frac{U_l}{u_{mb}}$$

(16.93)

in which U_l is the average velocity of the longshore current, u_{mb} is the maximum velocity of water particles due to wave breaking,

$$u_{mb} = \left(\frac{1}{4} g \frac{H_b^2}{h_b} \right)^{1/2}$$

(16.94)

Fig. 16.43 is a comparison of data from two beaches with Eq. (16.93), the coefficient K'' is approximately equal to the constant, 0.28 [63].

16.4.2.3 Unified approach to wave dynamics and wave energy

To unify these two approaches, one can substitute Eq. (16.30) into Eq. (16.93), and obtain

$$G_l' = 0.55 \frac{J}{c_f} \left(\frac{1}{8} \gamma H_b^2 c_{gb} \right) \sin \alpha_b \cos \alpha_b$$

(16.95)

The equations Eq. (16.87) and Eq. (16.95) for the longshore sediment transport

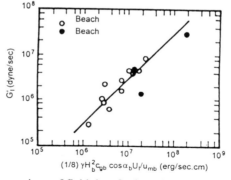

Fig. 16.43 Comparison of field data for longshore sediment transport rate
with Eq. (16.93) (after Komar, P.D., and D.L. Inman)

rates are almost equivalent if J/c_f is a constant; Komar verified theoretically that this is approximately the case [65]. The data for various beach slopes are combined in Fig. 16.42, showing that the longshore sediment transport rate is independent of J.

16.5 MOVEMENT OF SEDIMENT PARTICLES NORMAL TO THE SHORE AND BEACH PROFILE FORMATION

16.5.1 Movement induced by mass transport

According to the Longuet-Higgins theory, the velocity of mass transport near the bed u_{i0} is expressed by Eq. (16.24). If Eq. (16.9) is substituted into Eq. (16.24), one obtains

$$u_{io} = \frac{5}{4} \frac{u_m^2}{c} \tag{16.96}$$

For long waves, $h/L<<1$, so that Eq. (16 .9) can be simplified,

$$u_m = \frac{H}{2h} c$$

Substituting this equation into Eq. (19.96), one obtains

$$u_{io} = \frac{5}{4} \left(\frac{H}{2h}\right)^2 c \tag{16.97}$$

Substituting u_{i0} for U_α in Eq. (16.92), one obtains the landward sediment transport rate near the bed due to mass transport [59]

$$g_i' = K''W \frac{H}{2h} \tag{16.98}$$

The coefficient, K'', may decrease as the wave length decreases, but systematic experimental data concerning this trend are not available.

16.5.2 Direction of movement of sediment particles — neutral line concept

Sediment particles on a beach can move in either direction. The landward movement is due to wave action, and the seaward movement is due to the action of gravity along the sloping beach. The direction of sediment movement in a specific locality depends on the direction of the resultant of these two forces. Where they are just balanced, sediment particles will have no net movement in either direction, but will only move back and forth around their original position. Fig. 16.44 shows the direction and velocity of 3 mm glass balls on a 1:15 smooth slope under waves with a steepness of 0.0237 and a period of 1.68 sec; landward velocity is positive and seaward velocity negative [39]. At station 48, the net movement of sediment particles is zero. Further seaward, the sediment particles gradually move seaward; the farther from the coast, the weaker the wave action is. If the instantaneous effective force during the passing of the wave crest are not strong enough to move the sediment particles shoreward, sediment movement takes place only during the half period of seaward movement. Further out, in deep water, the movement of sediment ceases. Inside the point of equilibrium, sediment particles move landward due to wave action. In the breaker zone, sediment movement is retarded by the reverse flow due to the waves, and once again equilibrium is reached. The landward movement and seaward movement of sediment particles offset each other at the neutral line; on either side of it the sediment moves away from this line. The position of the neutral line depends on the wave characteristics, particle size, bottom slope, etc.

The concept of a neutral line is a simplified model of a natural phenomenon that is quite complex. For example, if suspended particles rise to a certain distance from the bed, they will be carried seaward. Therefore, a local balance of incoming and outgoing sediment fluxes exists, and it may not occur just at the neutral line. After storms at sea, large quantities of sediment particles are suspended by the turbulence caused by waves breaking; turbidity currents may even form and carry sediment particles along the bottom slope to the sea[1]. Experiments showed that weak unidirectional flows of water other than the mass transport may also change the direction of movement of sediment particles [66].

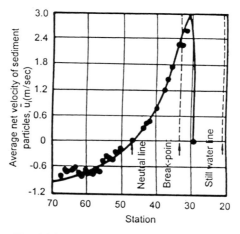

Fig. 16.44 Velocity and direction of glass balls due to wave action on a smooth 1:15 slope (after Eagleson, P.S., and R.G.Dean)

16.5.3 Formation of equilibrium profile of beaches

The movement of sediment on the two sides of the neutral line affects the form of a beach profile. Sediment particles moving landward inside the neutral line deposit near the coastline, forming shore banks, and the particles moving seaward outside the neutral line deposit near the foot of the underwater slope. This process continues until all the particles on the slope reach a state of equilibrium for their oscillating movements. At this stage, a dynamic equilibrium beach profile is reached instead of a neutral line. At each point on the slope, the drag along the slope due to wave action is equal to the component of the weight of a sediment particle along the slope. Nearer the shoreline, the forces acting on particles due to waves are larger; consequently, the slope of the equilibrium profile is steeper, and the sediment particles on the slope are coarser.

Actually, the equilibrium profile adjusts itself constantly to variations of the dynamic conditions in the sea. Obvious seasonal changes take place. In winter, storms are strong and they erode the foreshore area. The eroded sediment particles move seaward and deposit in regions farther offshore. There, they form bars in the breaker zone. In summer, storms are weak and sediment particles move landward, depositing in the foreshore area. The breaker zone bars disappear. These typical

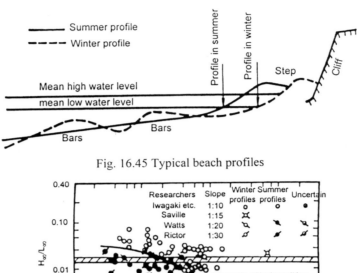

Fig. 16.45 Typical beach profiles

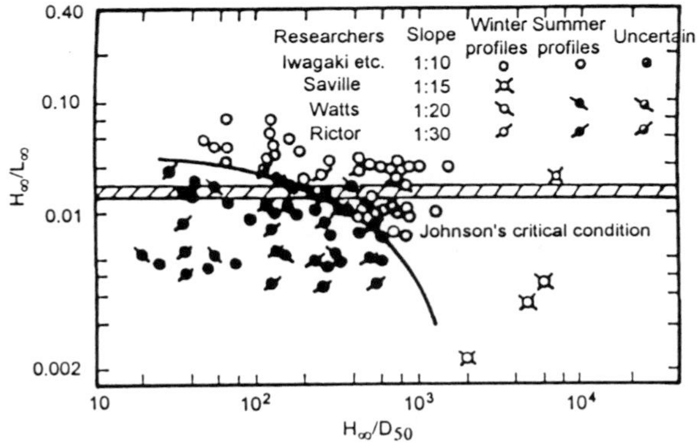

Fig. 16.46 The conditions for which winter and summer profiles form

beach profiles are shown in Fig. 16.45; the former is called the "winter profile" or "storm profile"; and the latter the "summer profile" or "normal profile."

The formation of the two seasonal profiles is closely related to wave steepness (H/L). During storms, waves are steep; sediment particles outside the breakers move landward, and sediment particles inside the breakers move seaward (Fig. 16.17). These two sediment fluxes meet at the breaker zone and form breaker bars there. In summer, waves are flat and sediment particles on both sides of the breakers move seaward so as to form steps along the shore. Johnson stated that a summer profile is created if H_∞ / L_∞ is smaller than 0.025, while a winter profile is created if H_∞ / L_∞ is larger than 0.030 [67]. As pointed out in more recent studies, the critical condition is also affected by the characteristics of the sediment particles on the beaches. Iwagaki and Noda reviewed experimental data from many researchers and concluded that the critical condition is related to both H_∞ / L_∞ and H_∞ / D_{50}, as shown in Fig. 16.46 [19], D_{50} is the median diameter of sediment particles on a beach. Johnson's critical condition is also shown in the figure, and it appears to be suitable for small values of H_∞ / D_{50}.

Dean suggested that the movement of suspended sediment can affect the transition between winter and summer profiles [68]. Sediment particles fall to the bed after they are suspended by wave action; meanwhile, they are acted upon by the oscillatory motion of the water. If the time period for settling is shorter than the wave period, the sediment particles are affected mainly by the landward velocity. If the time period is longer, the particles tend to be transported seaward. From such a consideration, Dean derived a relationship for the critical wave steepness in the form

$$\frac{H_\infty}{L_\infty} = 0.85 \frac{\omega}{c_\infty} \tag{16.99}$$

in which ω is the settling velocity of the D_{50} particle. Eq. (16.99) can be transformed into

$$\frac{H_\infty}{\omega T} = 0.85 \tag{16.100}$$

From Fig. 16.47 the critical value of $H_\infty/(\omega T)$ depends on the wave steepness in deep water, and it varies between 0.7 and 2 [69].

It is believed that $H_\infty/(\omega T)$ has a relationship with the angle of the slope, θ, in view of the importance of $H_\infty/(\omega T)$ to the beach profile. Fig. 16.48 shows such a relationship based on experimental data from many researchers [70].

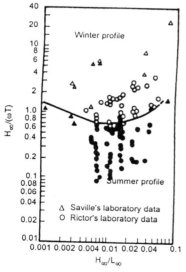

Fig. 16.47 Critical condition for two kinds of beach profiles (after Coastal Engineering Research Center)

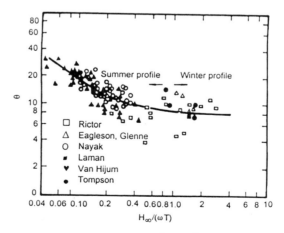

Fig. 16.48 θ vs. $H_\infty / \omega T$
(after Dalrymple, R.A., and W.W.Thompson)

779

16.6 MUD MOVEMENT DUE TO WAVE ACTION

If sediment particles carried by rivers deposit in estuarine and coastal areas, the consolidation process for particles smaller than fine silt is slow. The initial unit weight of such deposits is small, and they have a degree of fluidity. Such particles comprise what is called mud. Mud is easily transported by waves and tidal flows and may flow in the form of turbidity currents caused by their own weight. Dredging in the Thames River estuary in England for the maintenance of navigation amounted to 2 million m³/year. Most of the dredged material was placed on a beach 50 km south of the estuary. Later studies showed that the waves did not carry the dredged mud to the sea. Instead, it was transported back to the estuary in the form of turbidity currents, so that the dredging was an ongoing task that would never end [71]. Deposition caused by mud movement is also an important factor in the Xingang Harbor, Tianjin, China [70].

16.6.1 Basic phenomena of mud movement

The median diameter of mud particles is generally smaller than 0.005 mm. Turbid water carrying mud is generally non-Newtonian in nature.

If a layer of unconsolidated mud forms on the sea bottom, a clear interface exists between the mud and the clear water. The movements of this mud layer are affected by wave action. Observations have shown that the basic phenomena of mud movement are closely related to its unit weight [56,72,73].

Mud with a unit weight smaller than 1.10 has a fluidity that is so strong that it can move with a periodic oscillation as does water subjected to wave action. For such mud, the stronger the waves, the larger are both the amplitudes of the particle motion and the thickness of the moving layer of mud. In addition to the oscillations, mud particles also move landward as do water particles close to the bed; thus the mud moves in the direction of wave advance. Also, the mud layer may adjust its surface slope and even form an adverse slope against the direction of wave advance to resist the wave action by gravity. If the waves reach a certain strength, the interface becomes unstable and sediment particles are lifted from the mud layer into the clear water, creating a distribution of sediment concentration in the vertical direction. As the mud layer moves, because of the surface waves, the relative velocity of water particles near the mud layer is small. Therefore, hardly any turbulence is created there. For mud with a small unit weight, sediment particles of the mud layer may be entrained by the clear water and move in suspension, but only if the waves are strong enough [56].

For mud with a unit weight larger than 1.2, the oscillatory movement of mud is weak, perhaps even non-existent; thus, suspension of mud particles is the principal mode of movement. Along with the intensification of waves, flocs of mud particles may appear on the mud surface, moving to and from, then smoky looking sediment

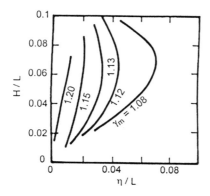

Fig. 16.49 Relationship between amplitudes of wave motions at mud layer surface and at water surface

sheets move upward and diffuse into the clear water, thus forming a layer of turbid water above the mud layer. If the waves intensify still more, sediment particles near the bottom diffuse upward into the clear water zone, so that the water is turbid from top to bottom.

For mud with a unit weight between 1.1 and 1.2, the mode of mud movement is transitional between these two patterns, i.e., movement within the mud layer and those due to suspension of mud particles coexist.

In the following discussion only the movement of mud with a unit weight smaller than 1.2 is discussed.

16.6.2 Mud flow near bed

Fluctuations within the mud layer caused by wave action can be quantified. Fig. 16.49 shows the relationship between the amplitude of mud layer surface, η, and the amplitude of the surface wave, H, for mud of various unit weights. The amplitude for the mud layer tends to increase with that of the surface wave. However, if the surface wave reaches a certain limiting height, further increase of the surface wave causes the amplitude at the mud layer surface to decrease instead. The smaller the unit weight of mud, the larger the amplitude at the mud layer surface. Zhao et al. derived an expression for the amplitude of the mud layer surface based on the assumption that the clear water is an ideal fluid and mud is a viscous one [74],

$$\eta = \frac{H}{2} \; \frac{\sinh kh - \dfrac{k}{m}\sinh mh_0}{\sinh k(h+h_0) - \sinh kh \cosh mh_0 - \dfrac{k}{m}\sinh mh_0 \cosh kh} \qquad (16.101)$$

in which h is the water depth, h_0, the thickness of moving mud,

$$k = 2\pi / L$$

$$m = \sqrt{\frac{2\rho + 3\rho_m}{2\rho - \rho_m}}\, k$$

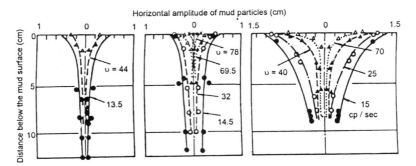

Fig. 16.50 Horizontal amplitude of mud particles with various coefficients
of kinematic viscosity due to wave action [75] (after Lhermitte, P.)

ρ, ρ_m are the densities of clear water and mud, respectively.

Like water particles, particles of mud move periodically with an elliptic orbit.
Fig. 16.50 shows the vertical distribution of the horizontal amplitude of mud particle
movement with various coefficients of kinematic viscosity[1] [75]. If the coefficient of
kinematic viscosity of mud is 40 cp/sec, the horizontal amplitude of mud particles
near the mud layer surface is only seven-tenths of that of water particles just above
the mud surface.

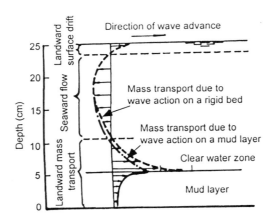

Fig. 16.51 Vertical distribution of velocity of mass transport due to wave action

---

[1] The viscosity of mud varies with it unit weight. Therefore, either quantities can be used to
identify the characteristics of mud.

Sediment particles on mud beaches also tend to move landward due to wave action and move seaward due to gravity. The beach slope reaches equilibrium if the two forces balance each other. Table 16.5 shows laboratory results for the reduction of the mud discharge along with the adjustment of the interface slope from a small positive value to various negative ones (the interface slope inclining downward in the direction of wave advance is positive) [56]. Obviously, the larger the mud discharge, without the effect of the slope of the mud layer surface, the steeper the final negative slope of the mud layer surface will be.

Fig. 16.51 shows the vertical distribution of the velocity of mass transport over the entire depth, including the mud layer. If mud movement exists, the velocity gradient at the bottom edge of the clear water is smaller than that without mud movement. The further it is down to the surface of the mud layer, the smaller the velocity. The velocity of mud flow varies with viscosity, as shown in Fig. 16.52 [75,76]. Experiments conducted at Tianjin University showed that, if the mud layer surface has a large adverse slope, the direction of mud movement in the upper mud layer is the same as the direction of wave advance, while it is opposite to the direction of wave advance in the lower mud layer, as shown in Fig. 16.53 [74]. The velocity gradients in Zhao's experiments were much smaller than those of Lhermitte.

Table 16.5 Mud discharge for various slopes of mud layer surface

Duration of wave action(sec)	0~25	25~60	60~90	90~120	120~180	180~360
Slope of mud layer surface	+0.001	-0.0006	-0.0018	-0.002	-0.0026	-0.003
Unit mud discharge (m²/s)	1.26	0.357	0.533	0.25	0.208	0.067

From experimental results, Lhermitte showed that mud on a sea bed 40 m deep might be induced to move landward by waves with a period of 12 sec and a wave height of 4 m. The mud moving along a beach slope generally does not reach the shoreline because strong turbulence in the breaker zone entrains the mud in suspension and takes it back out to sea. Only at sheltered beaches can mud reach the shoreline and forms a special mud beach.

16.6.3 Suspension of mud

When waves reach a certain intensity, sediment particles in a mud layer are entrained and carried into the clear water zone. There, they move as suspended load. The vertical distribution of concentration of suspended mud follows the usual formula (Section 16.3). In one study, Liu assumed that the mixing length and vertical fluctuating velocity due to wave action are proportional to the vertical amplitude of water particles and to their average horizontal velocity during half a wave period. Then, using their mean values along the vertical, he obtained the following vertical distribution for the mud concentration [56].

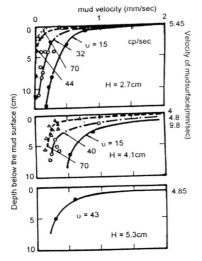

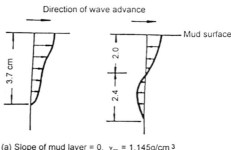

(a) Slope of mud layer = 0, γ_m = 1.145g/cm^3
L = 87cm, H = 2cm

(b) Slope of mud layer = -0.0025, γ_m = 1.145g/cm^3
L = 87cm, H = 3cm

Fig. 16.52 Relationship between mud velocity due to wave action and coefficient of kinematic viscosity of mud

Fig. 16.53 Velocity distribution of mud flow on two slopes of mud layer surface

$$\frac{S_\upsilon}{S_{\upsilon a}} = \exp\left(-\frac{100\omega}{HU}(y+a)\right) \tag{16.102}$$

in which U is the horizontal velocity of water particles during a half wave period averaged over the depth.

Although no systematic studies of the critical condition for the suspension of mud flow due to wave action have been made, some studies give the critical velocity for the instability of the mud interface in steady open channel flow, and they show that it is related to both the Reynolds number and Froude number just as for the critical condition of the stability of the interface of a density current [73,77].

16.6.4 Damping effect of mud on waves

Under certain conditions, mud may gather in some part of a bay and form a mud beach. For example, areas of unconsolidated mud exist in the bays on both sides of the Yellow River estuary. Even with strong waves at sea, the wave heights in those bay areas are significantly less, so that fishing vessels often use them as shelters. The same phenomenon occurs along the coast of Guiana [78].

The presence of mud near a coast greatly increases the energy dissipation of waves as they travel towards the shore. The pressure waves at the bed caused by the surface waves induce an oscillatory movement in part of the mud layer, and the large viscosity of the mud causes a phase lag between the pressure wave and the response of the mud. Therefore, a part of the energy propagates into the mud layer and is dissipated there by internal friction. The energy transmitted per unit area of mud surface and per unit time (averaged over a wave period) is [79]

$$E_m = \frac{\pi \gamma M H^2 \sin \phi}{4 T \cosh^2 kh} \qquad (16.103)$$

in which M is a constant of proportionally relating the amplitude of the mud wave to that of the pressure wave; its value was 0.039 for tests made in the Mississippi River estuary; ϕ is the phase lag between pressure wave and the mud wave.

E_m is at least one order of magnitude larger than the energy loss due to bed friction and infiltration during wave propagation. Fig. 16.54 shows the measured and calculated wave heights at various locations on a profile perpendicular to the coast in the Mississippi River estuary [79]. The wave height over a mud bed is much smaller than that for one without mud. Moreover, the effect of bed friction occurs only in shallow water areas, while the effect of mud on energy dissipation manifests itself in regions of moderate depth as well. Hence, the existence of mud beaches is favourable for oil exploitation in shallow regions.

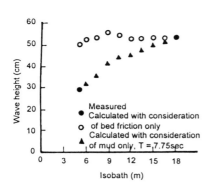

Fig. 16.54 Comparison of damping effect due to mud layer and to bed friction (after Tubman, N.W., and J.N. Suhayda)

16.7 FORMATION AND DEVELOPMENT OF RIPPLES ON SEA BEDS

As for both river beds and wind-blown sand, wave action can cause ripples on sea beds. This section treats the mechanism of formation and the morphological characteristics of ripples on sea beds and submarine bars (longshore bars).

16.7.1 Ripples

16.7.1.1 Formation of ripples on sea beds

Bagnold identified three types of ripples that form under waves during laboratory experiments. Sediment particles start to move by rolling to and from once the wave intensity exceeds the criterion for incipient motion of sediment; before long, they form a pattern of parallel transverse crests. Part of the bed surface sheltered by the lee side of the ripple is not affected by the water movement. Once this sheltering effect exerted by one ridge extends to the next, sediment particles in the troughs of ripples no longer move. As no more sediment particles are supplied to the crests, the ripples then form a stable pattern. The height of the ripples is in the range from a few particles diameters to more than 10. These stable ripples oscillate to and fro due to wave action, but their position can either remain stationary over long periods of time, or move in either direction. Bagnold called them 'rolling grain ripples'.

Kaneko and Honji spread glass balls with diameters of 0.2 to 0.86 mm on a rigid bed and found that, with wave action, the glass balls gradually formed crests perpendicular to the direction of wave advance [80]. They suggested that the mechanism of formation of "rolling grain ripples" is similar to the mechanism of two sediment particles settling one after the other. Because of the effect of viscosity, the trailing particle gradually catches up the leading one, and the two particles then settle together. Rolling grain ripples are often the "predecessor" of normal ripples on sea beds.

As the wave action intensifies, the height of the ripples gradually increases. When the maximum oscillatory velocity of water particles near the bed surface is more than twice the velocity for incipient motion, vortexes suddenly appear on the bed. The flow patterns near the bed at various stages are shown in Fig. 16.28. Dynamic equilibrium is achieved through the vortex action, and the ripples maintain a definite pattern on the bed. Bagnold named this type of ripples "vortex ripples"; they have almost the same shape as do the rolling grain ripples, but their characteristics are different.

If the amplitude of the water particle oscillation then reduces to about one-third of the length of the ripple, the vortex ripple changes into a new type. Transverse bridges form between two crests and their pattern resembles that of a brick wall, as shown in Fig. 16.55. Bagnold called this type of ripples "brick wall ripples." The same phenomenon took place in Manohar's experiment with fine sediment particles (smaller than 0.28 mm) and small horizontal amplitudes for the water waves.

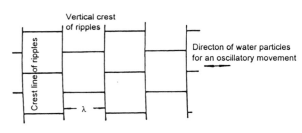

Fig. 16.55 diagram for "brick wall pples"

If the oscillatory movement of waves increases further, large amounts of sediment go into suspension, the height of the ripples diminishes, and their length increases. Finally, the ripples disappear and the bed surface becomes flat. The alteration of ripples and a flat bed is like that for alluvial rivers.

Experiments show that ripples only appear if the boundary layer is turbulent. Before the laminar layer in deep water loses its stability, either no ripples exist or they develop incompletely and have irregular shapes. Hence, the formation of ripples on sea beds may have some connection with the stability of boundary layer.

Until a complete understanding of the mechanism of formation of ripples on sea beds is achieved, the wave parameters controlling bed load movement on sea beds can be used to correlate the processes in the development of sea-bed ripples. The same flow parameters control both ripple development and bed load movement in alluvial rivers (as discussed in Chapter 6). For example, Manohar found from his experiments that a parameter

$$\psi_w = \left(\frac{\rho_s - \rho}{\rho} g \right)^{0.4} \frac{v^{0.2} D^{0.2}}{Na_0} \tag{16.48}$$

is related to both the initiation of movement of sediment particles and the intensity of bed load motion due to wave action. Thus, the various stages of ripple development may correspond to the various values of ψ_w , and these have been determined from experiments, as shown in Table 16.6 [9].

Table 16.6 Values of that correspond to various stages of ripple development

Stages of ripple development	ψ_w
First emergence of ripples	0.054
Gradual increase of ripple steepness	0.040-0.054
Gradual decrease of ripple steepness	0.024-0.040
Disappear of ripples	0.024

Ripple crests are generally parallel to the crests of the wave that cause them. If the longshore current is strong, however, it can create longitudinal ripples with crests that are perpendicular to the shore [61].

16.7.1.2 Characteristics of ripples on sea beds

Divers working on the California coast in the US observed ripples on the sandy sea bed. When the horizontal velocity of oscillatory movement near the bed was larger than 10 cm/sec, ripples occurred even at places where the water was as deep as 50 m. Observations along the Black Sea coast in the former USSR revealed that ripples were 90 to 110 cm long and 7 to 10 cm high in water 14 m deep. If the deposition of fine material in the troughs had been excluded, their height would have been 15 cm.

As a preliminary finding, the wave length of ripples, λ, is likely to be correlated with the horizontal amplitude of the water particles near bed, a_0. Fig. 16.56 shows the observed relationship between λ and 2 a_0[82]. It indicates that results for laboratory and field data are different. The wave periods obtained in laboratories correspond to those at lake shore areas or in sheltered coastal areas where the wind fetch is limited and the water depth is small. In those areas

$$\lambda = 1.6a_0 \qquad\qquad (16.104)$$

Nearer the shore, the depth is smaller and the values of a_0 are larger. In these regions, the ripples are longer and their wave length is greater. From field studies of waves with long periods, the data for different groups of particles follow different trends. Beyond some limiting values of λ , the wave length decreases with

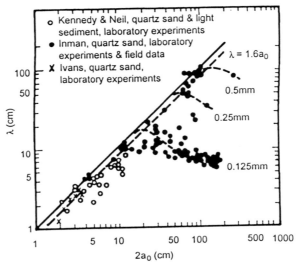

Fig. 16.56 Relationship between wave length of ripples and horizontal amplitude of water
(after Inman, D.L.)

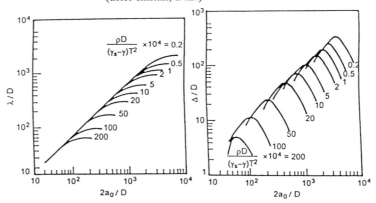

Fig. 16.57 Relationship of ripple forms and the parameters
$\rho D/(\gamma_s-\gamma)T^2$, $2a_0/D$ (after Mogridge, G.R., and J.W. Kamphuis)

788

increasing a_0. Finally, if a_0 is quite large and the bed shear stress exceeds some limit, the ripples disappear.

According to Mogridge and Kamphuis, the form of ripples depends on four dimensionless parameters [83]:

$$\frac{\lambda}{D}, \frac{\Delta}{D} = f\left(\frac{\gamma_s D^3}{\rho v^2}, \frac{\rho D}{(\gamma_s - \gamma)T^2}, \frac{\rho_s}{\rho}, \frac{2a_0}{D}\right) \qquad (16.105)$$

Viscosity probably has little influence. Ripples formed by light materials are longer and higher than those formed by natural sediment. However, the steepness of ripples, Δ/λ, is not likely to depend on the specific weight of the sediment. Thus, the principal parameters are $\rho D/(\gamma_s-\gamma)T^2$ and a_0/D. The former reflects the effect of wave period, and the latter the relative importance of the frictional and inertia forces in the oscillatory movement [83]. Fig. 16.57 shows the relationship between λ/D (Δ/D) and a_0/D for various values of $\rho D/(\gamma_s-\gamma)T^2$ [83].

Further analysis shows that the two parameters in Fig. 16.57 can be combined

$$\pi^2 \left(\frac{2a_0}{D}\right)^2 \frac{\rho D}{(\gamma_s - \gamma)T^2} = \frac{\rho u_m^2}{(\gamma_s - \gamma)D}$$

The discussion of the initiation of motion for sediment particles and of bed load movement indicates that this parameter reflects the intensity of the water movement. Fig. 16.58 shows the relationship between the steepness of ripples, Δ/λ, and the parameter, $(\rho u_m^2)/((\gamma_s-\gamma)D)$ [84]. If the latter is smaller than 40, Δ/λ has a constant value of 0.15. As it increases above this value, Δ/λ decreases sharply; if it is larger than 240, ripples do not form.

16.7.1.3 Movement of ripples

In addition to the oscillatory movement associated with the wave oscillation, unidirectional movement of ripples also occurs in shallow water. Scott observed in wave tank studies that null points or neutral lines also exist for the movement of ripples. On the sea side of the neutral line, ripples move slowly towards the shore because the amount of sediment transported landward across a ridge as the wave crest passes is larger than that of the suspended sediment transported seaward as the wave trough passes. The ripples are asymmetrical with a nearshore slope that is larger than the offshore one, as shown in Fig. 16.30 [47]. Outside the neutral line the situation is just opposite. Only in the vicinity of the neutral line are the ripples nearly symmetrical. Inside the neutral line, the speed of ripple movement increases landward, reaching its maximum near the breaker zone. Inside the breakers, ripples do not form. Outside the neutral point the speed of ripple movement increases at first

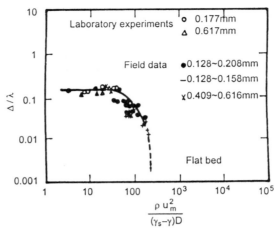

Fig. 16.58 Relationship between the steepness of ripples on beds and the parameter $(\rho u_m^2)/((\gamma_s-\gamma)D)$ (after Dingler, J.R., and D.L. Inman)

in the seaward direction and, then diminishes gradually to zero. According to limited data, the speed of ripple movement is probably proportional to the velocity of mass transport near the bed (Eq. 16.96).

Wave-formed ripples differ in some respects from ripples that are formed in unidirectional flows. Therefore, bed load rates calculated on the basis of the volume and speed of movement for ripples may lead to results that are quite incorrect. For example, observations in wave tanks show that even though ripples move landward continuously, the beach profile remains unchanged; hence, the net landward transport rate is zero. During one wave period, sediment moves landward part of the time and seaward the rest of the time. Although the amount of landward movement is larger than the seaward movement, and the ripples seem to move landward, the displacement of landward movement is much smaller than that of seaward movement (in one set of Scott's experiments, the ratio was only 1:25). Thus, the product of volume and speed of sediment transport in the two directions can be equal.

16.7.2 Submarine bars

16.7.2.1 Processes of formation of submarine bars

Due to wave action, one to three parallel bars with troughs between them can form along the coast as shown in Fig. 16.16. Such submarine bars are often located a little way outside the breaker zone.

Explanations of the causes for the formation of submarine bars differ. Their formation may be closely related to plunging breakers [84]. Two factors are involved. The water depth over the bar crest is closely related to the wave height (this point is discussed in section 16.7.2.2). When strong storms occur on the sea and short period waves reach the shoreline, submarine bars are destroyed. They can even disappear as the breaking waves change from plunging breakers to spilling breakers.

Plunging breakers scour a deep trough on the sea bed. Part of the sediment from the trough is transported along the coast by longshore currents, and part is transported seaward. At the same time, outside the breaker zone, the sediment on the sea beds moves landward. Hence, near the breakers, sediment being transported in the two opposite directions meets and forms the submarine bar. If the breaking waves are strong enough to move sediment from the open sea, the submarine bar reaches its limiting height.

On steep pebble beaches, the breakers form near the shore; then the space in the zone of breaking waves is not large enough for the full development of a longshore current. Consequently, the development of the submarine bar and trough is also restricted. Comparison of such a beach profile with a more typical profile is shown in Fig. 16.59. Only if the waves are steep enough, can submarine bars occur on steep pebbly beaches. For example, submarine bars appear on beaches with a slope of 1 on 20 if the wave is

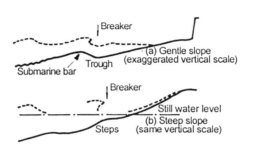

Fig. 16.59 Comparison of beach profiles with different slopes (after King, C.A.M.)

steeper than 0.0115; however, if the slope is 1 on 5, submarine bars appear only if the wave is steeper than 0.034 [86].

The formation of a series of submarine bars can be caused by various factors. If waves reform after breaking and then break again nearer the shore (Fig. 16.16), a series of submarine bars and troughs form at each of the breaker zones. Nearer the shore, the breaker height is smaller; therefore, the submarine bars and troughs are smaller. When the sea level rises during flood tides, the breaker zone moves landward; as it moves, a series of submarine bars can form. Sometimes one submarine bar is formed in the deeper area when strong storms occur; subsequently ordinary waves will break at places closer to the coast and a corresponding submarine bar forms there. At the same time, the submarine bar formed during the strong storm can still exist because ordinary waves are not strong enough to destroy it. In this way, two parallel submarine bars with different sizes can exist simultaneously.

The formation of submarine bars is closely related to wave breaking; hence waves passing over a submarine bar during summer often do not break because the wave height is small. For this situation, the flow near the bed is landward,

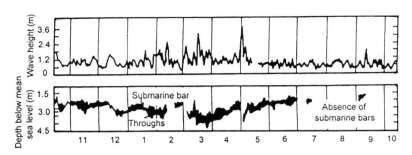

Fig. 16.60 Measured heights of waves and submarine bars along the coast of California, U.S. (after King, C.A.M.)

and it gradually destroys the submarine bar. Thus, the beach profile is gradually transformed from a winter profile to a summer profile (Fig. 16.45).

16.7.2.2 Characteristics of submarine bars

As the location of submarine bars depends directly on the location of the breakers, the submarine bar shifts seaward with the displacement of the breakers when the wave height increases; i.e., the water depth over the submarine bar should be proportional to the wave height there. Fig. 16.60 is a hydrograph of the wave height and the height of submarine bars and troughs observed along the coast of California, USA, and they tend to support this explanation.

Keulegan studied the characteristics of submarine bars in wave tanks and obtained the following simple geometrical relationships [87]:

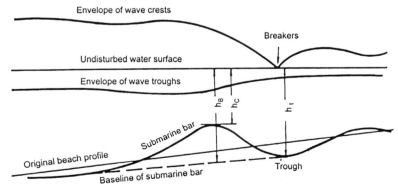

Fig. 16.61 Sketch of submarine bar and trough (after Keulegan, G.H.)

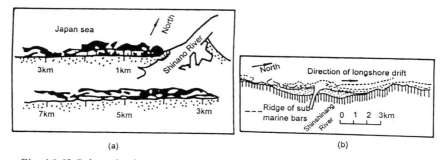

(a)   (b)

Fig. 16.62 Submarine bars in plan (after Masashi, Hom-ma, and C. Sonu)

$$\frac{h_c}{h_B} = 0.58$$

$$\frac{h_t}{h_c} = 1.69$$

These depths are defined in Fig. 16.61. Shepard analysed a large quantity of field data and obtained the values

$$h_c / h_B = 0.63$$

$$h_t/h_c = 1.30$$

The mean sea level was used as the datum. Shepard's values are smaller than the laboratory ones because submarine bars and troughs in nature are much larger than those in wave tanks [85].

Two basic patterns exist for the arrangement of submarine bars. In one pattern the bars lie parallel to the coast and form in a continuous line over long stretches. In the other, the bars form in a series of crescents. Both are shown in Fig. 16.62a. The crescent bars only form on gentle beaches in shallow water near a source of sediment. Along the coast of Japan, crescent bars exist on beaches with slopes flatter than 0.02. Straight submarine bars often form downstream of a sediment source if strong unidirectional longshore drifts occur. Crescent bars often occur upstream of a sediment source if the sediment moves within a closed system so that no sediment is carried outside this area, as shown in Fig. 16.62 [88].

16.7.2.3 Movement of submarine bars

The submarine bar is a stable accumulation of sediment. If the flow conditions remain constant, the location and morphological characteristics of submarine bars also remain unchanged. If the characteristics of the waves change, the location and morphological characteristics of submarine bars adjust in ways that correspond to changes in the breakers.

If the wave period remains constant and the wave height increases, the newly formed waves break over the crest of the submarine bar and cause it to decrease in size. After a period of time, a new series of submarine bars gradually forms further out to sea as shown in Fig. 16.63a. Along the coast of North Africa, a submarine bar shifted from 170 m to 205 m off the coast in two days due to the action of strong waves. If, on the contrary, the wave height reduces significantly, the original submarine bar moves landward and fills in the deep trough; at the same time, a new lower submarine bar forms inshore as shown in Fig. 16.63b. Fig. 16.64 shows the variation of Crane beach profiles on the east coast of USA from May to September 1967; it also illustrates the formation and displacement of submarine bars over a period of several months [89].

Along tidal sea coasts, the effect of fluctuations in the water stage on submarine bars is like that due to changes in wave height. Submarine bars formed during periods of high tides are scoured and shifted seaward in periods of low tides. If the

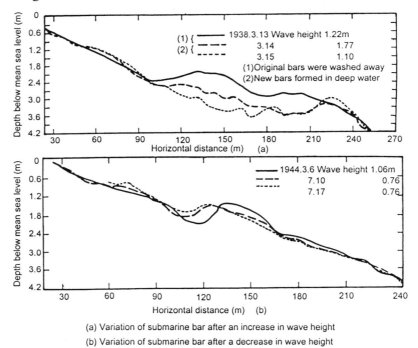

(a) Variation of submarine bar after an increase in wave height

(b) Variation of submarine bar after a decrease in wave height

Fig. 16.63 Variation of location and characteristics of submarines bars along with changes in waves

water stage fluctuates constantly and significantly, typical submarine bars and troughs hardly form at all. A viewpoint held in the past was that submarine bars were the accumulation of sediment along tideless sea coasts only, and that they did not exist along sea coasts with tides. Such a viewpoint is not correct. Along tidal coasts,

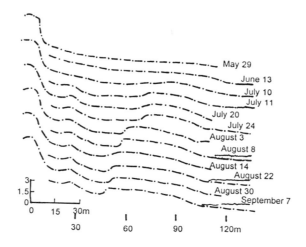

Fig. 16.64 Variation of Crane beach profile, Massachusetts U.S.,
from May to September 1967 (after Hayes, M.O.)

such as along the aforementioned California coast, typical submarine bars do occur; however, their shapes are not as protrusive as are those on tideless coasts. Moreover, parts of the troughs may be exposed during neap tides and concentrated flows can then occur in the trough; in this way, breaches may be eroded in the submarine bar so that it becomes broken in plan.

In addition to ripples and submarine bars, even more complex sediment accumulations can be induced by waves. Details of such occurrences are not included herein.

REFERENCES

[1] Vincent, G. E. "Contribution to the Study of Sediment Transport on A Horizontal Bed Due to Wave Action." *Proceedings, 6th Conference of Coastal Engineering*, 1958, pp. 326-355.
[2] Li, Yucheng. "Wave Flow Structure and Its Influence on Sediment Movement in Shallow Water." *Journal of Dalian Institute of Technology*, No. 2, 1963, pp. 31-43. (in Chinese).
[3] Mogridge, C. R. "Testing Sediment Movement Due to Wave Action." *Journal of Hydraulic Division, Proceedings*, ASCE, Vol. 96, HY7, 1970, pp. 1587-1606.
[4] Silvester, R. "Engineering Aspects of Coastal Sediment Movement." *Journal, Waterways and Harbor Division, Proceeding*, ASCE, Vol. 85, No. WW3 Pt.1, 1959, pp.11-39.
[5] Munk, W. H. "The Solitary Wave Theory and Its Application to Surf Problems." *Annals of the New York Academy of Science*, Vol. 51, 1949.
[6] Daily, J. W., and S. C. Stephan, Jr. "Characteristics of the Solitary Wave." *Transaction*, ASCE, Vol. 118, 1953, pp. 575-587.
[7] Bagnold, R.A. "Motion of Waves in Shallow Water and Interaction Between Waves and Sand Bottom." *Proceedings, Royal Society*, London, Series A, Vol. 187, 1946, pp. 1-15.

[8] Li, Huon. "Stability of Oscillatory Laminar Flow Along A Wall." *Tech. Mem. No.47*, U. S. Beach Erosion Board, 1954, p. 48.

[9] Manohar, M. "Mechanics of Bottom Sediment Movement Due to Wave Action." *Tech. Mem. No.75*, U. S. Beach Erosion Board, 1955.

[10] Collins, J. J. "Inception of Turbulence at the Bed Under Periodic Gravity Waves." *Journal Geophys. Res.*, Vol. 68, No. 21, 1963, pp. 6007-6014.

[11] Kalkanis, G. "Transportation of Bed Material Due to Wave Action." *Tech. Mem. No.2*, U. S. Coastal Engineering Research Center, 1964, p. 38.

[12] Jonson, I. G. "Waves Boundary Layers and Friction Factors." *Proceedings, 10th Congress of Coastal Engineering*, Vol. 1, 1966, pp. 127-148.

[13] Kamphuis, J. W. "Friction Factor Under Oscillatory Waves." *Journal, Waterways and Harbor Division, Proceedings*, ASCE, Vol. 101, No. WW2, 1975, pp. 135-144.

[14] Einstein, H. A. "Sediment Transport by Wave Action." *Proceedings, 13th Congress of Coastal Engineering*, Vol. 2 1972, pp. 933-952.

[15] Teleki, P.G. "Wave Boundary Layers and Their Relation to Sediment Transport." In *Shelf Sediment Transport, Process and Pattern*. Edited by D. I. J. Swift, D. B. Duane and O.H. Pilkey, Dowden, Hutchinson and Ross, Inc. 1972, pp. 21-59.

[16] Kajiura, K. "A Model of the Bottom Boundary Layer in Water Waves." *Bulletin*, Earthquake Research Institute, Vol. 46, 1968, pp. 75-123.

[17] Putnam, J. A., and J. W. Johnson. "The Dissipation of Wave Energy by Bottom Friction." *Transaction*, Amer. Geophys. Union, Vol. 30, No. 1, 1949, pp. 67-74.

[18] Putman, J. A. "Loss of Wave Energy Due to Percolation in A Permeable Sea Bottom." *Transaction*, Amer. Geophys. Union, Vol. 30, No. 3, 1949, pp. 349-356.

[19] Iwgaki, Y., and H. Noda. "Laboratory Study of Scale Effects in Two Dimensional Beach Processes." *Proceedings, 8th Congress on Coastal Engineering*, 1963, pp. 194-210.

[20] McGowan, J. "On the Highest Wave of Permanent Type." *The London, Edinburgh and Dublin Phil, Magazine and J. Sci.*, Vol. 38, 1894, p. 351.

[21] Michell, J. H. "The Highest Waves in Water." *Phil. Mag.*, Vol. 36, 1893, pp. 430-437.

[22] Suquet, F. "Experimental Study of Braking Waves." *La Houille Blanche*, Vol. 5, No. 3.

[23] Russell, R. C. H., and D. H. MacMillan. *Waves and Tides*. Phil. Library, N. Y., 1953, p. 348.

[24] Dyhr-Nielsen, M., and T. Sorensen. "Some Sand Transport Phenomena on Coasts with Bars." *Proc., 12th Congress on Coastal Engineering*, Vol. 2, 1970, pp. 855-865.

[25] Komar, P.D. *Beach Processes and Sedimentation*. Pretice-Hall Inc., 1976, p. 429.

[26] Bagnold, R. A. "Sand Movement by Waves, Some Small Scale Experiments with Sand of Very Low Density." *J. Inst. Civil Engrs.*, Vol. 27, No. 4, 1947, pp. 447-469.

[27] Longuet-Higgins, M.S. "Mass Transport in Water Waves." *Phil. Trans. Royal Soc. London*, Ser.A, Vol. 245, No. 903, 1953, pp. 535-581.

[28] Russell, R. C. H., and J. D. C. Osoro. "An Experimental Investigation of Drift Profiles in A Closed Channel." *Proc. 6th Congress on Coastal Engineering*, 1958, pp. 171-183.

[29] Longuet-Higgins, M. S. "The Mechanics of the Boundary Layer Near the Bottom in A Progressive Wave." *Proc., 6th Congress on Coastal Engineering*, 1958, pp. 184-193.

[30] Bijker, E. W., J. P. Th. Kalkwijk, and T. Pieeters. "Mass Transport on Gravity Waves." *Proc., 14th Congress on Coastal Engineering*, Vol. 1, 1974, pp. 447-465.

[31] Galvin, C. J. "Longshore Current Velocity, A Review of Theory and Data." *Rev. in Geophys.*, Vol. 5, No. 3, 1967, pp. 287-304.

[32] Putman, J. A., W. H. Munk, and M. A, Taylor. "The Prediction of Longshore Currents." *Trans.*, Amer. Geophys. Union, Vol. 30, No. 3, 1949, pp. 337-345.

[33] Longuet-Higgins, M. S. "Longshore Currents Generated by Obliquely Incident Sea Waves." *J. Geophys. Res.*, Vol. 75, No. 33, 1970, pp. 6775-6789.

[34] Bowen, A. J. "Rip Currents, 1. Theoretical Investigation." *J. Geophys. Res.*, Vol. 74, 1969, pp. 5467-5478.

[35] Bowen, A. J., and D. L. Inman. "Rip Currents, 2. Laboratory and Field Observations." *J. Geophys*, Res., Vol. 74, 1969, pp. 5479-5490.

[36] Shepard, F. P., and D. L. Inman. "Nearshore Circulation Related to Bottom Topography and Wave Refraction." *Trans.*, Amer. Geophys. Union, Vol. 31, No. 2, 1950, pp. 196-212.

[37] U. S. Army Coastal Engineering Research Center. *Shore Protection Manual*. 2nd Ed., Vol. 1, 1975, p. 495.

[38] Taylor, G. "Note on R. A. Bagnold's Empirical Formula for the Critical Water Motion Corresponding with the First Disturbance of Grains on A Flat Surface." *Proc., Royal Soc. London*, Ser. A, Vol. 187, 1946, pp. 16-18.

[39] Eagleson, P. S., and R. G. Dean. "Wave Induced Motion of Bottom Sediment Particles." *J. Hyd. Div., Proc.*, ASCE, Vol. 85, No. HY10, 1959, pp. 53-79.

[40] Eagleson, P. S., B. Glenne, and J. A. Dracup. "Equilibrium Characteristics of Sand Beaches." *J. Hyd. Div., Proc.*, ASCE, Vol. 89, No. HY1, 1963, pp. 35-57.

[41] Goddet, J. "Etude du Debit d'Entrainment des Materiaux Mobiles sous l'Action de la Houle." *La Houille Blanche*, No. 2, 1960, pp. 122-135.

[42] Rance, P. J., and N. F. Warren. "The Threshold Movement of Coarse Material in Oscillatory Flow." *Proc. 11th Congress on Coastal Engineering*, 1968, pp. 487-491.

[43] Komar, P. D., and M. C. Miller. "Sediment Threshold Under Oscillatory Waves." *Proc., 14th Congress on Coastal Engineering*, Vol. 2, 1947, pp. 756-775.

[44] Sato, S., T. Ijima, and N. Tanake. "A Study of Critical Depth and Mode of Sand Movement Using Radioactive Glass Sand." *Proc., 8th Congress Coastal Engineering*, 1964, pp. 304-323.

[45] Hammond, T. M., and M. B, Collins. "On the Threshold of Transport of Sand-Sized Sediment under the Combined Influence of Unidirectional and Oscillatory Flow." *Sedimentology*, Vol. 26, No. 6, 1979, pp. 795-812.

[46] Bagnold, R. A. "Beach Formation by Waves, Some Model Experiments in A Wave Tank." *J. Inst. Civil Engrs.*, No. 1, 1940, pp. 27-52.

[47] Scott. T. "Sand Movement by Waves." *Tech. Mem. No. 48*, U. S. Beach Erosion Board, 1954, p. 37.

[48] Einstein, H. A. "A Basic Description of Sediment Transport on Beaches." In *Waves on Beaches and Resulting Sediment Transport*. Edited by R. E. Meyer, Academic Press, 1972, pp. 53-93.

[49] Brown, C. B. "Sediment Transportation." Chapter 12, *Engineering Hydraulics*. Edited by H. Rouse, John Wiley & Sons, Inc., 1950.

[50] Madsen, O. S., and W. D. Grant. "Quantitative Description of Sediment Transport by Waves." *Proc., 15th Congress on Coastal Engineering*, Vol. 2, 1976, pp. 1093-1112.

[51] Skeath, J. F. A. "Measurements of Bed Load in Oscillatory Flow." *J. Waterways and Harbor Div., Proc.*, ASCE, Vol. 104, No. WW3, 1978, pp. 291-307.

[52] Nakato, T. et al. "Wave Entrainment of Sediment From Rippled Bed." *J. Waterways and Harbor Div., Proc.*, ASCE, Vol. 103, No. WW1, 1977, pp. 83-99.

[53] Hom-ma, M., K. Horikawa, and R. Kajima. "A Study on Suspended Sediment Due to Wave Action." *Coastal Engineering in Japan*, Vol. 8, 1965, pp. 85-103.

[54] Hattori, M. "The Mechanics of Suspended Sediment Due to Standing Waves." *Coastal Engineering in Japan*, Vol. 12, 1969.

[55] Kennedy, J. F., and F. A. Locher. "Sediment Suspension by Water Waves." In *Waves on Beaches and Resulting Sediment Transport*, Edited by R. E. Meyer, Academic Press, 1972, pp. 249-295.

[56] Liu, Jiaju. "Experimental Study of Movement of Xingang Mud Due to Wave Action." *J. Xingang Harbor Deposition Study*, No. 1, 1963, pp. 12-27. (in Chinese)

[57] Fleming, C. A., and J. N. Hunt. "Application of A Sediment Transport Model." *Proc., 15th Congress on Coastal Engineering*, Vol. 2, 1976, pp. 1184-1202.

[58] Das, M. M. "Suspended Sediment and Longshore Sediment Transport Data Review." *Proc., 13th Congress on Coastal Engineering*, Vol. 2, 1972, pp. 1027-1048.

[59] Bagnold. R. A., and D. L. Inman. "Beach and Nearshore Precesses." In *The Sea*. Edited by M. N. Hill, Vol. 3, Interscience Publisher, 1963, pp. 507-553.

[60] Huang, Jianwei. "Review of Calculation of Longshore Drift along Sandy Beaches." *Information on Water Resources and Water Transportation*, No. 2 1976, pp. 68-89. (in Chinese)

[61] Zenkovitch, B. P. "Study of Nearshore Sediment Movement." *Oceanologia et Limnologia Sinica*, Vol.1, No.2, 1955, pp.185-208. (in Chinese)

[62] Brenninkmeyer, B. M. "Mode and Period of Sand Transport in the Surf Zone." *Proc., 14th Congress on Coastal Engineering*, 1974, pp.812-827.

[63] Komar, P. D. and L. Inman, "Longshore Sand Transport on Beaches." *J. Geophys. Res.*, Vol.75, No.30, 1970, pp.5914-5927.

[64] Longuet-Higgins, M. S. "Recent Progress in the Study of Longshore Currents." in *Waves on Beaches and Resulting Sediment Transport, edited* by R. E. Meyer. Academic Press, 1972, pp.203-248.

[65] Komar, P. D. "The Mechanics of Sand Transport on Beaches." *J. Geophys. Res.*, Vol.76, No.3, 1971, pp. 713-721.

[66] Inman, D. L., and A. J. Bowen. "Flume Experiments on Sand Transport by Waves and Currents." *Proc., 8th Congress on Coastal Engineering*, 1963, pp.137-150.

[67] Johnson, J. W. "Scale Effects in Hydraulic Models Involving Wave Motion." *Trans., Amer. Geophys. Union*, Vol.30, No.4, 1949, pp.517-525.

[68] Dean, R. G. "Heuristic Models of Sand Transport in the Surf Zone." *Conf. on Engin. Dynamics in the Surf Zone*, Sidney, Australia, 1973, p.7.

[69] Coastal Engineering Research Center. *Shore Protection Manual*. Vol.1, 1975, pp. 4-78 to 4-84.

[70] Dalrymple, R. A., and W. W. Thompson. "Study of Equilibrium Beach Profiles." *Proc., 15th Congress on Coastal Engineering*, Vol.2, 1976, pp.1277-1296.

[71] Inglis, C. C., and F. H. Allen, "The Regimen of the Thames Estuary as Affected by Currents, Salinities and River Flow." *Proc., Inst. Civil Engrs.*, Vol.7, 1957, pp.827-876.

[72] Huang Jianwei, and Yuling Feng. "Preliminary Study of Data on Mud in Navigation Channel of Xingang Harbor." *J. Xingang Harbor Deposition Study*, No.1, 1963, pp.74-84, (in Chinese).

[73] Zhao Jinsheng, and Quan Chen. "Mud Movement and the Stability of Its Interface Under Combined Action of Waves and Gravity." *Thesis of Graduate Student*, Tianjin University, 1964, p.62. (in Chinese)

[74] Zhao Jinsheng, Zidan Zhao, and Zude Cao. "Preliminary Study of Mud Movement Due to Wave Action in Xingang Harbor." *J. Xingang Harbor Depostion Study*, No.1, 1963, pp.40-59. (in Chinese)

[75] Lhermitte, P. "Mouvement des Materiaux de Fond sous l'Action de la Houle." *Annals des Ponts et Chaussees*, 131 Annde, No. 3, 1961, pp.357-412.

[76] Migniot, C. "Action des Courants de la Houle et du Vent sur les Sediments." *La Houille Blanche*, No., 1977, pp.9-97.

[77] Li, Haolin. "Experimental Study of Characteristics of Mud Movement in Xingang Harbor." *J. Xingang Harbor Deposition Study*, No.1, 1963, pp.28-39, (in Chinese).

[78] Delft Hydraulics Laboratory. *Demerara Coastal Investigation.* The Netherlands, 1962, p.240.

[79] Tubman, N. W., and J. N. Suhayda. "Wave Action and Bottom Movements in Fine Sediments." *Proc., 13th Congress on Coastal Engineering*, Vol.2, 1976, pp.1168-1183.

[80] Kaneko, A., and H. Honji. "Initiation of Ripple Marks Under Oscillating Water." *Sedimentology*, Vol.26, No.1, 1979, pp.101-113.

[81] Vasseur, P., and R. G. Cox. "The Lateral Migration of Spherical Particles Sedimenting in A Stagnant Bounded Fluid." *J. Fluid Mech.*, Vol.80, 1977, pp.561-591.

[82] Inman, D. L. "Wave Generated Ripples in Nearshore Sands." *Tech. Mem. No. 100*, U. S. Beach Erosion Board, 1957, p.67.

[83] Mogridge, G, R., and J. W. Kamphuis. "Experiment on Bed Form Generation by Wave Action." Proc., *13th Congress on Coastal Engineering*, Vol.2, 1972, pp.1123-1142.

[84] Dringler, J. R., and D. L. Inman. "Wave Formed Ripples in Nearshore Sands." *Proc. 15th Congress on Coastal Engineering*, Vol.2, 1976, pp.2109-2126.

[85] Shepard, F. P. "Longshore Bars and Longshore Troughs." *Tech. Mem. No.15*, U. S. Beach Erosion Board, 1950, p.32.

[86] King, C. A. M. *Beaches and Coasts*, Edward Arnold Publisher, 1959, p.403.

[87] Keulegan, G. H. "An Experimental Study of Submarine Sand Bars." *Tech. Report No. 3*, U. S. Beach Erosion Board, 1948, p.40.

[88] Masashi, Hom-ma and C. Sonu. "Rhythmic Pattern of Longshore Bars Related to Sediment Characteristics." *Proc.*, *8th Congress on Coastal Engineering*, 1963, pp.248-278.

[89] Hayes, M. O. "Forms of Sediment Accumulation in the Beach Zone." in *Waves on Beaches and Resulting Sediment Transport*, edited by R. E. Meyer. Academic Press, 1972, pp.297-356.

CHAPTER 17

HYDROTRANSPORT OF SOLID MATERIAL IN PIPELINES

In recent years, hydrotransport of solid material in pipelines has been widely used in industry. The materials transported include sediment, coal, mining ore, syrup, paper pulp, sewage, and raw chemicals. Distances of transport can exceed 1000 m. Some examples of hydrotransport stations throughout the world are listed in Table 17.0 [1]. Hydrotransport is not only economical, it also causes little pollution. Therefore its range of application is becoming ever wider; it may well become one of the primary transport processes of the future.

Table 17.0 Large hydrotransport systems throughout the world (after Wasp, E.J.; J.P. Kenny, and B.L. Gandhi)

Name	Length(km)	Diameter(mm)	Transport capacity (million ton/year)	Year of operation
Coal				
Consolidation	174	254	1.3	1975
Black Messa	440	457	4.8	1970
ETSI	1640	965	25	1979
Orton	290	610	10	1981
Fine iron ore				
Savage River	85	229	2.3	1967
Pena Colorada	45	203	1.8	1974
Sierra, Grande	32	203	2.1	1976
Samarco	410	508	12	1977
Fine copper ore				
Bougainville	27	152	1.0	1972
West Irian	111	102	0.3	1972
Limestone				
Rubey	92	254	1.7	1964
Calaveras	27	178	1.5	1971

The operation of hydrotransport systems has evolved steadily with the accumulation of practical experience. In its early days, low concentrations of solid material were used to avoid blockage in pipelines. Later, it was found that coarse particles can be transported at high concentrations if a certain amount of fine particles are added to form a homogeneous slurry with a high viscosity. In the 1970's, a further discovery was that coarse particles can be transported at high concentrations even if no fine particles are added. In order to reduce the resistance to the flow, various substances have been added to the flows to test their effectiveness in drag reduction. This chapter presents the mechanics of hydrotransport. Some aspects of the technology involved in the construction of pumping stations used for this purpose are outside the scope of this book.

17.1 MODES OF SEDIMENT MOTION IN PIPELINES AND THE CLASSIFICATION OF FLOW PATTERNS

17.1.1 Basic modes of sediment motion

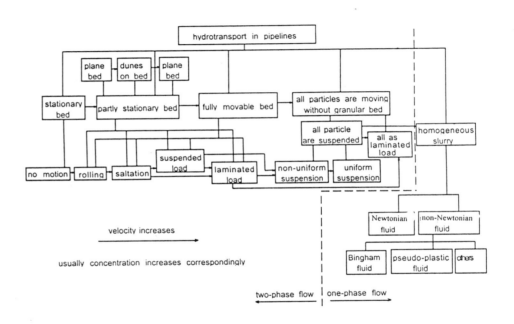

Fig. 17.1 Basic modes of sediment motion in pipelines

Basic modes of sediment motion in pipelines at different velocities and concentrations are shown in Fig.17.1. The velocities and concentrations for those on the left side of the diagram are lower than for those on the right. Like sediment particles moving in open channels, solid particles in pipelines move by rolling, saltation, and suspension, and as laminated load. If the concentrations are high, particularly those of fine particles (<0.01 mm), no sorting occurs even if the mixture is not moving, as discussed in Chapter 13. These mixtures are often homogeneous slurries that sometimes behave as Newtonian fluids, but in most cases they are non-Newtonian fluids. Usually a slurry is either a Bingham fluid or a pseudo-plastic fluid.

17.1.2 Classification of modes for heterogeneous flows in pipelines

With velocities changing from high to low, sediment motion in pipelines changes as follows:

1. Uniform suspension

With high velocities, sediment particles are uniformly distributed over a cross section because of the mixing due to intense turbulence. A nearly uniform suspension is called a pseudo-uniform suspension.

2. Non-uniform suspension

With lower velocities and correspondingly weaker turbulence, sediment particles are not uniformly distributed over the cross section; the concentration then increases toward the bed.

3. Bed load in contact with the bottom

With somewhat lower velocities, sediment particles move as bed load, sliding, rolling, or jumping along the bottom. Under certain conditions, laminated load may occur.

4. Deposition and stationary bed

If the velocities are lower than some critical value, particles settle out and form a stationary bed. At the beginning the settlement is unstable; sometimes particles deposit on the bed and form a series of ripples and sometimes they are washed away by the flow. If the velocities are still lower, the deposits on the bed have a stable configuration. The flow becomes one with a movable bed, and it is similar to an alluvial bed in open channel flow. ·The bed surface may be either plane or with ripples.

5. Blockage of pipeline

If the velocities are still lower, deposits accumulate on the bed, and finally they block the pipeline.

Extremely high concentrations entirely composed of coarse particles tend to damp turbulence, and the weight of the particles is then supported by a dispersive force caused by shear. In such cases particles are rather uniformly distributed over the cross section. The result is the laminated load described in Chapter 5. Such a phenomenon was not classified as an independent pattern until the late 1970s when it was observed in some experiments.

The conditions for the various flow patterns for the hydrotransport of gravel in pipes with diameters of 25 mm and 152 mm are shown in Fig. 17.2a. That for the transport of glass beads with a diameter of 0.06 mm is shown in Fig. 17.2b [2]. Both parts of the figure are schematic; laminated load is not indicated on them as an independent flow pattern. The critical conditions for the transitions from one pattern to another are discussed in sections 17.2 and 17.3.

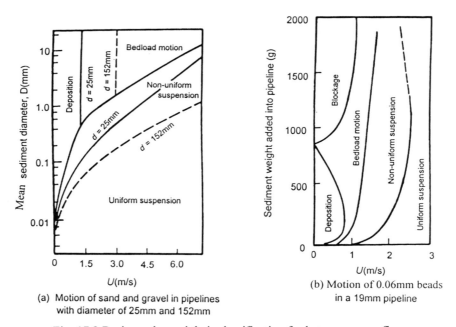

Fig. 17.2 Basic modes and their classification for heterogeneous flows
(after Newitt, D.M.; J.F. Richardson; M. Abbott; and R.B. Turtle)

17.1.3 A comparison of heterogeneous flows in pipelines and in open channels

Although some mechanical features for heterogeneous flows are common to both pipelines and open channels, their boundary conditions differ considerably.

1. For heterogeneous flows in pipelines, deposits on the bottom are usually disadvantageous; therefore the modes of sediment motion in pipes are usually the second and the third ones described in the preceding section, i.e., sediment motions over rigid beds. In contrast, sediment in fluvial rivers usually moves over granular and erodible beds. In pipes, all sediment particles can be suspended so that no particle moves as bed load.

2. In open channels, the driving force is the component of gravity acting in the direction of flow. Sediment increases the potential energy of the flow on the one hand, and it consumes part of the energy on the other. If the increase in energy is equal to that consumed, the existence of sediment does not increase the energy dissipation of the flow, and the flow is called an "automatic suspension." In pipes, the driving force

is the pressure gradient. The transport of sediment always increases the energy consumed by the flow in this case. The change of flow structure caused by the existence of sediment and its consequences introduces another kind of problem.

3. Because the flow in pipes is pressurized, phenomena related to a free surface cannot occur. For example, anti-dunes and sand waves cannot occur in a pipe, even if sediment does deposit on the bottom.

4. The scales of open channel flows, excepting these in small mountain streams, are usually much larger than the sediment they carry. However, the diameters of pipes are sometimes only a small multiple of the diameters of the particles they transport. Thus, large-scale turbulence in pipelines can be restricted to some degree.

5. The solid material transported in pipelines is often processed rather than natural. It may be a homogeneous slurry consisting only of fine particles. Such a case rarely occurs in a natural river.

In studies of the hydrotransport of solid materials, the basic concepts and theories described in other chapters should be applied. Besides, the characteristics of pipe flow should be considered.

17.2 MOTION OF HOMOGENEOUS SLURRIES IN PIPELINES

A homogeneous slurry is a kind of liquid that differs from water in both specific weight and viscosity. It may be either a Newtonian fluid or a non-Newtonian fluid. For a non-Newtonian fluid, the solid particles in it are not discrete but tend to form a distinctive structure. Such structured particles often settle at extremely low velocities, as described in Chapter 3. In an extreme case, the medium (water) moves through numerous tortuous capillaries among these structures. Hence, solid particles in a slurry will not deposit only if turbulence near the bed and a vertical exchange exist. Thomas conducted experiments that fully substantiated this phenomenon [3]. Thus, the critical velocity for the transition from laminar to turbulent flow possesses a significant meaning in the design of pipeline.

In this section the head loss for laminar and turbulent flow of a non-Newtonian slurry and the critical condition for the transition between them are discussed.

17.2.1 Laminar flow

17.2.1.1 Head loss [4]

For pipe flow, the rheological equation can be written in a general form

$$-\frac{du}{dr} = f(\tau) \tag{17.1}$$

in which τ is the shear stress at a distance r from the center of the pipe. The shear stress varies linearly from the center to the boundary, and one can write the following equations

$$\left.\begin{array}{c} \tau = \dfrac{r\Delta P}{2L} \\[2mm] \tau_w = \dfrac{R\Delta P}{2L} \end{array}\right\} \qquad (17.2)$$

in which τ_ω is the shear stress at the boundary, ΔP the pressure difference for two points a distance L apart, and R the radius of the pipe. Hence,

$$-\frac{du}{dr} = f(\tau_\omega \frac{r}{R})$$

Integrating, one obtains

$$u(r) = \int_r^R f(\tau_\omega \frac{r}{R}) dr$$

Under the action of the pressure difference ΔP, the discharge passing through the pipe Q is

$$Q = \int_0^R 2\pi r u(r) \, dr$$

or

$$Q = \pi \int_0^R r^2 f\left(\tau_\omega \frac{r}{R}\right) dr$$

Substituting $r = R\tau/\tau_\omega$ into the preceding equation, one obtains

$$\frac{Q}{\pi R^3} = \frac{1}{\tau_\omega^3} \int_0^{\tau_0} \tau^2 f(\tau) \, dr \qquad (17.3)$$

The function $f(\tau)$ can be determined from rheological measurements. Substituting $f(\tau)$ into Eq. (17.3), integrating numerically and using Eq. (17.2), one obtains a relationship between Q and ΔP. If the function $f(\tau)$ is simple enough, one can integrate Eq. (17.3) functionally.

The various rheological types of water-sediment mixture are described in Chapter 2. For each type, Eq. (17.3) is integrated in one of the following ways.

1. Newtonian fluid

The rheological equation for a Newtonian fluid is

$$f(\tau) = \frac{\tau}{\mu}$$

Substituting it into Eq. (17.3) and integrating it, one obtains

$$Q = \frac{\pi R^3 \tau_w}{4\mu}$$

Substituting Eq. (17.2) into the preceding equation, one obtains the conventional Poiseuille formula:

$$Q = \frac{\pi R^4 \Delta P}{8\mu L} \tag{17.4}$$

2. Bingham fluid

The rheological function for a Bingham fluid is a discontinuous function with the following form (Table 2.8)

$$\left.\begin{array}{ll} f(\tau) = 0 & 0 < \tau < \tau_B \\[2mm] f(\tau) = \dfrac{\tau - \tau_B}{\eta} & \tau_B < \tau < \tau_w \end{array}\right\} \tag{17.5}$$

The profiles for the shear stress and for the velocity are shown in Fig. 17.3. In the region $r < r_p$, the shear stress is less than the Bingham yield stress, no relative motion exists, and the fluid moves as a solid plug with a velocity u_p.

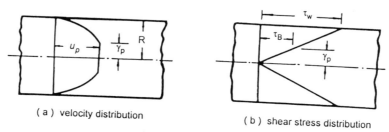

(a) velocity distribution   (b) shear stress distribution

Fig. 17.3 The profiles of shear stress and velocity for a Bingham fluid

Substituting Eq. (17.5) into Eq. (17.3), one obtains

$$\frac{Q}{\pi R^3} = \frac{1}{\eta \tau_w^3} \int_{\tau_B}^{\tau_w} \tau^2 (\tau - \tau_B) d\tau$$

Integrating it and substituting Eq. (17.2) into the resulting equation, one gets

$$Q = \frac{\pi R^4 \Delta P}{8L\eta}\left[1 - \frac{4}{3}\left(\frac{2L\tau_B}{R\Delta P}\right) + \frac{1}{3}\left(\frac{2L\tau_B}{R\Delta P}\right)^4\right] \tag{17.6}$$

807

Eq. (17.6) can be solved by trial-and-error. To simplify the calculation, Hedstrom worked out a diagram for Eq. (17.6) using dimensional analysis[5]. The pressure difference for a Bingham fluid in a pipe depends on the following factors:

$$\Delta P = f(\rho, \eta, \tau_B, d, U, L)$$

in which ρ is the density of the fluid, d diameter of the pipe, and U average velocity. From dimensional analysis, one obtains

$$f = \frac{d\Delta P / L}{\rho U^2 / 2} = f\left(\frac{\rho U d}{\eta}, \frac{\tau_B \rho d^2}{\eta^2}\right) \qquad (17.7)$$

in which f is the Darcy-Weisbach friction factor; it is four times the Fanning friction factor c_f (or λ) widely used in the chemical industry. Eq. (17.7) indicates that the friction factor is a function of two dimensionless parameters: the commonly used Reynolds number $Re = Ud/\nu$ and the Hedstrom number $\tau_B \rho d^2 / \eta^2$, usually written as He. Eq. (17.6) can be rewritten in the following dimensionless form:

$$\frac{1}{Re} = \frac{f}{64} - \frac{1}{6}\frac{He}{Re^2} + \frac{64}{3f^3}\frac{He^4}{Re^8} \qquad (17.8)$$

The derivation of Eq. (17.7) is not restricted to laminar flow, so that can also be used for turbulent flow. In the latter case, the relationship between the friction factor and the parameters Re and He cannot be obtained theoretically, but must be determined experimentally. Fig. 17.4 is a graphical representation of Eq. (17.8). It

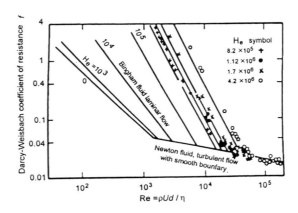

Fig. 17.4 Friction factor for a homogeneous slurry (Bingham fluid) flowing in pipelines

shows that because the Bingham yield stress is significant if the shear rate is low, a large Reynolds number means that the shear rate is high and the parameter He involving τ_B is not significant. Data from experiments by Dai et al., which were conducted in pipes of 100 mm and 200 mm carrying loess slurries (median diameter 0.0095 and 0.013 mm respectively), are also plotted in Fig. 17.4 [6]. The data follow closely the theoretical curves.

Some researchers replaced the rigidity of a Bingham fluid by an apparent viscosity and used it to obtain a new type of Reynolds number. Then they established the relationship between their Reynolds number and the friction factor. Shishenko's work is typical [7]. The apparent viscosity of a Bingham fluid is

$$\mu_a = \eta + \frac{\tau_B}{du / dr}$$

Shishenko first obtained the average velocity over the region for which $r_p < r < R$, within which the velocity varies,

$$\left. \frac{du}{dr} \right|_{ave} = -\frac{\Delta P}{4 L \eta}(R - r_p)$$

In the equation, r_p is the radius of the plug (Fig.17.3). Then, the average apparent viscosity over this region is

$$\mu_{a\,ave} = \eta + \frac{\tau_B}{\left| \dfrac{du}{dr} \right|_{ave}} = \eta \left[1 + \frac{4 \tau_B L}{\Delta P(R - r_p)} \right] = \eta \frac{R + r_p}{R - r_p} \tag{17.9}$$

The Reynolds number proposed by Shishenko has the following form

$$Re_1 = \frac{\rho U(d - 2r_p)}{\mu_{aave}}$$

Transforming it, one can show that

$$Re_1 = Re \frac{(A - 1)^2}{A(A + 1)} \tag{17.10}$$

in which

$$Re = \frac{\rho U d}{\eta} \tag{17.11}$$

A is the ratio of the pipe radius to the plug radius,

$$A = R / r_p \tag{17.12}$$

809

As a first approximation, Eq. (17.11) can be simplified and written in the form

$$Re_2 = Re \frac{A-1}{A+1} \qquad (17.13)$$

Fig. 17.5 is the relationship between the friction factor and Re_2 for a clay slurry and a peat slurry. For laminar flow

$$f = 64 / Re_2 \qquad (17.14)$$

so that the form of the formula is the same as for laminar flow of a Newtonian fluid; the only difference is that the commonly used Reynolds number is replaced by Re_2.

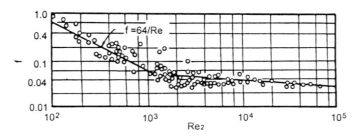

Fig. 17.5 f vs. Re_2 for Bingham fluid.

3. Pseudo-plastic fluid

The rheological equation of a pseudo-plastic fluid (Table 2.8) is

$$f(\tau) = \left(\frac{\tau}{K}\right)^{1/m} \qquad (17.15)$$

Integrating it, one obtains

$$Q = \frac{m\pi R^3}{3m+1}\left(\frac{R\Delta P}{2LK}\right)^{1/m} \qquad (17.16)$$

From Eq. (17.16),

$$\Delta P \sim LQ^m / R^{3m+1}$$

Wilkinson drew the following conclusions that are important in practice [4]:

(1) For a Newtonian fluid,

$$m = 1 \qquad \Delta P \sim R^{-4}$$

Thus, a slight increase of pipe size causes a great reduction in the pressure difference. In contrast, for a pseudo-plastic fluid with a small m-value,

$$m \to 0 \qquad \Delta P \sim R^{-1}$$

The pressure difference is inversely proportional to the pipe radius. Therefore, enlarging the pipe is not an effective way to reduce the power requirement.

(2) For a pseudo-plastic fluid with small m-value, the pressure difference hardly changes with the discharge. Hence, only by increasing the rotating speed of the pump, can one increase the discharge capacity of the pipeline.

(3) For a pseudo-plastic fluid with a small m-value, the discharge cannot be effectively determined by measuring the pressure drop over a given length of pipe.

Incidentally, the shear stress at the boundary τ_ω can be deduced from a balance of forces

$$\tau_w = \frac{d\Delta P}{4L}$$

so that the shear rate at the boundary is proportional to $8U/d$. Thus the rheological curve (the relationship between the shear stress and the shear rate) obtained by a measurement with a capillary viscometer is actually the relationship between $d\Delta P/4L$ and $8U/d$. The following equation can be deduced from Eq. (17.16).

$$\frac{8U}{d} = \frac{4m}{(3m+1)K^{1/m}} \left(\frac{d\Delta P}{4L} \right)^{1/m} \qquad (17.17)$$

That is, for a pseudo-plastic fluid, the relationship between $d\Delta P/4L$ and $8U/d$ depends only on the rheological properties of the fluid; thus, it is independent of the diameter of the pipe. It follows that the head loss for laminar flow of a pseudo-plastic fluid in a pipe can be deduced directly from the rheological data obtained from a measurement with a capillary viscometer. Behn and Shane proved this point from experiments with slurries flowing in pipes [6].

4. The unified method for various fluids

Depending on the properties of the transported sediments, the homogeneous slurry may have various rheological properties. A general method that can be applied to various kinds of fluids is preferable, even if the method is empirical. Metzner and Reed worked out such a method [9] .

Around 1930, Rabinowitsch and Mooney, working independently, found that for time-independent fluids, regardless of the fluid properties, the velocity gradient at the boundary can be expressed by the following equation:

$$-\left(\frac{du}{dr}\right)_0 = 3\left(\frac{Q}{\pi R^3}\right) + \frac{R\Delta P}{4L}\frac{d(Q/\pi R^3)}{d(R\Delta P/2L)}$$ (17.18)

Because of continuity

$$U = \frac{Q}{\pi R^2}$$

the preceding equation can be written in the form

$$-\left(\frac{du}{dr}\right)_0 = \frac{3}{4}\left(\frac{8U}{d}\right) + \frac{1}{4}\frac{8U}{d}\frac{d\ln(8U/d)}{d\ln(d\Delta P/4L)}$$

If one introduces m' from

$$m' = \frac{d\ln(d\Delta P/4L)}{d\ln(8U/d)}$$ (17.19)

the preceding equation can be written as

$$-\left(\frac{du}{dr}\right)_0 = \frac{3m'+1}{4m'}\frac{8U}{d}$$ (17.20)

Eqs. (17.19) and (17.20) can be used for pipes of various sizes, and also for Newtonian, Bingham, pseudo-plastic and dilatant fluids. From Eq. (17.19)

$$\tau_w = \frac{d\Delta P}{4L} = K'\left(\frac{8U}{d}\right)^{m'}$$ (17.21)

The procedure is as follows: one first obtains the relationship between $d\Delta P/4L$ and $8U/d$ experimentally; for each τ_ω, a pair of values for $8U/d$ and m' can then be determined; if these two values are substituted into Eq. (17.20), one can calculate the corresponding shear rate. Repeating this procedure for each τ_ω, one finally obtains the relationship between shear stress and shear rate for the fluid.

Eq. (17.21) and Eq. (17.15) are similar in form, but different in substance. The relationship between discharge and pressure difference, Eq. (17.16), is obtained by integrating Eq. (17.15). It requires that m remain constant within the region of integration. In contrast, Eq. (17.21) is a relationship between ΔP and U (and Q); no

812

integration is performed. Therefore, the latter equation, representing a sounder theoretical relationship, is tenable under any condition. It is valid so long as m' does not vary with the shear stress; even if m' does vary with the shear stress, Eq. (17.21) still gives an acceptable approximation that can be used in engineering design. For instance, starting from a relationship between $d\Delta P/4L$ and $8U/d$ obtained from a rheological measurement, one can choose values of K' and m' for the $8U/d$ that are suitable for a particular pipeline, and use them as the rheological properties of that pipeline.

A relationship exists between m and m', and also between K and K':

$$m = \frac{m'}{1 - \frac{1}{3m'+1}\left(\frac{dm'}{d\ln\tau}\right)} \tag{17.22}$$

$$K' = K\left(\frac{3m+1}{4m}\right)^m \tag{17.23}$$

If the relationship between $d\Delta P/4L$ and $8U/d$ forms a straight line on a log-log plot, i.e., if m' does not vary with τ_ω, then $m = m'$. The conditions $m = 1$ and $K = K'$ are for a Newtonian fluid.

Based on the characteristics mentioned above, Metzner proposed the use of the

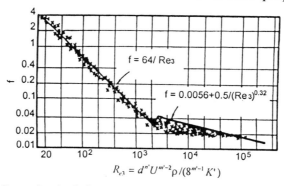

Fig. 17.6 Generalized relationship between f and Re_3 for non-Newtonian fluids
(after Metzner, A.B., and J.C. Reed)

following Reynolds number

$$Re_3 = \frac{d^{m'}U^{2-m'}\rho}{8^{m'-1}K'} \tag{17.24}$$

for establishing a relationship between friction factor and a suitable Reynolds number. Fig. 17.6 shows that data for various fluids are well distributed along the curve with no obvious grouping. It validates the method. In the laminar flow region

$$f = 64/Re_3 \tag{17.25}$$

Thus, it has the same form as that for a Newtonian fluid in the region of laminar flow.

17.2.1.2 Velocity profile

The basic relationships

$$\tau = r\frac{\Delta P}{2L}$$

and

$$-\frac{du}{dr} = f(\tau)$$

are known. Therefore,

$$-\int_{u}^{0} du = \int_{r}^{R} f\left(r\frac{\Delta P}{2L}\right) dr \tag{17.26}$$

If the function $f(\tau)$ is known, the velocity profile can be deduced from Eq. (17.26). Table 17.1 contains the equations describing the various velocity profiles for laminar flow of Newtonian, Bingham, pseudo-plastics and dilatant fluids in pipes. Two of them, the profiles for pseudo-plastic and dilatant fluids with different values of m, are shown in Fig. 17.7. The velocity profiles for pseudo-plastic fluids are more uniform than those for Newtonian fluid ($m = 1$), and those for dilatant fluids are less uniform. The larger the value of m, the less uniform the velocity profiles are. The profile approaches a triangle as the limit as m approaches infinity. The differences are particularly

Table 17.1 Velocity distributions for laminar flows of various fluids
(after Wasp, E.J.; J.P. Kenney; and B.L. Gandhi)

Fluid	$f(\tau)$	Velocity profile	Remark
Newton fluid	$f(\tau)=\dfrac{\tau}{\mu}$	$u = \dfrac{\Delta P}{4\mu L}(R^2 - r^2)$ $u = 2U(1-\dfrac{r^2}{R^2})$	r is the distance to pipe center, U is average velocity over cross section
Binghamm fluid	$f(\tau)=0,\ 0<\tau<\tau_B$ $f(\tau)=\dfrac{\tau-\tau_B}{\eta}$ $\tau_B<\tau<\tau_\omega$	$u = \dfrac{1}{\eta}\left[\dfrac{(R^2-r^2)\Delta P}{4L} - \tau_B(R-r)\right]$ $u_p = \dfrac{\Delta P}{4L\eta}(R-r_p)^2$ $r_p = 2L\tau_B/\Delta p$	r_p is distance of plug u_p is velocity of plug
Pseudo-plastic or dilatant fluid	$f(\tau)=(\tau/K)^{1/m}$	$u = (\dfrac{m}{m+1})(\dfrac{\Delta p}{2Lk})^{1/m}(R^{\frac{m+1}{m}} - r^{\frac{m+1}{m}})$ $u = U(\dfrac{3m+1}{m+1})\left[1-(\dfrac{r}{R})^{\frac{m+1}{m}}\right]$	—

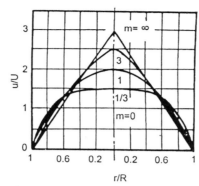

Fig. 17.7 Velocity profiles for laminar flows of Newtonian and non-Newtonian fluids

pronounced in the central part of the pipe. For a Bingham fluid, the fluid near the center of a pipe or near the water surface in a channel, all moves with the same velocity; i.e., it is a plug flow. For a pseudo-plastic fluid with a small value of m, as for a Bingham fluid, the fluid in those regions also moves at almost the same velocity. That is, "plug" flow also occurs for a pseudo-plastic fluid.

17.2.2 The transition from laminar to turbulent flow

For both Newtonian and non-Newtonian fluids, the relationship between the friction factor and the Reynolds number is the same if the flow is laminar. But usually the viscosity of non-Newtonian fluids is much higher than that of a Newtonian fluid; consequently, if laminar flow at a certain discharge passes through a pipe of a given diameter, the resistance for a non-Newtonian fluid is much larger than that for a Newtonian fluid. Therefore, it is not economical to transport a homogeneous non-Newtonian fluid with laminar flow. However, a small disturbance can keep solid particles in suspension in a slurry; too high a velocity would cause more energy consumption by the pump. From the view point of energy consumption, maintaining the flow at a rate just above the critical state between laminar flow and turbulent flow would be effective.

The critical condition for the transition from laminar to turbulent flow can be obtained from the relationships between the friction factor and the Reynolds number, that is, from Figs. 17.4 to 17.6. For instance, Fig. 17.4 indicates that if the rigidity is taken as the term for viscosity in the Reynolds number for a Bingham fluid, the critical Reynolds number for the transition from laminar to turbulent flow increases with increasing Hedstrum number. Hanks gave the expression for the critical Reynolds number [10].

$$(Re)_c = \frac{He}{8x_c}\left(1 - \frac{4}{3}x_c + \frac{1}{3}x_c^4\right) \qquad (17.27)$$

815

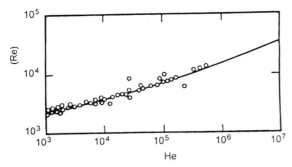

Fig. 17.8 Critical Reynolds number for the transition from
laminar to turbulent flow for Bingham fluids
(after Hanks, R.W.)

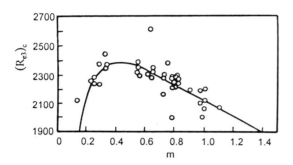

Fig. 17.9 Critical Reynolds number for the transition from
laminar to turbulent flow for exponential fluids
(after Hanks, R.W., and D.R. Pratt)

in which

$$x_c = \frac{\tau_B}{(\tau_w)_c} \qquad (17.28)$$

The relationship between x_c and Hedstrum number is as follows

$$\frac{x_c}{(1-x_c)^3} = \frac{He}{16,800} \qquad (17.29)$$

Solving Eq. (17.27) and Eq. (17.28), one obtains the relationship between the critical Reynolds number and the Hedstrum number, shown in Fig. 17.8.

The critical value of generalized Reynolds number [Eq. (17.24)], for the transition is in the range 2,000 to 2,500 (Fig.1 7.6). Hanks analysed data from a large number of tests with exponential fluids (pseudo-plastic and dilatant fluids), and he then established a relationship between $(Re_3)_c$ and m (Fig. 17.9) [11]. In designing a hydrotransport system for conveying oil shale powder, Faddick suggested that $(Re_3)_c = 4,000$. The corresponding critical velocity is [12]

$$U'_L = \left(\frac{4000 K' 8^{m'-1}}{d^{m'} \rho} \right)^{\frac{1}{2-m'}} \qquad (17.30)$$

17.2.3 Turbulent flow

17.2.3.1 Resistance to turbulent flow

1. Expression for resistance to turbulent flow with a smooth boundary

To reduce the power required for hydrotransport, smooth pipes are generally used. For turbulent flow, the head loss depends on the Reynolds number:

$$f = \frac{a}{(Re)^b} \qquad (17.31)$$

816

If the Blasius resistance formula for turbulent flow in smooth conditions is used, and if the Reynolds number has the form of Eq. (17.31) [7], Fig. 17.5 shows that for a Bingham fluid

$$a = 0.72$$

$$b = 1/6$$

If the Reynolds number takes the form of Eq. (17.24), the values of a and b vary with m', as shown in Table 17.2.

Table 17.2 Relationships between coefficients, exponents,
and values of m' for various fluids

m'	0.2	0.3	0.4	0.6	0.8	1.0	1.4	2.0
a	0.258	0.274	0.285	0.296	0.304	0.312	0.322	0.330
b	0.349	0.325	0.307	0.281	0.263	0.250	0.231	0.213

Another expression for the resistance to smooth turbulent flow of a Newtonian fluid is:

$$f = 0.0056 + \frac{0.5}{(\text{Re})^{0.32}} \tag{17.32}$$

Fig. 17.6 shows that for Reynolds numbers less than 70,000, the measured value of resistance for turbulent flow of a non-Newtonian fluid are always smaller than the corresponding value for a Newtonian fluid.

$$\sqrt{\frac{1}{f}} = 2\log\left(\frac{\text{Re}\sqrt{f}}{2 \cdot}\right) - 0.2 \tag{17.33}$$

Starting from the Prandtl-Karman logarithmic formula for resistance, Dodge and Metzner drew the following conclusion: for a non-Newtonian fluid obeying the exponential law (exponential fluid), formulas for resistance of turbulent flow in smooth conduits should have the following form [13]:

$$\sqrt{\frac{1}{f}} = A\log\left[\text{Re}_3\left(\frac{f}{4}\right)^{1-\frac{m'}{2}}\right] + B \tag{17.34}$$

The coefficients A and B, determined experimentally, are as follows:

$$A = 2 / (m')^{0.75} \\ B = -0.2 / (m')^{1.2} \Bigg\} \qquad (17.35)$$

Fig. 17.10 contains diagrams of resistance that include Eq. (17.25) and Eq. (17.35). Although the relationships are for turbulent flow of an exponential fluid, experience shows that the results also apply to non-exponential fluids provided that m' and K' are determined in accordance with the real shear stress τ_ω, at the boundary. Fig. 17.10 clearly shows that, in the turbulent flow region, the resistance factor for a Newtonian fluid is quite different from that for non-Newtonian fluids. For a given Reynolds number, the friction factor for psuedo-plastic fluids is smaller than that for Newtonian fluids, and the friction factor for dilatant fluids is larger.

Other empirical or semi-empirical formulas for resistance of non-Newtonian fluids in the smooth turbulent-flow region have been developed. Details are given in reference [14]; they are not discussed further herein.

2. Applicability of data on resistance from small pipes to flows in large pipes

In studies of hydrotransport, experiments are usually conducted in the laboratory using small pipes, and the results thus obtained are then used in the design of larger systems. The modeling law to be used in extending the small-pipe data to a larger scale has not been fully developed.

For resistance to turbulent flow with smooth boundaries, the Blasius formula can be transformed into

$$d^b \frac{d\Delta P}{L} = kU^{(2-b)}$$

in which values of the exponent b are given in Table 17.2 for various fluids. Thus, if experimental data from small pipes are used to obtain a relationship between

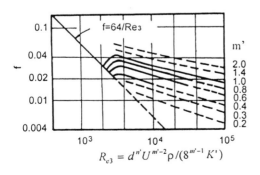

Fig. 17.10 The relationship between friction factor and Reynolds number, with various kinds of fluids and for both laminar and turbulent flow (after Dodge, D.W., and A.B. Metzner)

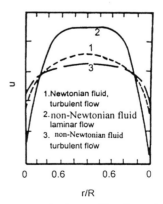

Fig. 17.11 Velocity profiles for turbulent flow of non-Newtonian fluids

$d^b(d\Delta P/L)$ and U in logarithm scale, and the points actually follow along a straight line, the straight line can be extended and the resistance in the larger prototype pipes can be determined with it. Based on a similar method of extension, suggested by Bowen in 1961 [15], Kenchington extended the results for resistance to flow of a cement slurry from small pipes and obtained satisfactory results.

17.2.3.2 Velocity profile

Many studies have been made for velocity profiles of turbulent non-Newtonian flows with smooth boundaries [14]. With reference to the logarithmic velocity profile of Newtonian flow, Dodge and Metzner proposed a velocity profile formula for exponential fluids flowing in smooth pipes, given in Table 17.3. Fig. 17.11 includes the velocity profile for $m' = 0.377$ and $Re_3 = 4,876$. The velocity profile of a non-Newtonian fluid is more nearly uniform than that of a Newtonian fluid at the same Reynolds number. In addition, Pai solved the Reynolds equation and deduced that the velocity profile formula is applicable to flows both in the main flow region and near the boundary [17]. Brodkey et al. applied this method to non-Newtonian fluids and obtained a formula for the velocity profile in smooth pipes [18]. These formulas are also listed in Table 17.3.

Table 17.3 Formulas for velocity profiles of turbulent non-Newtonian flows in smooth pipes (after Faddick, R.R.)

Newtonian fluid		Non-Newtonian fluid	
Author	Formula	Author	Formula
Prandtl-Karman	$\dfrac{u}{U_*} = 5.75\log\dfrac{yU_*\rho}{\mu} + 5.5$ (in main flow) $\dfrac{u}{U_*} = \dfrac{yU_*\rho}{\mu}$ (in laminar sub-layer) $\dfrac{u-U}{U_*} = 5.75\log\dfrac{y}{R} + 3.75$ $\dfrac{u_{max}-u}{U_*} = 5.75\log\dfrac{R}{y}$	Dodge-Metzner	$\dfrac{u}{U_*} = \dfrac{5.66}{(m')^{0.75}}\log\dfrac{y^m(U_*)^{2-m\rho}}{K} - \dfrac{0.4}{(m')^{1.2}} +$ $\dfrac{2.458}{(m')^{0.75}}\left[1.96 + 1.255m' - 1.628m'\log\left(3+\dfrac{1}{m'}\right)\right]$ (in main flow) $\dfrac{u}{U_*} = \left[\dfrac{y^m(U_*)^{2-m\rho}}{K}\right]^{1/m}$ (in laminar sub-layer) $\dfrac{u-U}{U_*} = 3.686(m')^{0.25} + 5.66(m')^{0.25}\log\dfrac{y}{R}$ $\dfrac{u_{max}-u}{U_*} = 5.66(m')^{0.25}\log\dfrac{R}{y}$
Pai Shih-I	$\dfrac{u}{u_{max}} = 1 + \dfrac{s-\alpha}{\alpha-1}\left(\dfrac{R-y}{R}\right)^2 + \dfrac{1-s}{\alpha-1}\left(\dfrac{R-y}{R}\right)^{2\alpha}$ in which $s = \dfrac{U_*R\rho/\mu}{2u_{max}/U_*}$, α is determined by experiment.	Brodkey-Lee-Chase	$\dfrac{u}{u_{max}} = 1 + \dfrac{s-\alpha}{\alpha - \dfrac{m+1}{2m}}\left(\dfrac{R-y}{R}\right)^{\frac{m+1}{m}}$ $+ \dfrac{\dfrac{m+1}{2m}-s}{\alpha - \dfrac{m+1}{2m}}\left(\dfrac{R-y}{R}\right)^{2\alpha}$ in which $s = \left[R^m(U_*)^{2-m\rho}/K\right]^{1/m}/2u_{max}/U_*$, α is determined by experiment.

Velocity profile formulas for non-Newtonian fluids flowing in rough pipes are rather scarce. In such flows, vortices cannot easily form behind protrusions on the boundary, and even if they do, the viscosity is extremely high in the central part of the vortices. Therefore, the energy dissipation rate is not high. As a result, roughness has less effect on the velocity profile than it does for Newtonian fluids.

17.3 TWO-PHASE FLOW IN HORIZONTAL PIPELINES

In the early stage of hydrotransport, water was used as the transporting medium. Later it was discovered that coarse particles could be transported at higher concentrations if the transporting medium was a homogeneous slurry consisting of water and fine particles. In some literature the phenomenon of sediment transport at high concentrations is called a state of hyperconcentration. The two-phase flows with water and with homogeneous slurry as the transporting mediums are discussed in the following sections.

17.3.1 Motion of two phase flow consisting of water and solid material

17.3.1.1 Profiles of concentration and velocity

Newitt studied concentration profiles of different types of sediment and the

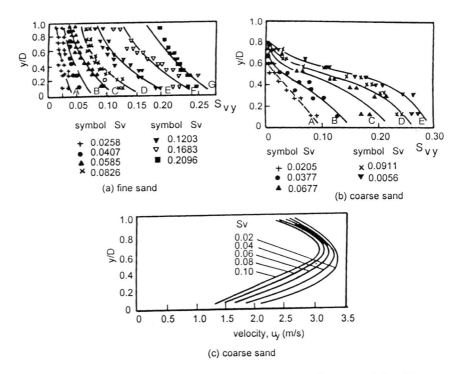

Fig. 17.12 Concentration and velocity profiles for two-phase flow (diameter of pipe 25 mm, pressure gradient 0.28) (after Newitt, D.M.; J. Richardson; and C.A. Shook)

820

corresponding velocity profiles in a pipeline 25 mm in diameter with a pressure gradient of 0.28 (head loss measured in terms of a clear water column) [19]. Fig. 17.12a and Fig. 17.12b are vertical profiles of the concentration for fine sand ($D = 0.15$ mm) and for coarse sand ($D = 1.55$ mm). The former yields a non-uniform suspension; the latter indicates some bed load motion. Fig. 17.12c presents velocity profiles for the motion of coarse sand, and they are obviously not symmetric. The higher the sediment concentration, the higher is the location of the point of maximum velocity. Fig. 17.13 contains velocity profiles measured by Japanese scientists for three different conditions: (a) non-uniform suspension, (b) sliding of material on pipe bottom, (c) deposition on pipe bottom [20]. The material used in their experiments was glass spheres. The velocity profile for clear water flow is also shown for comparison. If large amounts of sediment move along the bed, the velocity profile is markedly affected.

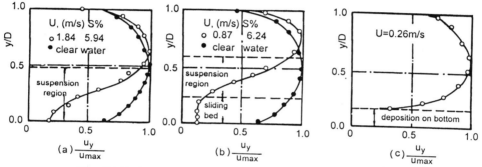

Fig. 17.13 Comparison of vertical velocity profiles for two-phase
flow in a pipeline with those for clear water

From their experimental data, Newitt et al. proposed the following formula for y', the distance between the points of maximum velocity and the center of the pipe [19]:

$$y' - \frac{d}{2} = 15.2 \left(S_v \sqrt{\frac{\tau_w}{\rho}} \right)^{1.81} \tag{17.36}$$

in which τ_ω is the shear stress at the wall. In the region $y > y'$ the velocity profile follows an exponential formula, exponent 1/7; in the region $y < y'$, the profile is described by the following equation:

$$\frac{(u_y / u_{max})_w - (u_y / u_{max})_m}{(u_y / u_{max})_w} \frac{\tau_w}{\rho} = (0.118 - 0.054 S_v) \left(1 - \frac{y}{y'} \right) \tag{17.37}$$

The subscripts m and ω in the ratio of u_y/u_{max}, the velocity at y to the maximum velocity, denote sand-water mixture and clear water, respectively.

The concentrations used in hydrotransport are usually rather high; hence the existence of sediment affects the velocity profile. The process is quite complex, as indicated in a preliminary way in Chapter 12. Because their experimental results covered only a limited range, the analytical results obtained by Newitt et al. may not be universally applicable. The study of the velocity and concentration profiles in a pipeline has not attracted enough attention; only in recent years have fragmentary data become attainable through improvements of measurement techniques, and these few data have not been systematically analyzed. Actually, there is close relationship between velocity field and shear stress field; therefore, once the velocity profile is known, the resistance to the flow can be determined. Shook and Daniel made a preliminary attempt to perform this operation [21].

17.3.1.2 Resistance

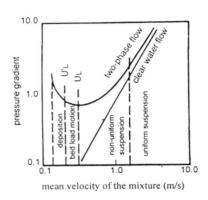

Fig. 17.14 The relationship between head loss and velocity for two-phase flow in a pipeline

Fig. 17.14 is the relationship between head loss and velocity in a pipeline transporting a sediment-water mixture with a given concentration of solid particles. The abscissa is the average velocity. It is the usual mean velocity of the sediment-water mixture for the condition of a rigid bed (no deposition), and a virtual velocity if deposition does occur and reduces the cross-sectional area of the flow. The ordinate is the head loss per unit length; it is that for a clear water column in this case, but it can be that of a column of the mixture in other cases.

The results shown in Fig. 17.14 divide into several distinct regions according to the magnitude of velocity. Variations of resistance in the various regions are different. For a homogeneous suspension, the $J_m \sim U$ relationship for two-phase flow is approximately parallel to that for clear water flow. For a heterogeneous suspension, the decrease of velocity causes the pressure gradient for a sediment-laden flow to become increasingly larger than that for clear water flow, and the curve becomes concave with a minimum value at "o". At this point the pressure gradient is a minimum, and the corresponding velocity is usually called the critical velocity-- that at which the operation of the pump is most economical. In most cases bed load motion is initiated at this critical velocity. To the left of point "o," the pressure gradient is larger and consequently bed material moves along the bottom. Some experimental data indicate that the pressure gradient remains constant for this type of flow. The above phenomenon may have different causes. In the design of a pipeline, the discharge divided by the cross section of the pipe is used as the average velocity. Whenever deposits form on the bottom, so the flow section is less than the cross section of the pipe, the real velocity is higher than the average velocity used in Fig. 17.14. If one replaces the average velocity by the real velocity as the abscissa, the left part of the curve shifts a little toward the right. A sediment-laden flow can adjust itself automatically. As the discharge decreases, the deposits on the bottom become larger and larger, and the flow section smaller and smaller. Consequently, the average velocity might remain unchanged and the pressure gradient could stay constant too. If a laminated load fills the pipe, the pressure gradient could remain constant over a certain range of velocity. This situation is discussed in more detail in the latter part of this section.

Formulas for the pressure gradients for the different flow conditions differ in the following ways.

1. Pressure gradient for uniform suspension of sediment

If the concentration is not high and the sediment is uniformly suspended, such a heterogeneous flow possesses the same viscosity but a larger specific weight than does water. In this situation, the pressure gradient in terms of a clear water column is

$$J_m = J_w s_m = J_w \left[1 + S_v \left(\frac{\rho_s - \rho}{\rho} \right) \right]$$

or

$$\frac{J_m - J_w}{J_w S_v} = \frac{\rho_s}{\rho} - 1 \tag{17.38}$$

in which J_m is the pressure gradient in terms of a clear water column for the transport of a water-sediment mixture, J_ω is the pressure gradient for clear water transported at the same velocity, and s_m is the specific weight of the water-sediment mixture. Eq. (17.38) indicates that a pressure gradient expressed in terms of a column of water-sediment mixture is the same as that for clear water flow at the same velocity and in the same pipe. If the head loss is expressed in terms of a clear water column, the $J_m{\sim}U$ relationship for sediment-laden flow is as shown in Fig. 17.14. It follows a straight line that is parallel to that for clear water flow but higher by the amount $S_v (\rho_s / \rho - 1)$.

If the concentration is high enough, the viscosity of the sediment-laden flow differs from that of clear water; the sediment also affects the structure of the turbulence, and the combined result of these two effects is rather complex. The pressure gradient (in terms of a column of water-sediment mixture) for a sediment-laden flow might be larger than, equal to, or less than that for a clear water flow, as discussed in Chapter 12. Because all the sediment particles can be suspended in pipe flows (no bed load), it is beneficial to study the effect of suspended load on flow resistance in experiments conducted in pipes. Quite a lot data of this type have been accumulated.

Part of these experimental data are shown in Table 17.4; in the experiments all sediment particles were suspended. It shows that there are three different types for the comparison between the head loss for a sediment-laden flow and that for a clear water flow with the same velocity:

(1) The head loss per unit distance for sediment-laden flow Δh_m (in terms of a column of water-sediment mixture) is equal to that for clear water flow Δh_ω (in terms of the column of clear water);

(2) The head loss per unit length for a sediment-laden flow is equal to that for clear water flow — both in terms of a column of clear water;

(3) The head loss per unit length for a sediment-laden flow (in terms of a column of water –sediment mixture) is less than that for clear water flow (in terms of a column of clear water).

Among these various situations, data of the (1) type indicate that the suspended sediment did not affect the energy dissipation of the flow, but the data of the (2) and (3) types indicate that the suspended sediment reduces the energy dissipation. No obvious differences in the experimental conditions for these data from those listed in Table 17.4 could be detected.

The problem of energy dissipation of clear water and sediment laden flows is discussed in more detail in Chapter 12. Apparently, all three situations are possible, but the conditions under which they occur are not yet well established. Thomas studied the head loss for heterogeneous flows for uniform suspensions [22]. He stated that the different situations are affected by the sizes of sediment. If the ratio of sediment size to thickness of laminar sublayer is relatively small (D/δ <5), the presence of the sediment changes the viscosity and the thickness of the laminar sublayer, and the head loss for a sediment-laden flow can be larger than that for clear water flow. If D/δ is larger (D/δ > 5), the sediment has little effect on the viscosity of the flow, and the head loss for sediment-laden flow is close to that for clear water flow. The thickness of the laminar sublayer used by Thomas is

$$\delta = \frac{5v}{u_*}$$

The experimental data he used are rather limited; therefore the generality of the preceding conclusion needs further investigation.

2. Head loss for flow with a non-uniform suspension and bed load

One cannot readily distinguish between flows with (a) a non-uniform suspension and (b) having bed load. In some experiments bed load occurred along with a non-uniform suspension. Therefore, in studies of head loss, many researchers may not have

Table 17.4 Comparison of head losses for sediment-laden and clear water flows

No.	Author	Pipe		Sediment				Concentration			U_c (m/s)
		material	d (mm)	material	D (mm)	γ_s	ω (cm/s)	γ_m	S_v	S_w	
(a) data of experiments in which the head loss in column of water-sand mixture of a sediment-laden flow is close to that of a clean water											
1	B.C. Knopos	iron	250	mine tailing	0.066	2.65	0.86	-	-	17~38	1.72~1.97
2	N.S. Blatchi	brass	25	sand	0.20	2.65	2.42	-	-	9~35	≈2.50
3	R.A. Smith	-	76	sand	0.24	2.60	-	1.01~1.26	-	-	1.49~2.84
4	A.П. Ufin	-	250	sand	0.25	2.60	-	1.01~1.26	-	-	2.45~5.10
		-	300	sand	0.25	2.60	-	1.01~1.26	-	-	3.25~5.54
		-	300	sand	0.32	2.60	-	1.07~1.44	-	-	3.96~4.53
5	G.W. Howard	-	102	sand	0.38	2.60	-	1.11	-	-	4.06
6	A.H. Krementov	iron	255	sand	0.79	2.65	8.70	-	-	20~90	4.35~5.35
7		iron	51	sand	1.6~9.1	2.65	-	-	<20	-	1.25~2.60
8	M.P. O'Brien R.G. Folsom	iron	50	Sacramento sand	-	2.65	-	-	-	10~30	1.37~3.95
		iron	76	Montreaol sand	-	2.65	-	-	-	5~15	4.27~6.70
		iron	38	sand brass minging	-	-	-	-	-	3.8~11.3	1.07~4.83
9		-	102	lime-stone	0.056	2.00	-	1.03	-	-	0.61~0.76
10	J.C. Williams	-	102	coal ash	0.15	2.00	-	1.21	-	-	2.07
11	D.M. Newitt et al.	brass	25	plastic	0.585~1.98	1.18	2.42~6.70	-	-	-	≈1.82~2.73
		brass	25	coal powder	1.32~5.98	1.40	6.8~14.8	-	-	-	≈2.73~3.35
(b) data of experiments in which the head loss in column of clear water of sediment-laden flow is close to that of clean water flow											
1	R. Durand	steel	40.6 &150	Sena sand	0.20	2.65	-	-	-	-	5.00
		steel	150	Ruwall sand	0.44	2.65	-	-	1.2	-	5.50
				plastic beads	0.44	1.60	-	-	-	-	>4.50
				emery	0.44	3.95	-	-	-	-	>4.50
(c) data of experiments in which the head loss in column of water-sand mixture of a sediment-laden flow is less than that of a clean water flow											
1	A.H. Krimentov	iron	306	sand	0.10	2.65	0.99	-	-	5~65	-
2	A.П. Polchevskii	iron	130	sand	0.47	2.65	5.90	-	-	<20~30	-
3	N.S. Blatchi	zinc-plating iron	25	sand	0.70	2.65	7.40	-	-	≈30~60	-
		brass	25	sand	0.70	2.65	7.40	-	-	15~65	-
4	M.P. O'Brien R.G. Folsom	iron	76	Sacramento sand	-	-	-	-	-	6~25	-
				Sandiego sand	-	-	-	-	-	2~26	-
5	Nampeko	-	440	peat	-	-	-	-	2.2~2.7	-	-

Note: d—pipe diameter; D—average diameter; γ_s—specific weight; ω—fall velocity; γ_m—specific weight of water—sand mixture; S_v—volumetric; S_w—in weight percentage; U_c—critical velocity at which head loss of a sediment-laden flow equals that of clear water flow.

distinguished between these two types of flow. Herein, the head loss for these two types of flow is also discussed as a whole in the following section. Some of the results relating to only one type of flow will be explained as it is introduced. The head loss for uniform sediment is discussed first, and then that for non-uniform sediment.

(1) Prediction of head loss

At present, predictions of the head loss for heterogeneous pipe flows are based on the assumption that resistances can be added; that is, the head loss for a sediment-laden flow consists of two parts: the head loss for clear water flow and the additional head loss caused by the sediment:

$$J_m = J_w + J_s \qquad (17.39)$$

in which J_m is the energy slope for sediment-laden flow, J_w the energy slope for a clear water flow with the same velocity, and J_s the additional energy slope caused by the sediment.

Usually J_s is determined from experiments. A few semi-theoretical results have been presented, but some of the basic assumptions used in deriving them are questionable.

Results of Nyrpic Hydraulics Laboratory and the revised Durand formula

In the early days, the work done at the Nyrpic Hydraulics Laboratory gave the most systematic results for hydrotransport. In 1953, Durand made a preliminary study of the topic [23], and in 1960, Gibert presented a more comprehensive summary [24]. In the literature, the result of their work came to be called the Durand formula.

Durand found for sediment of a certain size being carried without deposits in a horizontal pipeline, a functional relationship between the two dimensionless parameters: $\dfrac{J_m - J_w}{S_v J_w}$ and $\dfrac{U}{\sqrt{gd}}$, in which the energy slope is measured in terms of a water column, U is the average velocity of a sediment-laden flow, d is the diameter of the pipe. Fig. 17.15 shows such relationships for fine sand ($D = 0.44$ mm) and coarse sand ($D = 2.04$ mm); the results were obtained using pipes of various diameters. Thus, for each sediment there is a corresponding curve. Further study revealed that for sediment coarser

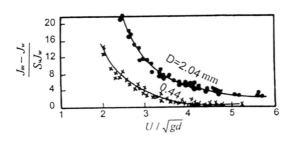

Fig. 17.15 Experimental relationships for coarse and fine sand (after Durand, R.)

826

than 2 mm, all of the data fall along a single curve; i. e., sediment diameter is no longer a factor (Fig. 17.16). Such experiments support the concept that the factor playing an important role in hydrotransport is not the diameter of the sediment itself, but a parameter closely related to the diameter. This parameter is simply the drag coefficient C_D, shown in Fig. 3.1.

Having introduced the parameter C_D, Durand got a unique relationship for natural sediments of all sizes (Fig.17.17); his empirical relationship is as follows:

$$\frac{J_m - J_w}{S_v J_w} = 180 \left[\frac{U^2}{gd} \sqrt{C_D} \right]^{-3/2}$$ (17.40)

At high velocities, the pressure gradient for sediment-laden flow is close to that for clear water flow.

Durand carried out additional experiments with particles of various specific weights, like plastic beads, and obtained a more generalized formula [25] :

$$\frac{J_m - J_\omega}{J_\omega} = 121 S_v \left[\frac{U^2}{gd \frac{\rho_s - \rho}{\rho}} \frac{\sqrt{\frac{\rho_s - \rho}{\rho} gD}}{\omega} \right]^{-3/2}$$ (17.41)

in which ω is the fall velocity of the sediment particles.

Although Eq. (17.41) is empirical, it is based on data that cover wide ranges of the significant variables:

diameter of pipe, d	40-580 mm
diameter of sediment, D	0.2-25 mm
specific weight of sediment, γ_s	1.5-3.95
concentration of sediment, S	50-600 kg/m^3

Therefore, the result is rather general.

The Durand formula is widely used at present. Some researchers have found that it can be applied to even higher concentrations. For example, Smith conducted experiments in pipes with diameters of 50 mm and 75 mm and with sands with diameters in the range of 0.22~1.4 mm, and he found Eq. (17.41) to be valid for volumetric concentrations as high as 33% (S=857 kg/m^3) [26]. Weidenroth carried out experiments with sand of 0.085~1.10 mm and reached the same conclusion [27].

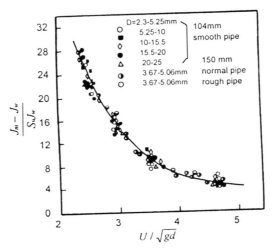

Fig. 17.16 Experimental relationship for sand coarser than 2 mm (after Durand, R.)

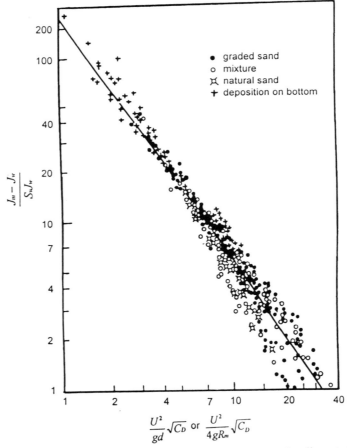

Fig. 17.17 Unified relationship of pressure gradients for various natural sediments (after Durand, R.)

Nevertheless, some laboratory data do not coincide with the Durand formula, and modifications to it have been proposed. For example, in 1967, Zandi and Govatos analyzed a set of 2,549 data points. They found that the Durand formula is not applicable if particles are moving along the bed in saltation. Excluding these data and data with $S_v < 5\%$, they obtained the following formula for the 1,452 data points remained [28]:

If

$$\frac{gd\dfrac{\rho_s - \rho}{\rho}}{U^2\sqrt{C_D}} < 0.1, \qquad \frac{J_m - J_w}{S_v J_w} = 6.3\left[\frac{U^2\sqrt{C_D}}{gd\dfrac{\rho_s - \rho}{\rho}}\right]^{-0.354} \qquad (17.42)$$

If

$$\frac{gd\dfrac{\rho_s - \rho}{\rho}}{U^2\sqrt{C_D}} > 0.1, \qquad \frac{J_m - J_w}{S_v J_w} = 280\left[\frac{U^2\sqrt{C_D}}{gd\dfrac{\rho_s - \rho}{\rho}}\right]^{-1.93} \qquad (17.43)$$

Hayden and Stelson proposed the following formula [29]:

$$\frac{J_m - J_w}{S_v J_w} = 100\left[\frac{U^2}{gd\dfrac{\rho_s - \rho}{\rho}}\frac{\sqrt{\dfrac{\rho_s - \rho}{\rho}gD}}{\omega}\right]^{-1.30} \qquad (17.44)$$

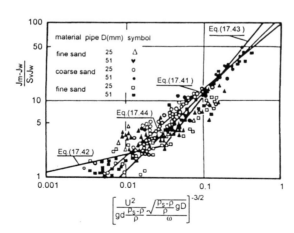

Fig. 17.18 A comparison of the Durand, Zandi-Govatos, and Hayden-Stelson formulas

Fig. 17.18 is a comparison of these formulas. Data in the figure are taken from the Hayden and Stelson experiments conducted in pipes with diameters of 25 mm and 50 mm carrying fine sand, coarse sand, and pebbles. The data deviate from the Zandi-Govatos formula. Bain and Bonnington stated that the data used by Durand include some with bed load, and that

829

the Durand formula is valid only within the range of the data from which the formula was derived [30].

Semi-empirical analysis by Newitt

Newitt and others attempted a theoretical analysis of the head loss for hydrotransport with suspended load and bed load [2]. The basic assumptions adopted by them are questionable, and the conclusions reached are therefore semi-empirical.

(i) Head loss for flow with sediment in suspension

Existing analyses of the head loss with suspended sediment are based on the assumption that the suspension of sediment takes a part of the energy from the mean flow, and consequently, that it causes additional energy dissipation. Such an assumption is questionable. As discussed in Chapters 10 and 12, the energy for sediment suspension comes from the kinetic energy of turbulence, not directly from the mean flow.

In the former Soviet Union, some researchers applied the gravity theory proposed by Velikanov to pipe flow [31]. According to the gravity theory, the suspension of sediment takes a part of the energy from the mean flow. In Canada, early in 1942, Wilson put forward the similar concept, and it was opposed by Einstein [32]. Newitt's analysis followed this concept, and it was used until the 1970s [33].

The energy per unit volume in a unit time required to suspend sediment is

$$E_1 = S_v(\rho_s - \rho)gK_1\omega \qquad (17.45)$$

in which ω is the fall velocity of an individual particle. In a sediment-laden flow the real fall velocity of the particles deviates somewhat from ω, and therefore a correction factor K_1 is introduced in Eq. (17.45). Because of the existence of sediment, a unit volume of sediment-laden flow requires additional energy

$$E_2 = J_{sm}U\rho_m \qquad (17.46)$$

in which ρ_m is the density of the water-sediment mixture, J_{sm} the additional head loss per unit distance (in terms of a column of water-sediment mixture). Since

$$J_{sm}\rho_m = J_s\rho$$

in which J_s is the additional head loss per unit distance in terms of the column of clear water, ρ is the density of clear water. Eq. (17.46) can be written as

$$E_2 = J_sU\rho \qquad (17.47)$$

If the additional head loss of a sediment-laden flow is required for suspending sediment,

$$E_1 = E_2$$

i.e.,

$$J_s = K_1 S_v \left(\frac{\rho_s - \rho}{\rho} \right) \frac{\omega}{U} \tag{17.48}$$

Substituting Eq. (17.48) into Eq. (17.39) and letting

$$J_w = f_w \frac{U^2}{2gd} \tag{17.49}$$

in which f_w is the friction factor for clear water flow at velocity U, one obtains

$$\frac{J_m - J_w}{S_v J_w} = \frac{2}{f_w} K_1 \left(\frac{\rho_s - \rho}{\rho} \right) \frac{\omega}{U} \frac{gd}{U^2}$$

By considering the fact that f_w varies slightly, the preceding equation can be rewritten in the form

$$\frac{J_m - J_w}{S_v J_w} = K_2 \left(\frac{\rho_s - \rho}{\rho} \right) \frac{\omega}{U} \frac{gd}{U^2} \tag{17.50}$$

From experimental data for transport of plastic beads, coal powder, and sand in a 25 mm pipe, Newitt obtained the value of 1,100 for K_2.

Eq.(17.50) can be compared with the empirical formula obtained by Durand [Eq.(17.41)]. At the velocity usually adopted in hydrotransport, and if all particles are suspended and their settling velocities follow Stokes law, then

$$\omega \sim \frac{\rho_s - \rho}{\rho} D^2$$

Substituting this value into Eq. (17.41) and Eq. (17.50), one can transform the Durand formula into

$$\frac{J_m - J_w}{S_v J_w} \sim \left(\frac{\rho_s - \rho}{\rho} \right)^{9/4} \frac{D^{9/4} d^{3/2}}{U^3} \tag{17.51}$$

and the Newitt formula into

$$\frac{J_m - J_w}{S_v J_w} \sim \left(\frac{\rho_s - \rho}{\rho} \right)^2 \frac{D^2 d}{U^3} \tag{17.52}$$

Both formulas show that $(J_m\text{-}J_w) / (S_v J_w)$ is inversely proportional to the third power of velocity, and the additional head loss caused by the existence of sediment rapidly decreases toward zero as the velocity increases. Relationships between $(J_m - J_w)/(S_v J_w)$ and the three other quantities D, d, and $(\rho_s\text{-}\rho)/\rho$ in these two formulas differ. Those for $(\rho_s\text{-}\rho)/\rho$ and D differ only a little. The effect of the pipe diameter is greater in the Durand formula than it is in the Newitt formula. Results of experiments with the transport of 0.4 mm sand in four different pipes are compared with these two formulas in Fig. 17.19a. The Durand formula appears to be more reliable than the Newitt formula.

(ii) Head loss if some sediment moves as bed load

If sediment moves as bed load, it has to take energy from the flow because of the relative motion between the bed load and the mean flow. The additional head loss thus caused is due to the friction between the bed load and the boundary. Because of this extra resistance, Newitt deduced that the additional energy consumption caused by bed load is [2]

$$J_s = K_3 S_v \left(\tfrac{\rho_x - \rho}{\rho} \right) \tag{17.53}$$

He conducted experiments with pebbles, crushed coal particles and particles of manganese dioxide in a 25 mm pipeline, and obtained a value of 0.8 for K_3. The value was independent of the velocity. As J_s does not vary with U, the relationship between pressure gradient and velocity for a sediment-laden flow should be parallel to that for clear water flow. For his data Newitt expressed $(J_m\text{-}J_w) / (S_v J_w)$ in the form:

$$\frac{J_m - J_w}{S_v J_w} = 66 \left(\tfrac{\rho_x - \rho}{\rho} \right) \tfrac{gd}{U^2} \tag{17.54}$$

Babcock conducted experiments with coarse sand and steel beads in a 25 mm pipeline; he verified the structure of Eq. (17.54), but proposed that the coefficient should be 60.6 [34]. However, he obtained a much smaller coefficient for the experimental data from tests in a 152 mm pipeline. As a preliminary estimate, this coefficient can be taken as $4000/d^{1.27}$, in which d is the pipe diameter in mm.

According to the classification of two-phase flow suggested by Durand, sediment moves as bed load if it is coarser than 2 mm. For such particles, the drag coefficient C_D approaches a constant, and Eq. (17.41) can be simplified to

$$\frac{J_m - J_w}{S_v J_w} = 42 \left[\tfrac{gd}{U^2} \left(\tfrac{\rho_s - \rho}{\rho} \right) \right]^{3/2} \tag{17.55}$$

A comparison of Eq. (17.54) and Eq. (17.55) shows that in both equations the ratio $(J_m\text{-}J_w)/(S_v J_w)$ does not vary with sediment size; however, the effects of pipe diameter, density, and average velocity are different in these two equations. Results of experiments with sand and crushed coal in several pipes are compared with Eq. (17.54) and Eq. (17.55) in Fig. 17.19b. Once again, the Durand formula fits the data better.

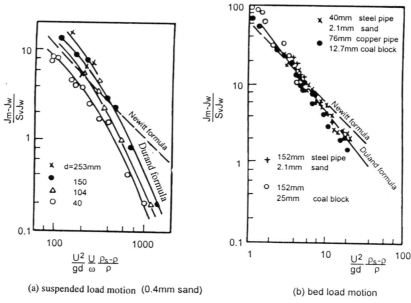

(a) suspended load motion (0.4mm sand)   (b) bed load motion

Fig. 17.19 Comparison between Durand formula and Newitt formula

Worster Study [35]

Worster conducted experiments with gravel and particles of coal coarser than 2 mm and obtained the relationship shown in Fig. 17.20. Such coarse particles usually move as bed load. As f_w varies little, according to Eq. (17.49), the abscissa in Fig. 17.20 is actually equivalent to the parameter proposed by Durand,

$$\frac{U^2}{gd\dfrac{\rho_s - \rho}{\rho}}$$

Hence, if the velocity is high

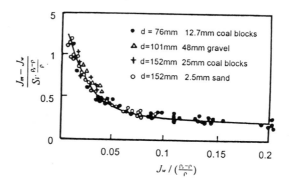

Fig. 17.20 The pressure gradient of flow carrying coarse particles (larger than 2 mm) of various densities (after Worster, R.C., and D.F. Denny)

833

$$\frac{J_m - J_w}{S_v \dfrac{\rho_s - \rho}{\rho}} \sim 0.25 \qquad (17.56)$$

The preceding equation is similar in form to Eq. (17.53) but the coefficient is quite different. In fact, $J_m - J_w$ in Eq. (17.56) and J_s in Eq. (17.53) correspond to the shear between solid particles suggested by Bagnold for a shearing sediment-laden flow, and $S_v(\rho_s - \rho)/\rho$ is the effective weight of sediment in a unit volume of sediment-laden flow. If the sediment moves as bed load, the effective weight equals the dispersive force among particles. That is, the coefficient in Eq. (17.53) and Eq. (17.56) corresponds to the ratio between the shear and the dispersive force among particles. From Bagnold's experiments (Fig. 9.3), this ratio varies within the range of 0.37 to 0.75.

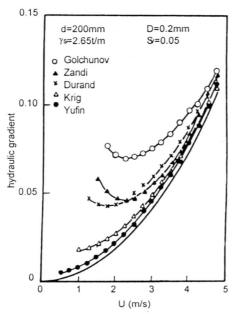

Fig. 17.21 A comparison of different
formulas for head loss
(after Sasic, M., and P. Marjamovic)

Other formulas have been proposed for non-uniform suspensions and for flow with bed load. The differences among them are rather large. Fig. 17.21 is a comparison of the relationship between pressure gradient and velocity obtained from five of these formulas, all for the case of sediment-laden flow with a volume concentration of 5% of fine sand ($D = 0.2$ mm) in a 200 mm pipe [36]. Among them, the Durand and Zandi formulas are close to each other, but the rest deviate considerably.

(2) Factors affecting head loss

As already discussed, the head loss depends on the velocity, viscosity of the fluid, size distribution and density of the solid material, its concentration, pipe size, etc. The effects of some of these factors are included in these formulas. Because the differences among the formulas are rather large, some researchers studied the effects of these factors on the head loss by varying one parameter at a time. Results from this approach show considerable scatter, and no systematic summary has been made.

Effect of the density of sediment

Duckworth et al. conducted experiments with solid materials both heavier and lighter than water. They obtained the following relationship between the friction factor for a sediment-laden flow, f_m, and that for a clear water flow, f_w [37]:

$$f_m = f_\omega + 13.8 M \left(\frac{U^2}{gd}\right)^{-1.24} f(\rho_s / \rho) \qquad (17.57)$$

in which M is the mass discharge of a heterogeneous flow; the function $f(\rho_s / \rho)$ is shown in Fig.17.22. For particles that are heavier than water, the head loss increases as the density of the solid particles increases, but for particles lighter than water, the head loss increases as the density of the particles decreases.

Effect of the size distribution

Shook et al. studied the effect of the size distribution on the head loss of heterogeneous flows through experiments for sediment particles of two median sizes — 0.2 mm and 0.5 mm, each with three different coefficients of non-uniformity. The experiments were

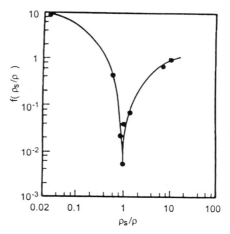

Fig. 17.22 Correction factor for the density of solid particles in Eq. (17.57) (after Duckworth, R.A., and G. Argyros)

conducted in pipes with diameters of 57 mm and 107 mm. The transporting medium had a specific weight of 1~1.35, viscosity of 0.0045~0.38 poise, and volumetric concentrations of 0.0045~0.38 [38]. The results indicate that the head loss was larger for the more uniform sediments. But in these experiments, a small amount of particles finer than 0.01 mm were included in the two least uniform sediments. Therefore, the following question arises: does the size distribution cause the reduction of the head loss or the presence of fine particles?

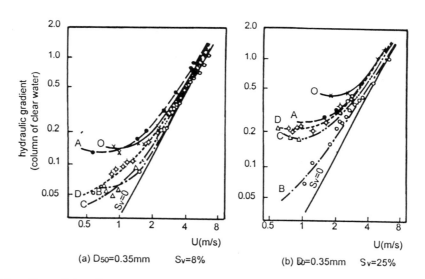

(a) $D_{50}=0.35$mm $S_V=8\%$

(b) $D_0=0.35$mm $S_V=25\%$

Fig. 17.23 The effect of size distribution and content of fine particles on head loss (Dai experiments)

835

Dai studied the head losses for two sediments with the following size distributions

Series	D_{50} (mm)	D_{65}/D_{10}
O	0.35	1.44
A	0.35	2.74

His experiments were conducted in a 25 mm pipeline. Fig. 17.23a and Fig. 17.23b are relationships between the head loss and the velocity for sediment concentrations of 8% and 25% [39]. If the concentration was low, the head loss for set A (the less uniform sediment) was slightly larger than that for set o; if the concentration was high, the head loss for set A was significantly lower than that for set o.

Effect of grain size

If sediment particles move as bed load, both the Durand formula [Eq.(17.55)], and the Newitt formula, Eq. (17.54), indicate that the head loss is independent of grain size. Fig. 17.24 shows the average velocities u_t of particles of various sizes at different average flow velocities U in the same pipeline [40]. For the particular conditions of Fig. 17.24, particles with $D/d < 0.025$ moved as suspended load and therefore had the same velocity as did the flow. The coarser particles lagged behind the flow, and u_t/U decreased as D/d increased. But if $D/d > 0.10$, the particles moving along the bed were more exposed to the flow and the drag force exerted by the flow was larger; consequently, the relative velocity between the solid particles and the flow was less. Thus, in a given pipe, the transport of large particles sometimes consumes even less energy than does the transport of particles of median size. However, the effect of sediment size on the head loss of heterogeneous flow with bed load is small compared to other factors.

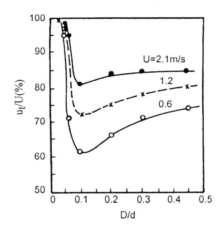

Fig. 17.24 Average velocities of particles with different sizes in pipeline

In addition, although all the existing formulas indicate that the additional head loss caused by the existence of sediment is proportional to the sediment concentration, some results indicate that this relationship holds only for low concentrations; at high concentrations, the situation may be different.

(3) Head loss for non-uniform sediments

This problem is already touched upon when the effect of size distribution on head loss is discussed. The relationships involved are particularly complex if a certain amount of fine particles are present. Further detail is included in section 17.5.4. If the amount of fine particles is not great and if the intent is to calculate the head loss by using the formula for uniform sediment, one must determine the proper diameter to represent a non-uniform sediment. Based on experimental results of three non-uniform sediments, Smith suggested the following representative diameter for non-uniform sediment:

$$D_e = \frac{\sum p}{\sum p / D}$$

(17.58)

in which p is the weight percentage of particles with diameter D. Eq. (17.58) means that the representative diameter of a non-uniform sediment D_e should be the diameter of that particle. The ratio of its surface area to its volume equals to the ratio of the total surface area of the non-uniform sediment to the total volume of the non-uniform sediment.

Condolios and Chapus [42] suggested a representative drag coefficient as follows:

$$(C_D)_e = \left[\frac{\sum p\sqrt{C_D}}{\sum p}\right]^2$$

(17.59)

Shen [43] chose to modify the Durand formula so that the representative drag coefficient would be

$$(C_D)_e = \left[\frac{\sum pC_D^{0.75}}{\sum p}\right]^{1.33}$$

(17.60)

3. Sediment motion and head loss if deposits occur

Once sediment is deposited on the bottom, the flow is one with a movable bed rather than a rigid one. Such flows are not essentially different from flow in a river; therefore, most of the research on sediment-laden flow in open channels is directly applicable to this case.

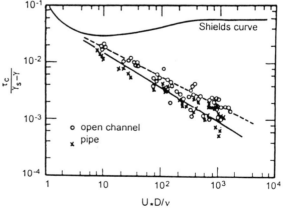

Fig. 17.25 Threshold condition for particles on a rigid bed (after Novak, P., and C. Nalluri)

837

(1) Threshold motion of sediment

With a movable bed on the bottom of pipe, a critical velocity for incipient motion exists much as for open channel flows. According to the experiments of Shen et al., if the flow is turbulent, the threshold condition can be determined from the Shields curve (Fig. 8.6) [44]. Some researchers have studied the threshold conditions for particles on a rigid bed in pipes with free surface flow (Fig. 17.25) [45]. Such particles can be moved by a much smaller drag force than can those on a movable bed in a flow with the same grain Reynolds number. Only at small Reynolds number are the threshold conditions for the two cases about the same.

(2) Bed load motion

Wilson conducted experiments in a square conduit, side length 94 mm, and in pipes with diameters of 89 mm and 53 mm. He obtained the following formula for the bed load transported by flow over a movable bed [46]

$$g_b = k \frac{\rho}{\rho_s - \rho}(\tau_0 - \tau_c)^{3/2}$$

in which g_b is the unit transport rate of bed load, in dry weight, τ_0 and τ_c are the shear stress on the bed and the threshold shear stress, respectively, and k is a coefficient. The preceding equation can be simplified to the form,

$$\Phi = \left(\frac{5.28}{\psi} - 0.248\right)^{3/2} \tag{17.61}$$

in which ϕ is the bed load function and and $1/\psi$ the flow intensity function. Eq. (17.61) and the modified Meyer-Peter formula [Eq. (9.65)] have exactly the same structure.

(3) Suspended load

In the discussion of the diffusion theory and its connection to the vertical distribution of sediment concentration (Chapter 15), the results of experiments conducted in a rectangular conduit, 24.2 cm by 7.6 cm, were described [47]. The experimental results indicate that the vertical distribution of suspended load fits with the following equation

$$\frac{S_v}{S_{va}} = \exp\left[-\frac{\omega}{\varepsilon_y}(y - a)\right] \tag{10.8}$$

In 1970, Sharp and O'Neill obtained the same result from experiments conducted in a pipe 51 mm in diameter [48].

(4) Total load

Graf and Acaroglu applied the concept suggested by Bagnold for the total load, but included both bed load and suspended load [49]. They expressed the energy conservation law in the form

$$(\gamma_s - \gamma)S_v U \frac{A}{p} \frac{U}{\omega} = \tau_0 U e_T \qquad (17.62)$$

in which A is the area of the flow section, p the perimeter of that part of the pipe with sediment deposits, ω the fall velocity of the sediment particles, e_T the efficiency of transporting the total load by the flow. From a comparison of Eq. (9.16) and Eq. (10.71), one can deduce that Graf et al. simply took the average velocity of the flow as the average velocity of the particles (total load), and transformed the submerged weight into an agitating force that maintains sediment motion by simply multiplying U/ω. All of the coefficients of proportionality appearing in their deduction are absorbed in e_T. Only if all particles move mainly as suspended load, is the preceding treatment logically correct. They assumed that the efficiency for the transport of sediment by the flow depends on the flow parameter ψ suggested by Einstein. Using experimental data obtained from both pipes and open channels and field data measured in natural rivers, they deduced the following equation:

$$\frac{S_v U R_m}{\sqrt{\dfrac{\rho_s - \rho}{\rho} g D^3}} = 10.39 \Psi^{-2.52} \qquad (17.63)$$

in which R_m is the hydraulic radius.

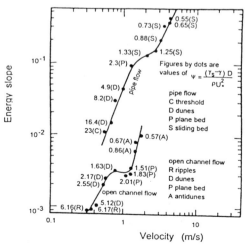

Fig. 17.26 Similarity between pipe and open
channel flows with granular beds
(after Acaroglu, E.R., and W.H. Graf)

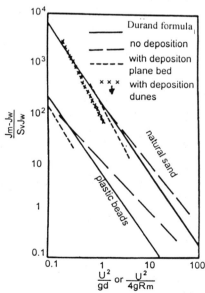

Fig. 17.27 Comparison of measured head
losses and the Durand formula
(after Wilson, K.C., and A. Brebner)

(5) Head loss

If deposits occur on the bottom, dunes may form and the head loss is then greatly affected by the bed configuration. Fig. 17.26 contains relationships between the energy slopes and the velocities in pipes and in open channels [50]; the data for open channels were taken from experiments carried out in a channel 0.61 mm wide and with a sediment size of 0.33 mm. Data for pipe flows were taken from experiments in pipes 76.2 mm in diameter with a sediment size of 2 mm. The similarity between them is striking; even the values of ψ at which bed configuration appears are of the same order. The only difference is that anti-dunes cannot occur in pipes; at high flow intensities the entire granular bed slides along the bottom of the pipe.

A method for predicting the head loss of a pipe flow with sediment deposits has been developed. If the velocity in a pipe is reduced to a certain value, sediment begins to deposit on the bottom. This velocity is called the critical silting velocity and is denoted by U_L'. For lower velocities, the deposits become thicker. Hence, the flow section reduces, and the velocity rises accordingly. Thus, a kind of restraining effect takes place. Even though the velocity is less than the critical silting velocity, siltation cannot increase indefinitely; as the deposits reach a certain thickness, the velocity rises enough that once again it reaches U_L''. In this way, a new equilibrium is established. Condolios, Chapus, and Gibert proposed the following equation [24,42]:

$$\frac{U_L'}{\sqrt{gd}} = \frac{U_L''}{\sqrt{4gR_m}} \qquad (17.64)$$

That is, the Durand formula is applied for flow over a granular bed (i.e., with deposits on the bed) and the pipe diameter is replaced by four times hydraulic radius (Fig.17.17). But this result is in question. Based on his own experiments, Wicks believes that Eq. (17.64) is not valid. Wilson et al. conducted experiments with sand of median size 0.69 mm and plastic beads 3.89 mm in diameter (specific weight 1.138); the experimental data indicate that the results of experiments with a rigid bed deviate from those with deposited sediment. The results are shown in Fig.17.27, with d replaced by $4R_m$ [51].

For sediment particles that all move as bed load, Shook and Daniel obtained the following equation using the Bagnold theory [52]:

$$J_m = \frac{f}{8gR_m}U^2 + \frac{\rho_s - \rho}{\rho}S_v \tan\alpha \qquad (17.65)$$

in which S_v is the concentration of bed load, α the dynamic friction angle for particles that shear and collide with each other. They conducted a series of experiments in a pipe 50.8 mm in diameter, and in a rectangular conduit 10×24.6 mm with four different sands (median diameters within the range 0.15~1.57 mm), nickel beads (median diameter 0.15 mm, specific weight 8.9) and lead beads (median diameter 2.03 mm, specific weight 10.87) and verified Eq. (17.65). In fact, Eq. (17.65), the Newitt formula [Eq. (17.53)] and the Worster formula [Eq. (17.56)] are identical.

4. Head loss if all sediment moves as laminated load

By the 1970s, research workers had learned that coarse particles without any fines can move at rather low velocities even if the concentration is high. Batin and Streat conducted experiments in horizontal and vertical pipes, 27 mm in diameter, using crushed glass particles (median diameter 0.4 mm and size varying within the range of 0.21~0.70 mm) [53]. In the experiments, the average velocity varied from 0.03 m/s to 1.44 m/s, the volumetric concentration varied from 57% to 64% and had a mean value of 59%, being slightly higher at low velocity. At such high concentrations, the ratio of free voids is less than zero (Table 13.2), and the particles are in direct contact with each other as they move. Under such conditions, turbulence no longer exists, and the suspended load in the common meaning of the term could not occur. Although the sediment particles were rather uniformly distributed in the pipe, they were not the homogeneous slurry of a non-Newtonian fluid mentioned in section 5.2.4. Fig. 17.28a presents distributions of particle velocities in horizontal pipelines, and no distinct plug can be detected [54]. In fact, the concentrations maintained in these experiments were close to the limiting concentration, a quantity that does not vary with flow conditions but depends rather on the minimum void clearance required to maintain relative motion between particles (discussed in Chapter 13). All these findings indicate that the sediment motion was laminated load, as defined by Bagnold and discussed in Chapter 5.

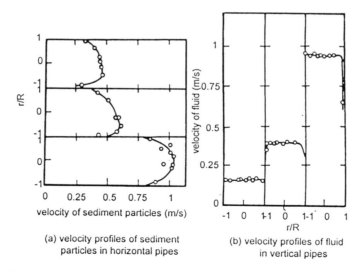

(a) velocity profiles of sediment particles in horizontal pipes

(b) velocity profiles of fluid in vertical pipes

Fig. 17.28 Motion of coarse particles at high concentrations (after Bantin, R.A., and M. Street)

The head loss for this type of flow is shown in Fig. 17.29 [54]. The most striking feature is that the head loss remains nearly constant ($J_m \approx 0.5$) over a rather wide range of velocity. As discussed in Chapter 13, the viscous shear stress caused by the deformation of fluid among solid particles at extremely high concentrations is negligible, and the agitating force along the flow direction due to the pressure gradient is balanced by shear stresses caused by the collisions among shearing solid particles.

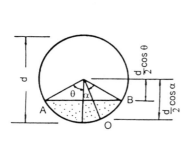

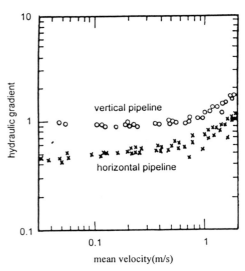

Fig. 17.30 Sketch for laminated load

Such a shear stress among the particles is finally transferred to the boundary by the friction force between particles and the boundary. Because this type of friction occurs between solid materials, the angle

Fig. 17.29 The hydraulic gradient for hyperconcentrated transport of coarse particles

of friction depends only on the properties of the friction surfaces; hence, it does not vary with the relative velocity of shear. Consequently, the hydraulic gradients are constant over a sizable range of velocity.

If the particles in a laminated bed load occupy the shaded area in Fig. 17.30 and the pressure in this region is hydrostatic, the pressure on the bed load at point o is

$$p = (\gamma_s - \gamma) S_v \frac{d}{2} (\cos\alpha - \cos\theta)$$

If f_s denotes the friction coefficient between solid particles and the boundary, the total friction force acting on a unit length of the affected boundary (the perimeter AOB) is

$$2 f_s \int_0^\theta p \frac{d}{2} d\alpha = f_s (\gamma_s - \gamma) S_v \frac{d^2}{2} (\sin\theta - \theta \cos\theta)$$

If all the particles in the pipe move as laminated load so that $\theta = 180^\circ$, the total friction force between sediment particles and the boundary is

$$2 f_s (\gamma_s - \gamma) S_v \frac{\pi d^2}{4}$$

This friction force should be equal to the difference of the two pressure forces acting on the ends of the isolated section, a unit distance apart; i.e.,

$$\gamma \, J_m \frac{\pi d^2}{4} = 2 f_s (\gamma_s - \gamma) S_v \frac{\pi d^2}{4}$$

After simplifying, one obtains

$$J_m = 2 f_s \left(\frac{\rho_s - \rho}{\rho} \right) S_v \tag{17.66}$$

Wilson conducted simple experiments with the sediment used by Batin and Streat, and found that $f_s = 0.29$. Substituting this into Eq. (17.66) and setting $S_v = 0.59$ and $\rho_s/\rho = 2.5$, one obtains 0.51 for J_m. This result agrees well with the measured data, as shown in Fig. 17.29 [55].

17.3.1.3. Determination of the critical condition

In heterogeneous pipe flows, several critical conditions occur, including those for the transitions from one flow type to another and the critical velocity U_L corresponding to the lowest point of the $J_m \sim U$ relationship.

1. Critical condition for the transition from uniform suspension to non-uniform suspension

According to the diffusion theory, the degree of uniformity of the vertical distribution of suspended particles depends on the parameter

$$z = \frac{\omega}{k U_*} \tag{10.17}$$

Fig. 10.10 shows that the vertical distribution of suspended particles is rather uniform if $z < 0.25$. For the data of the Newitt experiment, Shen obtained a critical z value for the transition from uniform suspension to non-uniform suspension of 0.19 [43]. Stevens and Charles [56] used 0.13, and Wasp [57] 0.11 for ω/U_* as the critical value. If the Karman constant is taken as 0.4, the corresponding values of z are 0.325 and 0.275.

2. Critical condition for the initial suspension of sediment particles.

According to the vertical distribution of suspended particles in Fig. 10.10, sediment particles move essentially as bed load if

$$z = \frac{\omega}{k U_*} \geq 5 \tag{10.20}$$

sediment is suspended if $z < 5$.

Wilson and Watt found that large-scale turbulence is reduced if the diameter of pipe is not quite large in comparison with the sediment size. Then the factor D/d can affect sediment suspension and should be considered [68]. From experimental data, they obtained the following empirical formula for the critical velocity, U_s,

$$U_s = 0.6\omega\sqrt{\frac{8}{f_w}}e^{0.45D/d} \qquad (17.67)$$

If $D/d = 0.002$, $e^{0.45D/d} = 1.094$, i.e., the error caused by neglecting the effect of D/d is less than 10%. This result together with the relationship

$$\frac{U}{U_*} = \sqrt{\frac{8}{f}}$$

one can transform Eq. (17.67) into

$$\omega/U_* = 1.667$$

For $\kappa = 0.4$, the critical condition for the initial suspension of particles is

$$z = \frac{\omega}{kU_*} = 4.17$$

a value close to that in Eq. (10.20). If one uses the friction factor for sediment-laden flow in place of that for clear water, the critical value of z is larger and thus closer to that in Eq. (10.20).

3. *Critical velocity U_L corresponding to the minimum value of the $J_m \sim U$ relationship*

In many papers the lowest point of the $J_m \sim U$ relationship is taken as the critical condition corresponding to the initiation of deposits; thus, no distinction is made between the critical velocity U_L and the critical silting velocity U_L'. Fig. 17.31 shows that although these two critical velocities differ somewhat, the difference is small enough to be neglected.

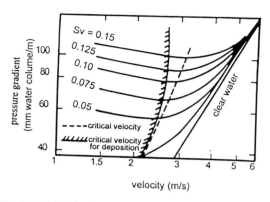

Fig. 17.31 Critical velocity and critical silting velocity

The critical velocity can be obtained directly by differentiating the known relationship between head loss and velocity. For instance, Shook used the Durand formula and obtained the following equation

$$\frac{J_m - J_w}{S_v J_w} = k \left[\frac{U^2 \sqrt{C_D}}{gd\left(\dfrac{\rho_s - \rho}{\rho}\right)} \right]^{-3/2}$$

in which he used the value $\kappa = 81$, proposed by Bantin, and set $dJ_m/dU = 0$. He obtained [55]

$$U_L = 2.43 \frac{S_v^{1/3} \sqrt{2gd\dfrac{\rho_s - \rho}{\rho}}}{C_D^{1/4}} \qquad (17.68)$$

4. Critical silting velocity

Usually deposits are not allowed to occur on the pipe bottom in order to reduce the risk of blockage of the pipeline; therefore many studies focus on the critical silting velocity. It is determined either by direct observation or by locating the minimum of $J_m \sim U$ relationship.

(1) Expression for the critical silting velocity

If the sediment is coarse and the concentration is low, the critical silting velocity actually corresponds to the threshold velocity for sediment particles on a rigid bed, shown in Fig. 17.25, and it is independent of the concentration. If the concentration has a certain value, the critical silting velocity corresponds to the transition from a non-saturated sediment transport state to a saturated one; in such a case, the concentration is an important factor. From an abundance of experimental data, Durand found [23]

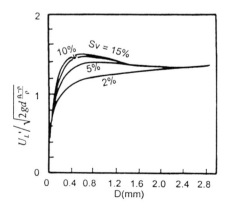

Fig. 17.32 Relationship between critical silting velocity and sediment properties

845

$$\frac{U'_L}{\sqrt{2gd\dfrac{\rho_s - \rho}{\rho}}} = F_L \qquad (17.69)$$

in which F_L is a function of concentration and grain size, as shown in Fig. 17.32. If sediment is smaller than 1 mm, F_L is a function of both concentration and grain size; if it is larger than about 1 mm, F_L approaches a constant value of 1.34, and the critical silting velocity is then independent of these variables. A number of experiments on hydrotransport in pipes were conducted at the Francis Soil Mechanics and Hydraulic Institute, Hannover College (HIHC). Their results and the Durand formula are compared in Fig. 17.33 for volumetric concentrations higher than 15% [60].

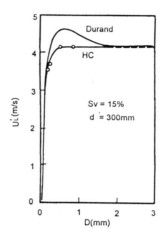

Fig. 17.33 A comparison of Hannover College experimental results and the Durand formula

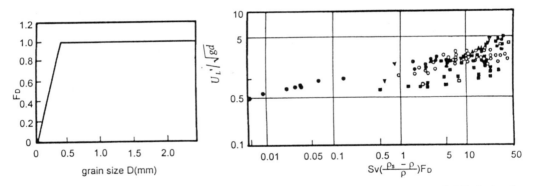

Fig. 17.34 Critical silting velocity, by Hayden and Stelsen (after Hayden, J.W., and T.E. Stelsen)

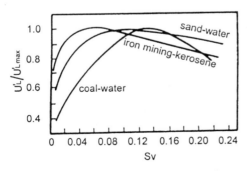

Fig. 17.35 Critical silting velocity, by Sinclair (after Sinclair, C.G.)

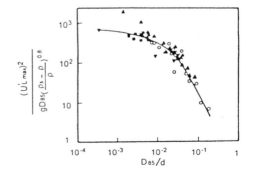

Fig. 17.36 Maximum critical silting velocity, by Sinclair (after Sinclair, C.G.)

846

Data from the Hayden and Stelsen experiments are shown in a plot of $U_L'/\sqrt{gd}$ vs. $S_v((\rho_s-\rho)/\rho)\,F_D$ in Fig. 17.34 [29], in which F_D is a function of grain size. The points scatter quite a lot. Sinclair carried out systematic experiments with sand, coal powder, and iron tailing (grain sizes vary from 0.03 mm to 2 mm) in small pipes ($d = $ 12.7, 19.1 and 25.4 mm). He found that the critical silting velocity reaches a maximum U'_{Lmax} for values of the concentrations S_v in the range of 0.05~0.20. Relationships between $U'_L/\,U'_{Lmax}$ and the concentration are shown in Fig. 17.35. U'_{Lmax} can be determined from relationship given in Fig. 17.36 [61].

(2) Comparison of various formulas

The preceding discussion includes a few of the existing formulas for critical velocity. Many more empirical or semi-empirical formulas have been proposed. Weidenroth and Kirchner compared the critical silting velocities determined from 15 various formulas for five different sands (D_{50}=0.20~5.95 mm) [62]; the formulas originated as follows:

No.	Author	No.	Author
1	Wilson (1942)	8	Sinclair (1962)
2	Durand (1953)	9	Brauer (1965)
3	Newitt (1955)	10	Yufin (1966)
4	Spells (1955)	11	Zandi (1967)
5	Cairns (1960)	12	Weidenroth (1967)
6	Hughmark (1961)	13	Rose (1969)
7	Yufin (1949)	14	Badcock (1970)
		15	Bain (1970)

The results are compared in Fig. 17.37 and Fig. 17.38, and the conditions for the comparison are as follows:

Figure	Pipe diameter mm	Sediment diameter mm	Volumetric concentration (%)
17.37(a)	100	varies	10
17.37(b)	varies	0.20	10
17.38(a)	100	0.20	varies
17.38(b)	300	0.20	varies

The differences among them are clearly quite large. They arise partly because of the different definitions of critical conditions adopted by different authors. Also, some formulas are applied in a range for which they are not suitable. For instance, formula 4 applies only in the range 0.05~0.5 mm and formula 5 for 0.24~0.38 mm. But the more important reasons are limitations on the scope of the various formulas.

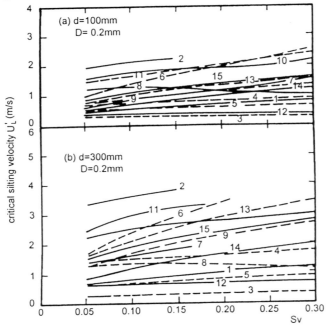

If a few extreme cases are excluded, the results determined from the rest of the formulas display some consistency. If the sediment size is larger than 1 mm, the grain size has no marked effect on the critical silting velocity. If the concentration is higher than 0.05, the critical silting velocity does not increase much with increase of concentration. As most formulas are derived from experiments with small pipes, large differences may occur if they are applied to large pipes, as is clearly shown in the comparisons in Fig. 17.37a and Fig. 17.38a.

Fig. 17.38 A comparison of the critical silting velocities with varying concentration and pipe diameter according to various formulas (after Wiedenroth, W., and M. Kirchner)

(3) Factors affecting the critical silting velocity

To obtain further verification of the effects of various factors on the critical silting

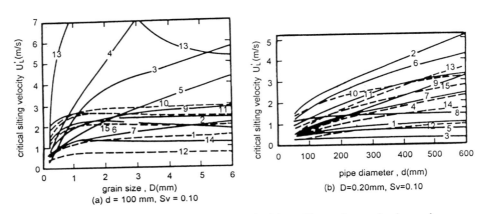

Fig. 17.37 A comparison of the critical silting velocities with varying grain size and pipe diameter according to various formulas (after Wiedenroth, W., and M. Kirchner)

848

velocity, one can conduct experiments by isolating the effects of one variable at a time. A series of experiments on heterogeneous flows in pipes was carried out at Karlsruhe University, Germany. The ranges and values for the relevant variables are as follows:

sediment size 0.1~10 mm

specific weight of sediment 1.10, 1.53, 2.65 and 4.55

volumetric concentration 0.05~0.25

pipe diameter 40, 80, 150, 200 mm

Part of the experimental data for relationships between two factors are shown in Fig. 17.39[63]. If the sediment is larger than 1 mm, the critical silting velocity indeed does not vary with the sediment size. But experimental results obtained for pipes of different sizes differ considerably, and the relationships between the critical silting velocity and the specific weight also show an effect of pipe diameter.

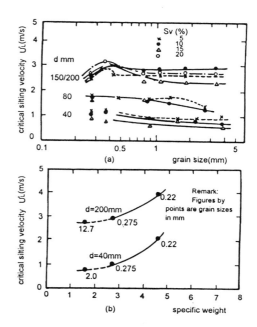

Fig. 17.39 Effect of grain size, specific weight, concentration, and pipe diameter on the critical silting velocity (after Weber, M., and E. Godde)

BMC and Jingxi Mining Bureau conducted experiments on the effect of concentration on critical velocity with local coal ($D < 25$ mm) in a steel pipe 100 mm in diameter. Their results are shown in Fig. 17.40; the concentration is in terms of the specific weight of the coal slurry [64]. Within the range of 1:1.7 to 1:1.2 (volumetric ratio of coal to water), the critical fall velocity decreases sharply as the ratio increases. The cause is that the settling velocity of coal particles decreases with increasing concentration. However, the range of concentration was not wide enough for the effect of the concentration on fall velocity to be fully displayed. If the concentration increases further, the slurry may possess non-Newtonian properties, and the head loss will then increase if the flow becomes laminar. Kazanskij analyzed the experimental data of various researchers and obtained the result shown in Fig. 17.41[65]; in it, p is the percentage of coarse particles with

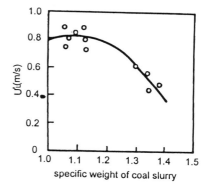

Fig. 17.40 U'_L vs. specific weight for coal particles

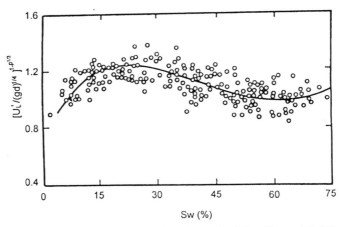

Fig. 17.41 Critical silting velocity U'_{Li} vs. S_w (after Kazanskij, I.B.)

$$Fr = \frac{\omega}{\sqrt{gD}} > 0.4$$

The curve for the relationship between the critical silting velocity and the concentration displays both a maximum and a minimum. This situation is much like that for hyperconcentrated flow in open channels (Fig.13.19).

5. Critical condition for pipe blockage

If a heterogeneous flow blocks the pipe, the intervals between particles are so small that sediment particles cannot move. One could expect that the free porosity ratio should then approach a constant smaller than zero. Without more experimental data, the value of the constant cannot be determined.

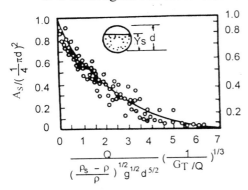

Fig. 17.42 Thickness of deposits vs. flow and sediment parameters (after Vanoni, V.A.)

From an analysis of experimental data for heterogeneous flow with deposits at Iowa University, Laursen obtained the relationship shown in Fig. 17.42, in which y_s is the thickness of the deposit and A_s the area occupied by deposit [43]. If one takes some value for $A_s/(0.25\pi d^2)$ (for example, 0.75) as the critical condition for pipe blockage, then the equation is

$$\frac{Q}{\left(\dfrac{\rho_s - \rho}{\rho}\right)^{1/2} g^{1/2} d^{5/2}} \left(\frac{1}{G_T / Q}\right)^{1/3} = 1 \qquad (17.70)$$

in which G_T is the transport rate (in percent by volume) of the total load, Q the discharge of the water-sediment mixture.

6. Unified standard for classification of flow regions

Various critical conditions for different situations are discussed in the foregoing. Some researchers, for example Turian and Yuan, would like to adopt an unified standard for classifying the different types of flow [66]. They collected the experimental data available and conducted supplementary experiments at rather large scale with the following ranges for the variables

$$d=12.6\sim699 \text{ mm}$$

$$D=0.03\sim38 \text{ mm}$$

$$\gamma_s=1.16\sim11.3 \text{ t/m}^3$$

$$S_v=0.006\sim42 \text{ \%}$$

$$U=0.009\sim6.7 \text{ m/s}$$

The total number of experimental data they collected was 3848. From dimensional analysis, they proposed a basic form for the friction factor f_m for heterogeneous flows in pipes

$$f_m - f_w = kS_v^a f_w^\beta C_D^r \left[\frac{U^2}{gd\left(\frac{\rho_s - \rho}{\rho}\right)} \right]^\delta \tag{17.71}$$

in which f_w reflects the effect of the Reynolds number, C_D the effect of sediment size. The coefficient and the exponents in Fig. 17.71 vary for different types of flow as shown in Table 17.5.

Table 17.5 Coefficient and exponents in Eq. (17.71) for different types of flow (after Turian, R.M., and Tran Fu Yuan)

Flow region	Description	α	β	γ	δ	κ	numbers of data points
0	deposition on bottom	0.739	0.772	-0.405	-1.096	0.404	361
1	bed load on bottom	1.108	1.046	-0.421	-1.354	0.986	1230
2	non-uniform suspension	0.869	1.200	-0.168	-0.694	0.551	493
3	uniform suspension	0.502	1.428	-0.152	-0.353	0.844	645

The critical condition for the transition from type a to type b is that the head losses calculated for these two types should be equal, thus

$$\frac{U^2}{gd\left(\frac{\rho_s - \rho}{\rho}\right)} = k'S_v^{a'} f_w^{\beta'} C_D^{r'} \tag{17.72}$$

in which

$$
\left.
\begin{array}{l}
k' = \left(\dfrac{k_b}{k_a}\right)^{\frac{1}{\delta_b - \delta_a}} \\[12pt]
\alpha' = \dfrac{\alpha_b - \alpha_a}{\delta_a - \delta_b} \\[12pt]
\beta'' = \dfrac{\beta_b - \beta_a}{\delta_a - \delta_b} \\[12pt]
\gamma' = \dfrac{\gamma_b - \gamma_a}{\delta_a - \delta_b}
\end{array}
\right\}
\qquad (17.73)
$$

The coefficients and exponents in Eq. (17.72) corresponding to the various critical conditions are listed in Table 17.6.

Table 17.6 Coefficients and exponents in Eq. (17.72) for various critical conditions
(after Turian, R.M., and Tran Fu Yuan)

Critical condition	κ'	α'	β'	γ'
$0 \rightarrow 1$	31.93	1.08	1.06	-0.06
$0 \rightarrow 2$	0.46	-0.32	-1.07	-0.59
$0 \rightarrow 3$	0.37	0.32	-0.88	-0.75
$0 \rightarrow 4$	2.41	0.23	-0.23	-0.38
$0 \rightarrow 5$	1.17	0.52	-0.38	-0.57
$0 \rightarrow 6$	0.29	1.08	-0.67	-0.94

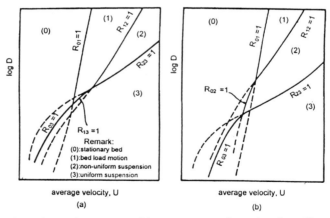

Fig. 17.43 Classifications for various types of heterogeneous flows in pipes (for given fluid, solid material, concentration, and pipe diameter) (after Turian, R.M., and Tran Fu Yuan)

Fig.17.43 shows a classification of the different types of flow for a given fluid, solid material, concentration, and pipe diameter. In the figure, $R_{ab}=1$ corresponds to the critical condition for the transition from type a to type b. In Fig. 17.44, lines of demarcation are shown for flows transporting solid material with a specific weight of 3 and a concentration of 5% in pipes with two different diameters. It shows that the lines of demarcation move to the right as the pipe diameter increases.

More studies of this type are available[67]; no further discussion is presented herein.

17.3.2 Two-phase flow in pipelines with homogeneous slurry as transporting medium

The preceding section indicates that maintaining the critical condition for deposition in two-phase pressure flow might be economical. In practice, however, some quantities are difficult to estimate; hence, to avoid blockage of a pipeline, the velocity adopted is usually some 20% to 30% higher than the critical velocity for deposition, and the sediment particles are thus mostly suspended. As pointed out in Chapter 3, the fall velocity of sediment particles reduces significantly with the increase of sediment concentration, and does so particularly if some of the particles are finer than 0.01 mm. The reduction of fall velocity favors the suspension of sediment particles. As a result, more attention has been paid in recent times to hyperconcentrated transport in pipeline. Based on their experimental results, Charles et al. achieved a saving of energy of 8% for the following conditions: conveyance of one million tons of 0.22 mm sand per annum, at a concentration of 56% (by weight), in a pipeline 200 mm in diameter, using a clay slurry at a concentration of 19% (by weight). The comparison is with clear water flow under the same conditions [68].

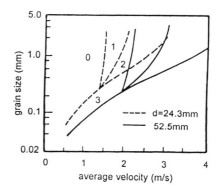

Fig. 17.44 Classifications of various types of flow for two different pipes (after Turian, R.M., and Tran Fu Yuan)

This section treats the calculation of head loss of hydrotransport of solid material by a homogeneous slurry. The drag reduction and the proper concentration of the slurry is discussed in section 5.

17.3.2.1 Early calculation

In the early 1960s, Wasp et al. suggested the following approach for calculating head loss in hydrotransport of coal provided that both coal particles and coal powder are carried in a pipeline [69]:

1. Divide the solid material into two parts: homogeneously suspended and heterogeneously suspended.

2. Calculate the head loss for the homogeneously suspended slurry according to the usual relationship for friction factor and Reynolds number. In the determination of unit weight and viscosity both fine particles and coarse particles are taken into consideration. For the reduction of head loss caused by the existence of coarse particles, Wasp et al. multiplied the friction factor obtained from the ordinary f — Re curve by 0.85.

3. Calculate the head loss of heterogeneously suspended sediment particles according to the Durand formula, using clear water as the carrying medium.

4. Add these two components of head loss to obtain the total head loss.

The preceding method of calculation is not a perfect one, although in their paper published in 1971, Wasp et al. stated that they used it successfully to calculate the head loss for the hydrotransport of sand (D = 0.26~1.23 mm, S_V = 5~25%) by bentonite slurry (S_V = 0 ~ 4.1%) [57].

17.3.2.2 Recent developments

More recently, the method of calculation introduced in the preceding section has been widely adopted for the calculation of head loss in hyperconcentrated pipe flow; however, instead of clear-water, a homogeneous slurry consisting of water and fine particles is taken as the transporting medium, and the solid particles, excluding the fine ones, are taken as the solid phase.

After a study of the transport of crushed stones (D = 3.34 mm, specific weight 2.687) and plastic beads (D = 3.80 mm, specific weight 1.304) with a bentonite slurry in a pipeline 41.2 mm in diameter, Japanese scientists obtained the following formula [70].

$$\frac{J_m - J_f}{S_v J_f} = \frac{2}{f_f}\left(\frac{U^2}{gd}\frac{\rho_f}{\rho_s - \rho_f}\right)^{-1.25} \tag{17.74}$$

in which subscript f denotes slurry. If figures relating to slurry in the preceding formula are replaced by the corresponding figures relating to clear water, they found that Eq. (17.74) can also be applied to the experimental data of two-phase pipe flow of clear water in the experiments conducted by Durand, Newitt, Babcock, and others.

Kenchington studied the transport of 0.75 mm sand (S_w = 5, 10, 15, 20%) by a clay slurry (S_w = 40%) in pipelines of 13, 25, and 51 mm [71]. With laminar flow, sand moved in the form of bed load; in such a case, the resistance consists of two parts: (1) resistance to slurry, which follows the law for the flow of an exponential fluid; (2) resistance caused by bed load motion. These two parts of the resistance can be calculated according to corresponding formulas and then added together. For turbulent flow, the resistance is essentially equal to the resistance for the slurry flow. In a pipeline 13 mm in diameter, the resistance to two-phase flow with a sand concentration of 20% (by weight) was even less than that for pure slurry without sand.

Hisamitsu et al. [72] studied the motion of four kinds of sand in a pipeline 78.8 mm in diameter:

sand	D_{50} (mm)
A	0.15
B	0.20
C	0.45
D	0.57~0.80

In their experiments, the liquid phase consisted either of a slurry of water and limestone powder or of water and clay. With concentrations of fine particles (percent of volume) not exceeding 8 to 10%, the slurries had the properties of a Newtonian fluid. The experiments showed that two-phase flow in a pipeline with clear water follows a slightly modified Durand formula:

$$\frac{J_m - J_w}{S_v J_w} = 120 \left(\frac{U^2}{gd\left(\frac{\rho_s - \rho}{\rho}\right)} \sqrt{C_D} \right)^{-1.5} + \left(\sqrt{\frac{\rho_s}{\rho}} - 1 \right) \tag{17.75}$$

and two-phase flow with a slurry follows the following formula:

$$\frac{J_m - J_f}{S_v J_f} = k \left(\frac{U^2}{gd\left(\frac{\rho_s - \rho_f}{\rho_f}\right)} \sqrt{C_D} \right)^{-1.5} + \left(\sqrt{\frac{\rho_s}{\rho_f}} - 1 \right) \tag{17.76}$$

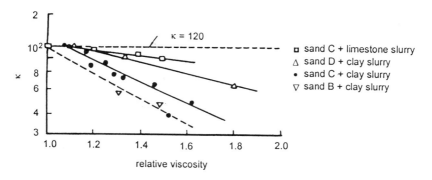

Fig. 17.45 Relationship between κ in Eq. (17.76) and the relative viscosity of the slurry
(after Hisamitsu, N.; Y. Shoji; and S. Kosugi)

855

Herein, C_D is the drag coefficient for sediment particles settling in slurry, and it can be determined from the material presented in Chapter 3. The coefficient κ depends on the relative viscosity of the slurry, Fig. 17.45.

17.4 HYDROTRANSPORT OF SOLID MATERIAL IN VERTICAL AND INCLINED PIPELINES

The text has so far treated only the hydrotransport of solid material in horizontal pipelines. The following sections present the changes in head loss if the pipeline is inclined to the horizontal.

17.4.1 Head loss in vertical pipelines

In vertical pipelines, the flow and the settling of sediment particles move in the same direction and no bed load motion (in the conventional sense) exists. Experience shows that the relative motion between sediment and water can be neglected. The head loss per unit length J_m (column of clear water) can be obtained from the following equation:

$$J_m = J_w \pm S_v \left(\frac{\rho_s - \rho}{\rho} \right) \tag{17.77}$$

The positive sign is used if the flow moves upward, and the negative if downward. That is, if the effect of the sediment on static pressure has been included, the head loss of two phase pipe flow does not different from that for clear water flow.

Eq. (17.77) is valid if the fall velocity of sediment particles is much smaller than the average velocity of water-sediment mixture. If these two velocities are of the same order of magnitude, Newitt suggested a modified form of Eq. (17.77) [73]:

$$J_m = J_w \pm S_v \left(\frac{\rho_s - \rho}{\rho} \right) \frac{U}{U_w} \tag{17.78}$$

in which U_w is the average velocity of the liquid phase.

17.4.2 Head loss in inclined pipelines

Worster and Denny recommended that the head loss of a flow in a pipeline inclined at an angle θ be taken as the vector sum of the head loss of a vertical pipeline and that of a horizontal pipeline in accordance with the inclination of the pipe [35], that is:

$$J_m(\theta) = J_w + J_s(0)\cos\theta + S_v \left(\frac{\rho_s - \rho}{\rho} \right) \sin\theta \tag{17.79}$$

in which,

$J_m(\theta)$ — energy slope for the two-phase flow in an inclined pipeline, column of clear water;

J_w — energy slope of clear-water flow;

$J_s(0)$ — the additional head loss per unit length due to the existence of sediment particles in horizontal two-phase pipe flow.

Experience accumulated at the NYRPIC Laboratory in France shows that the second term on the right side can be taken as [28]

$$J_m(\theta) - J_w - S_v\left(\frac{\rho_s - \rho}{\rho}\right)\sin\theta = 180 S_v J_w \left[\frac{U^2}{gd}\frac{\sqrt{C_D}}{\cos\theta}\right]^{-3/2} \qquad (17.80)$$

Experimental results obtained for flow in a 150 mm pipeline with inclinations to the horizontal of 0, 15, 30, and 45 degrees are shown in Fig. 17.46.

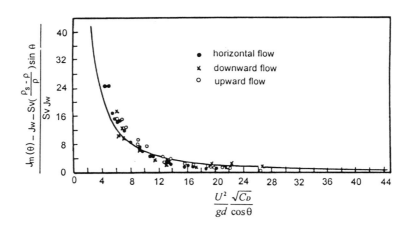

Fig. 17.46 Head loss for two-phase flow in inclined pipelines (after Zandi, I., and G. Govatos)

In Eq. (17.79) and Eq. (17.80), $\cos\theta$ does not change sign as the sign of θ changes. If the head loss for two-phase flow in an inclined pipeline is measured with a column of slurry, the additional head loss due to the sediment is the same whether the flow runs upward or downward. Experimental results in Fig. 17.46 appear to verify this result. However, the situation is different if significant quantities of sediment move in the form of bed load.

Chapter 9, in which the mechanism of bed load is treated, clearly indicates that as particles fall onto the bed, they jump once again from the bed and re-enter the main flow; hence, the flow must exert a force on them to accelerate them as they regain the momentum they have lost. In this way, the existence of bed load increases the energy dissipated by the flow. In upward pipe flow the acceleration is against the direction of gravity. In downward pipe flow the acceleration is in the direction of gravity. That is, if bed load constitutes an important part of the sediment motion, flow in a pipeline inclined upward consumes more energy than does flow in a horizontal pipe, and the flow in one inclined downward consumes less energy.

A series of experiments using crushed walnut shells (D_{50} = 1.50 mm, ρ_s= 1.35 g/cm^3), glass beads (D_{50} = 0.06 mm, ρ_s =.48 g/cm^3), and powdered coal (D_{50} = 1.40 mm, ρ_s= 1.76 g/cm^3) were conducted in pipelines inclined at various angles. The experimental results for the glass beads and powdered coal are shown in Fig. 17.47 [74]. The figure shows that the head loss for flow in pipeline inclined upward is larger than that for a similar flow in a horizontal pipeline; the loss for a flow in pipelines inclined downward is smaller. The larger the density difference between the solid and liquid phases is, the larger the difference in head loss will be. In a vertical pipeline, with the head loss measured by a column of water-solid particle mixture, the head loss of a two-phase flow is equal to that of a clear water flow; therefore, if the inclination angle is larger than some critical value, the head loss of an upward flow in a pipeline might be smaller than that of a flow in a corresponding horizontal pipeline. This deduction, drawn by inference from Fig. 17.47, has not been proved experimentally.

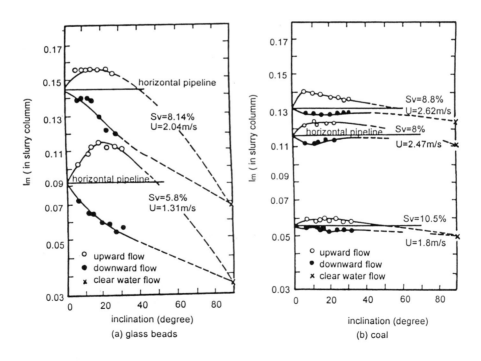

Fig. 17.47 Hydraulic gradient for flow with various concentrations of solid particles in pipelines at different inclinations (after Kao, D.T.Y., and L.Y. Hwang)

17.5 DRAG REDUCTION

In recent years many studies on drag reduction have been conducted for the purpose of saving energy. Several international symposiums on the topic have been held. Drag reduction at present consists of four different procedures: (1) adding a polymer or solution, (2) adding fibrous material, (3) adding air bubbles, (4) adding fine sediment. The topic can not be discussed fully herein; the following general outline of the studies emphasizes drag reduction achieved by adding fine particles.

17.5.1 Drag reduction by polymer addition

Early in 1948, Toms found that the surface friction occurring along the surface of a solid object in a fluid flow can be effectively reduced by adding a polymer to the fluid [75]. This method of drag reduction came to be called the "Toms effect." In 1972, Hoyt made a preliminary summary of the existing references [76], and Virk made another in 1975 [77]. Two years later, when Hoyt summarized the results of studies during 1975 and 1976, the number of papers cited was 171 [78]. BHRA Fluid Engineering published a literature index [79] including more than one thousand papers. These figures show how active research in this field became.

17.5.1.1 Polymers that cause drag reduction

Polymers possessing the following four conditions can have an effect on drag reduction:

1. long chain structures and few or no sub-chains;

2. large molecular weight($= 10^6$ or more);

3. flexible;

4. soluble.

Porch conducted experiments in a pipe with 41 kinds of polymers and studied their drag reduction effects and their processes of deterioration [80].

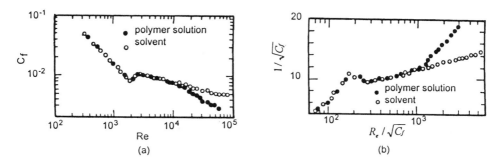

Fig. 17.48 Drag reduction by polymer addition (after Virk, P.S.)

17.5.1.2 Main characteristics of drag reduction

1. Fig. 17.48 presents the relationship between the Fanning friction factor c_f and the Reynolds number for both distilled water and an oxidized polyethylene solution (concentration by weight 0.03%, molecular weight 0.57×10^6) in a pipe 8.45 mm in diameter [77]. It shows that the addition of a polymer does not reduce the drag if the flow is laminar. Only for turbulent flow with a Reynolds number that is sufficient large (in Fig. 17.48, $Re > 12,000$) does the drag reduction increase significantly with increasing discharge. Within the range of the experiments, the discharge was 38% greater for a given shear stress on the boundary.

2. The critical condition for which drag reduction begins depends on the shear stress on the boundary τ_w and the properties of the polymer. τ_w reflects the flow characteristics. For given solution and polymer, the critical condition is a function of the rotating radius of the polymer, and it is independent of pipe diameter, viscosity of the solution and polymer concentration.

3. For a given concentration of polymer the smaller the pipe, the more the drag is reduced. This result can affect the applicability of data obtained from small pipes to flow in large pipes.

4. If the concentration of the polymer is low, the drag reduction increases with increasing concentration. If the concentration exceeds a certain value, however, further increase of the polymer concentration does not cause further decrease in drag; still, because of the increase of viscosity, complex situations can arise. Fig. 17.49 presents a comparison of the friction factors for flow of clear water and of a 0.15% concentration (by weight) of polyox-N750 (a kind of polymer) [78]. The Reynolds number used in the figure is based on the viscosity of clear water. The figure shows that, depending on the Reynolds number, the friction factor for flow with polymer can be either larger or smaller than that for clear water flow.

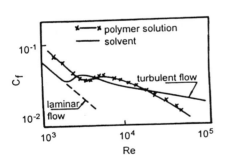

Fig. 17.49 A comparison of the values of c_f with and without high concentrations of polymers (after White, A., and Hemmings J.A.G.)

5. For a given Reynolds number, the effect of polymer on drag reduction has a maximum value. In Fig. 17.50, the critical condition under which the drag reduction starts to appear varies with the pipe diameter and the polymer used, but finally the friction factor can be described by the following equation; it is independent of the polymer properties and the pipe diameter:

$$\frac{1}{\sqrt{c_f}} = 19 \, Log \, Re \sqrt{c_f} - 32.4 \qquad (17.80)$$

6. The preceding experimental results were obtained for flow in hydraulically smooth pipes. For rough pipes, adding a polymer makes the laminar sublayer thicker and the relative roughness of the boundary less; consequently, the drag is reduced. The data from the Virk experiments indicated that both the critical condition at which the drag reduction occurs and the maximum drag reduction are the same as those for smooth pipes [81].

7. The effectiveness of polymers decreases with time, particularly for high rates of shear. The drag reduction of the polymers is caused mainly by their long-chain structure, and high rates of shear often break the long chains. The larger the molecular weight of the polymer is, the faster the polymer deteriorates. Consequently, many researchers have suggested the replacement of the polymer, by a soap solution, which has a rather complex chemical structure. Such solutions form a flocculation structure due to binding forces. Although the structure is destroyed if the shear rate is high and the drag reduction becomes less, both the structure and the drag reduction are recovered if the shear rate reduces. If a soap solution is used to reduce the drag, better results are obtained if the concentration is high. In experiments by Poreh et al. experiments with heterogeneous flows in a pipe 20 mm in diameter (median sand of 0.9 mm and phenolic moulding powder with a specific weight of 1.43 and a diameter of 1 mm), the head loss was reduced 70% by the addition of LTAB-1-naphthol, concentration 0.18% [82].

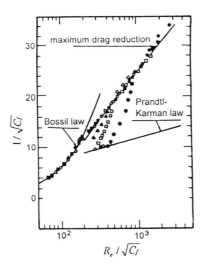

Fig. 17.50 Maximum effect on drag reduction by addition of polymer
(after White, A., and J.A.G. Hemmings)

861

17.5.1.3 Mechanism of drag reduction

No one has developed a generalized theory about the mechanism of drag reduction by adding polymers. Some researchers have suggested that the addition of polymers causes visco-elasticity in the sublayer, and that it reduces the transfer of eddies from the boundary to the central part of the pipe; in this way, it reduces the resistance to turbulence flow. Working with the basic properties of visco-elastic flow, Dou studied both the turbulence structure in a flow with drag reduction and the mechanism of the reduction [83]. However, drag reduction also occurs in two-phase air flows carrying solids [84] , and they do not possess visco-elasticity. Therefore, Dou's explanation does not fit this situation. Observations showed that the bursting phenomenon near the bed changes substantially if a polymer is added. The intervals between the formation of bursting lumps increases and the periods of bursting are prolonged. These actions decrease the rate of generation of turbulence [85].

Both of these explanations indicate that the addition of polymers interferes with the equilibrium among the supply, transfer, and consumption of energy within the flow. A part of the energy is absorbed by the polymer as energy is transferred from large eddies to small eddies. Thus the supply of energy to the small eddies is less, and if a polymer is added, small eddies with high frequencies are fewer or do not occur. Apparently, the smaller the small eddies and the higher their frequencies in the structure of turbulence, the more effective is the polymer in increasing drag reduction.

As discussed in Chapter 4, the Kolmogorov theory indicates that the size of the smallest eddy l_k is given by

$$l_k = (v^3 / \varepsilon)^{1/4} \tag{4.7}$$

and the turbulence velocity by

$$\sqrt{\overline{u_k'}^2} = (v\varepsilon)^{1/4} \tag{4.8}$$

If both sides of Eq. (4.7) are divided by both sides of Eq. (4.8), the result is a quantity with the dimension of time

$$t_k = (v / \varepsilon)^{1/2} \tag{17.81}$$

in which ε is the energy disippation rate per unit mass of water. Table 17.7 shows values of l_k and t_k for ordinary flows in pipelines of various diameters [79]. The table shows that the size of the smallest possible eddies increases rapidly and the frequency of turbulence decreases sharply with an increase of the pipe diameter. Consequently, the drag reduction from polymer addition is not as effective for large pipelines as it is for small ones.

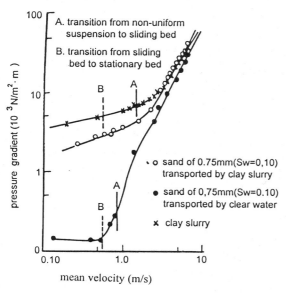

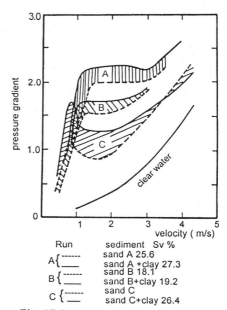

Fig. 17.54 Pressure gradient for the flow of clay slurry carrying coarse sediment particles (after Condolios, E., and E.E. Chapus)

Fig. 17.55 Drag reduction by addition of clay to two-phase flows (after Kazanskij, I.; H. Bruhl; and J. Hinsch)

The Dai experiments further reveal that there is a concentration of fine particles that optimizes the reduction of drag [39]. The head losses for two concentrations of the following five sets of sediments are shown in Fig. 17.23.

Sample	Median diameter (mm)	Average diameter (mm)	D_{65}/D_{10}	Content of particles finer than 0.01 mm (%)
O	0.35	0.355	1.44	0
A	0.35	0.392	2.74	0
B	0.35	0.404	4.23	4.3
C	0.35	0.363	24.41	9.0
D	0.35	0.371	62.86	13.5

The figure shows that the resistance reached a minimum for sample B (the content of fine particles was 4.3%); the minimum resistance for sample B was more clearly defined for a sediment concentration of 25%.

Sakamoto studied the hydrotransport of limestone in a pipe 50 mm in diameter [90]. The size of the limestone particles ranged widely. Particles finer than 0.01 mm constituted 8%; the coarsest particles were 16 mm. Sakamoto considered particles finer than 0.05 mm to be fine and particles larger than that to be coarse. For two different concentrations of coarse particles, he added different amounts of fine particles and measured the head losses. The experimental results reveal that the head loss was a minimum if the volume concentration of particles finer than 0.05 mm was 14.7% and the corresponding volumetric concentration of particles finer than 0.01 mm was 6.5%. Significantly, the limiting yield stress occurs at just this concentration. That is, the concentration of fine particles for optimum drag reduction corresponds to the maximum concentration of fine particles for the slurry to remain a Newtonian fluid.

17.6 PROBLEMS IN THE STUDY OF TWO-PHASE FLOW IN PIPELINES AND POSSIBLE IMPROVEMENTS

From the view point of its mechanics, two-phase flow in pipelines is much simpler than is flow in fluvial rivers. On the one hand, in hydrotransport the bed is usually rigid so that effects of movable beds are usually not present. On the other hand, only the resistance is important for two-phase flow in a pipeline; the more complicated study of sediment-carrying capacity is not a concern.

Although an abundance of experimental data and knowledge have been accumulated and considerable progresses has been achieved in the latest decades, still most of the resulting treatments are empirical, the formulas of different researchers differ greatly and many uncertainties arise in applying these formulas to flow in large pipes. Over many years of experience, the concepts and theories developed in the field of mechanics of sediment motion in open channels have rarely been applied to the field of two-phase flow in pipelines. Only in the decade of 1970s, did Shook, Wilson, and others begin to apply these theories to pipe flow.

As pointed out in Chapter 5, bed load affects flow resistance in a different way than does suspended load. Although different points of view have been presented as to the effect of suspended load on energy consumption, in contrast to the effect of bed load, the effect of suspended load is secondary and can be neglected only if the concentration of suspended load is not high. The existence of bed load, however, increases the consumption of energy.

For hydrotransport in pipelines, the resistance depends on the various types of flow. Strictly speaking, the present manner of subdivision of flow types is not scientific. Gaessler analyzed the existing data and gave the percentages of bed load in the total sediment load for the different types of flow, as shown in Table 17.8 [91]. The table shows that in homogeneous suspensions, all sediment particles move in the form of suspension; in all other types, various amounts of bed load occur. The percentages of bed load in the total load are different for experiments carried out by different researchers because their experimental conditions varied, and some differences occurred even if the experiments were for similar types of flow. Therefore, the experimental data are difficult to unify.

Table 17.8 The percentages of bed load in various flow regimes (after H. Gaessler)

Flow region	Percentage of the bed load in the total moving particles (%)
Uniform suspension	0
Non-uniform suspension	0-30
Bed load on bottom	30-80
Deposition on bottom	80-100

If the amount of sediment moving in the form of bed load is known, the energy required to maintain the motion of the sediment particles can be calculated. Wilson calculated the hydraulic gradient required for the transport of 1.5 mm coarse sand (S_v = 0.18) in a pipe 400 mm in diameter. In the calculation, he considered the following three cases: (1) all the particles moving as suspended load; (2) all the particles moving as bed load; (3) part moving as suspended load and part moving as bed load. The results are shown in Fig. 17.56 [92]. Although the method for calculating the hydraulic gradient suggested by Wilson may need refinement, the direction he has indicated is no doubt correct.

Fig. 17.56 reveals only that the hydraulic gradients caused by the different types of sediment motion are different. In fact, if the pipe diameter and sediment properties are known, the percentage of bed load in the total sediment load can be determined. Wilson proposed a function of the type

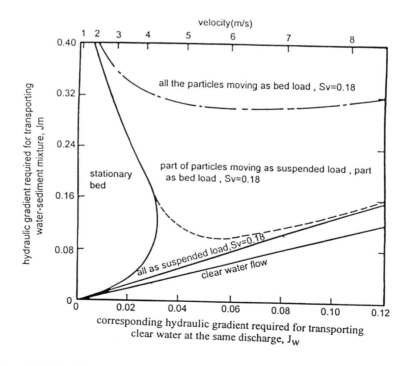

Fig. 17.56 Head losses for bed load and for suspended load (after Wilson, K.C.)

$$\frac{S_{vb}}{S_v} = \left(\frac{U_S}{U}\right)^a \tag{17.83}$$

in which S_v is the concentration of total load and S_{vb} is the part of the total that moves as bed load, U_s the critical velocity at which sediment particles start being suspended (Eq.17.67); the exponent α is slightly less than 2. This formula needs verification if it is to be applied under a wider range of conditions.

As discussed in Chapter 10, all sediment particles move as bed load if the suspension index z in the diffusion theory is larger than 5; if z is smaller than 0.25, not only are all sediment particles suspended, they are also nearly uniformly distributed. Therefore, if the concentration is not high, the ratio of the bed load to the total load should be a function of z. The ratio is one if z is larger than 5, and it approaches zero if z becomes small enough. The present need is to determine this function.

The experiments of Shook et al. serve to illustrate the preceding concepts [93]. They conducted experiments in a conduit with a rectangular cross section, 24.7×101 mm^2. The properties of sediment and the conditions of flow are listed in Table 17.9.

Table 17.9 Conditions for experiments on two-phase pipe flow, Shook et al.
(after Shook, C.A., et al.)

Set	Sediment	Median diameter (mm)	$\omega/U*$	$z=\omega_0/(\kappa U*)$	S_V	U (m/s)
A	sand	0.153	0.11-0.23	0.28-0.58	0.04-0.19	1.31-3.70
B	nickle filings	0.135	0.32-0.33	0.80-0.83	0.02-0.15	2.16-2.76
C	sand	0.35	0.22-0.57	0.55-1.43	0.03-0.20	1.29-3.84
D	sand	0.51	0.43-0.83	1.08-2.08	0.03-0.28	1.40-3.79

Note: ω_0 is the falling velocity of a single sediment particle in clear water, $k=0.4$.

Fig. 17.57a and Fig. 17.57b show the vertical concentration profiles for sets B and D, respectively. The solid lines are vertical concentration profiles obtained from the equation of diffusion theory,

$$\varepsilon_y \frac{dS_v}{dy} + S_v \omega = 0 \tag{10.7}$$

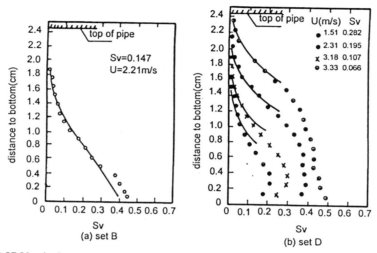

Fig. 17.57 Vertical concentration profiles for Sets B and D (after Shook, C.A., et al.)

in which the coefficient of sediment exchange is taken to be the same as the momentum exchange coefficient, and all concentrations are referred to that at a specific elevation. The figure shows that if z is 0.80, the vertical concentration profiles, except for a few points near the bed, tend to follow the diffusion theory; if $z = 1.5$, concentration profiles in the range from one-third to one-half of the distance from the bed deviate from the theoretical curves in that they are much more nearly uniform. The authors suggested that the deviation from the theoretical curves is due to the fact that part of the sediment moves as bed load. However, according to Einstein and Chien's experiments, described in Chapter 12, the vertical concentration profile of suspended load deviates from Eq. (10.7) also, and the distribution is more uniform if sediment particles are coarse and the density gradient near the bed is large. In other words, a part of the sediment Shook believed to be bed load was actually suspended load. Nonetheless, their experiments showed, to some degree, that (1) bed load occurs if z is larger than a certain value, and (2) the percentage of bed load in the total load is a function of z. If this function can be determined, it can be used to classify the types of flow, and the head loss can then be calculated. The study of hydrotransport in pipelines would then be made more rational, at least for flow at low concentrations; if the concentration is high and turbulence is fully damped, z as a physical parameter loses its meaning.

REFERENCES

[1] Wasp, E.J., J.P. Kenny, and B.L. Gandhi. "Solid-Liquid Flow Slurry Pipeline." *Transportation, Trans. Tech. Pub.*, 1997.

[2] Newitt, D.M., J.F. Richardson, M. Abbott, and R.B. Turtle. "Hydraulic Conveying of Solids in Horizontal Pipes." *Trans., Inst. Chem. Engin.*, Vol. 33, No. 4, 1995, pp. 93-113.

[3] Thomas, D.G.. "Transport Characteristics of Suspensions: 2. Minimum Transport Velocity for Flocculated Suspensions in Horizontal Pipes." *J. Amer. Inst. Chem. Engrs.*, Vol. 7, No. 3, 1961, pp. 423-430.

[4] Wilkinson, W.L. *Non-Newtonian Fluids: Fluid Mechanics, Mixing, and Heat Transfer.* Pergamon Press, 1960, p. 138.

[5] Hedstrom, B.O.A. "Flow of Plastic Materials in Pipes." *Indus. Engin. Chem.*, Vol. 44, 1952, pp. 651-656

[6] Dai, Jilan, Zhaohui Wan, et al. 1980, "An Experimental Study of Slurry Transport in Pipes." *Proc. of The International Symposium on River Sedimentation*, Vol.1, 1980, pp.195-204.

[7] Shisenko, R.E. *Experimental Study of Hydrotransport of Slurrry in Pipelines*, Oil Industry Press, 1951. (in Rusian)

[8] Behn, V.C., and R.M. Shane. "Capillary vs. Pipeline in Determining Sludge Flow Behavior", *Water, and Sewage Works*, Vol. 110, 1963, pp. 272-274.

[9] Metzner, A.B., and J.C. Reed. "Flow of Non-Newtonian Fluids--Correlation of the Laminar, Transition, and Turbulent Flow Regions." *J. Amer. Inst. Chem. Engrs.*, Vol. 1, 1955,pp. 434-440.

[10] Hanks, R.W. "The Laminar Turbulent Transition for Fluids with a Yield Stress." *J. Amer. Inst. Chem. Engrs.*, Vol. 9, 1963, p. 306.

[11] Hanks, R.W., and D.R. Pratt. "On the Flow of Bingham Plastic Slurries in Pipes and Between Parallel Plates." *J. Soc. Petroleum Engrs.*, 1967, pp. 342--346.

[12] Faddick, R.R. "Head Losses for Fine Oil Shale Slurries." *Proc., Hydrotransport 5*, 1978, pp.C4-49 to 60.

[13] Dodge, D.W., and A.B. Metzner. 'Turbulent Flow of Non-Newtonian Systems." *J. Amer. Inst. Chem. Engrs.*, Vol. 5, No. 21, 1959, pp. 189-203.

[14] Govier, G.W., and K. Aziz. *The Flow of Complex Mixtures in Pipes.* Van Nostrand Reinhold Co., 1972, p.792.

[15] Bowen, R.C. "Scale-Up for Non-Newtonian Fluid Flow." *Chem. Engin.*, Vol. 68, 1961.

[16] Kenchington, J.M. "The Design of Large Pipelines for Cement Slurries." *Proc., Hydrotransport 2*, 1972, pp.C4-41 to 60.

[17] Pai, Shih-I. *Viscous Flow Theory, 2.Turbulent Flow.* D. Van Nostrand Co., 1957, pp. 41-45.

[18] Brodkey, R.S., Jon Lee, and R.C. Chase. "A Generalized Velocity Distribution for Non-Newtonian Fluids." *J. Amer. Inst. Chem. Engrs.*, Vol. 7, No. 3, 1961, pp. 392-393.

[19] Newitt, D.M., J. Richardson, and C.A. Shook. "Hydraulic Conveying of Solids in Horizontal Pipes; Part 2, Distribution of Particles and Slip Velocities." *Proc., Symp. Interaction Between Fluids And Particles, Inst. Chem. Engrs.*, 1962, p.87.

[20] Toda, M., et al. "Hydraulic Conveying of Solids Through Horizontal and Vertical Pipes." *Chem. Engin. Japan*, Vol. 33, No. 1, 1969.

[21] Shook, C.A., and S.M. Daniel. "A Variable-Density Model of the Pipeline Flow of Suspensions." *Canadian J. Chem. Engin.*, Vol. 47, 1969, pp.196-200.

[22] Thomas, A.D. "Particle Size Effects in Turbulent Pipe Flow of Solid-Liquid Suspensions." *Proc., 6th Australian Hyd. and Fluid Mech. Conf.*, 1977.

[23] Durand, R. "Basic Relationships of the Transportation of Solids in Pipes--Experimental Research." *Proc., Minnesota Intern. Hyd. Conv.*, 1953, pp. 89-103.

[24] Gibert, R. "Transport Hydraulique et Refoulement des Mixtures en Conduites." *Annales des Ponts et Chaussees*, 130 Annee, No. 3 and 4, 1960, pp. 307-373, 437-494.

[25] Govier, G.W., and M.E. Charles. "The Hydraulics of the Pipeline Flow of Solid-Liquid Mixtures." *The Engin. J.*, Vol. 44, No. 8, 1961, pp. 50-57.

[26] Smith, R.A. "Experiments on the Flow of Sand-Water Slurries in Horizontal Pipes." *Trans., Inst. Chem. Engrs.*, Vol. 331 No. 2, 1955.

[27] Graf, W.H. *Hydraulics of Sediment Transport.* McGraw-Hill Book Co., 1971, pp. 425-502.

[28] Zandi, I. and G. Govatos. "Heterogeneous Flow of Solids in Pipeline." *J. Hyd. Div., Proc., Amer. Sec. Civil Engrs.*, Vol. 93, No. HY3, 1967, pp. 145-159.

[29] Hayden, J.W., and T.E. Stelsen. "Hydraulic Conveyance of Solids in Pipes." *Proc., Intern. Symp. on Solid-Liquid Flow in Pipes,* Univ. Pennsylvania, 1068.

[30] Pain, A.G., and S.T. Bonnington. *The Hydraulic Transport of Solids by Pipeline,* Pergamon Press, 1970, p.251.

[31] Coal Excavating Dept. of Beijing Coal College. "Hydrofilling and Hydrotransport." *Collection of Papers of Coal Mining Science,* vol.1, 1957, p.130.

[32] Wilson, W.E. "Mechanies of Flow, with Noncolloidal Inert Solids." *Trans., Amer. Soc. Civil Engrs.*, Vol. 107, 1942, pp. 1576-1594.

[33] Vocadlo, J.J., and M.E. Charles. "Prediction of Pressure Gradient for the Horizontal Turbulent Flow of Slurries." *Proc., Hydrotransport 21,* 1972, pp.C1-1 to 14.

[34] Babcock, H.A. "Heterogeneous Flow of Heterogeneous Solids." *Proc., Intern. Symp. on Solid-Liquid Flow in Pipes,* 1968.

[35] Worster, R.C., and D.F. Denny. "The Hydraulic Transport of Solid Material in Pipes." *Paper presented at a General Meeting of the Inst. Mech. Engrs.*, London, 1955, p. 12.

[36] Sasic, M., and P. Marjamovic. "On the Method for Calculation of Hydraulic Transport and Their Reliability in Practice." *Proc., Hydrotransport 5,* 1978, pp A5-61 to 76.

[37] Duckworth, R.A., and G. Argyros. "Influence of Density Ratio on the Pressure Gradient in Pipe Conveying Suspensions of Solids in Liquids." *Proc., Hydrotransport 2,* 1972, pp.D1-l to 11.

[38] Shook, C.A., et al. "Some Experimental Studies of the Effect of Particle and Fluid Properties Upon the Pressure Drop for Slurry Flow." *Proc., Hydrotransport 2,* 1972, pp.D2-13 to 22.

[39] Dai, Jilan. *The Effect of Size Composition on Hydraulic Properties of Two-Phase Flow,* Doctoral Dissertation, Tsinghua University, 1980, p. 44.

[40] Machinery Dept. of Beijing Coal College. *Preliminary Study of Hydrotransport of Coal, Teaching Material for Fluid Machines,* vol.1.

[41] Smith, R.A. "Experiments on the Flow of Sand-Water Slurries in Horizontal Pipes." *Trans., Inst. Chem. Engin.*, Vol. 33, No 4, 1955, pp. 85-92.

[42] Condolios, E. and E.E. Chapus. 'Transporting Solid Materials in Pipelines", *Chem. Engin.*, June 1963, pp. 93-98.

[43] Vanoni, V.A. (ed). *Sedimentation Engineering,* Amer. Soc. Civil Engrs., 1975, pp.245-278.

[44] Shen, H.W., and J.S. Wang. "Incipient Motion and Limiting Deposit Conditions of Solid-Liquid Pipe Flow." *Proc., Hydrotransport 1,* 1970, pp. H3-37 to 52.

[45] Novak, P., and C. Nalluri. "Correlation of Sediment Incipient Motion and Deposition in Pipes and Open Channels with Fixed Smooth Beds." *Proc., Hydrotransport 3,* 1974, pp.E4-45 to 56.

[46] Wilson, K.C. "Bed Load Transport at High Shear Stress." *J. Hyd. Div., Proc., Amer. Soc. Civil Engrs.*, Vol. 92, No. HY6, 1966, pp. 49-59.

[47] Ismail, H. "Turbulent Transfer Mechanism and Suspended Sediment in Closed Channels." *Trans., Amer. Soc. Civil Engrs.*, Vol. 117, 19521 pp. 409-446.

[48] Sharp, B.B. and I.C. O'Neill. "A Study of the Behavior of Large Light Particles in Turbulent Pipe Flow." *Proc., Hydrotransport 1,* 1970, pp.F4-45 to 56.

[49] Graf, W.H., and E.R. Acaroglu. "Sediment Transport in Conveyance System, Part 1." *Bull., Intern Assoc. Sci. Hydrology,* Vol. 13, No. 2, 1968.

[50] Acaroglu, E.R., and W.H. Graf. "The Effect of Bed Forms on the Hydraulic Resistance." *Proc. 13th Cong. Intern. Assoc. Hyd. Res.,* Vol.2, 1969.

[51] Wilson, K.C., and A. Brebner. "On Two-Phase Pressurized and Unpressurized Flow: Behavior Near Deposition Points." in *Advances in Solid-Liquid Flow in Pipes and Its Application.* Edited by I. Zandi, Pergamon Press, 1971, pp. 175-186.

[52] Shook, C.A., and S.M. Daniel. "Flow of Suspensions of Solids in Pipelines, Part 1: Flow with a Stable Stationary Deposit", *Cabadian J. Chem. Engin.*, Vol. 43, No. 2, 1965, pp. 56-61.

[53] Bantin, R.A., and M. Streat. "Dense-Phase Flow of Solid-Water Mixtures in Pipelines." *Proc., Hydratransport 1*, 1970, pp. G1-l to 24.

[54] Bantin, R.A., and M. Street. "Mechanism of Hydraulic Conveying At High Concentration in Vertical and Horizontal Pipes." *Proc., Hydrotransport 2*, 1972, pp. B2-12 to 24.

[55] Wilson, K.C., M. Street, and K.A. Cantin. "Slip Model Correlation of Dense Two Phase Flow." *Proc., Hydrotransport 2*, 1972, pp.M1-l to 10.

[56] Stevens, G.S., and M.E. Charles. "The Pipeline Flow of Slurries: Transition Velocities." *Proc., Hydrotransport 2*, 1972, pp.E3-37 to 62.

[57] Wasp, E.J., et al. "Hetero-Homogeneous Solids/Liquid Flow in the Turbulent Regime." In *Advances in Solid-Liquid Flow in Pipes and Its Application.* Edited by I. Zandi. Pergamon Press, 1971, pp. 199-210.

[58] Wilson, K.C., and W.E. Watt. "Influence of Particle Diameter on the Turbulent Support of Solids in Pipeline Flow." *Proc., Hydrotransport 3*, 1974, pp.D1-l to 9.

[59] Shook, C.A. "Pipelining Solids: The Design of Short Distance Pipelines." *Proc., Symp. on Pipeline Transport of Solids*, Canadian Soc. Chem. Engin., 1969.

[60] Fuhrboter, A.. "Uber die Forderung von Sand-Wasser-Gemischen in Rohrleitungen." *Mitt. des Franzius-Instituts fur Grund-und Wasserbau der Technischen Hochschule Hannover*, Heft 19, 1961, pp. 1-152.

[61] Sinclair, C.G. "The Limit Deposit Velocity of Heterogeneous Suspensions." *Proc., Symp.* Interaction *Between Fluids and Particles*, Inst. Chem. Engrs., 1962.

[62] Wiedenroth, W., and M. Kirchner. "A Summary and Comparison of Known Calculations of Critical Velocity of Solid Water Mixtures and Some Aspects of the Optimisation of Pipelines." *Proc., Hydrotransport 2*, 1972, pp. E1-1 to 22.

[63] Weber, M., and E. Godde, "Critical Velocity As Optimum Operating Velocity in Solids Pipelining." *Proc., Hydrotransport 4*, 1976, pp. D2-17 to 30.

[64] Mine Transport & Mining Equipment Dept. of Beijing Coal College and Jingxi Mining Bureau. *Hydrotransport of Dense Coal Slurry, Journal of Beijing Coal College*, Vol.1, 1959, pp. 18-25.

[65] Kazanskij, I.B. "Critical Velocity of Depositions for Fine Slurries-New Results", *Proc., Hydrotransport 6*, 1980, pp. 43-56.

[66] Turian, R.M., and Tran-Fu Yuan. "Flow of Slurries in Pipelines." *J. Amer. Inst. Chem. Angrs.*, Vol. 23, No. 3, 1977, pp. 232-242.

[67] Lazarus, J.H., and I.D. Neilson. "A Generalized Correlation for Friction Head Losses of Settling Mixtures in Horizontal Smooth Pipelines." *Proc., Hydrotransport 5*, 1978, pp.B1-l to 32.

[68] Charles, M.F., and R.A. Charles. "The Use of Heavy Media in the Pipeline Transport of Particulate Solids", In *Advances in Solid-Liquid Flow in Pipes and Its Application*, Edited by I. Zandi. Pergamon Press, 1971, pp. 187-197.

[69] Wasp, E.J., et al. "Cross Country Coal Pipeline Hydraulics." *Pipeline News*, 1963, pp.20-28.

[70] Masuyama, T., T. Kawashima, and K. Koctir. "Pressure Loss of Pseudo-Plastic Flow Containing Coarse Particles in a Pipe." *Proc., Hydrotransport 5*, 1978, pp.D1-l to 14.

[73] Kenchington, I.M.. "Prediction of Pressure Gradient in Dense Phase Conveying." *Proc., Hydrotransport 5*, 1978, pp.D7-91 to 102.

[72] Hisamitsu, N., Y. Shoji, and S. Kosugi. "Effect of Adding Particles in Flow Properties of Settling Slurries." *Proc., Hydrotransport 5*, 1978, pp.D3-29 to 50.

[73] Newitt, D.M. et al., "Hydraulic Conveying of Solids in Vertical pipes." *Trans., Inst. Chem. Engrs.*, Vol. 39, No.2, 1961, pp. 93-100.

[74] Kao, D.T.Y., and L.Y. Hwang. "Critical Slope for Slurry Pipeline Transporting Coal and Other Solid Particles." *Proc., Hydrotransport 6*, 1979, pp. 57-74.

[75] Toms, B.A. "Some Observations of the Flow of Linear Polymer Solutions Through Straight Tubes at Large Reynolds Numbers." *Proc., 1st Intern. Cong. on Rheology*, Vol.2, 1918, pp. 135-141.

[76] Hoyt, J.W. "The Effect of Additives on Fluid Friction." *J. Basic Engin.*, Amer. Sec. Mech. Engin., Vol. 94, 1972, p. 258.

[77] Virk, P.S. "Drag Reduction Fundamentals," *J. Amer. Inst. Chem. Engrs.*, Vol. 21, No. 4, 1975, pp. 625-656.

[78] Hoyt, J.V. "Polymer Drag Reduction--A Literature Review, 1975-1976", *Proc. 2nd Conf. On Drag Reduction*, 1977, pp. Al-l to 19.

[79] White, A., and Hemmings, J.A.G. "Drag Reduction by Additives-Review and Bibliography." *BHRA Fluid Engin.*, 1976, p. 181.

[80] Poreh, M., H. Rabin, and C. Elata. "Comparative Drag Reduction and Degradation Tests." *Israel Inst. Tech., Pub. No. 126*, Ch. 5, 1969, pp. 98-109.

[81] Virk, P.S. "Drag Reduction in Rough Pipes." *J. Fluid Mech.*, Vol.45, No. 2, 1971, pp.225-246.

[82] Poreh, M., et al. "Drag Reduction in Hydraulic Transport of Solids." *J. Hyd. Div., Proc., Amer. Soc. Civil Engrs.*, Vol. 96, No. HY4, 1970, pp. 903-909.

[83] Dou, Gouren. "Turbulence Structure for Drag Reduction by Addition of Polymers." *Study of Water Resources and Navigation*, Vol. 1, p.1-11.

[84] Pfeffer, R., and R.S. Kane. "A Review of Drag Reduction in Dilute Gas-Solid Suspension Flow in Tubes." *Proc., 1st Intern. Conf. on Drag Reduction*, BHRA Fluid Engin., 1974, pp.F1-l to 21.

[85] Achia, B.U., and D.W. Thompson. "Laser Holographic Measurement of Wall Turbulence Structures in Drag-Reducing Pipe Flow." *Proc., 1st Intern. Conf. on Drag Reduction*, BHRA Fluid Engin., 1974, pp. A2-l to 18.

[86] Radin, I.., J.L. Zakin, and G.K. Patterson. "Drag Reduction in Solid-Fluid Systems." *I. Amer.Inst. Chem. Engrs.*, Vol. 21, No. 2, 1975, pp. 358-371.

[87] Heywood, N.I., and J.F. Richardson. "Head Loss Prediction by Gas Injection for Highly Shear-Thinning Suspensions in Horizontal Pipe Flow." *Proc., Hydrotransport 5*, 1978, pp.C1-l to 22.

[88] Kenchington, J.M. "Prediction of Critical Conditions for Pipeline Flow of Settling Particles in A Heavy Medium." *Proc., Hydrotransport 4*, 1976, pp.D3-31 to 48.

[89] Kazanskij, I.., H. Bruhl and J. Hinsch, "Influence of Added Fine Particles on the Flow Structure and Pressure Losses of Sand-Water Mixtures." *Proc., Hydrotransport 3*, 1974, pp.D2-11 to 21.

[90] Sakamoto, M., et al. "A Hydraulic Transport Study of Coarse Material Including Fine Particles with Hydrohoist", *Proc., Hydrotransport 5*, 1978, pp.D6-79 to 90.

[91] Gaessler, H. "Experimentelle und Theoretische Untersuchungen uber die Stromungsvorgange beim Transport von Feststoffen in Flussigkeiten durch Horizontale Rǫhrieitungen." *Ph. D. Dissertation*, Tech. Univ. Karisruhe, 1967.

[92] Wilson, K.C. "A Unified Physically-Based Analysis of Solid-Liquid Pipeline Flow." *Proc., Hydrotransport 4*, 1976, p. 13.

[93] Shook, C.A., et al. "Flow of Suspensions in Pipelines, Part 2; Two Mechanism of Particle Suspension." *Canadian J. Chem. Engin.*, Vol. 46, No. 4, 1968, pp. 238-244.

CONCLUDING REMARKS

The modern study of the mechanics of sediment transport has been going on for more than 70 years since its commencement in the 1920s. The results of research introduced in the text cover a broad scope and convey the fruits of the many related studies. A review of the long history of development indicates the credits that are due for the progress and achievements made so far. In the prospects for the future, there remain, however, many links between the parts of this field that need further development and more penetrating study. The process of development in this discipline of science indicates the problems that remain to be solved, the nature of the theoretical and experimental research that should be focused on pertinent points and the data that must be collected and processed.

1. THEORETICAL STUDY

Sediment movement consists of the movement of large numbers of particles and it follows inevitably both the laws of mechanics and the random processes of statistics. In the 1940s, Einstein opened up an approach to the study of sediment movement by combining these two theories. However, not enough attention was paid to his approach. Only recently, partly under pressure from the study of environmental pollution problems, have research workers begun to follow this approach; and at last, real progress is being made. Along this pathway, many theoretical processes related to incipient motion, bed load transport, sediment diffusion, and the like remain to be clarified in fundamental ways.

In modern fluid mechanics, problems related to turbulence are receiving the study that may bring about a breakthrough. Existence of sediment alters turbulence structure in ways that make it more complicated than clear water flow. At present, not only are many ideas about fundamental concepts uncertain, but also experimental studies produce contradictory results. Thus, we do not yet have enough reliable data to allow us to develop rational knowledge about these phenomena. Still, due to advances in the field of measurement and observation techniques, researchers are obtaining new knowledge on the bursting phenomena of the turbulent shear flow and starting to apply it to the study of sediment movement.[1] Those working in this field must study closely of the various links with basic theory.

In the several chapters, the authors tried deliberately to present the various concepts presented in discussions of sediment movement under different dynamic actions and for many kinds of boundary conditions. However, the individual researchers may not be adequately aware of the common elements in the study of specific sediment problem while they follow their own lines of approach. For instance, common points arise in the mechanisms in open channel flow with hyper-concentration of sediment, in pipe flow with hyper-concentrations, in off-shore movement of heavily sediment-laden turbidity flows, in debris flow in mountain creeks and in wind-blown sediment movement in the desert. If we were all able to obtain an awareness of the common elements in these diverse studies of sediment movement and to coordinate the efforts of those working in different disciplines, theoretical studies would be greatly assisted and breakthroughs could be made earlier.

2. EXPERIMENTAL STUDY

According to preliminary information, more than 4,000 sets of flume experiments have been conducted in studies of sediment movement throughout the world. Further duplication of such flume tests within the range of prior experiments is probably not necessary. However, the range of variation of the data from flume experiments is still not broad enough to represent what occurs in natural rivers (Fig. 11.15).[2] Only 2% of the data in flume experiments is for values of h/D_{50} greater than 10^3, whereas, this value in natural rivers can have values 100 times larger. Even for specific values of the ratio h/D_{50}, the range of variation of the actual D_{50} used in the flume experiment appears to be too narrow. To overcome this shortcoming, additional experiments are needed with larger values of h/D_{50} and with a larger variety of sediment sizes. Also, the collection of more reliable data from field measurements is indispensable.

In order to answer questions not yet clearly understood, additional objective-oriented experiments should be conducted; these should include the period of stay, step length, jump height, and speed of movement of bed load particles; resistance loss and influence on turbulence structure of bed load and suspended load; relationship between the speed of bed load and flow intensity on fixed bed; and others. The proposed experiments do not require large equipment, elaborate installations or highly precise data so long as the experiments are intelligently planned and conducted with sound insights. A key to the success of such experiments is the further development of instrumentation and experimental techniques.

3. DATA PROCESSING OF EXISTING WORKS

The status of study of the mechanics of sediment transport at present is rather similar to the study of hydraulics in the later years of the 19th century. At that time, resistance formulas of many types were in use. Advancement in technique and verification in practice over several decades deleted all but three or four these formulas. Similarly, on many topics, the resistance of movable beds, criteria for sandwaves, incipient motion concepts, bed load transport rate, resistance loss for the two-phase flow in pipes and critical velocity for various stages of motion, the opinions of researchers vary widely and numerous formulas and hypotheses have been proposed. It is difficult for practicing engineers to select the one that is appropriate for their use. Systematic processing of the existing data and the screening out of the several more useful formulas for use in practice should proceed with the goal of "retaining the true and deleting the false" and "retaining the fine and deleting the crude". More effort just to create and propose more empirical formulas may only lead to more confusion in practice. Further research should be planned and organized on the basis of a deepen understanding of the existing problems. In this way, the study of the mechanics of sediment transport can be promoted and surer progress can be made toward a better solution of the sediment problems in China and throughout the world.

REFERENCES

[1] Sumer, B.M., and R. Deigaard. "Experimental Investigation of Motions of Suspended Heavy Particles and the Bursting Process." *Ser. Paper 23*, Inst. Hydrodynamics and Hyd. Eng., Tech. Univ. Denmark, 1979, p.106.

[2] Copper, R.H., A. W.Peterson, and T. Blench. "Critical Review of Sediment Transport Experiments." *J. Hyd. Div.*, Proc. Amer. Soc. Civil Engrs., Vol. 98, No. HY5, 1972, pp. 827-843.

NEW REFERENCES

Abdelhadi, M.L. "Environment and Socio-economic Impacts of Erosion and Sedimentation in North African Countries." *Proc. 6th International Symposium on River Sedimentation*, Central Board of Irrigation and Power, New Dehli, 1995.

Aguirre-Pe, J., and Ramon Fueutes. "Resistance to Flow in Steep Rough Streams." *J. Hydraulic Engineering*, ASCE, Vol. 116, No. 11, 1990, pp. 1374-1387.

Akiyama, J., and H. Stefan. "Turbidity Current with Erosion and Deposition." *J. Hydraulic Engineering*, ASCE, Vol. 111, No. 12, 1985, pp. 1473-1496.

Altinakar, M.S., W.H. Graf, and E.J. Hopfinger. "Weakly Depositing Turbidity Current on a Small Slope." *J. Hydraulic Research*, Vol. 28, No. 1, 1990, pp. 55-80.

Altinakar, M.S., W.H. Graf, and E.J. Hopfinger. "Flow Structure in Turbidity Flow." *J. Hydraulic Research*, Vol. 34, No. 5, 1996, pp. 713-718.

Annandale, G.W. *Reservoir Sedimentation*. Elsevier, Amsterdam, 1987.

Annandale, G.W. "Erodibility." *J. Hydraulic Research*, IAHR, Vol. 33, No. 5, 1995, pp. 471-494.

Armaly, B.F., F. Durst, J.C.F. Pereira, and B. Schonung. "Experimental and Theoretical Investigation of Backward-Facing Step Flow." *J. Fluid Mechanics*, Vol. 127, 1983, pp. 473-496.

Armanini, A.(Ed.) *Fluvial Hydraulics in Mountain Regions*. Springer Verlag, Berlin, 1991.

Asaeda, T., Nakai, M. Manandhar, S.K., and N. Tamai, "Sediment Entrainment in Channels with Rippled Bed." *J. Hydraulic Engineering*, ASCE, Vol. 115, No. 3, 1989, pp. 327-339.

Ashida, K. & Mofjeld, "Study on Hydraulic Resistance and Bed-Load Transport Rate in Alluvial Streams." *Trans.*, JSCE, Vol. 206, 1991, pp. 59-69.

ASCE Task Committee on Relations Between Morphology of Small Streams and Sediment Yield of the Committee on Sedimentation of the Hydraulics Division. "Relationships Between Morphology of Small Streams and Sediment Yields." *J. Hydraulic Engineering*, ASCE, Vol. 108, No.HY 11, Proceeding Paper 17450, 1982, pp. 1328-1365.

Bathurst, J.C. "Flow Resistance Equation in Mountains." *J. Hydraulic Engineering*, ASCE, Vol. 111, No. 4, 1985, pp. 1103-1122.

Becheteler, W. *Transport of Suspended Solids in Open Channels*. A.A. Balkema, Rotterdam, 1986.

Blanton, J.O. *Procedures for Monitoring Reservoir Sedimentation*. U.S. Bureau of Reclamation, Denver, Colorado, 1982.

Bonnecaza, R.T. et al. "Particle Driven Gravity Currents." *J. Fluid Mechanics*, Vol. 250, 1993, pp. 339-369.

Bo Pedersen, F. *Environmental Hydraulics: Stratified Flows*. Springer Verlag, 1986.

Bouvard, M. *Mobile Barrages and Intakes on Sediment Transporting Rivers*. IAHR Monograph, A.A. Balkema, Rotterdam, 1992.

Bray, D.I., and K.S. Davar. "Resistance to Flow in Gravel Bed Rivers." *Canadian J. Civil Engineering*, Vol. 14, No. 2, 1987.

Brereton, G.J., W.C. Reynolds, and R. Jayaraman. "Response of a Turbulent Boundary Layer to Sinusoidal Free-Stream Unsteadiness. *J. Fluid Mechanics*, Vol. 221, 1990, pp.131-159.

Brownlie, W.R. "Flow Depth in Sand-Bed Channels." *J. Hydraulic Engineering*, ASCE, Vol. 109, HY 7, 1983, pp. 959-990.

Celik, I., and Rodi, W. "Suspended Sediment Transport Capacity for Open Flow." *J. Hydraulic Engineering*, ASCE, Vol. 117, No. 2, 1991, pp. 191-204.

Chang, H.H. "Energy Expenditure in Curved Open Channels." *J. Hydraulic Engineering*, ASCE, Vol. 109, No. 7, 1983, pp. 1012-1022.

Chang, H.H. "Regular Meander Path Model." *J. Hydraulic Engineering*, ASCE, Vol. 110, No. 10, 1984, pp. 1398-1411.

Chang, H.H. "Variation of Flow Resistance Through Curved Channels." *J. Hydraulic Engineering*, ASCE, Vol. 110, No. 12, 1984, pp. 1772-1782.

Chang, H.H. "Formation of Alternate Bars." *J. Hydraulic Engineering*, ASCE, Vol. 111, No. 1, 1985, pp. 1412-1420.

Chang, H.H. "River Morphology and Thresholds." *J. Hydraulic Engineering*, ASCE, Vol. 111, No. 3, 1985, pp. 503-519.

Chang, H.H. "Water and Sediment Routing Through Curved Channels." *J. Hydraulic Engineering*, ASCE, Vol. 111, No. 4, 1985, pp. 644-658.

Chang, H.H. "River Channel Changes: Adjustments of Equilibrium." *J. Hydraulic Engineering*, ASCE, Vol. 112, No. 1, 1986, pp. 43-55.

Chang, H.H. *Fluvial Processes in River Engineering*. John Wiley & Sons, Inc, 1988.

Chang, H.H. "Selection of Gravel-Transport Formula for Stream Modeling." *J. Hydraulic Engineering*, ASCE, Vol. 120, No. 5, 1994, pp. 646-651.

Chen, C.L. "General Solutions for Viscoplastic Debris Flow." *J. Hydraulic Engineering*, ASCE, Vol. 114, No. 3, 1988, pp. 259-282.

Cheng, K.J. "An Integrated Suspended Load Equation for Non-equilibrium Transport of Non-uniform Sediment." *J. Hydrology*, Vol. 79, 1985, pp. 359-364.

Cheremisinoff. N.P. (ed.) *Theory of Minimum Energy and Energy Dissipation Rate, Encyclopedia of Fluid Mechanics*. Vol. 1, Gulf Publishing Company, 1984.

Chien N., R. Zhang, and Z. Zhou. *River Fluvial Mechanics*. Science Press, Beijing, 1987. (in Chinese)

Chien, N. *Movement of Hyperconcentrated flow*. Tsinghua University Press, 1989. (in Chinese)

Chikita, K. "Sedimentation by River-Induced Turbidity Currents: Field Measurements and Interpretation." *Sedimentalogy*, Vol. 37, 1990, pp. 891-905.

Chikita, K., and Y. Okumura. "Dynamics of Turbidity Currents Measured in Katsurazawa Reservoir." *J. Hydrology*, Vol. 117, 1990, pp. 323-338.

Christodoulou, G.C. "Interfacial Mixing in Stratified Flows." *J. Hydraulic Research.*, Vol. 24, No. 2, 1986, pp.77-92.

Clifford, N.J. et al. *Turbulence: Perpectives on Flow and Sediment Transport*. Wiley Chichester, 1993.

Colemann, N.L. "Velocity Profiles with Suspended Sediment." *J. Hydraulic Research*, Vol. 19, No. 3, 1981, pp. 211-230.

Coleman, N.L., and C.V.Alonso. "Two-Dimensional Channel Flows over Rough Surface." *J. Hydraulic Engineering*, ASCE, Vol. 109, No. 11, 1983, pp. 175-188.

Coleman, N. L. "Effects of Suspended Sediment on the Open Channel Velocity Distribution." *Water Resources Research*, Vol. 22, No. 10, 1986.

Cui, Yantao, G. Parker, and C. Paola. "Numerical Simulation of Aggradation and Downstream Fining." *J. Hydraulic Research*, Vol. 34, No. 2, 1996, pp. 185-204.

Davis, T.R.H. "Large Debris Flows: A Macro-Viscous Phenomenon." *Acta Mechanics*, Vol. 63, 1986, pp.161-178.

Dawdy, D.R., and V.A. Vanoni. "Modeling Alluvial Channels." *Water Resources Research*, Vol. 22, No. 9, 1986, pp. 71S-81S.

Dietrich, W. "Settling Velocities of Natural Particles." *Water Resources Research*, Vol. 18, 1982, pp. 1615-1626.

Fan, J., and G.L. Morris. "Reservoir Sedimentation. I: Delta and Density Current Deposits." *J. Hydraulic Engineering*, ASCE, Vol. 118, No. 3, 1992, pp. 354-369.

Fan, J., and G.L. Morris. "Reservoir Sedimentation. II. Reservoir Desiltation and Long-term Storage Capacity." *J. Hydraulic Engineering*, ASCE, Vol. 118, No. 3, 1992, pp. 370-384.

Fang, Hongwei. "Compostition and Competent Velocity of Non-uniform Bed Load." *J. of Hydraulic Engineering*, Vol. 4, 1994, pp.43-49. (in Chinese)

Ferro, V., and G. Giordano. "Experimental Study of Flow Resistance in Gravel-Bed Rivers." *J. Hydraulic Engineering*, ASCE, Vol. 117, No. 1, 1991, pp. 1239-1246.

Garcia, M. "Hydraulic Jumps in Sediment-Driven Bottom Currents." *J. Hydraulic Engineering*, ASCE, Vol. 119, No. 10, 1993, pp. 1094-1117.

Garcia, M. "Experiments on the Entrainment of Sediment into Suspension by a Dense Bottom Current." *J. Geophysical Research*, Vol. 98, No. C3, 1993, pp. 4793-4807.

Grant, W.D. "Movable Bed Roughness in Unsteady Oscillatory Flow." *J. Geophysical Research*, Vol. 87, 1982, pp. 469-481.

Griffitlis, G.S. "Form Resistance in Gravel Channels with Mobile Bed." *J. Hydraulic Engineering*, ASCE, Vol. 115, No. 3, 1989, pp. 340-355.

Gyr, A. "Towards a Better Definition of the Three Types od Sediment Transport." *J. Hydraulic Research*, Vol. 21, 1983, pp.1-15.

Han, Qiwei, and Mingmin He. *Statistic Theory of Sediment Motion*. Science Press, 1984. (in Chinese)

Haque, M.I., and Mahmood, K. "Geometry of Ripples and Dunes." *J. Hydraulic Engineering*, ASCE, Vol. 111, No. 1, 1985, pp. 48-63.

He, Mingmin, and Qiwei Han. "The Determination of the Size Distribution of Carrying Capacity and of Effective Bed Material." *J. Hydraulic Engineering*, No. 3, 1990, pp. 1-12. (in Chinese)

Herbich, J.B. (ed.) *Handbook of Dredging Engineering*. McGraw-Hill Co., New York, 1992.

Hey, R.D., J.C. Bathurst, and C. Thorne. (ed.) *Gravel-bed Rivers, Fluvial Processes, Engineering and Management*. John Wiley & Sons, Inc, 1982.

Hey, R.D. "Bar Form Resistance in Gravel-Bed Rivers." *J. Hydraulic Engineering*, ASCE, Vol. 114, No. 12, 1988, pp. 1498-1508.

Hino, M., M.Kashiwayanagi, A. Nakayama, and T. Hara. "Experiments on the Turbulence Statistics and Structure of a Reciprocating Oscillatary Flow." *J. Fluid Mechanics*, Vol. 131, 1983, pp. 363-400.

Holtorff, G. "Steady Bed Material Transport in Alluvial Channels." *J. Hydraulic Engineering*, ASCE, Vol. 109, N0. 3, 1983, pp.368-384.

Hu, Chunhong, and Yujia Hui. "Bed-Load Transport. I: Mechanical Characteristics." *Proc.*, ASCE, J. Hydraulic Engineering, Vol. 122, No. 5, 1996, pp. 245-254.

Hu, Chunhong, and Yujia Hui. "Bed-Load Transport. II: Stochastic Characteristics." *Proc.*, ASCE, J. Hydraulic Engineering, Vol. 122, No. 5, 1996, pp. 255-261.

Hui, Yujia, and Chunhong Hu. "Kinematic Characteristics of Saltation of Solid Grain in Flowing Water." *J. Hydraulic Engineering*, No. 12, 1991, pp. 59-64. (in Chinese)

Hunt, B. "Newtonian Fluid Mechanics Treatment of Debris Flows and Avalanches." *J. Hydraulic Engineering*, ASCE, Vol. 120, 1994, pp. 1350-1365.

Hussain, A.K.M.F. "Coherent Structure-Realty and Myth." *Phys. Fluids*, Vol. 26, No. 10, 1983, pp. 2816-2850.

Ikeda, S. "Incipient Motion of Sand Particles on Side Slopes." *J. Hydraulic Engineering*, ASCE, Vol. 108, 1982, pp. 95-114.

Ikeda, S., and T. Asaeda. "Sediment Suspension with Rippled Bed." *J. Hydraulic Engineering*, ASCE, Vol. 109, N0. 3, 1983, pp. 409-423.

Ikeda, S. "Prediction of Alternate Bar Wavelength and Height." *J. Hydraulic Engineering*, ASCE, Vol. 110, No. 4, 1984, pp. 371-186.

Ikeda, S., and T. Nishimura. "Flow and Bed Profile in Meandering Sand-Silt Rivers." *J. Hydraulic Engineering*, ASCE, Vol. 112, No. 7, 1986, pp. 562-579.

Jaeggi, M.N.R. "Formation and Effects of Alternate Bars." *J. Hydraulic Engineering*, ASCE, Vol. 110, No. 2, 1984, pp. 142-156.

Julien, P.Y. *Erosion and Sedimentation*. Cambridge University Press, Cambridge, U.K., 1995.

Karim, M.F., and J.F. Kennedy. "Means of Coupled Velocity and Sediment-Discharge Relations for Rivers." *J. Hydraulic Engineering*, ASCE, Vol. 116, No. 8, 1990, pp. 1350-1360.

Knight, D. W., and J. D. Demetrion. "Flood Plain and Main Channel Flow Interaction." *J. Hydraulic Engineering*, ASCE, Vol. 109, No. 8, 1983, pp. 1073-1092.

Knight, D. W., and M.E. Hamed. "Boundary Shear in Symmetrical Compound Channels." *J. Hydraulic Engineering*, ASCE, Vol. 110, No. 10, 1984, pp. 1412-1430.

Knight, D. W., and H. S. Patel. "Boundary Shear in Smooth Rectangular Ducts." *J. Hydraulic Engineering*, ASCE, Vol. 111, No. 1, 1985, pp. 29-47.

Kobayashi, N., and N. Seo. "Fluid and Sediment Interaction over a Plane Bed." *J. Hydraulic Engineering*, ASCE, Vol. 111, No. 6, 1985, pp.903-921.

Krogstad, P.A., R.A. Atonia, and L.W.B. Browne. "Comparison between Rough- and Smooth-Wall Turbulent Boundary Layer." *J. Fluid Mechanics*, Vol. 245, 1992, pp.599-617.

Kuhnle, R.A. "Incipient Motion of Sand-Gravel Sediment Mixtures." *J. Hydraulic Engineering*, ASCE, Vol. 119, 1993, pp. 1400-1415.

Lau, Y. Lam. "Hydraulic Resistance of Ripples." *J. Hydraulic Engineering*, ASCE, Vol. 114, No. 10, 1988.

Lai, J.S., and H.W. Shen. "Flushing Sediment through Reservoirs." *J. Hydraulic Research*, Vol. 34, No.2, 1996, pp. 237-255.

Lavelle, J.W., and H.O. Mofjeld. "Do Critical Stresses for Incipient Motion and Erosion Really Exist?." *J. Hydraulic Engineering*, ASCE, Vol. 113, 1987, pp. 370-385.

Lavelle, J.W., and H.O. Mofjeld. "Bibliography on Sediment Threshold Velocity." *J. Hydraulic Engineering*, ASCE, Vol. 113, 1987, pp. 389-393.

Lee, H-Y., and Odgaard, A.J. "Simulation of Bed Armoring in Alluvial Channels." *J. Hydraulic Engineering*, ASCE, Vol. 112, No. 9, 1986, pp. 794-801.

Lin, B., G. Dou, J. Xie, D. Dai, J. Chen, R. Tang, and R. Zhang. "On Some Key Sedimentation Problems of Three Gorge Project (TGP) in the Light of Recent Findings." *International Journal of Sediment Research*, Vol. 4, No. 1, 1993, pp. 57-74.

Liu, Dayou. *Dynamics of Two-Phase Flow*. High-Education Press, 1993. (in Chinese)

Long, Y. "Manual on Operational Methods for the Measurement of Sediment Transport." *Operational Hydrology Report No. 29*, World Meteological Organization, Geneva, 1989.

Mehrota, S.C. "Permissible Velocity Correction Factors." *J. Hydraulic Engineering*, ASCE, Vol. 109, No. 2, 1983, pp.305-308.

Middleton, G.V. "Sediment Deposition from Turbidity Currents." *Annual Review of Earth Planetary Science*, Vol. 21, 1993, pp. 89-114.

Misri, R. L., R.J.Garde, and K.G. Ranga Raju. "Bed Load Transport of Coarse Nonuniform Sediment." *J. of Hyd. Eng.*, ASCE, Vol. 110, No. 3, 1984, pp.312-328.

Morris, L.M., and Jiahua Fan. *Reservoir Sedimentation Handbook*. McGraw-Hill, New York, 1997.

Muller, A., and A. Gyr. "On the Vortex Formation in the Mixing Layer behind Dunes." *J. Hydraulic Research*, Vol. 24, 1986, pp. 359-375.

Nakagawa, H., and I. Nezu. "Experimental Investigation on Turbulent Structure of Backward-Facing Step Flow in an Open Channel." *J. Hydraulic Research*, Vol. 25, 1987, pp. 67-88.

Nakato, T. "Test of Selected Sediment-Transport Formulas." *J. Hydraulic Engineering*, ASCE, Vol. 116, No. 3, 1990, pp. 362-379.

National Academy of Sciences. *An Evaluation of Flood Level Prediction Using Alluvial River Models*. Committee on Hydrodynamic Computer Models for flood Insurance Studies, Advisory Board on the Built Environment, National Research Council, National Academy Press, Washington, D.C., 1983.

Nezu I., and H. Nakagawa. *Turbulence in Open-Channel Flows*. IAHR Monograph, A.A. Balkema, Rotterdam, 1993.

Ni, Jinren, Guangqian Wang, and Hongwu Zhang. *Basic Theory of Solid-Liquid Two Phase Flow and Its New Application*. Science Press, 1991. (in Chinese).

Nokes, R.L., and I.R. Wood. "Turbulent Dispersion of a Steady Discharge of Positively or Negatively Buoyant Particles in Two Dimensions." *J. Hydraulic Research*, Vol.25, No. 1, 1987, pp.103-122.

Odgaard, A.J. "Meander Flow Model I: Development." *J. Hydraulic Engineering*, ASCE, Vol. 112, No. 12, pp. 1117-1136.

Offen, G.R., and S.J. Kline. "A Proposed Model of the Bursting Process in Turbulent Boundary." *J. Fluid Mechanics*, Vol. 70, 1995, pp. 209-228.

Otsubo, K., and K. Muraoka. "Critical Shear Stress of Cohesive Bottom Sediments." *J. Hydraulic Engineering*, ASCE, Vol. 114 No. 10, 1988, pp.1241-1256.

Parker, G., P.C. Kingeman, and D,G. McLean. "Bed Load and Size Distribution in Paved Gravel-Bed Streams." *J. Hydraulic Engineering*, ASCE, Vol. 108, 1982, pp. 544-571.

Parker, G., P. Diplas, and J. Akiyama. "Meander Bends of High Amplitude." *J. Hydraulic Engineering*, ASCE, Vol. 109, No. 10, 1983, pp. 1323-1337.

Parker, G., and E.D. Andrews. "Sorting of Bedload Sediment by Flow in Meande Bends." *Water Resour. Res.*, Vol. 21, No.9, 1985, pp. 1361-1373.

Parker, G., and Coleman, N.L. "Simple Model of Sediment-Laden Flows." *J. Hydraulic Engineering*, ASCE, Vol. 112, No.5, 1986, pp. 356-375.

Parker, G. "Surface-Based Bedload Transport Relationship for Gravel Rivers." *J. Hydraulic Research*, Vol. 28, No. 4, 1990, pp.417-436.

Parker, G. "Selective Sorting and Abrasion of River Gravel. I: Theory." *J. Hydraulic Engineering*, ASCE, Vol. 117, No.2, 1991, pp. 131-149.

Parker, G. "Selective Sorting and Abrasion of River Gravel. II:Applications." *J. Hydraulic Engineering*, ASCE, Vol. 117, No.2, 1991, pp. 150-171.

Petersen, M.S. *River Engineering*. Prentice-Hall, Englewood Cliffs, New Jersey, 1986.

Prestegaard, K. "Bar Resistance in Gravel Bed Streams at Bankful Stage." *Water Resources Research*, Vol. 19, 1983, pp. 472-476.

Proffitt, G. J., and Sutherland, A. J. "Transport of Nonunform Sediment." *Journal of Hydraulic Research*, Vol. 21, No. 1, 1983.

Raudkivi, A.J. *Loose Boundary Hydraulics(3rd Edition)*. Pergamon Press, 1990.

Raudkivi, A.J. "Sedimentation: Exclusion and Removal of Sediment from Diverted Water." *IAHR Hydraulic Structures Design Manual No. 6.*, A.A. Balkema, Rotterdam, 1993.

Ikeda, S., and G. Parker. "River Meandering." *Proceedings of the Conference Rivers '83*, New Orleans, Louisiana, October, 1983.

Robinson, S.K. "Coherent Motion in the Turbulent Boundary Layer." *Annual Review of Fluid Mechanics*, Vol. 23, 1991, pp.601-639.

Rosso, M., M. Schiara, and J. Berlamorit. "Flow Stability and Friction Factor in Rough Channels." *J. Hydraulic Engineering*, ASCE, Vol. 116, No. 9, 1990, pp. 1109-1118.

Rutherford, J.C. *River Mixing*. J. Wiley & Sons, Ltd., Chichester, 1994.

Samaga, B.R., K.R. Raju, and R.J. Garde. "Bed Load Transport of Sediment Mixtures." *J. Hydraulic Engineering*, ASCE, Vol. 112, No. 11, 1986, pp. 1003-1018.

Samaga, B. R., K. G. Ranga Raju, and R.J. Garde. "Suspended Load Transport of Sediment Mixtures." *J. Hydraulic Engineering*, ASCE, Vol. 112, No. 11, 1986, pp. 1019-1035.

Savage, S.B., and K. Hutter. "The Motion of a Finite Mass of Granular Material down a Rough Incline." *J. of Fluid Mechanics*, Vol. 199, 1989, pp. 177-215.

Savage, S.B. and C.K.K. Lun. "Particle Size Segregation in Inclined Chute Flow of Dry Cohesionless Granular Solids." *J. of Fluid Mechanics*, Vol. 189, 1988, pp. 311-335.

Schwab, G.O., D.D. Fangmeir, W.J. Elliot and R.K. Frevert. *Soil and Water Conservation Engineering, 4th ed.* John Wiley and Sons, New York, 1993.

Sedimentation Committee, Chinese Society of Hydraulic Engineering. *Handbook of Sedimentation Engineering*. Environment Science Press, 1989. (in Chinese)

Sekine, M., and H. Kikkawa. "Mechanics of Saltation Grains." *J. Hydraulic Engineering*, ASCE, Vol. 118, No. 4, 1992, 1992, pp. 536-558.

Shao, Xuejun, Zhenhuan Xia. "The Distribution of Solid Particles Suspended in a Turbulent Flows: A Stochstic Approach." *ACTA Mechanica Sinica*, Vol. 23, No. 1, 1991, pp. 28-35. (in Chinese)

Shen, H.W., and C.S. Hung. "Remodified Einstein Procedure for Sediment Load." *J. Hydraulic Engineering*, ASCE, Vol. 109, No. 4, 1983, pp. 565-578.

Shen, H.W., Jau-yan Lu. "Developing and Prediction of Bed Armoring." *J. Hydraulic Engineering*, ASCE, Vol. 109, No Hy4, 1983, pp.611-629.

Simons, D.B., and F. Senturk. *Sediment Transport Technology (2nd Edition)*. Water Resources Publications, Fort Collins, Colorado, 1992.

Song, Tiancheng, Zhaohui Wan, and Ning Chien. "The Effect of Fine Particles on the Two-Phase Flow with Hyperconcentration of Coarse Particles." *J. of Hydraulic Engineering*, No. 4, 1986, pp. 1-9. (in Chinese)

Song, Tiancheng, and Zhaohui Wan. "Experimental Study on the Resistance Relation in Hydrotransport." *J. of Sediment Research*, No. 2, 1987, pp.30-41. (in Chinese)

Song, Tiancheng, and W.H. Graf. "Velocity and Turbulence Distribution in Unsteady Open-Channel Flow." *J. Hydraulic Engineering*, ASCE, Vol. 122, No. 3, 1996, pp. 143-154.

Strand, R.I., and E.L. Pemberton. *Reservoir Sedimentation*. U.S. Bureau of Reclamation, Denver, Colorado, 1982.

Sumer, B.M., and A. Muller. *Mechanics of Sediment Transport*, A.A. Balkema, 1983.

Sumer, B. M. "Lift Forces on Moving Particles Near Boundary." *J. Hydraulic Engineering*, ASCE, Vol. 110, No. 9, 1984, pp. 1272-1278.

Swamee. P.K., and Ojha, C.S.P. "Bed-Load and Suspended Transport of Coarse Non-uniform Sediment." *J. Hydraulic Engineering*, ASCE, Vol. 117, No. 6, 1991, pp. 774-787.

Takahashi, T. "Mechanics and Existence Criteria of Various Type Flows during Massive Sediment Transport." *Proc. International Workshop on Fluvial Hydraulics of Mountain Regions*, Trent, Italy, 1989, pp.119-130.

Takahashi, T. "Debris Flow Initiation and Termination in a Gully." *Proc. Hydraulic Engineering '93*, Vol. 2, 1993, pp. 1756-1761.

Thorne, C. et al. (Ed.) *Sediment Transport in Gravel-bed Rivers*, J. Wiley & Sons, Ltd, Chichester.

Tu, Haizhou, and W.H. Graf. "Friction in Unsteady Open-Channel Flow over Gravel Bed." *J. Hydraulic Research*, Vol. 31, No. 1, 1993, pp. 99-110.

UNESCO. *Methods of Computing Sedimentation in Lakes and Reservoirs*, Paris, France, 1985.

Utami, T., and T. Ueno. "Experimental Study on the Coherent Structure of Turbulent Open-Channel Flow Using Visualization and Picture Processing." *J. Fluid Mechanics*, Vol. 174, 1987, pp. 399-440.

Van Rijn, L.C. "Equivalent Roughness of Alluvial Bed." *J. Hydraulic Engineering*, ASCE, Vol. 108, No.10, 1982, pp. 1215-1218.

Van Rijn, L.C. "Sediment Transport, Part I: Bed Load Transport." *J. Hydraulic Engineering*, ASCE, Vol. 110, No.10, 1984, pp. 1431-1456.

Van Rijn, L.C. "Sediment Transport, Part II: Suspended Load Transport." *J. Hydraulic Engineering*, ASCE, Vol. 110, No.11, 1984, pp. 1613-1641.

Van Rijn, L.C. "Sediment Transport, Part III: Bed Forms and Alluvial Roughness." *J. Hydraulic Engineering*, ASCE, Vol. 110, No.12, 1984, pp. 1733-1754.

Wan, Zhaohui. "Bed Material Movement in Hyperconcentrated Flow." *J. Hydraulic Engineering*, ASCE, Vol. 111, No. 6, 1985, pp.987-1004.

Wan, Zhaohui, and Tiancheng Song. "The Effect of Fine Particles on the Vertical Concentration Distribution and Transport Rate of Coarse Particles." *J. of Hydraulic Engineering*, 1987, pp.20-31. (in Chinese)

Wan, Zhaohui, and Zhaoyin Wang. *Hyperconcentrated Flow*. IAHR Monograph, A.A. Balkema, Rotterdam, 1994.

Wang, Guangqian, and Jinren Ni. "Kinetic Theory for Particle Concentration Distribution in Two-Phase Flow." *J. Mechanic Engineering*, ASCE, Vol. 116, No. 12, 1990.

Wang, Guangqian, and Jinren Ni. "The Kinetic Theory for Dilute Solid / Liquid Two-Phase Flows." *Inter. J. of Multi-Phase Flow*, Vol. 17, No. 2, 1991, pp. 273-281.

Wang, Shiqiang, and Ren Zhang. "Experimental Study on Transport Rate of Graded Sediment." *Proceeding of International Conference on River Flood Hydraulics*, U.K., 1990.

Wang, Shiqiang, and W.R. White. "Alluvial Resistance in Transition Regime." *J. Hydraulic Engineering*, ASCE, Vol. 119, No. 6, 1993, pp.725-741.

Wang, Shiqiang, Ren Zhang, and Yujia Hui. "New Equations of Sediment Transport Rate." *International J. Sediment Research*, Vol. 10, No. 3, 1995, pp.1-18.

Wang, Xingkui, and Ning Chien. "Turbulence Characteristics of Sediment-laden Flow." *J. Hydraulic Engineering*, ASCE, Vol. 115, No. 6, 1989, pp. 781-800.

Wang, Zhaoyin. "Experimental Studies of the Hyperconcentrated Flows." *Ph. D. Dissertation*, Institute of Water Conservansy and Hydroelectric Power Research, Beijing, 1984. (in Chinese)

Wang, Zhaoyin, Xingkui Wang, and Yumin Ren. "Statistic Properties and Spatial Density Distribution of Bingham Fluid in Open Channel Flow." *J. of Hydraulic Engineering*, No. 4, 1993, pp. 12-22. (in Chinese)

Whiting, P.J., and W.E. Dietrich. "Boundary Shear Stress and Roughness over Mobile Alluvial Beds." *J. Hydraulic Engineering*, ASCE, Vol. 116, No.12, 1990, pp. 1495-1990.

Wiberg & Smith. "Calculations of the Critical Shear Stress for Motion of Uniform and Heterogeneous Sediments." *Water Resour. Res.*, Vol. 23, 1987, pp.1471-1480.

Wilcock, P.R. "Critical Shear Stress of Natural Sediments." *J. Hydraulic Engineering*, ASCE, Vol. 119, 1993, pp. 491-505.

Wilson, K.C. "Mobile-Bed Friction at High Shear Stress." *J. Hydraulic Engineering*, ASCE, Vol. 115, No. 6, 1989, pp. 825-830.

WMO (World Meteorological Organization). "Measurements of River Sediments." *Operation Hydrology Report No. 16*, WMO No.561, Geneva, 1981.

Yalin, M.S. *River Mechanics*. Pergamon Press, 1992.

Yang C.T., and A. Molinas. "Sediment Transport and Unit Stream Power Equation." *J. of Hydraulic Division*, ASCE, Vol. 108, No. HY6, 1982, pp.776-793.

Yang. C.T. "Unit Stream Power Equation for Gravel." *J. Hydraulic Engineering*, ASCE, Vol. 110, No. 12, 1984, pp. 1783-1798.

Yang C.T., and S. Wan. "Comparison of Selected Bed-Material Load Formulas." *J. Hydraulic Engineering*, ASCE, Vol. 117, No. 8, 1991, pp. 973-989.

Yang, C.T. *Sediment Transport---Theory and Practice*. McGraw-Hill Companies, 1996.

Yang, C.T., A. Molinas, and Baosheng Wu. "Sediment Transport in the Yellow River." *J. Hydraulic Engineering*, ASCE, Vol. 122 No. 5, 1996, pp. 237-244.

Yen, B.C. *Channel Flow Resistance*. Water Resources Publications, 1991.

Zhou, Zhide. "Resistance to Flat-Bed Flows in Fine Sand Channels." *J. Hydraulic Engineering*, No. 5, 1983, pp.58-64. (in Chinese)

Zyserman, J.A., and J. Fredsøe. "Data Analysis of Bed Concentration of Suspended Sediment." *J. Hydraulic Engineering*, ASCE, Vol. 120, No. 9, 1994, pp. 1021-1042.

LIST OF SYMBOLS

A area; half amplitude of surface wave

a constant; length of longest axis of particle; half amplitude of sand wave; amplitude of movement of water particles in horizontal direction under wave action

a_o amplitude in horizontal direction of water particles in vicinity of bed under wave action

B width; average channel width

b length of intermediate axis of particle; amplitude in vertical direction of water particle under wave action; subscript pertaining to river bed, bed load or break wave zone

C cohesion; Chezy coefficient; coefficient in expression of thickness of laminar layer in vicinity of flow boundary

C_D drag coefficient

C_L lift coefficient

c length of shortest axis of particle; average velocity of air molecules; wave velocity

c_f Manning's friction factor; (in Chapters 16 and 17)

c_x speed of propagation of wave groups

D diameter of sediment (sphere, granular material)

D_A diameter of a sphere having same surface area as that of a particle

D_m mean diameter

D_n nominal diameter

D_r degree of dispersion

D_s diameter of a sphere enclosing a particle

D_{50} median size of sediment of which 50 % by weight is finer

d diameter of pipe; thickness of loose deposits below earth surface

E energy; wave energy per unit width; coefficient of dispersion; ratio of suspension energy to potential energy provided by flow in unit time

E_l energy of wave propagating along the coast per unit of coast length in unit time

E_m energy transferred through unit area of mud in unit time

e porosity ratio; density of distribution of sand waves (rock pieces) expressed by the projected area in flow direction

e' density of distribution of vegetation (grass or wood); density of distribution of sand waves (rock pieces) expressed in cross sectional area perpendicular to the flow

e_b efficiency of bed load transport by flow

e_s efficiency of suspended load transport by flow

F force; resistance; flocculation factor; probability density; complex potential

F_D tractive force

F_L lift force

F_r Froude number

F_r'' effective Froude number related to density current

f function; friction coefficient; linear frequency; Darcy-Weisbach coefficient of resistance

f_b resistance coefficient related to bed resistance

f_b' resistance coefficient related to grain resistance

f_b'' resistance coefficient related to resistance of sand waves

f_e coefficient of head loss at abrupt channel expansion

f_s friction coefficient between sediment and boundary in hydrotransport of solid

G_l' transport rate of littoral current in terms of submerged weight

g gravitational acceleration

g' gravitational acceleration in density current

g_b bed load transport rate per unit width in dry weight

g_b' bed load transport rate per unit width in submerged weight

g_{bv} bed load transport rate per unit width in volume

g_i' transport rate per unit width of bottom current toward coast induced by mass flow

g_s suspended load transport rate per unit width in dry weight

g_T total load transport rate per unit width in dry weight

H depth in reservoir; height of surface wave

H_b wave height of breaking waves

He Hedström number

h water depth

h' thickness of density current

h_a height of water column corresponding to atmospheric pressure

h_c critical depth

h_f head loss

h_f' thickness of the frontal part of density current

h_i distance of the interface of density current from the sill of outlet

h_L distance of the surface of flowing layer from the sill of outlet in selective withdrawal

h_o water depth at plunging point of density current in reservoir

I_1, I_2 parameters used in numerical integration of suspended load transport rate

i_b percentage of sediment with a size D in bed load

i_o percentage of sediment with a size D in bed material

i_s percentage of sediment with a size D in suspended load

i_T percentage of sediment with a size D in total load

J slope; gradient

J' energy gradient required to overcome grain resistance

J'' energy gradient required to overcome bed form resistance

J_e energy gradient

J_m energy gradient for pipe flow with suspended particles

J_s additional energy gradient for transporting water and sediment mixtures in pipes due to existence of sediment

J_w energy gradient for pipe flow

j relative phase difference in analysis of stability of sand waves

K coefficient of thickness in fluids following exponential law; correction coefficient of resistance; coefficient of permeability

K' number of waves per unit length

K_b resistance coefficient of river bed

K_b' coefficient of grain resistance

K_s dimension of roughness

K_u Keulegan number

$K(x)$ function of self-correlation

k coefficient; coefficient of permeability; correction coefficient for the influence of impingement of saltation particles on uplift force in eolian sediment motion; height of vegetative cover; number of waves

L length; distance of a single step in bed load movement; wave length of surface wave

L_x dimension of macro eddies

L' range reached by disturbance of flow

l mixing length; dimension of eddies; average free travelling distance of gas molecules

l_k dimension of smallest eddies

log   logarithm based on 10

ln   logarithm based on e

M parameter of mobility of sediment; percent of discharge that could be measured by depth integration

m exponent; exponent of plasticity of fluids following exponential law; exponent in exponential velocity distribution formula; mass

m' mass of sediment under water

N number; additional pressure of thin film water between grains; angular velocity in oscillation of wave movement

N_c angular velocity of waves at incipient motion

n exponent; frequency; Manning's roughness coefficient

o subscript pertaining to bed

P dispersive force among particles

PE potential energy

pH   symbol denotes logarithm of the reciprocal of the concentration of hydrogen cations

PI plastic index

p pressure; wetted perimeter; probability of sediment in movement

$p_a(D)$ frequency of occurrence of sediment with size D in the armored bed layer

$p_o(D)$ frequency of occurrence of sediment with size D in original bed material

Q discharge

Q_b discharge passing through an area A_b related to grain resistance

q discharge per unit width; probability of sediment remaining in static state on bed

q_o discharge per unit width at incipient motion

R radius of a circle, hydraulic radius; radius of pipe

R_b hydraulic radius related to bed resistance

R_b' hydraulic radius related to grain resistance

R_b'' hydraulic radius related to bed form resistance

R_c radius of bend

R_e Reynolds number

R_{e*} grain Reynolds number

R_i Richardson number

R_m hydraulic radius (Chapter 17)

R_w hydraulic radius related to bank resistance

R_ζ correlation of longitudinal fluctuating velocities at x and x+ζ

R correlation of vertical fluctuating velocities at t and t+ζ

r radius; radius of curvature; ratio of the product of viscosity and fluid density in upper and lower layers of a density current

S sediment concentration in kg/m³

S_c critical concentration of forming a homogeneous slurry at hyperconcentration

$S(f)$ frequency distribution

S_o sorting coefficient; concentration of density current at outlet

S_v volummetric sediment concentration

S_{va} volummetric sediment concentration of suspended load at a height "a" from the bottom

S_{vb} volummetric concentration of bed load

S_{vm} average volummetric sediment concentration in a vertical

S_{vo} content of static particles on bed surface expressed by volume ratio; saturated content of solid materials in non-viscous debris flow (water debris flow)

S_{v*} content of loose deposits expressed by volume

S_w sediment concentration expressed in weight

S_{wb} content of bed load expressed in weight

S_{wp} periodical variation of sediment concentration expressed in weight induced by oscillation of flow

S_{wT} sediment concentration of total load in weight

s distance between two adjacent particles

T time; period; shear stress between particles

t time

t_o time lag between movement of suspended sediment and water particles

U average velocity in a vertical or in a cross section in flow direction

U' velocity of density current

U_c incipient velocity; velocity of the front of debris flow; relative velocity of adjacent layers of fluid; velocity of outflow through orifice

U'_f velocity of the front of density current

U_l average velocity of littoral current

U_L average pipe flow velocity in solid hydrotransport in which head loss is minimum

U_L' non-silting velocity; average pipe flow velocity in solid hydro-transport when deposits start to accumulate at bottom

U_L'' critical pipe flow velocity of transition from laminar to turbulent flow

U_* friction velocity

U'_* friction velocity corresponding to grain resistance

U_{*c} friction velocity at incipient motion

u flow velocity in longitudinal direction at a point or at certain height

u' instantaneous turbulence fluctuation of u at a point

u'' difference of time-averaged velocity at a point with the cross-sectional average velocity

u_b velocity of bed load movement

u_D velocity exerting upon a particle on bed under wave action

u_i velocity of mass flow induced by waves

u_m horizontal velocity of water particles in vicinity of bed under wave action

u_o velocity exerting upon a particle on bed under action of flowing water

u_p velocity of the plug in Bingham fluid

u_r rotary component of velocity of eddies; relative velocity of solid and fluid; ratio of velocity and friction velocity at a point

u_s velocity of suspended sediment

u_t velocity of forward and backward movement of sediment under oscillating flow

u_{max} maximum velocity in a vertical

V volume

v flow velocity perpendicular to flow direction at a point or at a certain height

v' instantaneous turbulence fluctuation of v at a point

W gravitational force

W' submerged weight of sediment

W_b' submerged weight of bed load in a water column of unit bed area

W_c power required for incipient motion of sediment

W_s' submerged weight of suspended load in a water column of unit bed area

w transverse velocity at a point or at certain height; subscript pertaining to river bank or clear water

w' transverse fluctuating velocity

w_b energy provided by unit water body in unit time

w_d energy dissipated in unit water body in unit time for overcoming density gradient

w_s energy lost in unit water body in unit time for overcoming local resistance

w_t energy of unit water body transferred to river bottom in unit time

X displacement of sediment particle in x direction; maximum grain size subject to hiding effect in a sediment mixture; parameter of sediment size

x direction of flow

Y displacement of sediment particle in y direction; dimensionless parameter used in stability analysis of sand waves; correction coefficient for lift force in movement of sediment mixture; parameter in sediment transport

y direction perpendicular to flow

y_n effective height in saltation

Z x+iy; relative smoothness of bed (=h/D)

z transverse direction; exponent in diffusion equation

z_1 measured value of exponent in diffusion equation

α exponent, coefficient; dynamic angle of friction of shearing of particles in collision; angle between flow direction and horizontal axis on inclined plane; ratio of resistance of the bed and that of the interface in density current

α_b angle between wave line and coast line

β exponent; coefficient; dynamic coefficient of friction; ratio of coefficient of sediment transfer and that of momentum transfer; angle

γ specific weight of water

γ_b specific weight of bed material considering porosity

γ_f specific weight of fluid

γ_m specific weight of water sediment mixture

γ_s specific weight of sediment grains

Δ height of sand wave; boundary roughness

δ phase difference in stability analysis of sand waves; thickness of laminar layer in vicinity of boundary; scope of reservoir from which the afflux from an outlet comes in selective withdrawal

ε proportional coefficient; rate of energy dissipation in unit mass of water body; number of waves Yalin's criterion for incipient motion

ε_m coefficient of momentum transfer

$\varepsilon_{x,y,z}$ coefficient of diffusion in x, y, z direction

ζ coefficient; elliptical function

η coefficient; coefficient of rigidity of Bingham fluid; eddy viscosity; distance between the bed and the reference plane; fluctuations of lift force exerting on sediment particle; relative height (= y/h); height of mud surface wave

Θ parameter of flow intensity

Θ' parameter of effective flow intensity

θ angle; dip angle of beach or inclined plane; correction coefficient of lift for fine particles in laminar state in movement of large range of sediment mixture; density ratio of top and bottom layers in density current

θ_1 coefficient of correction in computing sediment transport rate per unit width from measured data by points

θ_2 coefficient of correction in computing sediment transport rate per unit width from measured data by depth integration

κ Karman constant

Λ sphericity of sediment particle

λ ratio of grain diameter to distance between particles; ratio of step length to grain size in bed load movement; wave length; Fanning's friction factor (Chapter 14)

λ_c composite coefficient of resistance including resistance at interface and that at reservoir bottom in density flow

λ_i coefficient of resistance at the interface of density currents

λ_L coefficient of local resistance in density current

λ_0 coefficient of resistance at reservoir bottom of density current

λ_x dimension of small scale eddies; distance of formation and collapse of laminar layer near boundary in x direction in bursting phenomenon

λ_z distance between low flow zones uprising from the bed in z direction in bursting phenomenon

μ viscosity

μ_a apparent viscosity

μ_o viscosity of clear water

μ_r relative viscosity

ν kinetic viscosity

ξ proportional coefficient; distance of water surface from a reference plane; parameter reflecting the hiding effect in movement of sediment mixtures; parameter characterizing density current

Π sphericity of sediment particle

σ standard deviation; geometric root mean square of sediment size distribution; degree of scattering in sediment composition

τ shear stress in flow

τ' shear stress in fluid among particles

τ_b shear stress on bed of channel (bed resistance)

τ_b' grain resistance on bed surface

τ_b' form resistance of bed configurations

τ_B Bingham yield stress

τ_c shear stress at incipient motion

τ_e effective shear stress of flow regarding sediment movement

τ_i shear stress at interface in stratified flow

τ_L resistance among solid particles in loose deposits

τ_o shear stress on perimeters (if it is the river bed, then $\tau_o = \tau_b$)

τ_o' shear stress of fluid among particles on bed surface

τ_w shear stress on channel bank or pipe wall

Φ dimension expressing sediment size; potential function; parameter of bed load transport intensity

Φ_T parameter of intensity of total load transport

ϕ internal angle of friction of sediment; angle of repose

ρ density of water

ρ_f density of fluid

ρ_s density of sediment (solid, sphere)

$\Delta\rho$ density difference of top and lower layers

χ parameter reflecting transition from smooth to rough boundary in velocity formula

ψ parameter of flow related to sediment movement

ψ_o electric potential in thermodynamics

ω fall velocity of particle (sphere); angular frequency;

ω_o fall (settling) velocity of single particle in infinitely body of clear water

ω_F settling velocity of floc

∞ subscript pertaining to characteristics of deep water wave

Authors Index

Affiliation Index

Subject Index